DIGITAL
PRINTED CIRCUIT DESIGN & DRAFTING

BY DARRYL LINDSEY

Founder of the renowned Masters Design & Technical Center

Published by

The Innovators
Bishop Graphics, Inc.
Westlake Village, California 91359

FIRST PRINTING—1986

Printed in the United States of America.

Distributed by Prentice-Hall, Inc.
A division of Simon & Schuster
Englewood Cliffs, NJ 07632

Prentice-Hall International (UK) Limited, *London*
Prentice-Hall of Australia Pty. Limited, *Sydney*
Editora Prentice-Hall do Brasil, Ltda., *Rio de Janeiro*
Prentice-Hall Canada Inc., *Toronto*
Prentice-Hall Hispanoamericana, S.A., *Mexico*
Prentice-Hall of India Private Limited, *New Delhi*
Prentice-Hall of Japan, Inc., *Tokyo*
Prentice-Hall of Southeast Asia Pte. Ltd., *Singapore*

ISBN 0-13-214009-8 025

Contents

Preface

This instruction book on analog printed circuit board design contains the necessary information to prepare you for immediate entrance into the ever expanding industry.[1] This guide begins with the basics: (basic printed circuit (PC) Design and basic electronic components), and builds from there. The information assumes you know absolutely nothing about printed circuit boards[2] and electronics. *"YOU DON'T NEED TO BE AN ELECTRONIC EXPERT."* No college degree is needed, to excel in this field. A large percentage of the students at Masters PC Design school attend while working in careers unrelated to electronics. They have the same desire: to get into a dynamic field that pays well and has always had many job openings, good working conditions, and chances for advancements. *"YOU WILL HAVE JOB SECURITY."* A few have language problems. But they discover quickly that they can learn to communicate in the language of the PC designer that will be rapidly mastered from the pages of this book. *"A PC DESIGNER DOESN'T NEED FANCY WORDS."*

Through your reading assignments and designing projects you will gradually develop a proficiency—a certain touch, a certain art—in printed circuit designing. You will be shown exact step-by-step methods. Then your next project builds from the previous project with more information and bigger challenges.

All projects must be completed. Practice is very important. Check your work against the examples. Read the discussions in each section. *"COMPLETE ALL PROJECTS."* They teach you how to do each PC design project.

[1]Electronics deals with television and radio, pocket calculators, computers, satellites, etc.

[2]Printed circuit boards are the devices on which electronic circuits have been printed or etched. The circuits are printed onto an insulating board or base board. The board provides both support for components and an interconnecting material such as copper.

Acknowledgment

The Author would like to express his thanks for all the assistance that has been provided for the various editions of this book. I am indebted to former students, colleagues and friends for their time and efforts in helping prepare this book. I am indebted especially to Charles T. Walling. And finally I am indebted to my wife, Priscilla Lindsey, for her time and patience in helping prepare the manuscript.

Introduction

To develop, perfect, and produce an electronic device, requires the efforts of many. Not least among those concerned is the PC designer—especially in the design and first production stages of a product.

Printed circuit (PC) design is a relatively new and dynamic profession that is finding greater demands and increasing opportunities. The building block used by the electronic engineer of today is the printed circuit board (PCB)*. PC design is essential to the development and production of electronic equipment. Everyday, industry introduces innovative instruments, new developments, and progressive concepts. The aircraft industry is striving for smaller, more compact, and lighter gear; military electronic requirements are increasing. Microminiaturized equipment for our space satellites places greater demands on the PC designer. Almost every factory in America uses control, safety, and monitoring devices that contain PC boards. Home alarm, safety, and lighting controls, pocket radios, hand-held calculators, digital clocks, and even automobiles have PC boards.

The great electronic centers located in every major U.S. city use thousands of PC boards in their large computer systems, desk top calculators, minicomputers, control terminals, recording and memory units, and home entertainment centers. There are test sets, and signal and system analyzers which provide service, maintenance, and troubleshooting capabilities for all the electronic products invented. These all contain PC boards. The demand for qualified PC designers is increasing.

This instruction guide helps someone who is entering the PC design field—whether he or she has absolutely no background in electronics or is an electronic engineer. This guide begins with the basics of digital circuit design. For analog design, see the first volume in this series: *Analog Printed Circuit Design and Drafting.* The first volume in this series takes the designer deeper into the field of analog designing.

Although this instruction guide encompasses the full scope of printed circuit designing, only the bare essentials are explained about the electronic components and circuits. To do PC design, it is not necessary to understand fully the operation of the electronic circuits or components. You will, however, become acquainted with the basic components that make up digital circuits. You will be able to recognize these parts, even though their operational characteristics might remain foreign to you. Learn to recognize them and how they physically fit into an electronic package. Chapters 2 and 3 provide a fundamental understanding of components. You need to know almost everything about their physical appearance and almost nothing about their electronic function.

*Also referred to as circuit card assembly (CCA), printed circuit card (PCC), printed wiring board (PWB), printed circuit assembly (PCA), and other associated and similar titles.

How To Use This Book

This instruction guide uses the building block technique. A subject is first introduced with fundamental information. And then the following sections build from there.

(Information expands as your knowledge grows.) For example, the section schematics gives basic information. You will be able to then read the first schematic in Chapter Two. On the second schematic, there are additional features. You will receive the information you need before reading the more advanced schematic. Then, with each new section on schematics, you will receive more details until you have sufficient knowledge to read any schematic with confidence.

The material in this guide comes to you much the same as an eight course dinner—not all at once, but one course at a time. You begin with the wine and appetizer, soup and salad—the basics. Then later you have meat and potatoes. The big difference with this analogy is that, with PC designing, you continue using the basics along with the advanced information.

Often, a chapter containing additional information about a subject includes references to previous chapters. These references are for your benefit in case you wish to review the previous material.

Instruction Guide Format

This instruction guide is divided into chapters. Each chapter contains an introduction, the objectives of the chapter, and the discussion or main body of the chapter. The introduction and objectives define what is contained in the chapter and what you can expect to learn. The discussion body of the chapter contains a formal presentation of the subject material together with supporting references, diagrams, pictures, layouts, drawings, etc.

Examinations

A quiz at the end of each chapter tests you on the skills defined in the objectives. This quiz gives you a chance to monitor your short-range progress. If you fail to understand a concept in a Chapter, it is easy to review the material in that chapter immediately. *(Check your own progress).* More comprehensive tests are given in chapters 8, 9, 10, and 11. The test after chapter 7 includes material from chapters 1, 2, 3, 4, 5, 6, and 7.

Remove the quiz from the guide. Then, with all PC design books and material out-of-reach, complete the test.

The answers to all tests and quizzes are in the back of this guide. Store the tests and answers in a binder for quick reference. The page number or numbers to each question refer you to the text that discusses the subject of the questions. *(Store your tests in a convenient location).*

1

THE PRINTED CIRCUIT BOARD

As a PC designer, you will design and then layout the artwork for a unit that will become a printed circuit board. Since PC boards will be your main concern, this chapter is an introduction. It contains background on the evolution of printed circuitry and PC boards, the involvement you will have in the designing and manufacturing of the boards, and a general discussion on how PC boards are fabricated.

In Table 1-1, a printed circuit board or a PC board is defined as an insulating material onto which an electronic circuit has been printed or etched. (A PC board is shown in Figure 1-1). The PC board is an efficient use of modules and modular design. *(PC BOARDS ARE MODULES)*. Usually, one PC board contains one complete function. (More than one function, however, may be contained on one schematic diagram). Should some part of the PC board fail, an electronic technician could quickly pinpoint the fault. The fault can be quickly repaired with a spare board. The PC board slides in and out, which makes replacement quick and simple.

During WWII and into the Korean War, such modular design was not used. If a synchronizer unit failed on a radar set, the entire radar was inoperative until the single faulty part that caused the breakdown was located and replaced. Sometimes this process could take days or even weeks. (If a spare synchronizer was available, of course it could be exchanged, but not quick-and-easy like a PC board. Usually spare units were not available). *Shorter downtimes are possible using PC boards.*

The units during that time were constructed by hand wiring and by point-to-point soldering. The wires needed to be stripped of their insulation, tined, and soldered. Each discrete component was installed by hand electrically and mechanically. The units were large, awkward, and bulky. Few electronic units were available for tight places such as planes, small ships, or home emergency use—places where electronic sophistication could have saved lives. The need for smaller more compact electronic equipment was a long accepted fact, but it took time for the design engineers to supply the need. They realized the need to decrease the time between unit failure and repair.

Figure 1-1. Printed Circuit Boards

A natural evolution took place. Smaller components were invented. Modular design became the state-of-the-art. Soon the printed circuit concept was born.

The printed circuit was developed from an etching process practiced years ago. *(Printed circuits are made from an etching process).* The artisan deposited wax (or other acid resistant material) over the design that was retained on the metal. Then, after the acid was applied and the wax was removed, the metallic design would remain. (It is also analogous to the batik art which is a method of dyeing designs on cloth by waxing the sections that are not to be dyed). For the etching process, four basic items were needed: the (1) base material upon which the (2) base metal was placed (bonded), (3) the resistant material, and (4) the etchant. There was, of course, a fifth ingredient—the ability of the artisan.

The PC board of today is also developed by an etching process. And the copper traces that remain are the conductive strips that take the place of insulated wire conductors. *(Wire conductors are replaced by printed circuit traces).*

Today the PC designer is the artisan. He designs the artwork of the electronic circuit and then lays it out on a piece of plastic film or other suitable transparent material. This becomes the artmaster.

The remainder of the process, however, is completed by other professionals when the artmaster goes to a photographer. Photographs are made of the artmaster. The photographs then go to a fabrication shop that produces the pattern for making the printed circuits on the PC board. *(Photos are made of the artmaster).* The printed circuit board itself is fabricated from laminated plastic or from fiberglass impregnated with an epoxy filler. The laminated fiberglass forms the insulating base material which provides the mechanical support for the components, traces, hardware, etc. The conductive material (usually copper foil) is bonded to one or both sides of the base material. *(Laminated plastics make the base material).*

The photos are used in the process of applying the etchant resistant material (called resist) to the copper foil on the board. One such method used is shown in Figure 1-2. When the fabrication (fab) shop has completed their job, the printed circuit board (with the printed circuits as copper foil in the exact likeness of the artmaster) is ready to have components and hardware loaded. (An example of a board, as it comes from a fab shop, is in Figure 1-3).

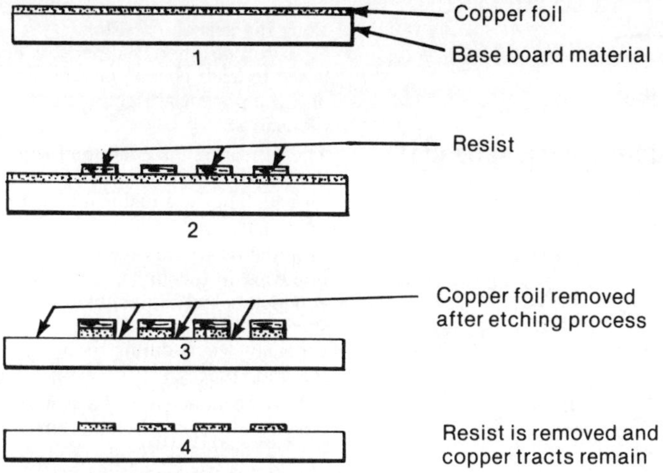

Step 1 The process begins with copper foil bonded to an insulating base board material (referred to as copper clad base material). The bonding occurs through an electroplating operation.

Step 2 The resist or etchant resistant material is applied to the copper foil as part of the photoprinting or photoengraving process. The resist is a material such as mask or ink which prevents those parts of the copper foil from etching away. The resist is deposited on the copper cladding in the exact pattern of the art master. (The resist mask is also applied by a screen printing process in much the same way as signs or reproductions are made in silkscreen printing.)

Step 3 After etching, only the portions protected by the resist have copper foil remaining. The rest of the copper has been removed.

Step 4 The resist is removed by washing, only the printed circuits remain.

Figure 1-2. Etching Process

When the PC board is loaded, it is ready for further tests and evaluations, and is plugged into the equipment where it performs its special function. (Do you suppose your part in this drama is finished? Just hang on a bit longer).

As often happens, something wrong is discovered by the engineer or the test technician that makes certain "modifications" necessary. These changes may be only minor. However, it is not uncommon to hear the engineer say to you one day, "Well, your design was good but *they* want some bigger components, an extra control circuit and some additional filtering...plus a few changes I made to make it work".

And so you start again: new layout, new tape-up, new art master, etc. You have real job security. You are a vital part of a team. You interface (work closely) with electronic and mechanical engineers, technicians, research and development (R&D) lab personnel, production lines, machine shops, and photo labs to name a few. You make up the parts lists, fabrication and assembly drawings, and silkscreen drawings (if required). Your work doesn't end here. As the product goes on the market, there are production and service engineers who find improvements to the original design. These improvements, when made, still need your expertise for implementation. Then the sales force and marketing engineers want new options on the product, greater versatility, and more streamlined capabilities. This all means that the engineers will come back to you for more PC design work.

Many times engineers may become stumped on the details of incorporating an improvement. It will be you, the PC designer who will show them how it can be done.

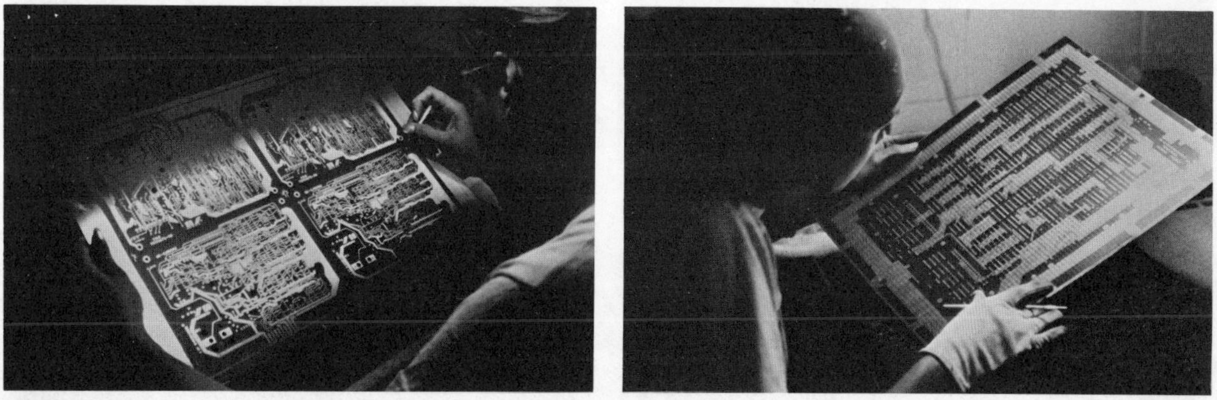

Figure 1-3. A board from a fabrication shop

Terms and Definitions

These are the initial terms and definitions needed in the preliminary stages of learning PC designing. (See Table 1-1). Additional terms are added to this list, as well as expanded definitions, as you progress through each section. The beginning PC designer should become acquainted with these terms and definitions because a working knowledge of them gives quicker insights into this vital profession. A complete list of all Terms and Definitions pertaining to PC designing is given in Appendix A. *(FOR FASTER PROGRESS IN PC DESIGNING LEARN THESE TERMS AND DEFINITIONS).*

Table 1-1. Terms and Definitions Applicable to Printed Circuit Boards

AIR GAP:	The non-conductive air space between current carrying conductors such as traces, pads, ground planes, etc.
ANALOG CIRCUIT	A circuit comprised primarily of individual (discrete) components, such as resistors, capacitors, diodes, transistors, etc. The circuit output is a continuous electrical signal that varies in frequency, amplitude, etc. as a function of the input. The magnitude is represented by physical variables such as voltage, current, resistance, rotation, etc.
ANNULAR RING:	The width of the conducting surface (material) surrounding a hole. Generally referred to as the "minimum annular ring" since it concerns the narrowest width of the conductor ring around the hole.
ARTWORK:	An accurately scaled configuration used to produce a master pattern. The artwork is prepared by a PC designer usually at an enlarged scale of 2-1 (2:1) or 4 to 1 (4:1). Then it is reduced photographically to produce the artwork master.
ARTWORK MASTER	An accurately scaled pattern (usually 1:1) which is used to produce the production master.
AWG:	American Wire Gage. A method of specifying wire diameter. The higher the number, the smaller the diameter.
AXIAL LEADS:	Leads coming out the ends and along the axis of a resistor, capacitor, or other axial part. As opposed to leads coming out the side of a component at right angles to the axis.
BASE MATERIAL:	The insulating material upon which the conductive pattern may be formed. The base material may be rigid or flexible.
BRIDGING:	A condition where excess solder builds up in the air gap between conductors and causes them to short together. Solder bridging generally occurs during the wave soldering process.
BUS:	A heavy conductive metal strip or trace on the printed circuit board used to distribute voltage, grounds, etc., to smaller branch traces.
BYPASS CAPACITOR:	A capacitor used for providing a comparatively low impedance AC path around a circuit element.
CONFORMAL COAT:	A coating that is generally sprayed, dipped, or brushed on to protect the completed printed circuit board from fungus, moisture and debris.
CONNECTOR TONGUE:	A series of contacts on the edge of the printed circuit board that mate with a receptacle to provide an electrical connection between the printed circuit board and other circuitry. (Also called edge board contacts).
DIGITAL CIRCUIT:	A circuit comprised mostly of integrated circuits and operates like a switch (i.e., it is either "ON" or "OFF").
DIP SOLDERING:	A process whereby printed boards are brought in contact with the surface of a static pool of molten solder for the purpose of soldering the entire exposed conductive pattern in one operation.
DISCRETE COMPONENT:	A component which has been fabricated prior to its installation (i.e., resistors, capacitors, diodes, transistors, etc).
DOUBLE-SIDED PRINTED BOARD:	A printed board with conductive patterns on both sides.
FEED-THRU:	A plated-thru hole in a printed circuit board that is used to provide electrical connection between a trace on one side of the printed circuit board to a trace on the other side. (Also referred to as a plated-through hole because connection is made through the hole by plating or depositing metal on the wall of the hole).
FLOW SOLDERING:	Also call wave soldering. A method of soldering printed circuit boards by moving them over a flowing wave of molten solder in a solder bath.

GLASS EPOXY: A material used to fabricate printed circuit boards. The base material (fiberglass) is impregnated with an epoxy filler which then must have copper laminated to its outer surface to form the material required to manufacture printed circuit board.

GRID: A two-dimensional network that consists of two sets of equally spaced, parallel lines superimposed on each other so that the lines of one set are perpendicular to the lines of the other, to form squares. This orthogonal (mutually perpendicular) network is used for locating exact points on the PC board.

GROUND PLANE: A conductor layer, or portion of a conductor layer, used as a common reference point for circuit returns, shielding, or heat sinking.

HEAT SINK: Any device that absorbs and draws off heat from a a hot object, radiating it into the surrounding atmosphere.

MASTER PATTERN: An accurately scaled pattern which is used to produce the printed circuit within the accuracy specified on the master drawing.

MOTHERBOARD: A relatively large printed circuit board (PCB) on which components, modules, receptacles (PCB socket contacts), and other electronic subassemblies are mounted. Other PC boards are inserted into the receptacles (sockets) of the motherboard. Since the receptacle contacts interface with each other through the traces or conductor strips on the motherboard, the motherboard provides electrical connections between the inserted PC boards. (The motherboard is also called a back plane or matrix board).

MULTILAYER PRINTED BOARD: A printed board consisting of alternate layers of conductive patterns and insulating materials bonded together, with conductive patterns in more than two layers, and with the conductive patterns interconnected as required.

PAD: A portion of the conductive area to which components, terminals, traces, etc., are mechanically attached. (ALso called land area).

PLATING: A uniform coating of conductive metal upon the base material of the printed circuit board.

PRINTED CIRCUIT: A circuit where the interconnections between components, terminals, subassemblies, etc., are made by conductive strips (traces) that have been printed or etched onto an insulating board.

PRINTED CIRCUIT ASSEMBLY: A printed circuit board to which discrete components, hardware, and other electronic devices have been attached to form a complete operating unit.

PRINTED CIRCUIT BOARD (PCB): An insulating material onto which an electronic circuit has been printed or etched. (See printed circuit). Also referred to as circuit card assembly (CCA), printed circuit card (PCC), printed wiring board (PWB), printed circuit assembly (PCA), and other associated and similar titles.

PRODUCTION MASTER A 1 to 1 (1:1) scale pattern which is used to produce one or more printed boards (rigid or flexible) within the accuracy specified on the master drawing. There are two principal types of production masters; a single image and multiple image. A single-image production master is used in the process of making a single printed board. A multiple-image production master is used in the process of making two or more printed boards simultaneously.

RADIAL LEAD: A lead extending out the side of a component, rather than from the end.

SCHEMATIC DIAGRAM A drawing that shows, by means of graphics symbols, the electrical connections, components, and functions of a specific electronic circuit arrangement.

SOLDER MASK: A type of conformal coat that is placed on the side of the board to be soldered. This coating prevents solder bridges from forming between traces and pads during the dip soldering or wave soldering process.

TRACES: The metallic conductive strips that provide connections between components, terminals, etc., on printed circuits.

WAVE SOLDERING: See flow soldering.

Design Inputs

As a PC designer, you need certain inputs and information before you can do your design work. This section on design inputs contains a description of the information you must have to begin your design task and a "check list" to ensure that you have all necessary information to complete the project.

Before you begin any PC design project, you should receive a design package. This package is supplied by either the design or project engineer. (They may be one and the same person in smaller companies). Essentially the design package consists of a schematic diagram, a parts list, some data sheets, some miscellaneous design features concerned with board size and shape, components types and ratings, or other requirements, guidelines, and restrictions. The most important item in this design package is the schematic diagram. It may be a very rough copy as seen in Figure 2-6 or as neatly drawn as the one in Figure 2-7. (YOU START WITH A DESIGN PACKAGE).

Schematic Diagram

A circuit is a combination of electrical and electronic components connected in a particular way to produce a certain desired function. An example is an amplifier circuit in a stereo or a light circuit in the home. A schematic is a drawing or diagram of a specific electrical or electronic circuit showing both the components as graphic symbols and the interconnections of the components that produce the desired function. The schematic diagram received by a PC designer in preparation for laying out a PCB may contain several circuits. Each circuit may be composed of dozens of individual discrete electrical components such as transistors, diodes, resistors, etc., together with many compact electronic devices called integrated circuits (ICs). The schematic may also have switches, relays, and adjustments.

Circuit Design Features

All of these component parts have been fitted together by design to produce a certain result in a system. The engineers can visualize this end result. They know what is required. You, the PC designer, are given exact guidelines, tolerances, and restrictions so your PCB will produce the required result. This helps to make the overall system operate within the exact specifications placed on the designers. For example, let us assume that a system has a noise problem and the PC board you are about to design is believed to be the central cause of that noise. Therefore, the engineers will specify certain guidelines and restrictions on your design. Maybe these restrictions are concerned with the length of the traces, or special shielding is required, or with the design of the ground plane. Perhaps new devices with low RFI emission characteristics are to be tried. Any design inputs you receive from the design engineer, either written or by word-of-mouth, must be incorporated into your PC layout.

It is common, however, for an important design requirement to be overlooked. A PC designer may complete the assignment only to discover that all transistors in one circuit should have been shielded. This requires an almost complete redesign of the PC board, thereby delaying the project.

NOTE: Blaming the "crummy" system or an engineer for the oversight or criticizing the company for lack of proper coordination, will not recover lost time and delay in the design schedule. Pinning the blame on someone will not help. But it is to your advantage as a PC designer to ensure that time-consuming oversights never happen.

But how will I know if a unique feature or special function is to be designed into my board? To ensure that you have all necessary information to complete the project, a "checklist" is provided (much like a pilot's checklist used before take-off and landing). Table 1-2 lists questions to ask the design engineer. These questions and the answers ensure you have all the information before beginning your project. Also, as you proceed through this guide, these questions will become more refined and meaningful to you. As your knowledge of PC designing expands, so will your awareness of what is required. The list in Table 1-2 will have increased significance. *(IF YOU ARE CONFUSED OR STUMPED, ASK THE ENGINEER).*

Included with the schematic and the other material in the design package, may be notes, comments, and the schematic diagram made by the design engineer. *(LOOK FOR EXTRA INFORMATION THAT MAY BE ON ENGINEER'S SCHEMATIC).* Some of the information you can expect on the schematic includes:

— Notes
— IC (Integrated circuit) table
— REF DES (Reference designation) table
— Component values, ratings, compositions, sizes, manufacturer's numbers, tolerances, etc.
— Capacitor polarization
— CW (clockwise) rotation of adjustments

Table 1-2. Questions to Ask Design Engineer

1. Check to see if there are any special size components, or if all the components are standard.
2. How many decoupling capacitors are there for each IC?
3. Are there any critical leads on the board?
4. Are there any special size decoupling capacitors required on the input voltage pins?
5. Should there by any spare IC gates?
6. By using the maximum size of all the components, will all the components fit on the board? (This is your responsibility).
7. Are there pre-assigned pins on the connectors?
8. Are there any variable resistors or capacitors and where should they be placed on the board?
9. Are there any test points? Are there placement restrictions? (Try to place all test points on the edge).
10. Do any components require heat sinks?
11. Will boards have keying pins?

*The one thing you can expect the engineers to do on the schematic is be inconsistent. This is not intentional nor are these mistakes on their part. Because there is such a massive amount of information that needs to be conveyed, there is bound to be oversights. It is the PC designer's responsibility to locate these oversights.

Schematic Notes

These are located at conspicuous places on the diagram. They are a list of "inclusive" information about the components on the drawing because they concern every part referenced. For example, two notes may read: "Resistors in ohms; capacitors in uF". The complete meaning is: all resistor values are to be read in ohms; all capacitor values are to read in uF (microfarads). With these notes as guides, the engineer may now write, in an abbreviated form a single number such as 100 (read as 100 ohms) by the resistors and a number such as 10 (read as 10 uF) by the capacitors. A more important note would be: resistors are ¼W, 5% (W is their wattage rating and 5% is the tolerance of their value in ohms). The complete meaning is: all resistors are rated at ¼watt with a tolerance of ±5% unless otherwise noted. With this note, the engineer does not have to write anything by the resistors unless their ratings are something other than ¼W or have a tolerance different than ±5%. Again, this note saves the engineer much writing time. Other such notes as might appear are shown below:

— All capacitors are ceramic disc.
— All transistors are TO-18 case.
— All ICs are 7400.
— All diodes are 1N914.
— All switches are DPDT.
— ETC.

Many of the titles and words used in these discussions may be unfamiliar or foreign to you now but in due time each will be fully explained.

IC (Integrated circuit) and REF DES (Reference designator) tables are important. The IC table usually contains the reference designation for each IC and its type. Consider U9-74L04. (This information, as you will learn later, becomes invaluable when you begin making connections to U9).

Information about components is important. Any information that is written at or near the component will usually be of help to you during the design of the PCB; however, the information may not be complete. Even if the information is explicit and seems to be everything you will need, many times valuable data is missing which you will need to ask the engineer.

If there are two or more components used for the same function, the engineer may only identify one part completely and give only partial information on the other parts. It may seem safe to assume that all parts are the same, and then design your board on this assumption. But any assumption of this nature will invariably lead the PC designer into making costly mistakes. The only safe course is to check out each component. This can be done quickly and easily, if the components are numbered on the schematic. Using the parts list, compare the descriptions and part numbers of each part. If the components are not numbered or if there is no parts list available, then ask the engineer. It is poor practice to begin designing a PC board by assuming any information. Get the full and complete facts about the schematic and the board and then begin your design with confidence. *(DO NOT ASSUME ANY INFORMATION).*

Summary

The design of electronic equipment involves many technical requirements and areas of consideration. Some areas that must be considered in the design phase are electrical, mechanical, environmental, serviceability, function, and application.

The electrical and electronic design requirements include choice of circuitry, electrical power available, circuit functions, unit size, interconnections, component availability, to name a few. Mechanical engineering must consider weight, size, shape, ease of fabrication, mounting needs, equipment markings, temperature extremes, etc. The environment or where the equipment is to be used must be determined. Equipment used onboard a submarine or atmosphere would have different design factors than one used solely on dry land. One to be used in Alaska has different temperature requirements than one used on the Mojave desert. The function of the equipment may dictate weight, size, and shape which, in turn, would determine some mechanical and electrical design features. Depending on its function, it may need certain types of displays or none at all. Always its maintainability, serviceability, accessibility, which are all interrelated, must be taken into account. Is it to be stationary or provided a service manual for field service? Many of these design factors will not come under the PC designer's area of responsibility, but it is important for him/her to be aware of how they are interrelated and how they affect the final design.

Quiz - Chapter 1

1. Printed circuit boards provide a quick-and easy exchange for faulty electronic equipment. T F
2. The printed cicuit board uses an ____________________ process to remove un-wanted metallic.
3. Before you can start a new project, you should have a "check list" to ensure that you have covered all necessary information. T F
4. The least important part of the engineering design package is the schematic. T F
5. How will you know if a unique feature or special function is to be designed into your board? ________ ____________________
6. Schematic notes are used by the engineer only. T F
7. Schematic's information about components will normally be written at or near the component. T F
8. Schematic-all information about components will be complete if it is written near the component. T F
9. It is a poor practice to begin designing a P.C. board by assuming. T F
10. The engineer's schematic could be inconsistent. "Not intentionally" but because there is such a massive amount of information that oversights can occur. T F

The answers for each question can be found on the pages listed below:

Question 1. Pg. 2
Question 2. Pg. 2
Question 3. Pg. 5
Question 4. Pg. 6
Question 5. Pg. 6
Question 6. Pg. 7
Question 7. Pg. 7
Question 8. Pg. 7
Question 9. Pg. 7, 8
Question 10. Pg. 7

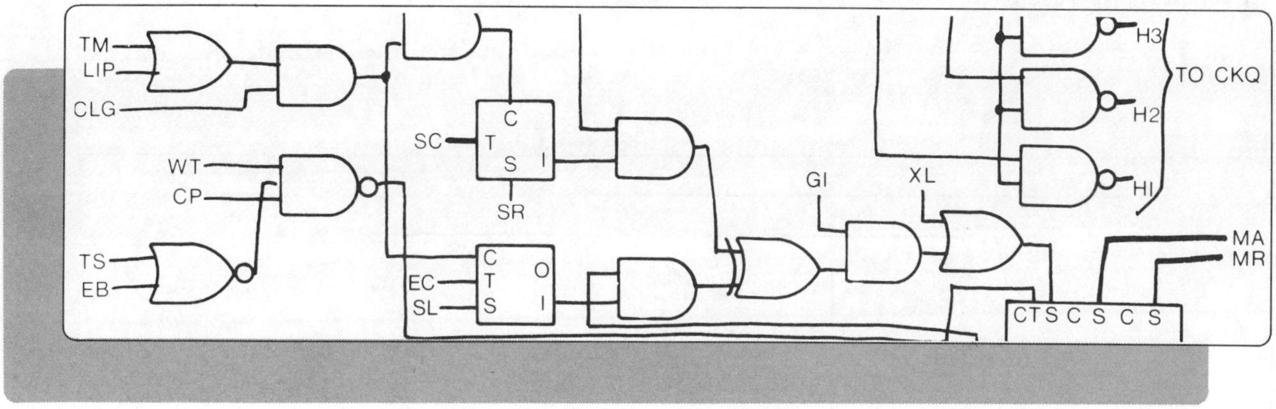

2

SCHEMATIC DIAGRAMS

A schematic diagram is a system of graphic symbols that represent electronic, electrical, and electro-mechanical components. The components are connected electrically as shown by the interconnecting lines (Figure 2-1). The schematic is the first step in a new or revised electronic design because it displays and identifies the components that make up the equipment. Your principal concern as a PC designer is the conversion of the schematic diagram into a layout (which would be transformed into an artmaster). Therefore, learning to read a schematic is of primary importance. This chapter contains the description of all the components parts on the schematic diagram in Figure 2-1. It presents the schematic symbols called logic gates or logic diagrams. Digital applies to the term logic.

Figure 2-1 demonstrates the basic parts of the schematic: each symbol, its reference designation, its fundamental purpose, and an introduction to the basic schematic layout. After studying the material in this chapter, you should be familiar with all the essentials of a schematic: particularly the logic symbols and the discrete symbols shown in Figure 2-1. You should be able to:

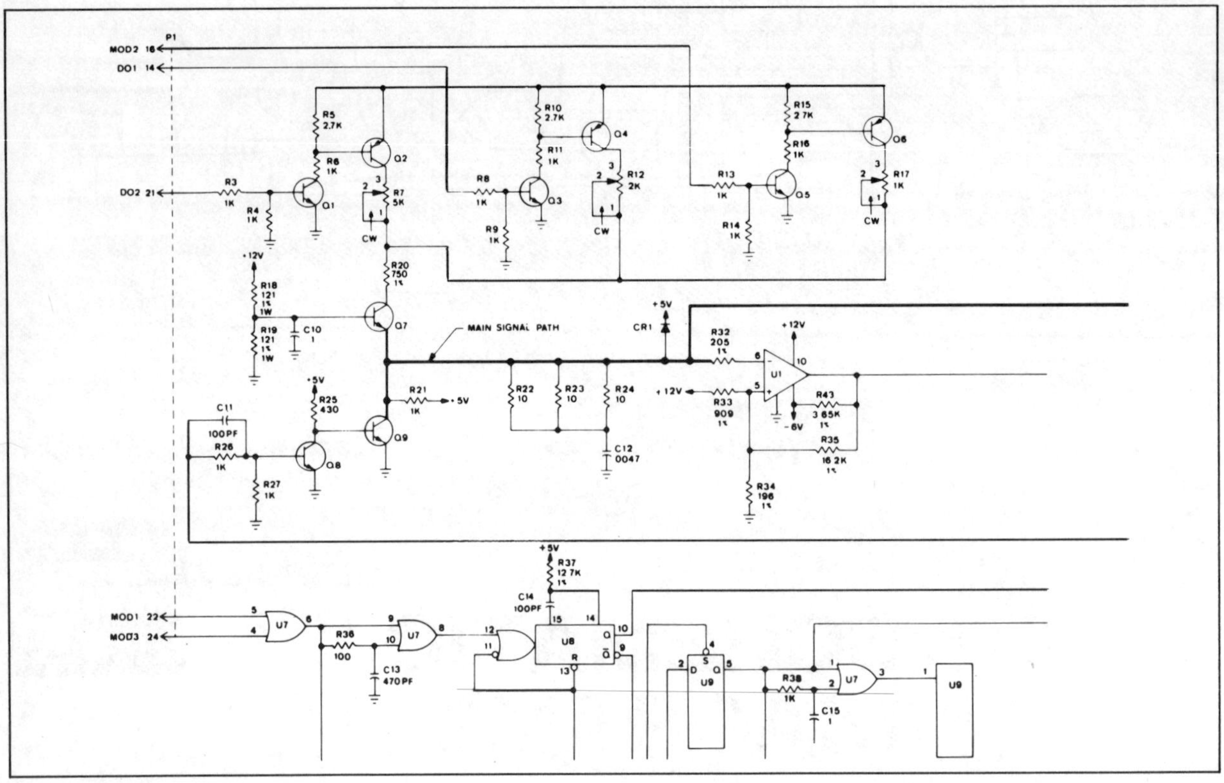

Figure 2-1. A Schematic Diagram

1. Recognize the schematic symbols presented in this section.
2. Recognize the reference designations.
3. Understand the basic definitions of the logic symbols.
4. Know the fundamentals of a logic layout.
5. Define a logic diagram.
6. Know the fundamentals of reading a logic diagram and a basic schematic.

Schematic Symbols and Reference Designators

The following is a description of each circuit component symbol in Figure 2-1. A brief description of their application and electrical operation is also given.

Schematic Diagram Reference Designators

Each component is identified by an alphanumeric reference designation such as: C1, R1, Q1, etc. (C for capacitor, R for resistor, and Q for transistor, etc). The number shows the difference between electrical components of the same type (C1, C2, C3, etc.). The use of reference designators provides an easily recognizable identification for every component part on the schematic. See Table 2-1 for reference designations. See Table 2-3 for a schematic check list.

Schematic Diagram Symbols

To simplify the schematic diagrams, symbols represent the various circuit components. In most cases, the symbol represents what the component actually does in the circuit or how it physically constructed. Many companies have standards of their own that are usually based on a combination of military and commercial practices. The military uses Mil-Std (standard) 275 (with the latest revision letter); industry uses ASA (for American Standards Association) Y14.5.

It would be impractical to include all of the military and industry standards. Only the most commonly used symbols in schematics for printed circuit boards are described below.

Table 2-1. Reference Designations

A	assembly	MG	motor-generator
AT	attenuator; isolator termination	MK	microphone
B	fan; motor	MP	miscellaneous mechanical part
BT	battery	P	electrical connector (movable portion); plug
C	capacitor		
CP	coupler		
CR	diode; diode thyristor: varactor	PS	power supply
DC	directional coupler	Q	transistor: SCR: triode thyristor
DL	delay line	R	resistor
DS	annunciator; signaling device (audible or visual); lamp; LED	RT	thermistor
		S	switch
		T	transformer
E	miscellaneous electrical part	TB	terminal board
		TC	thermocouple
F	fuse	TP	test point
FL	filter	U	integrated circuit; microcircuit
G	generator		
H	hardward	V	electron tube
HY	circulator	VR	voltage regulator; breakdown diode
J	electrical connector (stationary portion); jack	W	cable transmission socket
K	relay	Y	crystal unit (piezoelectric or quartz)
L	Coil; inductor		
LS	meter	Z	tuned cavity; tuned circuit

Logic Symbols

Standard symbols used in logic are prescribed in standards issued by the IEEE, ANSI and by the military. The military standard graphic symbol for logic diagrams is Mil. Standard 80-6B. This has been superseded by the IEEE standard graphic symbols for logic diagrams. The IEEE standard is 91-1973, and the ANSI standard is Y32.14. These standards allow both rectangular and distinctive shapes for logic element symbols. We recommend the use of distinctive shapes (See Figure 2-2). Equivalent symbols for the AND, NAND, OR, NOR, EXCLUSIVE OR, EXCLUSIVE NOR, BUFFER and INVERTER are shown in Figures 2-3A and 2-3B. These symbols are equivalent and may be used on the same schematic drawing. The choice of these symbols must be decided by the electrical engineer.

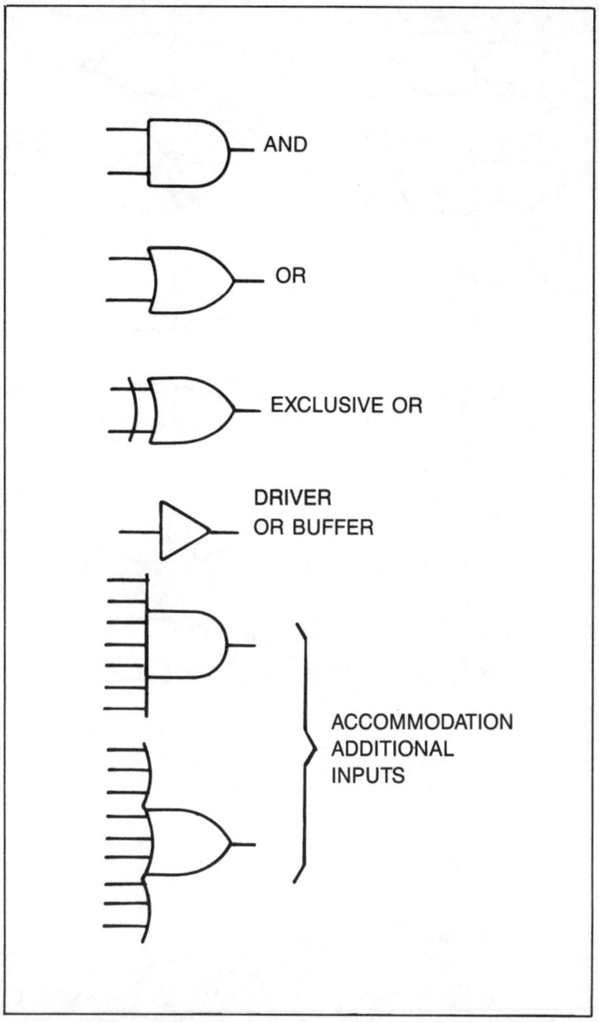

Figure 2-2. Standard Logic Symbols

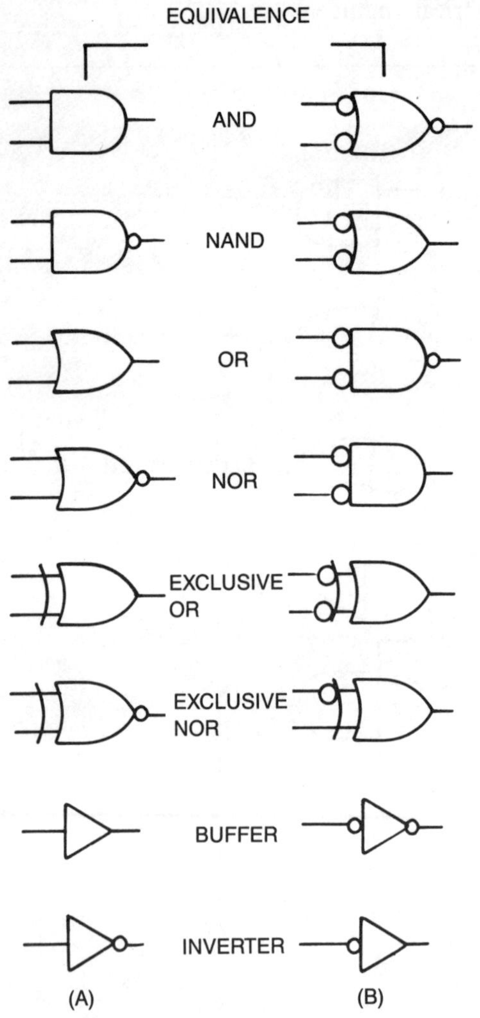

Two symbols, one device. Any physical logic circuit is correctly represented by either of two standard symbols.

Figure 2-3. Equivalent Logic Symbols

Table 2-2. Schematic Check list.

1. Before the schematic is drawn, the production engineer reviews the schematic for flow and arrangement of the logic of components.
2. Primary signal lines on all schematics will be drawn heavier.
3. All test point should have a name.
4. The schematic should be drawn with "-" voltages down and " + " voltages up.
5. If possible show the PC board components or controls with their related functions.
6. Do not draw power and gorund lines on the schemati unless it is very easy to see. Use the symbols and codes.
7. All inputs and outputs on the connectors should show the signal directions and be assigned a name.
8. Switches hsould be marked N.C. (normally closed) or N.O. (normally open).
9. IC's use reference designation - "U".
10. "OP-AMPS" eliminate all sash lines around the symbol.
11. IC's with gates that have internal connections should have abox drawn around the logic.
12. Decoupling logic capacitors should be placed in the notes only.
13. Coding inputs and outputs from gate to gate use anumber "(3)" to indicate the number of places it terminates. A note will be required to denote (3).
14. IC's lists all input and output coding.
15. When possible, show internal logic on all the blocks drawn on the schematic.
16. Dash lines shown on the schematic must be defined in the notes.
17. Resistor and capacitor values should be complete. Example: 200K, ¼W, 5% (or in the notes)
18. Zener diodes should have the voltage rating and tolerance.
19. Spare gates should be shown.
20. Transformers - the normal operating voltage on all output terminals should be listed.

Schematic Name____________________

Schematic Number____________________

Capacitor

Reference designation: C Electronic Symbol:

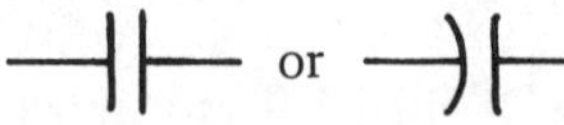

This is a device consisting of two conducting surfaces separated by an insulating material or dielectric such as air, paper, mica, glass, plastic, film, or oil.

A capacitor stores electrical energy, blocks the flow of direct current, and permits the flow of alternating current, to a degree, dependent upon the capacitance and the frequency.

a. 

b. C1

c. 1uf

d. 25V

e. ± 10%

Value Expressed IN:

Farads (f) - most general applications (especially in PC boards) are in microfarads (mf) or picofarads (pf). (See item C).

Tolerance:

The tolerance reflects the amount that the capacitor is allowed to deviate from the value shown on the schematic. Most common tolerances for PC board applications are ± 10% and ± 20%. (See item E).

Rating:

Capacitors are rated by the amount of voltage that they are able to withstand safely. Capacitor physical sizes (of the same material) are controlled by their voltage ratings, with most common values within a voltage rating all being the same size. The most common voltage rating for PC board applications ranges from approximately four volts to several thousand volts. (See item D).

Material:

The majority of capacitors used for PC board mounting are most always constructed externally using metal or paper. Internally, these capacitors are commonly mica, foil, and plastic. There are various other methods of construction.

Connector Pin

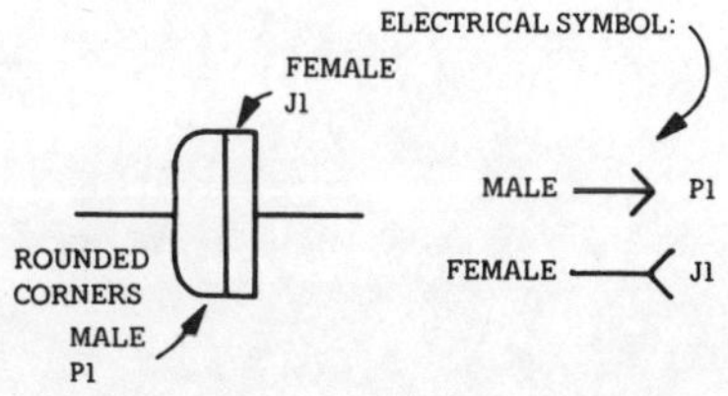

Crystal

Reference Designation: Y Electronic Symbol:

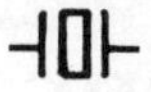

This is generally used to generate a signal with an accurate frequency output. It can also be used to convert pressure to an electrical signal, as can be done with a piezo electrical crystal transducer.

Diode

Reference Designation: CR Electronic Symbol:

This is a two-terminal device which conducts electricity more easily in one direction than in the other.

Fuse

Reference Designation: F Electronic Symbol:

 or 

Two types of fuse symbols could be used. The first symbol represents the fuse element only. The second represents the complete fuse.

Chassis Ground

Electronic Symbol:

This indicates a connection made to the chassis.

Earth Ground
Common Ground

Electronic Symbol:

This indicates a connection to ground.

Inductor

Reference Designation: L   Electronic Symbol:

This is used to introduce inductance into an electrical circuit. The inductor is wound into a spiral, or coil to increase its inductive intensity. It may use an iron or air core.

Resistor

Reference Designation: R   Electronic Symbol:

This is a device which resists the flow of electric current in a circuit. There are two main types: fixed and variable.

a. 

b. R1

c. 200 Ohms

d. ¼W

e. ± 5%

Electronic Symbol: (See Item a)

Reference Designation: R (See Item b)

Value Expressed In: Ohms (See Item c)

Tolerance:

The tolerance reflects the amount that the resistor value is allowed to deviate from that shown on the schematic. Most common tolerances used in PC board applications are ±2%, ±3%, ±5% and ±10%. (See Item e).

Rating:

Resistors are rated by the amount of power they are able to dissipate in a circuit. Resistor sizes (of the same material) are controlled by their wattage or power dissipation. The most common rating for PC board applications are 1/10W, 1/8W, 1/4W, 1/2W, and 1W. (See Item d).

Material:

Resistors are made from a variety of materials, but the most commonly used on PC boards are carbon, wire-wound, and metal film.

Relay Symbol

Reference Designation: K   Electronic Symbol:

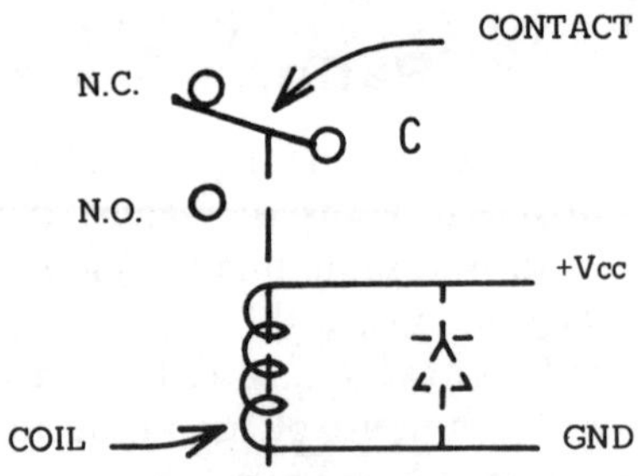

This is an electromechanical device with contacts that are opened and closed by a magnetic field, derived from a permanent magnet or electro-magnet.

Rating:

Relays are rated by (a) the amount of current that the contacts are designed to carry and (b) the voltage range at which the coil is designed to operate (i.e., 5V, 10V, 15V, and 100V).

NOTE: To protect the voltage source to the coil (especially in transistor supply), a diode should be connected across the coil of the relay and physically located as near the coil as possible.

Transformer

Reference Designation: T Electronic Symbol:

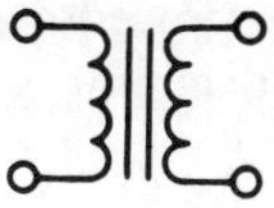

This is an electrical device which, by electromagnetic induction, transforms energy from one or more circuits to one or more other circuits.

Transistor

Reference Designation: Q Electronic Symbol:

This is a semiconductor device usually made of silicon or germanium having three or more electrodes. It may be used as an amplifier signal generator, or to perform logical functions.

P-N-P Transistor
(Positive-Negative-Positive)

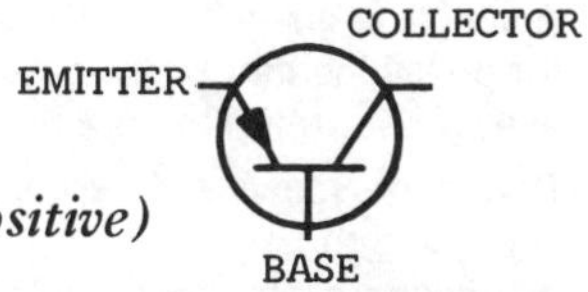

N-P-N Transistor
(Negative-Positive-Negative)

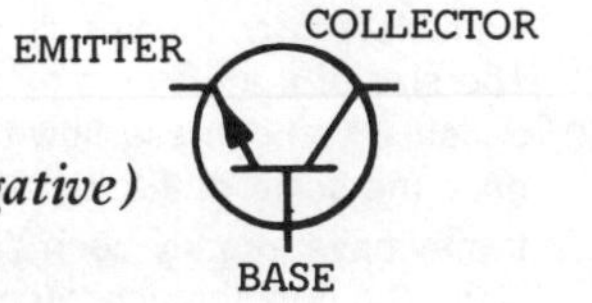

FET
(Field Effect Transistor)

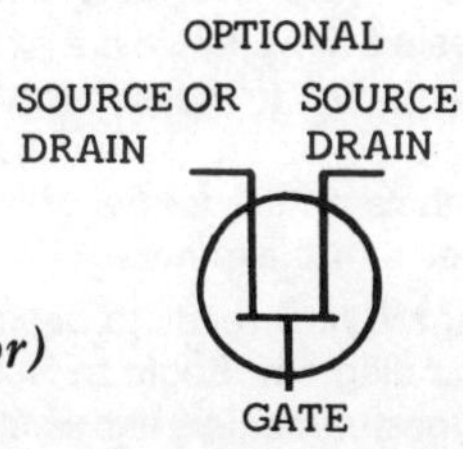

Variable Resistor

Reference Designation: R Electronic Symbol:

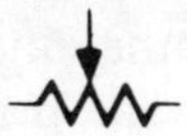

Zener Diode

Reference Designation: VR Electronic Symbol:

This is a device that has a sudden rise in current above a certain reverse voltage (the Zener value). if forward-biased, the diode is an ordinary rectifier.

Operational Amplifier Symbol

Reference Designation: U12U2 Electronic Symbol:

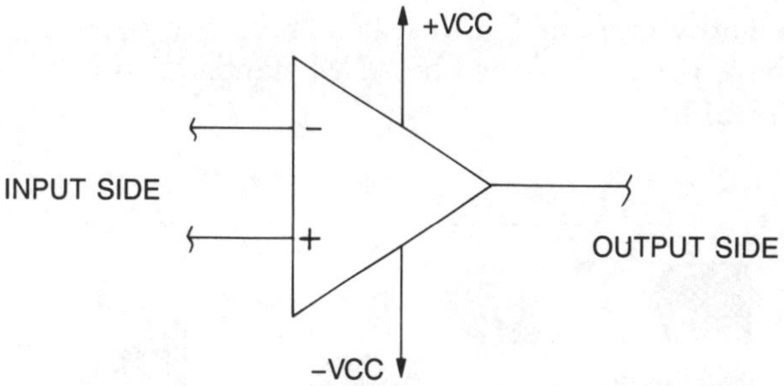

The operational amplifier (sometimes called computing amplifier) puts out an enlargement of an electrical waveform. It not only amplifies but also performs certain operations such as addition and mathematical sign reversal. However, its functions are not limited to arithmetic computation.

The abbreviation is OP AMP. This is an extremely stable (usually direct coupled) linear amplifier, originally intended for performing mathematical operations. OP AMP's, especially the integrated-circuit variety, have a wide range of applications outside of computation.

Logic Diagrams

The logic diagram should be arranged to describe the logic functions of the entire diagram with clarity and simplicity. The primary inputs should be placed on the left side of the schematic diagram. Secondary inputs should also be placed on the left side of the diagram. However, secondary inputs may enter from the right side of the schematic diagram when required for polarity.

Primary outputs should be located on the right side of the schematic diagram. Secondary outputs may be placed on the left side of the schematic diagram for clarity if necessary. Arrowheads may be used on the conductor lines to indicate the signal direction (Figure 2-11). Logic function symbols are drawn with templates (Figure 2-4).

Logic schematic diagrams are not drawn to a specific scale. However, there are five basic sizes used for logic symbols. The templates come in five different sizes: Full, three-quarter, half, three-eighths, and one-fourth scale. These five sizes have been established by ANSI standard; — Y32.14.

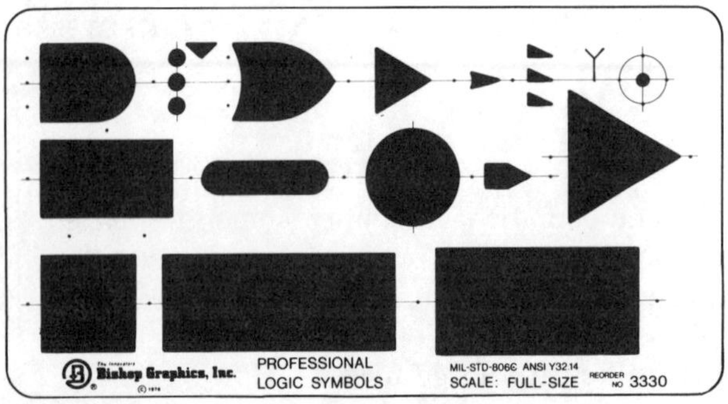

Figure 2-4. Logic Symbol Templete

Drawing Logic Diagrams

The electronic draftsperson must be familiar with the numerous logic symbols and the proper technique used to draw them. Certain procedures should be followed to become consistent and proficient in drawing logic diagrams. These procedures typically vary from company to company. However, the basics should remain the same. These are the basic procedures to follow for the creation of a logic diagram:

1. Obtain the engineer's sketch. Normally the information that the engineer provides the draftsperson is in a sketch form. See Figure 2-5 for a typical sketch. The sketch could be routed through the drafting supervisor in a large company. In small companies the engineer communicates directly with the draftsperson.
2. Check for completeness and accuracy of the engineer's sketch before you begin to draw the schematic logic diagram. The draftsperson will need to examine and ask questions of the engineer to make certain that the information is complete and accurate.
3. Plan the layout. When interpreting the engineer's sketch, the draftsperson searches for areas of the logic diagram which need to be improved i.e., overcrowded areas, undefined symbols, etc. The draftsperson also looks for alternate routes for the conducting signal paths to simplify the logic diagram. For some logic diagrams, the draftsperson may find it necessary to draw several sketches to determine the best placement. The draftsperson must be careful not to change the electrical meaning by routing the conductor paths improperly. See Figure 2-6 for questions that would be asked of the engineer's sketch shown in Figure 2-5.
4. Determine space and logic symbol sizes. What scale are you going to use for your logic symbols? Using the engineer's sketch as a guide, the spacing of the logic symbols and conductor signal lines must be determined. The size of the drawing may now be established because you can see how much area will be required once the scale of the logic has been determined.
   It may have already been dictated by the drafting supervisor. If this is the case, the determination must be made at this time as to whether it will fit on the requested drawing size. At this point, the draftsperson will make notes to himself on the engineer's sketch as to where he would place major sections or areas of logic symbols on the formal drawing. See Figure 2-7 for examples of these notes.
5. You are now ready to begin the formal layout of the logic diagram. Begin by using the sketch as a guide as suggested earlier, and lightly draw the major sections and connector areas. Continue by lightly drawing the required symbols and conductor lines in each major section.

When all the required symbols, conductor lines, and connector pins are drawn on the schematic diagram, the draftsperson should examine the layout for accuracy and proper spacing. The draftsperson should also try to simplify the logic diagram. If any modifications are necessary, they should be made at this time. Also, the engineer should be consulted before the formal drawing is darkened in.

6. Complete the formal logic diagram after determining the location for all the required symbols, conductor lines, and lettering. The logic diagram is ready for finalizing after the engineer has been consulted and after all questions pertaining to the completeness and accuracy have been answered. At this point, you are ready to letter all the required designations. You must darken the logic symbols and conductor lines, and then place the required lettering on the drawing. See figure 2-8 for completed logic diagram of Figure 2-5.

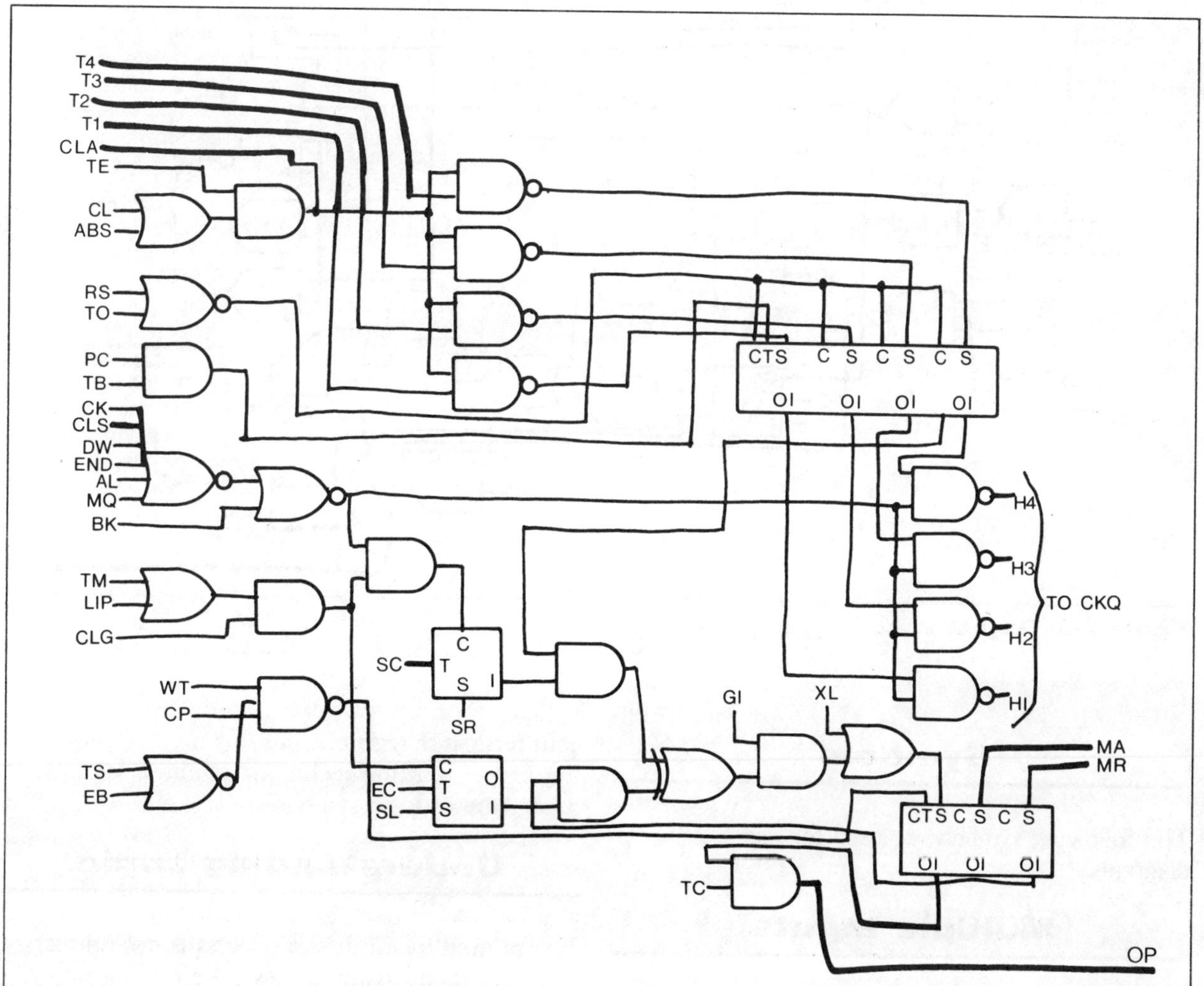

Figure 2-5. Engineer's Sketch of Logic Diagram

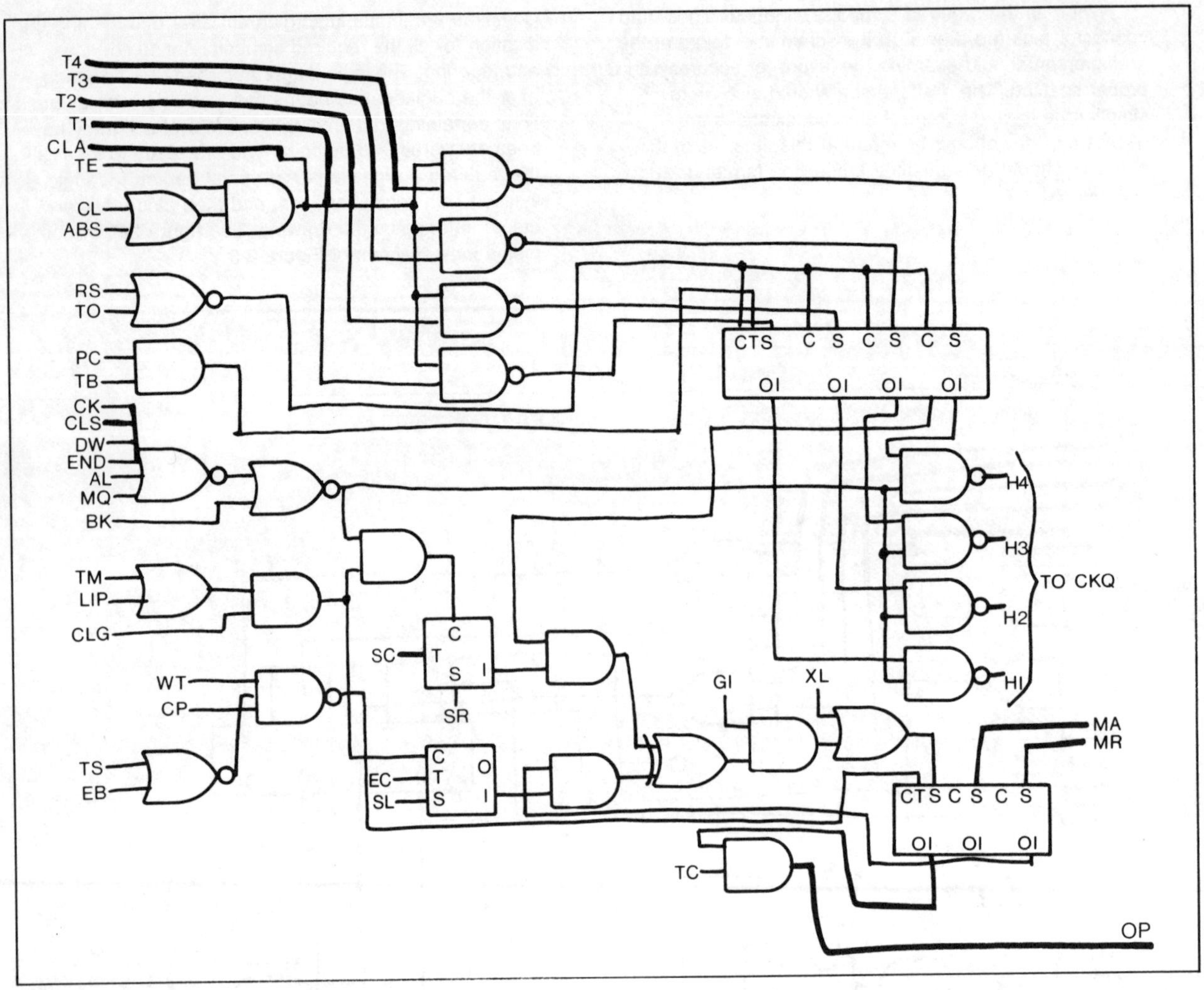

Figure 2-6. Rough Schematic

Symbols

The following symbols are used for logic diagrams.

Multiple Inputs

To accommodate more than three inputs to one logic gate symbol, the input side of the symbol is extended. Figure 2-9 illustrates the acceptable method of extending a logic symbol to provide for several inputs. Figure 2-9A is for ANDS and NANDS. Figure 2-9B is for ORS and NORS. There are other logic symbols to represent more complicated logic devices such as flip-flops, counters, shift registers, etc. Figure 2-10 shows the proper drafting technique required for this other type of logic symbol.

Understanding Logic

The printed board design process is evolutionary and consists of many phases. These phases may or may not be accomplished in a specific order. All, however, will ultimately be required and accomplished through necessity. Digital logic is one of the evolutionary phases. Since you may not even know the meaning of the words "digital logic", this chapter describes the printed circuit board design process.

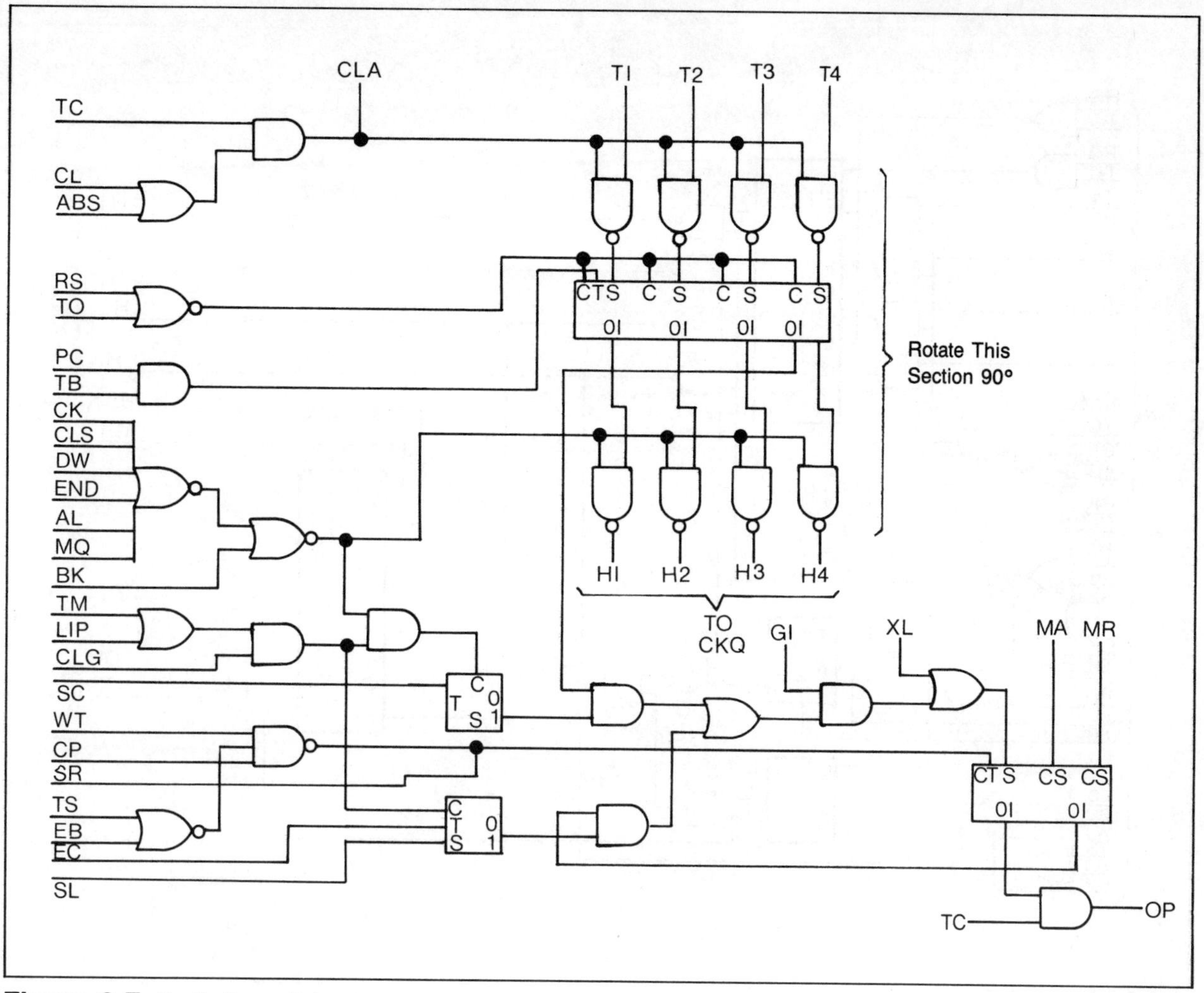

Figure 2-7. Neatly Drawn Schematic

Component combining and grouping which are the real secrets to the successful PC board layout, is a secondary intent of this chapter. Grouping will be discussed following the digital logic discussion. An understanding of digital logic is a must before component grouping can be accomplished successfully. Chapter 3 will provide exercises on the practicing of component grouping. This chapter will not make you an electrical technician or an electronics engineer. Instead you will have sufficient knowledge of digital logic to form your PC board designing tasks more effectively and efficiently. Experience has shown that understanding digital logic and applying this knowledge can solve many problems in the initial design stage rather than at the scheduled completion time. Problem solving creates an understanding of digital logic which aids the PC design process.

The word, "digital", comes from the word digit, which means one, two, three, etc. However, in the science of modern electronics, certain words or jargon have taken on different meanings. Also, certain other words are assumed to be included in many electronic definitions. For instance, the word digital not only means digits but implies binary, which means a two-state system as opposed to decimal which means a ten-state system. By combining the two definitions, the word digital means a digit that can be in two-states (two-state electronics). These two-states are zero equals OFF or LOW, and one equals ON or HIGH.

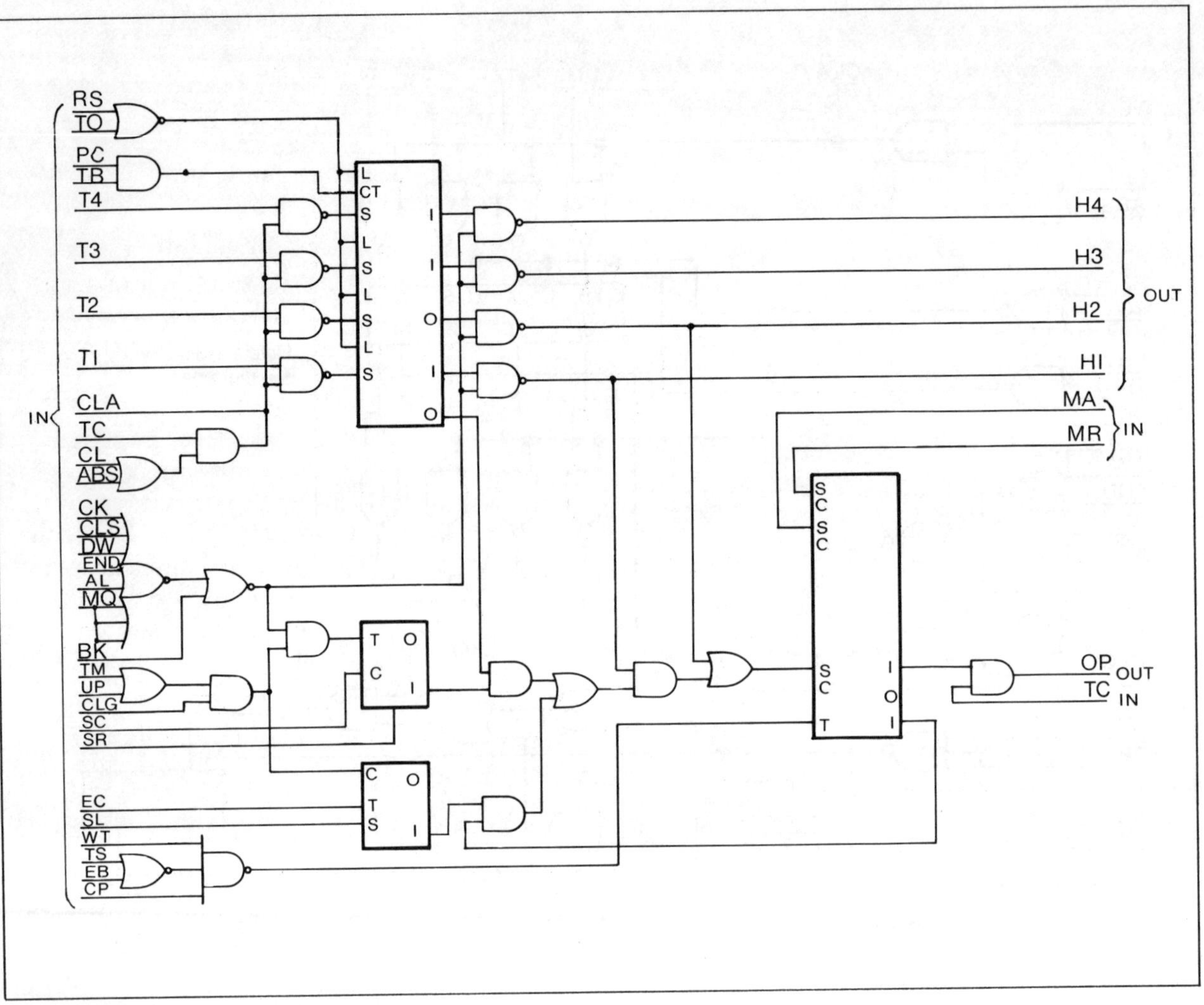

Figure 2-8. Completed Logic Diagram

Logic is the science of valid, efficient reasoning. In other words, it is a way of performing a given task in the most efficient manner available to you. As in the definition of digital, the word logic implies digital electronics The words "digital" and "logic" mean an efficient decision making method of using two-state electronics. It is relatively inexpensive as opposed to non-digital logical electronics and is most useful in the art of electronic design.

Digital "integrated circuits", or IC's are the hardware through which digital logic concepts are made possible. The IC is a relatively small, sealed package with pins protruding from the sides. The package types are: Dual in-line packages, (DIP), flat pack, and TO-5 cans. See Figures 2-12 and 2-13 for examples.

As the name implies, circuits (not just components) are integrated into the IC. There are four common families of integrated circuits available today. They are:

1. SSI (small scale integration)
2. MSI (medium scale integration)
3. LSI (large scale integration)
4. VLSI (very large scale integration)

The difference between these four is not the types of circuits contained within each package, but the density of the circuitry contained within each package. Each of these four families can be further subdivided into the type of circuitry used to fabricate the particular IC device. These subdivisions are:

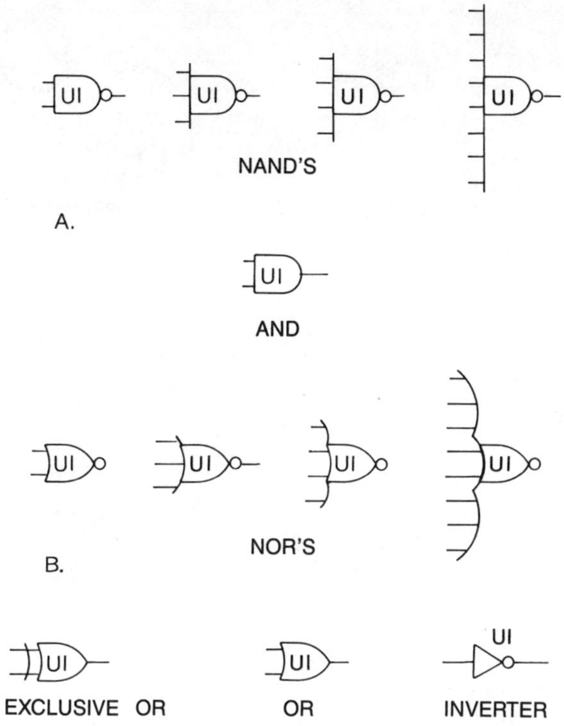

Figure 2-9. Drafting of Logic Symbols

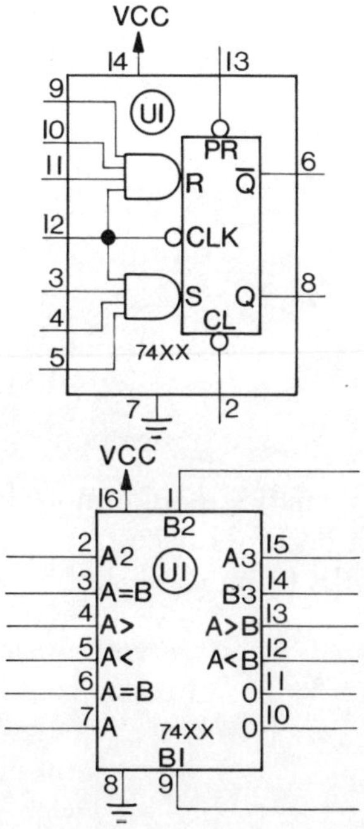

Figure 2-10. Proper Drafting Symbols (Flip-flop & Counts)

1. TTL (transistor-transistor logic)
2. ECL (emitter-coupled logic)
3. CMOS (complimentary metal-oxide, semiconductor)
4. DTL (diode-transistor logic)
5. RTL (resistor-transistor logic)
6. HTL (high-threshold logic)

This chapter points out some of the abbreviations used in the industry today...it does not define each one in detail. The differences among these types are switching speeds, power consumption, cost, isolation, and sensitivity.

A circuit is made up of many discrete components such as resistors, capacitors, transistors, and diodes that can perform some function. For example, it might count up or count down, multiply or divide. In fact, it might do all of these at the same time. The discrete component cannot perform a function readily without first being connected into a circuit.

Integrated circuits are usually formed on a silicon wafer consisting of many circuits. Various photographic and chemical etching processes are performed to produce the pre-designed electronic circuits. The individual circuits are separated from the wafer and are usually called die or chips (Figure 2-17 shows the IC (chip) building block). Gold wires are connected between the output pins (or the chip) and IC pins. Then the IC is sealed. This process is the same for all IC's made today (SSI, MSI, LSI and VLSI). See Figure 2-13 for an example of internal connections and gold wires connecting to the chip.

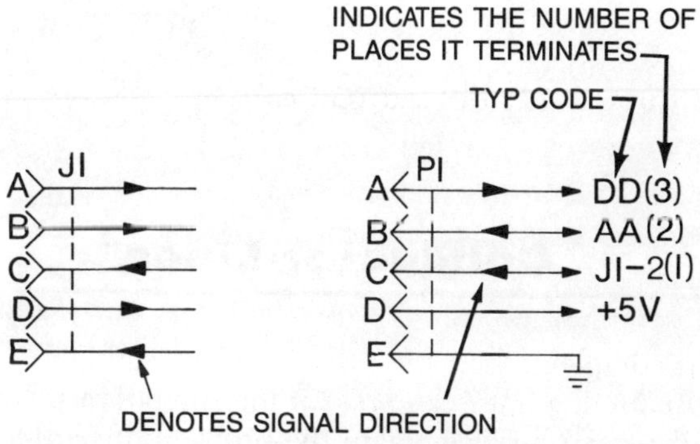

Figure 2-11. Male & Female Connector

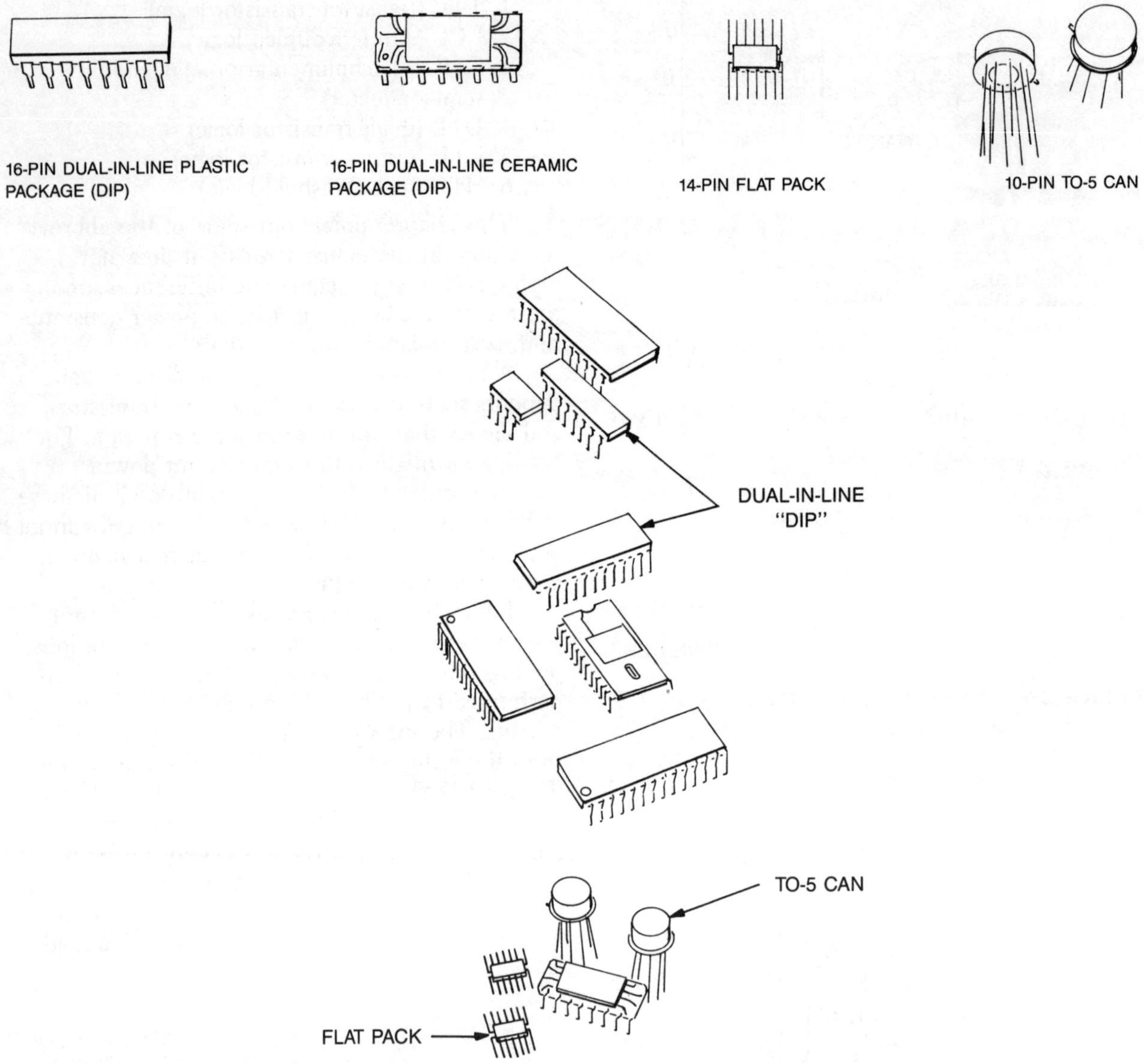

Figure 2-12. DIP and TO Can

Conductor Lines

In complex schematics, the interconnecting or conducting lines can present interpretation problems. To help eliminate the problem of hard-to-read schematics, a coding method is used. Figure 2-14 illustrates the use of the ground symbol to show interconnections to a common point.

Some schematics require more than one sheet. It then becomes necessary to identify the conductor paths to indicate continuity of the electrical paths from one sheet to another (Figure 2-15).

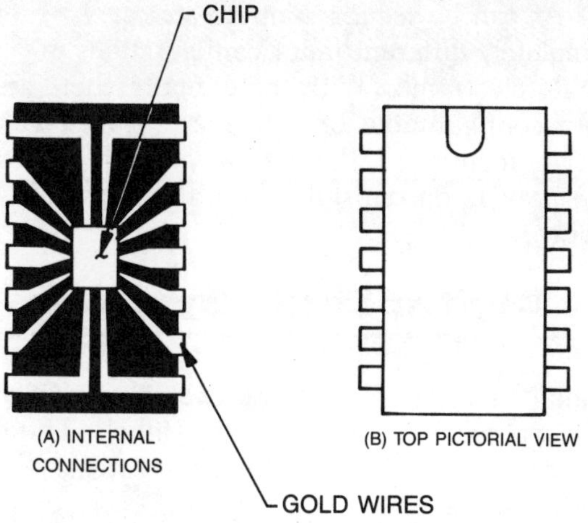

Figure 2-13. IC Internal Connections

Termination

In complex schematics, the interconnecting or conducting lines can present problems. Parallel interconnecting lines might be coded. See Figure 2-16.

Logic Rules

For this discussion, an ON or Logic 1 state is identified as +5 Vdc and an OFF or Logic 0 state as 0 Vdc. 0 Vdc is actually ground potential and is abbreviated GND. This arrangement of +5 Vdc = Logic 1 and 0 Vdc = Logic 0 is referred to as positive true logic, meaning the more positive voltage is a Logic 1. Negative true logic could be used where the less positive voltage is a Logic 1 and the +5 Vdc and 0 Vdc levels could be other values. These rules can be applied to an actual digital logic gate symbol. Figure 2-18 below shows the AND gate.

Leads A and B are inputs. Lead C is an output. There could have been many more inputs (designated as the fan-in), however, for this explanation only two are used. To make this gate function, we want the output (C) to change state when we make certain changes to the input leads (A and B). Now remember, only two states exist on any one lead whether it be input or output (Figure 2-18 describes the operation of Figure 2-19).

Most IC data sheets provide a truth table which shows all possible combinations of input conditions and the resulting output. For instance, the different conditions that exist for 2 input leads are shown in Figure 2-20.

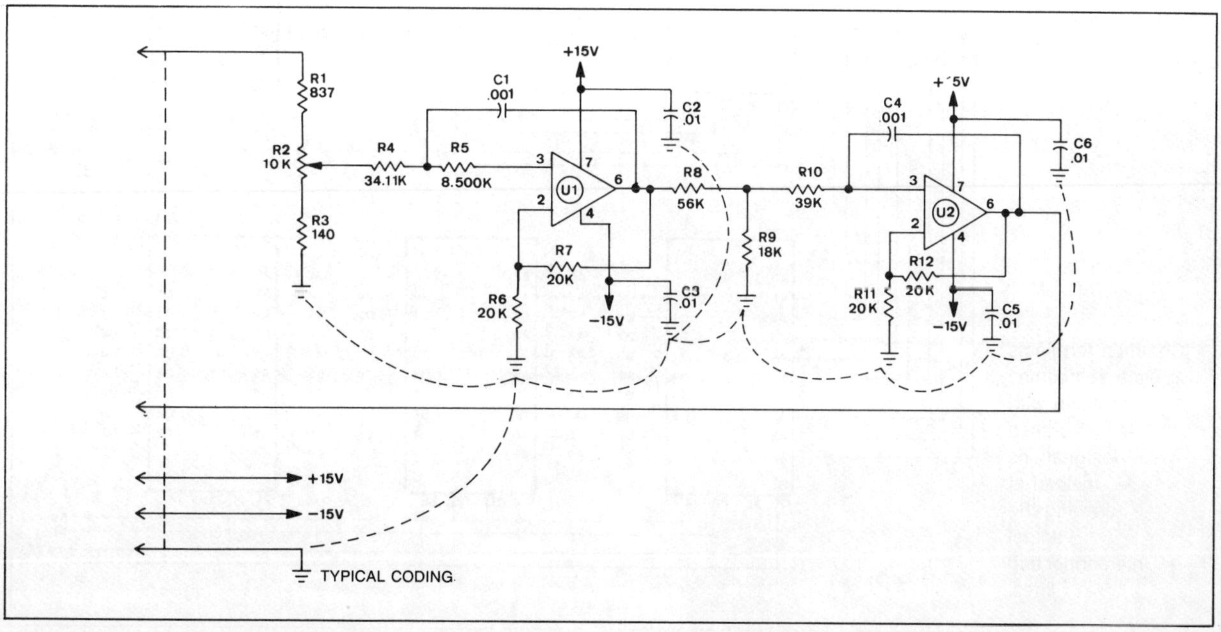

Figure 2-14. Symbol Coding Method

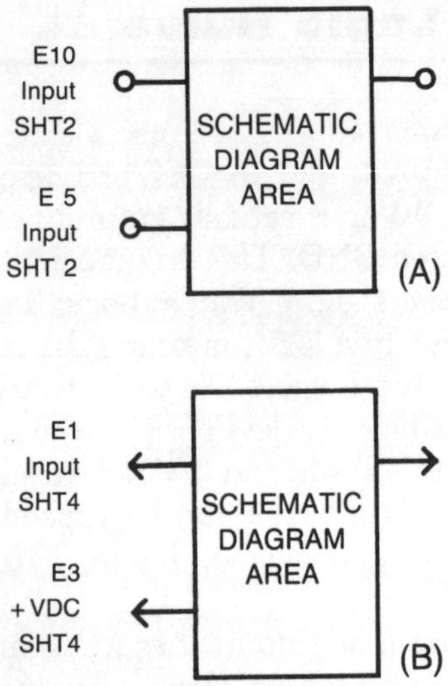

Figure 2-15. Electrical Path From One Sheet to Another

As can be seen, a 2-input gate can have four completely different input configurations in 2-state electronics. With three inputs, there are eight configurations or states, see Figure 2-21.

As mentioned, the data sheet truth table will also provide output states from the various inputs.

Digital Logic Systems

Digital logic systems fall into two categories: basic gate symbols and complex gate symbols. There are three gate symbols in the basic category. These are:

* AND Gate

* OR Gate

* INVERTER

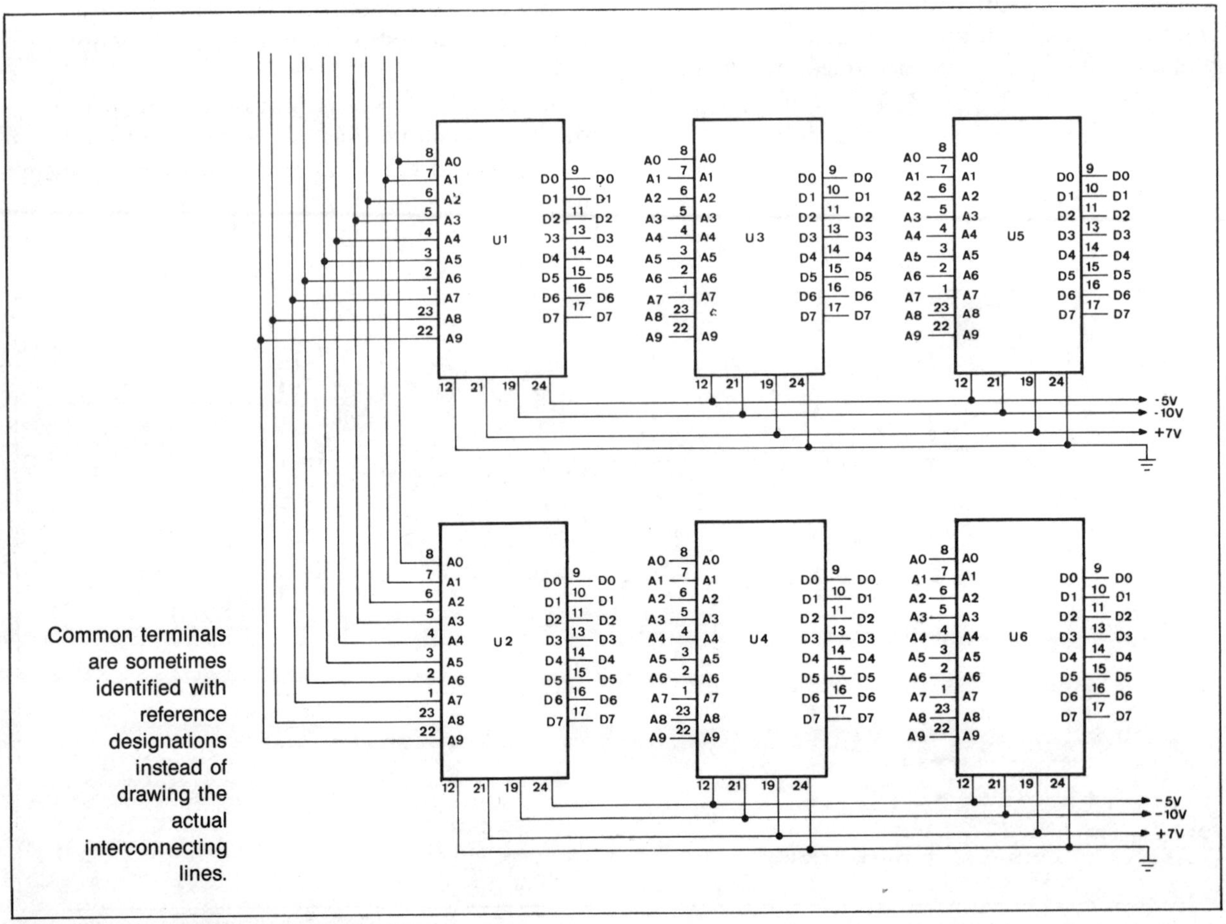

Figure 2-16. Parallel Interconnecting Lines

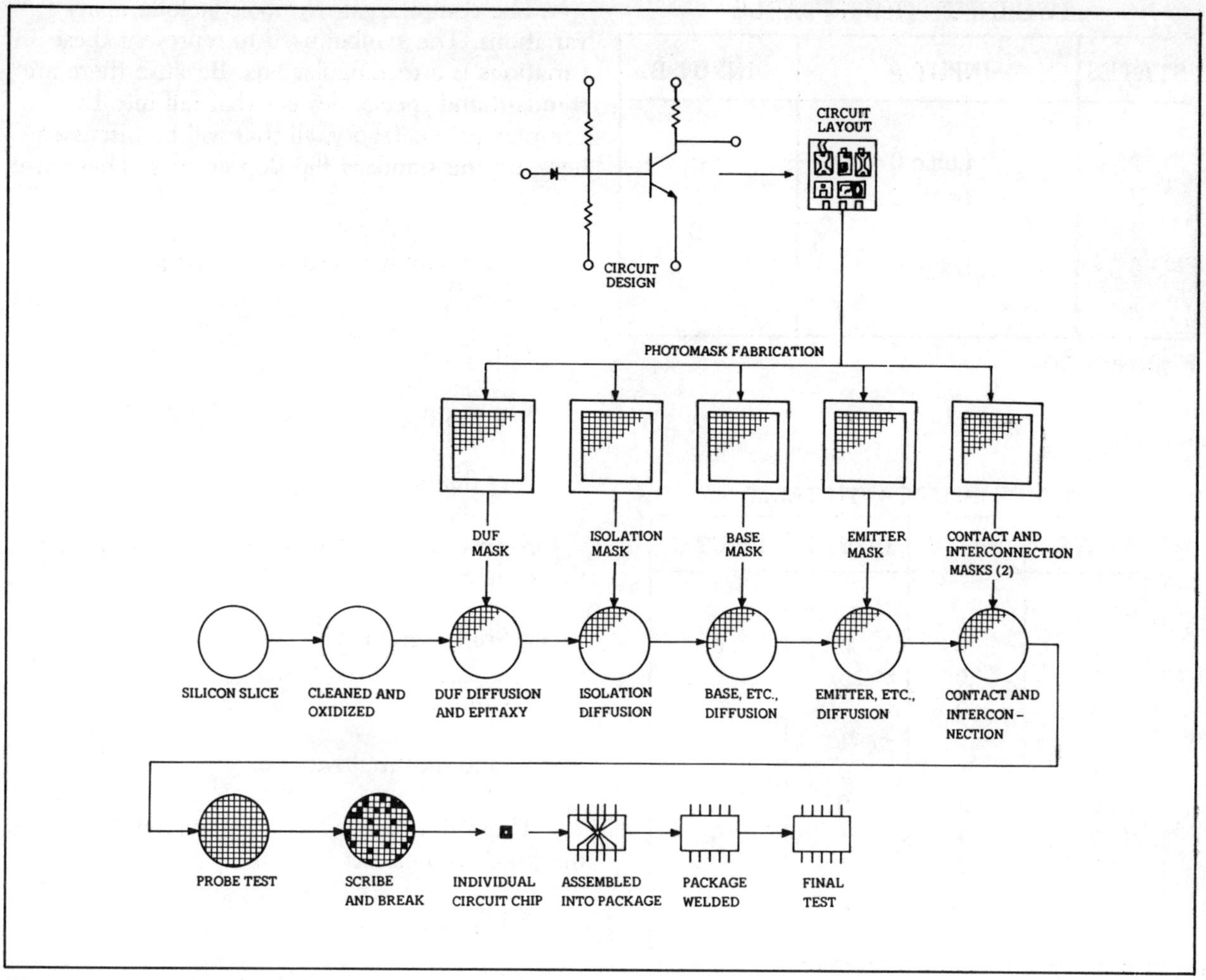

Figure 2-17. IC Building Block

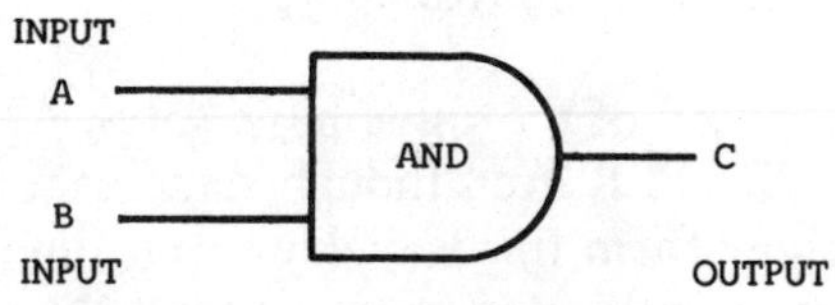

Figure 2-18. Basic Logic Gate

STATE	INPUTS		OUTPUT
	A	B	C
1	0	0	0
2	0	1	0
3	1	0	0
4	1	1	1

Where 1 = + 5Vdc or Logic 1
0 = 0Vdc or Logic 0

Figure 2-19. Logic Gate Truth Table

TWO INPUT TRUTH TABLE

STATES	INPUT A	INPUT B
1	Logic 0	0
2	Logic 0	1
3	Logic 1	0
4	Logic 1	1

Figure 2-20.

THREE INPUT TRUTH TABLE

STATE	INPUT A	INPUT B	INPUT C
1	0	0	0
2	0	0	1
3	0	1	0
4	0	1	1
5	1	0	0
6	1	0	1
7	1	1	0
8	1	1	1

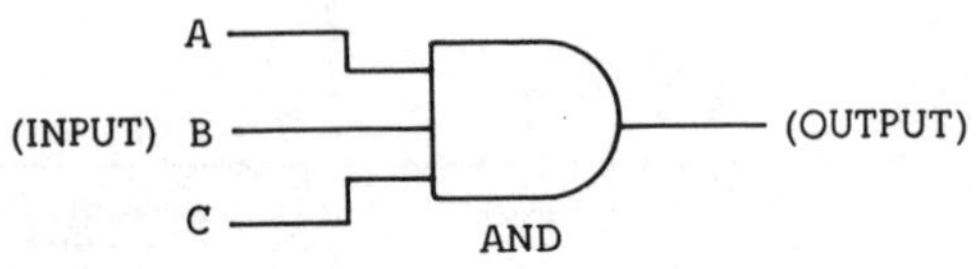

Figure 2-21.

The complex gate symbols include many variations. The symbol used to represent these variations is a rectangular box. Because there are standard and special devices that fall into the complex gate category, all that will be discussed here are the standard flip flop devices. These are:

* Basic flip-flop
  (Commonly known as the RS flip-flop or latch)

* JK flip-flop

* T flip-flop

* D flip-flop

Special devices that will not be discussed in this chapter include:

* Shift registers

* Counters

* One-shot multivibrator

These devices also use the rectangular box as the logic symbol. Actually, all complex devices;

* Flip-flop

* Shift registers

* Counters

* One-shot multivibrator

are composed of the three basic gates; AND, OR, and INVERTER even though some data sheets don't show them this way. However, flip-flop packs are widely used to understand their basic operation.

Before we attempt a discussion of logic gate generation, let's review some basic rules:

* There are three basic logic symbols: AND, OR, INVERTER

* All logic gates have at least one input lead

* Most logic gates have a single output lead

* The normal positive logic convention is:
  An ON state is a Logic 1, or high
  An OFF state is a Logic 0, or low

* The NAND gate (not AND) is the AND gate with an inverter on the output lead

* The NOR gate is the OR gate with an inverter on the output lead

* Logic symbols represent circuits within an IC package

* Each IC package contains at least one logic symbol (hence circuit) usually more

* The total number of inputs on a gate refers to fan-in

Basic Logic Symbol Definitions

The AND gate logic symbol is shown in Figure 2-22 along with a dual inline (DIP) IC that includes four 2-input AND gates (called a Quad 2 input AND).

The AND gate shown in Figure 2-22 is a device whose output pin is a Logic 1 when, and only when, all of the inputs are also a Logic 1 as shown in State 4 of the AND truth table. When any input goes to a Logic 0 state, the AND gate output also goes to a Logic 0 as shown in States 1 to 3 of the AND truth table. The AND gate can include up to 8 input leads. If only one input is ever used, the device no longer functions as an AND gate. Rather it functions as a buffer element. An analogy to the AND gate is a lamp with two (or more) switches required to turn the bulb ON. In Figure 2-23, Switch 1 and Switch 2 must be ON to energize the light bulb.

If either or both Switch 1 and 2 are OFF, the light bulb will also be OFF. Therefore, the AND gate is actually a series element. That is, it requires all input states to be ON before the output state can be ON.

The symbol for a 2-input OR gate is shown in Figure 2-24 along with a quad 2-input, DIP or gate.

The OR gate shown in Figure 2-24 is a device whose output pin is a Logic 1 when any or all inputs are a Logic 1, as shown in states 2-4 in the truth table. When, and only when, all inputs are a Logic 0, as shown in state 1 of the truth table, will the OR gate output be a Logic 0. The OR gate can include up to ten input leads. If only one input is ever used, the device no longer functions as an OR gate. Rather it functions as a buffer element. An analogy to the OR gate is a lamp with two (or more switches, where each switch can supply power directly to the lamp and light it. In Figure 2-25, Switch 1 or Switch 2 or both switches 1 and 2 being on will energize the bulb. The lamp will only go OFF when both switches are OFF.

Therefore, the OR gate is actually a parallel element. That is, it can be in the ON state by any input or inputs or all inputs being in the ON state.

The inverter of Figure 2-26 inverts as an incoming signal. Also shown in the Figure is a DIP, Hex inverter (six inverters).

The inverter shown in Figure 2-26 is a device whose output is a Logic 0 when the input is a Logic 1 as shown in State 2 of the inverter truth table. The inverter output is a Logic 1 when the input is a Logic 0 as shown in State 1 of the truth table. NAND gates and NOR gates as shown in Figure 2-27 (with all input leads tied together) also function as inverters.

An analogy to the inverter requires some reverse thinking. For example, when you turn your lamp switch ON you expect the lamp to go ON. If, however, the switch was wired backward the lamp would go OFF when the switch was turned on. The inverter functions this way, when the output state is reversed from the input state.

The NAND gate symbol of Figure 2-28 is similar to that of the AND gate, with a small circle at its output terminal.

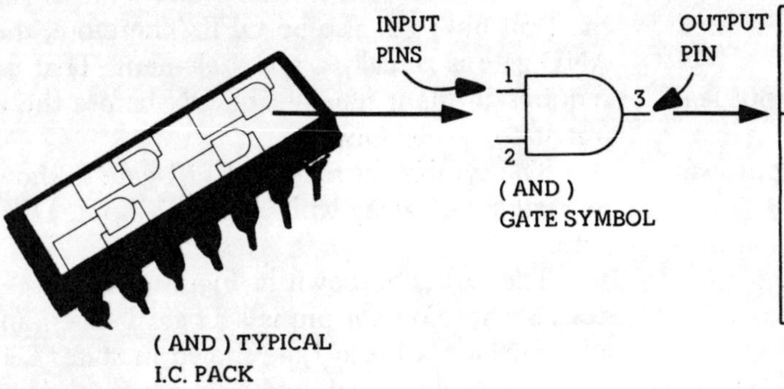

AND TRUTH TABLE

STATE	INPUT PIN		OUTPUT
	1	2	3
1	0	0	0
2	0	1	0
3	1	0	0
4	1	1	1

AND TRUTH TABLE

Figure 2-22. AND Gate

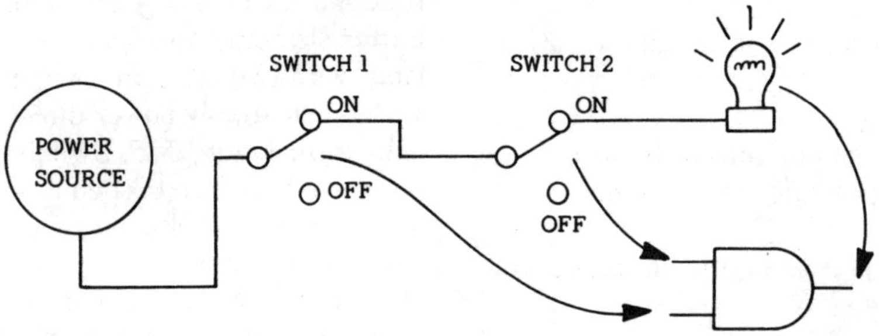

Figure 2-23. Analogy of the AND Gate

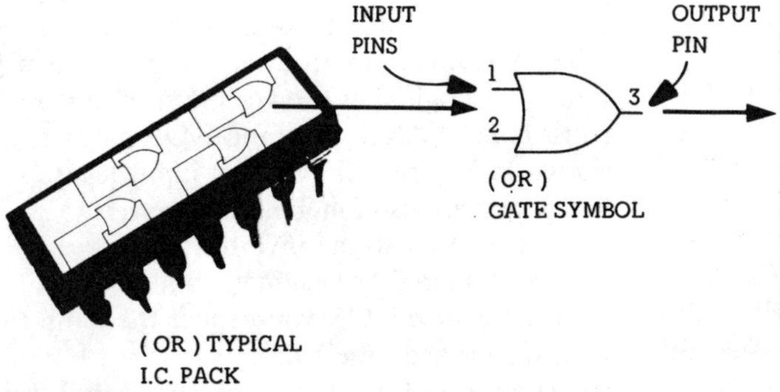

OR TRUTH TABLE

STATE	INPUT		OUTPUT
	1	2	3
1	0	0	0
2	0	1	1
3	1	0	1
4	1	1	1

Figure 2-24. OR Gate

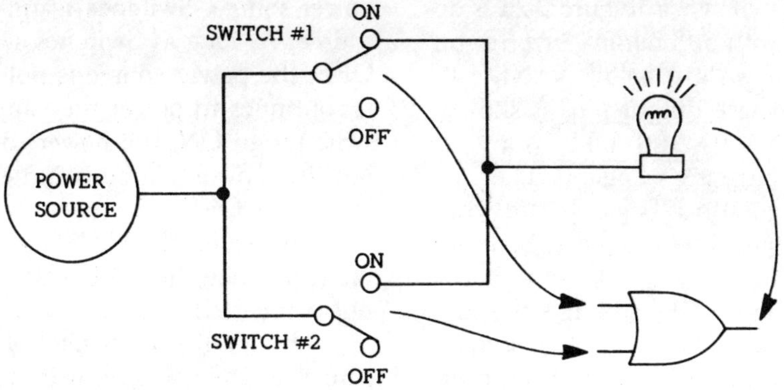

Figure 2-25. Analogy of the OR Gate

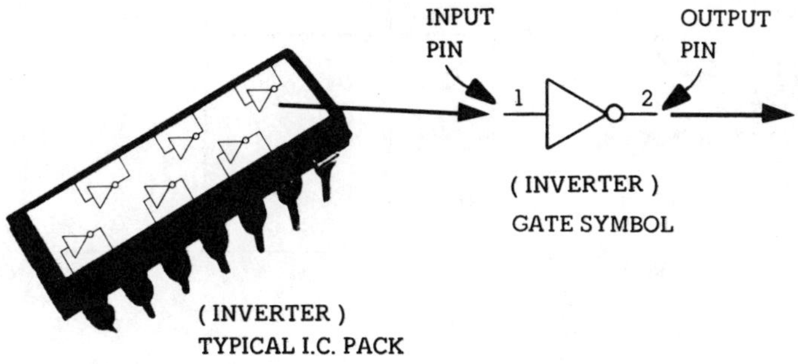

INVERTER TRUTH TABLE

STATE	INPUT	OUTPUT
	1	2
1	0	1
2	1	0

Figure 2-26. Inverter Gate

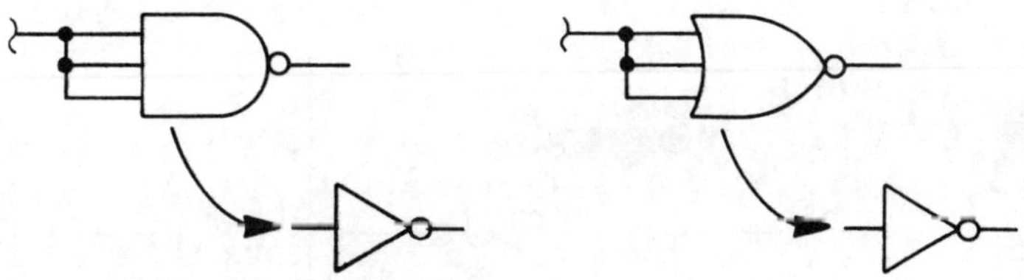

Figure 2-27. NAND And OR Gates Connected As Inverters

The NAND gate shown in Figure 2-28 is actually an AND gate with a "built-in" inverter on the output lead as shown just below NAND gate symbol above. The device output pin, as shown in States 1-3 of the NAND truth table, is a Logic 1 when any or all inputs are a Logic 0. The NAND gate output goes to a Logic 0 state when and only when all inputs are a Logic 1, State 4 on the truth table.

An analogy to the NAND gate, like the inverter, requires reverse thinking. An ON lamp is normally associated with pushing a button in or flipping a switch UP. In Figure 2-29 below, the lamp is always ON, powered directly from a power source. Switches 1 and 2 control the power source. As long as Switches 1 and 2 or both are OFF, the power source is not grounded out, and it continues to power the lamp. If, however, both switches go ON, the power source is grounded out (the power source is turned OFF) and the lamp goes OFF.

Therefore, the NAND gate is a serial element just like the AND gate, except that the output is inverted.

The NOR gate symbol of Figure 2-30 is similar to the OR gate with a small circle at the output.

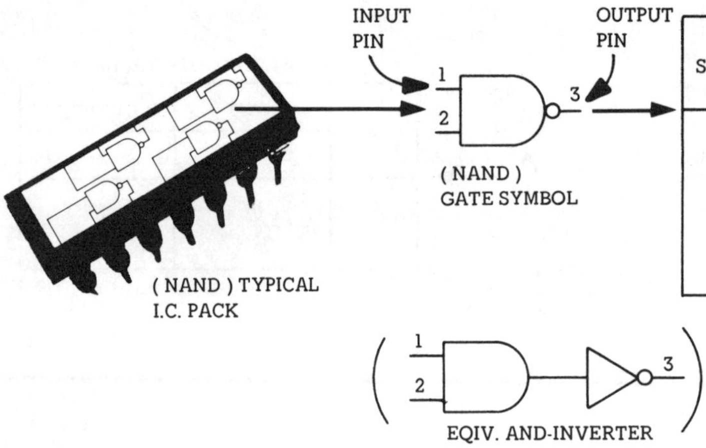

NAND | TRUTH TABLE

STATE	INPUT		OUTPUT
	1	2	3
1	0	0	1
2	0	1	1
3	1	0	1
4	1	1	0

Figure 2-28. NAND Gate

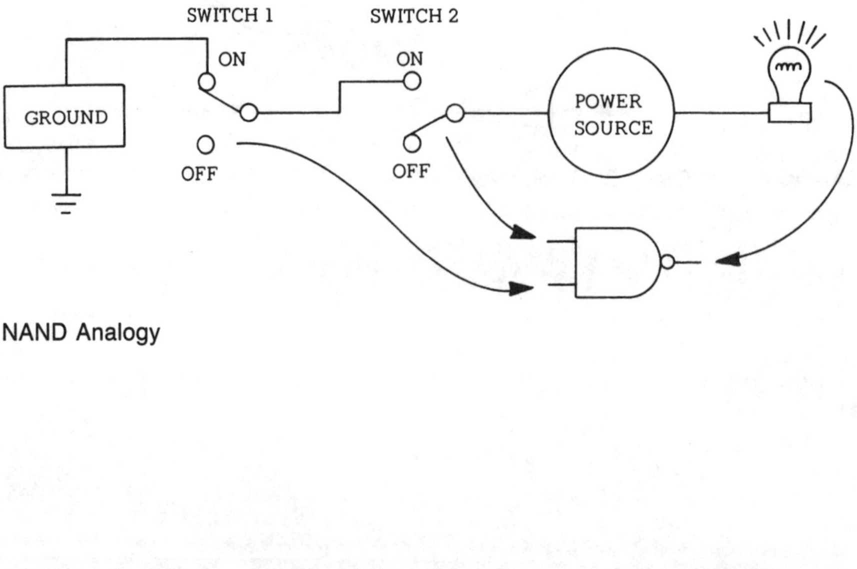

Figure 2-29. NAND Analogy

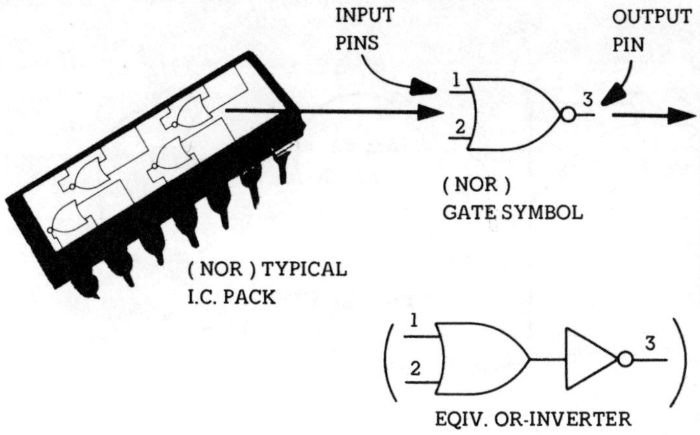

NOR TRUTH TABLE

STATE	INPUT		OUTPUT
	1	2	3
1	0	0	1
2	0	1	0
3	1	0	0
4	1	1	0

Figure 2-30. NOR Gate

The NOR gate shown in Figure 2-30 above, is actually an OR gate with a "built-in" inverter on the output lead, as shown just below NOR gate symbol above. The device output pin, as shown in State 1 of the NOR truth table, is logically a Logic 1 only when all inputs are in Logic 0 state. The NOR gate output pin is a Logic 0 when any or all inputs are in a Logic 1 state, as shown in States 2-4.

An analogy to the NOR gate, like the inverter, and NAND gate requires reverse thinking. As with the NAND gate analogy, an ON lamp is associated with pushing a button in or flipping a switch up. In Figure 2-31 the lamp is always ON, powered directly from a power source. The power source is controlled by Switches 1 and 2. If a ground is supplied to the power source, the lamp will go OFF. As can be seen, both Switch 1 and 2 are connected directly to a ground. If either switch individually or both are turned ON, the power source is grounded out (turned-off) and the lamp goes off. Therefore, the NOR gate is a parallel element just like the OR gate except that the output is inverted.

The exclusive OR gate shown in Figure 2-32 is similar to the OR gate. The exclusive OR gate output, however, is a Logic 1 when and only when one of the inputs is in a Logic 1 state, as shown in States 2 and 3 in the truth table below. When all inputs are either a Logic 0 or a Logic 1, the exclusive OR gate output is a Logic 0 as shown in States 1 and 4 above.

The exclusive OR gate is therefore a type of lock-out device. It locks out all Logic 0 states at input as well as more than one Logic 1 state.

The logic gates shown in Figure 2-33 are of the open- collecter type. These gates are used in timing circuits, as drivers and for wired AND functions. If a logic gate is connected to the output of the open-collector gate, it then functions just like a normal NAND gate or inverter. The manufacturing part number is the only means of identifying an open-collector gate from a non-open collector gate.

Complex Logic Symbol Definitions

As previously mentioned, the operation of various flip-flop types is important. Devices such as shift registers, counters, and single shot generators will not be discussed due to their vast variety. The flip-flop can perform the function of a memory device, a divider, and a counter all in one. It is adaptable to many designs requiring complex logic devices. Like its counterparts, the basic logic gates, the flip-flop is contained within an IC package. In most cases, at least two flip-flops are available in each package.

The R-S flip-flop is shown symbolically in Figure 2-34.

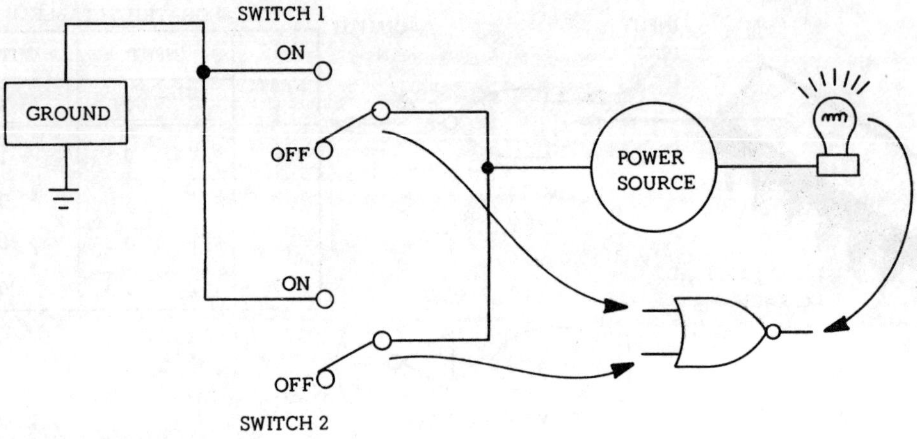

Figure 2-31. NOR Analogy

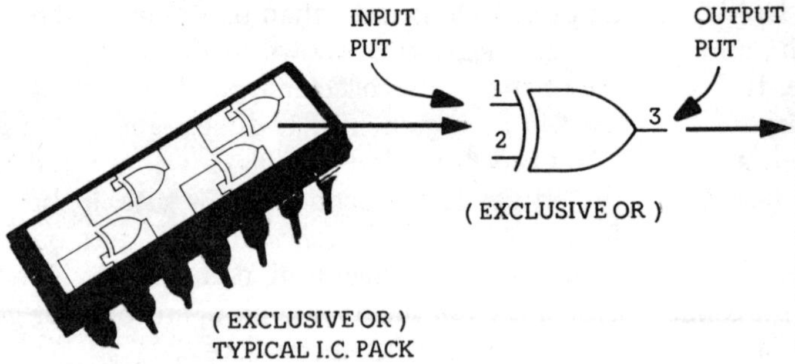

EXCLUSIVE OR TRUTH TABLE

STATE	INPUT		OUTPUT
	1	2	3
1	0	0	0
2	0	1	1
3	1	0	1
4	1	1	0

Figure 2-32. Exclusive OR Gate

Figure 2-33. Open Collector

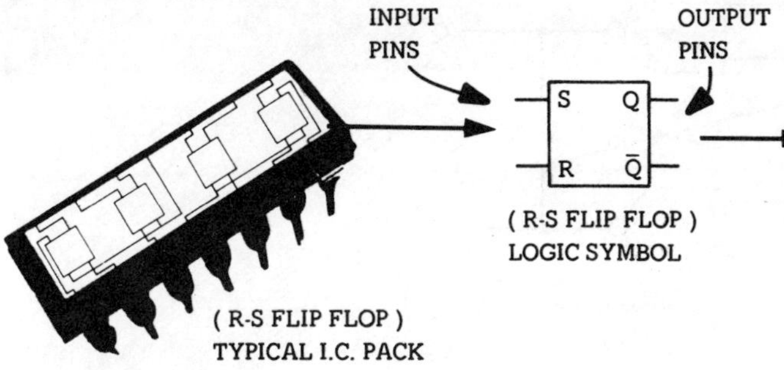

R-S FLIP FLOP TRUTH TABLE

STATE	INPUT		OUTPUT	
	R	S	Q	$\overline{Q}$
1	0	0	N/A	N/A
2	0	1	0	1
3	1	0	1	0
4	1	1		

NO CHANGE IN STATE

Figure 2-34. R-S Flip-Flop Logic Symbol

The R-S (reset set) flip-flop is the most basic of flip-flops and is generally referred to as a latch. Actually, most latches are a combination of two cross-connected NAND gates, as shown in Figure 2-35.

As shown in the R-S flip-flop truth table, when a Logic 1 is applied to the S input lead (State 2) the $\overline{Q}$ output lead goes to a Logic 1 and Q goes to a Logic 0. When the R input lead goes to a Logic 1 (State 3) the flip-flop changes state; $\overline{Q}$ goes to a Logic 0 and Q goes to a Logic 1. When both R and S inputs are in a Logic 2 state, there is no change in the flip-flop's output state. A Logic 0 state on both input leads is not allowed. Q and $\overline{Q}$ are never in the same state.

Another form of flip-flop is the "D" type shown in Figure 2-36.

The "T" (toggle) or "clocked" flip-flop changes state as a result of an input pulse called a trigger, clock pulse or toggle at the T lead. As shown in the truth table, state 1, after the first toggle occurs, the Q and $\overline{Q}$ states change. When the second toggle occurs, the flip-flop reverts back to the original state with Q a Logic 0 and $\overline{Q}$ a Logic 1. If a trigger is placed on the SD lead, the flip-flop goes to State 1. If a trigger is placed on the RD lead, the flip-flop goes to State 2. As long as trigger pulses occur at the "T" input, the flip-flop will change state back and forth. Q and $\overline{Q}$ states are never the same.

The J-K flip-flop shown in Figure 2-37 is an important element in most digital electronic designs. This flip-flop offers complete flexibility in that all possible controls are "built-in", but not all are required for the gate to function. For example, the SD and RD (set direct and reset direct) leads need never be used.

In addition, by trying the J and K input leads to a Logic 1 state, the flip-flop acts just like a "I" flip-flop (already discussed).

As shown in the truth table, the J-K flip-flop Q output goes to a Logic 1 when the J input is a Logic 1, as in State 3 of the truth table. A trigger occurs at T. The flip-flop $\overline{Q}$ output goes to a Logic 1, as in State 2 of the truth table, when the K input is a Logic 1 and a trigger occurs at T. Therefore, a Logic 1 on one of the input leads (while a trigger pulse is occuring) causes the J-K flip-flop output to change state. The J-K flip-flop may have direct set (SD) and direct reset (RD) inputs. In which case, a transition from a Logic 1 to a Logic 0 on either lead will cause the flip-flop to attain a certain state. When SD is triggered, the flip-flop reverts to a set state with Q at Logic 1, as in State 3 as the truth table. When RD is triggered, the flip-flop reverts to a reset state with Q a Logic 1, as in State 2 of the table. Q and $\overline{Q}$ are never in the same state.

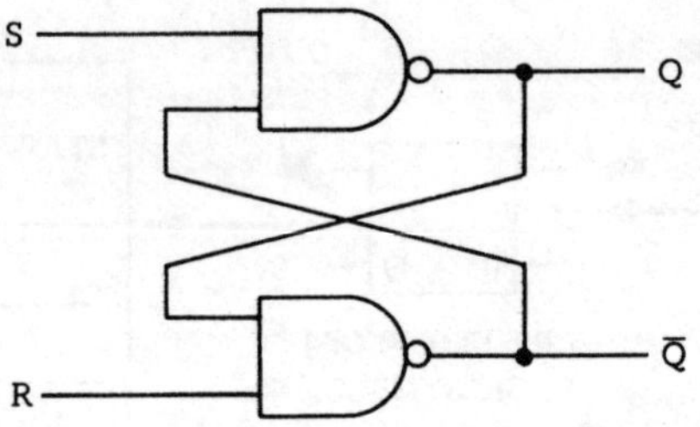

Figure 2-35. Schematic R-S Flip Flop

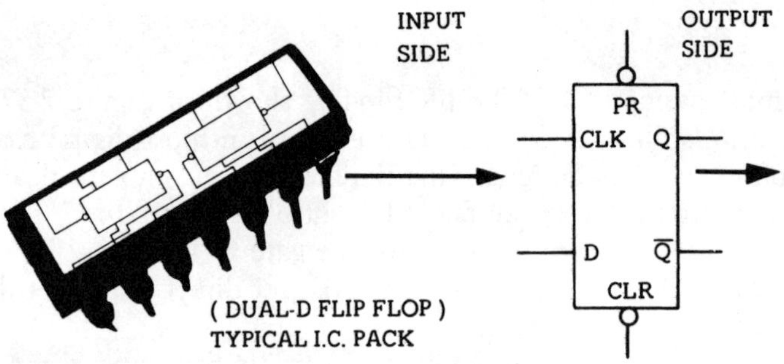

DUAL D - FLIP FLOP | TRUTH TABLE

STATE	INPUT	OUTPUT	
	D	Q	$\overline{Q}$
1	1	1	0
2	0	0	1

Figure 2-36. Dual-D Flip Flop

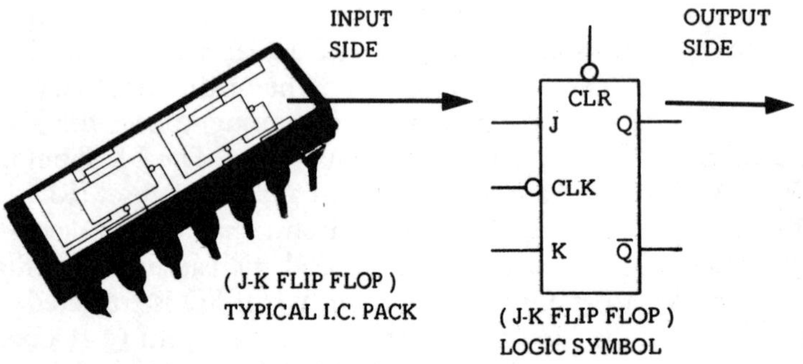

J-K FLIP FLOP | TRUTH TABLE

STATE	INPUT		OUTPUT
	J	K	Q
1	0	0	QN
2	0	1	0
3	1	0	1
4	1	1	$\overline{Q}$N

Figure 2-37. J-K Flip Flop

Combining

Combining describes methods for properly using a logic gate in applications other than what it was designed for. (Substitution might be a more appropriate word). It should be realized, however, that one gate cannot be arbitrarily substituted for another. The rules for combining logic gates (as well as some typical design methods used in practice today) will be described.

Without first considering the impact of combining or substituting might have on the fit, form, and function of your project (be it the repair of a drain pipe, bicycle, automobile, or a printed circuit board layout), you may end up with a real problem. On the other hand, good combining or substitution habits will undoubtedly result in the most economical and compact project possible.

A classic example of effects produced in substitution could be using a spark plug not rated for a car's engine. The negative effects in this case might be poor gas mileage and poor engine ignition. Of course the mere fact that an automobile is now running might be a more positive effect. Similar to this example the side effects of substituting logic gates will become obvious during the board test phase. If some very basic rules are followed however, the effects will be positive.

Before attempting logic gate combining, check with the design engineer and refer to the manufacturer's specifications. This practice will ensure that gate combining wasn't the cause for the circuit's unsuccessful operation.

Be aware of the following characteristics when combining and substituting logic gates:

* Open-collector gates vs. non-open collector gates (Totem pole logic)

* Input pins and output pin numbers are compatible

* Combining logic gates should be done under the Engineer's direction

To be more specific, combining is a process involving the numbers of logic gate input leads. Before going on, consider this example: If a structure must have four windows minimum to meet your needs having a fifth window would most likely be okay. If it weren't, you could "board-up" the extra window much easier than rebuilding the complete structure. Similarly, if a digital circuit requires a 2-input NAND gate and a 4-input NAND gate is available, the spare input leads can be ignored so the gate functions as a 2-input device (much better than adding a complete IC package). The whole point is to conserve space and save money by using what's available. (See Figure 2-38 through 2-41).

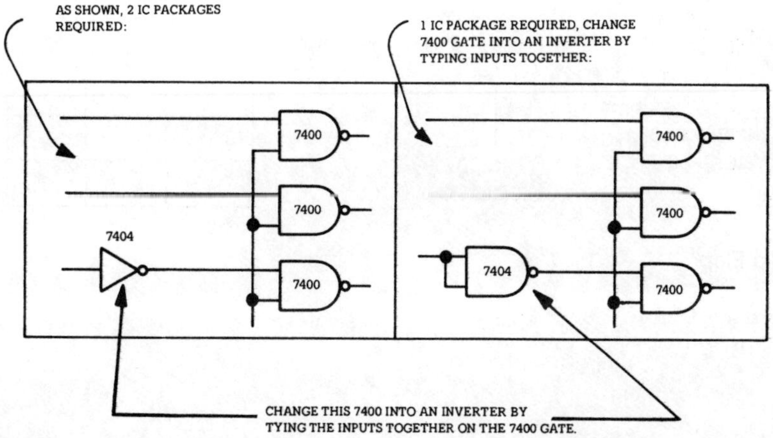

Figure 2-38. Combining

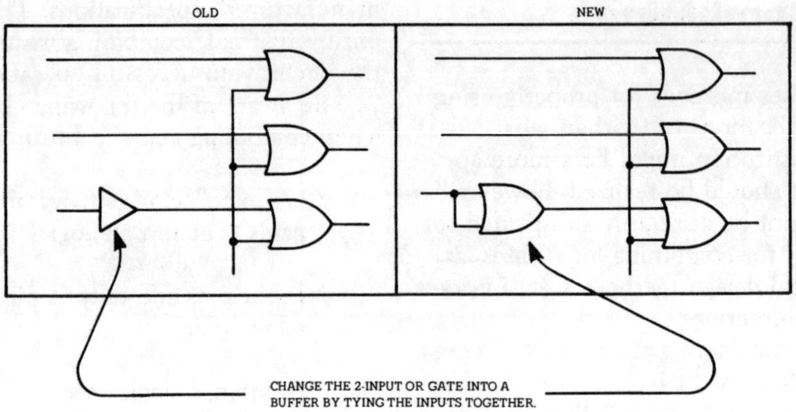

Figure 2-39. Combining

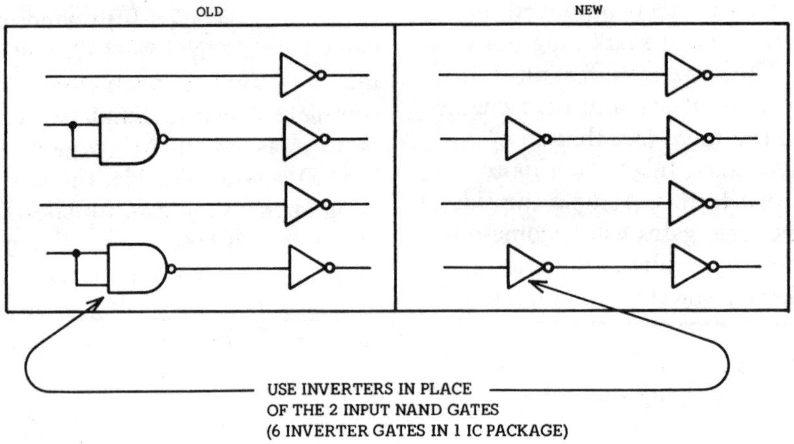

Figure 2-40. Combining

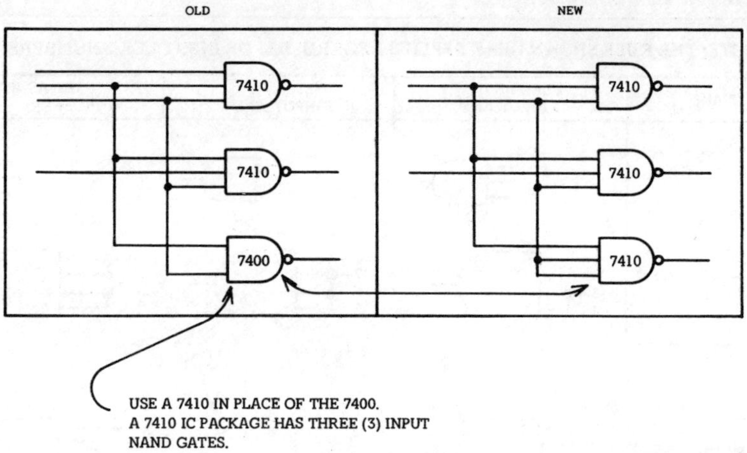

Figure 2-41. Combining

Some additional rules to remember follow:

* AND - When combining AND gates, the unused (spare) input lead(s) MUST BE connected HIGH* or to a functioning input lead (See Figure 2-42)

* NAND - When combining NAND gates, the unused (spare) input lead(s) MUST BE connected HIGH* or to a functioning input lead (See Figure 2-43)

* OR - When combining OR gates, the unused (spare) input lead(s) MUST BE connected LOW* or to a functioning input lead (See Figure 2-44)

* Exclusive OR - When combining exclusive OR gates, the unused input lead(s) MUST BE connected. As shown in Figure 2-45.

* NOR - When combining NOR gates, the unused input lead(s) MUST BE connected LOW* or to a functioning input lead (See Figure 2-46)

* Exclusive NOR - When combining exclusive NOR gates, the unused input lead(s) MUST BE connected. (See Figure 2-47)

Logic Notations

HIGH = VCC POSITIVE SUPPLY

HIGH = GND NEG. SUPPLY

LOW = GND POS. SUPPLY

LOW = VCC NEG. SUPPLY

In introducing the basic gates, truth tables, and logic equations, several different notations for the two binary states were used. For the gates, we used Logic 1 and Logic 0 and for the light bulb circuit we used Open and Closed. It is also common practice to use the terms Logic 1 and Logic 0 interchangeably with High and Low, or True and False, respectively.

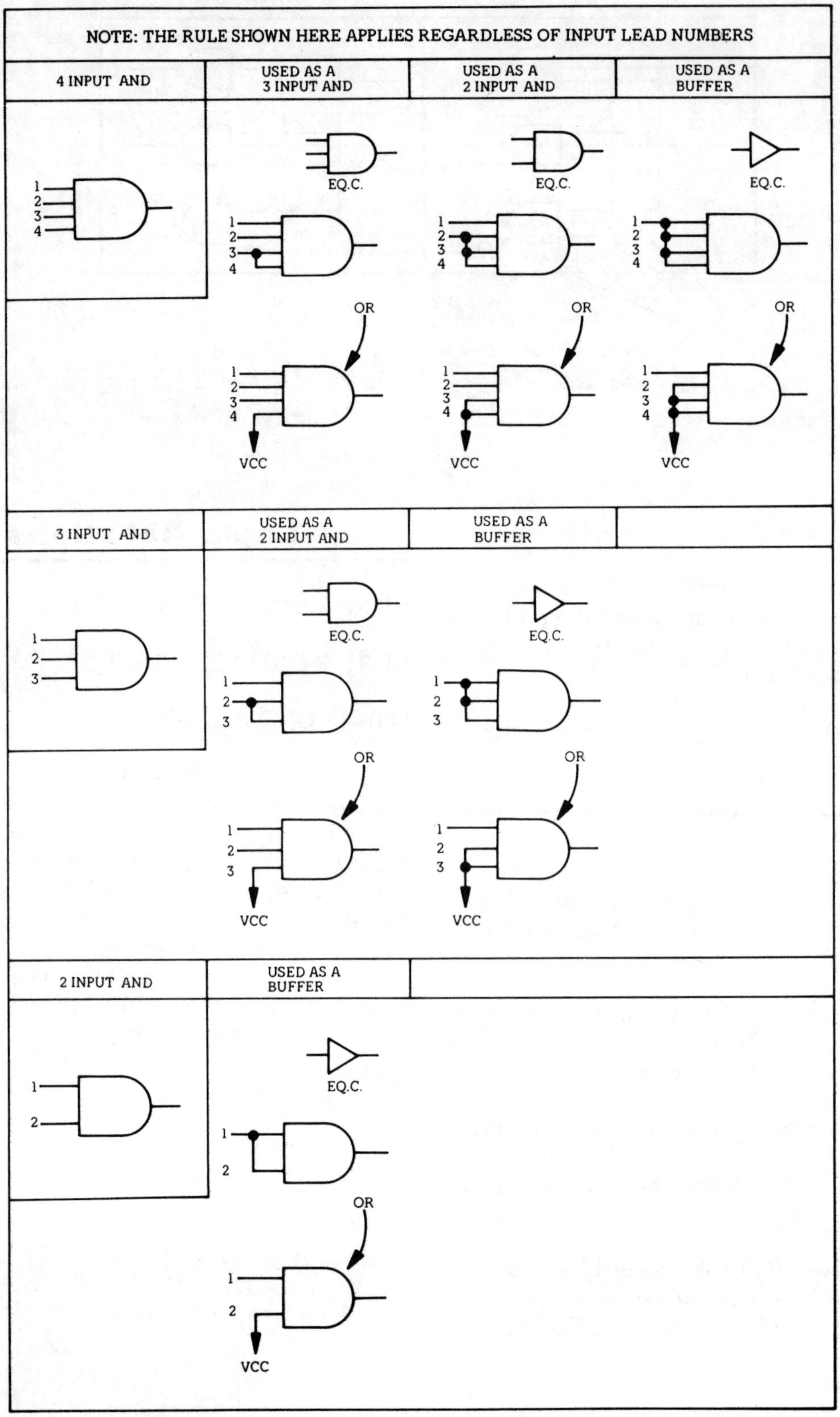

Figure 2-42. AND Gate Combining

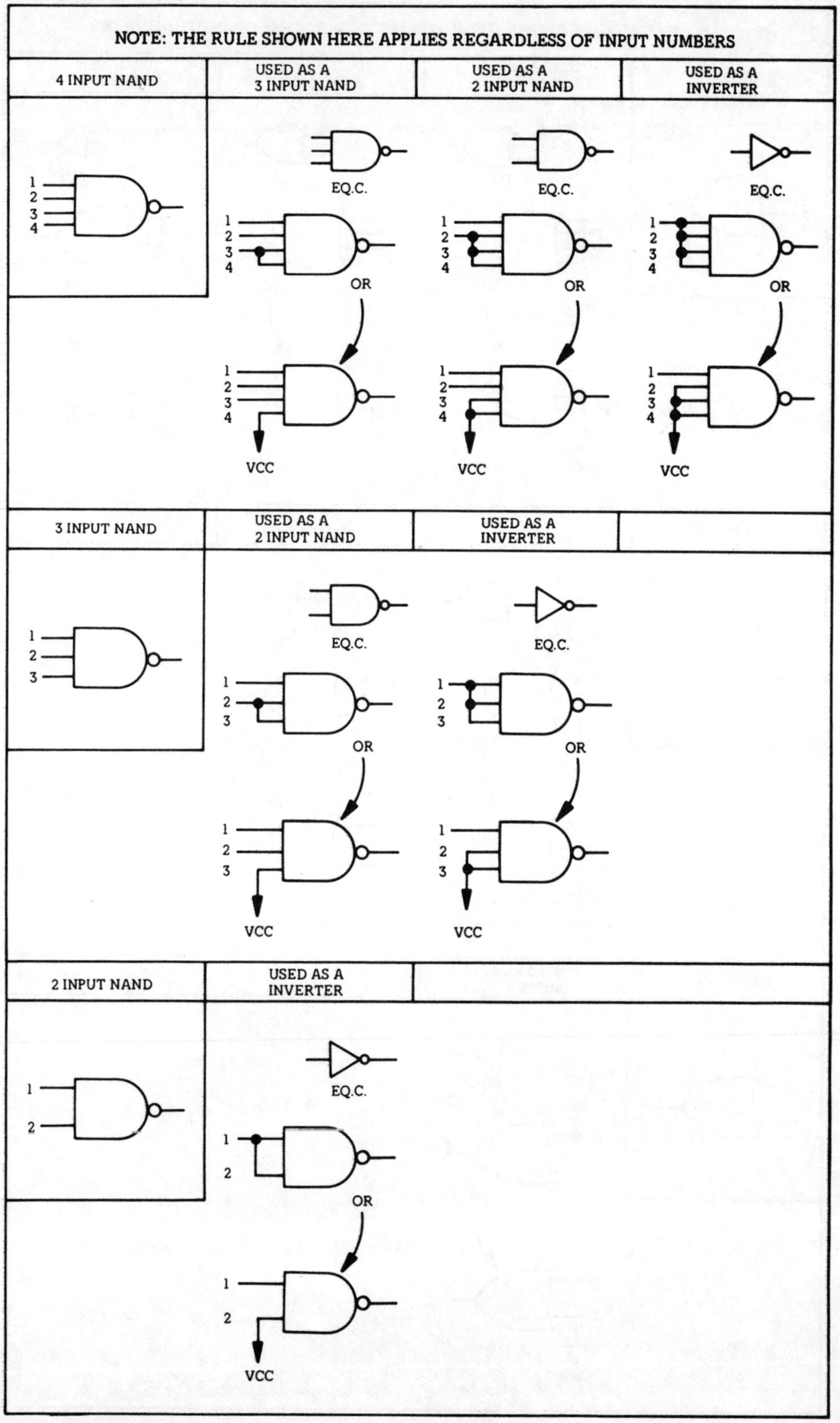

Figure 2-43. NAND Gate Combining

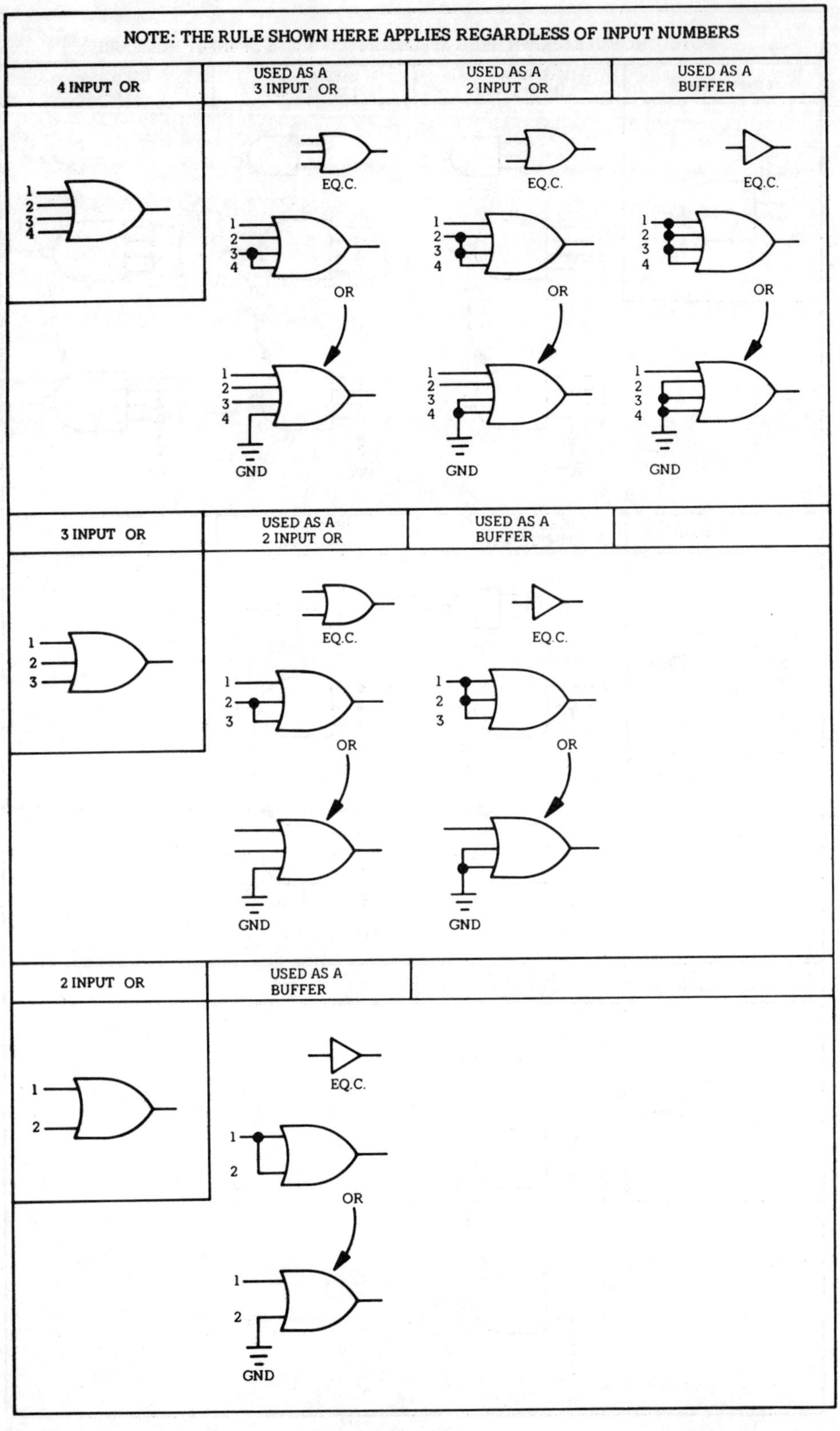

Figure 2-44. OR Gate Combining

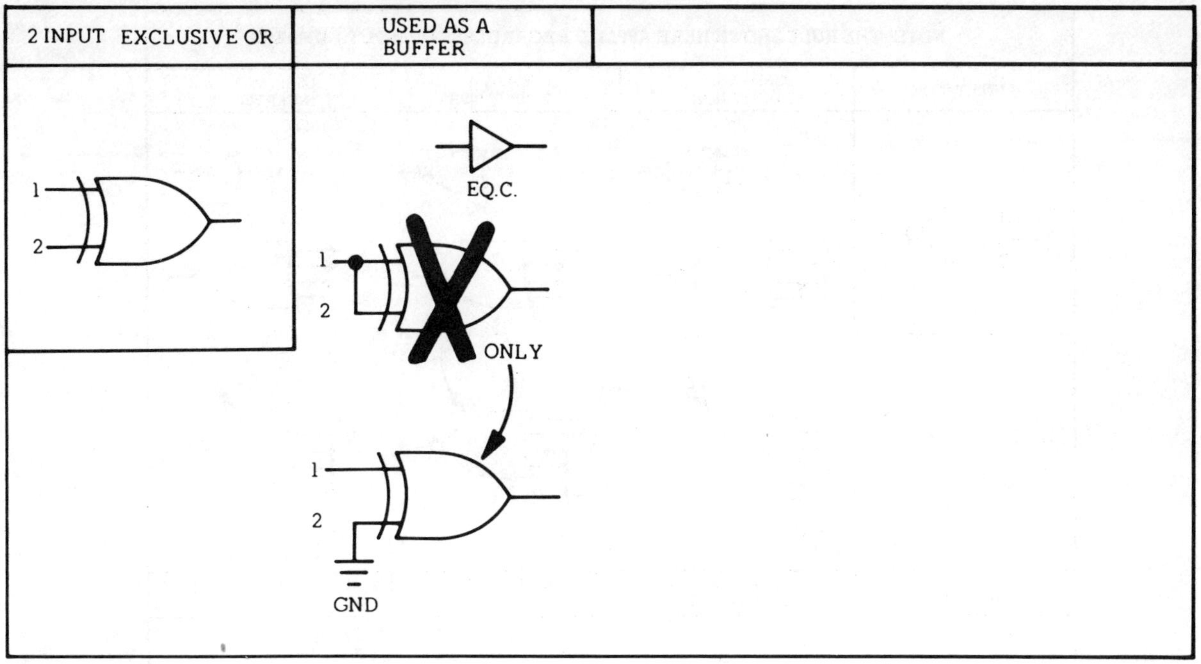

Figure 2-45. Exclusive OR (Buffer) Gate Combining

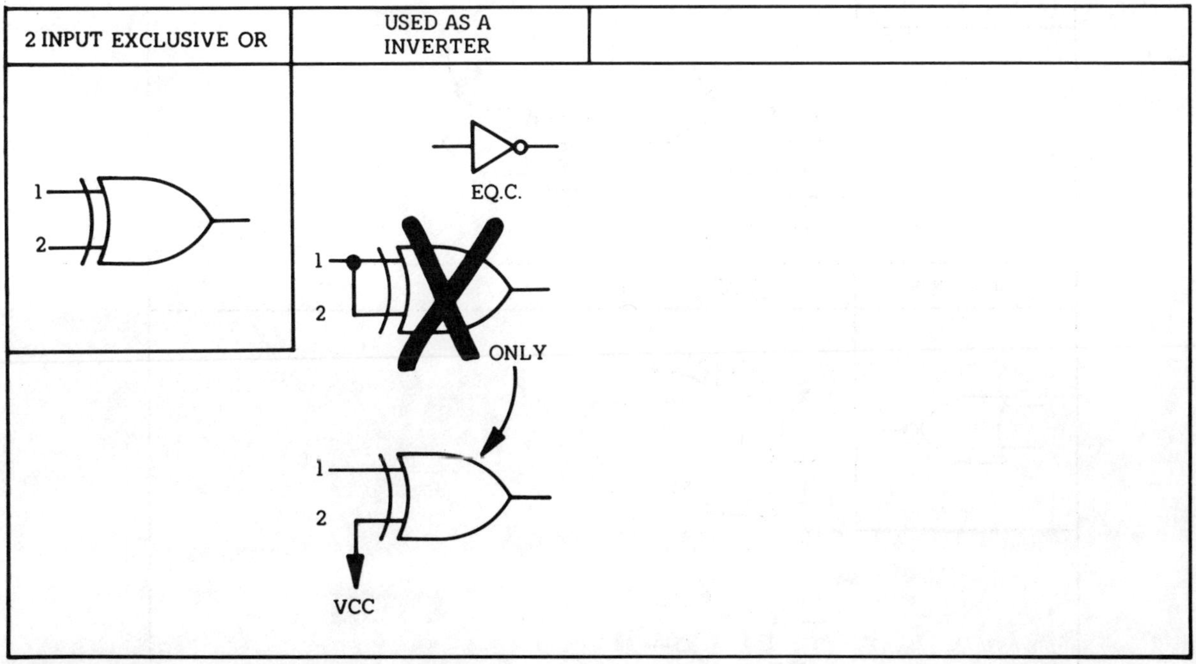

Figure 2-45A. Exclusive OR (Inverter) Gate Combining

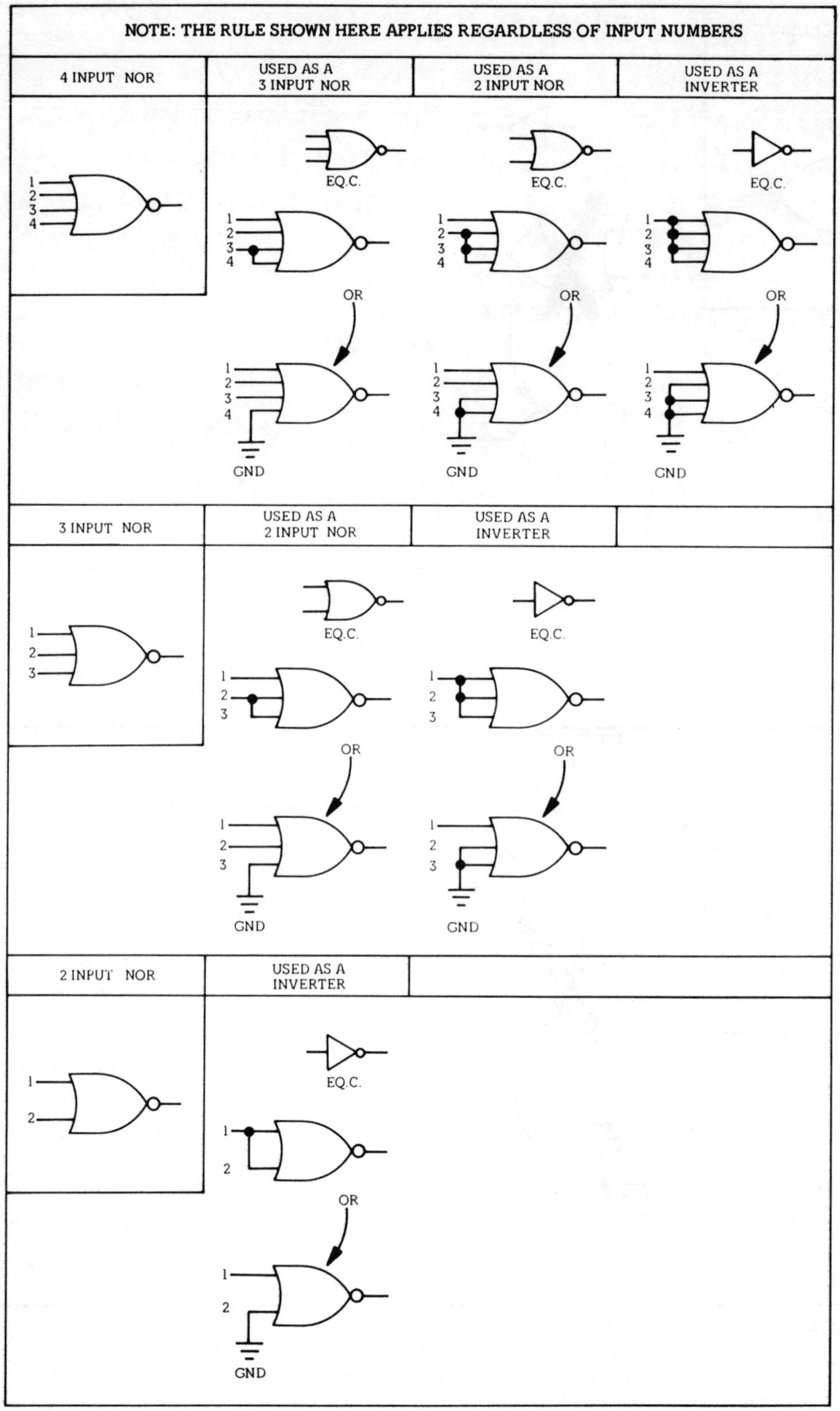

Figure 2-46. NOR Gate Combining

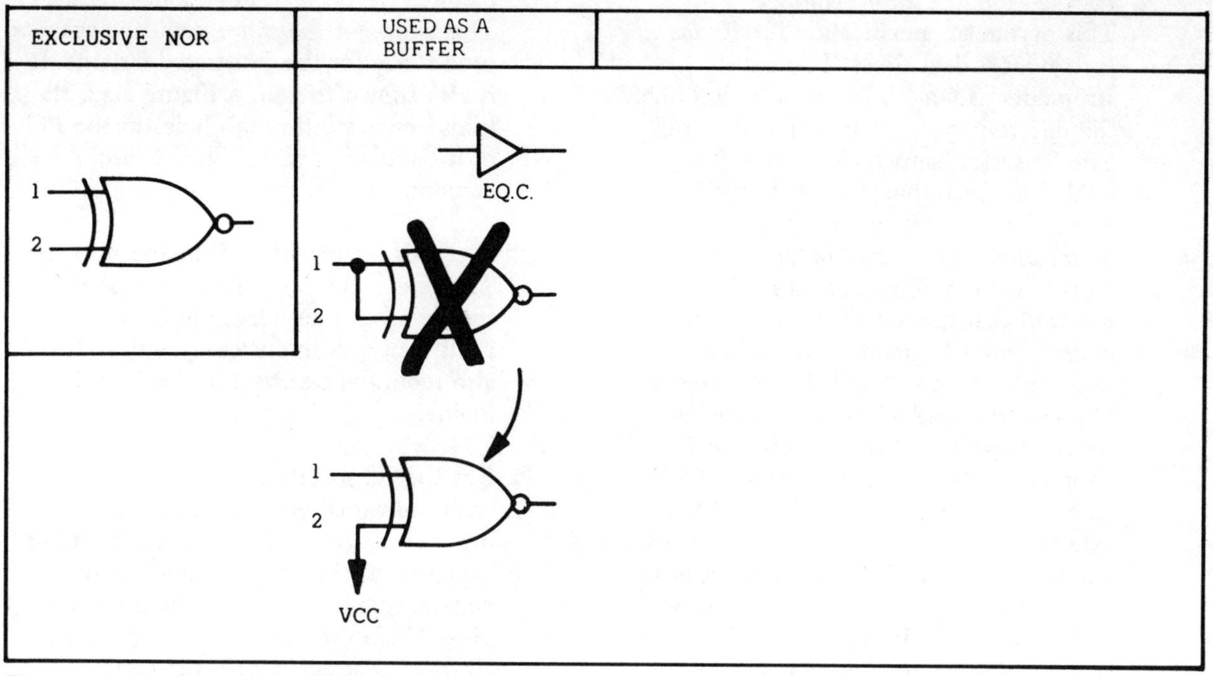

Figure 2-47. Exclusive NOR Gate Combining

Logic Combining and Grouping

The total number of IC devices required in the design are identified. After that, the gates are combined to use as many spares as possible. Then, space is allocated on the drawing representing the PC board for each integrated circuit package. It is important that IC grouping and combining be accomplished before the PC board layout has begun. You will learn what specific problems can be expected by doing otherwise when we work the example in Figure 2-49.

Note: Always verify combining of gates with the design engineer. As the PC board designer, you may recommend, but the engineer should make the final decision. We suggest that you combine and group gates together, then obtain the engineers final approval

Identify the logic gates by refering to the manufacturing specification sheets. A typical specification sheet for a 2-input NAND is shown in Figure 2-48. It shows the W package, and the J or N dual in-line package. We will be using the dual in-line package for our layout.

Most of the data that is called out on this specification sheet is not used by the PC designer. The most important information concerning the PC designer is in the block in the upper right hand corner described by the letter F. To acquaint you with the information that is provided on the typical specification sheet for logic gates, we have itemized each area: A through J provides a description of each segment of the specification sheet:

A. Package Identification Number
This particular specification is a Texas Instruments data sheet. Texas Instruments uses a 54/74 series to describe the different types of IC's that they sell. The 54 series happens to be a higher quality package than the 74. It must meet higher specifications. Typically the 54 series is sold to the military and the 74 is sold to the commercial field. You will find that this number varies from manufacturer to manufacturer. The last two digits on the 54 and the 74 series, in this case the double 00, is the identifier for the logic gate itself. In this case it happens to be a 2-input positive NAND gate. To find a 2-input positive NAND gate under another manufacturer (for example, Signetics), the package identification number may be an 8471. If you wish to buy Motorola, it might be an MC-3000 number. Intersil may place it in the 5500 series. Likewise, with National Semiconductor it would be another number. Each major manufacturer would have its own way of designating its IC's.

B. This is the identifier for the IC package itself describing what is inside the IC package. In this case, its called out as a quadruple, meaning that four, 2-input positive NAND gates are inside the IC package.

C. W = ceramic flat. This is a flat package previously shown in Figure 2-12. This particular package is not as popular today as the dual in-line package. The mounting configurations for the flat pack are quite a bit different. The leads are actually soldered or welded to the top of the component side of the PC board rather than the leads penetrating through the holes and being soldered to the non-component side as in a dual in-line.

D. J stands for ceramic dual in-line package. This is one of the more popular packages used today for the integrated circuits. It is also shown to you in Figure 2-12. Its leads penetrate through holes in the PC board and are soldered to the non-component side.

E. N stands for plastic dual in-line package. This is exactly the same case style as the J we have discussed. But the body itself is of a different composition. It is also mounted exactly like the J dual in-line.

F. F is the IC specification that identifies each individual gate, the input pin number assignment, output pin number assignment, the identification of the voltage supply VCC, and the ground pins. This information in F is primarily all the PC designer needs from the specification sheet. This is the information that the designer must have in order to complete the interconnections and assign the proper pin number on the schematic to be completed or routed onto the PC board.

G. This is a schematic of each individual gate represented in F. It shows the two input pins, A and B, and the output pin Y. Each individual gate will have a VCC* supply, internally connected with a ground return and to its respective voltage and ground pins.

H. Recommended Operating Conditions. The minimum and maximum voltage rating is given for the series 5400 and 7400 along with the fan-out which will be explained later. Also, the operating temperatures range is listed under H.

A

B

CIRCUIT TYPES SN5400, SN7400
QUADRUPLE 2-INPUT POSITIVE NAND GATES

schematic (each gate)

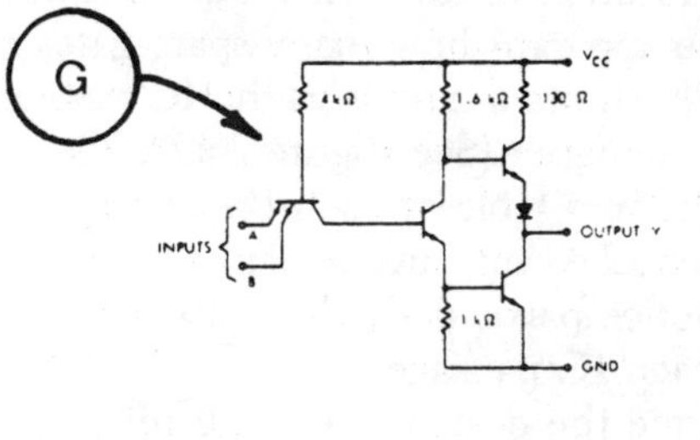

NOTE: Component values shown are nominal.

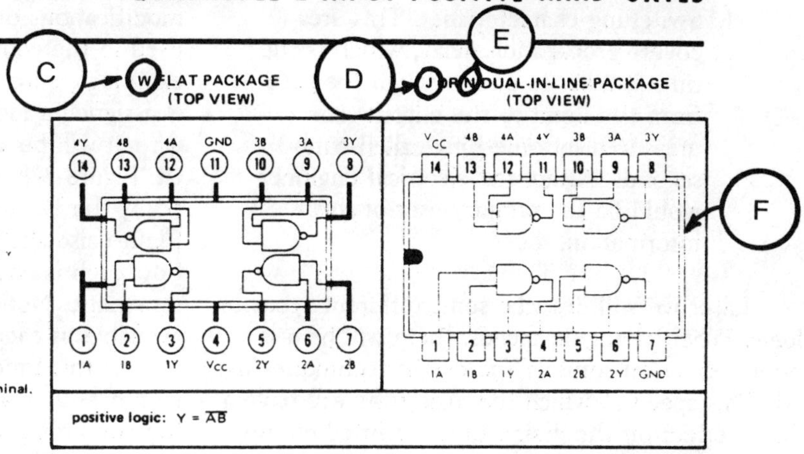

H

recommended operating conditions

		MIN	NOM	MAX	UNIT
Supply Voltage V_{CC}:	SN5400 Circuits	4.5	5	5.5	V
	SN7400 Circuits	4.75	5	5.25	V
Normalized Fan-Out From Each Output, N				10	
Operating Free-Air Temperature Range, T_A:	SN5400 Circuits	−55	25	125	°C
	SN7400 Circuits	0	25	70	°C

I

electrical characteristics over recommended operating free-air temperature (unless otherwise noted)

PARAMETER		TEST FIGURE	TEST CONDITIONS†			MIN	TYP‡	MAX	UNIT
$V_{in(1)}$	Logical 1 input voltage required at both input terminals to ensure logical 0 level at output	1				2			V
$V_{in(0)}$	Logical 0 input voltage required at either input terminal to ensure logical 1 level at output	2						0.8	V
$V_{out(1)}$	Logical 1 output voltage	2	V_{CC} = MIN, I_{load} = −400 μA	V_{in} = 0.8 V,		2.4	3.3		V
$V_{out(0)}$	Logical 0 output voltage	1	V_{CC} = MIN, I_{sink} = 16 mA	V_{in} = 2 V,			0.22	0.4	V
$I_{in(0)}$	Logical 0 level input current (each input)	3	V_{CC} = MAX,	V_{in} = 0.4 V				−1.6	mA
$I_{in(1)}$	Logical 1 level input current (each input)	4	V_{CC} = MAX,	V_{in} = 2.4 V				40	μA
			V_{CC} = MAX,	V_{in} = 5.5 V				1	mA
I_{OS}	Short-circuit output current§	5	V_{CC} = MAX		SN5400	−20		−55	mA
					SN7400	−18		−55	
$I_{CC(0)}$	Logical 0 level supply current	6	V_{CC} = MAX,	V_{in} = 5 V			12	22	mA
$I_{CC(1)}$	Logical 1 level supply current	6	V_{CC} = MAX,	V_{in} = 0			4	8	mA

J

switching characteristics, V_{CC} = 5 V, T_A = 25°C, N = 10

PARAMETER		TEST FIGURE	TEST CONDITIONS		MIN	TYP	MAX	UNIT
t_{pd0}	Propagation delay time to logical 0 level	65	C_L = 15 pF,	R_L = 400 Ω		7	15	ns
t_{pd1}	Propagation delay time to logical 1 level	65	C_L = 15 pF,	R_L = 400 Ω		11	22	ns

† For conditions shown as MIN or MAX, use the appropriate value specified under recommended operating conditions for the applicable device type.

‡ All typical values are at V_{CC} = 5 V, T_A = 25°C.

§ Not more than one output should be shorted at a time.

Figure 2-48. Typical Specification Sheet

I. Electrical characteristics over recommended operating free-air temperatures. These specifications are primarily used by the electrical engineer.

J. Switching characteristics. This area covers propagation delay, which is the time it takes to pass through the gate from the input to the output side measured in something called nanoseconds. Again, the electrical engineer would be the primary user of this information.

Later we will describe some different types of logic. When changing speeds, there will be a direct effect on some of the design techniques used. This speed at which the logic runs will have a direct effect on the design of the printed circuit board.

Integrated Circuit Combining and Grouping

The engineer's schematic should have all of the logic gates identified with their respective manufacturer's specification number such as 7400 (or perhaps just the 00 without the prefix number 74 or 54). The engineer should provide the specification sheets to accompany his schematic. However, if he does not, then it is the PC designer's responsibility to obtain the proper catalogues from the engineer and to look up each type of logic required for the engineer's schematic. Reference to these specification sheets identifies the total number of logic gates, the respective gate pin numbers, inputs and outputs, the VCC pin, and ground pin. These are all required to complete your PC board layout.

A small logic diagram Figure 2-49 will be used to introduce the grouping and combining techniques required on digital PC boards. This shows the use of logic specification sheets, and the assignment of IC identification numbers, and modifications of the gates. This information is used in the combining techniques. A table of each logic gate by IC package type is written as it is assigned a location on the schematic. In this way, it will be apparent how many spare gates exist. Figure 2-49 shows a circuit in the IC's assigned by the PC designer (See Figure 2-49.). The Figure also shows a table of each IC package and how each is checked off once assigned to the schematic. Notice particularly the spare gates available in each IC package.

At this time the designer is trying to eliminate as many IC gates as possible by using the combining technique shown in Figures 2-42 through 2-47. Let's itemize the procedures that would be required to group the logic gates together into IC packages. Consider the possible combining techniques that could be applied to Figure 2-49. Using the specification sheets listed below the schematic, you see that there are a total of seven IC's called out on the schematic. The one specification sheet that has been excluded is the 141 IC-2, mainly because for this exercise, you can not modify or combine this type of IC.

IC 5 is a 2-input NOR gate. There are two available 2-input NOR gates. Check the combining sheets (Figure 2-46). Notice that the 2-input NOR gates can be made to act like an in-

---

* Symbol for collector-voltage supply of a bi-polar transistor

---

verter gate either by tying the 2 inputs of the unused inputs low. We do have inverter gates available on the schematic to combine into IC 5. However, is there a need or an advantage to combining? Observing and studying our schematic, we can see that we do have an inverter package on the top of the schematic communicating directly to IC 5 with its output and the input coming from IC 2. This particular inverter gate could be combined into IC 5. This would shorten the lead or trace length on the PC board between the output of the inverter to the input of the 2-input NOR called IC 5. This is an option. Before we would make this a permanent change, we would always consult with the engineer. However, we do not recommend that you run over and question the engineer on every individual manuever. We would like you to make all your combining changes and then sit down with the engineer at one time and consult with him on the entire schematic.

IC 6 (an inverter package) has six gates. We have used five out of the six. Leave the inverter gates last because it is one of the easier gates to combine into other IC's. This eliminates the inverters on the PC board as an individual IC within itself. Do we have five available spares that can electrically act as inverters to virtually eliminate IC 6? Look at our spare count. Notice that IC 1 has no spare because IC 1 is a 2-input NAND gate, it could act as an inverter; but it has no spares. IC 5 has two available spare that could also act as inverters because it is a 2-input NOR.

IC 7 is a 3-input NOR called a 27. Examine the specification sheet. There are three in a package. The schematic shows us that there is only one required. Therefore, we have two spares. These 3 input NORS cannot act as inverters, however, IC's 5 and 7 seem to be the only two IC packages that could possibly eliminate IC 6. IC 5 and IC 7 have a total spare gate count of four; but the schematic requires five inverter gates. We know that IC 5 required only two 2-input NOR gates. Can you combine the two 2-input NORS (called IC 5) into IC 7 (called a 3-input NOR package that has two spares)?

Examine the combining sheet, Figure 2-46. Notice that a 3-input NOR called a 27, can be made electrically the same as a 2-input NOR. This is done by either tying the third unused input to one of the other two active inputs or tying the third unused input - low. Therefore, the answer is: yes. We can combine the 2-input NORS (called IC 5) into IC 7. This virtually eliminates IC 5 from the schematic giving us a total IC count of six IC's required to electrically satisfy our schematic on Figure 2-49.

If you do change the NOR package, IC 5, and replace it by combining into the NOR package, IC 7 , make your modifications by marking up the engineer's schematic. Then seek the engineer's approval before proceeding into the actual layout. In this case the design engineer may possibly want 3-input NORS on the board for future expansion. So, you must consult with the engineer before you make these changes on your own.

Also, discuss the tie off techniques for the unused inputs. Note at this time you have not assigned any gate pin numbers either to the input or to the output of the gates from the specification sheet. This will be when the interconnecting is made on the layout.

On the typical schematic, you would proceed with the combining techniques that we have shown you until you are convinced that the minimum of IC packages have been achieved.

The next exercise (called grouping exercise number #1) will give you a more in-depth understanding of grouping and combining logic gates.

Let's review the logic considerations before grouping the schematic.

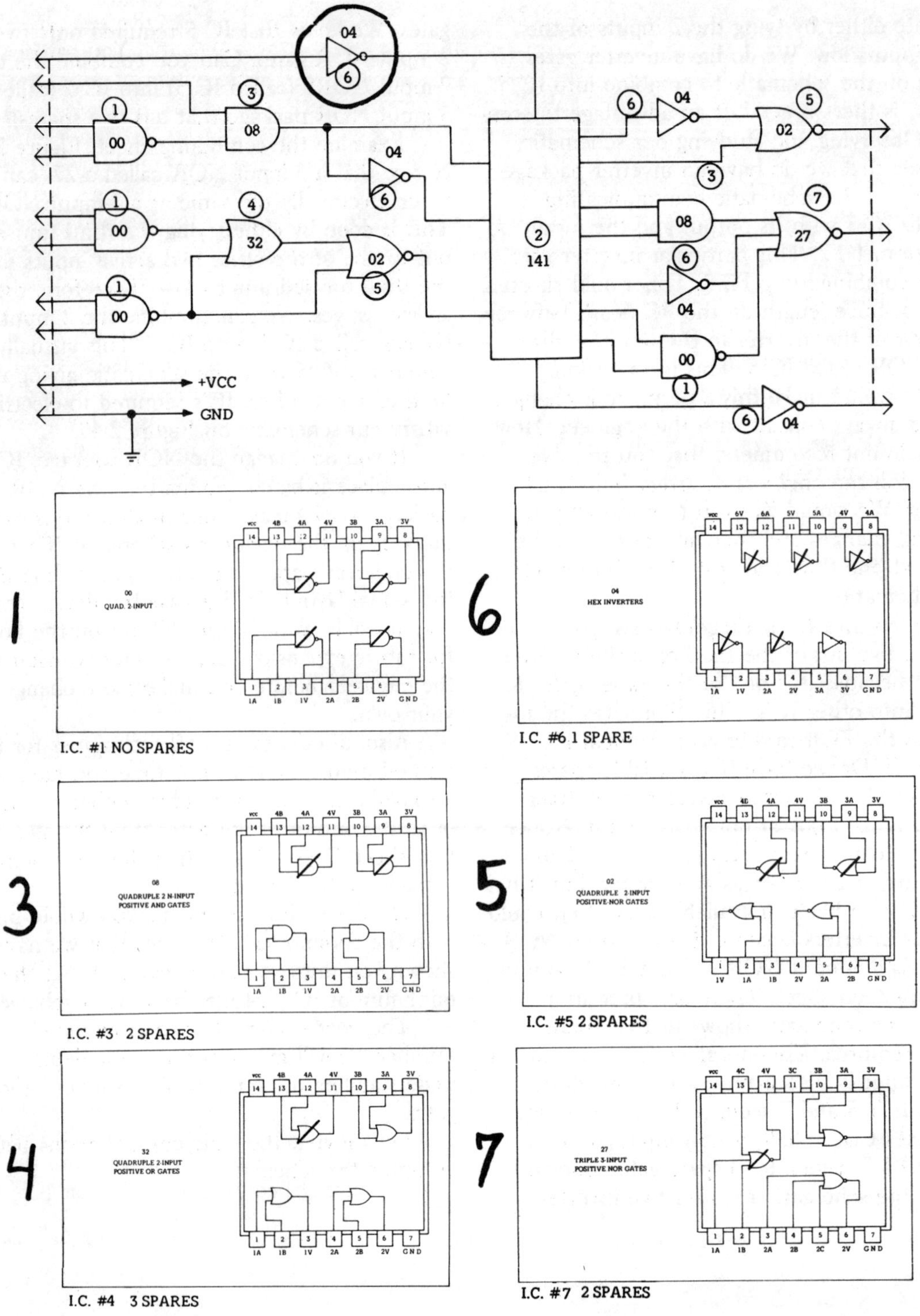

Figure 2-49. Integrated Circuit Grouping

GROUPING AND COMBINING: EXPLANATION FOR Figure 2-49

1. IC 1 was assigned to the 2-input NAND gate (the 00). Look at the specification sheet for IC 1 (the 00). Notice that there are four gates inside one IC package. Examine the schematic. There are a total of four gates required to be 2-input NAND gates. Three of the NAND gates are clustered together in the same area of the schematic. They communicate to the same common point called the connector. This makes it very attractive to put these three gates all inside the one IC package (IC 1). The fourth gate is not common to the connector, but it is the only 2-input NAND gate left on the schematic. Therefore, put that into the same package as the other three NAND gates.

2. IC 2 is a very easy choice because the designer did not have the same choice to use a combining technique as he would have with other individual gates.

3. The specification sheet shows us that IC 3 is identified as a 2-input AND gate. It has four of them in the package. On the schematic notice that 2 AND's are required which would leave us two spares in the IC package. Therefore on the specification sheet, IC 3 would have two gates crossed off with the remaining of two spare gates. These two AND gates can be other gates. Notice the combination techniques shown in Figure 2-42. This gives you the combining options that were described earlier in this chapter. But before we go into the combining, let's continue with pure identification of the existing gates that were given to us by the engineer.

4. IC 4 is a 2-input OR gate. Once again, there are four in a package. Examining the schematic, we can see that there is only one gate required on the entire schematic. This leaves a remainder of three spare 2-input OR's. The combining techniques shown in Figure 2-44 shows your options with the remaining spares for IC 4.

5. IC 5 is a 2-input NOR gate. Once again, look at the schematic. You can see that the total requirement of 2-input NORS is two. On the specification sheet for IC 5, check off two gates, leaving a remainder of two spares. The combining technique shown in Figure 2-46 shows the options available to you for the remaining spares.

6. IC 6 is called the inverter. There are a total of six gates in the package. Looking at our schematic we can see that we have a total count of five gates or inverters. This leaves one inverter as a spare. You have no combining options with this inverter gate. It can only be another inverter in the 04 package.

7. IC 7 is a 3-input NOR package. Specification sheet #27 shows there are 3 gates inside this package. The schematic shows that there is only one required. Therefore, there will be two spares in the IC 7 package. Figure 2-46 shows the combining options for the two remaining spares in IC 7. At this time, eliminate as many IC gates as possible by combining them into the available spares that were left on the specification sheets. This does two things for you: you can possibly reduce your IC count, or you can possibly place the gates (that are communicating together on the schematic) inside the same IC to reduce the trace lengths on the PC layout.

8. IC 1 cannot be combined because we have used all of the gates inside the IC package. IC 2 cannot be combined because it is the type of package that the PC designer cannot change. IC 3 has two available spares. We know from the combining sheet (Figure 2-42) that IC 3 has two available spares that can also act as a buffer gate besides being 2-input AND gates. However observing the schematic, we can see that there are no buffer gates required on the schematic. Therefore, no logical combining can be done with IC 3. IC 4 (the 2-input OR gate) has three spares available. Checking our combining sheet (Figure 2-44), we can see that the 2-input OR gates can also act as buffers

similar to IC 3. However, once again we can see that there are no buffer gates required on the schematic. Therefore, there are no combining options available for IC 4.

Logic Considerations

These are points to consider before grouping your logic schematic:

1. Schematics should be readable. All symbols and reference designations used by the engineer should be understood by the PC designer before proceeding.

2. Specification sheets are used to identify all logic symbols and logic families along with component case sizes for the discrete type of components on your logic diagram.

3. Are spare gates required by the engineer? If so, which types of gates are there and how many?

4. Are there crucial lead lengths for some of the logic gates as far as input or output connections are concerned? This would change from one logic family to another and will be explained later in the book.

5. Following proper combining rules, the following points should be considered. The points will be listed in order of their importance:

   A. Combine gates for reduction of total IC packages.

   B. Combine gates that intercommunicate to each other electrically.

   C. Choose spare gates that can be as adaptable as possible. What choice would you rather have; a 2-input NAND spare, or a 3-input NAND spare? The answer should be a 3-input NAND spare. It can be more things such as a 3-input NAND, a 2-input NAND, or a single input inverter gate.

6. Identify open collector gates versus totem pole/(non-open collector) gates. Open collector and non-open collector gates cannot be combined without the special instructions which will be given to you later in this book.

7. Do not take large numbered input gates and tie the inputs together to create a single input gate such as a buffer or an inverter gate. For example, taking a 4-input NAND gate and tying all four inputs together to create a single input inverter would be a poor design. It is more acceptable to take 2-input NAND gates and tie the inputs together to create a single input. There are times when it could be acceptable to take 3-input NAND gates and tie all three inputs together to create a single input (but rarely the 4-input NAND gate, and absolutely never an 8-input NAND gate).

8. When tying off inputs, check with the engineer about the technique to be used: tying the unused inputs high (voltage), or tying the unused input low (ground) or tying the unused input to another used input of the same gate.

Grouping Exercise 1

Using the logic consideration (points 1 through 8) and the engineer's schematic (Figure 2-50), group and combine to find the lowest number of IC's required on this schematic. At the same time, combine gates for the shortest trace connection by following the steps below:

1. Is the schematic readable? Are all the components and gate symbols identified? Examine the schematic (Figure 2-50). Notice that the gate symbols have not been labeled with an identification number such as a 7400 or a 7404, etc. This means that you must ask the engineer to identify all of the gates on the schematic. It is impossible for the designer to proceed with the grouping or plotting at this point without having the identification of each individual gate. There are two reasons. One is that you have different logic families such as open collector and non-open collector (Another way of saying non-open collector is totem pole. You cannot mix these two logics together). The other reason is that you must have the number to know which manufacturer that the engineer is using. Obtain the specification sheet from the engineer. Before proceeding to step 2, ask the engineer for the identification of all logic gates. See Figure 2-51 for the properly identified schematic.

2. Specification sheets are used to identify logic families and symbols. The specification sheet that we used in Figure 2-48 is not necessary to combine gates. Figure 2-52 gives you the minimum amount of information required to group and combine the schematic on Figure 2-51. What we have done for your convenience is to pull off or extract the minimum amount of information. We compiled it onto one sheet for each different type of logic gate that the engineer has assigned to the schematic.

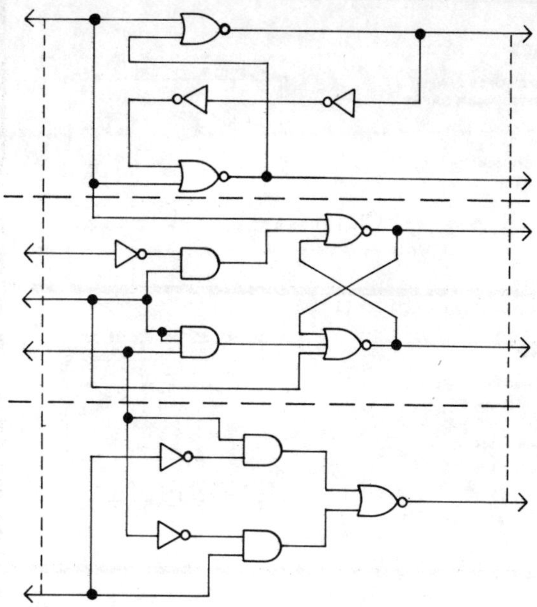

Figure 2-50. GROUPING EXERCISE NO. 1

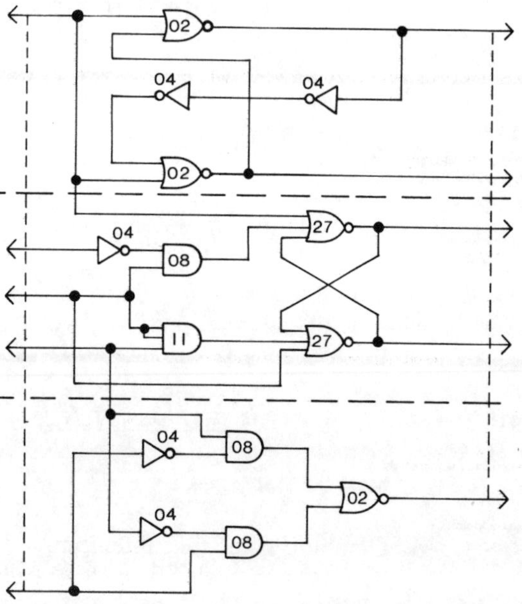

Figure 2-51. MASTERS GROUPING EXERCISE NO. 1

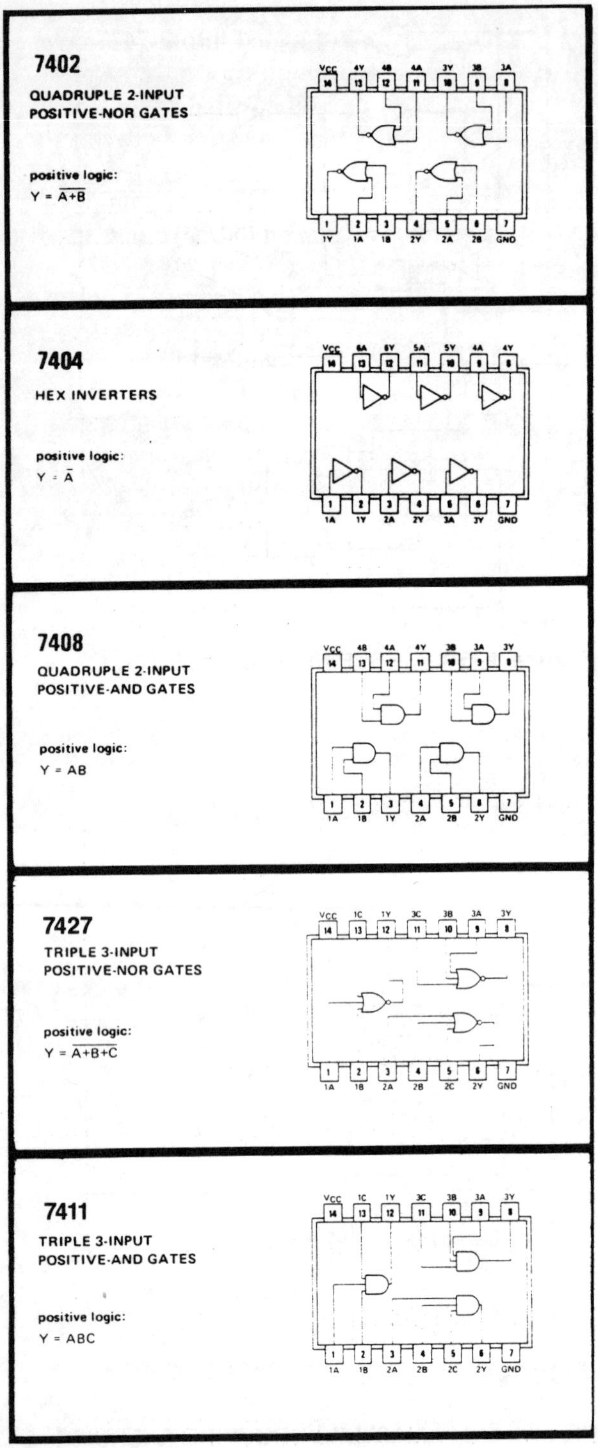

Figure 2-52. Min. amount of information for grouping

Figure 2-52 identifies each type of logic gate. It provides the picture for the total quantity of gates within each package ,the input and output assignments for each gate, and the voltage and ground pin number for each type package. It distinguishes between open collector and non-open collector. If we were dealing with an open collector gate, it would clearly call out in bold letters (under the identification number) "open collector".

3. Before we proceed with grouping, we must go back and ask the engineer whether spare gates are required. How many are needed of each type of logic? In this case, the engineer is not going to require any particular spares. This is not to say that we do not want any spares, it just means that it is not mandatory that you end up with spares of any particular type.

4. Before grouping or combining our schematic, ask the engineer if there are any critical leads or sensitive lead lengths on the schematic. This would affect your decision on which gates to be placed together into the same package. If there are critical leads, you would try to place those gates (that are intercommunicating with those critical leads) within the same package or very close to each other on the PC layout. The engineer in this particular case is not going to request any short lead lengths or specify any critical trace lengths on the PC layout.

5. Combining our gates. We are going to use the combining rules provided in this chapter, Figure 2-42 through 2-47.

6. Combine these gates using the technique called tying off unused inputs. For this particular schematic, tie off to another used input. On the job, you would seek the engineer's permission to do so. But this is a more popular technique than tying the unused input to a high or a low (meaning a voltage or ground) point.

7. Combine the gates that seem to be the simplest or the most obvious as far as communication to each other. Make the traces as short as possible. To do this, look at the schematic in Figure 2-53.

   We will give you an explanation of our ideas as we study this grouping and combining exercise. Look at the schematic in Figure 2-53. Notice the top section of the schematic, could all of the gates in that section be placed inside one package? The answer is yes. You would take the 02's, 2-input NOR gates. Look at the specification sheet in Figure 2-66. Notice that there are four 02's inside that one package. Using your combining rules sheet (Figure 2-46), you can see that the inverter with a single input and a single output could be substituted with a 2-input NOR called an 02. See the schematic Figure in 2-54 for this conversion along with the labeling of each IC and the changing of the picture of the gate on the schematic. Note that at the top of the schematic (Figure 2-54), an accurate count of all spare gates is being kept while the designer is grouping and combining. On more complicated schematics this technique is very valuable for tracking how many spare gates are in each package as you work through the schematic.

Note: IC 1 = 0 spares.

---

8. Continuing with grouping and combining, refer back to the schematic of Figure 2-53, Section B. Similar to the question in Step 7, can all of the gates in Section B be placed inside one package? Review the specification sheets (Figure 2-52), for the IC called 7427, a 7408 and a 7404. Notice that the 7427 has only three gates inside that one package (3-input positive NOR gates). The 7408 has 4 gates inside that one package. Using your combining rule sheet on the 3-input NOR (Figure 2-60), you can see that the AND gate, 7408 and the AND gate 7411 cannot be combined into a 7427. However, the 7404 inverter gate can be combined into a 7427. We prefer to hold the less obvious inverters until last. This means that there is no direct electrical connection between the 3 NOR inputs on the 7427 and the inverter, a 7407. Place the 7427's into one package called IC 2. IC 2 would have one spare at this time. The 7408 on Figure 2-52 shows four gates inside that package. Look at the schematic on Figure 2-53. Notice that the 7411 is not really required to be a 3-input AND gate. It can be reduced to a 2-input AND because two inputs are strapped together on the input side. Therefore, change the 11 to an 08; place both gates in Section B in the same package called IC 3, which would have two spares now. The 7408 inverter gate will be left until last.

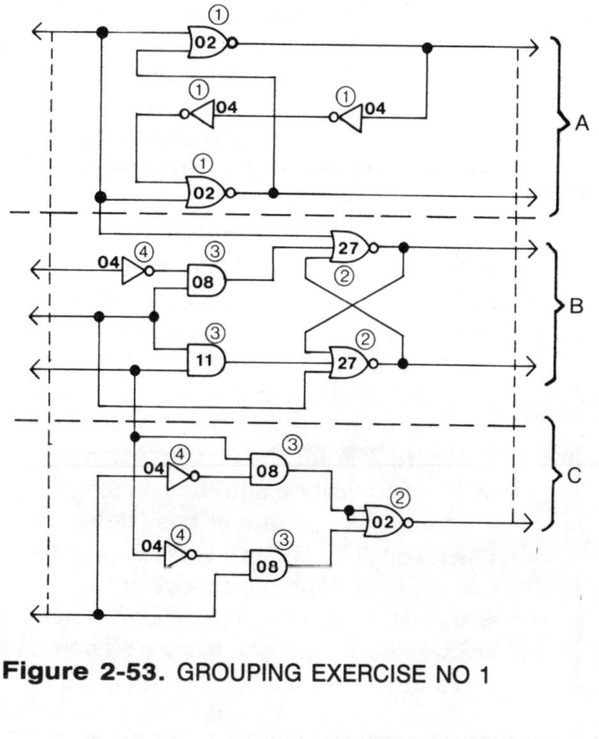

Figure 2-53. GROUPING EXERCISE NO 1

9. Examine Section C on schematic 2-53. Remember that we have one spare in IC 2 called a 3-input NOR 7427. There are two spares in IC 3 called 7408. Examine Section C. Can you place all of these gates inside one IC package? The answer is no. You cannot place AND gates and NOR gates in the same IC package. You can place inverters and NORS in the same IC package, and combine them. But in this case, there is no direct electrical connection between the inverters and the NOR gates. This means you are not necessarily trying to place the inverters with the NOR called 7402 in Section C. You might be looking more at the likelihood of Section B, where IC 2 communicates directly to IC 3. Consider communication to be important on the schematic and on the PC board itself when routing the traces. It is much easier to maintain short traces by having IC 3 talking to IC 2, or vice versa. Therefore, place the two 7408, 2-input AND gates in the same package alled IC 3 (which has two spares). Label these two gates IC 3. The spare count becomes zero. Can you change a 7402, or can you combine a 7402 into a 7427 package?

   Once again, examine the combining rules on Figure 2-46. It shows that a 2-input NOR can be placed or combined into a 3-input NOR package by tying off one used input. Therefore, label the 2-input NOR IC 2 and change the number of the gate to a 27. See the schematic in Figure 2-54 for the correct assignment of IC identification and the correct drafting of the picture of the gate on the schematic. The last two gates in Section C are 7404's. Once again look at the specification sheet in Figure 2-52. There are six inverters in one package. There are three inverters remaining on the schematic in Figure 2-53. Label those three inverters "IC 4" on Figure 2-67. Refer to the schematic on Figure 2-54 for a complete identification and conversion of all logic symbols. This is the finished grouping and combining schematic with a full description of all spare gates. IC 1 has 0 spares, IC 2 has 0 spares, IC 3 has 0 spares, and IC 4 (which was the last package to be combined and grouped) has 3 spares.

   Have you grouped the gates together? Have you achieved the optimum number of IC's along with reasonably short connections from one IC to the other? Examine these schematics once again. Section A will have very short connections. Section B has a mirror image communication to Section C (from IC 2 to IC 3 and IC 3 to IC 4). On the actual layout, IC 2 talks to IC 3; IC 3 talks to 4. This works out to be good intercommunication on the layout.

   IC 1 is fairly isolated, but this is not a problem on the layout.

Grouping Exercise 2

This grouping exercise is similar to grouping exercise 1 except that it is a more complex schematic. For example, we have an open collector and a non-open collector (called totem pole gates). There are unused inputs on the gates. The complexity of each grouping exercise will gradually increase. Examine the logic consideration (points 1 - 8) that was given to you in grouping exercise 1 and the engineer's schematic in Figure 2-55. Group and combine to find the lowest number of IC's required on this schematic. At the same time, combine the gates with the shortest trace connection by following the steps below.

1. Is the schematic readable? Are all the components and gate symbols identified? Examine the engineer's schematic in Figure 2-55. Notice that the gate symbols have been labeled with an identification number a 7400 or a 7404 etc. However, the prefix 54 has been dropped. The engineer has only put the last two or three digits in the symbol. Examine the schematic further. Note that there are flags or little tails on the output side of some of the gates such as 05 and 136. These gates are identified by the engineer as an open collector. Make sure that all of the gates are marked correctly as an open collector.

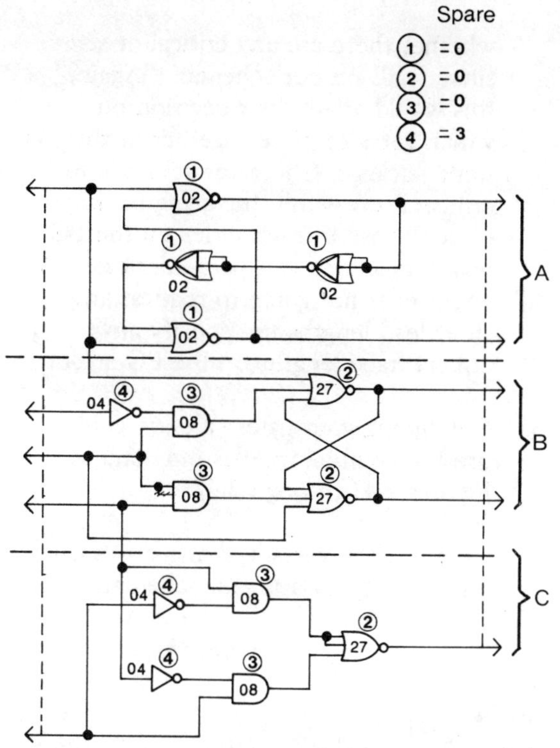

Figure 2-54. GROUPING EXERCISE NO 1. Correct Procedure

Make sure that all of the gates that have not been marked with an open collector are truly totem pole type gates. In order to do this, we must have the manufacturer's specification sheet for each logic family listed on the schematic. Therefore, we must obtain the specification sheets called the 7400, the 7402, 7404, 7408, 7410, 7411, 7405, and 74136. It is the engineer's responsibility to show you where to obtain these specification sheets and which manufacturer to buy them from. Before we proceed to Step 2, we should go ask the engineer for these specification sheets of all the logic gates on the schematic of Figure 2-53. You can then properly identify each logic gate.

2. Specification sheets are used to identify logic families and symbols. The entire specification sheet in Figure 2-48 is not necessary to combine your gates. Figure 2-57 provides the minimum amount of information to group and combine the schematic. Using the specification sheets, check each logic gate on the schematic of Figure 2-55 for correctness. Make sure all the inputs have been identified on the schematic. Remember that you cannot combine open collector gates with totem pole gates. See Figure 2-56 for corrections on the schematic. Example A shows that the engineer failed to identify all the open collector gates with a flag. This is one of the reasons why you must double check the schematic using the specification sheet number against the specification sheet shown in Figure 2-51. Example B shows the modification to a gate called an 11 which is the same as 7411. Looking at your specification sheet Figure 2-71, it calls out the 11 as a triple 3-input positive AND gate. We know looking at the schematic that the third input on the 7411 package was not needed because it was tied directly to a $ 5V. Refer to the logic rules, Figure 2-42. Tying off to a high (voltage) is the alternate way of deal-

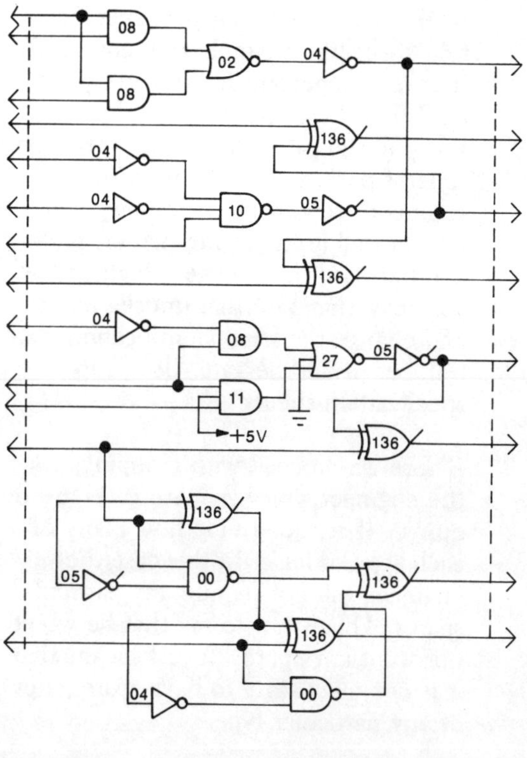

Figure 2-55. GROUPING EXERCISE NO. 2

ing with an unused input. Of course the other option was to tie that unused input to a used input of the same gate. Therefore, a change was made to bring the gate back to its minimum requirement. We have a better understanding of the true options available when combining gates. We eliminated the third input by eliminating the 7411 and changing it to an 08 package.

Example C is similar because there is a third unused input in a gate called a 27. Look at this specification sheet of Figure 2-57. A 27 is called out as a 3-input NOR gate. With Example B, reduce this to its minimum requirement. Using the logic rules in Figure 2-46, you can see that the unused input is being tied low, in this case to ground. This was one of the alternate techniques used on this unused input. The other one of course was to tie it to a used input of the same gate. Therefore, Example C was reduced to a 2-input NOR called a 7402 on your specification sheet (the example in Figure 2-57). It is very important for you to understand that these specification sheets show all the gates that come with each style of package called out by the engineer's schematic. If the engineer has unused inputs being tied off to another used input or to a high or low, you may eliminate that unused input and change the logic identification number inside the gate. Use your specification sheets.

3. Before we proceed with grouping, ask the engineer whether spare gates are required. If so, determine how many of each type of logic. In this case the engineer will not require any particular spares. This is not to say that he would not want any spares. It just means that it is not mandatory to have spare gates of any particular type.

4. The engineer must be asked (before we can group or combine our schematic) whether there are any critical or sensitive leads on our schematic. Again, this would affect your decision on which gates to place together in the same package. Check for sensitive or critical leads within the same package or very close to each other on the PC board layout. In this particular case, the engineer is not going to request any short lead lengths nor specify any critical trace lengths on the PC layout.

5. Combining your gates. We are going to use the combining rules provided in Figures 2-42 through 2-47.

6. Combine the gates by tying off unused inputs for this particular schematic.

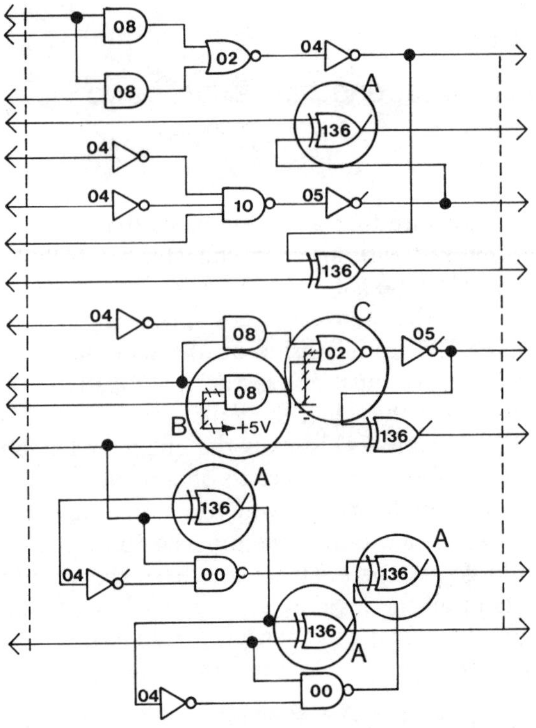

Figure 2-56. GROUPING EXERCISE NO. 2

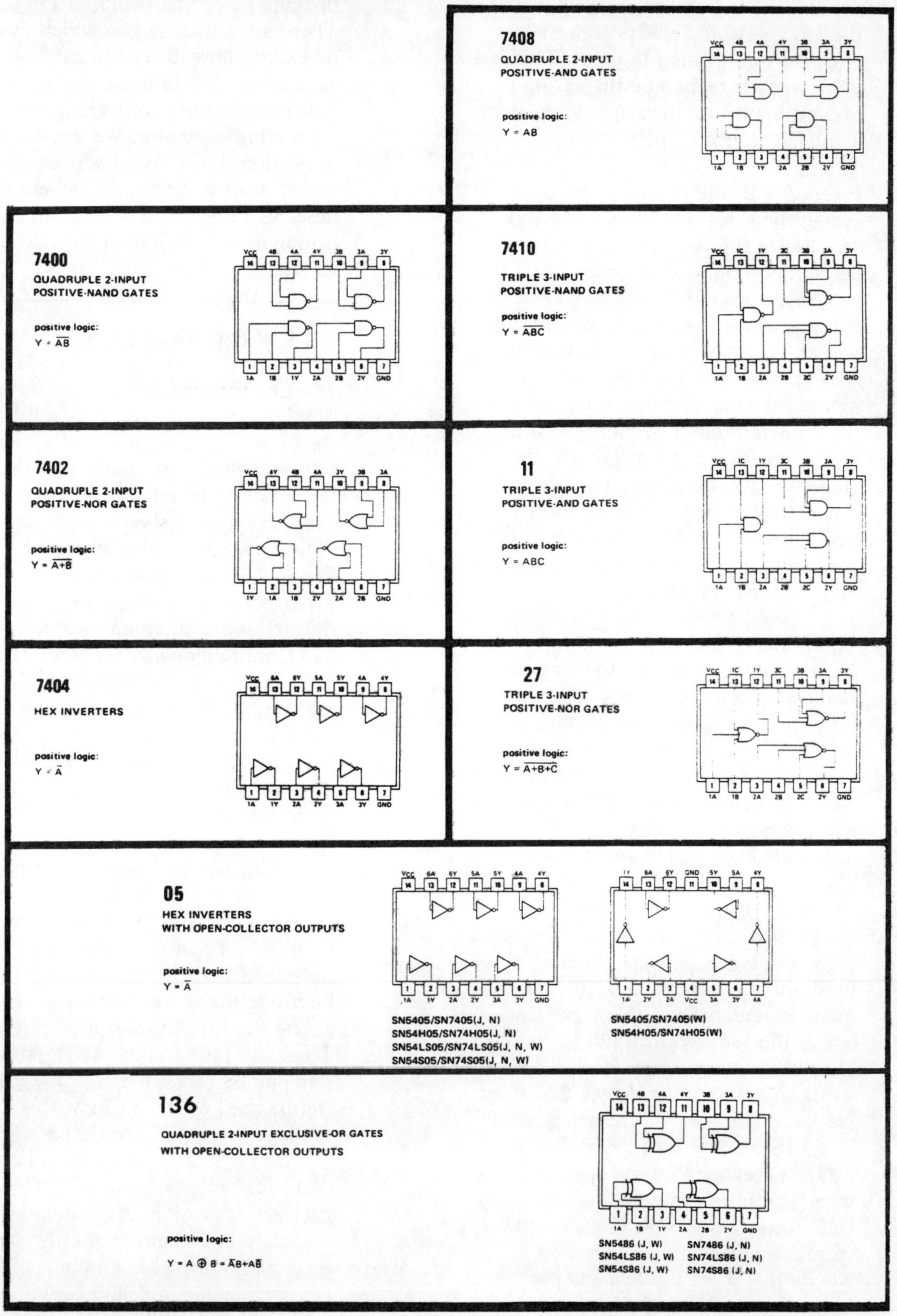

Figure 2-57. Specification Sheet

Tie off to another used input. On the job you would always seek the engineer's permission to do so. This is a more popular technique than tying the unused input to a high or low, meaning a voltage or ground point.

7. Proceed with the schematic by combining the gates that seem to be the simplest or the most obvious as far as communicating to each other. Make the traces as short as possible. Look at the schematic Figure 2-58. We will give you an explanation of our ideas as we walk through this grouping and combining exercise. The first thing we did was to determine if Section A was a logical breaking point as far as intercommunication of gates. The answer was yes. Most of the gates are intercommunicating with each other in that particular section. The same situation exists for Sections B and C.

   The next step is to find out what would be the easiest or the simplest combining technique on the schematic. Remember that we cannot combine open collector and totem pole logic gates in the same package. We must start with the open collector gates. Section C has three gates called 136's which are exclusive OR gates. These gates should be inside one package. This is called IC 1. Look at the specification sheets in Figure 2-57. Notice that there are four of these logic gates inside one package. Therefore, we have one more to pick up. At this point, determine how many gates are left of the 136 exclusive OR gates. We have labeled three gates IC 1. To take a quick count you would see that there are three more left on the schematic. Using the combining rules in Figure 2-45, we can see that the open collector inverter can be combined into an open collector exclusive OR. Examine the specification sheet in Figure 2-57. Notice that if we use two 136 packages, we have a total of eight logic gates called 2-input EX* ORS. Count the remaining gates that could be 136's. There are a total of five which would be exactly how many 136 gates we would have left to make two packages. Label one of these remaining gates "IC 1", leaving no spares. We chose the 136 in Section A because it was directly tied to IC 1 in Section C (where the other two EX* ORS were communicating directly to the open collector inverters, the 05's).

---

* EX = Exclusive

The 05's will be combined into a "136" 2-input EX OR. This will be shown to you in Figure 2-59, using the combining rules in Figure 2-45. Now set up an IC count column and a spare count column as shown in the upper right hand part of the schematic in Figure 2-58. This helps to keep track of what spares are available as we go through our combining technique.

Continuing with Section B, the 136 and the 05 will be placed in one package, IC 2. In Section A, the 136 and the 05 will be placed inside the same package called IC 2. We will label all four gates IC 2 at this point. In our spare column, show IC 2 as having no spares.

Using the philosophy of attacking the easiest and most obvious grouping combination first, the 08 gates was placed in IC 3 because it was obvious in Section B that these two gates should be inside the same IC package. Examine the specification in Figure 2-57. There are four 2-input AND gates inside one 08 package. All of the gates will be used for IC 3. Show this in your spare column, "IC 3 = 0 spares".

Remember the IC's should communicate with each other in the same package. If possible, IC's should communicate in a chain. Not only does IC 3 communicate with a 2-input NOR gate in Section A, it also communicates with a 2-input NOR gate in Section B.

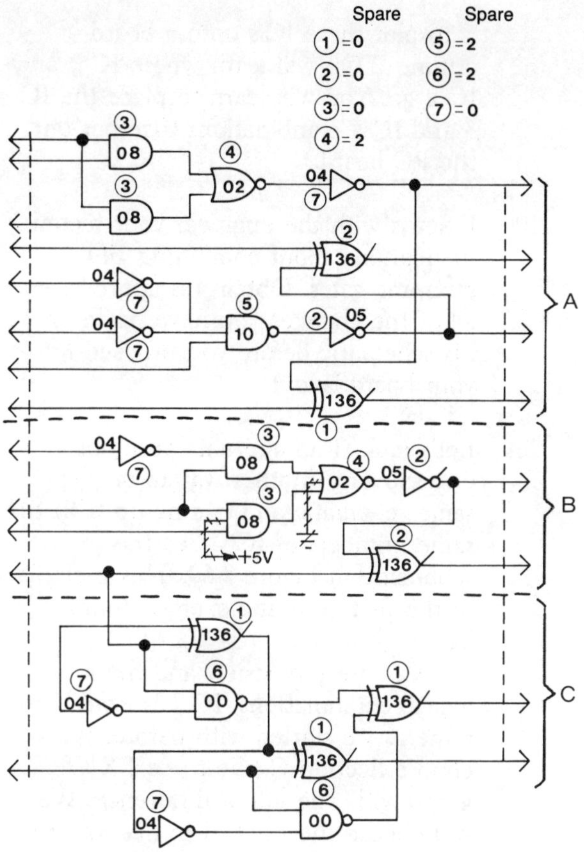

Figure 2-58. GROUPING EXERCISE NO. 2

Label IC 4 the 02, 2-input NOR. This gives us a total of two spares which will be listed in our spare column.

IC 5 was selected because there is one 3-input NAND on the schematic. Look at the specification sheet under 7410. Notice that there are three in a package. Therefore, IC 5 would have two spares.

The only two gates (7400's) were saved last, except for the open collector inverters. This is because inverters and buffers are the easiest gates to combine into other IC packages. You combine not only for shorter trace length, also for fewer IC's. The only two gates (other than the inverters) were the 7400's called the 2-input NANDS (IC 6). Review your specification sheet. You have two spares.

IC 7 is the 04 package. Review the specification sheet. There are four in the package making a total count on your schematic. The total required number of inverters (04's) is seven. Therefore, seven would have zero spares.

You have now combined all of the obvious gates into their respective packages. However, you have not considered total IC count (at this point, 7 IC's is the total).

8. Minimizing the total IC count, we will show the proper logic symbol for each combining technique, such as IC 2. Combine the open collector inverter 7405 into a 2-input EX OR gate 74136 (Schematic Figure 2-59). Sections A and B show conversion of the 05 inverter gate into a 2-input EX-OR package, tying the unused input high.

   Let's review some of the possibilities remaining to reduce the IC count. IC 6 is a 2-input NAND. We have used two of its four gates.

   Review the spare count on our schematic (Figure 2-58). IC 6 has two spares; IC 5 has two spares; and IC 5 is a 3-input NAND. Can we take IC 6 (a 2-input NAND), and combine those two gates into IC 5? The answer is yes.

   Review your combining rules in Figure 2-42. Tie one of the unused inputs on the 3-input gate called a 7410 to one of the used inputs. This changes the gate on our schematic (Figure 2-59) to a 7410 instead of a 00. The third input is tied off to a used input of the same gate, IC 5.

   We have now eliminated IC 6 from the schematic by using the two spares out of IC 5. We must show this in our spare column. The two spares are crossed out on IC 5, leaving zero spares. IC 6 would be totally eliminated. Now we have a total IC count on the schematic of 6 rather than 7 IC's.

The last combination does not reduce the IC count as much as it shortens the trace lengths. Look at the IC 4 called a 7402, a 2-input NOR package. Two spares are available. Using our combining rules (Figure 2-46), combine inverters into a 2-input NOR package by tying off one of the unused inputs to the same gate. Looking at IC 4 in Section A, we know that the inverter (IC 7) could be changed to IC 4. This changes the schematic to show a 2-input NOR with the 2-inputs tied together. Likewise with Section B, the 2-input NOR (IC 4) communicates with IC 3. This is a nice combination to pick up the inverter, giving an IC 3 to IC 4 communication. This gives a lot of intercommunication from IC to IC and keeps them close together.

The remaining inverters could stay as IC 7. This increases the spare count on IC 7 to two. This would be shown in your spare column. We could continue to give more possible combinations on this schematic, but we will increase your knowledge little-by-little using the next two grouping exercises of three and four.

9. Have you grouped the gates together to achieve the optimum number of IC's and reasonably short connections from IC to IC? Examine the schematic in Figure 2-59. Section A shows IC 3 communicating with IC 4. IC 2 communicates with IC 5 which communicates with IC 7. Section B has IC 3 and 4 communicating with each other. This is excellent because this is what we have in Section A. However, IC 2 does not share the same intercommunication to IC 5 that Section A did.

   In Section C IC 1 communicates with IC 5 which is communicating with IC 7. The IC 5 to IC 7 intercommunication is good because in Section A IC 5 and IC 7 are also communicating.

   One of the things that we will teach you is in the grouping exercises, is to plot these IC's onto a board outline. YOu will learn where IC 1 and IC 2 go. You will learn to place the IC 3 and IC 4 combinations together on the PC board.

10. Discuss with the engineer your recommendations about combining and grouping gates. Obtain his approval on all of the changes you have made on his schematic before you proceed into your board layout.

11. Let's look at an alternate grouping exercise showing another way to do the same schematic and ending up with the same numbers of IC's. See the schematic in Figure 2-60. This is similar to the first grouping exercise (number 2).

   Choose the easiest and the most logical combinations. IC 1 is exactly the same as we started with before. We will choose Section C (the three EX OR gates) with one minor difference. We will choose the EX OR in Section B. IC 2 is very different because we choose the 2-input NAND gates called the 00's. We know that only two were required on the schematic. We had two spares, and combined the inverters called 04's into 00's. Put the two inverter gates of Section C into IC 2. Section C now has all of IC 1 communicating with IC 2, which is a good combination. IC 3 is in Section B. We chose to attack the 2-input AND's just like we did in the first grouping exercise number two. We choose the 2-input AND's in Section A to accompany the 2-input ANDS in Section B. This used up all of the gates in IC 3, leaving no spares. In IC 4, we choose to pick up the remainder of the open collector gates. Changed the 05's once again to 2-input EX OR's by tying the unused input high and using all of the gates in IC 4.

   IC 5 is very different from the first exercise in number 2. We took IC 5 (a 3-input NAND having three gates). We combined the single input inverter 04

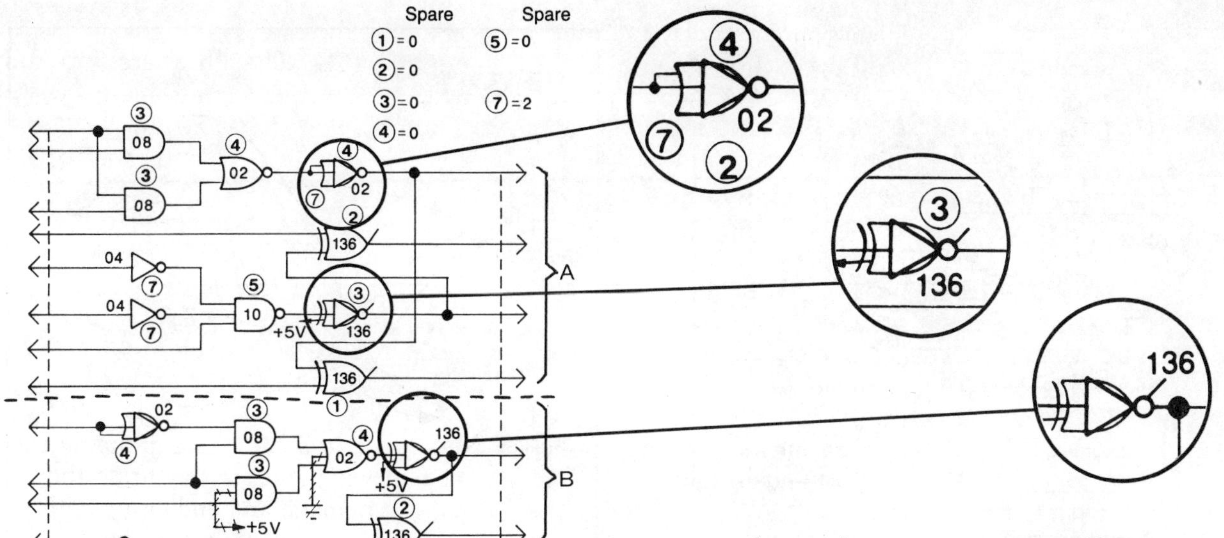

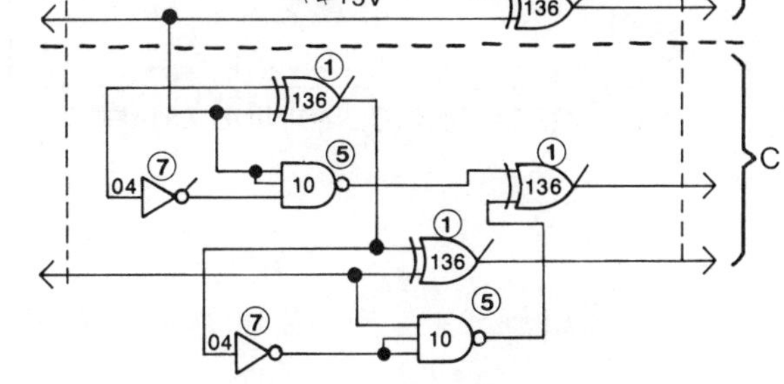

Figure 2-59. GROUPING EXERCISE NO. 2

into this 3-input NAND package. Using the two inverters in Section A, we used up all the spares in IC 5. However, the electrical connections in IC 4 and IC 5 in Section A are very good. IC 6 in the last IC is the 2-input NOR package. There are a total of four gates in this package. We only need 2-input NOR's. Examine your schematic. The remaining two gates that have not been accounted for are inverter gates. They can be combined into IC 6 by changing the inverter to look like a 2-input NOR and by tying the unused input to the used input, for a remainder of zero spares. The spare count column on this schematic now has no spares whatsoever. The inverters have virtually been eliminated. There is no need for an inverter package.

However, there are some marginal maneuvers in Section A of IC 5 7410. Changing those two spares into inverters by combining the inverters into that package was a questionable manuever. The change was not bad, it was just marginal. Perhaps it would have been better if we could have ended up with a spare 3-input NAND and with a slightly longer trace length on the PC board. The PC designer has to make the final decision as to whether this was a better grouping exercise than grouping exercise number 2.

Before we go to the next grouping exercise, let's examine what you have learned on the first grouping exercises. See our grouping and combining chart in Figure 2-61. Redraft this chart showing which gates can be combined together. For example, the AND gate (shown on the right hand side on the top of the chart) can be combined with any other AND. A 2-input AND can be combined into a 3-input AND package; or a 2-input AND can be combined into a 4-input AND package. The AND gate can also have a buffer combined into it. This means that a buffer gate with a single input and a single output can be used in an AND package that has two, three, or four inputs. We know that the 4-input would be an unlikely choice. With the color blue, we have drafted from the AND down to the buffer. That is the possible combination for the NAND. It can be combined with any other NAND, such as a 3-input or a 4-input. It can also be combined with the inverter. Draft a red line from the NAND over to the inverter. The OR is similar to both the AND and the NAND because it can be combined within its own family (the 2-input OR to the 3-input OR). It could also have the buffer. Draft in the color blue to show that the OR

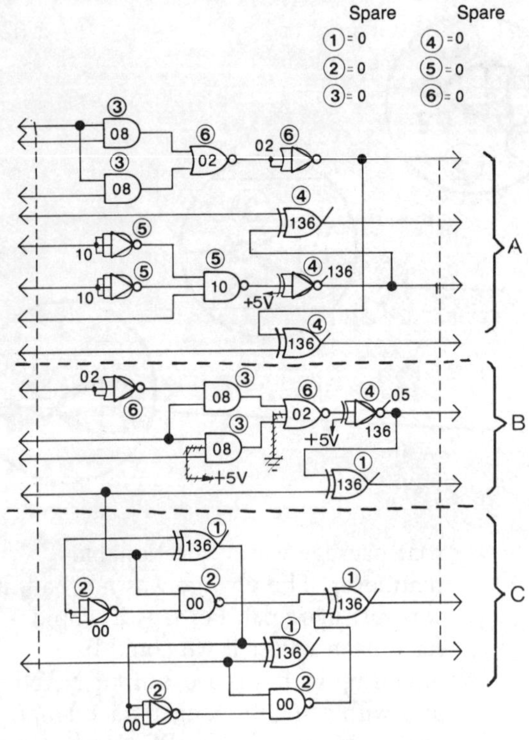

Figure 2-60. ALTERNATE GROUPING EXERCISE NO. 2

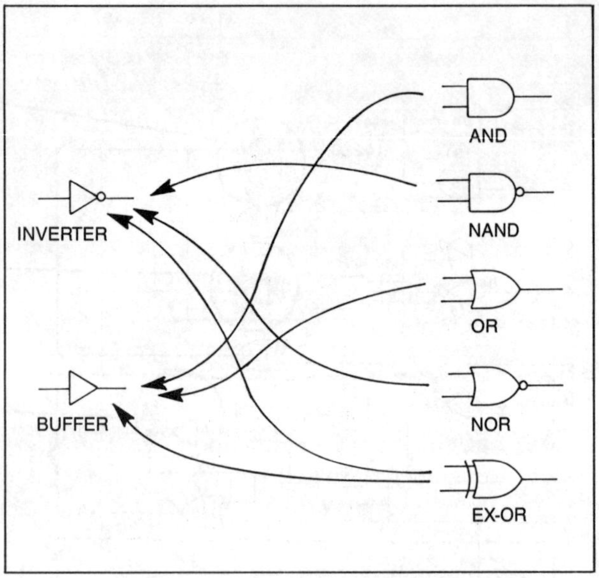

Figure 2-61. GROUPING/COMBINING CHART

could have a buffer combined into its package. The NOR is like the NAND in the sense that it can be combined within its own family. It can have the inverter combined into it. This makes a single input into a double input by tying the two inputs together. The EX OR* will be the gate that has the capability of combining both the inverter and the buffer. In red, draw from the EX OR up to the inverter. In blue, draw from the EX OR to the buffer. You cannot tie the EX OR inputs together as you can for the AND, the NAND or the NOR. We can not tie the inputs together to create an inverter or a buffer. For an inverter to use the 2-input EX OR and still have a spare input, the spare input must be tied high. In the case of the buffer, the spare input on the EX OR must be tied low. Use this chart (Figure 2-61) in grouping exercises 3 and 4 to decide which gates shall be combined together.

---

* Exclusive OR

Grouping Exercise 3

Using the logic consideration points (1 through 8) and the engineer's schematic (Figure 2-62) group and combine the lowest number of IC's required on this schematic.

1. Is the schematic readable and are all components and gate symbols identified? Examine the schematic in Figure 2-62. All the gate symbols have been labeled with the identification number (such as 7400 or 7404, etc.) You would have been given a set of specification sheets by the engineer to match each one of these numbers and any other discrete components that might be on the schematic.

2. Specification sheets identify logic families and symbols. At this time you must obtain all the specification sheets required to identify each type of logic family on your schematic (Figure 2-62). See Figure 2-63 for a complete set of specifications required for each logic family. At this time, identify all open collector types of gates versus the non-collector type. Make sure that the schematic indicates this. You can use the flag or tail from previous schematics. Examine the schematic on Figure 2-62. These are open collector gates. The 136 is a 2-input EX OR* open collector style. We also have an inverter open collector called an 05. Notice that the engineer has flagged each gate with a tail to indicate an open collector style of gate. You cannot intermix an open collector with a non-open collector (called a totem pole). At this time, identify all of the logic gates. The engineer has tied off unused inputs or terminated them high or low, and converted them to their lowest required state. See the example in Figure 2-64. There are two gates in Example A called 27's. The inputs are tied to used inputs. In this case, they are tied low. Change the gates to look like a 02 which is a 2-input NOR instead of a 3-input NOR. Example B shows 3-input ANDS (called 11's) that can be converted back to 2-input AND's (called 08's) by removing the unwanted input.

   In this case both are tied high. Remove them from the schematic. Change the identification of the gate to an 08. This manuever is the PC designer's responsibility. The engineer could have built a breadboard of the schematic. He might use available IC's from the stock room. He may not choose the best or the minimum required in Figure 2-64). As PC designers, we must choose the minimum number of inputs by grouping and combining. All this is done with the aid of the specification sheets (Figure 2-63).

3. Before grouping, ask the engineer whether spare gates are required, and, if so, how many of each type of logic. In this case the engineer will not require any spares.

4. Are there any critical leads or sensitive lead lengths on our schematic? Again, this would affect your decision on which gates should be placed together in the same package. Trying to place those gates (that are intercommunicating with critical leads) within the same package or very close to each other on the PC board layout. In this particular case, the engineer will not request any short lead lengths or specify any critical trace lengths on the PC layout.

5. Combine the gates using the combining rules provided in Figures 2-24 through 2-47.

6. When combining gates, use the technique of tying off unused inputs. Tie off to another used, obtaining the engineer's permission to do so. But this is a more popular technique than tying the unused input to a high or to a low (a voltage or ground point).

* EX OR: Exclusive OR

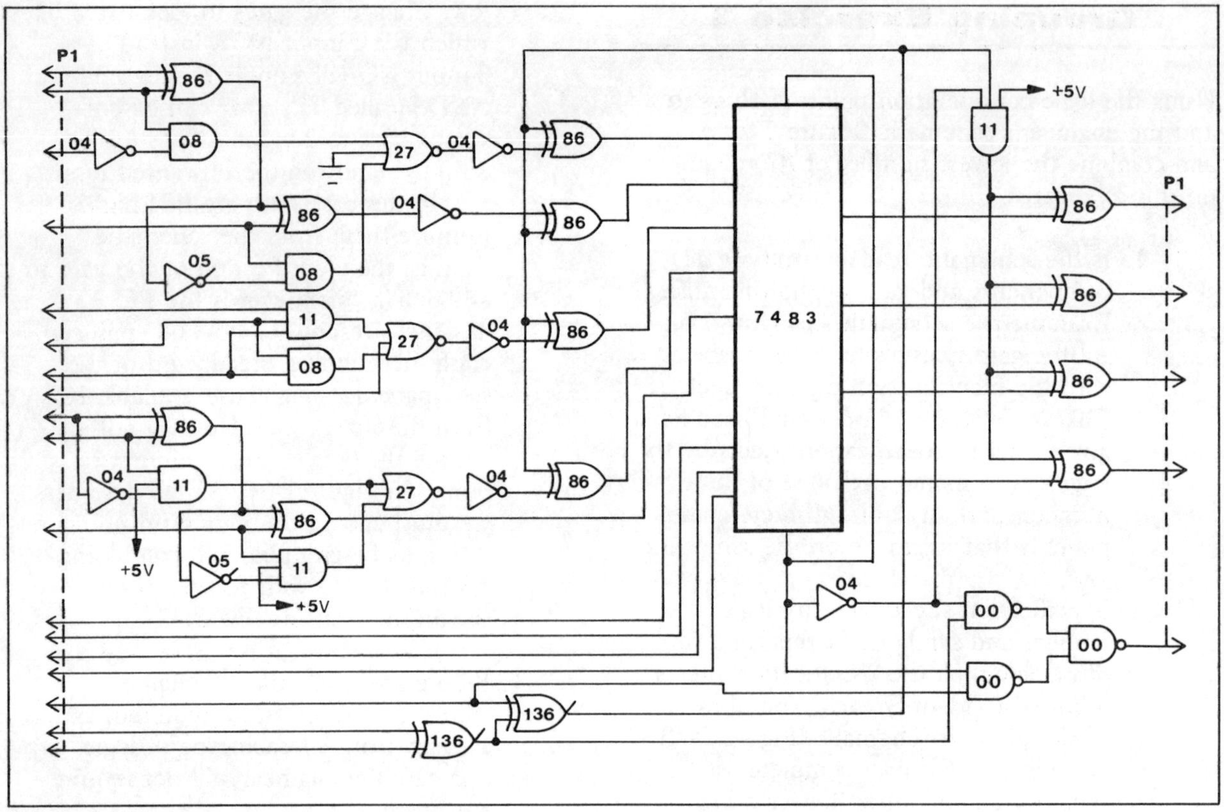

Figure 2-62. Engineer's Schematic

7. Combine the simplest gates that most obviously communicate with each other. Make the traces as short as possible.

Look at your schematic in Figure 2-65. Use your specification sheets in Figure 2-63. We placed IC 1 on the 7411 because you can see from your specification sheets that you cannot combine or modify the gates. Therefore, it is easy to start with this particular IC. List the IC's (assigned to the gates) on the side or the bottom of the schematic. Show the spare count. IC 1 now would have zero spares.

IC 2 was assigned to the 2-input EX OR* package called the 86. It is a non-open collector style EX OR. There are four of them in a package (according to the specification sheet). They are all communicating to IC 1 and to each other. Therefore put all four of the 2-input EX OR's in the same package called IC 2. List IC 2 at the bottom, with zero spares. IC 3 is highly communicative to IC 1 and to itself. Put all four of the 2-input EX OR's (called 86) into IC 3 giving you zero spares. Tying IC 2 to IC 1 to IC 3 is a logical flow.

Group these IC's together. When you plot them on the PC board, they will communicate to each other very efficiently.

Continue with the simplest or the most obvious combinations. IC 4 picks up the 2-input NAND gates called the 00. Look at your specification sheet. There are four in a package. At this

---

* EX OR: Exclusive OR

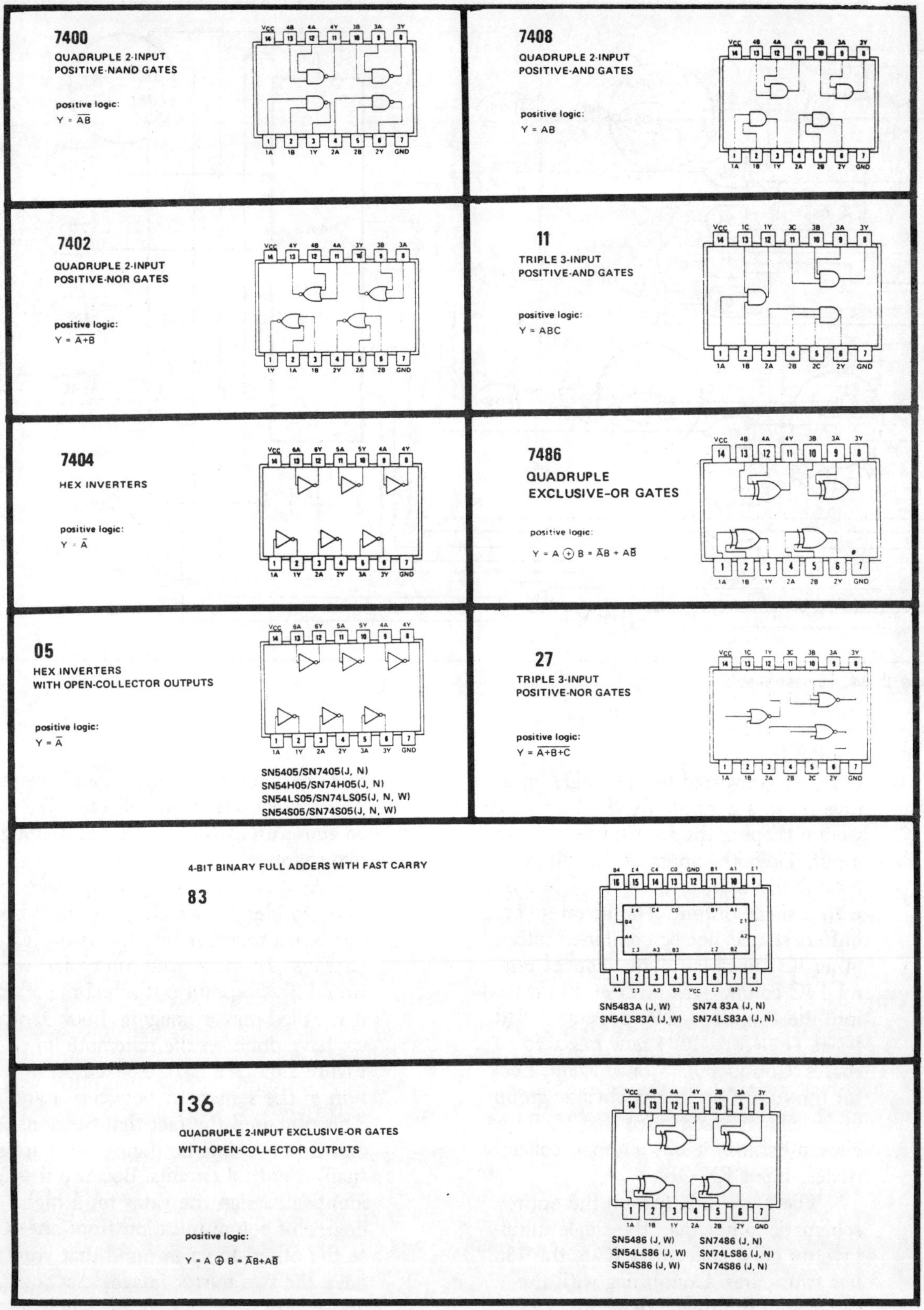

Figure 2-63. Spec Sheet for Fig. 2-62

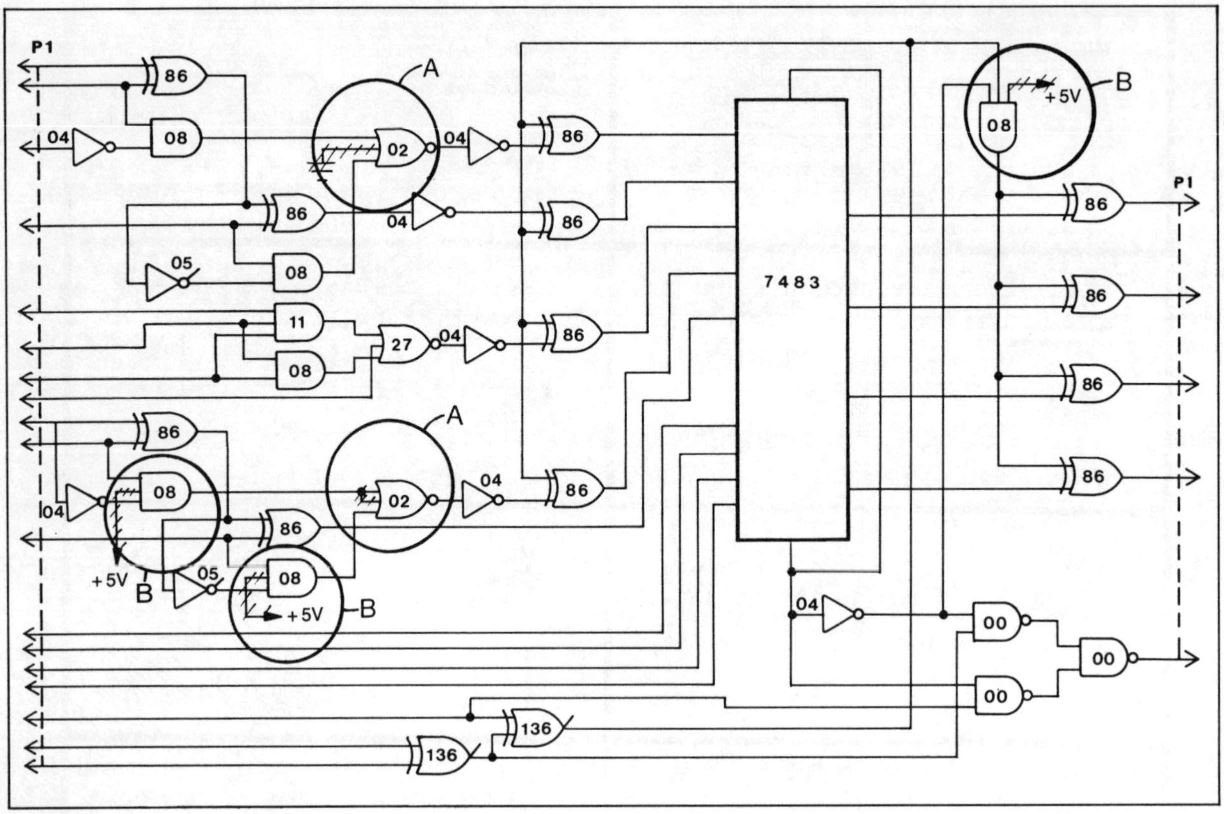

Figure 2-64. Unused Inputs

time, we converted the 04 single input inverter to a 2-input NAND. Electrically we are keeping the inverter as a single input. Tie both inputs of the 2-input NAND together, creating a single input with a single output. The inverters or buffers should not be combined into other IC's until last. This reduces our total IC count. This inverter was placed into the 2-input NAND package called IC 4. Therefore, IC 4 now has zero spares. Consider IC 5 once again. Look for the easiest and most obvious grouping combinations. Notice that IC 5 was chosen because it was an open collector (style 2-input EX OR).

There are only two on the entire schematic (other than the single input inverter open collectors). IC 5, the 136, has two spares. Continuing with the removal of the open collectors, notice that there are two single input inverters (05's), which are open collector. We picked those up as IC 6, with a remaining of two spares.

At this point, it is getting more difficult to identify which gates should be combined together into the same IC package. To make your job easier, we would like to point out a technique and it is called mirror imaging. Look at what we have done on the schematic in Figure 2-65. We have blocked out a section of the schematic between A and B. This obscured the fact that Sections A and B (on schematic Figure 2-66) are actually identical circuits. Because they are identical, assign the gates for a high degree of communication from one IC to the other. Keep in mind that you have the two mirror images, Sections A and B. Let's go back to the schematic in

Figure 2-65. Whatever you do in Section A, you will repeat in Section B. For example, IC 7 is an 86 (called a 2-input EX OR). There are two of them in Section A. Label both of them IC 7. Therefore, we will mirror image down to Section B, which also has two. IC 7 has zero spares.

IC 8 is an 08 (2-input AND package). There are four in the package. Two of them are required in Section A. Likewise two are required in Section B. Therefore, IC 8 has zero spares. We now have a good combination of gates talking to each other (called IC 7 and IC 8). There's a high degree of communication between both Section A and Section B.

IC 9 is called an 02 (2-input OR package). There are four in the package. There are only two required: one in Section A and one in Section B. Therefore, there are two spares for IC 9.

IC 10 is not a mirror image area. Look at the remaining gates to be assigned. There is a 27 (called a 3-input OR gate). There are three in a package. Because there is only one on the schematic, IC 10 would have two spares.

With IC 11 there are a few gates left. There are the AND gates called the 11 and the 08's. The only other ones remaining are the inverters. We will use a combining technique that we have used before. We will change the 08's (which are 2-input AND's) to 11's. Use all three of these AND gates as IC 11. This is because the 11 package on the specification sheet has three gates. If we do this, there would be zero spares for IC 11.

The only thing left is the inverters. Look at the schematic on Figure 2-65. There are six inverters. Look at the specification sheet for the 04. Notice there are six inverters in the package. Therefore, call each one of the remaining inverters IC 12 (which would have zero spares).

Combining permits intercommunication from one gate to another, or from one IC to another, or from sets of gates to sets of gates. Now we will combine to reduce the total count of IC's. See Step 9 for this technique.

9. Look at the schematic on Figure 2-67. We now have a total count of twelve IC's. We would like to reduce this if possible by using the combining techniques from the schematic in Figure 2-65. We will do two things at one time. Not only will we combine for intercommunication from IC to IC, but we will also reduce the total count of IC's. We know that the benefits of reducing the total count are:

   a) Conservation of space on the PC board
   b) Lowering the cost of purchasing the IC
   c) Easier logistics as far as handling, stocking and storing
   d) Easier assembly of the component

   All of these things reduce the overall cost. By reducing the total IC count, we can save many dollars in the end product of the assembled PC board. To reduce the total count, start with IC 1. Look at all the spares on IC 1 through IC 12. 1 through 4 have zero spares. We can do little to reduce the count on those IC's. IC 5 has been assigned as a 2-input EX OR with an open collector (called a 136). Looking at your specification sheet, notice that a 136 has four gates. Using your combining rules (Figure 2-65), notice that the open collector inverters (called 05's) can be combined into that IC 5, using the two spares as inverters. See the schematic of Figure 2-67. Notice that we would tie the unused input to high, in this case, to + 5V. This would virtually eliminate IC 6. You can cross it out on the schematic; IC 6 no longer exists. We have reduced our total count to eleven.

   IC 7 and IC 8 have zero spares. We cannot reduce those two IC's. IC 9 however does have two spares. It is an 02 which is a 2-input OR package with four gates. Should anything on the schematic be combined into IC 9? Using your combining rule sheet (Figure 2-44), notice there is nothing obvious to com-

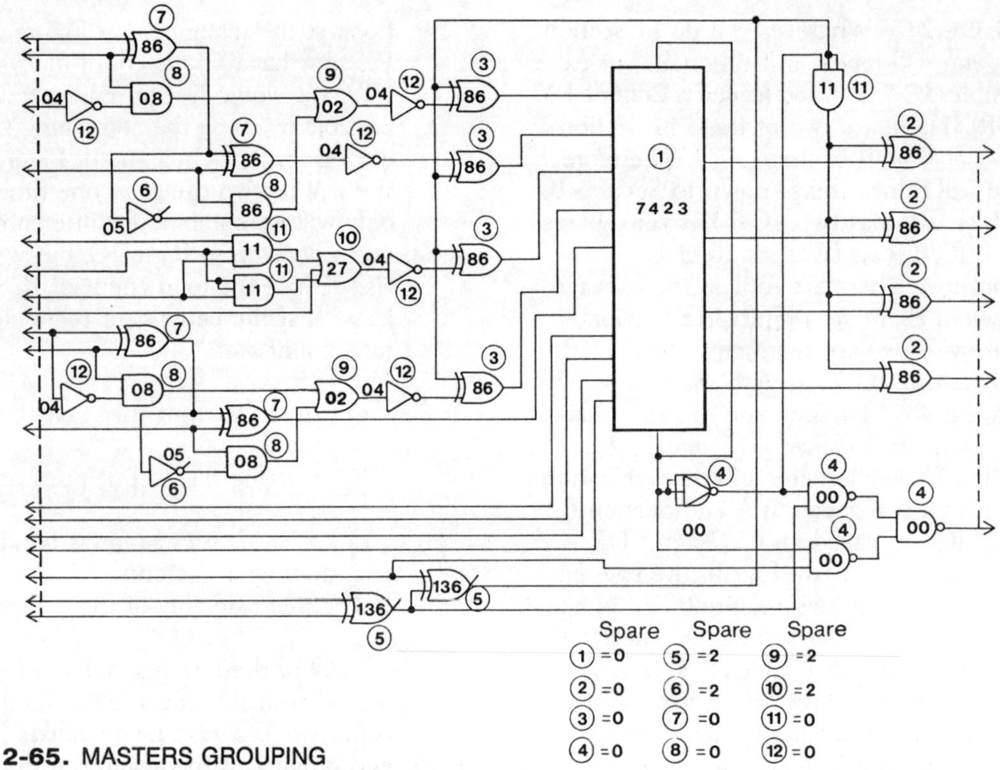

Figure 2-65. MASTERS GROUPING EXERCISE NO. 3

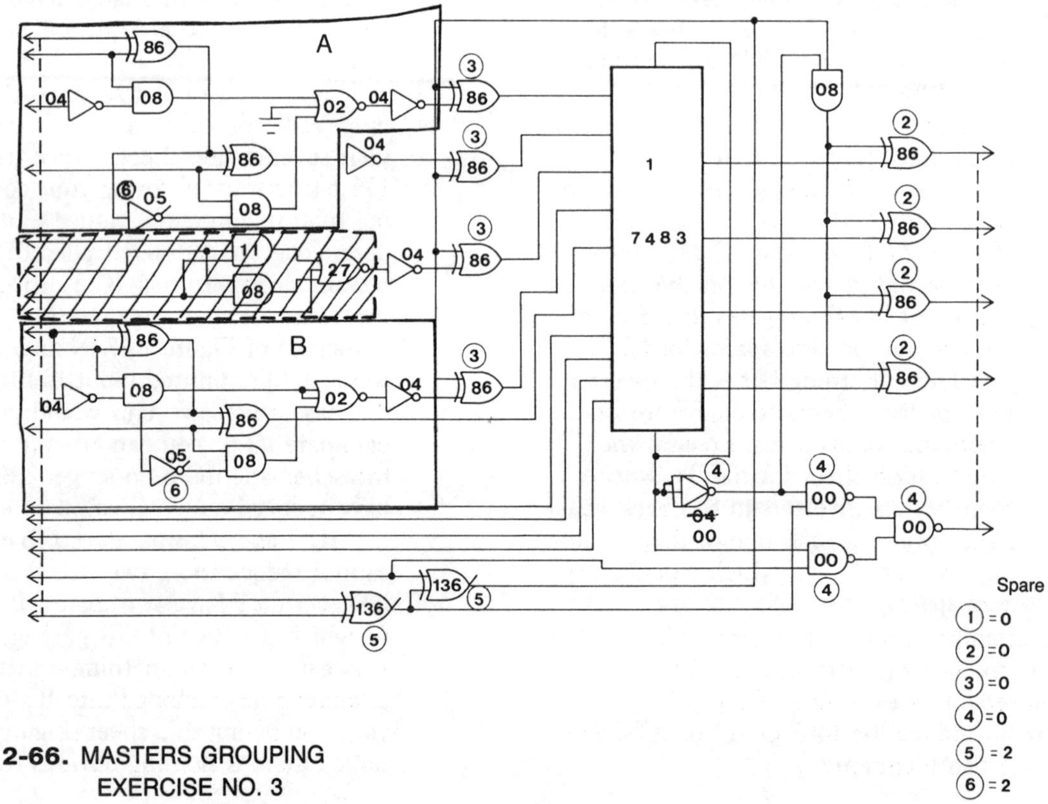

Figure 2-66. MASTERS GROUPING EXERCISE NO. 3

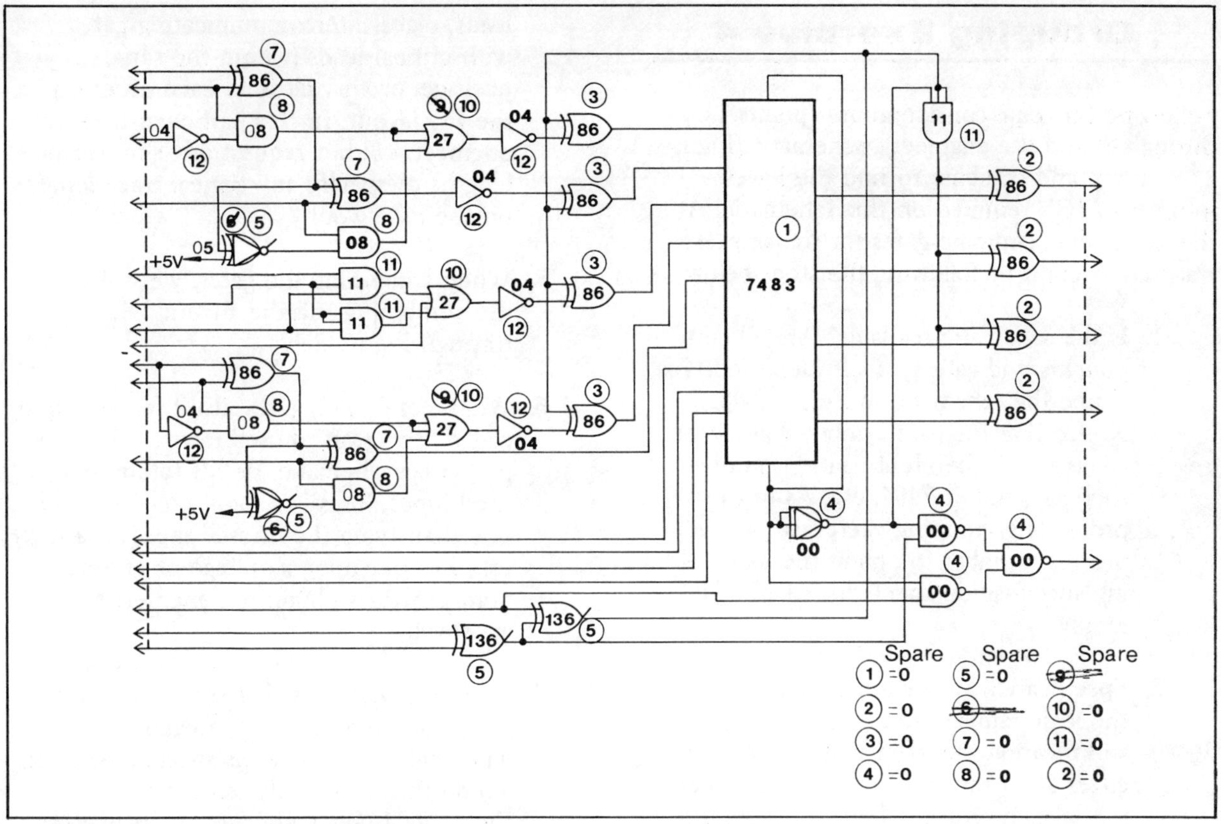

Figure 2-67. MASTERS GROUPING EXERCISE NO. 3

bine into IC 9. Let's move on to IC 10 which also has two spares. IC 10 is a 3-input OR package. Look at the combining rules on Figure 2-58 and your specification sheet IC. IC 10 is a 3-input OR with two remaining spares. The 3-input OR's can absorb the two 2-input OR's used in IC 9. This eliminates IC 9 as shown on schematic Figure 2-67. The IC 10 spare count would go to zero. IC 11 and 12 have no spares. Our total IC count has been reduced from twelve IC's to ten, by eliminating IC 6 and IC 9.

10. Have you grouped the gates together to achieve the optimum number of IC's along with achieving reasonably short connections from one IC to another? Examine the schematic in Figure 2-67. Notice that each gate has been properly identified and all unused inputs have been tyed off.

    You are now ready to take your schematic back to the electrical engineer and seek his approval for each combining decision that you have made. It is important that you consult with the engineer in this area. When in doubt about tying the unused inputs correctly or which gates can be combined, the electrical engineer should always be consulted before proceeding into the board layout.

Grouping exercise number 4 will have more gates. After you walk through this grouping exercise you will notice that the more gates clustered together in quantity, the easier it becomes to group and combine your schematic.

Grouping Exercise 4

Following the logic considerations (points 1 through 8) and the engineer's schematic (Figure 2-68) group and combine to find the lowest number of IC's required on this schematic. At the same time, combine gates for the shortest trace connection by following the steps below:

1. Is the schematic readable? Are all components and gate symbols identified? Examine the schematic in Figure 2-68. Notice that the gate symbols have been labeled with the identification number such as: 7400 or 7404, etc. As in the preceeding grouping exercises, you will use this number to obtain the manufacturing specification sheets for each logic family.

2. Specification sheets are used to identify the logic families and symbols. The specification sheets in Figure 2-69 will cover all of the gates that the engineer has labeled on the schematic in Figure 2-68. Using your specification sheets, identify all of the logic gates (open collector versus non-open collector/totem pole). Reduce the logic gates to the absolute minimum. This is done by removing unnecessary inputs, by changing the picture of the gate, and by reassigning the gate identification. See Figure 2-70, Example A. Also, Example B shows the identification for open collector gates (a flag on the output side).

3. Before grouping, ask the engineer whether spare gates are required. If so, determine how many of each type of logic. In this case the engineer is not going to require any particular spares. This is not to say that we do not want any spares. It just means that is is not mandatory.

4. Ask the engineer before grouping or combining the schematic: are there any critical leads or sensitive lead lengths? This affects your decision on which gates would be grouped together into the same package. If there were sensitive or critical leads, place intercommunicating gates with critical leads (within the same package) or very close to each other on the PC layout. In this particular case, the engineer will not request any short lead lengths or specify any critical trace lengths on the PC layout.

5. When combining the gates, we will use the combining rules in Figure 2-42 through Figure 2-47.

6. When combining gates, use the technique called "tying off unused inputs". For this particular schematic, tie off to another used input. This is a more popular technique than tying the unused input to a high or a low (meaning a voltage or ground point). Always obtain the engineer's approval.

7. Combine the gates that seem to have the most communication with each other. This makes the traces as short as possible. To do this, look at the schematic in Figure 2-71. We will give you an explanation as we go through this grouping and combining exercise.

   Using our specification sheets on Figure 2-69, we proceeded to assign IC 1 to the 74-73 called a JK flip-flop. We cannot combine any gates in this particular package. There are two flip-flops inside one IC package. Our schematic only calls for one. Therefore, we would call out IC 1 as having one spare. IC 2 was selected as the 7400 (the 2-input NAND gates). Reviewing your specification sheets, you can see that IC 2 has four gates inside one package.

   Our schematic requires three. After reducing the Example A in Figure 2-70 to an inverter, IC 2 would be listed with one spare. IC's 3, 4, 5 and 6 were a logical combination because there was intercommunication between the same types of gates (called the 7408 2-input AND gates). Reviewing the specification sheet, notice that there are four in a package. IC's 3, 4, 5 and 6 would have zero spares. IC 7 was selected as a 3-input AND gate called a

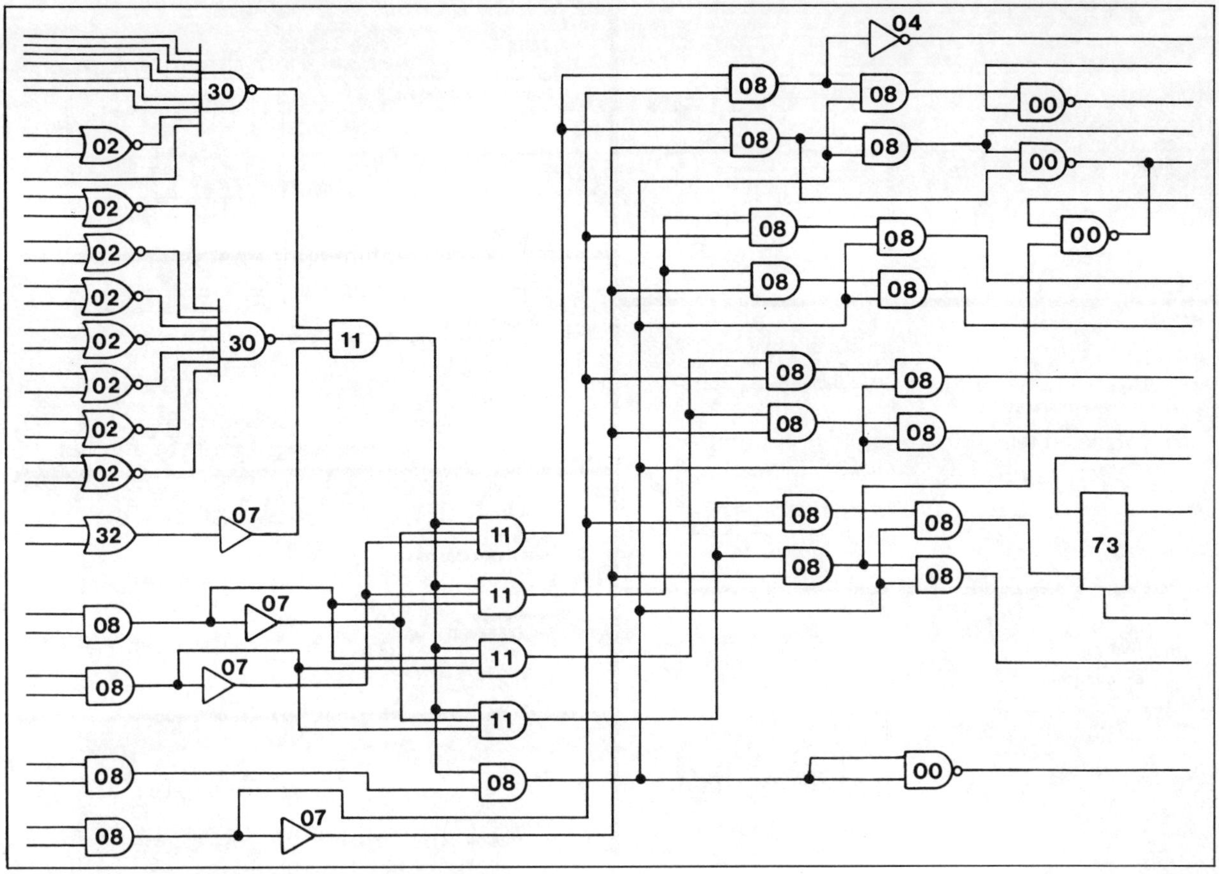

Figure 2-68. GROUPING EXERCISE NO. 4

7411. Reviewing your specification sheet, you can see that there are three in a package. These were placed together because there was intercommunication to each other and because there was only a total of five AND gates of the 3-input 7411 style. This results in zero spares for IC 7. IC 8 picked up the remaining two 7411's. This results in one spare. We changed the 2-input AND (called a 7408) to a 7411. This fills out the 7411 3-input AND package (called IC 8). IC 8 now has zero spares. IC 9 was selected as a buffer package, 7407. It was assigned as an open collector using the flag on the output side of the gate. Even though there are six in the package, our schematic only required four. We ended up with two spares in IC 9. IC 10 was a logical combination of gates (called the 7408). The specification sheet indicated that there were only four in a package; we used all four. This became a good combination because all four of them communicated with the connector on the input side. They all communicated with IC 9 on the output side. The layout now flows nicely from IC 9 to IC 10. IC 10 would be listed with zero spares. IC's 11 and 12 were easy to assign. Looking at your specification sheets, you can see that there is only one gate inside the package (an 8-input AND). Make sure that all of the inputs are accounted for. If you make a count on IC 11, you would find that we came up with one input short. We want to add one input to the schematic and tie it off to one of the used inputs for IC 11. IC's 11 and 12 would have zero spares. IC's 13 and 14 are 7402's (2-input NOR packages). According to your specification sheet, there are four gates inside a 7402. A total of 8 is re-

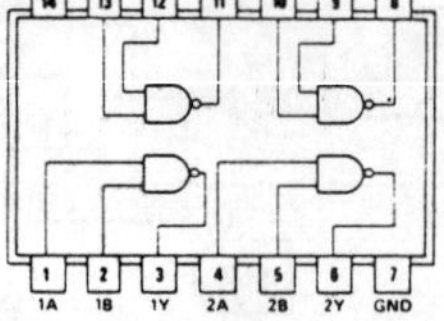

7402

QUADRUPLE 2-INPUT
POSITIVE-NOR GATES

positive logic:
$Y = \overline{A+B}$

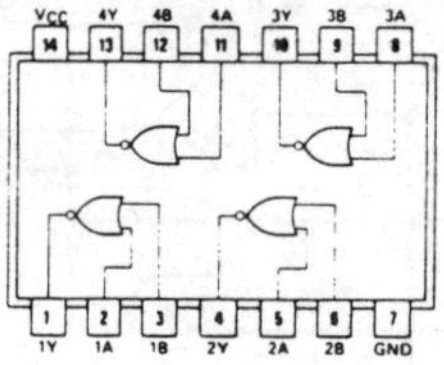

7404

HEX INVERTERS

positive logic:
$Y = \bar{A}$

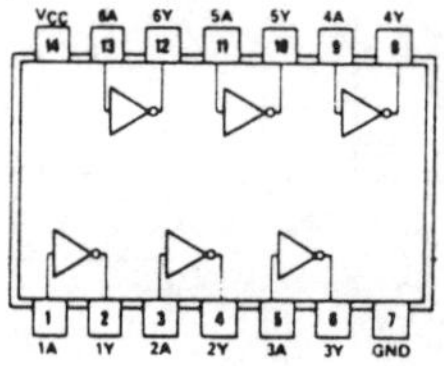

7408

QUADRUPLE 2-INPUT
POSITIVE-AND GATES

positive logic:
Y = AB

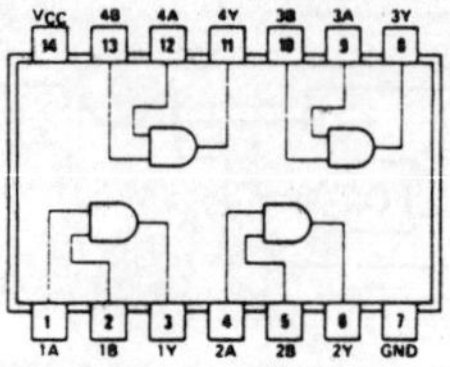

11

TRIPLE 3-INPUT
POSITIVE-AND GATES

positive logic:
Y = ABC

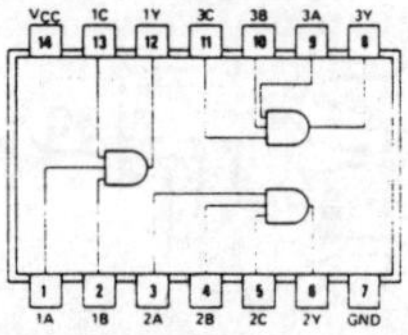

7430

8-INPUT
POSITIVE-NAND GATES

positive logic:
$Y = \overline{ABCDEFGH}$

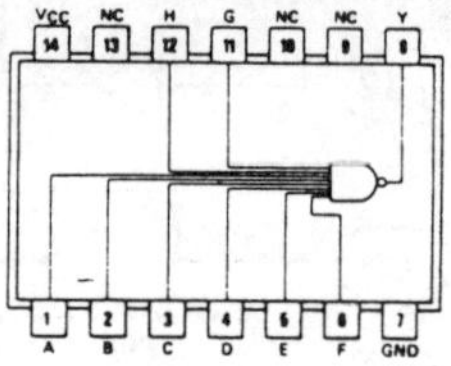

7432

QUADRUPLE 2-INPUT
POSITIVE-OR GATES

positive logic:
Y = A+B

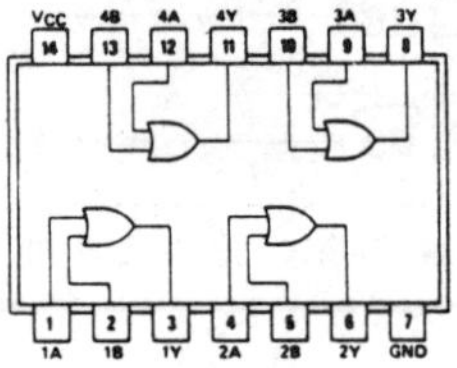

07

HEX BUFFERS/DRIVERS
WITH OPEN-COLLECTOR
HIGH-VOLTAGE OUTPUTS

positive logic:
Y = A

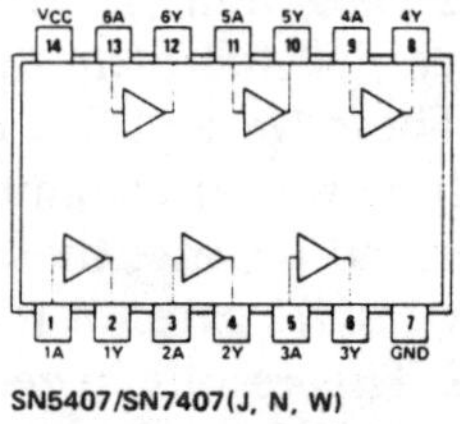

SN5407/SN7407(J, N, W)

DUAL J-K FLIP-FLOPS WITH CLEAR

73

'73, 'H73, 'L73
FUNCTION TABLE

INPUTS				OUTPUTS	
CLEAR	CLOCK	J	K	Q	$\bar{Q}$
L	X	X	X	L	H
H	⎍	L	L	Q_0	$\bar{Q}_0$
H	⎍	H	L	H	L
H	⎍	L	H	L	H
H	⎍	H	H	TOGGLE	

'LS73
FUNCTION TABLE

INPUTS				OUTPUTS	
CLEAR	CLOCK	J	K	Q	$\bar{Q}$
L	X	X	X	L	H
H	↓	L	L	Q_0	$\bar{Q}_0$
H	↓	H	L	H	L
H	↓	L	H	L	H
H	↓	H	H	TOGGLE	
H	H	X	X	Q_0	$\bar{Q}_0$

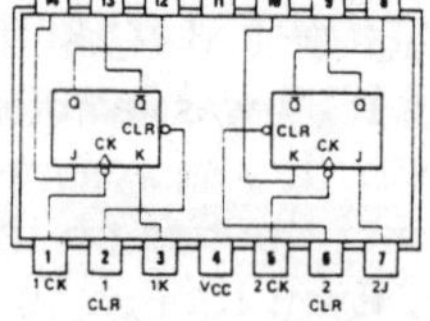

SN5473 (J, W) SN7473 (J, N)
SN54H73 (J, W) SN74H73 (J, N)
SN54L73 (J, T) SN74L73 (J, N)
SN54LS73 (J, W) SN74LS73 (J, N)

Figure 2-69.

quired on the schematic. Therefore, IC's 13 and 14 would have zero spares.

This becomes again a very nice flow for our PC layout. Most of the inputs on IC's 13 and 14 go to the connector; the outputs are common to IC's 11 and IC 12. IC 15 is a 7432 (a 2-input OR gate). There is only one required on the schematic. On the specification sheet you can see that there are four inside one package. Therefore, IC 15 would be listed with three spares. IC 16 is a 7404 inverter. Two are required on the schematic. The specification sheet states there are six in a package. If we assign these to the schematic, it would leave four spares. Once again, the inverter gates were last because we will combine them into another IC that can act as an inverter gate (such as 7400 called IC 2). Examine the schematic spare count list. Eliminate IC 15 and IC 16 if possible. Are there spares on the schematic that can eliminate both inverters on IC 16, or the single 2-input OR 7432 IC 50. Examine each IC package for a spare count.

Look at each gate to see if we can combine IC 16 inverters into our spares (listed at the bottom of the schematic). IC 1 has one spare. It cannot be used for IC 16 or IC 15. IC 2 has one spare. It can be used for IC 16. However, it cannot be used for IC 15. Can that one spare in IC 2 be used to eliminate one of the gates in IC 16? If there were two spares in IC 2 (7400), we would do that at this time. We would eliminate IC 16 by combining the inverters into the 7400. However, we only have one possible gate in IC 2 that can do that. We would not eliminate IC 16; we would still have a remaining inverter. Therefore, we will not attempt to combine. We will save the 2-input NAND spare in IC 2. This is because it is more valuable to use it as a 2-input NAND spare than it would be to pick up the fifth spare inverter in IC 16.

IC 9 has two spares as buffers. We cannot use those two spares to eliminate IC 16 nor 15. IC 15 has three spares that cannot be used to eliminate IC 16. Likewise IC 16 has four spares that cannot be used to eliminate IC 15.

At this point we have reduced our IC count to the lowest possible number (16). We have assigned spares for future design modifications of this PC board. They will be shown as spares on the final schematic. Their input and output pins will be labeled after the schematic has been assigned all of the input and output numbers at each gate.

8. Are the gates grouped together to achieve both the optimum number of IC's and reasonably short connections from one IC to another? Consult with the electrical engineer about the grouping of the gates and the assignment of the unused inputs (such as IC's 8, 11, and 16).

Alternate Grouping Exercise 4

The objective on schematic Figure 2-72 (an alternate grouping exercise to the schematic in Figure 2-71) is to show you that there is not just one way of grouping your schematic. This becomes a personal judgement. There are several correct ways to do it. Comparing schematic Figure 2-71 to 2-72, you can see that IC's 1 and IC 2 are relatively the same. Keep your spare count exactly the same as in the schematic in Figure 2-71. IC's 3, 4, 5, and 6 have the same spare count, but the grouping of the gates is different. Most likely, the trace lengths on the actual layout of the PC board would not be much different between the two schematics. However, it became a personal judgement as to which gates to place in IC's 3, 4, 5, and 6. IC's 7, 8, 9, 10, 11 and 12 are exactly the same on the two schematics. However, 13 is different. Let's suppose that the engineer requested that there be at least two 2-input NOR's (called 7402) as spare gates. Yet you did not want to increase your total IC count on the PC board. One way is to subdivide the 7402's. Instead of having eight of them on the schematic, create two more for a total of ten. Change the two inverters (called the 7404's) to 7402's. The sixteenth package becomes a 7402. Combine the two 7404's with one of the 7402's into that package. Now

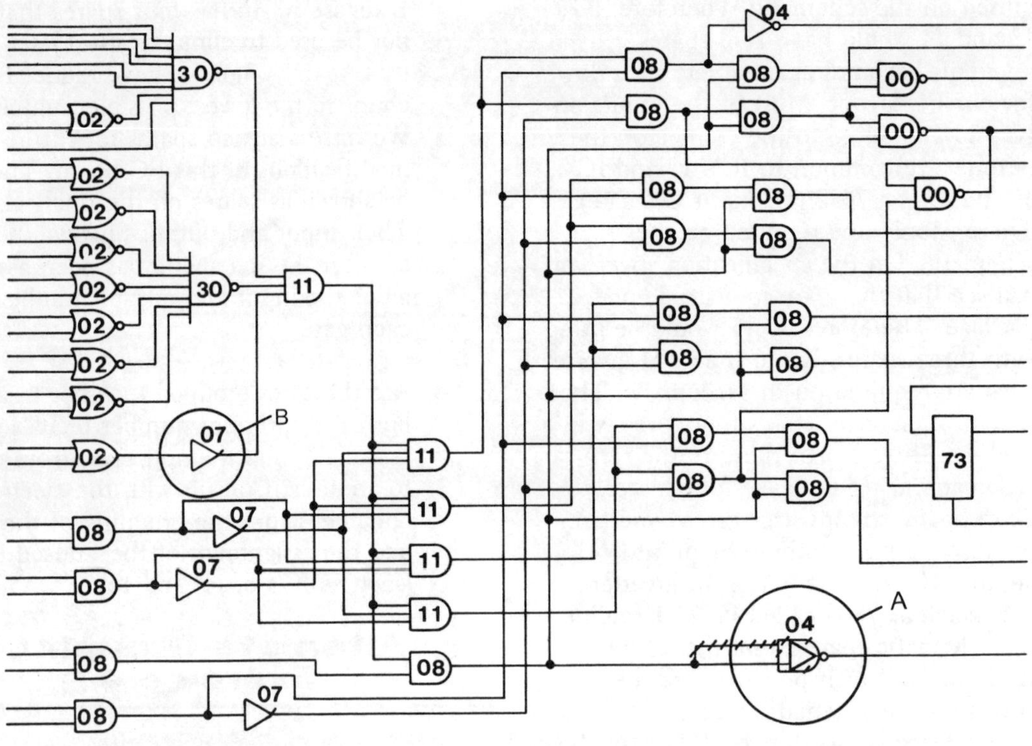

Figure 2-70. GROUPING EXERCISE NO. 4

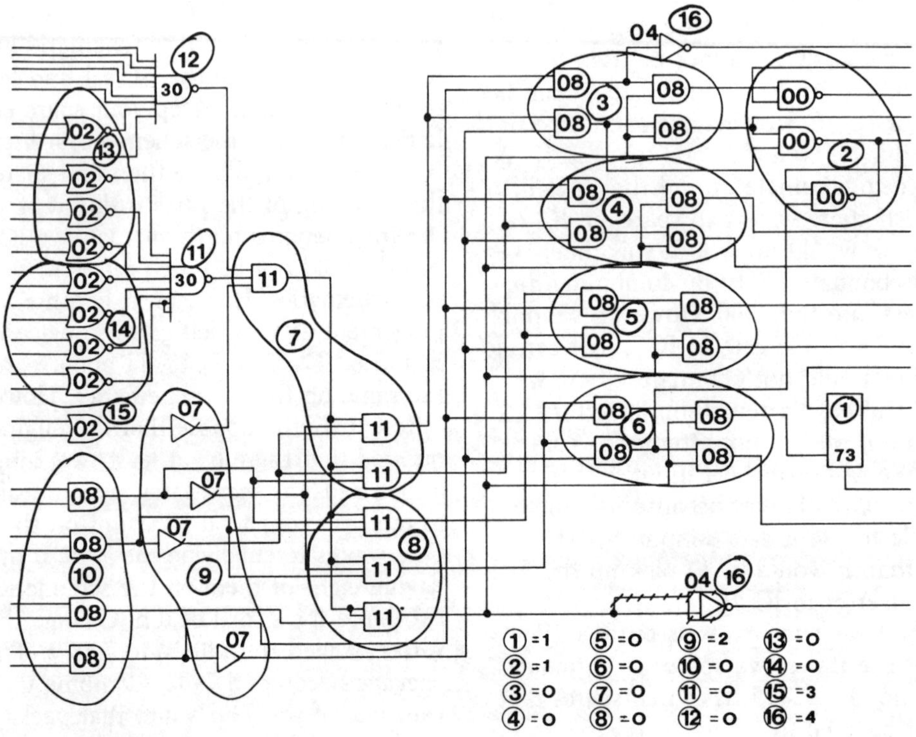

Figure 2-71. GROUPING EXERCISE NO. 4

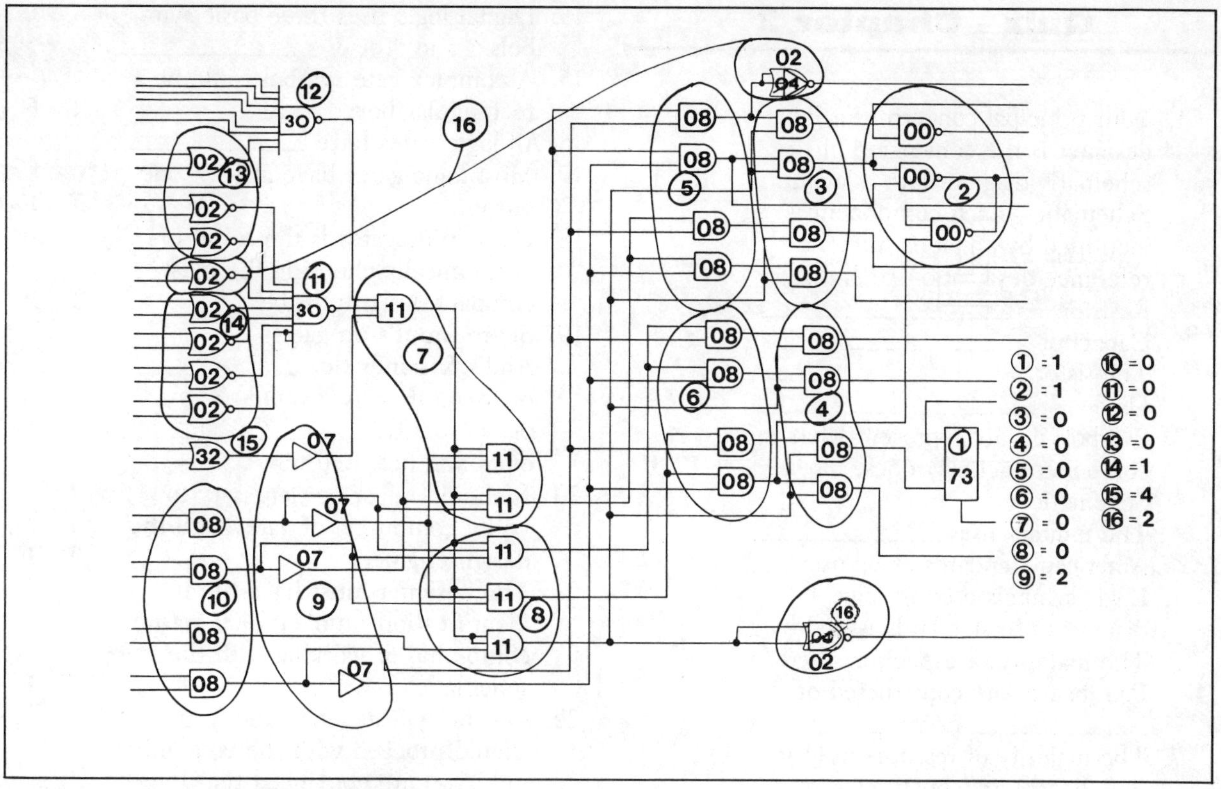

Figure 2-72. GROUPING EXERCISE NO. 4

there is one spare in IC 16 and one spare in IC 13. This satisfies the request for two spares by the engineer. Or we could have left all 7402's in IC 13 by converting the two inverters (called 7404's) into 7402's. This gives a remainder of two spares to satisfy the engineer's request of two spare 2-input NORS. IC's 14 and 15 were not changed.

These four grouping exercises should familiarize you with both the names of the gates and with the various combinations that are available to you to change the gates for a reduced IC count. The exercises acquainted you with the different styles of packages: how many gates are inside each different IC package, and how to prepare for the first layout by grouping and combining gates. There are more grouping exercises for the PC boards later on in this book. Refer to these exercises to review grouping techniques.

Quiz - Chapter 2

1. Your principal concern as a P.C. designer is the conversion of the schematic diagram into a layout. T F
2. Schematic—Each component is identified by an alpha-numeric reference designation such as:
   Resistor = ______________
   Capacitor = ______________
   Transistor = ______________
   Diode = ______________
3. Symbols do not represent what the component actually does. T F
4. Schematics-
   The military uses ______________
   American standards assoc uses ____
5. Logic symbols used in logic diagrams use standards issued by __
6. The majority of capacitors used for P.C. boards are constructed of ______________ or ______________.
7. The majority of resistors used for P.C. boards are constructed of____, ______________ or ______________
8. Name a semiconductor device usually made of silicon having three or more electrodes.
   ____
9. Schematics are drawn showing input line on the left side of the sheet and outputs on the right side. T F
10. Logic symbols are drawn using a template and have only one basic size. T F
11. Logic symbols - to accommodate multiple inputs the logic symbol must be drawn the same as a 2 input. T F
12. The word digital not only means digits but implies binary, which means a two state. T F
13. Name these common IC families
    SSI ______________
    MSI ______________
    LSI ______________
    VLSI ______________
14. Digital logic uses three basic symbols * and * or & ______________
15. A complex gate symbol could be a rectangular box. T F
16. All logic gates have ______________
17. Most logic gates have at least one output. T F
18. Combining gates is the result of economically choosing the most compact design possible. T F
19. Spare Input's on gates:
    And's & nands tie ______________
    or to another used input
    or & Nors tie______________
    or to another used input
20. It should never be necessary to verify combining of gates with the design engineer. T F
21. The designer must have the IC identification number 7400 before combining or working with the schematic. T F
22. On the typical schematic, you should proceed with the combining until you are convinced that the minimum number of IC packs have been achieved. T F
23. It would be poor practice to tie large numbered input gates together to create a single input gate. T F
24. On critical leads you would place those gates that are intercommunicating in different IC packages. T F
25. Open-collector gates can combine together. T F
26. Using the philosophy of attacking the easiest and most obvious grouping combinations first, the P.C.B. designer should look for gates that communicate with each other. T F
27. It is not the P.C. designers responsibility to obtain all logic specification sheets. T F
28. Combining - you cannot intermix an open collector gate with a totem pole/no open collector gate. T F

The answers for each question can be found on the pages listed below:

Question 1. Pg. 11
Question 2. Pg. 12, 14, 15
Question 3. Pg. 12
Question 4. Pg. 13
Question 5. Pg.
Question 6. Pg. 15
Question 7. Pg. 16
Question 8. Pg. 17
Question 9. Pg. 18
Question 10. Pg. 19
Question 11. Pg. 20
Question 12. Pg. 21
Question 13. Pg. 22
Question 14. Pg. 26
Question 15. Pg. 28
Question 16. Pg. 29
Question 17. Pg. 29
Question 18. Pg. 37
Question 19. Pg. 39
Question 20. Pg. 45
Question 21. Pg. 48
Question 22. Pg. 49
Question 23. Pg. 52
Question 24. Pg. 54
Question 25. Pg. 62
Question 26. Pg. 62
Question 27. Pg. 67
Question 28. Pg. 67

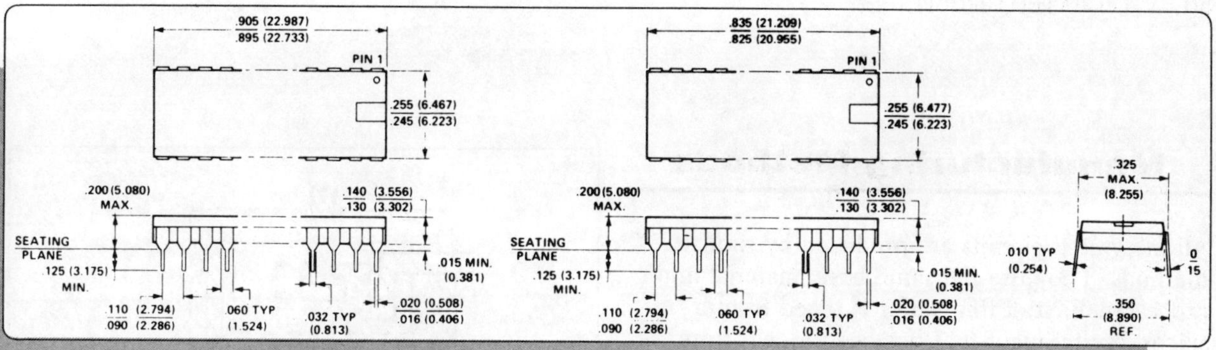

3

INTEGRATED CIRCUITS

So far in the study of digital logic, we have regarded a gate or any other logic element as a unit. We have only talked about the inputs, outputs, and communication to each other without mention of what is inside that makes the element or gate perform its function. The PC designer does not have to understand about gate function. However, it is informative to learn how integrated circuits are manufactured and how their internal circuits function. As you work more with logic gates, you will become interested in their behavior. You will learn of the different circuit conditions and the methods by which they can be recognized. You will learn about the internal structure of integrated circuits and the theory behind it.

This chapter discusses how integrated circuits are made and how their internal components (such as resistors, diodes, transistors, etc.) are interconnected into logic elements. We will discuss the classification of logic gates into their respective families and classify them according to their complexity. We will look at the physical dimensions of the different types of packages that the logic elements are placed into. We will study the electrical characteristics of each type of logic family, covering not only the IC itself but the voltage and ground bussing techniques and decoupling capacitors.

Manufacturing Methods

All integrated circuits are produced by similar methods. They use the same basic material and processes. First, a thin wafer is sliced from a silicon crystal ingot that has been grown from molten silicons. The wafer is polished and inspected under a microscope. It serves as a base called the substrate on which many integrated circuits will be formed. The wafer may be anywhere from three to six inches in diameter, providing space for several hundred integrated circuits. This process was shown in Figure 2-17.

The operation of an integrated circuit is dependent on the interreaction between a P-type (position) and N-type (negative) regions of a single piece of semiconductor (usually silicon). Both P-or N-type regions are made by heating the wafer in a closed, high-temperature oven in the presence of gaseous chemical. The high temperature causes atoms of the chemical to diffuse into the solid wafer silicon. (Figure 3-1). Impurities that change the electrical characteristics of the silicon in this region. By precise arrangement of layers and areas of P-and N-material transistors, diodes and resistors can be produced on the same piece of material. For example, an NPN transistor (Figure 3-1), is produced by diffusing an N-layer, a P-layer, and then another N-layer, each on top of the previous one. Transistors, diodes, or resistors are connected into a circuit by a pattern of aluminum lines (Figure 3-2) deposited on the surface of the wafer.

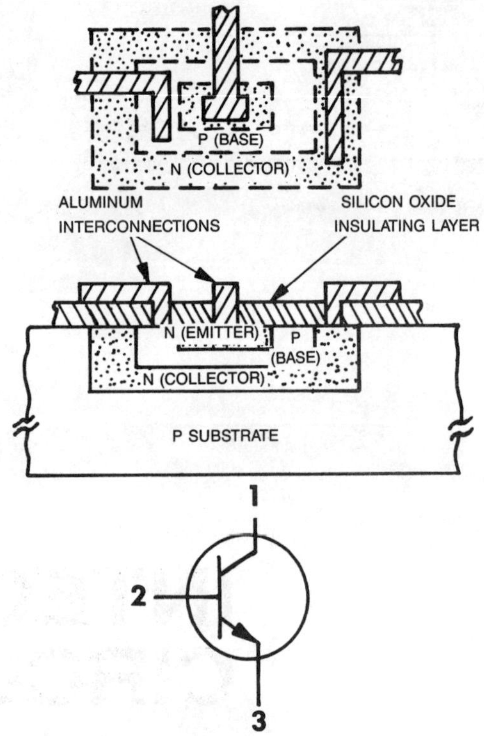

Figure 3-1. Cross section of substrate, showing P-type and N-type regions of an NPN bipolar transistor

Bipolar and MOS Integrated Circuits

There are two basic types of integrated circuits: bipolar and MOS (metal-oxide-semiconductor). The difference between the two is the type of transistor formed on the substrate bipolar junction transistors, whether NPN-or PNP-type are constructed from two layers of semiconductor material. For example, two N-type layers are interconnected by a P-type layer to form an NPN transistor (Figure 3-1). The current flow through the transistor is determined by the amount of current into the center layer (called the base of the transistor). MOS integrated circuits use the field effect transistor (FET).

This type of transistor consist of two areas of the same type of semiconductor, either N or P, interconnected by narrow channels. The current flow through the channel is controlled by voltage on a contact which is insulated from the channel by a thin layer of oxide. These integrated circuits are referred to as complementary MOS (called CMOS). Let us compare the individual characteristics of MOS versus the bipolar. Bipolar integrated circuits can only operate at one power supply voltage, generally consume more power, but they operate at higher speeds. On the other hand, MOS integrated circuits can operate at any power supply voltage, from 5V to 20V. They consume much less power, but are much slower in speed than bipolar.

Most standard logic families are bipolar circuits. The one exception is CMOS which is a logic family of its own. This has been developed and expanded in recent years.

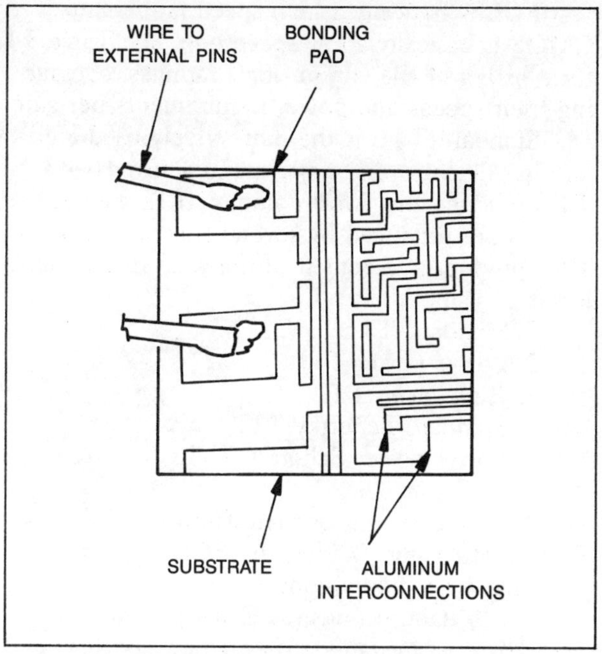

Figure 3-2. Enlarged section of the surface of a wafer, showing the aluminum connecting strips and bonding pads on one integrated circuit

IC Classification

In Chapter 2, we talked about four basic classifications of integrated circuits:

SSI = small scale integration
MSI = medium scale integration
LSI = large scale integration
VLSI = very large scale integration

Integrated circuits are commonly classified according to their complexity. In the early sixties, no more than one or two logic gates could be fabricated on an wafer. The technology has advanced to the point where now several hundred logic gates can be fabricated on a small area of chip. In the process, it has become possible to manufacturer whole, new classes of logic circuits on one integrated circuit chip. New names such as LSI (large scale integration) have been coined to indicate a level of complexity. The definition of these terms that are used today are not exact, but most manufacturers define their complexity of circuits in these four categories: SSI, MSI, LSI and VLSI.

All IC devices in a logic family are compatible with each other. Their logic levels are the same; they can operate from the same power supply voltage; and the output of one gate can supply the proper amount of power to drive the input of another gate. Because each logic family uses different logic element circuits, any two different logic families are not compatible even though they may have similar characteristics.

Six of the largest and best known logic families currently in use are:

TTL = transistor transistor logic
ECL = emitter couple logic
CMOS = complementary metal oxide semiconductor
DTL = diode transistor logic
RTL = resistor transistor logic
HTL = high threshold logic

Each logic family is designed for a different application. Each has its strengths and weaknesses because the basic operating circuit is different in each family. These characteristics determine how well a particular family will work in a given application. For example, circuits in a computer must switch or operate very fast. The logic family with gates that switch extremely fast will be the one selected for that circuit. On the other hand, if the circuit was designed to operate on a very low-power battery (such as a watch), then speed would not be as important as power consumption. Therefore, a different type of logic would be selected for the watch. To give you a comparison among logic families, there are commonly accepted performance characteristics that integrated circuit manufacturer's specify for their products. These specifications usually appear in the manufacturer's catalogue.

1. Propagation: Delay and Speed

Propagation delay is a measure of how rapidly logic level changes, from the input of a gate to its output of the same gate. Speed is directly related to propagation delay, which is typically measured in nanoseconds (ns.). Speed is measured at megahertz, abbreviated: MHz.

2. Power Dissipation

Power dissipation is a measure of both how much power a gate uses and how much heat it generates. It is commonly measured in miliwatts, abbreviated: mW.

3. Fan Out

This is a measure of how much load can be connected to the output of a gate. Because the load normally is another logic element, fan out is usually stated in terms of the number of logic gate inputs that can be connected to one output. If a logic gate has a fan out of ten, it means that ten inputs of other logic gates can be connected to one output.

4. Noise Immunity

This is a measure of how much noise a logic signal can have superimposed on it before the noise causes a gate to change output incorrectly. This immunity is usually specified in milli volts, abbreviated: mV.

There are other parameters that are equally important to the circuit designer or the electrical engineer when he is choosing a logic family for a particular system. For a typical PC design, these are the four areas of interest for digital printed circuit board design.

Logic Family Comparison

There is no single logic family that has all of the best characteristics. Usually a logic family is strong in some features and weaker in others. The nature of an electronic circuit often requires that one characteristic suffer if another is to be obtained. For example, increased speed normally increases power consumption. But this means higher power consumption, which could be a problem. The electrical engineer optimizes all the characteristics of a particular family. For example, standard TTL is a general purpose family. Whereas low power TTL (and perhaps the high-speed TTL) are subfamilies.

ECL is basically a high speed family and CMOS is basically a low speed one. See Table 3-1 for a listing of the various logic families, comparing their speeds and power requirements per gate.

Standard TTL is the most widely used logic family. Almost every major manufacturer has a TTL product line. Most common TTL integrated circuits are produced by several companies. The TTL product line consist of the following families listed in Table 3-1.

1. Standard TTL
2. Lower power TTL
3. High speed TTL
4. Standard Schottky TTL
5. Lower power Schottky TTL

Almost every major manufacturer markets an ECL product line. Within the ECL logic family there are four subdivisions.

1. 8 nanoseconds (ns ECL) gate propagation delay, 30 megahertz, frequency, speed.
2. 4 nanoseconds gate propagation delay, 75 megahertz, frequency, speed.
3. 2 nanoseconds gate propagation delay, 125 megahertz, frequency, speed.
4. 1 nanosecond gate propagation delay, 400 megahertz, frequency, speed.

Because ECL is so fast, it is commonly used in large computers or high density digital communication systems such as satellites—or space vehicles. However, high speed ECL creates problems.

Notice the asterisk on Table 3-1: "Critical leads can be improved by making the trace thicker or shorter". This is especially true to 1 and 2 nanoseconds ECL.

These two logics are running at such high speeds that they require special printed circuit board design techniques. Therefore, they must be segregated from logic like TTL. Also in Table 3-1, it says "if standard IC's and high speed on the same board—break up ground". You must check with the engineer to see if separate grounds are required. Later on in this book, we will give you an explanation of different design techniques required for ECL logic.

Table 3-1. Logic Speed

Name	Family	Speed	PWR Per Gate
Standard TTL	TTL	25 MHz	10 MW
Low Power TTL	TTL	3 MHz	1 MW
High Speed TTL	TTL	50 MHz	22 MW *
Low Power Schottky	TTL	25MHz	2 MW
Standard Schottky	TTl	125 MHz	19 MW *
ECL	ECL	30 MHz to 400 MHz	22 MW to 60 MW *
CMOS	MOS	5 MHz-10 MHz	.01 MW very low

* Critical leads can be improved by making the trace thicker or shorter.

* If standard IC's and high speed on the same board - break up ground.

CMOS

The complementary metal oxide semiconductor (CMOS) logic family is noted for its low power consumption. The CMOS logic family has some operating characteristics that are much different from the bipolar families that we have just finished describing.

1. Low power consumption per gate.
2. Power supply voltage can be varied between 3 and 18 volts.

Each of the above characteristics is usually an advantage over bipolar logic. The low power consumption CMOS is useful in battery powered digital circuits. A major disadvantage of CMOS is its relative slow operating speed.

DTL, RTL, and HTL

Diode transistor logic (DTL) is one of the oldest logic families, but in some cases is still popular. It has not diminished nearly as much perhaps as the RTL. There are several generations of DTL logic that has been introduced by various manufacturers. These logics have been mostly replaced by TTL.

High threshold logic (HTL) is a special purpose family. Its predominate feature is high noise immunity. Therefore, it is used in the industrial environment where electromechanical devices may produce large transient voltages.

Electrical Characteristics

It takes some form of power to operate most digital logic circuits. Typically digital circuits operate from a single +5V power supply. The +5V has a return path called ground. This voltage and ground path is generally drafted onto the PC board in the form of a bus structure. See Figure 3-3. This bus structure must be routed to the various IC's on the PC board. In order to route to the correct pin, use an IC specification sheet that identifies the voltage or the ground pin (Figure 3-4).

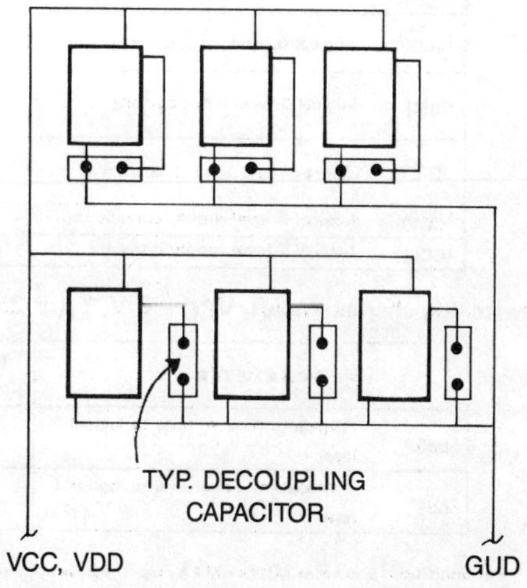

Figure 3-3. Acceptable bus structure

CIRCUIT TYPES SN5404, SN7404
HEX INVERTERS

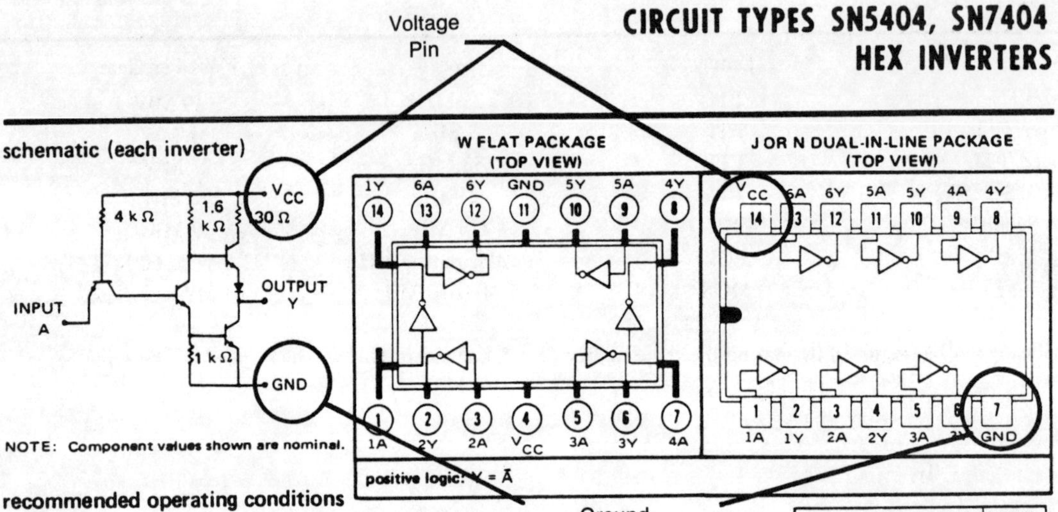

recommended operating conditions

	MIN	NOM	MAX	UNIT
Supply Voltage V_{CC}: SN5404 Circuits	4.5	5	5.5	V
SN7404 Circuits	4.75	5	5.25	V
Normalized Fan-Out From Each Output, N			10	
Operating Free-Air Temperature Range, T_A: SN5404 Circuits	−55	25	125	°C
SN7404 Circuits	0	25	70	°C

electrical characteristics over recommended operating free-air temperature range (unless otherwise noted)

PARAMETER		TEST FIGURE	TEST CONDITIONS†			MIN	TYP‡	MAX	UNIT
$V_{in(1)}$	Logical 1 input voltage required at input terminal to ensure logical 0 level at output	15				2			V
$V_{in(0)}$	Logical 0 input voltage required at any input terminal to ensure logical 1 level at output	16						0.8	V
$V_{out(1)}$	Logical 1 output voltage	16	V_{CC} = MIN, I_{load} = −400 μA	V_{in} = 0.8 V,		2.4	3.3		V
$V_{out(0)}$	Logical 0 output voltage	15	V_{CC} = MIN, I_{sink} = 16 mA	V_{in} = 2 V,			0.22	0.4	V
$I_{in(0)}$	Logical 0 level input current	18	V_{CC} = MAX,	V_{in} = 0.4 V				−1.6	mA
$I_{in(1)}$	Logical 1 level input current	18	V_{CC} = MAX,	V_{in} = 2.4 V				40	μA
			V_{CC} = MAX,	V_{in} = 5.5 V				1	mA
I_{OS}	Short-circuit output current§	19	V_{CC} = MAX		SN5404	−20		−55	mA
					SN7404	−18		−55	
$I_{CC(0)}$	Logical 0 level supply current	20	V_{CC} = MAX,	V_{in} = 5 V			18	33	mA
$I_{CC(1)}$	Logical 1 level supply current	20	V_{CC} = MAX,	V_{in} = 0			6	12	mA

switching characteristics, V_{CC} = 5 V, T_A = 25°C, N = 10

PARAMETER		TEST FIGURE	TEST CONDITIONS		MIN	TYP	MAX	UNIT
t_{pd0}	Propagation delay time to logical 0 level	65	C_L = 15 pF,	R_L = 400 Ω		8	15	ns
t_{pd1}	Propagation delay time to logical 1 level	65	C_L = 15 pF,	R_L = 400 Ω		12	22	ns

† For conditions shown as MIN or MAX, use the appropriate value specified under recommended operating conditions for the applicable device type.
‡ All typical values are at V_{CC} = 5 V, T_A = 25°C.
§ Not more than one output should be shorted at a time.

Figure 3-4. IC Specification Sheet

Note that the voltage pin is referred to as VCC (symbol for collector-voltage supply of a bipolar transistor) pin 14 and the ground pin is called out as GND 7. There are other symbols used for voltage pin descriptions such as VDD (drain-voltage supply). In the case of VCC, there are bipolar transistors inside the IC. In the case of the VDD, there are field effect transistors (FET). Likewise, it could be VEE (emitter-voltage supply). Again, a bipolar type of transistor is being used.

The example in Figure 3-4 for the voltage and ground pin is a 7405 open collector inverter gate (a 14 pin package). The total pin quantity on IC's can change. For example, Figure 3-5 shows a 16-pin package. In this case, the voltage pin (VCC) is pin 16 and the ground pin is 8. The voltage and ground pins can be located on different pins of the IC. On Figure 3-6, the voltage pin is now on the low numbering side of the IC pin 5. The ground is on the high numbering side of the IC pin 12. To determine where the voltage and ground pins will be bussed in Figure 3-3, we must identify every IC on the PC board with its voltage and ground.

Consider packaging the actual case size for integrated circuits. Figure 3-7 shows a dual-in-line package (DIP). It is a chip which has the logic elements located on it. The chip is encapsulated into the package. Leads interconnect from the chip to the pins on the outer edges of the package. The package can be either plastic or ceramic material. Figure 3-8 shows the various sizes of the DIP. The most popular package is shown in Figure 3-7 (a 14-pin dual-in-line). But, as you can see in Figure 3-8, it also comes in 16, 18, 20, 22, 24, 28, and 40 pins. There are two other packages not shown in Figure 3-8: a 4-pin and a 6-pin dual-in-line. All these different configurations have the same spacing (.100) from center of pin to center of pin along the edges. However, the spacing between rows of pins changes as the size of the package increases. For example: A 14-, 16-, 18-, 20-pin package is .300 from the center of one row of pins to the center of the other. But a 22-pin is .400 from center to center. A 24-pin is .600 from center-to-center: a 28- and a 40-pin are both .600.

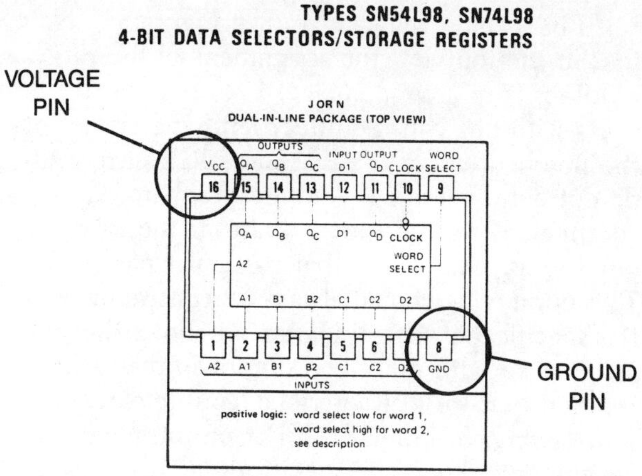

Figure 3-5. 16 Pin DIP Voltage Pins

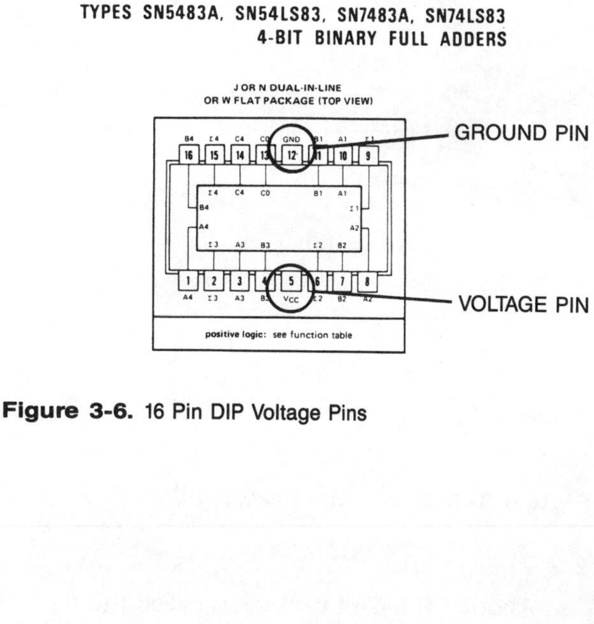

Figure 3-6. 16 Pin DIP Voltage Pins

Pin spacing will change. See Figure 3-7. Note that in the top view the assignment of the pin numbers is in a counter-clockwise rotation. It starts with pin 1 in the upper left-hand corner of the notched end of the IC. Some IC's have a little dot denoting pin 1. And are numbered in a counterclockwise direction whatever the total quantity of pins are on that particular package. The numbering should always be consistent with the specification sheet. Figure 3-4 shows the pins assigned to both input and output in the same rotation that we just described (counterclockwise: 1 through 7, 8 through 14). For pin number assignment, always use specification sheets to verify the pin numbers, not only for voltage and ground but also for the input pins and the output pins of each gate.

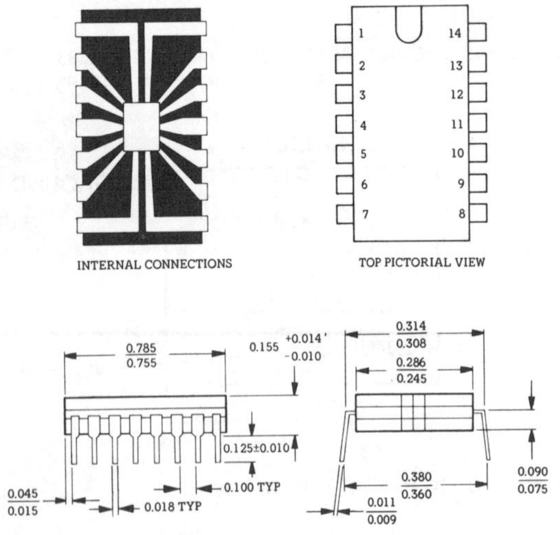

Figure 3-7. Dual-in-line package (DIP)

Another standard package is called the flat pack. A flat pack has the same parts, but its shape is different; the leads come out on the sides. The outer case is usually made of ceramic material. Figure 3-9 shows three basic configurations for the flat pack: 14-, 16-, and a 24-pin. Note that the flat pack has its leads on a much smaller more compact spacing than the dual-in-line. The leads are on .050 spacing center-to-center. DIP leads go straight down from the body. You would place the leads into holes on the PC board which allows wave-soldering, re-flowing, dip soldering or even hand soldering. Flat pack leads do not penetrate through the PC board. They are on top or on the component side of the PC board. For low quantities, you would hand solder each individual lead to the PC board. This makes the flat pack very impractical and expensive to assemble. Most companies today do not use the flat pack configuration. The dual-in-line is more popular because of assembly costs. For explanation of wave-soldering, dip-soldering, or re-flow, see the Analog Design book.

The pin number assignment on the flat pack is numbered identically to the DIP (counterclockwise from the pin in the upper right hand corner on the notched end of the IC).

See Figure 3-4. The flat pack is called out by a "W". The dual-in-line is called by a J or N. These two packages are not always pin compatible. Their voltages, ground input, or output numbers may vary according to the gates. Make sure never to interchange the pin numbers. On the specification sheet cross out which package you are not using. This helps you to avoid confusing pin number assignments.

Decoupling Capacitors

Figure 3-3 gave you an example of a bus structure for voltage and ground pins on IC's. It also gave you an example of strapping the decoupling capacitors across the voltage and ground pins.

Decoupling is the linking of two circuits or devices by electrostatic lines of force. Figure 3-10 gives us an example of a coupling capacitor. There are two basic styles of capacitors used on typical digital PC boards. There is a filtering input capacitor that is placed very close to the input connector pins. It allows AC (alternating current) to pass through, but block DC (direct current). There are two basic circuits used for this. One is called an RC (resistance-capacitance) filtering circuit that is a combination of a polarized capacitor with a fixed resistor (Figure 3-11). The other version of the same input filtering capacitor is called an LC (inductance-capacitance) filtering circuit. Figure 3-12 gives you an example of an LC circuit which has an inductor/coil and a polarized capacitor.

PACKAGING INFORMATION All dimensions in inches and (millimeters)

PLASTIC DUAL IN-LINE PACKAGE TYPE P

16-LEAD PLASTIC DUAL IN-LINE PACKAGE TYPE P

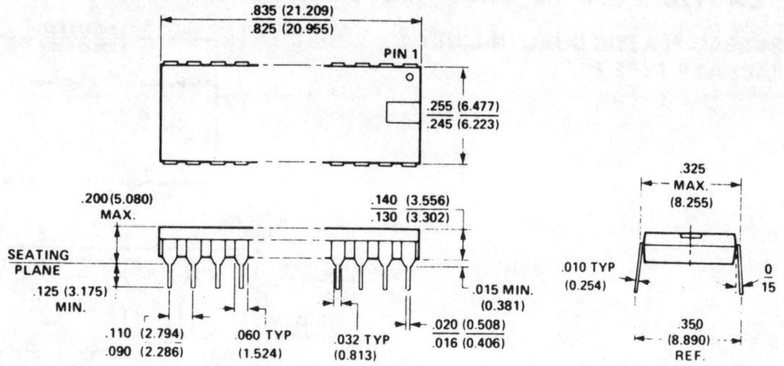

LEADS INSERT ON .100" x .300" CENTERS

18-LEAD PLASTIC DUAL IN-LINE PACKAGE TYPE P

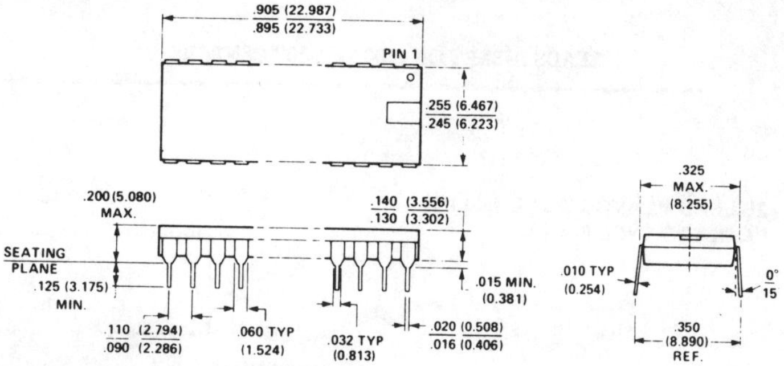

LEADS INSERT ON .100" x .300" CENTERS

20-LEAD PLASTIC DUAL IN-LINE PACKAGE TYPE P

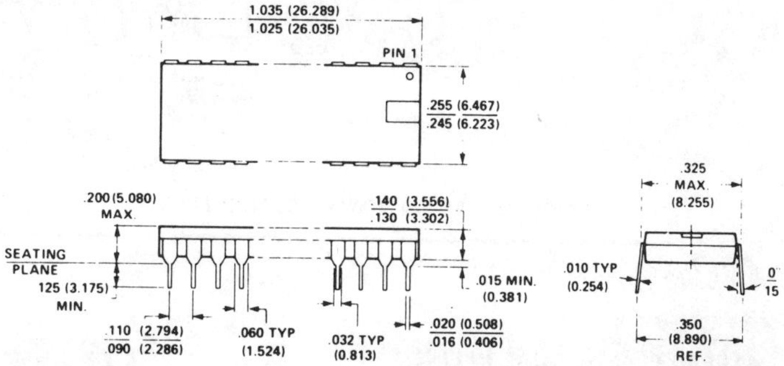

LEADS INSERT ON .100" x .300" CENTERS

22-LEAD PLASTIC DUAL IN-LINE PACKAGE TYPE P

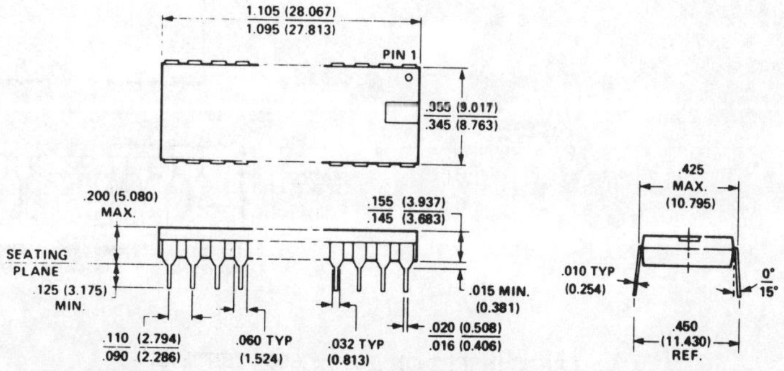

LEADS INSERT ON .100" x .400" CENTERS

Figure 3-8. Various size's of Dual-Inline-Package

PACKAGING INFORMATION All dimensions in inches and (millimeters)

PLASTIC DUAL IN-LINE PACKAGE TYPE P

24-LEAD PLASTIC DUAL IN-LINE PACKAGE TYPE P

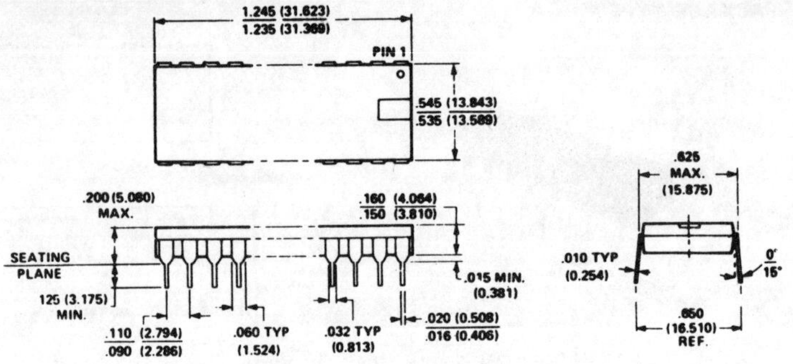

LEADS INSERT ON .100" x .600" CENTERS

28-LEAD PLASTIC DUAL IN-LINE PACKAGE TYPE P

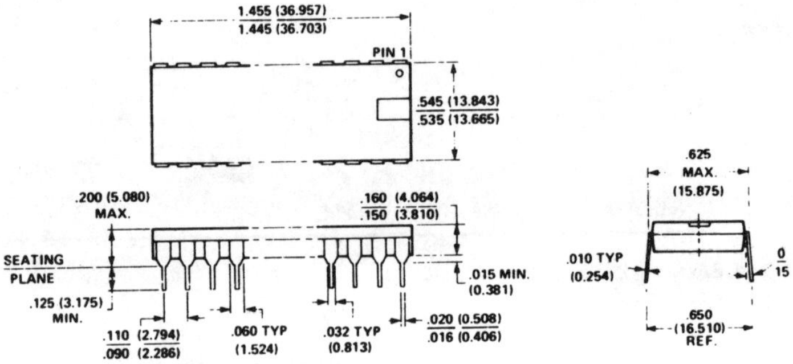

LEADS INSERT ON .100" x .600" CENTERS

40-LEAD PLASTIC DUAL IN-LINE PACKAGE TYPE P

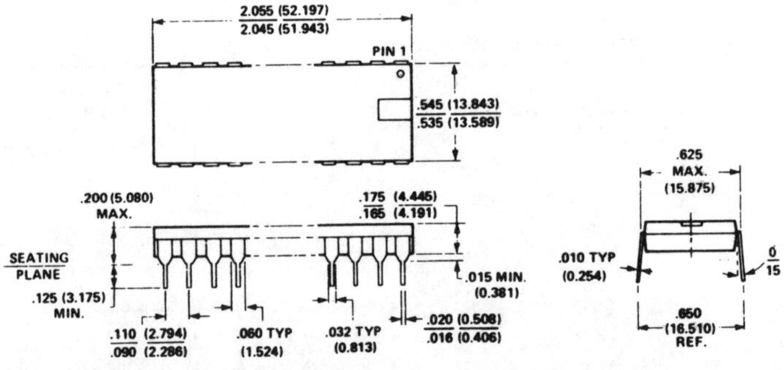

LEADS INSERT ON .100" x .600" CENTERS

Figure 3-8 continued

W ceramic flat packages (inch dimensions, see page 54 for metric dimensions)

These hermetically sealed flat packages consist of an electrically nonconductive ceramic base and cap, and a 14-, 16-, or 24-lead frame. Hermetic sealing is accomplished with glass. Tin-plated ("bright-dipped") leads (—00) require no additional cleaning or processing when used in soldered assembly.

14-PIN W CERAMIC FLAT PACKAGE OUTLINE

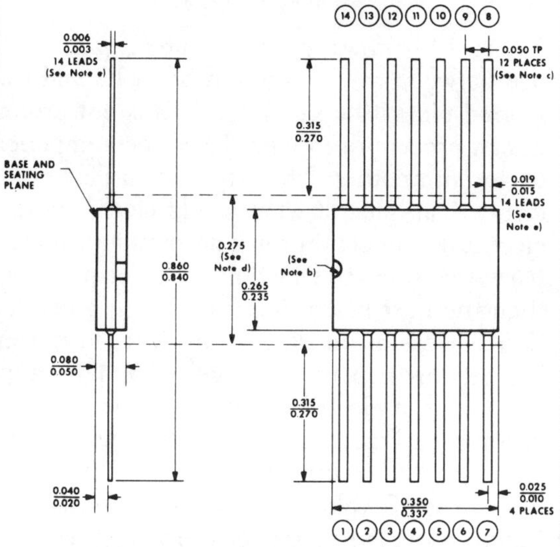

Falls Within
JEDEC MO-004AA Dimensions

16-PIN W CERAMIC FLAT PACKAGE OUTLINE

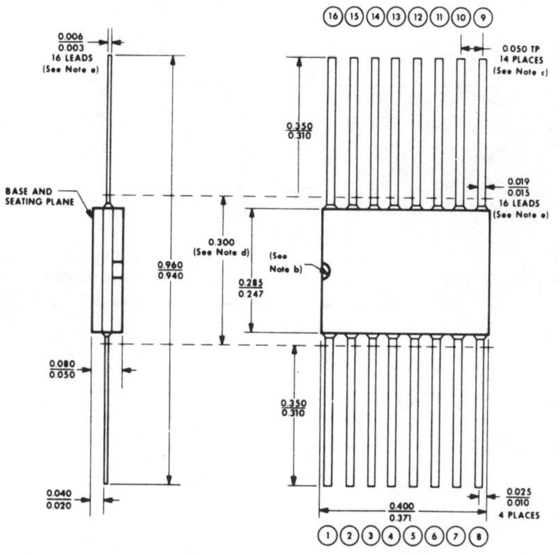

Falls Within
JEDEC MO-004AG Dimensions

24-PIN W CERAMIC FLAT PACKAGE OUTLINE

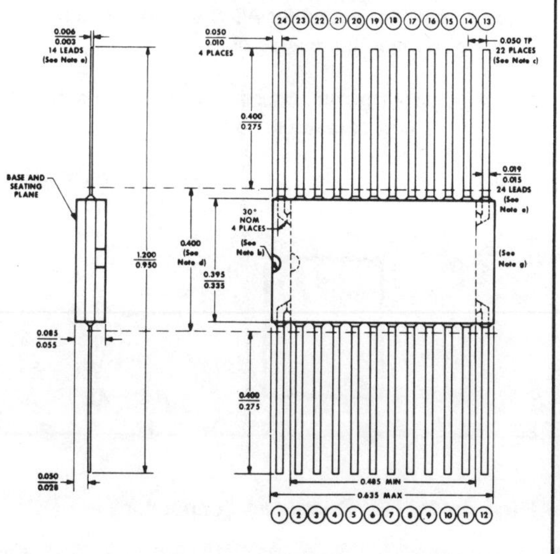

Falls Within
JEDEC MO-019AA Dimensions

NOTES:

a. All dimensions are in inches.
b. Index point is provided on cap for terminal identification only.
c. Leads are within 0.005 radius of true position (TP) at maximum material condition.
d. This dimension determines a zone within which all body and lead irregularities lie.
e. Not applicable for solder-dipped leads.
f. When solder-dipped leads are specified, dipped area extends from lead tip to within 0.050 of package body.
g. End configuration of 24-pin package is at the option of TI.

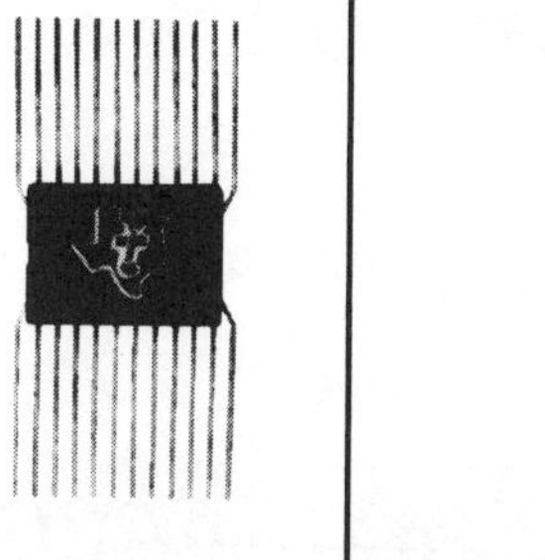

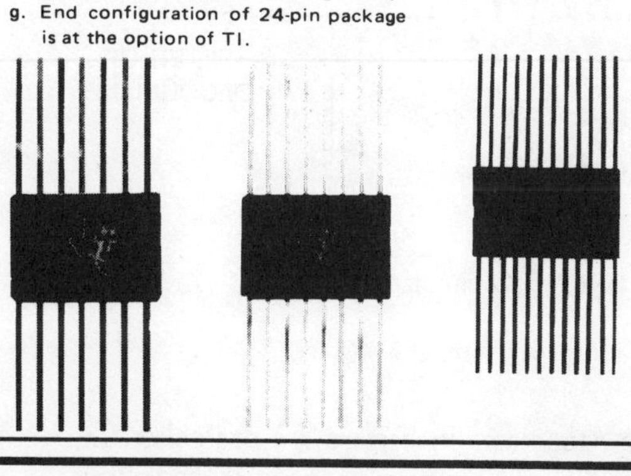

TEXAS INSTRUMENTS

Figure 3-9. TTL integrated circuits mechanical data

Coupling - The linking of two circuits or devices by electrostatic lines of force or by direction connection for the purpose of transferring energy from one to the other.

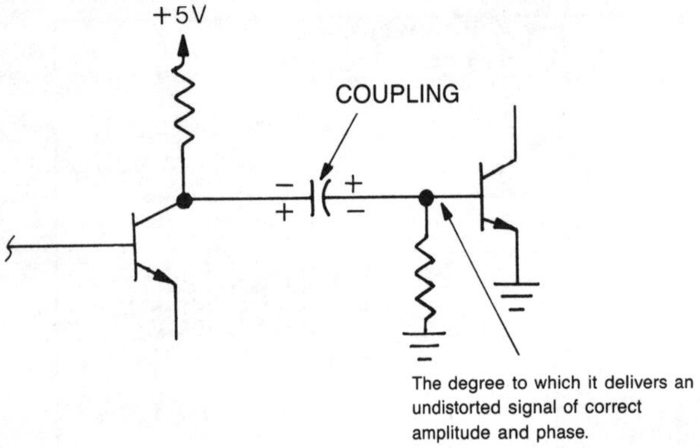

Figure 3-10. Coupling capacitor

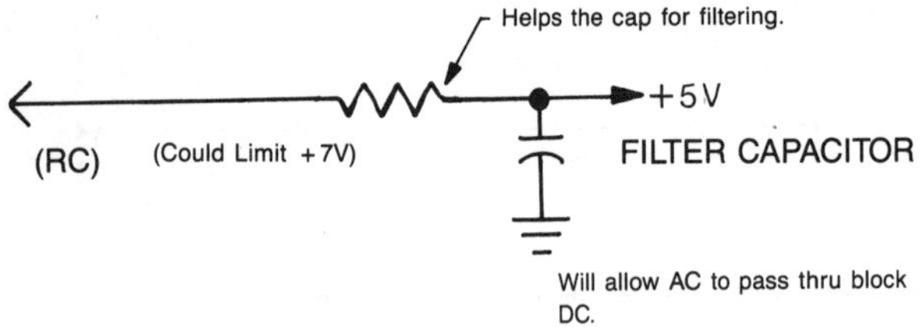

Figure 3-11. Polarized capacitor with fixed resistor

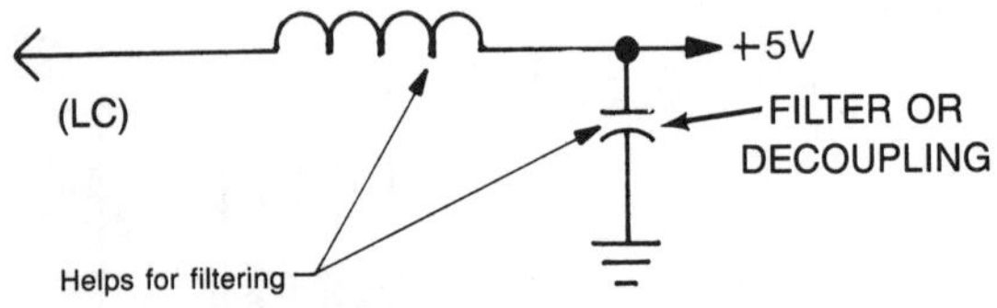

Figure 3-12. LC circuit

The capacitors in Figure 3-3 are called decoupling. Do not confuse them with coupling capacitors, which provide a low impedence path to ground for AC and also prevent common coupling between the stages of a circuit. Figure 3-13 shows that the noise would be removed from the voltage lines and terminated to the ground line through the aid of the decoupling capacitor.

Ideal Electrical Connections

Figure 3-14 shows the decoupling capacitor strapped between the voltage pin of an IC and the ground pin of the same IC. This is not proper design procedure to piggy-back one component on top of another. This was just an illustration to give you an idea of what would be the perfect electrical connection for a decoupling capacitor. However, since this technique is not possible, then the next best thing is to design a configuration as close as possible to this idea. For example, use the bus structure in Figure 3-3. The proper technique of electrical connection would be the example in Figure 3-15. If we change the bus structure (the routing of the voltage and ground bus to the IC voltage and ground pins as in Figure 3-16), the proper electrical connection for the decoupling capacitor on the bus structure of Figure 3-16 is the example in Figure 3-17.

Decoupling capacitor - A capacitor which provides a low impedance path to ground for AC & to prevent common coupling between the stages of a circuit.

* Impedance - The total opposition offered by a circuit or evice to the flow of (alternating current) AC

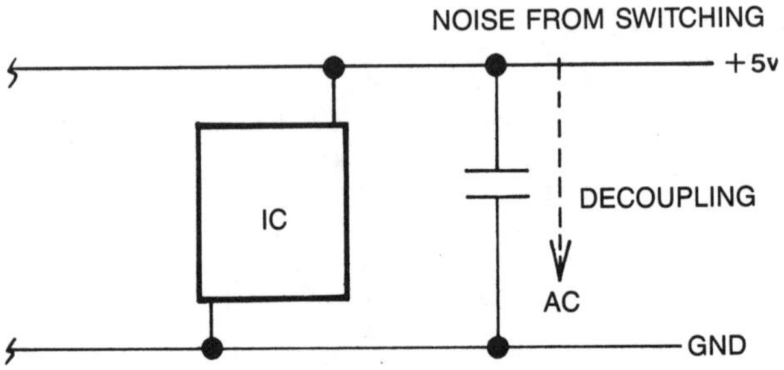

Figure 3-13. Decoupling capacitor

Improper Placement

See Figure 3-18. It is wrong both to strap the decoupling capacitor from the voltage line of one row of IC's and to terminate to the ground row of a completely separate row of IC's (Figure 3-18). It defeats the purpose of the decoupling capacitor used for an IC. The ideal is shown in Figure 3-15.

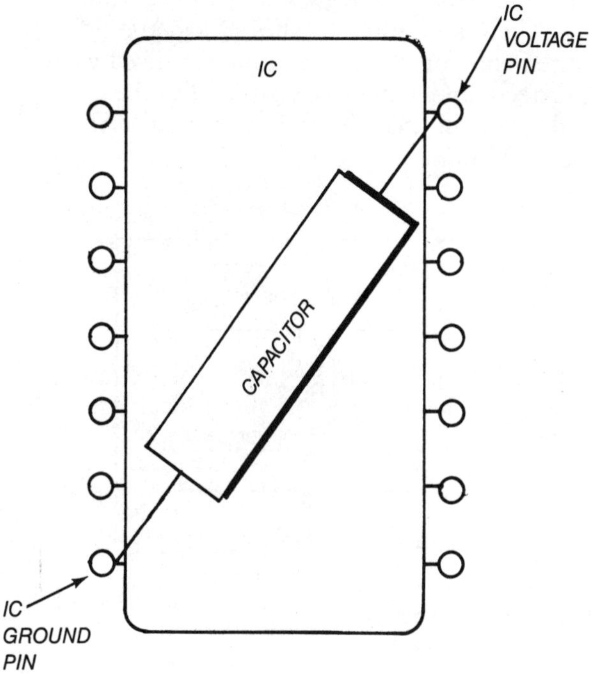

Figure 3-14. Decoupling capacitor between voltage pins

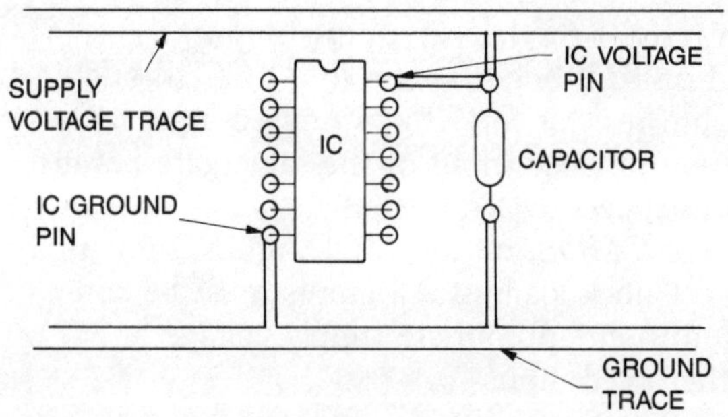

Figure 3-15. Proper technique

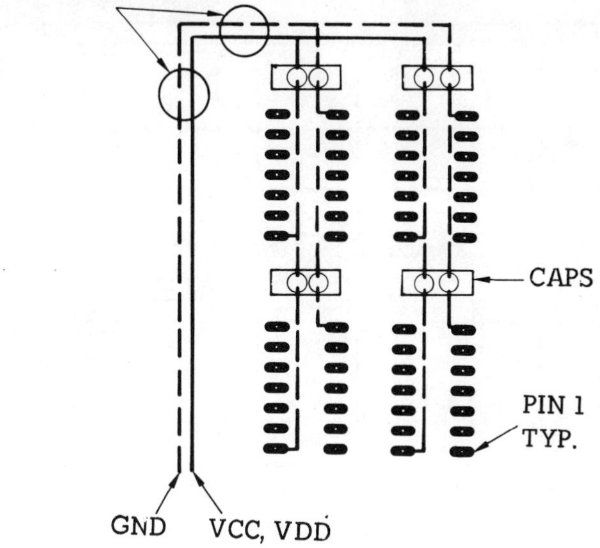

Figure 3-16. Acceptable bus structure

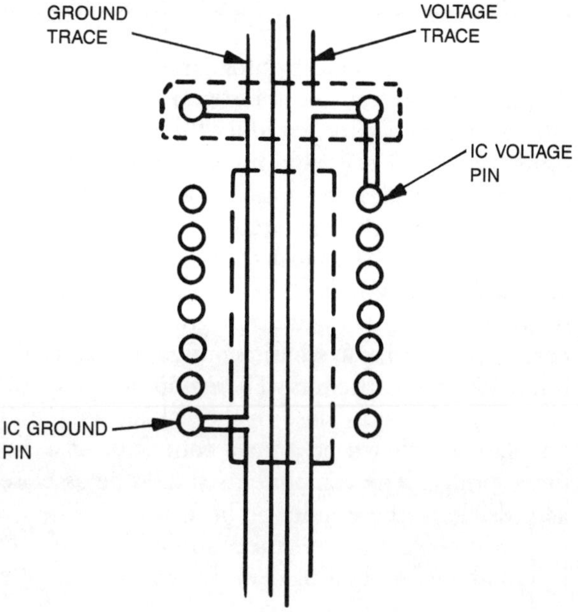

Figure 3-17. Proper placement of IC filter cap's

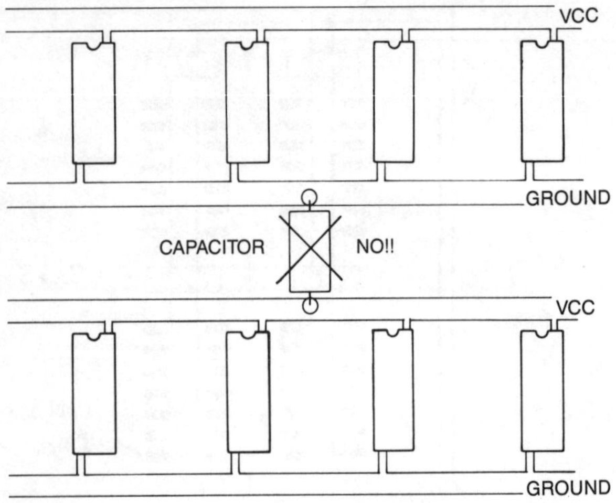

Figure 3-18. Improper placement of IC decoupling capacitor

Input Filtering Capacitors

When input filtering capacitors are installed (Figure 3-19), conducted interference is excluded. These filter capacitors conduct the desired currents, but reject the undesirable currents. The printed circuit designer must be able to recognize the basic input filtering circuits when they appear on a schematic. The most recognizable networks are those dealing with the circuit's supply or operating voltages. Figure 3-20 gives you a description of the five basic configurations, called input filtering capacitors. The components should always be placed as close to the voltage source as possible, whether it be a male connector or a diode bridge. The components should be as close as possible to these input-point locations. The responsible electrical engineer should be consulted about the best placement. Various capacitor materials are used as filters, depending on the frequency impinging on the circuit. The engineer decides the type of capacitors used for your input filtering circuit.

Integrated Circuit Families

Table 3-1 introduced the various logic families that the PC designer would be interfacing to on a typical PC board. TTL logic families were discussed in Chapter 2. We talked about grouping and combining gates to reduce total IC count. All of the rules in Chapter 2 about tying unused inputs together, tying high, or tying low were based on TTL logic. The handling of the unused inputs will differ from one logic family to the next.

There are different types of logic requirements for the unused inputs of used gates. Our recommendations are given with the understanding that the engineer would make all final decisions.

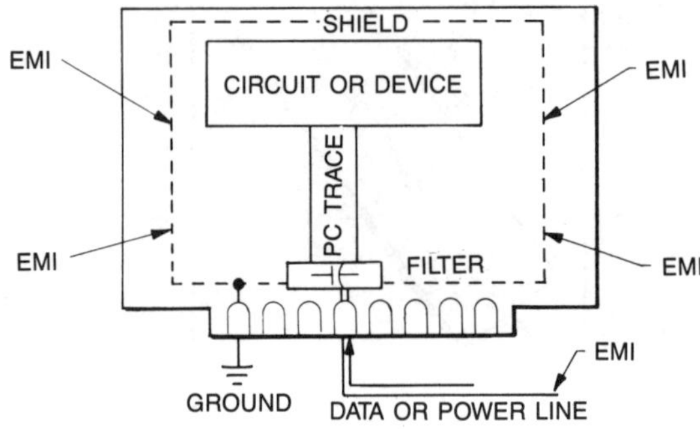

Figure 3-19. Input filtering

Unused Inputs of Used Gates

For TTL, tie to the used input of the same gate. Check loading. OR tie to a VCC through a 1K ohm resistor for 1 to 25 unused inputs.

For TTL Schottky, tie to a VCC through a 1K ohm resistor for 1 to 25 unused inputs. Do not tie to a used input of the same gate because it loses speed.

For CMOS, tie to a VDD, a GND for used input. Check loading. All inputs must be connected to an appropriate supply voltage or another used input.

For ECL, insure high speed operation. All unused inputs should be tied to VEE through a pull-down resistor.

These are the various techniques for tying off unused inputs of a used gates for the logic families. All logic rules are discussed in Figures 2-42 through 2-47. They are based on the TTL logic family.

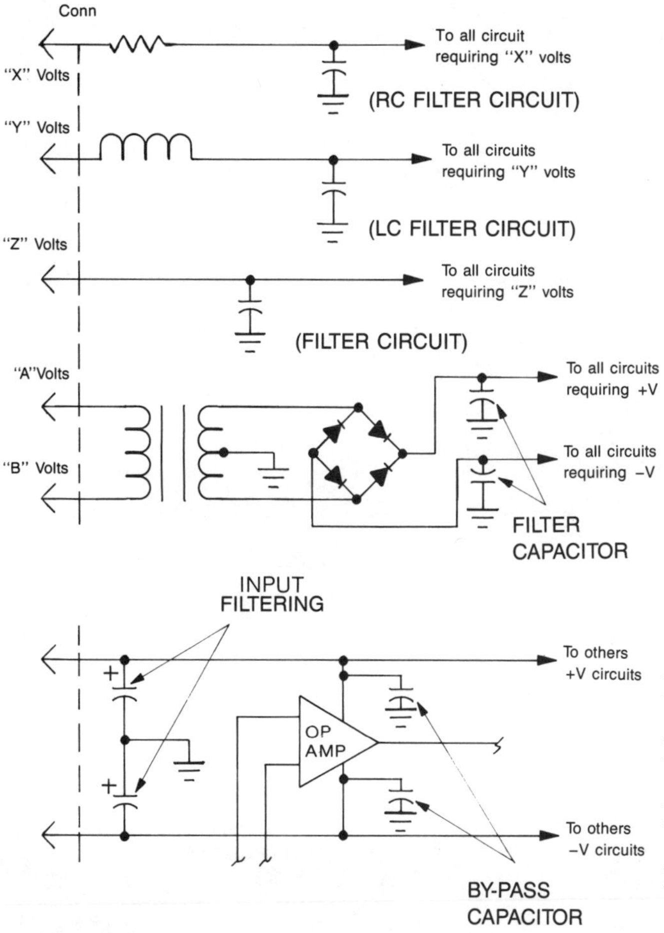

Figure 3-20. By-pass capacitor

Quiz - Chapter 3

1. All integrated circuits are produced by similar methods. They use the same basic material silicon crystal ingot. T F
2. There are two basic types of integrated circuits: Bipolar and ______
3. Name these logic families
   TTL: ______________
   ECL: ______________
   CMOS: ______________
4. Propagation delay is a measure of how rapidly logic level changes from input to output. T F
5. Power dissipation is a measure of how much heat a gate will dissipate. T F
6. Fanout - This is a meaure of how much load can be connected to the input of a gate. T F
7. Noise Immunity: This is a measure of how much noise a logic signal can have superimposed on it before it causes malfunction. T F
8. There is no single logic family that has all of the best characteristics. T F
9. ECL Logic is considered to be a slow speed family. T F
10. TTL logic is a general purpose family. T F
11. CMOS logic is basically a high speed family. T F
12. High speed logic creates problems for P.C. designers. T F
13. High speed & low speed logic if used on the same board should have separate grounds. T F
14. The CMOS logic family is noted for its low power consumption. T F
15. The voltage pin is referred to as VCC = collector voltage supply. There are other symbols such as __________ and __________
16. The IC package can be either plastic or ceramic material. T F
17. Dual-in-line can be purchased in package sizes starting from 6 pin, 8 pin, 14 pin, 16 pin, _____, _____, _____, _____ & _____

18. It is correct to strap a decoupling capacitor from the voltage line of one row of IC's to the 2nd row of a separate row of IC's. **T F**
19. Decoupling is the linking of two circuits or devices by electrostatic lines of force. **T F**
20. Input filtering capactitors should not be placed close to the input connector. **T F**

The answers for each question can be found on the pages listed below:

Question 1. Pg. 84
Question 2. Pg. 84
Question 3. Pg. 85
Question 4. Pg. 85
Question 5. Pg. 86
Question 6. Pg. 86
Question 7. Pg. 86
Question 8. Pg. 86
Question 9. Pg. 86
Question 10. Pg. 86
Question 11. Pg. 86
Question 12. Pg. 86
Question 13. Pg. 86
Question 14. Pg. 87
Question 15. Pg. 89
Question 16. Pg. 89
Question 17. Pg. 89
Question 18. Pg. 95
Question 19. Pg. 90
Question 20. Pg. 90

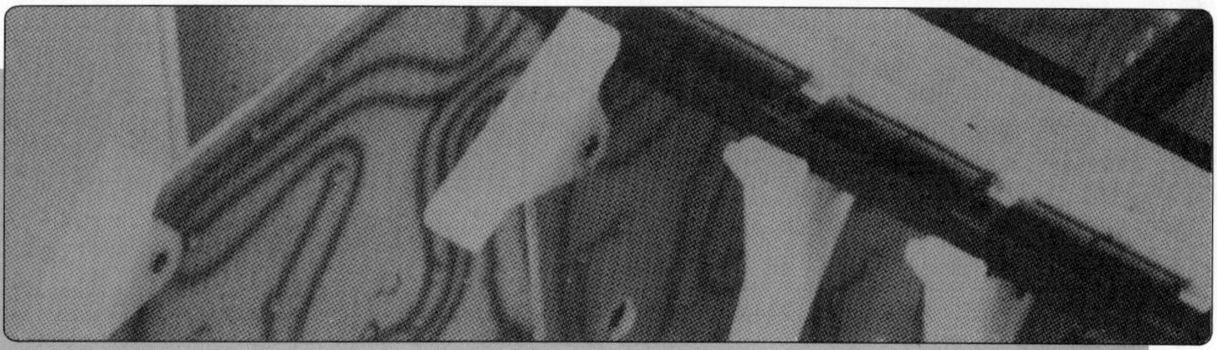

4 PACKAGING CONSIDERATIONS

Packaging considerations were also covered in the Analog Design book. The information in this chapter considers board sizes, environment, board interface, restrictions (mechanical and electrical), and the manufacturability of PC boards. Review the Analog Design book for more extensive coverage of these areas.

Multi-layer PC boards and packaging considerations specific to digital PC boards are introduced in addition to what was covered in the Analog Design book.

Board Outline and Size

Before a printed circuit board is started (locating the components and traces), several items must be defined. One of the first is the working board area as shows in Figure 4-1. The final choice of a printed board size and shape will probably be a compromise. Some factors indicate a large board, others the opposite. This is a review of some of the major factors which affect the selection of a board's size and shape.

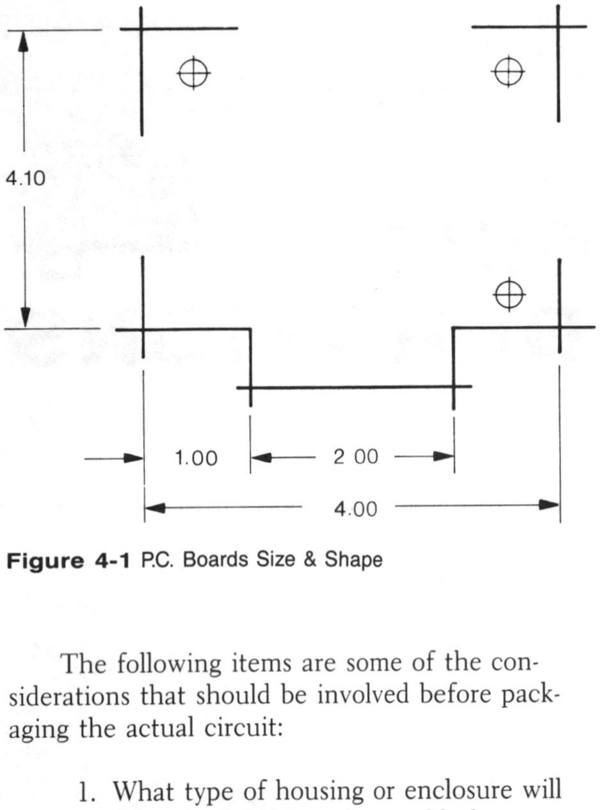

Figure 4-1 P.C. Boards Size & Shape

The following items are some of the considerations that should be involved before packaging the actual circuit:

1. What type of housing or enclosure will the completed board assembly be mounted in? See Figure 4-2.

2. What is the best method for fastening or mounting the PC board for this application (card guides, standoffs, etc)? See Figures 4-3 and 4-4.

3. What type of input/output terminations are available (connector, cable, wires etc)? See Figures 4-5 and 4-6.

4. Will the board require card extractors or special extractor tools to aid in the removal of the board from the enclosure? See Figure 4-7.

5. Is the mechanical integrity of the packaging design high? If the board is large or irregular in shape, are stiffeners or special braces required? See Figure 4-8.

6. What is the desired accessibility for adjustable components? See Figure 4-9.

7. What is the heat dissipation required?

8. What is the shielding required (board-to-board or on the board itself)? See Figure 4-10A & B.

9. What is the type of circuit and its relationship to other circuits (placement and area required).

10. What is the cost of manufacturing? Board manufacturing cost will vary with board size, tolerance etc. (Figure 4-12).

11. Are there tooling holes? See Figure 4-13.

12. What is the *board outline?* See Figure 4-14 & 4-15. Find the board dimensions including connector size and tolerance. (Only two decimal dimensioning).

These questions and others that are typical to any number of design requirements are covered in this chapter.

Some equipment might have to fit into an existing enclosure, which can limit the board size or leave the designer with only a choice from a small number of preferred board sizes. Unless there is a mandatory requirement to use a standard enclosure, the enclosure or case should be designed to fit the system to avoid forcing the system being forced into an enclosure size that may dictate the use of a non-standard or non-optimum board size.

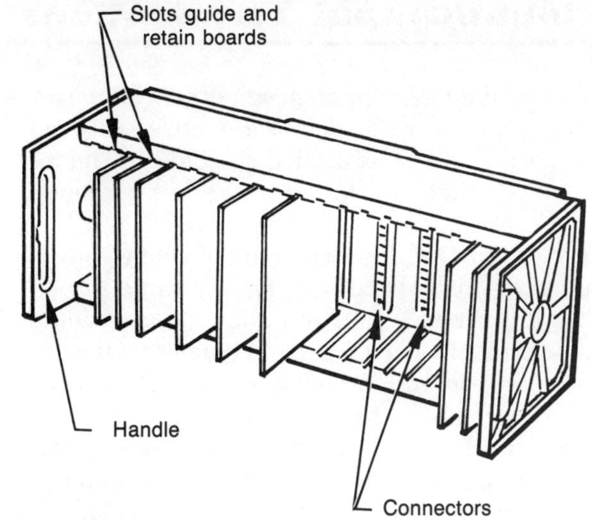

Figure 4-2. Housing or enclosure

Generally the board size is determined by mechanical black box designer via electronic and mechanical engineers.

Methods of Mounting Boards

If the type of electrical interface requires that the PC board card be inserted into a mating connector, then card guides offer the most convenient solution (Figure 4-3). They not only provide a quick "connect/disconnect" capability, but also a means of testing the board out of the unit by means of extender cards. One of the most economical ways to provide this type of interface is by the use of an edge connector. This method not only eliminates the cost of buying an additional connector, but saves manufacturing time by not having to assemble a second connector.

If the electrical interface is by means other than a connector, or a card guide is not practical, then mounting holes must be included on the board so that it can be installed with screws, standoffs or other mechanical fasteners.

Whatever method is chosen, sufficient clearance should be provided so that components or conductors will not interfere or short out to the mounting hardware.

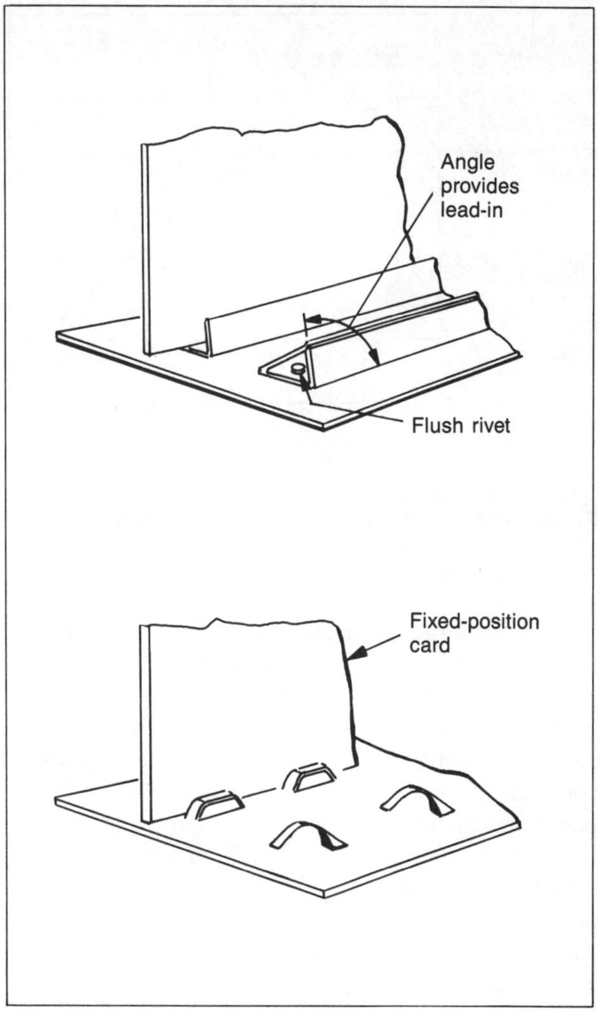

Figure 4-3 Fastening or Mounting

Method of Mounting The Board At The Next Assembly

There are various methods available to secure the board assembly to the next assembly. The application best suited for the particular board must be decided prior to starting the board layout. This is important because, if standoffs are used to mount the board, make sure that components or traces will not be in the position where a hole is drilled or a fastener is installed. In the case of card guides, there must be sufficient area clear of components to allow room for the card guide along the edge of the board.

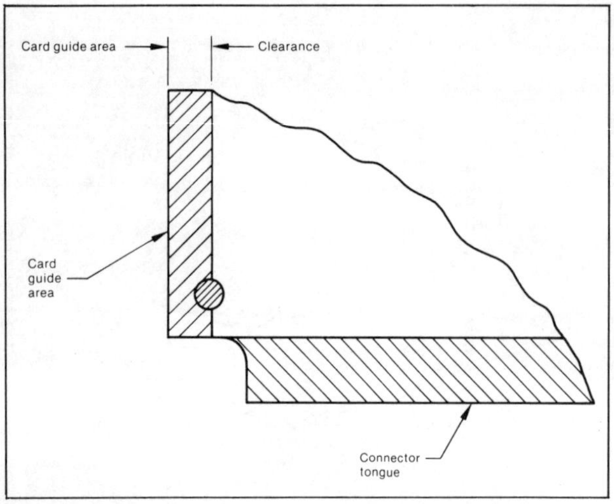

Figure 4-4. Items Associated With Board Mounting

Other items associated with the board mounting are items such as: mounting holes, stiffeners, and metallic or non-metallic card guides. (If card guides are non-metallic, traces may be routed right out to the .060" distance from the edge of the board.) One basic consideration on all printed circuit packaging is the leads that protrude through the non-component side of the printed circuit board. Adequate clearance must be provided for all leads, wires, hardware, etc. to ensure they will not short out to an adjacent board, sheet metal or object mounted underneath the board. The allowance for these leads does not generally exceed .060- to .100-in.: therefore, a minimum clearance of an additional .060- to .100-in. should be designed into the mechanical package. If the board is an exceptionally large one, or made of thinner material, it might require even additional clearance to ensure that flexing of the board under vibration or stress would not cause the leads to short out.

Input/Output Terminations

One of the basic requirements of any PC board is to provide means of electrical interface between the board and its associated equipment. The most common method is to use connectors, terminals, and cables.

1. CONNECTORS - The type of connector or interface that will be used for any particular board is generally chosen during the mechanical design of any given piece of equipment: this is necessary because not only does the PC board have to consider the type of connector or connection, but the associated mechanical package must provide space for the connector or connection that will mate with the printed circuit board. The choice of connectors that are available are too numerous to cover, but there are models and configurations to fit almost any requirement.

The following single part connector, double part connector, discrete connector and wire wrap are just a few of the type of terminations available today.

Single Part Connector

Single part printed circuit connectors are the most common type. They use one edge of the printed circuit board as the plug/connector (which has printed and plated conductors as the connector contacts).

The other half of the single part printed circuit board connector is usually an assembly of mating contacts in a plastic moulded receptacle. See Figure 4-5.

Double Part Connector

Double part connectors consist of self-contained multiple contact plug (male) and receptacle (female) assemblies. Usually the plug/male connector half mounts on the printed circuit board; the receptacle/female connector half mounts to an interconnecting board (motherboard) or back plane. See Figure 4-6.

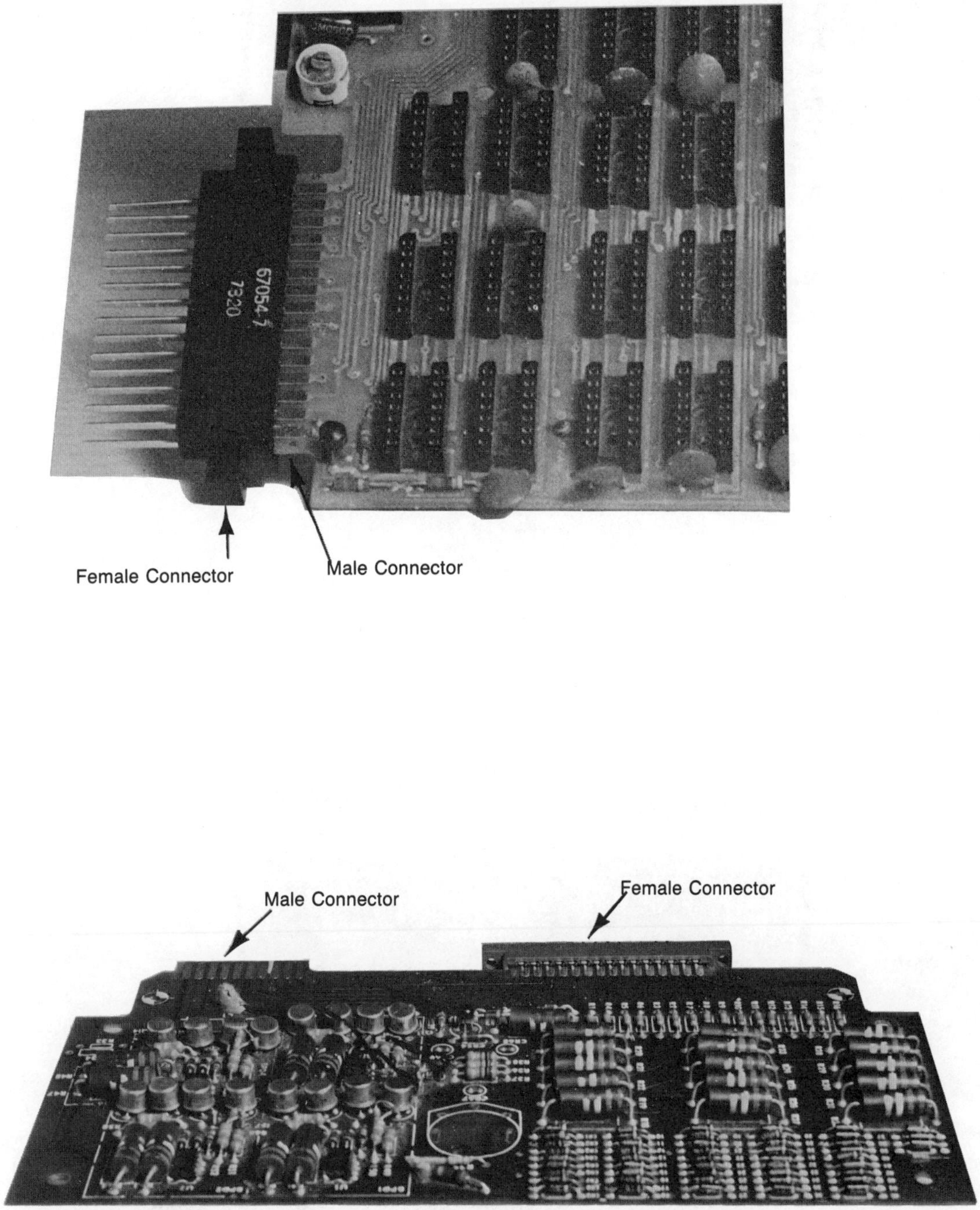

Figure 4-5. Single Part Connector

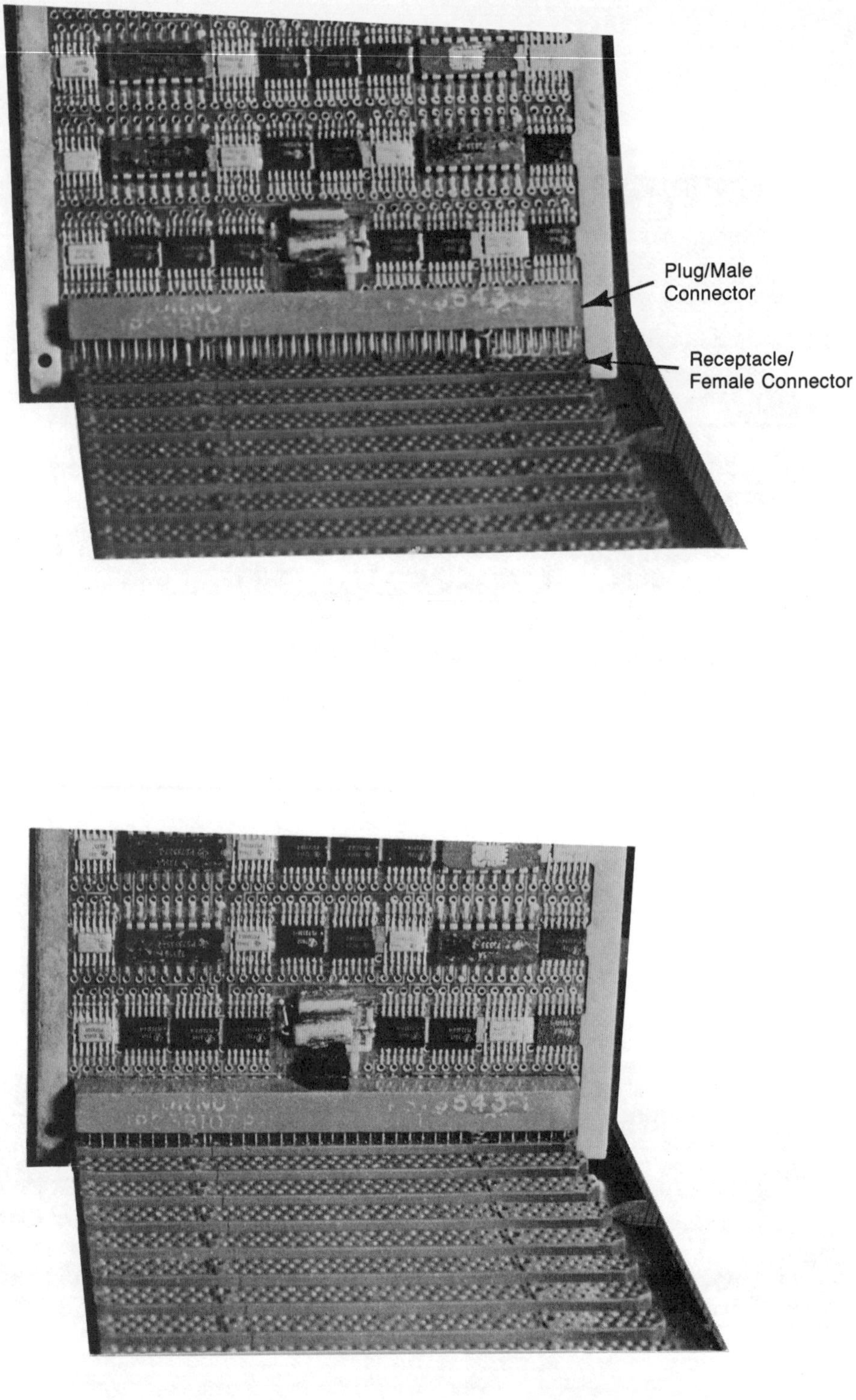

Figure 4-6. Double Part Connector

Some basic disadvantages of using single part and double part connectors are:

A. The connector location is usually limited to one edge of a printed circuit board.

B. The direction of insertion (plug-in) is usually limited.

C. The plug and receptacle halves are limited to a specific number of contacts and spacing.

Tools for Board Removal

There are many kinds of tools for printed circuit board extraction. Many use a minimum of board space, thereby maximizing the amount of board area available for components. They protect the board from damage during extraction. Below are only a few of the board extractors currently being used. There is the simple "Finger Hole" hook type which requires one finger hole in the exposed edge of the printed circuit board. There is a "Coat Hanger" hook type which requires two holes in the exposed edge. Most common are the "Side Grasping" or pin squeeze board extractors. See Figures 4-7A, B, C and D.

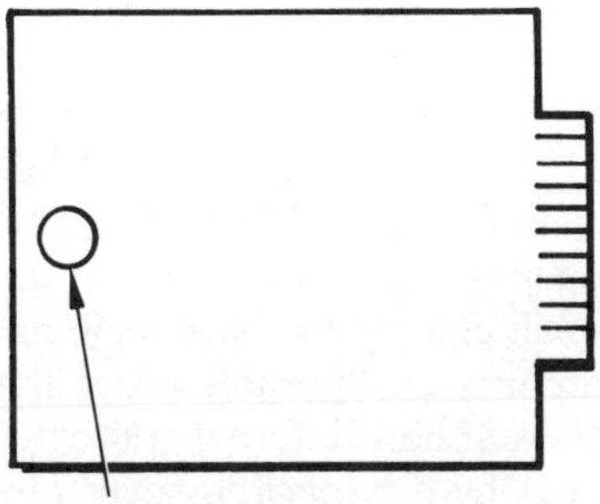
(A)
Finger Hole

Figure 4-7A.

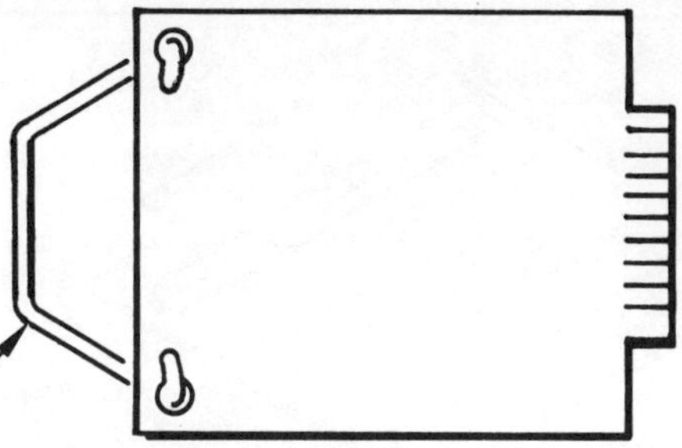
(B)
Hook Type

Figure 4-7B.

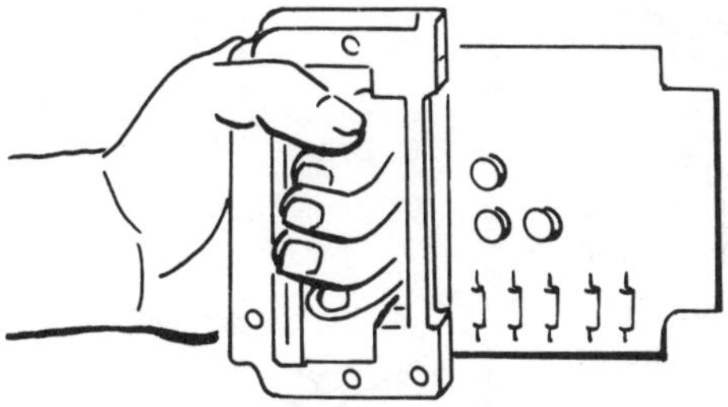
(C)
Cam-action Friction Board Extractor

Figure 4-7C. Tools for Board Removal

(D)
Pull-tab P.C. Board Ejector

Figure 4-7D.

Board Stiffeners and Braces

Stiffeners are used on PC boards to ensure that the board remains flat. One basic consideration is the leads that protrude through the non-component side of the board. Adequate clearance must be maintained for all leads, wires, hardware, etc. to ensure they will not short or damage an adjacent board, sheet metal wall, or object mounted beside or underneath the board. If the board is exceptionally large or made of thinner material, it might require bracing or stiffeners. This avoids the warpage or twisting that might set in the board from wave or flow soldering. See Figure 4-8.

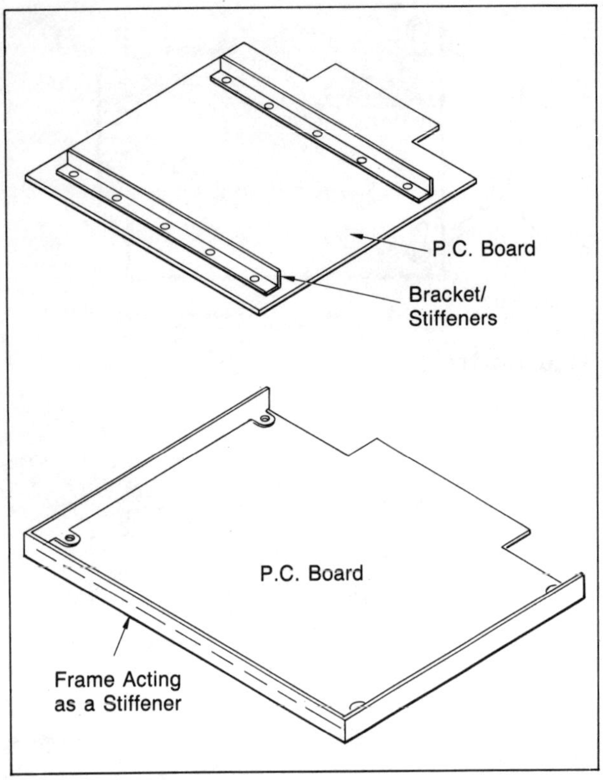

Figure 4-8. Stiffeners

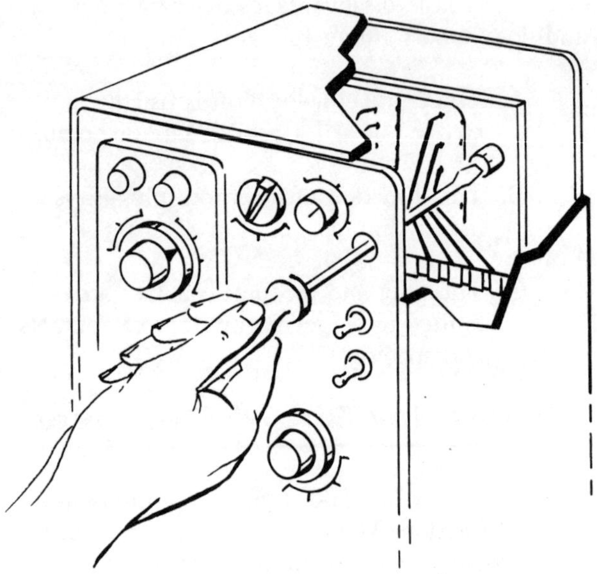

Figure 4-9. Accessing a Printed Circuit Board

Accessibility for Adjustable Components

Adjustable components (usually resistors) are common to many printed circuit board assemblies. Placement of variable components (resistor, pot) on the PC board should be considered for ease of access. While the assembled board is mounted inside the black box most of the time, the adjustment is made while the assembly is on electrically. See Figure 4-9.

Shielding

High-frequency technology is growing in the fields of communication, instrumentation, medicine, control and computation. Designers must consider the need for electromagnetic compatibility. This requirement is caused not only by the increased use of high-frequency techniques within equipment, but by also a vast increase in the number of users of the electromagnetic spectrum in general. Integrated circuit technology has forced equipment into smaller volumes requiring even stricter electromagnetic compatibility within designs.

The printed circuit designer is faced with the problem of shielding. A shield reduces the electromagnetic field strength. A shield contains (or excludes) electromagnetic energy by reflecting or absorbing the energy. Solid shields have discontinuities (cracks) at covers, doors, panels, ventilating openings, etc., many digital instruments as well as small computers, have logic and low-level analog circuitry on the same board that must be protected from both internal and external interference. The electrical engineer must advise the PC designer on shielding techniques. The examples below illustrate shielding techniques using bus bars for walls, containment shielding, and ground planes on the PC board. See Figures 4-10A, B, 4-11.

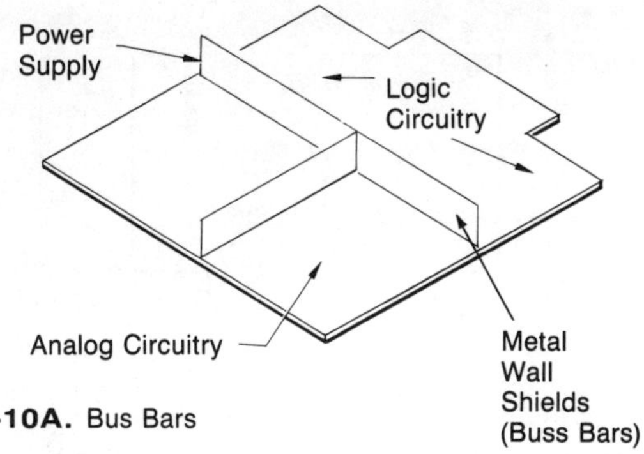

Figure 4-10A. Bus Bars

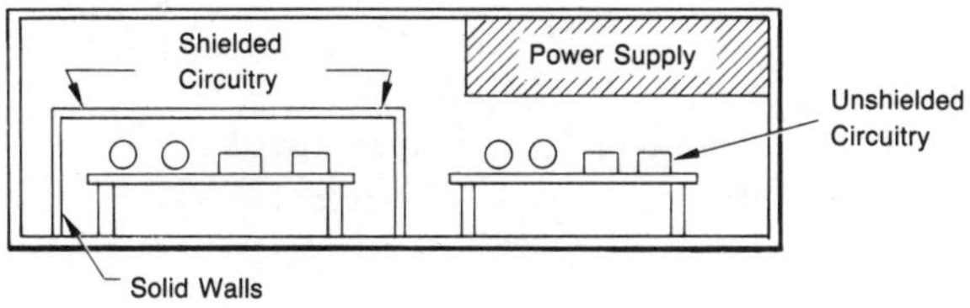

Figure 4-10B. Containment Shielding

Ground Plane

Ground planes are very common on digital PC boards and should be considered in the initial layout. Some components use ground patterns for electrostatic shielding.

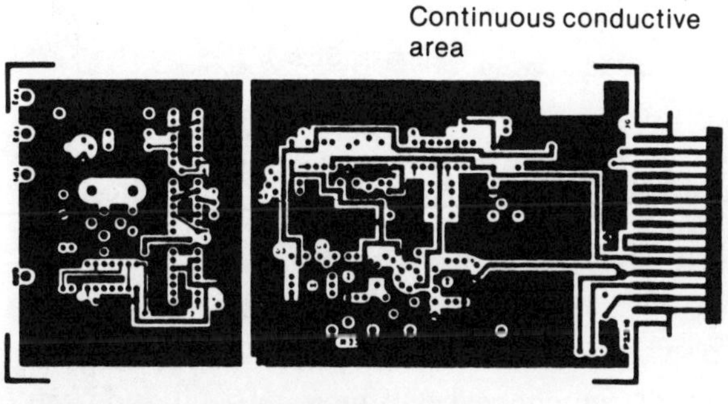

Figure 4-11. Ground Planes

Type of circuit and its relationship to other circuits (placement and area consideration).

1. Predetermine the location of all components influenced by special structural consideration.
2. Arrange the components to achieve the optimum shape for the PC board.
3. Identify the components with short traces or interconnects requirement (engineer will advise).
4. Identify the components that should not be placed near each other, to prevent magnetic or capacitive coupling (engineer will advise).
5. Increase the space between conductors with high potential differences.
6. Make sure that metal-cased components without insulator do not touch each other. The case could be electrically hot and their metal case or mounting hardware does not short circuit patterns on the PC board.
7. Identify where separate grounds must be used on the same schematic (engineer will advise).
8. Identify the placement of variable components on the PC board for ease of access.

Cost of Manufacturing

The board designer directly influences how a board will be manufactured. A design which is unnecessarily stringent, such as narrow conductors with tight airgaps, very small diameter holes (see hole formulas), and pad diameters with small annular rings. As the conductor density rises, yield goes down. As the conductor density decreases, yield increases. Similarly, the designer impacts the reliability of the product in field service. His specification of materials, plating thicknesses, metallic overcoatings, and assembly characteristics directly affects how reliably the assembly will operate. A board may match the fabrication drawing and still not be reliable, if it was underdesigned for its application.

Just as underdesigning can result in increased cost, the same applies to overdesign: increased fabrication cost, lower yields, schedule delays, and higher material cost (such as gold plating) etc. In some cases, particularly in government applications, the cost of added board reliability may be justified by reduced assembly problems. The manufacturability will be discussed in specific design parameters in this chapter, as well as material and fabrication costs.

The printed circuit designer needs to be aware not only of printed circuit fabrication, but also of the quality program for controlling the manufacturing. This evaluation is the final inspection of the product.

Control of design and manufacturing information is the first step toward control of the end product.

Tooling Holes

Tooling holes are used as PC board reference points upon which other dimensions are based. These are noted in Figure 4-12.

The manufacturer will "tool-up" or design his fixtures for drilling the PC board using these tooling holes as a reference point. It is, therefore, necessary that the tooling holes be located accurately according to the dimensions given on the fabrication drawing.

In cases of very dense boards, there may not be sufficient room on the board to locate the pads for the tooling holes. Therefore, locate them outside the perimeter of the board outline. They will appear on the final positive and negative film which the manufacturer will use to produce the board. The same accurate dimensioning would still apply. After the board is complete, they are trimmed with the other excess material.

A hole size of .125 inch diameter is recommended for board tooling holes. The hole tolerance should be held very close. Example .125 ± .001. Any shift of board position due to oversize holes directly affects the permissible component hole location.

To obtain the most accurately located holes, tooling holes should be used with the longest possible span between them, and be on the same edge, if possible.

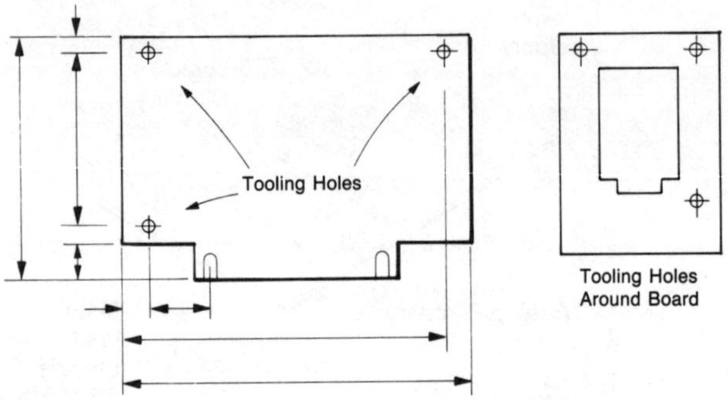

Figure 4-12. Tooling Holes/Board Dimensions

These tooling holes will establish the X and Y axis for fabrication. See Figure 4-13.

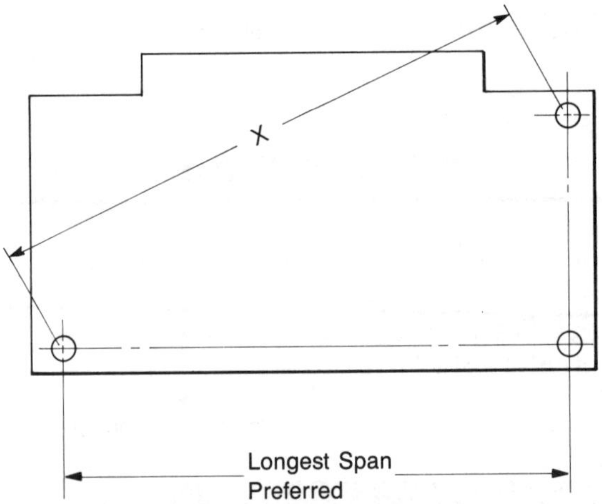

Figure 4-13. Tooling Hole

Although the greater span would be diagonal, such placement makes it more difficult and time consuming to remove the board and install components. Common hole sizes are .100, .125, and .150. The most common hole size is .125.

Board Outline

Select the shape and size of the board to accommodate all components. Fit the available space with the chassis or cabinet area.

The strength of a board is unlikely to limit its size; bowing, warping and the effects of vibration are far more likely to be restrictive. However, large boards (typically "12x12" and larger) are stiffened with metal stiffeners (Figure 4-8).

Accurate dimensions and tolerance are only as useful as a manufacturer's ability to work with them. Combine your design guidelines with the actual capabilities and limitations of the "real world" (know your PC board manufacturer).

All board outlines must be expressed to include some tolerance. Most common are ±.005 and ±.010 inch. The tolerance is important because calculating the actual working board area to its smallest size is the objective. In Figure 4-14, the board dimensions are 4.00 *x* 8.00 inches or a total of 32.00 square inches. The tolerance of .02 inch may not seem like much, but subtracting .02 from 4.00 and 8.00 inches 3.80 *x* 7.80 equals a total of 29.64 square inches, which is significant reduction in board area. The tolerance could be important in conductors and traces on very dense boards, where a trace runs within .040 inch of board edge. The tolerance could be significant, because in this case the trace could come within .02 of board edge, which is very poor design practice. This could damage the traces when cutting the board outline.

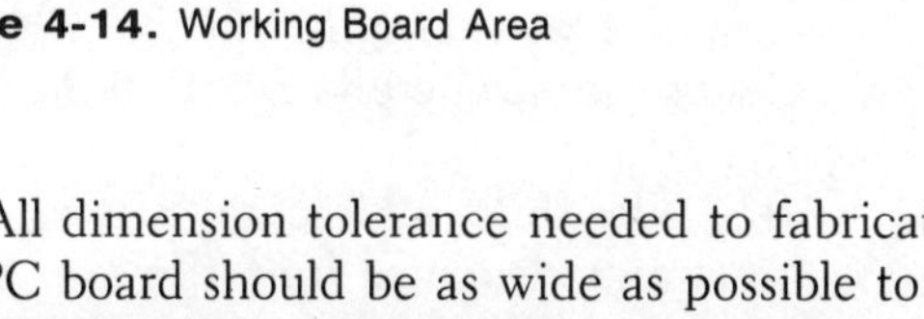

Figure 4-14. Working Board Area

All dimension tolerance needed to fabricate the PC board should be as wide as possible to reduce manufacturing cost. The two most common tolerances are ±.005 and ±.010 inch. They are cost effective because they are "real world" limitations. If the designer chooses a tolerance of ±.002 or ±.001, it is very difficult and costly to produce. Some board outlines have dimensions that are not as critical as others (Figure 4-15).

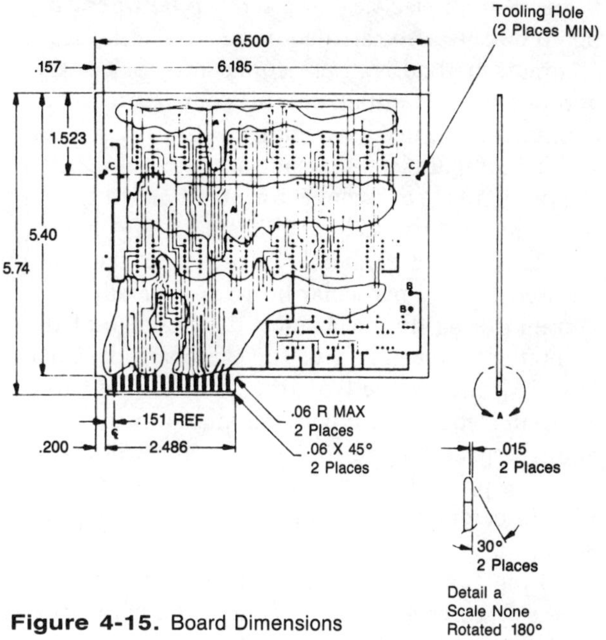

Figure 4-15. Board Dimensions

All dimensions that are shown as two places past the decimal, such as "5.74", "5.40", and ".06 R" maximum, have no need for a tight tolerance. In this case ±.02 would be appropriate. Note all the dimensions shown as three places past the decimal such as "6.500", "6.185", and ".157" etc. These dimensions are much more critical for the board outline to fit or mate with other components. An acceptable tolerance would be ±.005 or a preferred ±.010. All dimensions critical to the board configuration should be referenced from the tooling holes or register marks etched on the printed circuit board. Notice that in Figure 4-15, all board outline dimensions are taken from these points. All other features such as pads, traces, and reference markings shall be located by the film or photomaster and does not need to be dimensioned.

Input/Output Connectors:

One of the basic requirements of any PC board is to provide an electrical interference between the board and its associated equipment. The most common method is to use connectors. Select the appropriate I/O (input/output) connectors to accommodate all necessary input/output signals and still be compatible with space and environmental requirements. Your connector manufacturing representative can help in this area.

The type of connector or interface that will be used for any particular board is generally chosen during the mechanical design stage of the equipment. Not only does the PC board designer have to consider the type of connection, but the associated mechanical package must provide enough space for the connections that will mate with the printed circuit board, and the connector or connectors which also must mate. See Figure 4-16.

Problem: Using the connector specification sheets (Figures 4-16 and 4-17), find the connector board outline dimension.

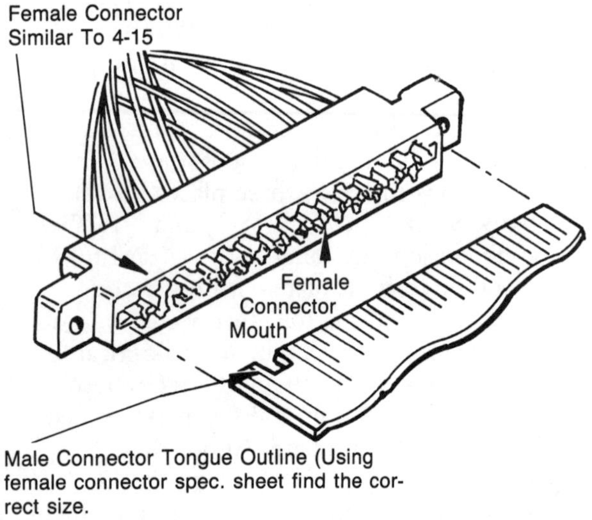

Figure 4-16. Female Connector

STEP 1 See the female connector specification sheet (Figure 4-16) for the dimension 2.504 ±.005 (the size of the female connector mouth). The first subtraction is the tolerance ±.005 from the dimension 2.504

EXAMPLE:

The connector dimension equals	2.504
Subtract the Neg tolerance	− .005
Step 1 answer	= 2.499

STEP 2 The male connector diagram (Figure 4-17) specifies that the unknown dimension will have a tolerance of ±.010. The second subtraction is the male connector tolerance from the answer of step 1.

EXAMPLE:

The answer from step 1 is	2.499
Subtract the mail connector Pos tolerance	− .010
Step 2 answer	= 2.489

STEP 3 At this point, we must consider the worst case fit condition for both the male and female connector. It is theoretically possible to have both male and female connectors in the exact size. We know two objects the same size do not fit together. To insure that this problem does not occur, subtraction 3 provides a minimum clearance. Perhaps an acceptable clearance would be from .003 to .010 inch (check with the design engineer). A .003 inch clearance would be absolute minimum, and the .010 inch clearance would be maximum. Choose somewhere between .003 and .010 in. which could be .006 or .007. The third subtraction is the minimum clearance from the answer of Step 2.

Edge connectors for 1/16″ (1.588 mm) boards

TRW CINCH CONNECTORS

.156″ (3.962 mm) contact center Cardcon* double and single readout as specified

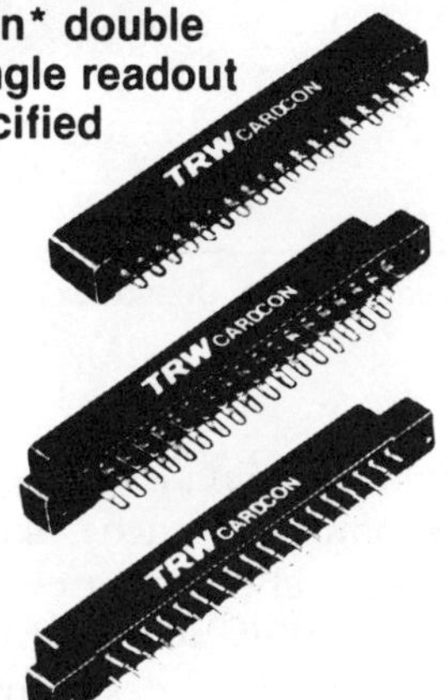

New cost effective methods of manufacture, selective plating and assembly permits TRW Cinch to price Cardcon 20% to 50% below most other P.C. Board connectors. Cardcon meets applicable performance requirements of MIL-C-21097.

Insulator: Blue glass-filled polyester—94V-O

Contact style: Semi-bellows—bifurcated

Contact material: Spring brass alloy 260

Contact plating: .000020″ minimum gold on contact area and gold flash on tail with .000050″ copper underplate.

Termination: Solder eyelet terminals; DIP solder terminals with either .200″ (3.56 mm) or .140″ (5.08 mm) between row spacing

Mounting: Available with mounting ears and .128″ holes or without mounting ears. For other type mounting consult factory.

Test voltage: 1800 RMS AC at sea level

Rated current: 5 AMPS.

Polarizing key: Between-contact type (P/N 50-PK-2) or In-contact type (P/N 50-PK-4). Order separately.

Card guide: Part No. 50-GP-1B. Order separately.

Part is blue to match Cardcon.

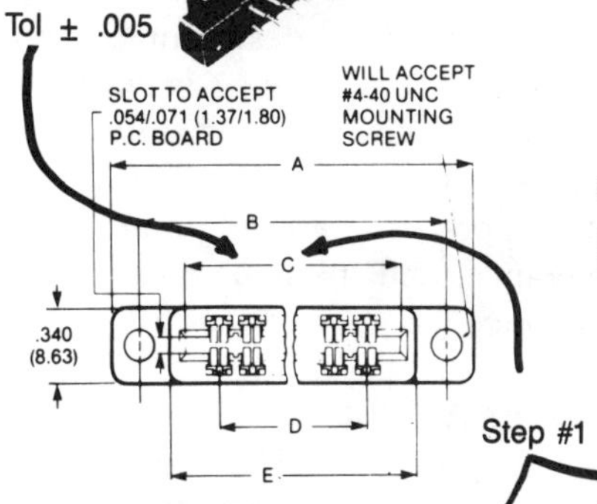

Solder eyelet terminals

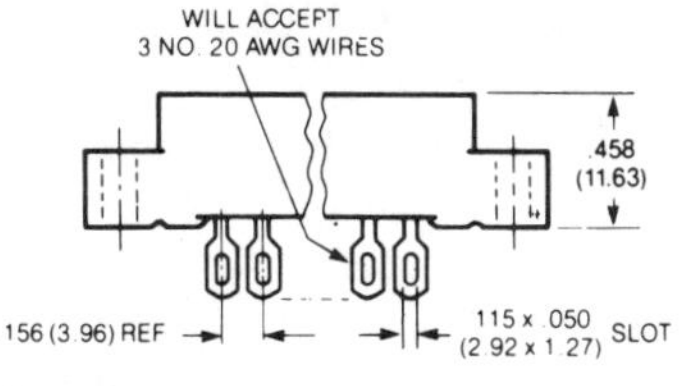

Double readout

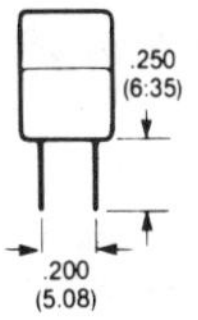

Single readout

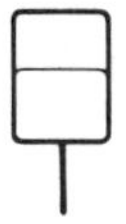

DIP solder terminals

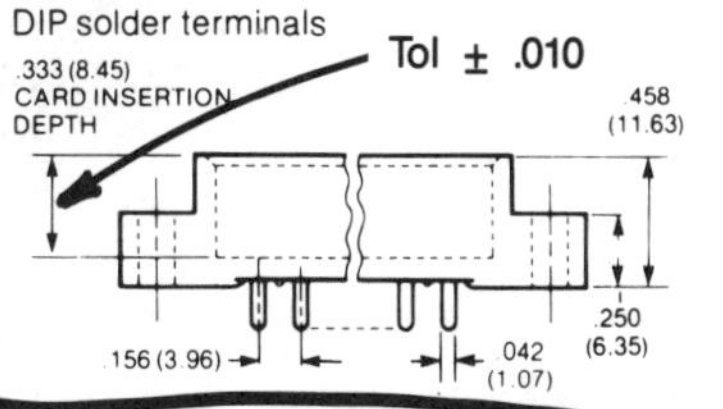

Double readout

STYLE 1 STYLE 2

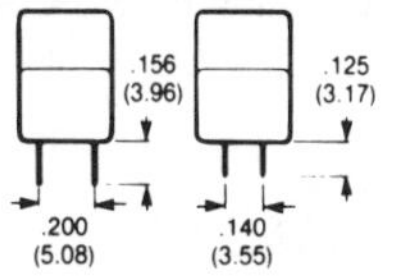

Single readout

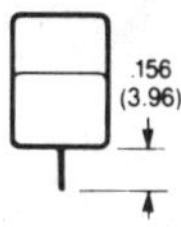

Board retention P.C. tails

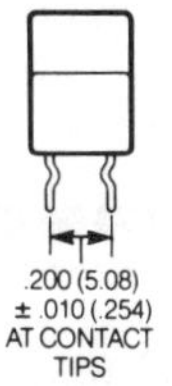

*Trademark of TRW, Inc.

Contact Positions	Dimensions									
	A		B		C		D		E	
	In.	MM	In.	MM	In.	MM	In.	MM	In.	MM
6	1.785	45.33	1.531	38.88	1.100	27.94	.780	19.81	1.239	31.47
10	2.410	61.21	2.156	54.76	1.724	43.78	1.404	35.66	1.864	47.34
12	2.723	69.16	2.469	62.71	2.036	51.71	1.716	43.58	2.177	55.29
15	3.191	81.05	2.937	74.59	2.504	63.60	2.184	55.47	2.645	67.18
18	3.660	92.96	3.406	86.51	2.972	75.48	2.652	67.36	3.114	79.09
22	4.285	108.83	4.031	102.38	3.596	91.33	3.276	83.21	3.739	94.97
24	4.597	116.76	4.343	110.31	3.909	99.28	3.588	91.13	4.051	102.89
25	4.754	120.75	4.500	114.30	4.067	103.30	3.744	95.09	4.208	106.88

Contact Positions	Distributer Part Numbers									
	Single Readout				Double Readout					
	Solder Eyelet		DIP Solder .156″ long		Solder Eyelet		DIP Solder .125″ long Style 2		*DIP Solder .156″ long Style 1	
							.140″ Between Rows		.200″ Between Rows	
	.128″ Mounting Hole	W.O. Mounting Ears	.128″ Mounting Hole	W.O. Mounting Ears	.128″ Mounting Hole	W.O. Mounting Ears	.128″ Mounting Hole	W.O. Mounting Ears	.128″ Mounting Hole	W.O. Mounting Ears
6	50-6SN-7	50-6SN-8	50-6SN-5	50-6SN-6	50-12SN-9	50-12SN-10	50-12SN-2	50-12SN-4	50-12SN-1	50-12SN-3
10	50-10SN-7	50-10SN-8	50-10SN-5	50-10SN-6	50-20SN-9	50-20SN-10	50-20SN-2	50-20SN-4	50-20SN-1	50-20SN-3
12	50-12SN-7	50-12SN-8	50-12SN-5	50-12SN-6	50-24SN-9	50-24SN-10	50-24SN-2	50-24SN-4	50-24SN-1	50-24SN-3
15	50-15SN-7	50-15SN-8	50-15SN-5	50-15SN-6	50-30SN-9	50-30SN-10	50-30SN-2	50-30SN-4	50-30SN-1	50-30SN-3
18	50-18SN-7	50-18SN-8	50-18SN-5	50-18SN-6	50-36SN-9	50-36SN-10	50-36SN-2	50-36SN-4	50-36SN-1	50-36SN-3
22	50-22SN-7	50-22SN-8	50-22SN-5	50-22SN-6	50-44SN-9	50-44SN-10	50-44SN-2	50-44SN-4	50-44SN-1	50-44SN-3
24	50-24SN-7	50-24SN-8	50-24SN-5	50-24SN-6	50-48SN-9	50-48SN-10	50-48SN-2	50-48SN-4	50-48SN-1	50-48SN-3
25	50-25SN-7	50-25SN-8	50-25SN-5	50-25SN-6	50-50SN-9	50-50SN-10	50-50SN-2	50-50SN-4	50-50SN-1	50-50SN-3

*For board retention P.C. tails add "R" to end of part number.

Figure 4-16 (cont.) Female connector spec sheet

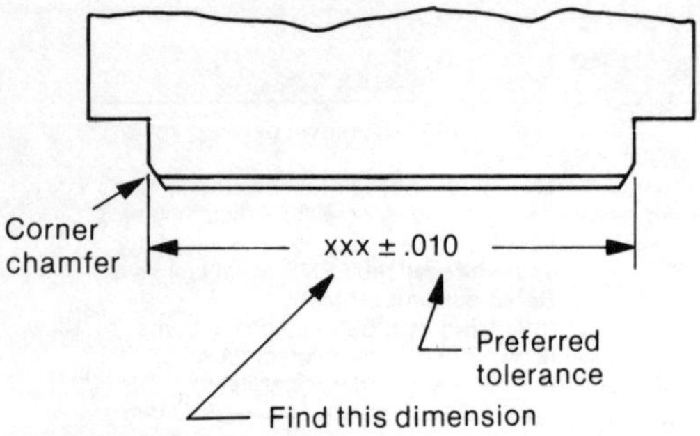

Male Connector Tongue
PC Board Male Connector Formula: There are three steps to remember - SUBTRACT, SUBTRACT, SUBTRACT

1ST SUBTRACTION - FEMALE CONNECTOR
Mouth Tolerance

2ND SUBTRACTION - MALE CONNECTOR
Tongue Tolerance

3RD SUBTRACTION - CLEARANCE
Minimum clearance between the female connector mouth and the male connector tongue

Figure 4-17. Connector outline dimensions

EXAMPLE:

The answer from step 2	2.489
Subtract the minimum clearance	− .006
In this case use .006	
Step 3 answer	= 2.483

If you choose to use the absolute minimum clearance, subtract .003 from the answer to step 2.

EXAMPLE:

The answer from step 2	2.489
Subtract the absolute	- .003
The minimum clearance	= 2.486

The correct dimension for Figure 4-17 is 2.486 ±.010, using the absolute minimum.

The correct dimension for Figure 4-17 would be 2.479 ±.010 using the maximum Step 3 clearance.

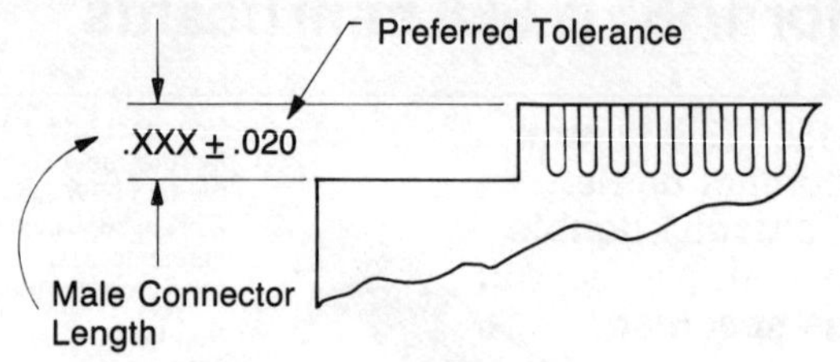

Figure 4-18. Male connector length and tolerance

STEP 4 To find the minimum length for the male connector tongue, see Figure 4-16; the section A-A view provides a dimension of .300 ±.010 called card depth. Good design practices ensure that the male connector tongue will have the maximum insertion depth at all times. Add the card depth, the female connector tolerance, and the male connector tolerance.

EXAMPLE:

Card depth	.300
Female connector tolerance	.010
Male connector tolerance	+.020
Minimum Answer	= .330

Chamfer Connector

The male PC board connector should be provided with a lead-in chamfer on the edge and corners to permit easy mating with the female connector. This prevents wear or damage. See Figure 4-19.

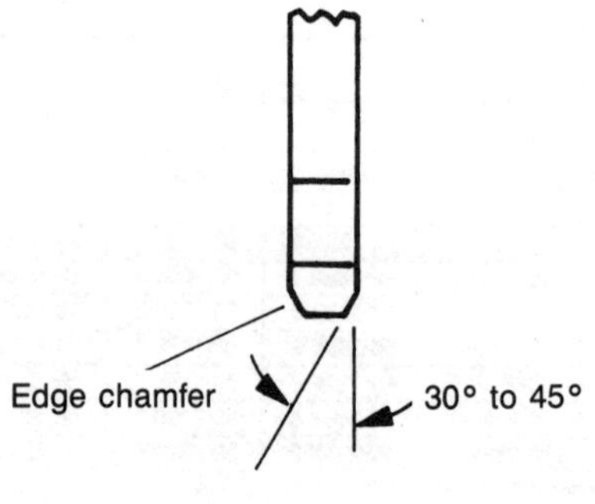

Figure 4-19. Mail Connector Tongue Edge View

Common contact spacings are: .100 inch, .125 inch, .150 inch, .156 inch, .200 inch. High density spacing includes .050 inch and .075 inch (See Figure 4-20).

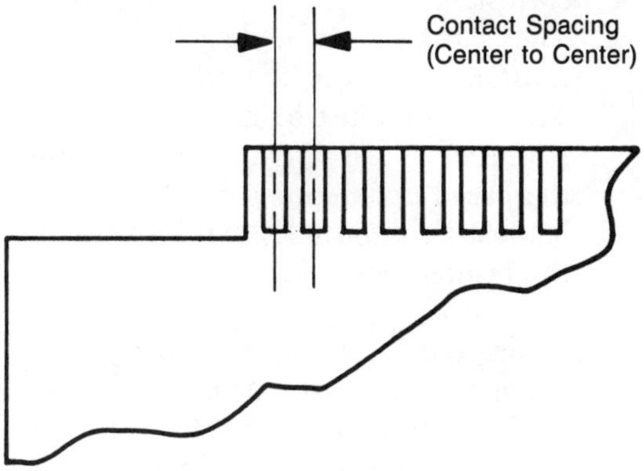

Figure 4-20. Contact Spacing

High density spacing of many contacts can cause difficulty with the insertion and removal of the male connector. The gripping of the female contacts creates a tremendous insertion or extraction force. Figure 4-21 illustrates a possible solution by chamfering the tongue (beviled edge) to insert a small number of contacts at one time. The chamfer typically would be approximately 1° or 2° maximum.

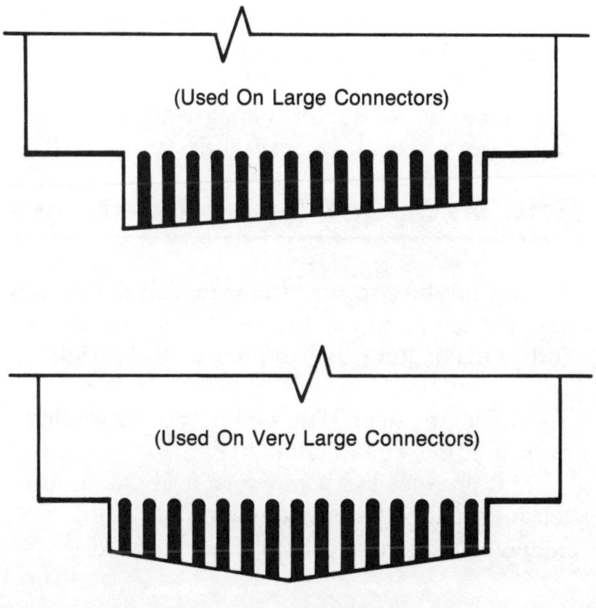

Figure 4-21. Chamfered Connectors

Keying Connectors

Whenever it is possible to install a connector on the printed circuit board using two different ways, or to install the board in the wrong connector, keying should be considered. Cut slots into the male printed circuit board with the keying plugs in the female connector. This helps proper installation. The width and length of the slot can be determined from the connector specification sheet (plug size). See Figure 4-22.

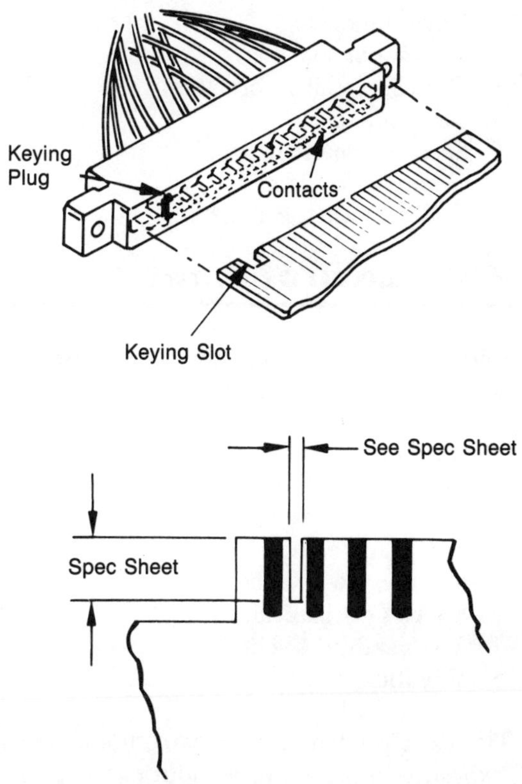

Figure 4-22. Key Connectors

Modern Connectors

Specialized electronic components involve many engineering decisions such as contact plating, mounting, connector profile, contact spacing, material etc. To achieve the optimum in performance, reliability, and economy, the designer and connector supplier must exchange engineering information.

Restrictions

The following are typical mechanical and electrical design restrictions and limitations.

1. Environment
2. Contamination
3. Critical circuit information
4. Mounting parts
5. Securing components
6. Component spacers
7. End product specifications

Environment

The following are typical environmental test requirements for printed circuit boards.

1. Altitude
2. Corrosion (Salt Spray)
3. High Temperature
4. Low Temperature
5. Moisture Resistance
6. Shock (Mechanical)
7. Temperature Cycling
8. Vibration

The basic printed circuit board should withstand exposure to the above without either physical damage or change in operating characteristics, such as cracking, and delamination.

Contamination

Protect printed circuit boards against dust, dirt, contamination, humidity, salt spray etc. There are many insulating compounds that can be applied as protective coatings: epoxies, polyurethanes, silicones, acrylics, polystyrenes, and varnishes.

Some of the more important considerations involved in the selection of printed circuit board coatings are condensed in the following list.

1. It must have the ability to prevent corrosion.
2. The coating must be flexible - its resistance to cracking during shock.
3. The coating must be easy to apply.
4. It must be transparent for the viewing of the board's component markings.
5. It must be removed easily for repairing the printed circuit board.

Conformal coating thickness (having the same shape as the board and its components) typically .003 inch minimum to .010 inch maximum. (Military applications could require greater thickness).

Critical Circuit Information

The electrical engineer usually supplies information relating to the traces or conductor configuration. This type of information comes with the schematic or logic diagram. The critical areas are either circled or in the form of a note. The critical area is usually either a lead or trace length between components or circuits, or the location of electrically radiating components such as transformers, inductors, etc.

Sometimes restrictions such as voltage and ground distribution are critical to the electrical design of the printed circuit board. Routing of these traces or conductors (along with their widths) can affect the electrical design.

End Product Specifications

The end product specifications (which define the operating environment for the PC board), is another major item that influences the design. This information is usually supplied by the responsible engineer. The end product specification should be well documented.

End product PC board specifications supply such information as environment, vibration, temperature range, electrical shielding, etc.

This particular aspect of the printed circuit board design is somewhat subjective. Often the PC board design is not proven until it is tested by the engineer. It is not a good practice either to overdesign or underdesign the board. Otherwise parts or leads will come off during vibration tests.

Examples of printed circuit board specifications are:

1. Military specifications (for example MIL-275)
2. ANSI standards (American National Standards Institute)
3. IPC (Industrial Process Control)
4. Good commercial practices

Printed Circuit Board Materials

Materials for PC boards generally fall within a couple of standard laminate grades with "G-10" and "FR-4" being the most common due to the availability of the material. Table 4-1 lists these laminates. Although the laminates range in a variety of material, the finished PC board material (laminate with copper applied) is generally treated the same as far as etching the final printed circuit board.

Manfacturability

Additive and subtractive processes:

This final section introduces the techniques of PC board manufacturer. Designers may avoid the disasterous consequences of designing a board that cannot be manufactured. Both subtractive and additive techniques are presented. Conventional subtractive processes are currently the most popular manufacturing techniques. This approach begins with a PC board which is entirely coated with copper. PC board artwork described in previous sections is used to construct an acid-resistent shadow mask on the copper. When submerged in acid, the copper is eaten away, leaving a pattern of conductive "wires" in the copper. This conductive pattern is the board's printed circuit.

The subtractive process may be augmented (or even replaced) by an additive process. Although more costly to use, additive processes eliminate the time-critical risks of acid etching, and allow manufacturers to improve PC board circuit density and product yield.

Table 4-1. Laminates

Grade	Composition	Remarks
XXXPC	paper/phenolic	High moisture resistance
FR-2	paper/phenolic	Similar to XXXPC but flame retardant
XXXP	paper/phenolic	Best mechanical characteristics of paper/phenolic grades
FR-3	paper/epoxy	High mechanical and electrical characteristics, flame retardant
FR-4	glass/expoxy	Fame retardant, chemical resistant, low water absorption
G-3	glass/phenolic	High flexural strength and dimensional stability
G-5	glass/melamine	High resistance, high impact strength
G-9	glass/melamine	Same as G-5 but better electrical characteristics
G-10	glass/epoxy	Same as FR-4 but not flame retardant
G-11	glass/epoxy	Same G-10 but higher flexural strength under heat
G-30	glass/p0lymide	High dimensional stability under heat, flame retardant
FR-5	glass/epoxy	Same as G-11 but flame retardant
GPO-1	glass/polyster	General purpose mechanical and electrical grade
GPO-2	glass/polyster	Similar to GPO-1 but lower flammability

Today there is a constant pressure throughout the entire electronics industry to provide more compact and denser PC boards. This pressure has brought about the use of solder masks which will be discussed further in this chapter.

The basic sequence for manufacturing PC boards by the conventional or subtractive process is:

Subtractive Process

Starting with panels of copper-clad base material, holes are fabricated by drilling or punching. The panels are sensitized and given an electroless copper flash to make the holes conductive for electroplating. A plating resist mask is applied by screening or photoprinting. Copper is electroplated with the minimum thickness in the holes, usually at least 0.001 inch. A second metal is electro-plated over the copper to act as an etch-resist and to provide corrosion protection for the finished circuit. The screened or photoprinted mask is stripped. The background copper is etched. A solder mask may be applied if specified by the end-user.

Table 4-2. Subtractive Process

A. Copper-clad base material.
B. Drill or pierce holes.
C. Clean and sensitize holes.
D. Electroless copper flash.
E. Print reverse pattern mask.
F. Electroplate copper.
G. Electroplate etch resist metal.
H. Strip mask.
I. Etch foil copper.
J. Apply solder mask (optional).

Additive Process

The additive process begins with panels of catalytic base material coated with a thin layer of catalytic adhesive. The base material can be any type normally used for conventional circuits. And the adhesive can be applied either by the laminate supplier or by the printed circuit manufacturer. After hole formation by drilling or punching, a reverse pattern is applied by screening or photoprinting. A chemical activation cycle prepares the adhesive for copper deposition to achieve bond strengths averaging 12 to 14 pounds per inch. When the panels are placed in the electroless copper bath, copper autocatalytically deposits to the required thickness. This thickness can range from a copper foil equivalent of to 2 ounces. If the plating resist is not a permanent epoxy mask, a stripping operation follows. A solder mask is applied to one or two sides if required. To ensure future solderability and prevent oxidation, the copper is protected with an inexpensive, organic *resin coating.*

* For a step-by-step process of the manufacturing techniques required to make a 2-sided PC board, see chapter 4 in the Analog Desgin book.

Table 4-3. Additive Process

A. Begin with an adhesive coated unclad laminate.
B. Drill or punch the holes.
C. Print the reverse pattern mask.
D. Activate.
E. Deposit electroless copper.
F. Strip the mask (optional).
G. Apply the solder mask (optional).
H. Apply the protective overcoat.

Component Assembly

The complexity and variety of components demand that proper component layout and placement procedures assure an economical production of PC board assemblies. These guidelines suggest parameters for locating attaching parts to the PC board. Provisions are made for designing PC boards manually or automated insertion.

When designing printed circuit boards with components to be assembled, the following suggestions should be considered.

1. Design for spacing and the use of standard tools and components.
2. Design with safety factors in mind
3. Consider the accessibility of parts. (Don't stack components.)

4. Design for maintenance and repair.
5. Select parts carefully. Parts to be mounted on the PC board must not only meet the storage and operating conditions specified for the equipment, but they must also maintain their integrity after having been subjected to heat during wave soldering. (See wave soldering)
6. Fasteners should be easily removed.
7. Consider adequate clearance between components and mounting brackets or card guides.
8. Set up both components and traces for wave soldering.
9. Use maximum dimensions of component for eliminating interference of two adjacent components.

Wave Soldering

Figure 4-23 illustrates wave soldering. Component leads are soldered to a printed circuit board by traveling at a predetermined rate of speed through a wave of molten solder.

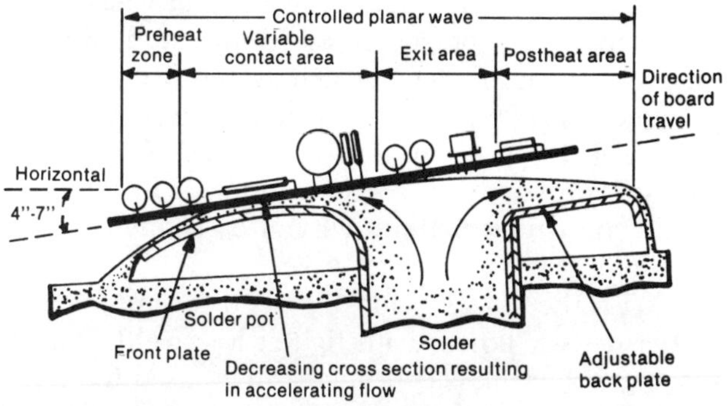

Figure 4-23 Wave Soldering

For the following reasons wave soldering is the most widely used method for soldering of components to printed circuit boards:

1. Cost saving over hand soldering.
2. There is reduced inspection time because of knowledge that every joint is soldered under identical conditions.
3. There is reduced scrap which would result from burning boards with soldering iron.
4. There is the reduced possibility of vapor, or flux entrapment.
5. The actual soldering time per solder joint is extremely short because of the excellent heat transfer conditions, the fast circulation of solder, and the preheating of the board as it travels through the solder (across the solder wave surface).

Basics

The basic machine consists of a drive conveyor that travels over a fluxer unit, a preheater unit, and the wave solder unit, (Figure 4-24).

Board Design Considerations

1. When passing over the preheat and solder wave, a board that is too thin to support the components will usually warp. The problem increases as the area increases, to the point where it is almost impossible to reach all of the terminal areas with a flat solder wave (on a warped board). The center of the board usually sags while the outer edges miss the wave completely. A minimum thickness of .06 (1/16") is standard for PC boards, without using special fixtures.

2. Consideration should be given to the board outline. Unconventional shapes contribute to hidden costs because of the special fixturing which needed to ensure reliable soldering.

3. Prevent board sagging by uniformly distributing heavy components on the printed circuit board.

4. Radial lead components placed on the board should have a bend or a loop on the lead. This prevents the component body from being buried by sitting too low in the wave solder.

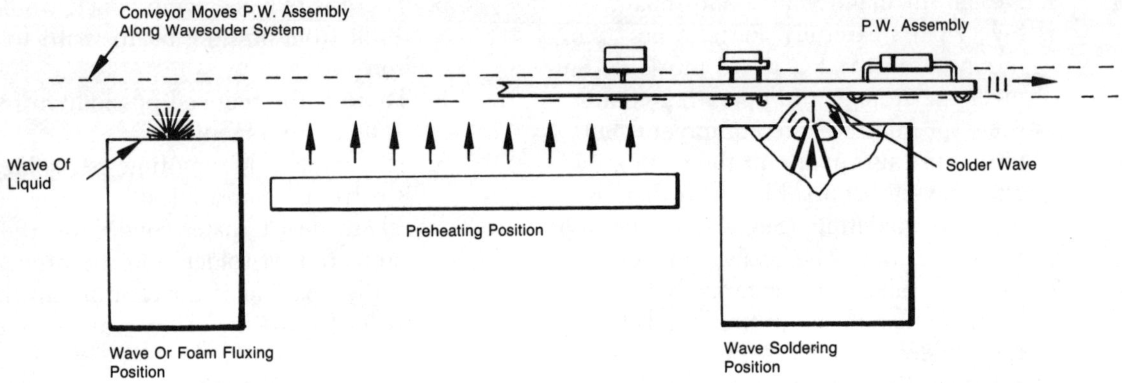

Figure 4-24. Wave Solder Unit

5. Mounting holes or holes for hardware, screws, etc. must not contain solder. Because all round holes have a tendency to collect solder, solder build-up requires removal which increases cost. To achieve solderless holes, plug or mask off all holes that are not to be filled with solder.

6. For reliability and consistency, the traces on the noncomponent side of the PC board should be routed in the direction that the board travels on the solder wave. The majority, but not necessarily all of the traces, should flow in the direction of the wave. A wash-board effect from the traces running against the wave causes bridging. The excess solder that builds up can short out adjacent conductors.

7. Trace or pad spacing must be as wide as possible. If space limitations require that conductor spacing be .010 inch or less, wave soldering becomes more difficult, even if these boards can be wave soldered satisfactorily with the aid of solder masking.

8. Solder masking is a screened on coat of epoxy resin, covering all areas of the PC board (except pads which require soldering). Clearance around these pads is typically .010 inch.

A wave soldering pallet is usually provided with a mask to protect printed circuit conductors (such as contact fingers), which are not to be soldered. The PC board must be mounted in the grooves in the pallet to protect the desired area. Unless the designer takes the connector into account, the circuit pattern will not be aligned in the direction of the solder wave.

With the current industry trend toward higher density boards with tighter line widths and spacing, solder masks are increasingly used to eliminate bridging between adjacent conductors during wave soldering.

Quiz - Chapter 4

1. The definition of working board area: ______________________________ ______________________________
2. Generally the board size is determined by the mechanical designer. T F
3. If the P.C. board inserts into a mating connector, the __________ ________________ offer the most convenient solution.
4. If the electrical interface is by means other than a connector, & card guides are not practical, then ______________________ holes must be included.
5. The application for mounting the P.C. board must be decided after starting the board layout. T F
6. The leads that protrude through the non-component side of the printed circuit board are not an important consideration. T F
7. The type of connector or interface that will be used is generally chosen during the mechanical design stage. T F
8. Single part P.C. connectors are the most common. They use one __________ of the printed circuit board as the plug.
9. Double part connectors are the most common type. T F
10. One of the basic disadvantages of using a single part connector is: the direction of insertion (plug-in) usually limited. T F
11. Tools used for the printed circuit board extraction, help protect the board from damage. T F
12. Stiffeners are used on P.C. boards to insure that the board remains: ______________________________
13. Board stiffeners are only used to avoid problems when wave or flow soldering is used. T F
14. Placement of adjustable components should be considered for __________________ of access.
15. Electromagnetic problems: is caused only by the increased use of high frequency techniques. T F
16. A shield reduces the electromagnetic field strength. T F
17. A shield excludes electromagnetic energy by reflecting or __________ the energy.
18. Many digital boards have analog circuitry on the same board that must be shielded from each other. T F
19. The P.C. designer must advise the electrical engineer on which shielding techniques to use. T F
20. Ground planes are not good for electrostatic shields. T F
21. Before the P.C. design layout is started, the P.C. designer should identify the components that should not be placed near each other. T F
22. Metal-cased components without an insulator can touch each other. T F
23. The P.C. board designer directly influences how a board will be manufactured. T F
24. If the conductor density rises, the board yield goes up. T F
25. A board may match the fabrication drawing and still not be reliable. T F
26. Boards size: warding, bowing, and the effects of ________________ are most likely to effect it's size.
27. It is important to combine your design guidelines with your local P.C. board manufacturer capabilites. T F
28. All board outlines must be expressed to include some tolerance: most common are __________ and __________.
29. The tolerance is important because calculating the actual working area to its smallest size is the objective. T F
30. The formula to find the P.C. board male connector size, there are three steps to remember __________ ______________, __________
31. To find the minimum length for the male connector tongue, you must find the max card __________

32. The female connector should be provided with a lead-in chamfer to permit easy mating. T F
33. A common contact spacing for male & female connectors is .125. T F
34. To help in proper installation of connectors or mating male to female. Keying is very common. T F
35. To achieve optimum performance in printed circuits environment contamination must be considered. T F
36. Name two of a possible 8 environmental test requirements for printed circuit boards __________ & __________
37. One very common coating used to protect printed circuit boards is __________ coating.
38. Critical leads or traces are not provided by the electrical engineer. T F
39. Voltage and ground distribution are never critical to an electrical design. T F
40. Often the printed circuit board design is not proven until it is tested by the engineer. T F
41. Printed circuit board specifications are: (name two)
    1. Military MIL-275
    2. __________
    3. __________
42. The most popular manufacturing technique for printed ciruict boards is the additive process. T F
43. Today there is a constant pressure throughout the entire P.C. board manufacturing industry to manufacture more compacted, denser P.C. boards. T F
44. One important component assembly consideration is to use maximum dimensions of components. T F
45. Maintenance and repair is not an important consideration for component assembly. T F
46. Wave soldering is the most widely used method for soldering of components to P.C. boards. T F
47. Wave soldering has no cost saving over hand soldering. T F
48. Board design consideration: prevent board sagging by __________ distributing heavy components on the P.C. board.
49. Board design consideration: radial lead components should have a __________ on the lead.
50. Board design consideration: solder masking is a screened on coat of epoxy resin, covering all areas of the P.C. board except __________ which require soldering.

The answers for each question can be found on the pages listed below:

Question 1. Pg. 102
Question 2. Pg. 102
Question 3. Pg. 103
Question 4. Pg. 103
Question 5. Pg. 103
Question 6. Pg. 104
Question 7. Pg. 104
Question 8. Pg. 104
Question 9. Pg. 104
Question 10. Pg. 107
Question 11. Pg. 107
Question 12. Pg. 107
Question 13. Pg. 107
Question 14. Pg. 108
Question 15. Pg. 108
Question 16. Pg.
Question 17. Pg. 108
Question 18. Pg. 108
Question 19. Pg. 108
Question 20. Pg. 109
Question 21. Pg. 109
Question 22. Pg. 109
Question 23. Pg. 109
Question 24. Pg. 109
Question 25. Pg. 109
Question 26. Pg. 111
Question 27. Pg. 111
Question 28. Pg.
Question 29. Pg. 111
Question 30. Pg. 114
Question 31. Pg. 114
Question 32. Pg. 114
Question 33. Pg. 115
Question 34. Pg. 115
Question 35. Pg. 116
Question 36. Pg. 116
Question 37. Pg. 116
Question 38. Pg. 116
Question 39. Pg. 116
Question 40. Pg. 117
Question 41. Pg. 117
Question 42. Pg. 117
Question 43. Pg. 118
Question 44. Pg. 119
Question 45. Pg. 119
Question 46. Pg. 119
Question 47. Pg. 119
Question 48. Pg. 119
Question 49. Pg. 119
Question 50. Pg. 120

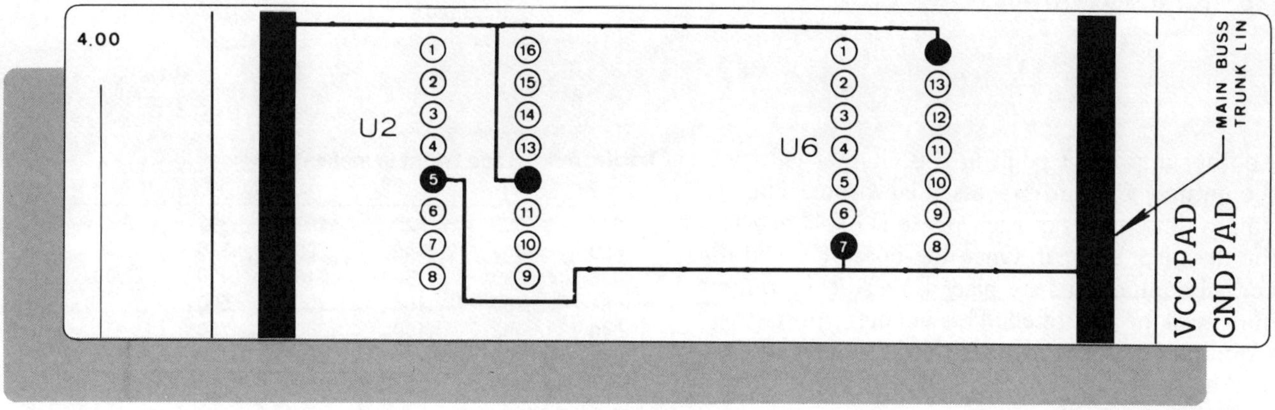

PC DESIGN LAYOUTS

5

This chapter provides the basic ground rules by which the designer/student will construct future boards. The beginning of this chapter addresses "traces" for signal paths as well as voltage and ground paths. Explanations are presented on how to code traces, paths, and pads on the layout.

This chapter discusses how to place sections of the schematic onto the PC board outline. Also, how to calculate the area required for all its components or subdivided sections is discussed. The designer/student should be able to determine (with a high degree of confidence) that the components will fit onto the board outline far in advance of actually drafting out the layout. This can save an enormous amount of design effort. Without this knowledge, one could possibly invest a lot of time only to find out that the components on the schematic will not fit onto the board outline.

Another important point in this chapter is how the entire PC board process is layed out. The time requirements of each phase is explained to the designer/student. Once the board is layed out and all components are placed, these components must now be identified. This chapter covers the requirements for component designations and markings.

Signal Traces

A signal trace is generally considered to be any trace on a board other than power and ground. These traces are generally low voltage and low current. They do not require a very wide trace or air gap.

Traces on the PCB are produced by special adhesive tape that is applied to the layout during the design process. The most commonly accepted tape width used in industry today is .030-.040 in. at a 2:1 scale with a .030-.040 in. air gap between adjacent traces or pads (see Table 5-1). In cases of extremely dense boards, tape-ups are used with .026- and even sometimes smaller traces. But as the trace and air gap decrease, the cost and unreliability of the board increase. The .040-in. trace and air gap combination provide a .020-in. conductor on the actual board. This combination provides ample material for conduction of electricity and also allows sufficient tolerance for the board manufacturing (from the initial tape-up through the photography, etching, plating and wave soldering processes). Each of these steps should be allowed an acceptable amount of tolerance so that cumulative errors do not affect the reliability of the completed board.

In cases where the trace width (tape) must go down to .026- or .031-in. widths, it becomes very critical to maintain at least a .026- .031 inch air gap with the same width as the tape illustrated in Figure 5-1. The air gap should be adequate to avoid solder bridging between traces during the wave soldering operation.

Table 5-1 Tape Width in Inches

.015	.050	.125	.187	.312	.800
.020	.062	.140	.200	.375	1.000
.026	.070	.150	.218	.400	2.000
.031	.080	.156	.250	.500	
.040	.093	.160	.280	.750	
.046	.100		.300		

Figure 5-1. Air Gap

Power/Ground Traces

The primary consideration in a power or ground trace is to provide a direct connection from the device to the power supply. Because this is not always possible, the next best step is to increase the width of the conductor trace to provide enough mass to accomplish essentially the same thing.

As shown in Figure 5-2 a nominal trace width for power and ground can be established at .100-in. (2:1 scale). As in the case of the signal trace, this can be reduced for some requirements. But be sure the electrical engineer is aware of this condition. A .100-in. trace is easy to plan as the design progresses. This is because it is an even grid width; the air gap could be the same.

Color coding can be used in three different methods to help the PC designer: conductor layer coding, voltage/ground pads, and input/output pads. Although each color coding method serves a separate purpose, they are all used during the course of a typical PC layout.

A standard colored pencil such as those available from office supply or drafting supply outlets can be used for each color. Choose a readily available color to represent a particular voltage, ground or trace. Become familiar with what it represents and do not change it. This avoids the confusion of what each color means on each board. After using these approaches for a while, you will automatically relate the color code to its function without referring to a color key.

Figure 5-2. Power and Signal Traces

- When wide conductors must be transferred from one conductive layer to another or when feed-thru's are necessary, use more than one plated-through hole as shown

- Conductor Spacing
  The conductor spacing shall be established for each printed circuit board design.
- Routine Designs: .025 min, voltage limitations 400 VDC or AC peak. (See safety requirements high voltage)
- High Density: The "preferred minimum" is .015, and the voltage limitation is 50 VDC or peak.
- The "minimum by exception" is .010.

Minimum's:

When the printed circuit board design requires the preferred "minimum" (and more importantly the "minimum by exception" conductor spacing), the printed board assembly must be conformally coated or a suitable solder mask must be applied.

NOTE With present state-of-the art capabilities it is possible to use smaller land diameters, annular rings and hole diameters. However, these smaller geometries significantly increase manufacturing cost and reduce manufacturability.

Conductor Layer Coding

This coding system indicates which side of the PCB a particular trace is located. Therefore, on a typical two-sided board, two different colors would be used, red traces indicate component side traces, blue traces indicate non-component side traces. The component body outlines would be drawn with yet another color such as black or some color that you would not use for any of the other color coding methods. If a board is intended to be single-sided (all conductors on one side of the board only), then all traces should be one color only. This is shown in Figure 5-3. See the example of the layout in Figure 5-4.

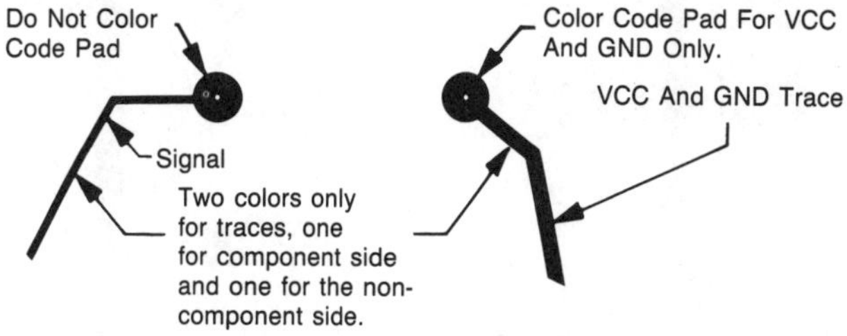

Figure 5-3. Color Coding

Voltages and Ground Coding

In this method, a different color is chosen to represent each voltage and ground that is used, red might be +5V, orange +12V, green −12V, and blue ground. Color code the component lead pad to be connected to the particular voltage or ground (pad only).

During the design stage, the colors stand out as flags to indicate where the nearest voltage or ground pins are located for the most direct bus hook-up. This allows the designer to route the voltage or ground bus without stopping and looking up each IC's correct voltage or ground pin. See Figure 5-4 for a typical color coded bussing.

During the taping operation, the color coded pad on the layout can be seen through the hole in the pad on the artwork. As shown in Figure 5-5, a general rule would tell the PCB taper that all color coded pads were to be .100 inch tape minimum regardless of the color. All other traces could be .050-in. wide or whatever size you choose for signal traces.

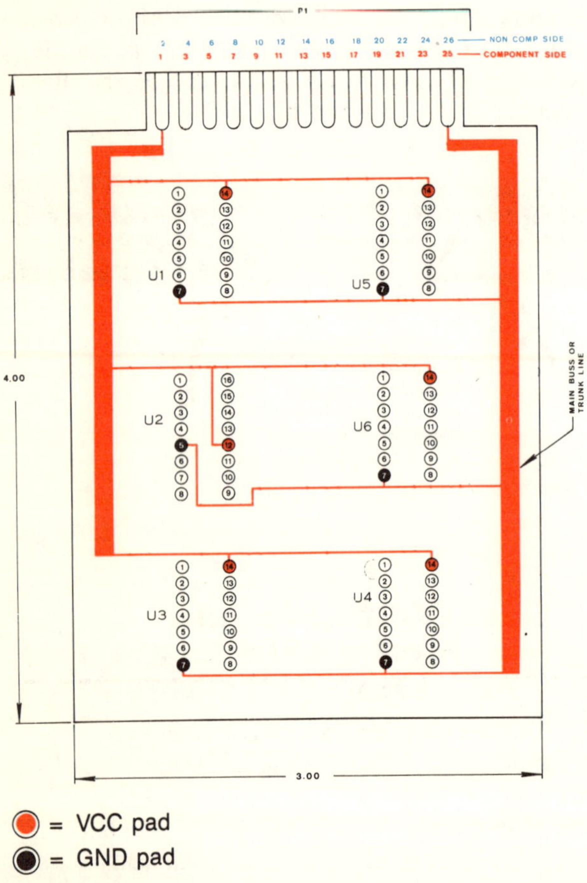

Figure 5-4. Figure Layout

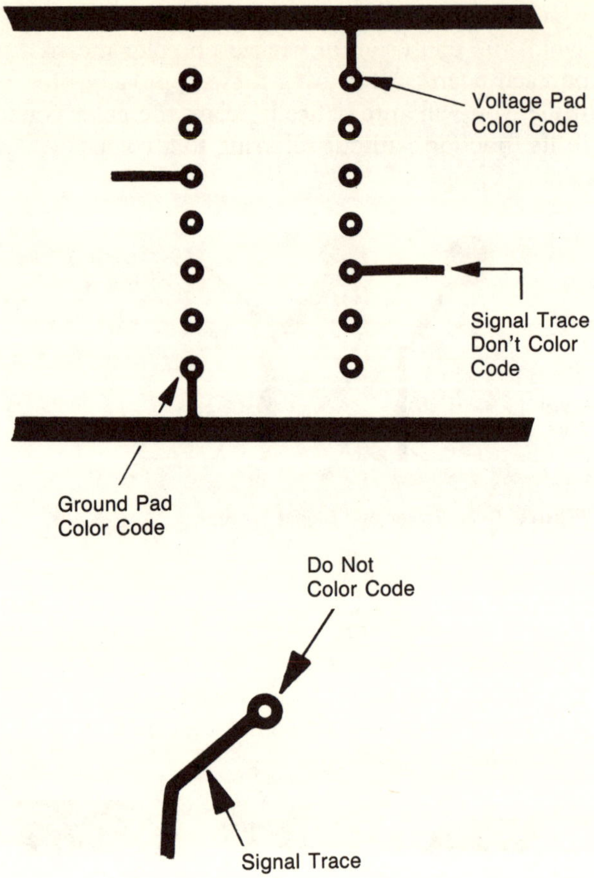

Figure 5-5. Color Coded Pads

Output Coding

Components (pads) that are to be connected to a connector(s) may also use a coding method. As the circuit is being layed out, the particular pad that is to go out to the connector can be coded with the connector pin number. It can be routed after the rest of the layout is completed (see Figure 5-6). The designer completes the connections between local related components and then routes the output connections (traces to the connector) in whatever path is most available and less crowded.

The following are examples of layouts for the three basic multiple lead component types: flatpacks, cans, and dual in lines. The examples show typical land area use for these three styles of IC packages.

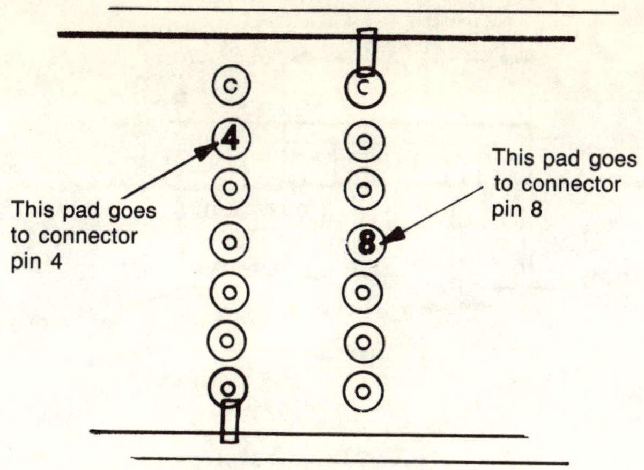

Figure 5-6. Connector Coding

Flat Packs

The flat pack is designed for the component side. It is either surface mounted (see Figure 5-7) or through-hole staggered mounted (see Figure 5-8).

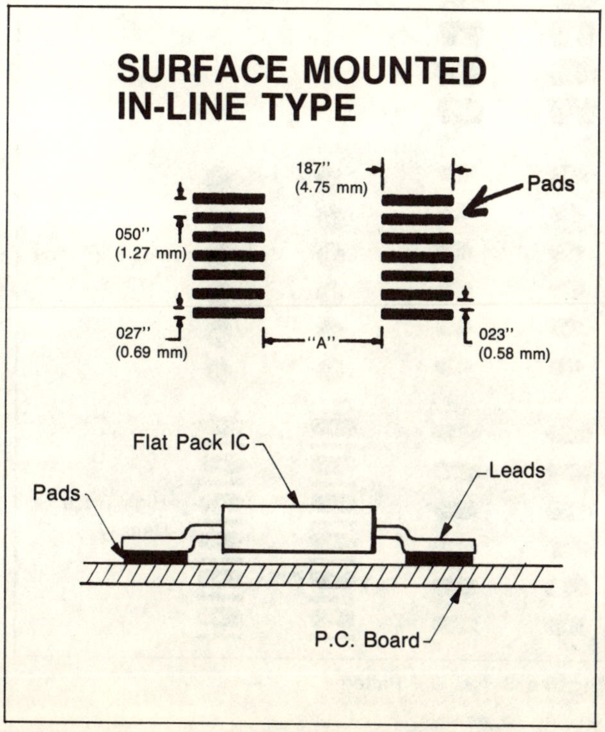

Figure 5-7. Component Side Surface Mounted

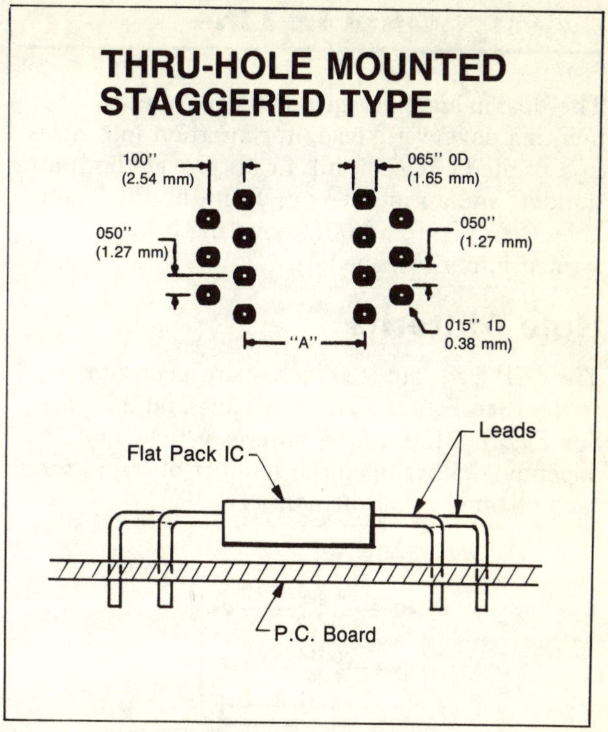

Figure 5-8. Staggered Hole Mounting

Multiple Lead Cans (Transistors)

Because the packaging technology was well established for transistors in metal "TO" cans, the rugged construction made the 10-or 12-lead IC a natural first package. See Figure 5-9.

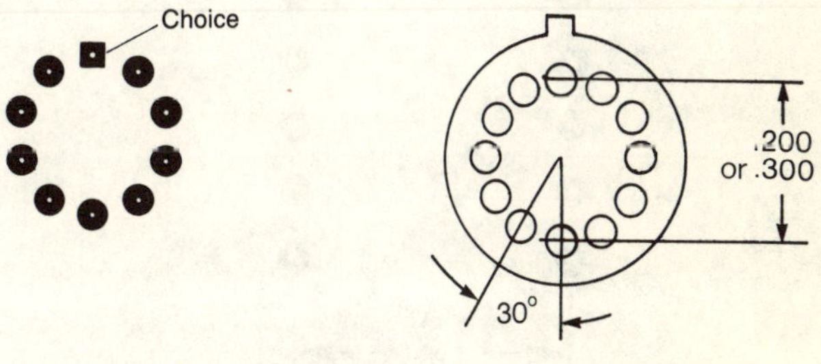

Figure 5-9. TO Pattern

Dual In Line

The dual-in-line package (DIP) has its leads pointing downward, ready for insertion into holes in a printed circuit board. Leads provide both a standoff and an interference fit in the PC board holes. See Figure 5-10. It shows the leads are formed into a V-shape.

Hole Patterns

The DIP leads are .100 inch apart, center-to-center. See Figure 5-11 for spacing and lead size. See Figure 5-12 for DIP patterns which vary depending either upon the number of traces for each channel or on the density.

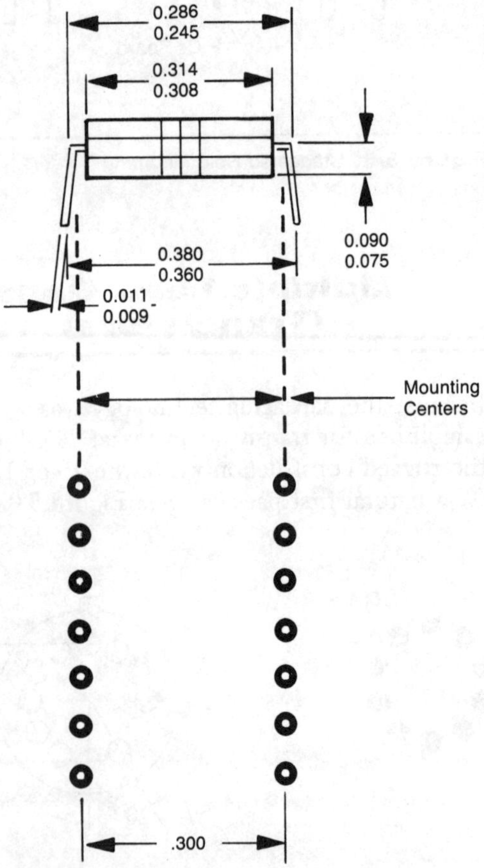

Figure 5-10. DIP Leads

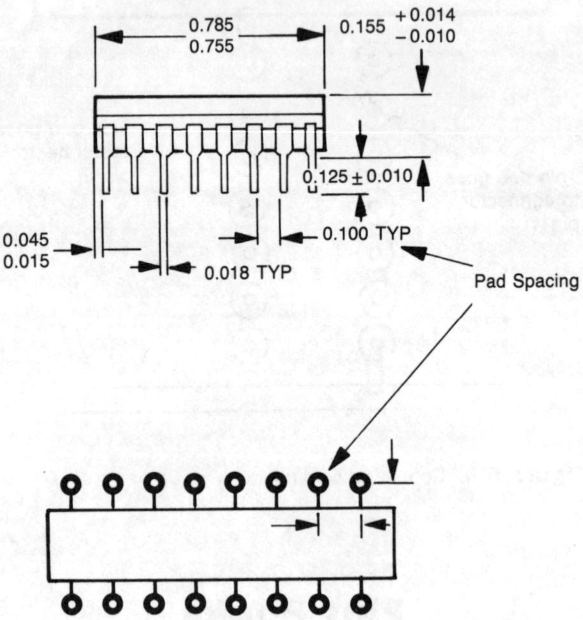

Figure 5-11. Spacing and Lead Size

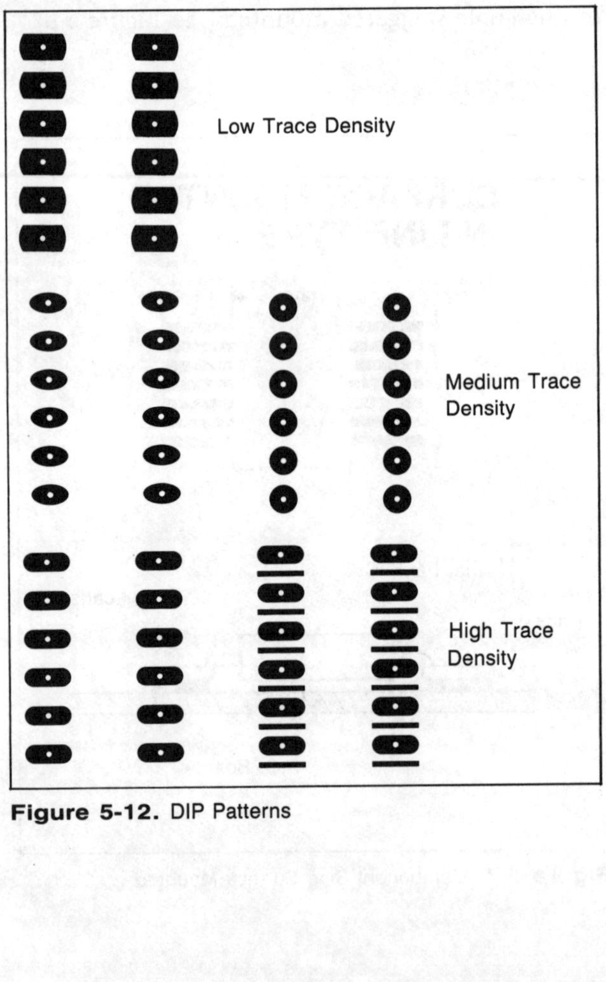

Figure 5-12. DIP Patterns

Templates

A standard component template is one of the most necessary pieces of equipment the designer uses in PC design. It not only saves valuable time but also assures that standard component lead spacing and body sizes are considered. Although several models and styles are available, pick one that will be compact, yet complete for the requirements of your particular type of design. Figure 5-13 shows a multipurpose template which has both analog components (discrete) and patterns for 8-, 14-, 16-, and 24-pin IC's.

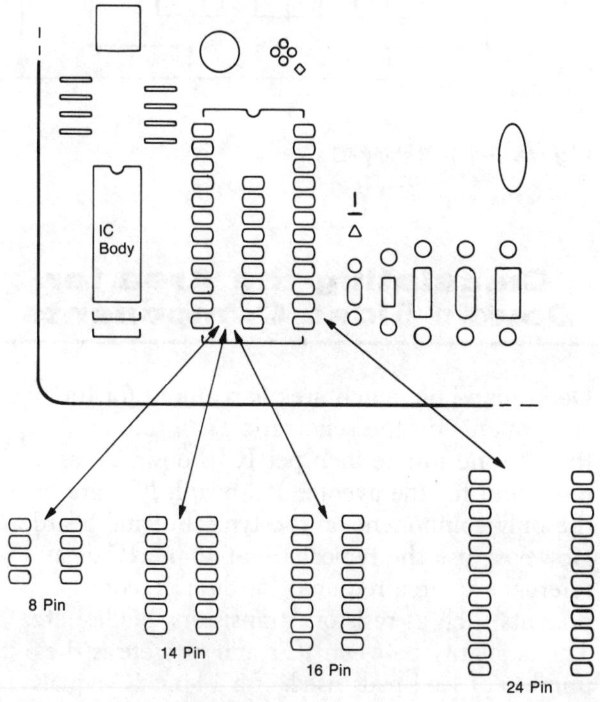

Figure 5-13. Multipurpose Template

IC Template

There are templates dedicated to IC's that provide a choice of pad styles. See Figure 5-14.

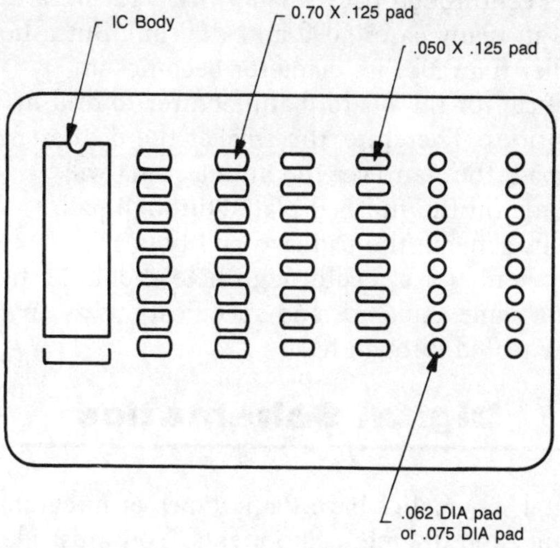

Figure 5-14. Dedicated IC Template
Template can be adjusted to accommodate 8-, 14-, 16-, 24-pin etc.

Feedthroughs

A feedthrough consists of a pad on both sides of the board with a plated-through hole connecting them. They connect traces on one side of the board to traces on the opposite side (when a component pad is not available for this purpose). Feedthrough holes are constructed like any other plated-through holes. But because no terminals or component leads are mounted in them, they are generally smaller than any other pad and plated-through hole. This avoids the consuming of excess space on the board needed for trace routing. Feedthroughs are especially necessary in digital layouts where vertical and horizontal traces must be kept on opposite sides as much as possible. Then a feedthrough pad is needed; the connection is continued on the opposite side of the board.

Feedthrough sizes vary depending on the density of the board. But on tape-ups (where practical) feedthrough pads should be .125-in. at 2:1. This results in a .062-in. pad on the actual board (which could have a .025- to .031-in. diameter hole drilled through it).

Feedthrough pads smaller than .125-in. are used in many cases (.090 and .100 in.). But a hole smaller than .021-in. diameter becomes impractical for the board manufacturer to drill in quantities. Therefore, the smaller the diameter of the pad, the narrower the annular ring will become on the finished plated-through pad.

In general, the pad size and hole size requirements for a feedthrough hole should be held to the same standards and specifications as any other plated-through hole.

Digital Schematics

Digital schematics have many types of integrated circuits and discrete components. You must identify each size. Before the initial layout, you must know if all the components on the schematic will fit on the board outline given to you.

In most cases, digital schematics are made up of discrete and logic symbols. These logic symbols must be grouped together in order to determine how many integrated circuits are going to be placed within the board outline. Also grouping determines the total count of all resistors, capacitors, diodes, transistors, etc. See Figure 5-24 for a typical digital schematic with these logic gates grouped. Notice that each grouping of logic gates is identified with its own IC number. Once you have chosen which are the logic gates (U1, U2, U3, etc.), assign the IC's to the proper areas within the board outline. See Figure 5-15 for plotting IC's on the board outline.

IC location is determined by which ICs provide easy electrical flow and interconnection (from one IC to the next). Obtain the shortest and most direct interconnects from IC to IC. Assign the IC's on the entire board before beginning the formal layout.

For analog boards, we had taught you sketched individual circuits before placing them down with a formal template. Digital boards have no sketching. Once you have plotted the IC's (as in Figure 5-15), you go directly into the formal template layout. Before drawing the first line down on a layout, calculate the total area required for the components to make sure that all the components on the schematic would fit onto the PC board outline. Many hours will be wasted on a layout if there is not enough room to connect all of the traces.

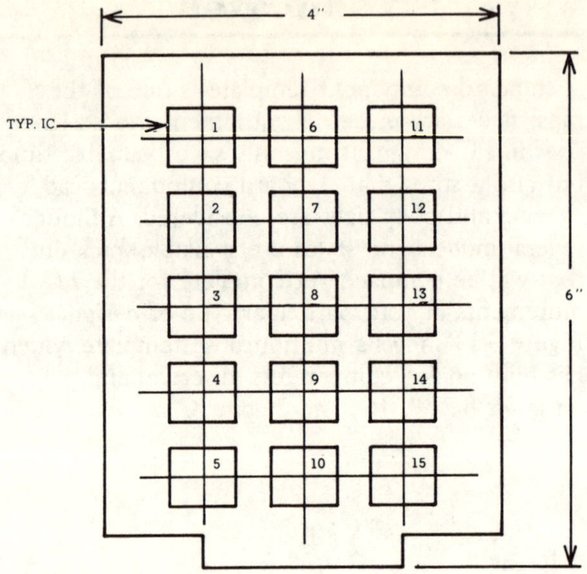

Figure 5-15. Plotting IC's

Calculating the Area for Double-Sided Components

Determine how much area is required for the components on the schematic. As a rule of thumb, one square inch per IC is desirable as a minimum for the average PC board. IC's are not the only component on the typical digital board. However, use the typical 14- or 16-pin IC as a reference to area required for discrete components such as resistors, transistors, diodes, etc. The similarity between IC's and discrete is the number of lead pads used. If a 14-pin IC equals one square inch, this represents the area required to connect all necessary pads for that IC. If seven discrete components (with two pads each for 14 pads total) equal one square inch, this also represents the area for the components and traces to connect all 14 pads. See Table 5-2 for a breakdown of discrete component density per square inch. The one square inch used in Table 5-2 does not mean that this is the maximum number of components per square inch that a designer can fit on the typical board. It means that one square inch is a comfortable area to allow for "X" number of discrete components.

Example: See Figure 5-18 for an example on how to calculate area required for discrete and integrated circuit components.

Using the digital schematic (Figure 5-18), calculate the number of square inches required to fit all of the components in this schematic on a PC board. See Table 5-2 and Figure 5-16.

STEPS

1. Determine the case size of all IC's and discrete components. For our working example, we will identify all the components for you. All IC's are equivalent to a 14-pin case size. All resistors = RL07; all capacitors = CK60. This identifies all the components on the digital schematic.

2. Using the component case size given in Step 1, find the case size in Table 5-2.

RESISTORS RL07 = 7 per sq. in.

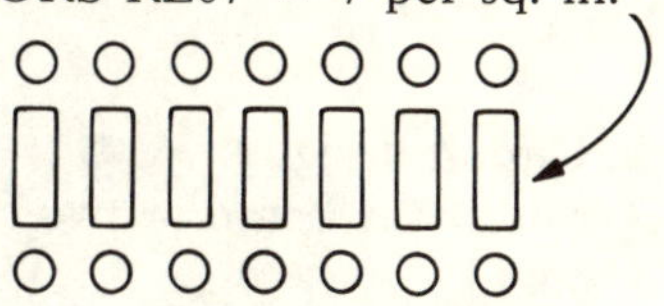

CAPACITOR CK60 = 7 per sq. in.

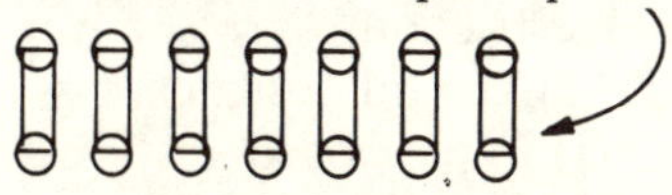

IC 14-Pin = 1 per sq. in.

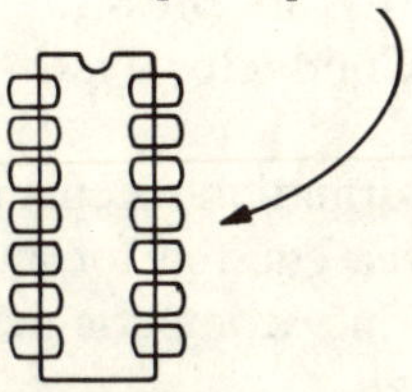

3. Using the quantity per square inch listed in Step 2, find the square inches required for schematic Figure 5-18. Count the total number of resistors on the schematic that are RL07. In this case all six resistors are RL07. Step 2 shows seven RL07 = 1 square inch. Six would = .86 square inches. You could call six RL07's equal to one quare inch. This is a rule of thumb. Six or seven RL07's could be rounded off to one square inch. List R1 through R6 = 1 square inch.

   Count the total number of capacitors on the schematic that are CK60. In this case, all three capacitors are CK60.

   Step 2 shows seven CK60s = one square inch. Three would equal .5 square inches, approximately. C1 through C3 = .5 square inches.

   Count the total number of IC's on the schematic. There are 14 IC's. All IC's are 14-pin case size. Step 2 shows that 14-pin IC's are equal to one square inch. IC 1 through IC 14 equals 14 square inches.

4. Add all the component areas together.

R1 through R6	= 1 (square inch)
C1 through C3	= .5
IC 1 - IC 14	= 14
TOTAL:	= 15.5 square inches

5. 15.5 square inches is not an absolute; it is only a guide. In many cases, the same

Table 5-2. Board Component Density

	Double Sided Boards	
Resistors:	RC05, RL07 RC06 RC07, RL07 RC08 RC12 RN50 RN55 RN60 RN65	7 Comp's per sq. in.
Capacitors:	CK05, CK06, CK60, CK62, CK12	= 7 Comp's per sq. in.
Transistors:	TO-5, TO-12, TO-18 TO-39, TO-46 TO-52, TO-107 TO-72, TO-17, TO-92	= 3-4 Comp's per sq. in.
Diodes	DO-7, DO-41, DO-13 DO-14	= 7 Comp's per sq. in.
IC'S	14 & 16 PIN	= 1 per sq. in.

COMPONENT COUNT (Equivalent 14- and 16-pin IC calculations)

	Total No. on Board		Equiv. 14/16 IC
14- and 16-pin IC	________	× 1.00 =	________
20 pin IC	________	× 1.25 =	________
24 pin IC	________	× 1.50 =	________
28 pin IC	________	× 2.00 =	________
40 pin IC	________	× 3.5 =	________
64 pin IC	________	× 6.5 =	________
8 pin IC	________	× .50 =	________
test points	________	÷ 10 =	________
jumper pads	________	÷ 10 =	________

Convert all components on the board to an equivalent 14- 16-pin IC count using the following conversion factors:

	Total No. on Board		Equiv. 14/16 IC
1/4 -watt resistors	________	÷ 7 =	________
1-watt resistors	________	÷ 4 =	________
disc type caps	________	÷ 7 =	________
large caps	________	÷ 4 =	________
3-pin components (Transistors, variable resistor, etc.) pre assigned	________	÷ 3 =	________
connector pins used	________	÷ 20 =	________
Other components (list)			
Total Equivalent 14 and 16-pin IC =			________

Figure 5-16. Component Equivalent To 14 And 16 Pin IC

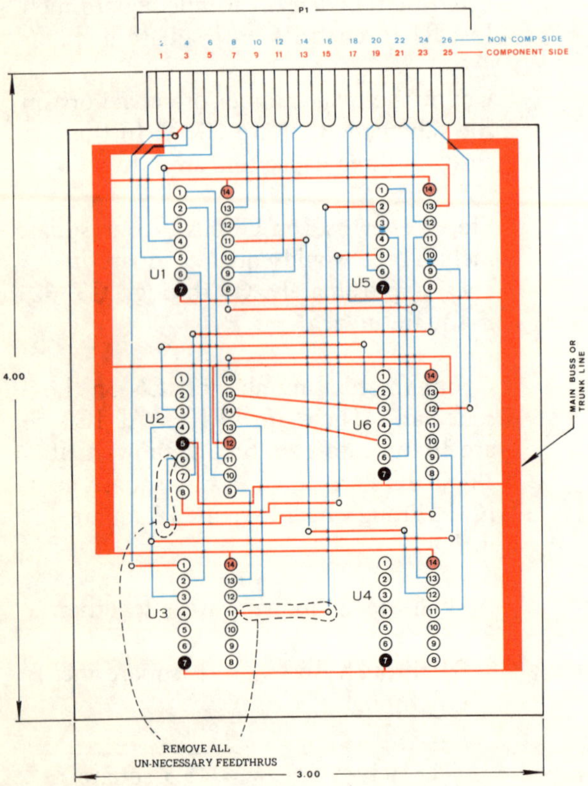

● = VCC pad
● = GND pad

Figure 5-17. Formal Digital Template Layout

schematic could be layed out in half this area. Some of the limiting factors are listed below:

A. Circuit design
B. Multilayer board or two sided board
C. Packaging considerations and restrictions
D. PC board designer
  (1) Experienced
  (2) Capabilities
  (3) Work pressure
E. Conductor width/spacing

6. This information calculates the approximate area equired for any given engineering schematic versus the given board size.

Calculating the Area for Multilayer Board Components

The use of multilayer boards is particularly important when the component densities of double-sided boards must be exceeded. Double-sided board density has an average of one square inch per IC. Multilayer boards have a density beyond

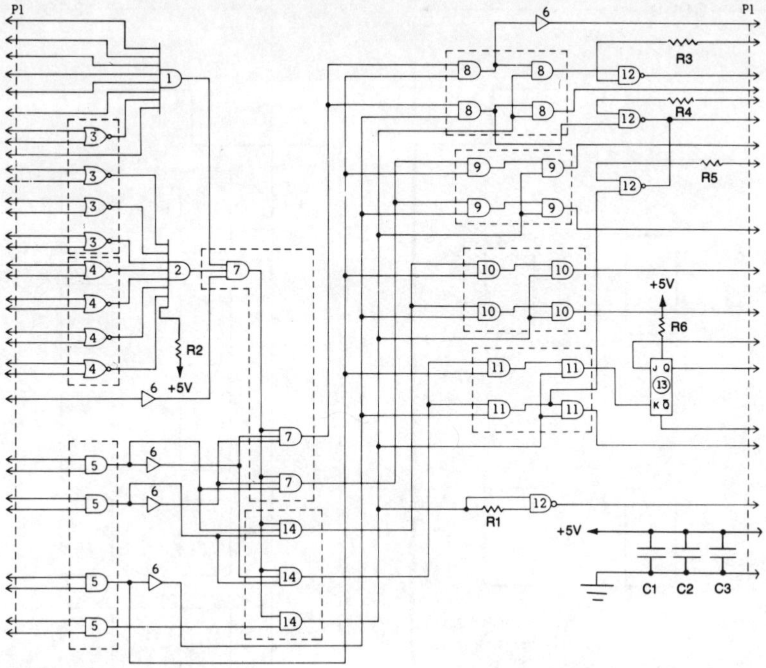

Figure 5-18. Digital Schematic

that of a typical double-sided board. The density can be increased to twice the amount. This can be increased to a maximum density of body to body. This method often requires more layers.

Need for Multilayer PC Boards

Many factors will influence the need for a multilayered printed circuit board. Some of these are as follows:

1. When the interconnections and the subassemblies (typically multilayer motherboards) are complex, require costly wiring for harnessing, can not be designed easily on two-sided PC board.
2. When a large percentage of traces or interconnections require coupling or shielding.
3. When frequency requirements call for careful control and uniformity of conductor impedence, with minimal distortion and signals (critical lead lengths).
4. When interconnections, size and weight are critical as in military, airborne, missile and space application.
5. When the quantity of components (discrete and integrated circuits) exceed the normal design area and close spacing is the absolute necessity.
6. When a large number of traces or interconnections are needed (motherboard or back plane).

Comparing Double-Sided to Multilayer Boards

Using the same size board (3 × 4 inches), compare the density factors in the three examples. All are 12 square inches. The first example in Figure 5-19 shows a density factor of 2 square inches for each IC (using a double-sided PC board). The 2 square inches for each IC gives a low density factor for routing conductor paths for voltage, ground, and signal lines.

Figure 5-20 shows a density factor of one square inch for each IC (a double-sided PC board). The increased density makes it more difficult to route conductor paths for voltage, ground, and signal lines.

Figure 5-21 shows a multilayer board with four layers: two signal layers, one voltage layer,

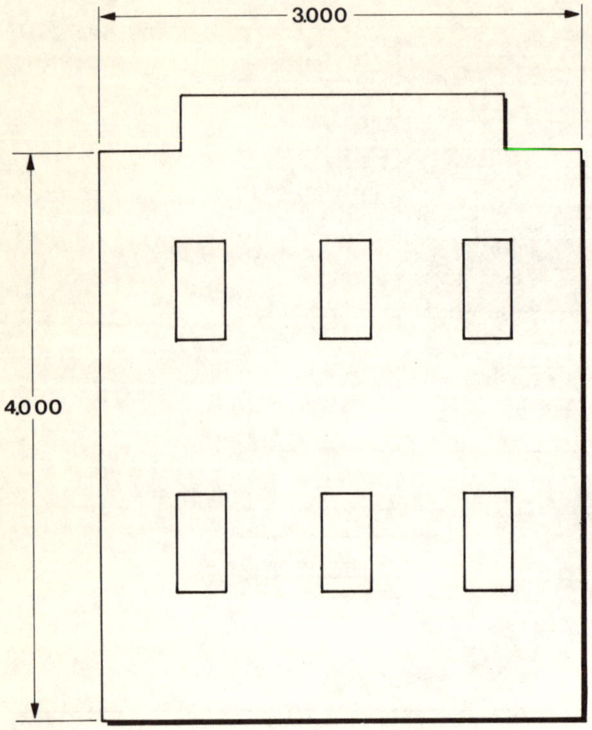

Figure 5-19. Double-Sided Board With Two Square Inches Per IC

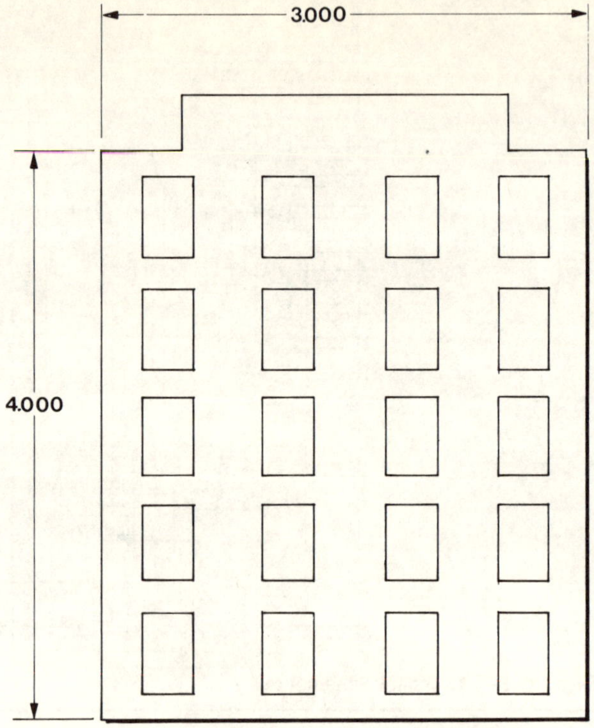

Figure 5-21. Multilayer Board (4-layer) With (.6) Square Inches Per IC

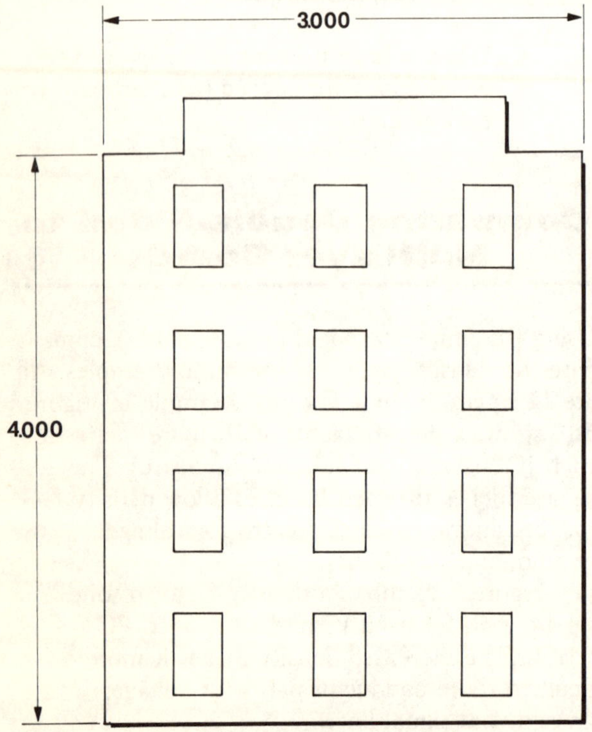

Figure 5-20. Double-Sided Board With 1 Square Inch Per IC

extremely difficult to interconnect all the conductor paths for voltage, ground, and signal lines. However, the four layers would be relatively easy to design.

Air Gaps, Traces and Pad Sizes

The board layout should be made to an enlarged scale: either 2:1 or 4:1. All component pads should be located at grid intersections (except feedthroughs). Feedthroughs are not mandatory on gridded intersections, but it is preferred. The pad spacing for discrete components should be determined by the maximum component body size. See the standard formula in chapter 3 of the analog design book. All components should be mounted on only one side of the board. Components should be placed or located so that any component may be removed without removing any other component (unless the assembly is not intended to be repaired). Only one component lead should be placed in a lead hole, except for and one ground layer. The density factor is .6 square inch for each IC. If this were the density factor on a double-sided PC board, it would be some commercial bus bars that are designed to fit in the same mounting hole as the integrated circuit lead. When a board design requires that jumpers be used (a jumper is an insulated wire used to substitute a trace on the board), they should be placed on the component side of the board. They should be constructed of solid insulated wire designed to be as straight and as short as possible. Whenever possible, avoid crossing over components. Those components requiring special holddowns because of shock, vibrations, or weight considerations, should be so indicated on the layout. The type of fastening should also be specified (clamping, bolting, rattails, adhesives, etc).

Conductor Widths

The preferred board layout is at a scale of 2:1. All traces, pads, air gaps, etc. have to be multiplied by a factor of 2. For conductor widths, establish a normal size that suits the majority of applications. Careful considerations must be given to the design of the conductive patterns. A good printed circuit board design includes: manufacturability, repairability/maintain-ability, and all functional requirements.

Primary Concerns

In the design phase, the primary concerns are the design widths of the conductors, conductor spacing, component holes, through connections (called Vias), and component mounting.

Component holes are determined from the maximum lead diameter of the component. Via holes are determined from the smallest hole to be drilled in the board. The formula for calculating a component hole size (which will be used to calculate the pad size) is covered in chapter 3 of the analog design book.

Connector Size and Board Outline

The formula to compute the connector size for a male PC connector, and to compute the board outline (including the tooling holes) is covered in chapter 4 of this digital design book.

Using the Grid Pattern

The grid pattern and the formula template layout is discussed in this chapter. Draw the layout as viewed from the component side of the printed circuit board. The grid pattern should be .050 or .100 inch spacing as shown in this chapter. See Figures 5-34 and 5-35.

PC Design Process Flow and Time Requirements

1. The layout of the printed circuit board begins with a schematic diagram, a rough bill of materials, complete information on special environmental and performance requirements, and all necessary specification sheets for integrated circuits and discrete components.

2. The board outline must be determined next. Besides the size of the board, determine what method of fastening or mounting is being used (card guides, screws, stand-off, etc). What type of electrical interface is required (connector, cable, wires, etc.)? The board size must be known in order to compute the total square inches available for the schematic given by the engineer.

3. Find out component case sizes. Determine whether the schematic (with all its components) will fit on the board outline.

4. Grouping is required at this time. Assign the gates that communicate together electrically on the schematic, they should be placed in the same IC package. Determine which gates will be IC 1, IC 2, etc. Grouping must be done in order to compute the total number of IC's required to fit within the assigned board outline. It is desirable to have one square inch minimum for each IC.

5. Plotting the grouped IC's onto the PC board is the next step. Assign IC numbers to locations on the board. The IC's that communicate to each other should be in line or in the same rows with each other.

6. The formal template layout is the next step. Bypass the rough sketch that would normally be done on an analog type board design. We will not sketch any part of the circuit. Conduct a formal template layout at this time. Establishing the conductor paths in an XY direction. This means that the majority of conductor paths on the component side must run in one direction, and the majority of the conductor paths on the non-component side must run in the opposite direction. The layout should be viewed from the component side of the printed circuit board.

7. This step will be used to establish the test point locations. Clean up all unnecessary via holes (called feedthroughs). In the formal layout we would have a tendency to place feedthroughs because of the XY system being used. Now it is time to remove the vias that were not necessary.

8. Part numbers (or other suitable identifications such as reference designations) from the schematic identify all items and components appearing on the layout.

9. All component lead holes are located at the intersections of grid lines. In locating components on the grid, the mounting length is determined by the maximum body size for discrete components. IC's on their mounting centers are determined from the body size (referenced by the number of pins; 14, 16, etc.). Component parts are mounted on only one side of the board.

10. Conductor widths on the finished PC boards are determined for voltage and ground traces from Table Figure 5-22 or from the electrical engineer.

11. After completion of the layout drawing, an artmaster is taped. The artmaster is made to an enlarged scale (generally 2:1 or 4:1). It is then photoreduced to provide a 1:1 scale (negative and positive). This is the photomaster (a tool for manufacturing the PC board). An alternative method is to have the layout digitized on CADD and photoplotted to create the 1:1 scale, negative and positive.

12. Also stemming from the layout:
    a. Fabrication detail drawing
    b. Assembly drawing and parts list
    c. Silkscreen drawing and photomaster

13. Each designer works differently. Table 5-3 is the recommended percentage of time spent on each phase of PC board design.

Determining Conductor Thickness and Width

Conductor patterns are shown in Figure 5-3. This takes into account applicable current, temperature, and copper thickness. Any conductor path carrying 1 A (ampere) of current (or less) shall be 0.010 in. (10 mils) minimum, with an air gap of 0.015 in. (15 mils) minimum.

Example of design criteria: Current = 10 A
Operating = 100° (T max above ambient)
Copper weight = 1 oz

a. Find 10 A in the left-top scale.
b. Draw a line to intersect the 100oC graph line.
c. Draw a vertical line from there down to intersect the 1-oz copper graph line.
d. Draw a horizontal line to the left from there to find the minimum conductor width - 0.070; therefore, it would require a tape width of 0.140 at 2:1 to carry a 10 A current on a board with 1 oz copper-clad at 100° rise above ambient (room temperature).

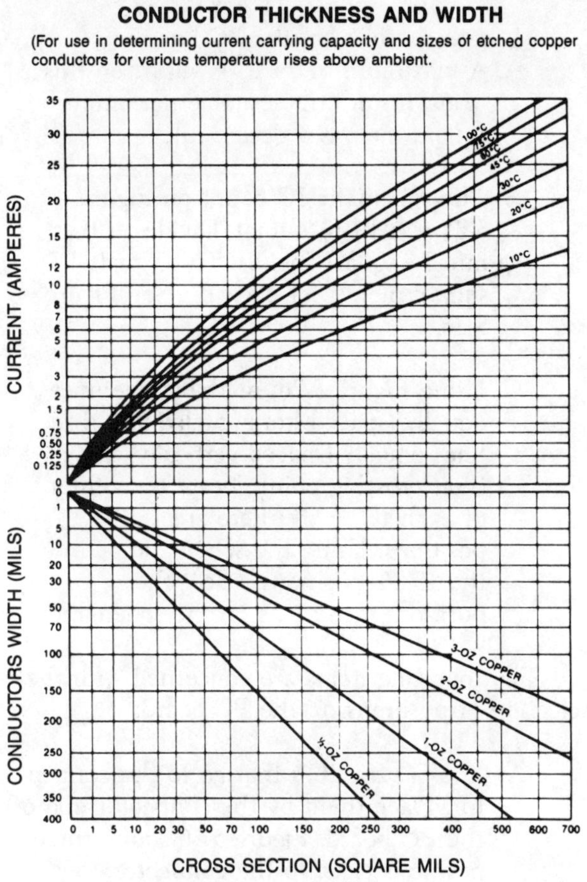

Figure 5-22. Conductor Thickness and Width

Table 5-3. Analog and Digital Board Time Requirements (layout only)

LAYOUT	TIME REQ'D
Schematic 1. Component Shape 2. Lead Length	10%
Sketch 1. Starting Point 2. Color Code 3. Shape 4. Board Outline	50%
Formal Layout 1. Plot Components Per Sketch 2. Draft Signal Traces 3. Draft Voltage & GND Traces 4. Final Check	40%
Layout = 60% of Total Effort **Tape-up** = 40% of Total Effort	

Table 5-3 is for layout only. When compared to the total effort required to produce a finished PC board, layout equals 60% of the total effort; tape-up equals 40%.

Integrated Circuits

Review the grouping of gates. Because this step is an integral part of plotting, Figure 5-23 shows how to assign IC packs by area. This assigning process is done to keep traces short. It helps eliminate unnecessary via holes called feed-throughs. This is where grouping plays its most important role. Successful grouping (as shown in Figure 5-23) places the grouped gates close to one another and to the connector pins. Grouping (which was covered in chapter 2) holds that the rule still applies, regardless of complexity. Some of the most important rules are as follows:

1. Determine the gate requirements or numbers.
2. Combine where possible (referring to manufacturer's specifications).
3. Group the gates by IC packs to enhance the mechanics of the upcoming design. Keep traces short, close to connectors, while minimizing feedthrough requirements.

Make sure that there is a minimum number of IC packs and that the majority of gates are grouped for shortness of lead lengths and communication to each other. Plot the IC pack on the board outline.

Board Outline

Using Figure 5-24, plot the grouped IC's onto a board outline. Look at the schematic. Notice the gates have been grouped to facilitate signal flow. This flow determines how the actual PC board is to be layed out. In Figure 5-24 the outputs from both IC 3 and IC 4 are connected directly into the input on IC 2. Likewise, IC 5 outputs are connected to the inputs of IC 6. Furthermore, IC's 3, 4, and 5 inputs come directly from the P1 connector. IC 1 and 2 outputs are connected to IC 7 inputs. IC 7 and IC 14 are connected to the inputs of IC 8, 9, 10 and 11, etc.

This is the flow in Figure 5-24. The PC board that we are plotting this schematic onto should flow in a similar manner. Undoubtedly gates will flow to the left or to the right and up and down, but these should be the exceptions. Isolated gates that do not fall into the mainstream cause no problem in the design of the layout as long as most of the schematic is consistent with the proper flow. (Keep the flow in mind and don't be concerned about gates that don't fit the overall left to right or up and down flow pattern).

Now that all logic is assigned and grouped, look at their placement on the layout (called plotting). Count the total number of IC packages on Figure 5-24. Divide the PC board square inches by that amount. If there were twelve IC's on a schematic and a board 4 in. x 6 in. (as shown in Figure 5-25), there are 4 inches wide by 6 inches high of useable board area.

Formula: Multiply the width times the length,

4 in. x 6 in.
= 24 square inches
Divide the number of IC's (in this case 12), into 24. This equals 2 sq. in. per IC.

Return to Figure 5-23. Assign the IC's on the board outline (Figure 5-26) using the following steps:

1. Count all the IC's on the schematic. Use the IC specification sheet. The total count of ICs is 14.

2. Figure the total square inches on the PC board (Figure 5-25). 4 in. × 6 in. equals 24 square inches.

3. As a rule of thumb, one square inch per IC is desirable for the average PC board. (Use this as a means of figuring the maximum or total number of IC's that can fit on any given PC board, without having to relax design standards).

4. Count the total number of inputs and outputs for the PC board connector. Make sure there are enough pins on the connector.

5. Locate the IC's on the PC board:
   A. A maximum of 24 IC's will fit on this board (using the desirable minimum square inch as a design rule).

   B. The schematic IC count on Figure 5-24 was 14 IC's total. Divide 14 IC's into 24 square inches. This equals 1.7 square inches for each IC. See Figure 5-26.

   C. Using Figure 5-26, mark off five lines equally (spaced horizontally) and three lines equally (spaced vertically). At their crossing point, locate the center of each IC. Note there are 15 positions. Using the schematic and the 15 IC positions, select the position of the IC's on the schematic.

The following items are concerned with the selection of positions on the PC board:

A. Look at the IC's that go to the connector. Place them by the connector end of the PC board. Figure 5-26 shows three positions close to the connector: (positions 5, 10, and 15 are the most preferable; positions 4, 9, and 14 are second in preference).

B. Since the first row of IC's by the connector positions (positions 5, 10, and 15) are selected first, the next positions to be selected are the ones that connect most to positions 5, 10, and 15. Select the IC that connects to position 5, using position 4. Then select the IC that connects or flows to position 4, using position 3 (position 2 and 3 likewise, positions 2 to position 3, and 1 to 2). Return to position 10 for the connections from position 9 to 10 and 8 to 9, etc. Note that the flow on the schematic should be running vertically. Start from the connector flowing back (in this case, to the top of the PC board).

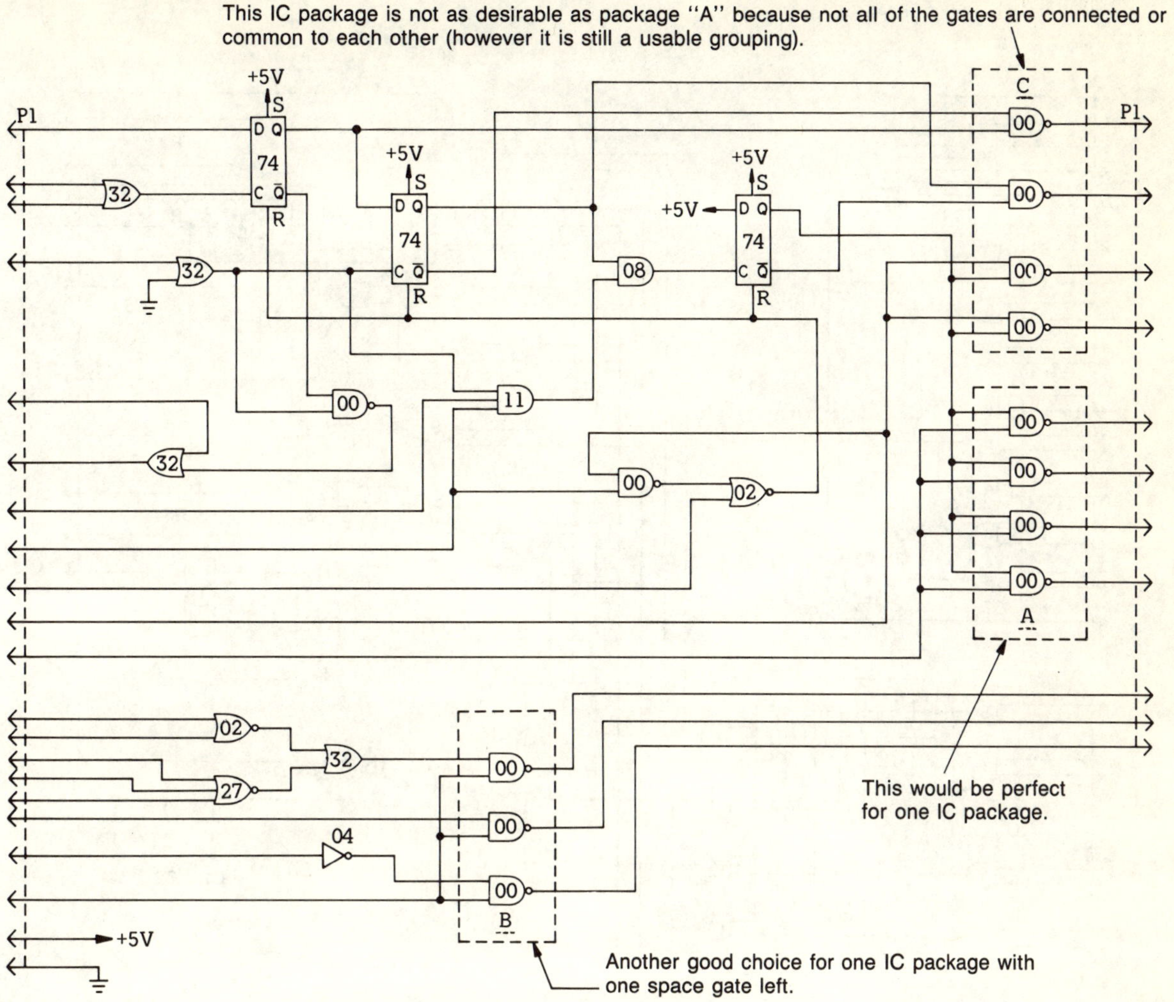

Figure 5-23. IC Grouping Before IC Plotting

Note: If there were not any more connections in row 10 that connected to positions 7 to 8 and 6 to 7, then stop on that row and go to the next row (positions 14 to 15, etc).

6. Return to Figure 5-24. Use the schematic and the IC count of 14 IC's to assign the IC positions on the PC oard. See Figure 5-27, following the steps below.

A. Look at Figure 5-24 (IC 3, IC 4, IC 5 and IC 1 are the IC's that connect to the connector the most). Select positions 5, 10 and 15. IC 3, 4, and 5 are placed in these positions respectively. See Figure 5-27. IC 1 should be close to the connector also. Put IC 1 in position 4 because it connects to IC 3 and the connector.

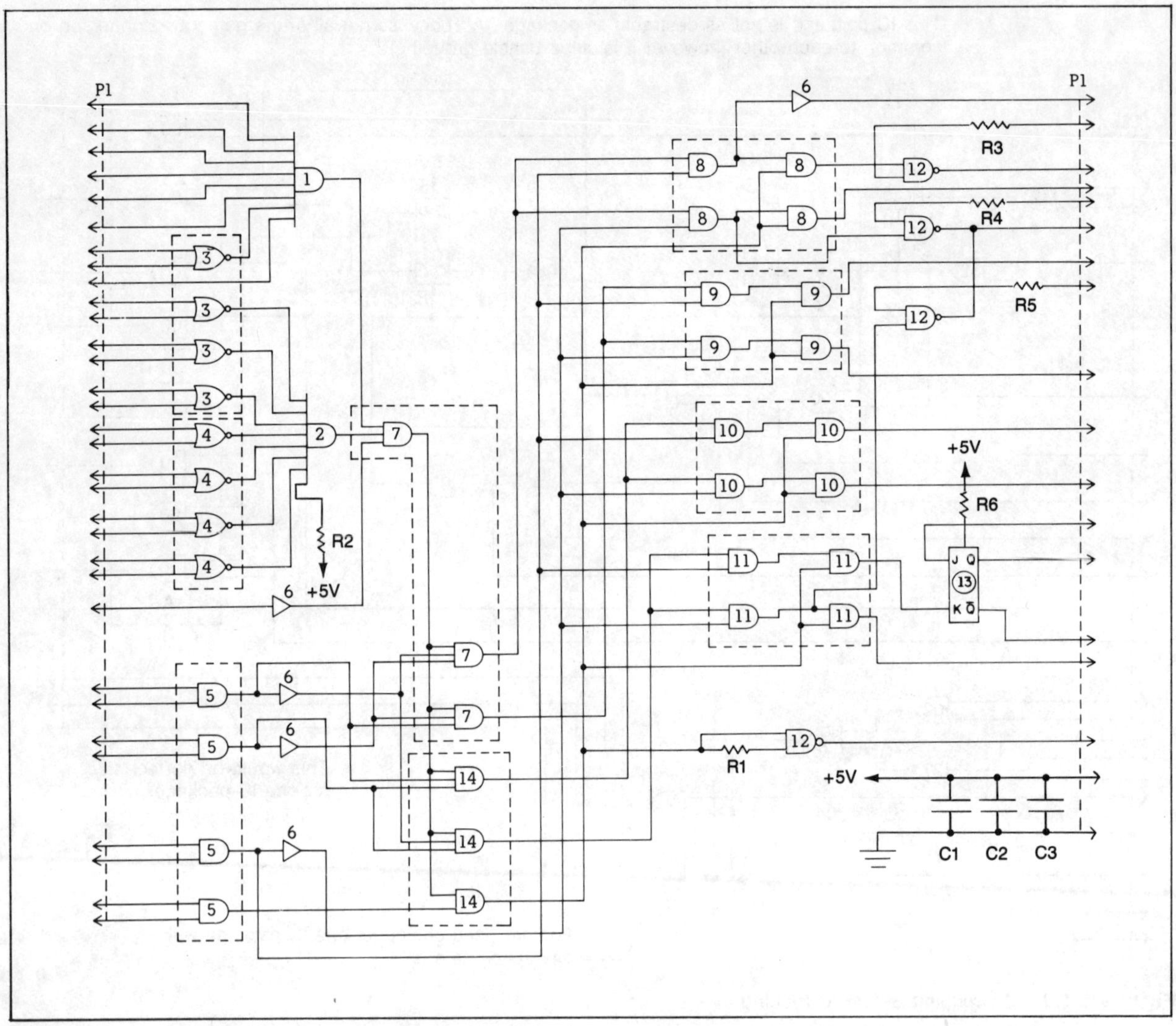

Figure 5-24. Grouping IC

B. Notice which IC's are connected to IC 4 for position 10. IC 2 is connected to IC 4, and IC 7 is connected to IC 2 on Figure 5-23. Therefore, select position 9 for IC 2 and position 8 for IC 7.

C. Return to Figure 5-24 (for IC 5 or position 15). IC 6 connects to IC 5. Therefore, position 14 should be assigned to IC 6. IC 6 also connects to IC 7 and IC 14. IC 7 is still close. Therefore, IC 14 will be assigned to position 13 to get a close connection to IC 6.

D. On the schematic, the next group of IC's that are common to each other are: IC 8, 9, 10 and 11. They should be close together, using positions 3, 2, 7, and 11.

E. The last IC's are out of the common flow. IC 12 and and IC 13 are the last two IC's which need positions assigned. IC 12 connects mostly to IC 8, 11 and IC 7. So, position 6 would be close to all three IC's. IC 13 should be close to IC 11. It would fit very nicely in position 12. Note position 1. Use this as a spare

NOTE: One square inch per IC would be desirable except on PC boards that relax their standards.

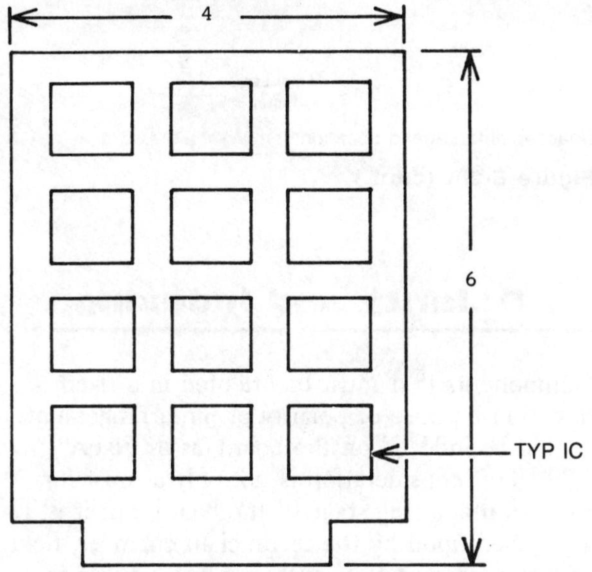

Figure 5-25. Plotting IC's

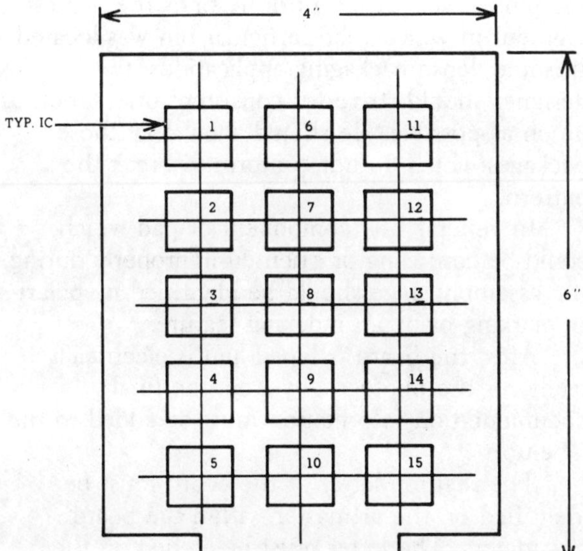

Figure 5-26. Plotting IC's Onto A Board Outline

IC position (perhaps providing the pads for a 16-pin IC that can also be adapted to a 14-pin IC position). This would be for future expansion of the PC board, since we do have the area to place the pads into. This is just one way to select positions or to plot the IC's. There are several ways and they would be correct. But this will give some idea how to start.

Plotting will be exercised on the board assignments at the back of this book.

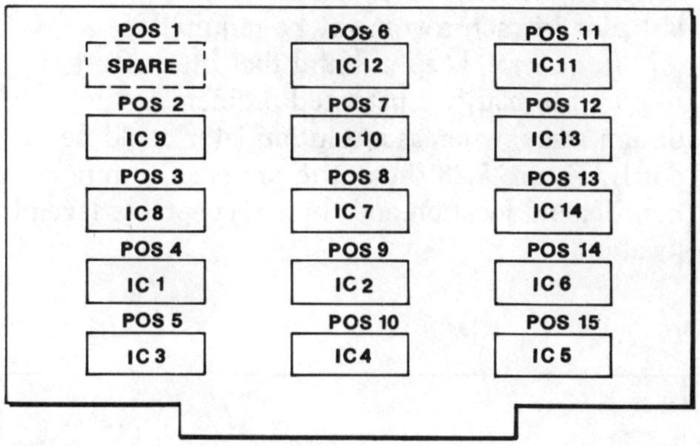

Figure 5-27. Plotting IC's Final Board Plot

Component Designation Marking

Some applications require that all reference designations of the electrical components either be etched or silk-screened on the PC board for easy identification. This aids technicians and field service repairmen to repair or trouble-shoot boards without the use of an assembly or loading diagram. Boards that have a high failure rate or have to be repaired by persons without proper access to a full documentation package rely on this information.

In cases where such rigid identification is not necessary, it is still important to identify such components as connectors, test points, switches, or any component that requires access for calibration or trouble-shooting. If a particular board requires that jumpers be installed to reflect different models or functions, then they should be marked on the board also.

Integrated circuit packages should be identified on the board. The IC's are not generally oriented in any numerical sequence on the schematic. Therefore, in order to relate a reference designation fro the schematic to the PC board, the integrated (IC) mechanical package must be identified by a reference designation number on the board for quick reference.

It is helpful to identify enough of the contacts on a connector or any other cluster of pins that are pre-assigned a number arrangement. Generally, the first and last pins are enough to establish this sequence; in other cases, the first or last pins in each row might be required.

In general, keep in mind that little effort is required to clarify a patented problem at the design stage, whereas a solution later could be costly. Figure 5-28 shows the preferred component legend location and the unacceptable legend location.

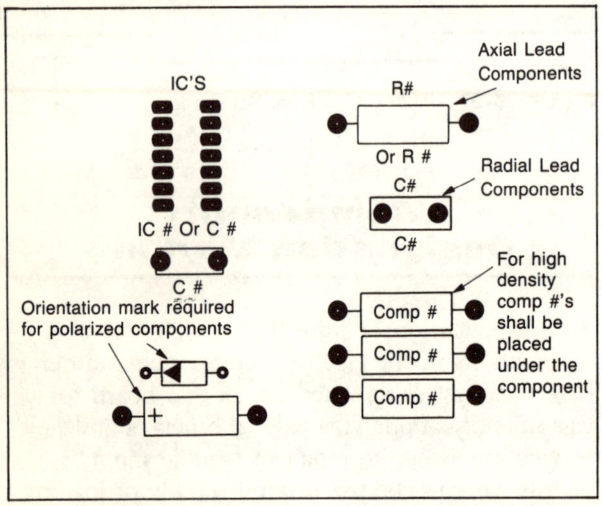

Figure 5-28. Preferred Component Legend

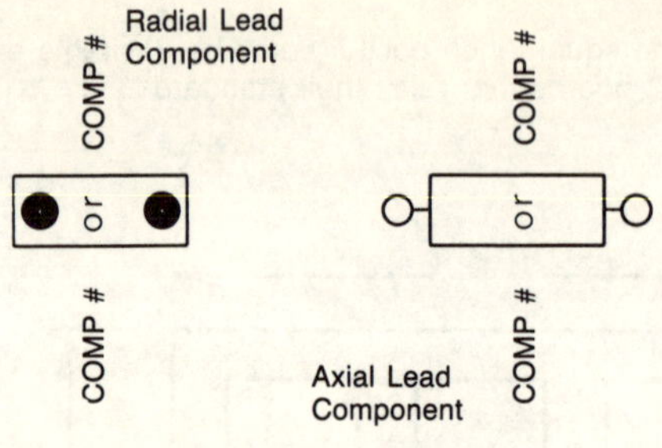

Unacceptable Legend Location

Figure 5-28. (cont.)

Polarity and Indexing

Components that must be oriented in a fixed direction because of polarity or pin arrangements should be indexed on the board (as in Figure 5-29). This consideration is not only a must for the manufacturing stage of the board, but it is also relied upon by the technician, engineer, field service representative, customer and any other person that would have need to trouble-shoot or follow the flow of the circuit on the PC board.

Index various pins on a connector to aid in finding a particular pin number. In addition to identifying the pin number (as described in Figure 5-30), every tenth pin could be indexed. In large connectors, this helps the technician to scan the pins in sets of ten until he finds the nearest segment in which that particular pin was located. In some dense packaging applications, the designer should strive for consistent orientation as much as possible. He should allow only those packages which do not conform to break the pattern.

In general, any component or pad which could be confusing or oriented improperly during the assembly stage should be identified by polarity marking or by an indexing feature.

After the board is taped and is electrically ready for the photography step, the final documentation information must be added to the tape-up.

The revision letter of the board must be identified on the artmaster within the board parameters. The letter must be etched on the

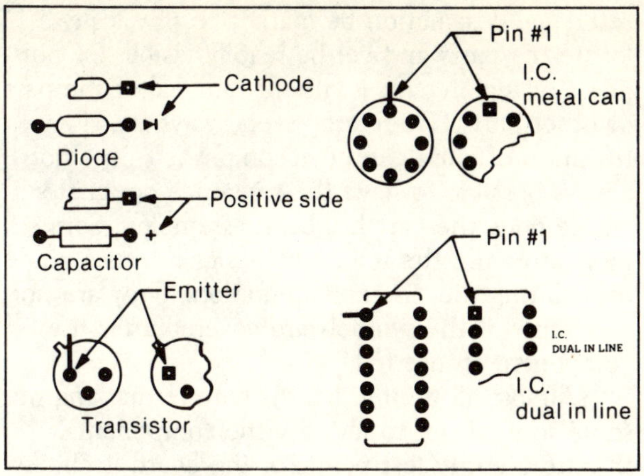

Figure 5-29. Component Polarity and Indexing

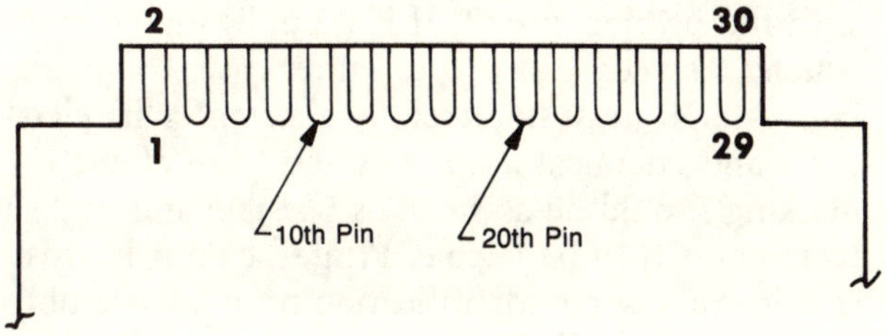

Figure 5-30. Connector Indexing

board to avoid confusion with any previous or later changes that might occur. This is generally done (Figure 5-31) on the noncomponent side of the board along with the PC board part number, if applicable.

The assembly number of the board should be identified in the same manner on the component side of the board. This helps anyone to find the proper assembly drawing and any other documentation involved in the manufacturing, procurement or sale of the board. No assembly revision letter should be etched on the board. This information should be added at whatever revision the assembly drawing is at the time the board is assembled (Figure 5-31).

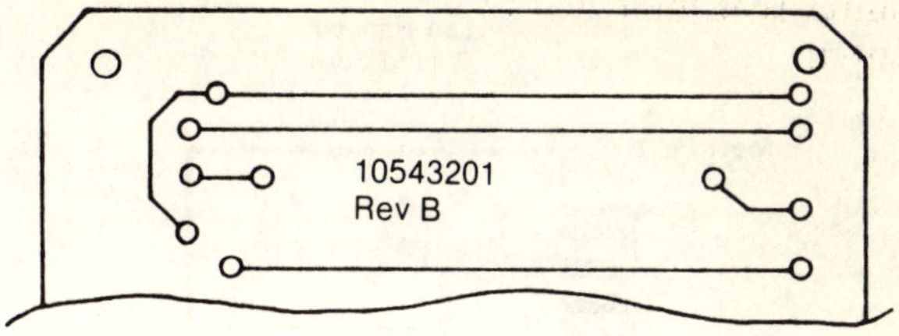

Figure 5-31. Revision Marking

Figure 5-32A/5-32B shows the correct component marking for left-handed viewing versus right-hand viewing of a printed circuit board.

Marking Requirements

A specification that clearly defines the requirement for component marking is the American National Standards Institute (ANSI) document Y32.16 on electrical and electronics reference designations. This document describes the methods for the assignment of reference designators to components and assemblies.

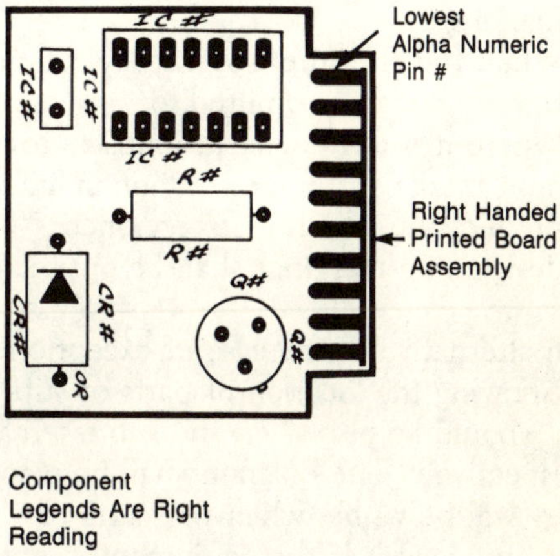

Figure 5-32A. Right-handed Printed Board Assembly

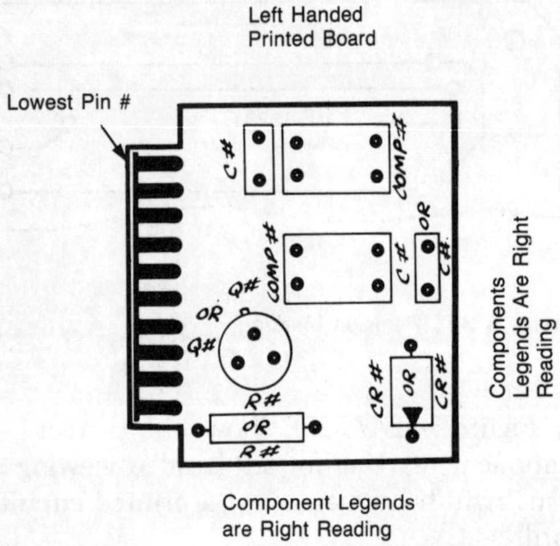

Figure 5-32B. Left-handed Printed Board Assembly

Reference Designations

Equipment should be permanently and legibly marked with the reference designations for each sub-assembly and part except in the following cases:

1. Where space limitations prevent such marking.
2. When the customer specifically requires that markings be omitted, or
3. Where it is customary in industry to omit markings on specific non-military products, in which case reference designation markings shall be optional.

As an alternative in the case of exception 1, diagrams showing the location of parts of sub-assemblies should be placed on the sub-assembly or unit respectively. The diagram shall be placed where they will be visible when the parts or assemblies are viewed. If the equipment is not completely marked, or if diagrams are not placed on the unit for any reason, part location diagrams shall be included in the maintenance data for the set.

As may be seen from this requirement, the decision to provide board marking is based primarily on the maintenance of that board. ANSI standard Y32.16 also requires that the reference designation be marked to physically locate the parts and yet be readily visible for purposes of maintenance without removal of the part or other parts. There are several ways that identification of parts can be accomplished. It is not always possible to have the reference designator visible once the part has been assembled. Some fabricators use the reference designator as a guide only during the assembly operation. They are not concerned if the symbols are covered after the component is inserted.

Almost all printed boards require marking of some sort - if not to identify the components, then to identify test points or the board itself. As a minimum, some form of marking should be applied that correlates to the orientation of polarized components. Marking may be applied after the board is etched, or it may be etched along with the printed circuit. If it is applied later, it is usually a screen process. In either case, the marking material must be carefully chosen for its electrical and chemical properties. The size of the marking should be as small as possible and yet large enough to be legible. Printed circuit boards should carry some identification prior to assembly of components. This identification is usually the printed board fabrication drawing number and assembly. The assembly number should be located on the component side where the assembling of the electrical components takes place. This number is usually preceded by the letters "ASSY".

In addition to identification, serialize the board primarily for traceability. Some companies also display their company logo, federal code identification, or UL number. Many of these can be etched at the same time that the conductors are processed. All nomenclature (identifying) information should appear on the PC artwork. Figure 5-33 shows an example of a finished marked board.

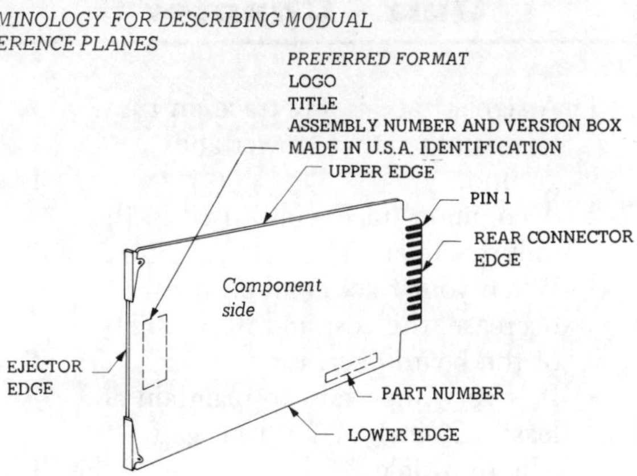

Figure 5-33. PCB Nomenclature

Types of Grid Patterns

The use of a grid pattern for printed circuit layouts has become common. Grid systems are used to place traces and to position pads. This also makes the locating of components in the specific pattern simpler. The grid can take the place of a scale. Simply count the grids. Grid systems consist primarily of .025 × .025, .050 × .050 and .100 × .100 inches. In order to arrive at the final size, divide the grid by the scale. For example, a layout using a .100 grid drawn at 4:1 would have the actual size of .025 inch. When considering the scale of the layout, the designer should use a grid that is equally divisable. It is not a good idea to use a 0.100 inch grid and prepare the layout at the 3:1; rather a 2:1 or 4:1 layout scale should be used.

Grid sheets come in several different types and materials as shown in Figure 5-34 and 5-35. Most are printed on a polyester based film that does not shrink or stretch. Precision black grids on polyester film are the most common. Layouts and artmasters are prepared by overlaying the grid with a clear piece of polyester film or other plastic material. For added dimensional stability, grids also are prepared on glass plates. Most glass grids are accurate to within ±.002 inches over 48 inches (with an accuracy of ±.0005 inch between any two lines).

Positive Line Grids (See Figure 5-34).

- Assures the placement of PC symbols and tapes
- Serves as an exacting measuring layout tool where maximum accuracy and stability are required
- Sizes up to 38‘ x 58‘ available ("off-the-shelf")

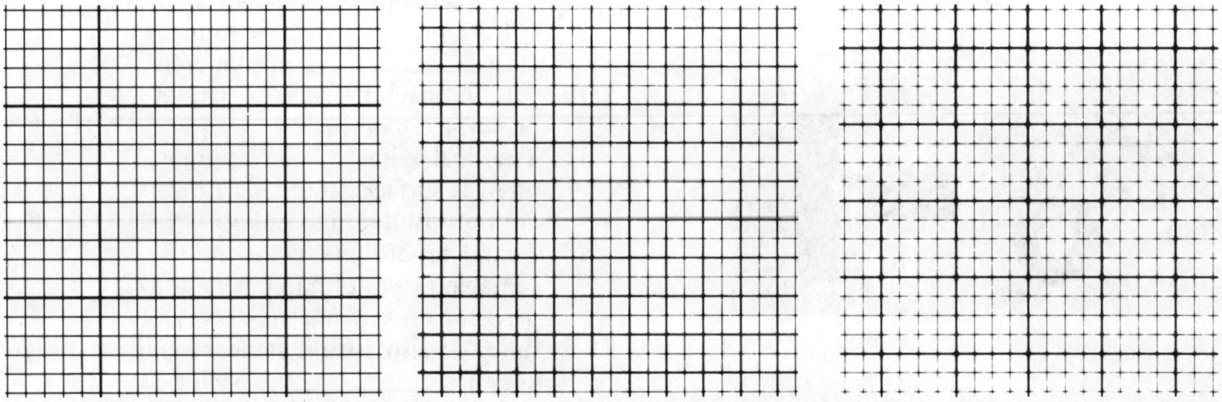

A	B	C
10 × 10 lines/inch (.100)	10 × 10 lines/inch (.100)	10 × 10 lines/inch (.100)
Heavy accent on the 10th line	Heavy accent on the 2nd line	Heavy accent on the 4th line
Colors are black, blue, and brown	Colors are black, blue, and brown	Colors are black, blue, and brown

Figure 5-34. Positive Line Grids

- Available in a variety of patterns and colors

Negative Line Grids (See Figure 5-35).

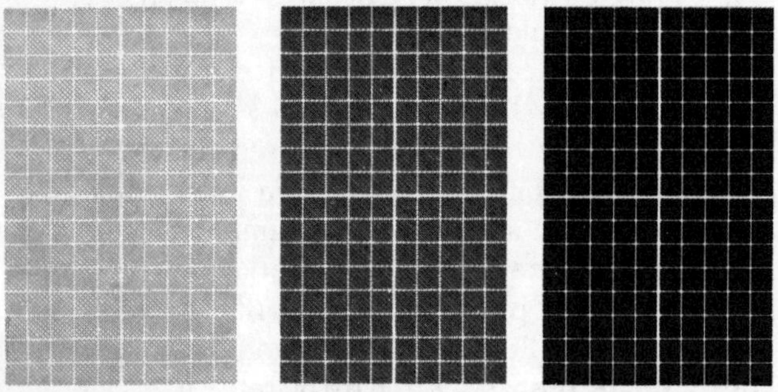

Light Medium Dark

Figure 5-35. Negative Line Grids

The negative line grids have the same characteristics as the precision positive line grids, except that the lines are clear instead of solid. Splitting a clear line is more accurate than splitting a solid black line, by some opinions (Figure 5-21). The translucent gray background reduces light table glare, with less eye fatigue. There are three shades of black: light, medium, and dark.

Example Placing Pads & Traces

Quiz - Chapter 5

1. A signal trace is any trace on the boards other than power and ground. T F
2. A common trace width used in the industry today is ______________
3. When your traces and air gaps decrease, the cost and unreliability of the board decrease. T F
4. It is very important to maintain at least .026 inch at 2 x 1 air gap when possible. T F
5. Color coding can be used for three different functions
   1. Color code component side & non-component traces.
   2. Color code voltage & ground pads
   3. ______________
6. It is common practice to use more than one plated through hole, when wide conductors must be transferred from one side to the other. T F
7. Conductor spacing: the minimum by exception is ______________
8. If the P.C. was intended to be single-sided, all conductors would be on one side & only one color for traces would be used. T F
9. During the taping operation the color coded pads on the layout will tell the taper which pads to place ______________ size tape on.
10. The flat pack IC package cannot be surface mounted. T F
11. The dual-in-line IC package most often is surface mounted. (Leads are not inserted into holes) T F
12. Templates are one of the most valuable pieces of equipment the designer uses. It not only saves time but also assures standard component ______________ spacing & body sizes.
13. A feedthrough consists of a pad on both sides of the board with a plated through hole connecting them. (Which has no component lead) T F

14. Feedthroughs are especially necessary on analog boards. T F
15. Digital schematics are made up of logic symbols only. T F
16. "Plotting" is the placement of IC on the printed circuit board. T F
17. IC location is determined by which IC's ________________ to each other.
18. Like analog all digital boards are sketched before completing a formal layout. T F
19. Before the layout begins the designer should calculate the total area required, to insure that all ________________ will fit on the board.
20. Calculating the area: the one square inch means that this is the maximum number of comp's per square inch. T F
21. The similarity between IC's and discrete components is the number of lead pads used. T F
22. A typical multiplayer P.C. board component density is ____________ the amount of a double-sided board.
23. Many factors will influence the need for a multiplayer P.C. board? Name two:
    1. ________________________
    2. ________________________
24. The board layout should be made to an enlarged scale: either________ or 4 x 1.
25. Component pads should be located on grid intersections. T F
26. Components may be mounted on both sides of the P.C. board. T F
27. Stacking components is acceptable, if necessary. T F
28. Typically only one component lead should be placed in a lead hole. T F
29. Some components require special holddowns because of ____________ or vibrations, or weight considerations.
30. Holddowns for components can be accomplished in many ways. Name two.
    1. ________________________
    2. ________________________
31. In the Design phase, the primary concerns are the design widths of conductors & spacing of conductors. T F
32. A typical grid pattern should be .050 or________________
33. When placing the IC packs on the P.C. board, you should plot for shortness of leads and communications. T F
34. Placement of IC's: it is important to place IC's that connect to the connector, as close to the ________ ______________ as possible.
35. Component marking: silkscreening component identification aids only the P.C. designer. T F
36. Component marking: on most P.C. boards it is important to identify test points and variable comp's. T F
37. IC's are not generally marked for idenification. T F
38. Components that have a fixed direction for polarity must be identified by indexing. T F
39. It is a good practice to identify or index every tenth pin on the connector. T F
40. Marking requirements: revision letters are marked on the board to avoid confusion with previous or new changes that might occur. T F
41. The assembly number & revision number are marked on the same side of the board. T F
42. In all cases equipment should be permanently and legibly marked with reference designations. T F
43. It is not always possible to have the reference designator visible once the part has been assembled. T F
44. Printed circuit boards should carry some indentification prior to assembly of components. This identification is usually the P.C. board ________________ drawing number.

45. Grids: negative line grids have the same characteristics as the precision positive line grids except that the lines are "clear." T F

The answers for each question can be found on the pages listed below:

Question 1. Pg. 126
Question 2. Pg. 126
Question 3. Pg. 126
Question 4. Pg. 126
Question 5. Pg. 126, 127
Question 6. Pg. 127
Question 7. Pg. 127
Question 8. Pg. 127
Question 9. Pg. 128
Question 10. Pg. 129
Question 11. Pg. 130
Question 12. Pg. 131
Question 13. Pg. 131
Question 14. Pg. 131
Question 15. Pg. 132
Question 16. Pg. 132
Question 17. Pg. 132
Question 18. Pg. 132
Question 19. Pg. 132
Question 20. Pg. 133
Question 21. Pg. 132
Question 22. Pg. 135
Question 23. Pg. 135
Question 24. Pg. 136
Question 25. Pg. 136
Question 26. Pg. 136
Question 27. Pg. 136
Question 28. Pg. 136
Question 29. Pg. 137
Question 30. Pg. 137
Question 31. Pg. 137
Question 32. Pg. 137
Question 33. Pg. 139
Question 34. Pg. 140
Question 35. Pg. 143
Question 36. Pg. 143
Question 37. Pg. 144
Question 38. Pg. 144
Question 39. Pg. 144
Question 40. Pg. 145
Question 41. Pg. 145
Question 42. Pg. 146
Question 43. Pg. 146
Question 44. Pg. 146
Question 45. Pg. 148

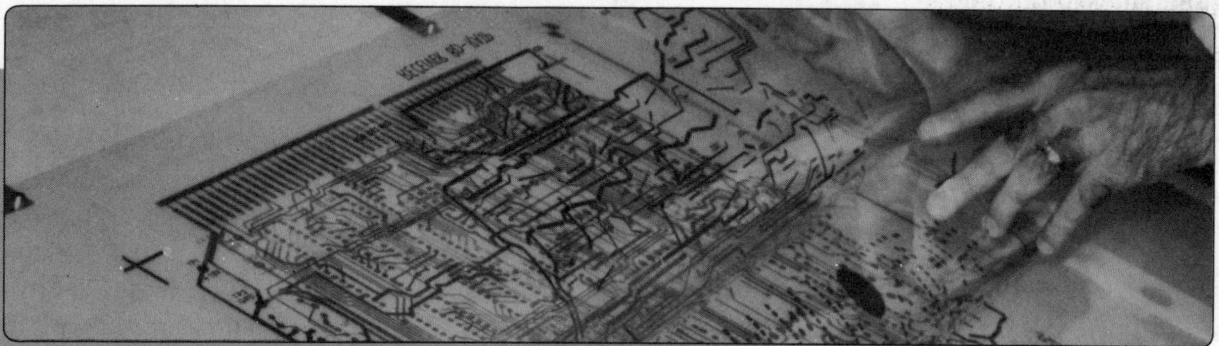

6 ARTMASTER TAPE-UP

This chapter discusses the basic materials for hand taping digital PC boards. Three methods are used:

1. 2-sheet method

2. 3-sheet method

3. Red and Blue taping techniques

This chapter covers basic taping standards for conductor widths, conductor spacing, and board edge clearance. Taping do's and don'ts are presented. Many basic techniques for hand taping are also covered in the Analog Design book such as:

1. Taping exercises
2. Applying Stik-on patterns
3. Applying connector strips
4. Applying pads
5. Applying tape
6. Component designation marking
7. Taping guidelines

The Analog Design book has a photo tour showing the production of the final 1:1 film for your PC board.

The artmaster, Figure 6-1, is a tool which is generally prepared at an enlarged scale used to photographically produce the master pattern. It is constructed by accurately placing precision width tapes and pads on a stable-based polyester film to represent conductors on the finished board. The tape widths, pad diameters and other conductor patterns must be accurately calculated to insure that the desired mass of conducting material will be the result on the completed project.

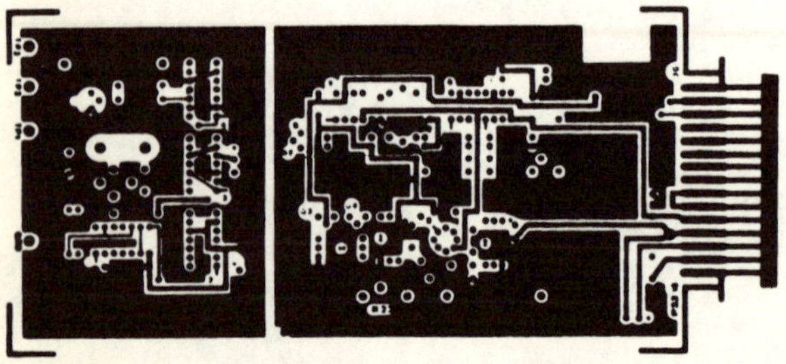

Figure 6-1. Artmaster

Scale

The most common artmaster scales are 2:1 and 4:1 although there are some companies that work at 3:1 and 5:1 scales. It must be noted that the two latter scales, 3:1 and 5:1, are not industry standards and should be avoided.

Base Material

Artwork base material should be .005" to .007" thick stable-based polyester film. This material is temperature and humidity stable, and provides a smooth nonporous surface to which tape and patterns will adhere best.

Tape and Pads

Black crepe tape, pads and patterns provide a material that is easy to work with and the traces can be routed either by the sweeping bend method or by the cut and bend method. The latter is preferred due to the fact that better proficiency in both time and appearance can be accomplished in a shorter learning time. With the black crepe method, a separate layer or sheet must be prepared for each side of the board. This can be accomplished by a 2-layer method of Figure 6-2 where the first layer has all the pads plus the traces from one side of the board and the second layer has the same pads spotted with the traces from the opposite side.

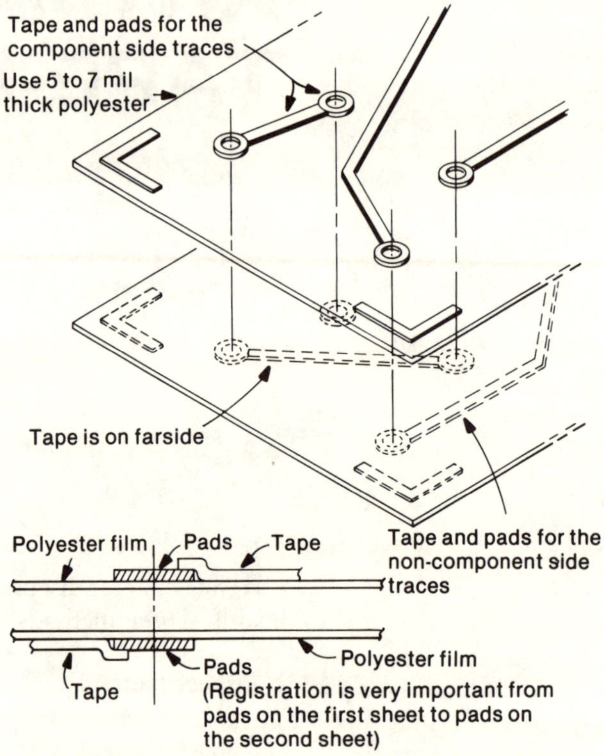

Figure 6-2. 2-Layer Method

The 3-layer method shown in Figure 6-3 is essentially the same except one layer is used for a "pad pattern", and then this is overlayed with a sheet for the noncomponent traces and a dif-

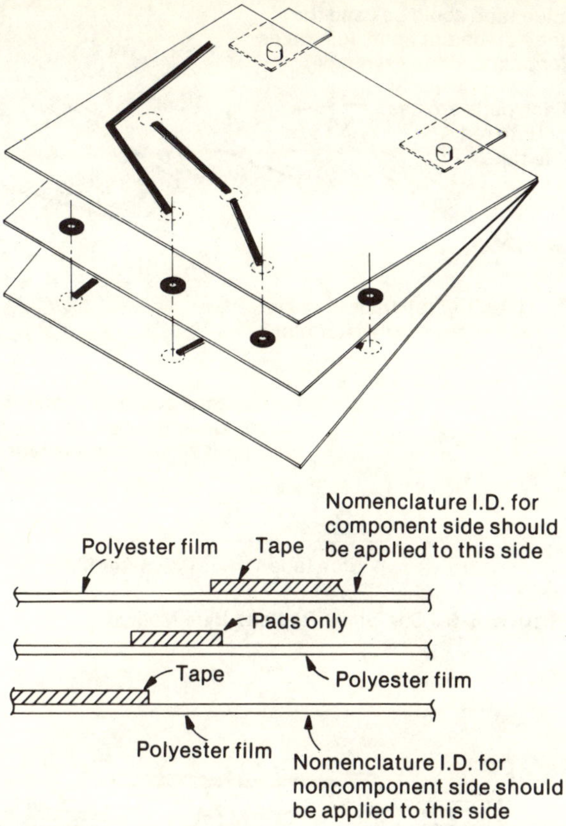

Figure 6-3. 3-Layer Method

ferent sheet for the component traces. This method saves time by not having to "spot" the second set of pads opposite the first set, but most important, insures perfect registration of all pads from component side to noncomponent side because the same pad pattern is used for both sides with the respective trace pattern superimposed over it. Figures 6-4 and 6-5 show the pad master. The red and blue material method illustrated in Figure 6-6 provides tapes in the same widths as black crepe except they can be taped on one sheet of polyester film with red on one side and blue on the other side.

The pads would be black crepe placed only on one side and, as in the 3-layer method, would be used in both the component side and noncomponent side.

This method allows the photographer, by the use of red and blue filters on his camera, to filter out the blue tape in one case and photograph only the red tape with the black pads for one side of the board. He then changes the filters, repeats the process and photographs only the blue tape with the black pads for the opposite side of the board. The tape-up does not have to be flipped over or moved during this entire process. The end result is absolute registration for 2-sided PC boards.

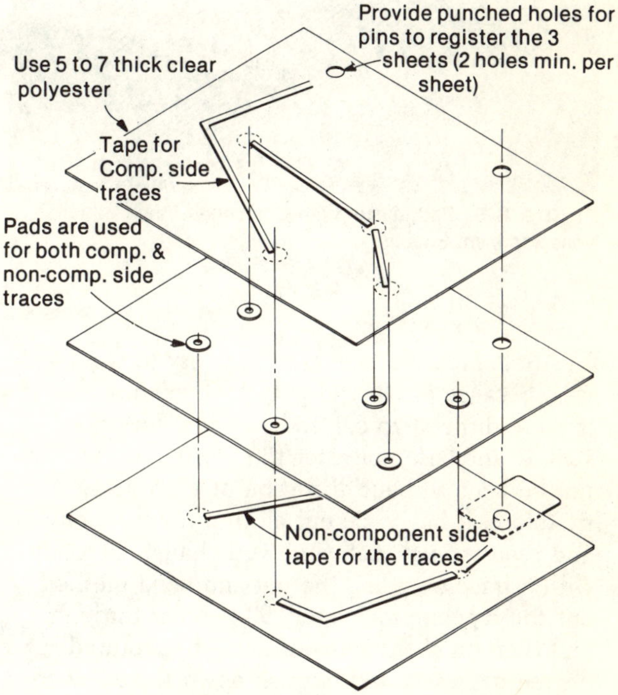

Figure 6-3a. 3-Layer Method

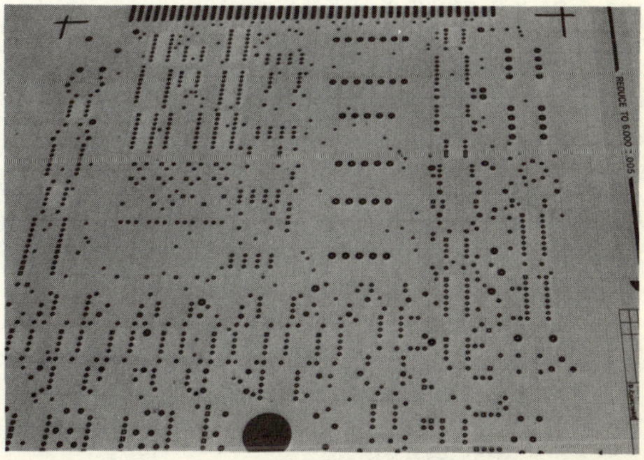

Figure 6-4. Pad Master

Figure 6-5. Pad Master With Component And Noncomponent Sheets

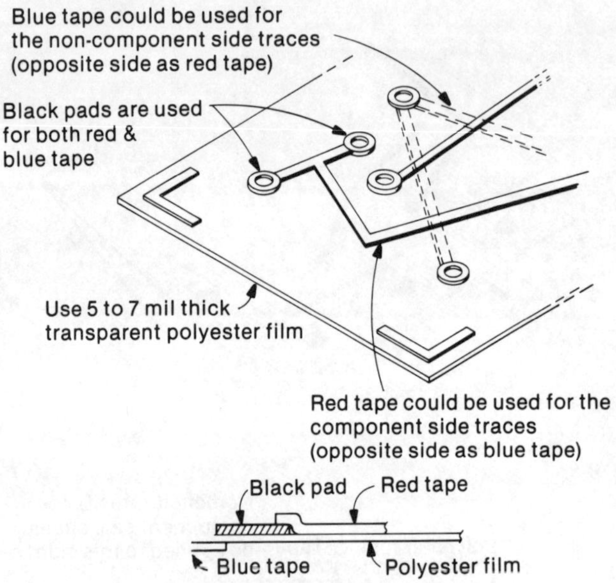

Figure 6-6. One Sheet Red and Blue Method

Red and blue tape is not as easy to work with because of its rigid polyvinyl composition. It is more difficult to cut and at times does not stick to the base polyester film. Red and blue cannot be bent without distortion of the traces; therefore, it has to be cut all the way through and repositioned each time you change direction with a trace or, using the cut and bend method, cut the tape approximately 90 percent through and then bend the remainder to the required degree necessary and burnish down the edges to keep the uncut edge from sticking up.

Taping Standards

Regardless of taping methods or type of material used, the physical act of taping must conform to some basic standards.

Conductor angles should be made at preferred angles of 30, 45 or 90 degrees. Although the traces could be directed at almost any angle, it is important that all traces running parallel be at the same angle for uniformity. The minimum angle that any trace should be placed at is 90 degrees. Angles less than 90 degrees provide a situation during the manufacturing process which could allow etching solution to build up on the inside angle and etch away excess material as illustrated in Figure 6-7.

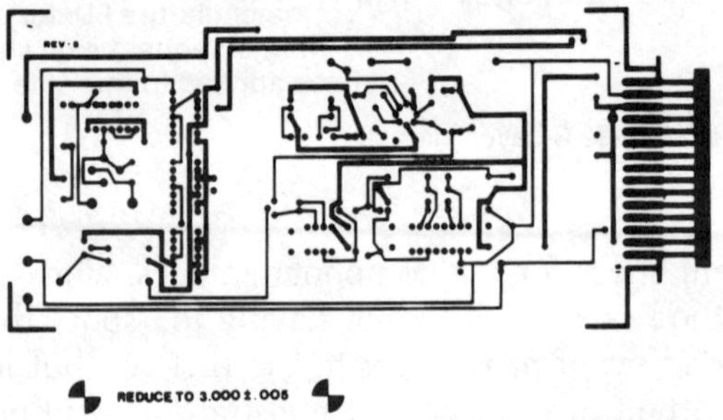

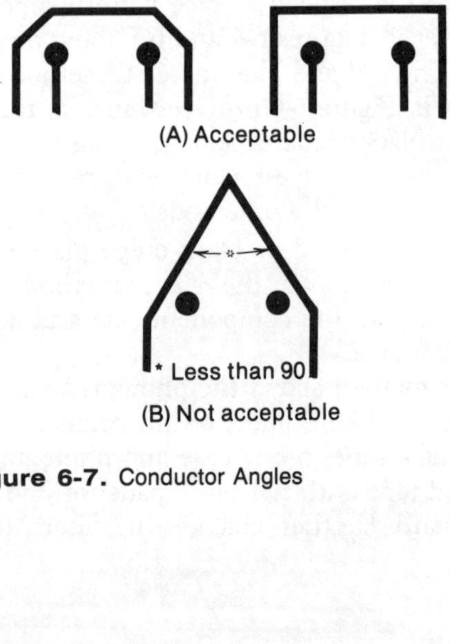

Figure 6-7. Conductor Angles

Conductor Routing

The same problems discussed under "Conductor Angles" also apply to trace routing. If the junction of traces coming off a pad form a pocket, the same overetching problem can occur as shown in Figure 6-8.

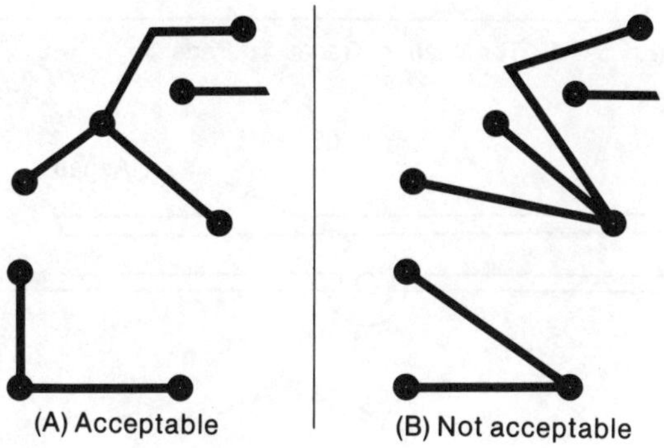

Figure 6-8. Conductor Routing

Conductor Spacing to Board Edge

Traces should not come any closer to the edge of the board than .050" at a scale of 1:1 inch and, if at all possible, at least .100" should be left. This gives the board manufacturer the necessary tolerance that is required to shear the board to the size specified. If less than the above mentioned air gap is left, a portion or all of the trace could get sheared off or shorted out at the next assembly.

If metal or electrically conductive card guides are used, then the gap from the edge of the card guide to the edge of the nearest conductor should be the standard required air gap (.050" or greater); see Figure 6-9.

Problem Areas

See Table 6-1. Do not overlap tape or cut the pads. Tape has a natural tendency to creep or return back to its original shape if it is bent or stretched too far. Even under the most careful conditions, this can become a problem. Care must be taken to avoid situations where tape

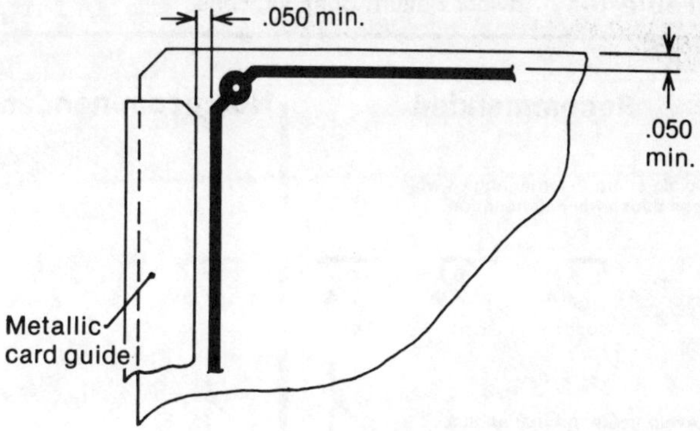

Figure 6-9. Spacing for Conductive Card Guides

could distort and cause both electrical and manufacturing problems. See Figure 6-10 for examples of poor taping techniques.

Terminating traces to pads should be very deliberate. Trying to connect a trace with a con-

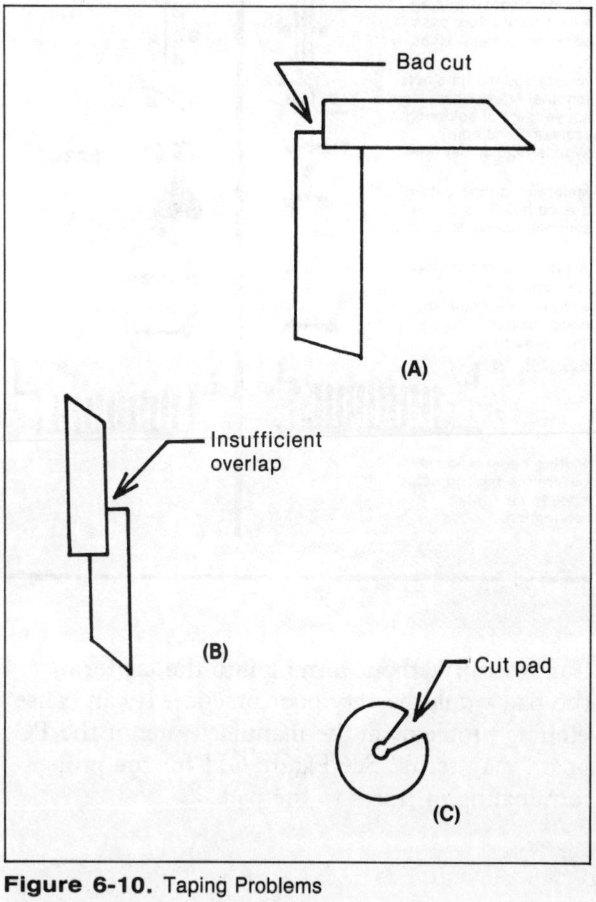

Figure 6-10. Taping Problems

Table 6-1. Artwork Pattern Configurations

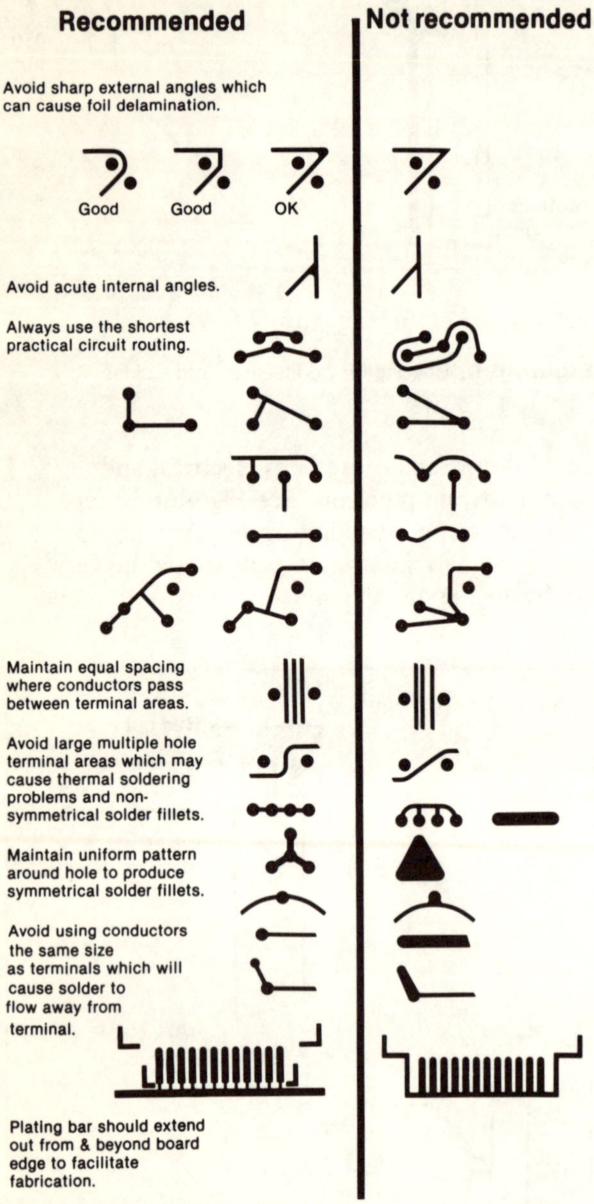

tinuous run without turning into the center of the pad would be very poor practice. It can cause etching problems in the manufacturing of the PC board may result. See Figure 6-11 for the proper terminating of traces to the pad.

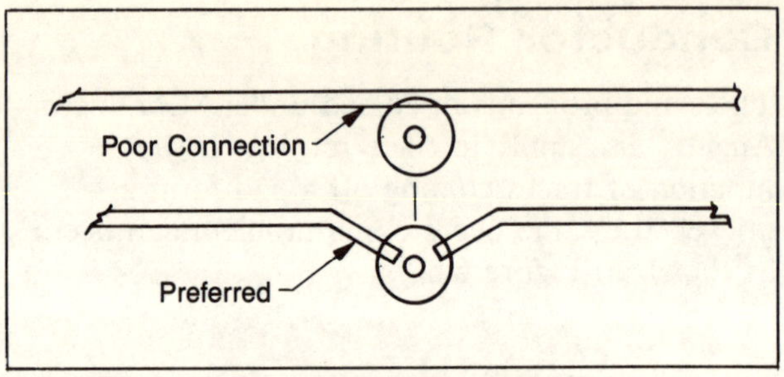

Figure 6-11. Terminating Traces To Pads

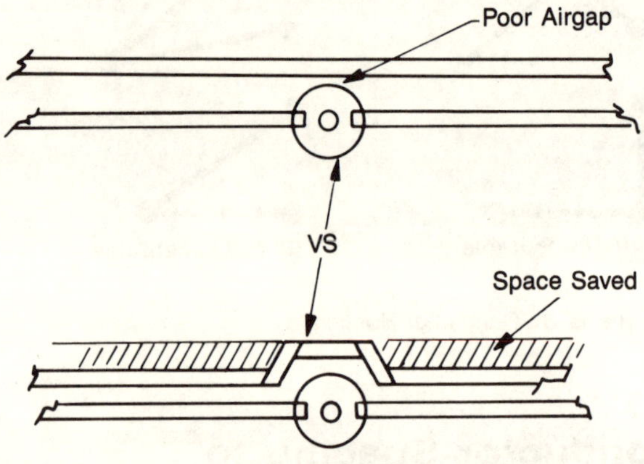

Figure 6-12. Routing Traces

On very dense digital PC boards, routing traces with tight air gaps can be a problem. To conserve space, we will step around the holes, or feedthrough pads. Figure 6-12 shows the technique to conserve space and to maintain proper air gap as you work around the feedthrough pads.

Terminating traces to pads can be aesthetically pleasing, especially for a series of connections in-line. Figure 6-13 shows four traces turning on the same angle into a row of pads rather than the same traces terminating to the same row of pads at different angles. Consistency on turning the traces is aesthetically more pleasing than the inconsistent angles shown in Figure 6-13.

When terminating traces into pads, the angle must avoid creating poor air gaps. Figure 6-14 demonstrates a 45 degree turn into a pad. This leaves sufficient air gap on the turn. A much tighter angle creates a very poor air gap. This could cause a short on the finished PC board.

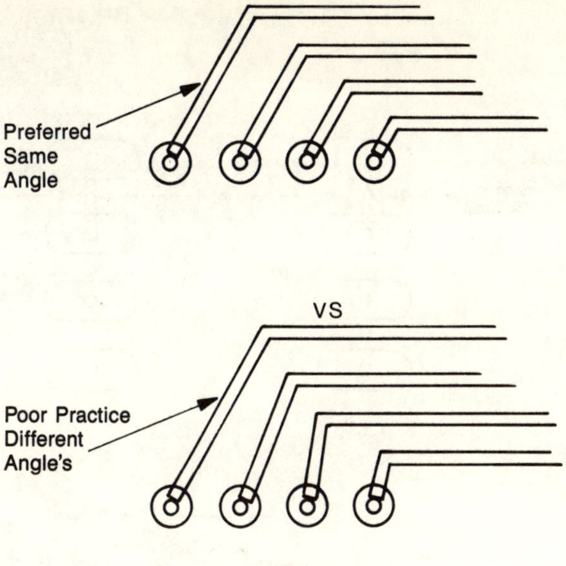

Figure 6-13. Trace Terminations

Routing traces thru the IC pad on digital PC boards is a common practice. Use very thin tape, typically .012 to .015 at 1:1. Figure 6-15A demonstrates a termination of a trace coming through an IC pattern using a tee. This would be poor practice. The preferred way is to turn the tape into the pad, terminate that end, and start on the opposite end of the pad with a new routed trace. The small piece of tape creating a tee is poor because it can be moved or kicked off of the artwork very easily. Figure 6-15B shows tape being routed through an IC pattern, changing channels as it routes through. Be careful when the designer changes directions or channels, he may turn the tape too close to the pad, creating a poor air gap. Move the turn out into the center. Create as much air gap as possible around the turn clearing the pads.

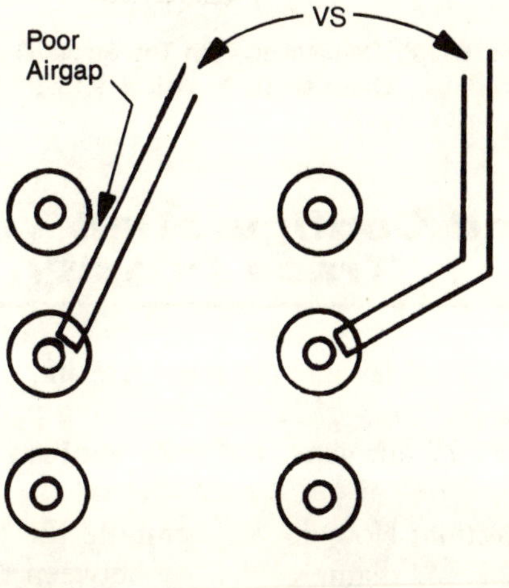

Figure 6-14. 45 Degree Trace Turn

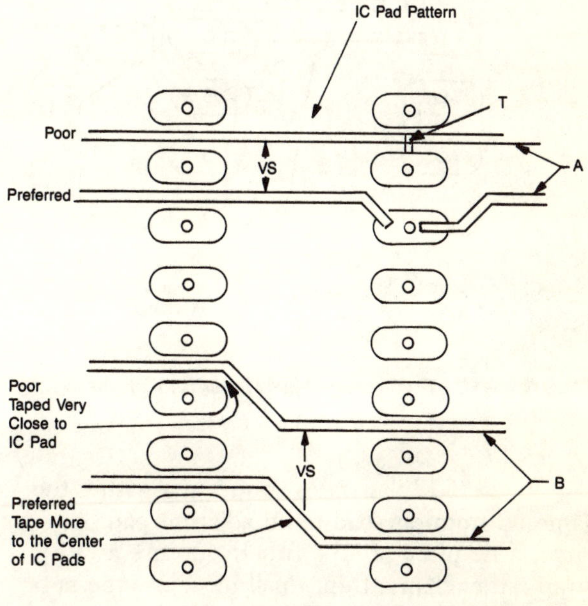

Figure 6-15. Trace Termination Through An IC Pattern

Figure 6-16 shows a typical IC pattern. It illustrates a connection called a feedthrough hole under an IC pattern. Avoid terminations using feedthroughs under an IC pattern. Either remove the feedthrough completely or reroute the feedthrough so that it is exposed and not covered by any component body. If there is poor plating in the hole, there is no way to repair the feed-through hole without totally removing the IC body from the PC board.

Figure 6-16B shows a typical IC pattern. It illustrates the routing connection within the IC itself, using a channel going through two IC pads. Avoid this practice if possible, by routing inside the IC pattern. Everytime that you route these traces through IC patterns, shorts may occur because of the extremely tight air gaps that are being created.

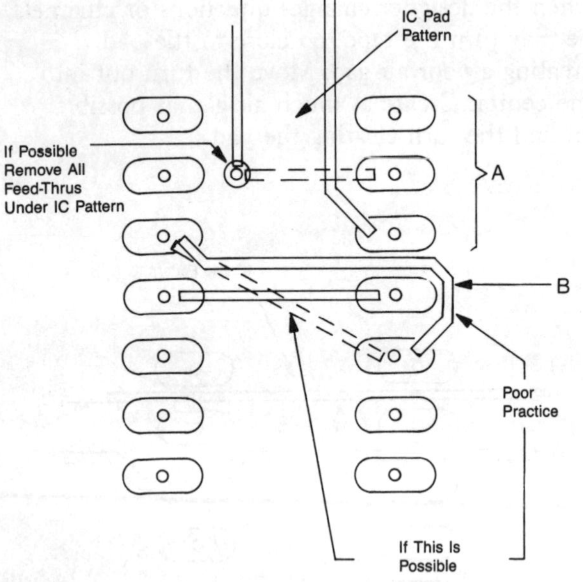

Figure 6-16. Feedthrough Hole Under An IC Pattern

Figure 6-17 illustrates connecting within the same IC from one pad to an adjacent pad. A large, wide piece of tape is better to use as a strap rather than a thin, small piece of tape not because of the electrical connection, but because of the mechanical connection. When drilling holes into the IC pads, the hole may be misregistered. It could cause both a disconnect from the strap to the adjacent pad and the tape could fall off.

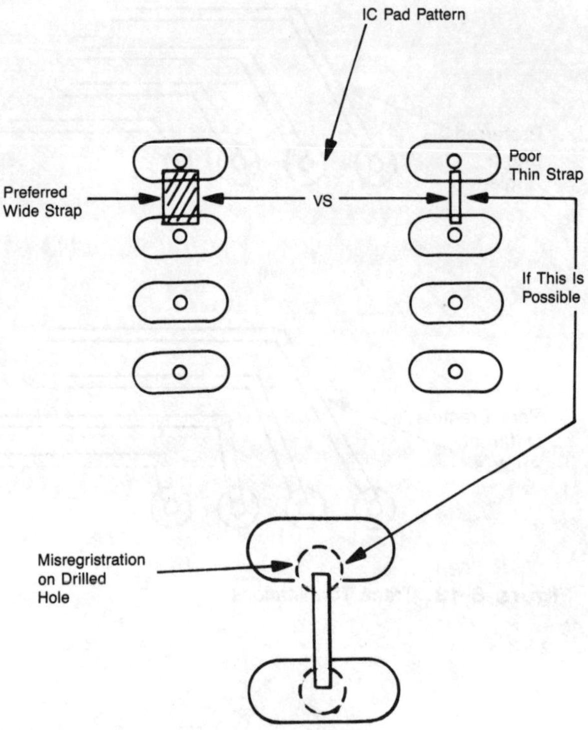

Figure 6-17. Connecting With The Same IC From One Pad To An Adjacent Pad

Pad Configuration Versus Trace Density

Figure 6-18 demonstrates the packing density between the rows of IC's by changing pad widths (diameters), air gaps, and trace widths.

Question: How do you compute the trace diameter density between rows of IC's?

Answer: IC pad spacing minus one pad diameter divided by the number of traces and air gaps = trace density.

Figure 6-19 shows an example of how to compute the number of traces that can fit between rows of an IC pattern. The typical 14-or 16-pin IC pattern (pad size) is .300. Find the area between the two rows. Subtract one pad diameter, in this case, a .075 diameter pad. This leaves .225 between the two rows of pads. The air

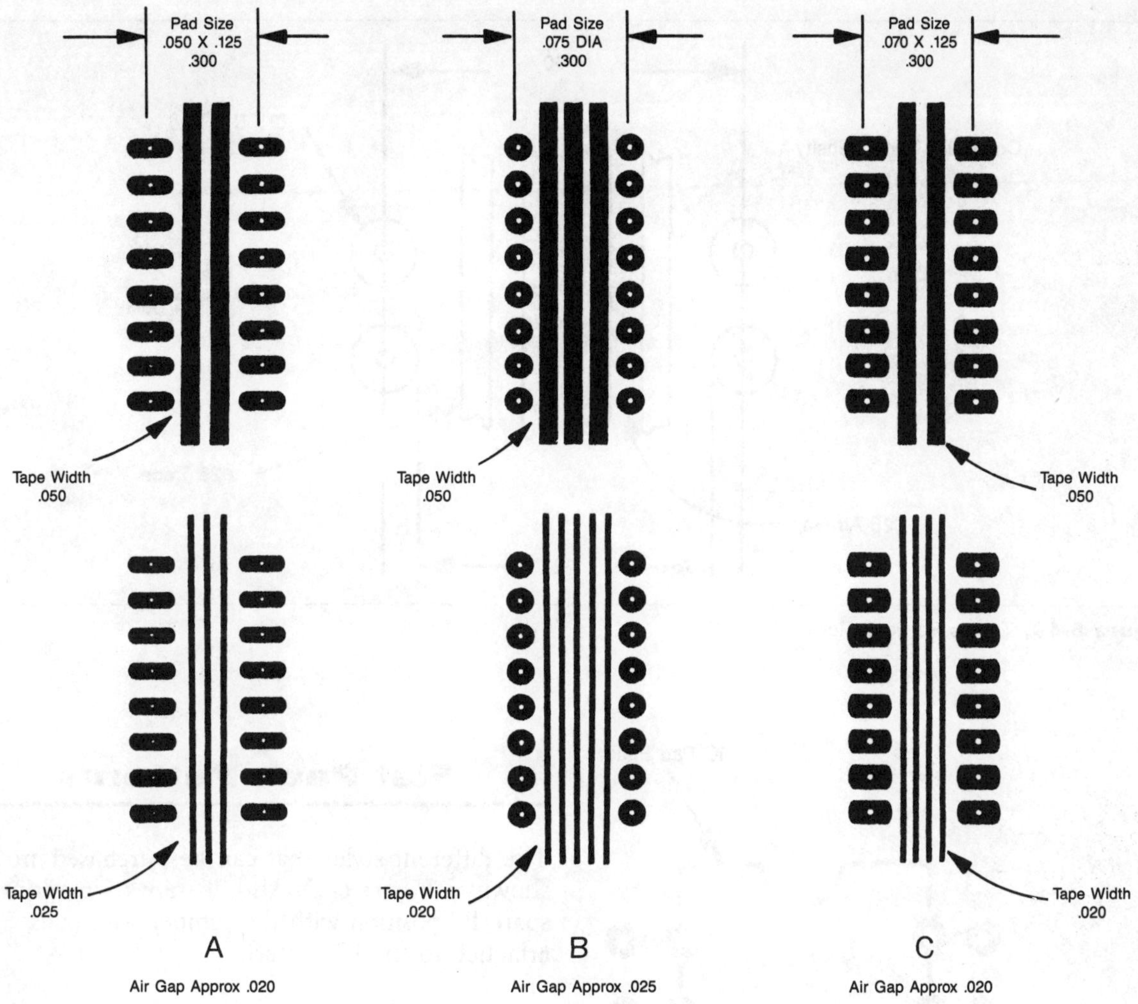

Figure 6-18. Packing of trace density

gap and the trace width in this particular case is exactly the same which is .025. Dividing .025 into .225 equals 9 (5 air gaps and 4 traces) that will fit between the two rows of pads. This means a .025 trace with a .025 air gap.

Spare IC Positions

Most printed circuit boards require spare IC positions by the electrical engineer, especially in the prototype stage or in the early production stage (Figure 6-20). If spare IC positions are used, place down a pattern that would accommodate a 14-or 16-pin IC. If it is a 14-pin IC fitting into a 16-pin pattern, one of the sets of pads would be unused. Use a second pad strapped to the IC pattern. This provides a jumper wire pad so that the engineer can solder a wire to the pad without attaching to the leg of the IC (See Figure 6-20).

Dual In-Line Patterns

Figure 6-21 shows some of the more popular configurations for dual in-line patterns. The figure provides a short explanation about the advantages of each. This is a small sampling of the different styles, sizes, and configurations that can be purchased.

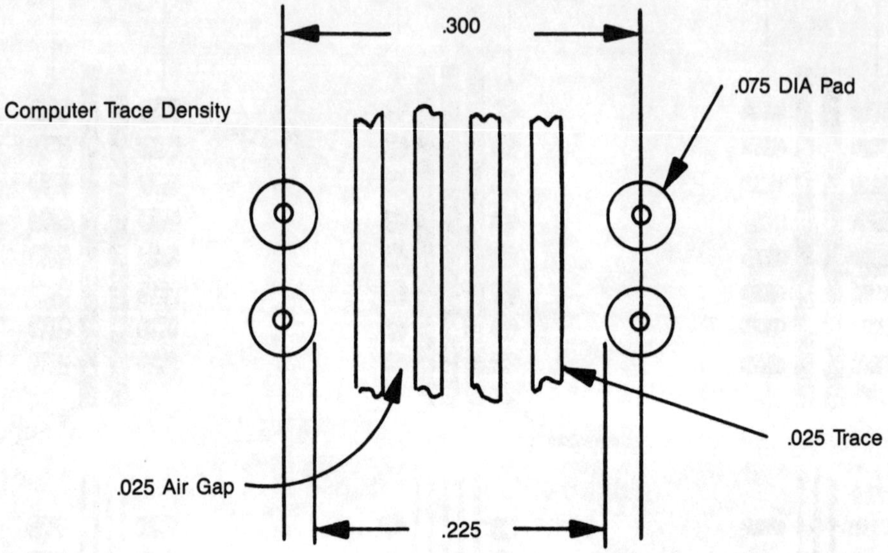

Figure 6-19. Traces Between Rows

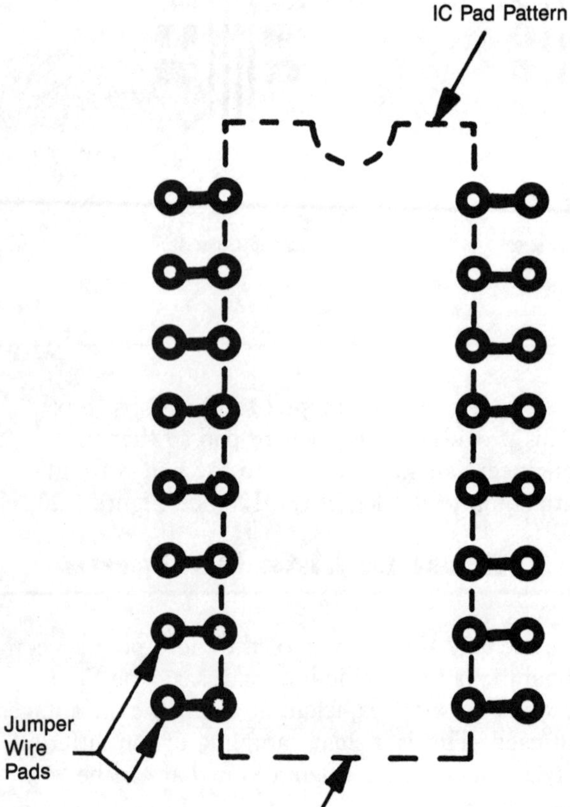

Figure 6-20. Spare IC Positions 14-Or 16-Pin IC

Flat Pack Patterns

The different styles that can be purchased are shown in Figure 6-22. Also, it shows a preprinted spare IC position with the jumper wire pads attached to the IC pattern.

Component Designation Marking

Some applications require all reference designations of the electrical components either be etched or silk-screened on the PC board for easy identification. This aids technicians and field service repairmen to repair or trouble-shoot boards without the use of an assembly or loading diagram. Boards that have a high failure rate or have to be repaired by persons without proper access to a full documentation package rely on this information.

In cases where such rigid identification is not necessary, it is still important to identify such components as connectors, test points, switches or any component that will require access for calibration or trouble-shooting. If a particular board requires that jumpers be installed to reflect different models or functions, then they should be marked on the board also.

DUAL IN-LINE (DIP) PATTERNS

A

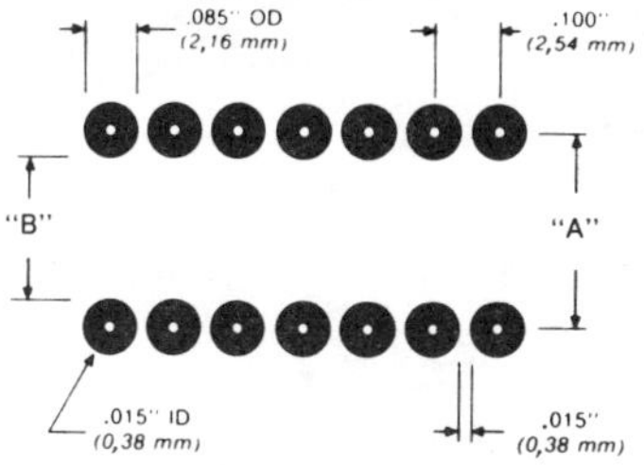

Round Pad

Basic all purpose terminal area.

B

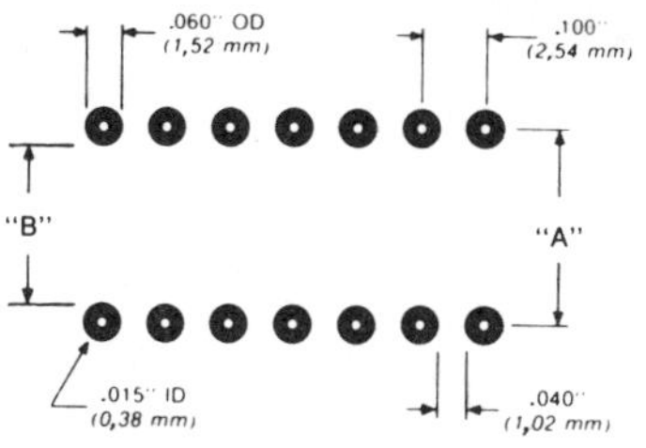

Round Pad

Basic all purpose terminal area.

C

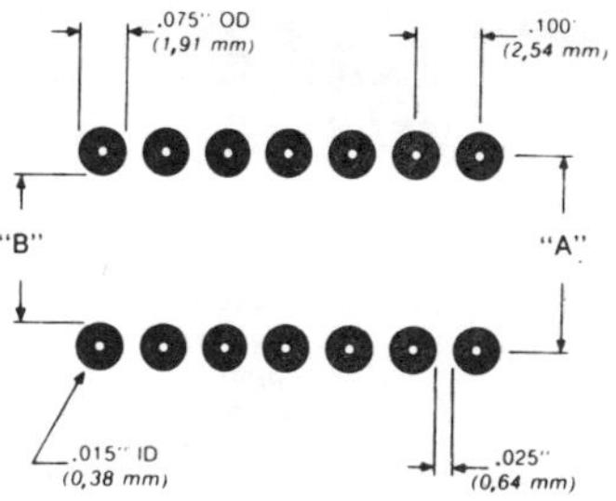

Cut Pad

Basic large terminal area.

D

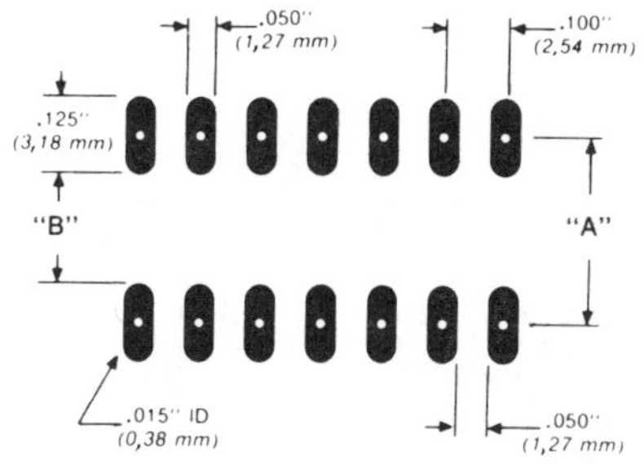

Oval Pad

Use where a larger terminal area is desired.

E

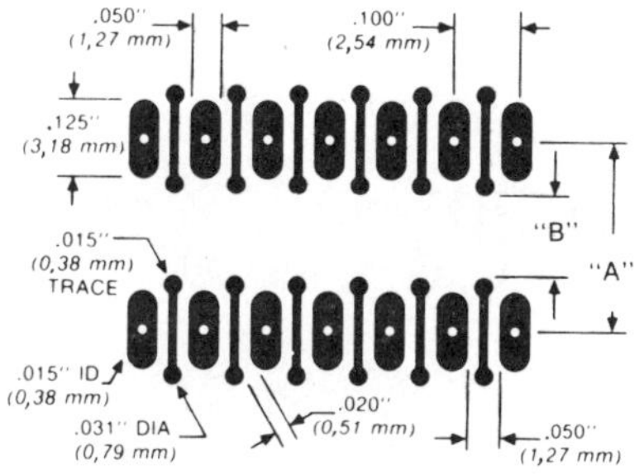

Oval Pad With Reduced Conductors

Use for a larger term area with reduced

F

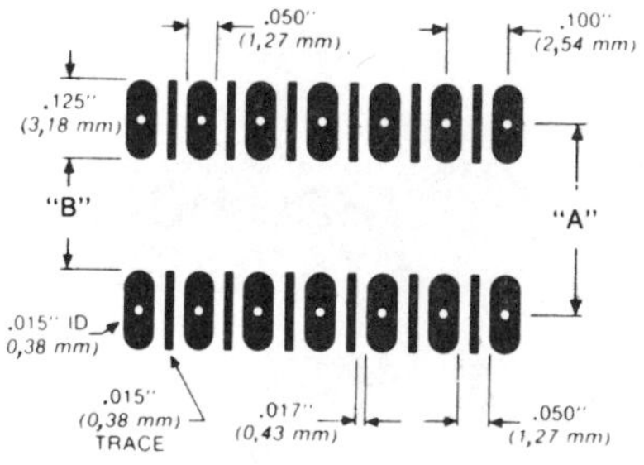

Oval Pad With Conductors

Use where a larger terminal area is desired, with conductors between terminals.

G

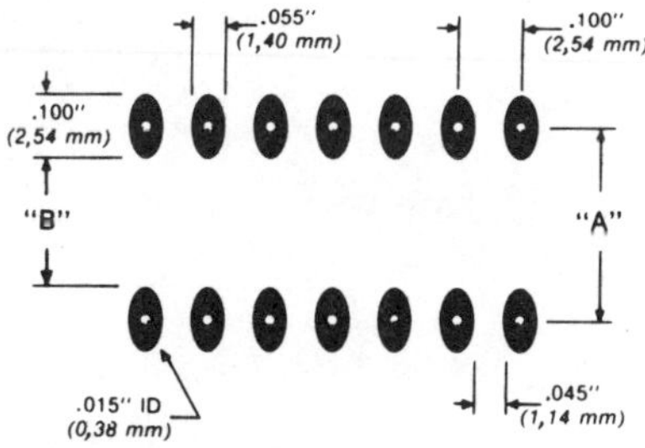

Elliptical Pad

Use where a larger terminal area is desired in wave solder applications (Solder has less tendency to bridge).

H

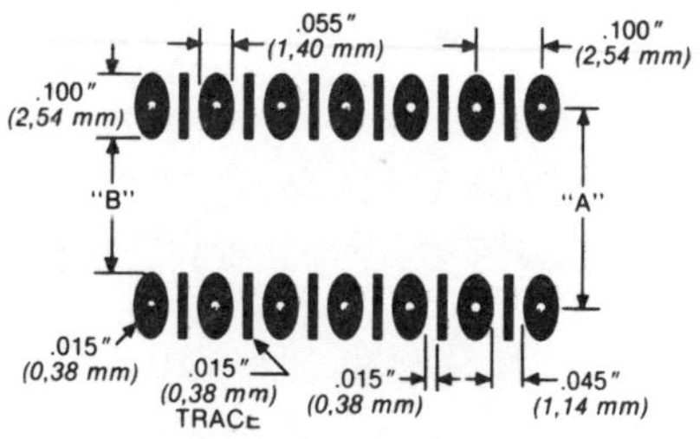

Elliptical Pad With Conductors

Use where a larger terminal area is desired with conductors between terminals in wave solder applications. (Solder has less tendency to bridge.

I

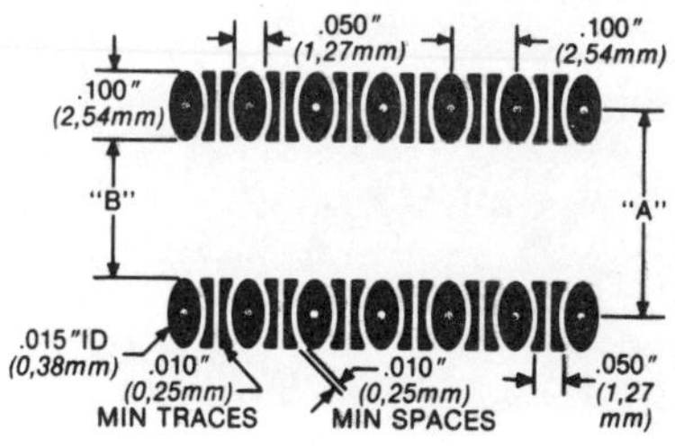

Elliptical Pad With Dual Conductors

Use where a larger terminal area is desired, with 2 conductors between terminals.

Figure 6-21. Dual In-Line (DIP) Patterns

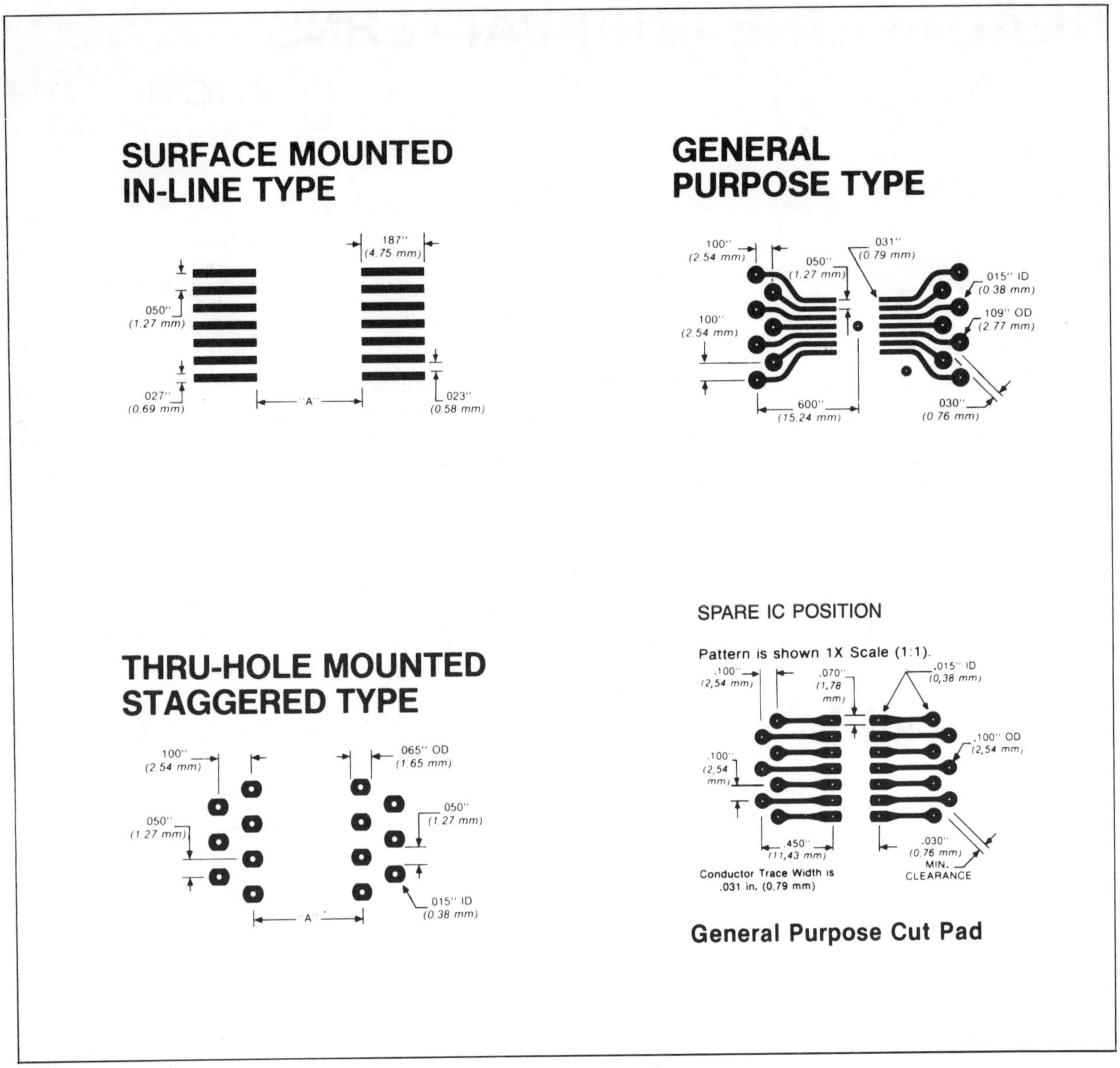

Figure 6-22. Flat Pack Patterns

Integrated circuit packages should be identified on the board. The IC's are not generally oriented in any numerical sequence on the schematic. Therefore, in order to relate a reference designation from the schematic to the PC board, the integrated circuit (IC) mechanical package must be identified by a reference designation number on the board for quick reference. See Figure 6-23.

In general, keep in mind that little effort is required to clarify a patented problem at the design stage, whereas downstream a solution could be costly.

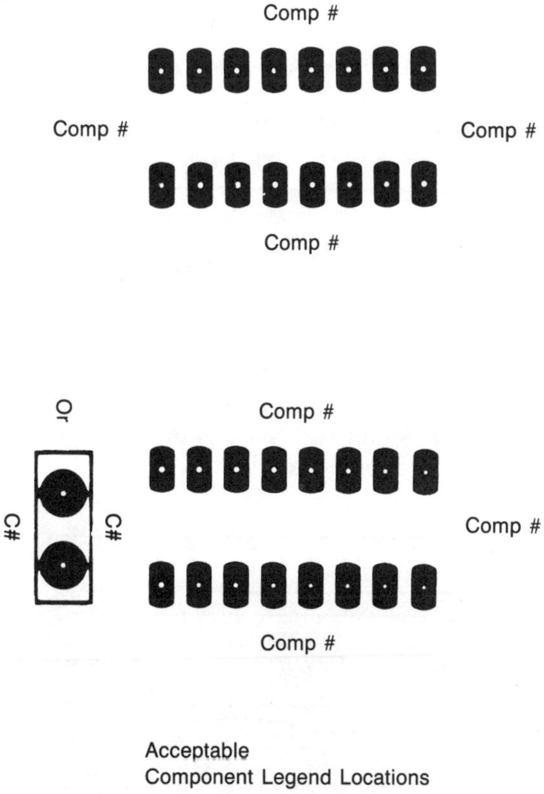

Figure 6-23. Board Reference Designations

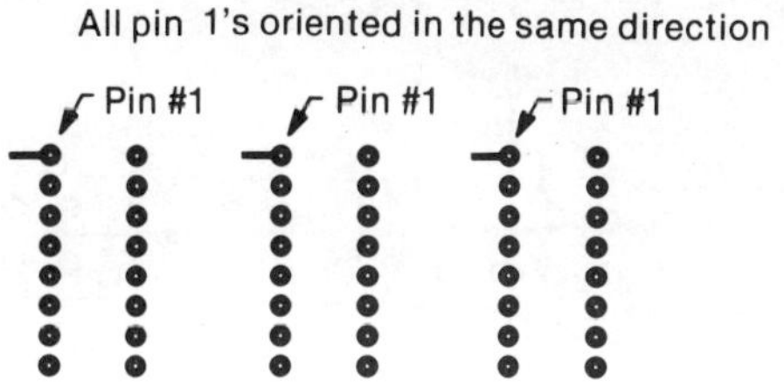

Figure 6-24. Component Orientation

Polarity and Indexing

Components that must be oriented in a fixed direction due to polarity or pin arrangements should be indexed on the board (as in Figure 6-24). This consideration is not only a must for the manufacturing stage of the board, but it is also relied upon by the technician, engineer, field service representative, customer and any other person that would have need to trouble-shoot or follow the flow of the circuit on the PC board.

It is very helpful to index various pins on a connector to aid in finding a particular pin number. In addition to identifying the pin number as described in "Board Reference Designations" Figure 6-25, every tenth pin could be indexed. In large connectors, this would allow the technician to scan the pins in sets of ten until he found the nearest segment which the particular pin was located.

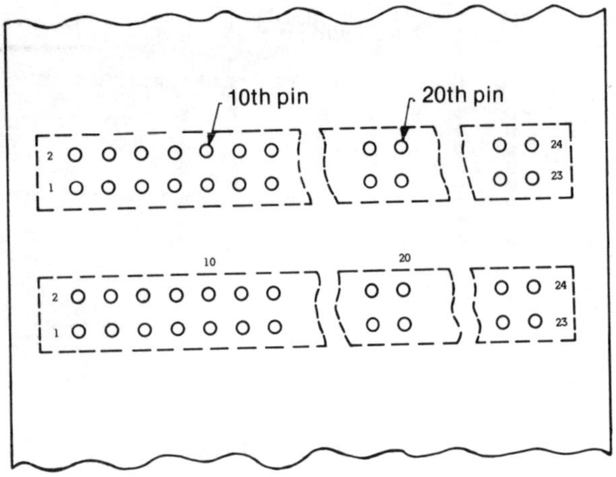

Figure 6-25. Connector Indexing

As shown in Figure 6-26 integrated circuits should be placed on the board so that all are oriented in the same direction in addition to having pin "1" indexed.

This practice aids the technician in finding a particular pin on an integrated circuit without first turning the board over to see how that particular package was oriented, pin "1" top or bottom.

In some dense packaging applications, due to the flow of the circuit and the pin assignments of some integrated circuits (i.e., counters), it is not possible to follow this pin orientation process. The designer should still strive for consistent orientation as much as possible allowing only those packages which won't conform to break the pattern.

In general, any component or pad which could be confusing or oriented improperly during the assembly stage should be identified by polarity marking or an indexing feature. (See Figure 6-26A)

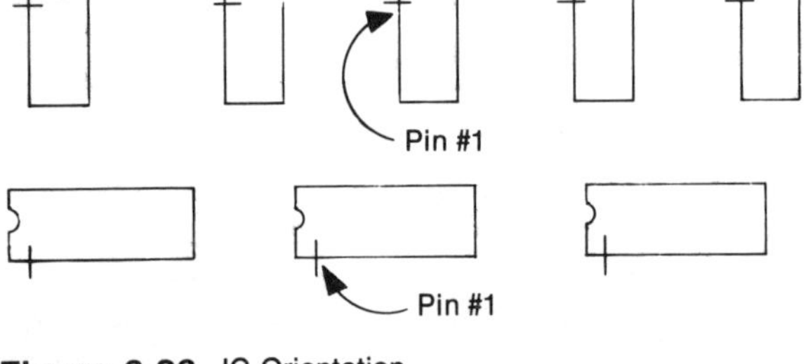

Figure 6-26. IC Orientation

After the board is taped and is electrically ready for the photography step, the final documentation information must be added to the tape-up.

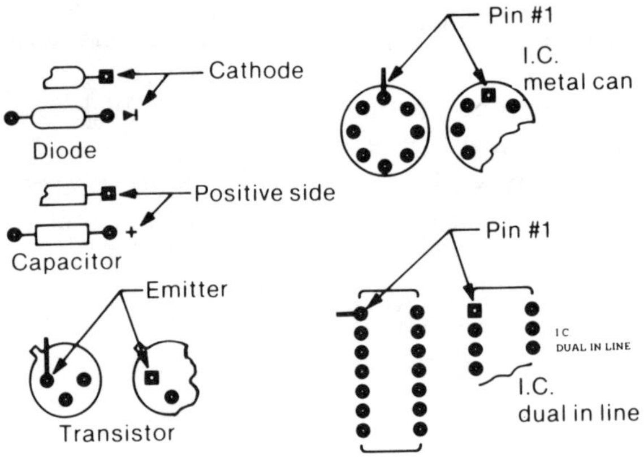

Figure 26A. Polarity Marking or Indexing

Board Indexing

The revision letter of the board must be identified on the artmaster somewhere within the board parameters. This allows for it to be actually etched on the board to avoid confusion with any previous or later changes that might occur. This is generally done as in Figure 6-27 on the non-component side of the board along with the PC board part number, if applicable.

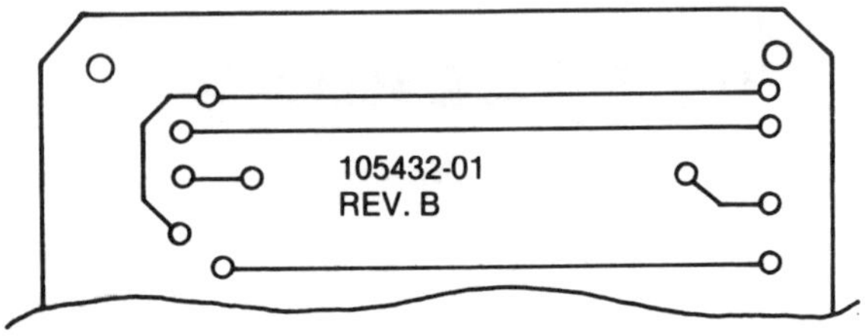

Figure 6-27. Revision Markings

The assembly number of the board should be identified in the same manner on the component side of the board. This gives quick reference to anyone in finding the proper assembly drawing and any other documentation involved in the manufacturing, procurement or sale of the board.

No assembly revision letter should be etched on the board; this information should be added at the actual stage at whatever revision the assembly drawing is at the time the board is assembled.

Artwork Master

The master pattern number refers to the artmaster itself and will document the number and revision letter of the finished tool to which the board is made. This number and revision letter is important for traceability purposes in case failure of the final product should result. This identification marking should be outside the parameter of the board as it will not be part of the physical board once it is manufactured, and it should be entered in the documentation system.

The reduction targets and dimensions provide the necessary information to the photographer to aid him in the proper reduction of the artmaster. Targets should be carefully and accurately placed so that they will appear on all copies of the master pattern. Reduction marking should extend along one vertical side and/or along one horizontal side of the tape-up.

Examples of an artwork master containing the master pattern number and reduction targets are shown in Figure 6-28. See Table 6-2 for taping guidelines. See Figures 6-29 and 6-30 for examples of poor and good taping techniques.

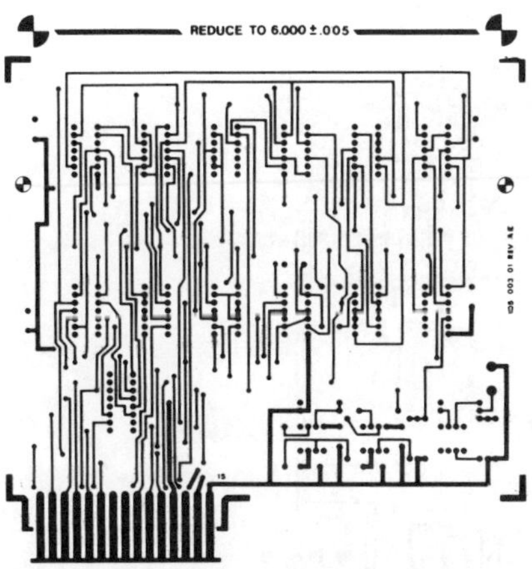

Figure 6-28. Artwork Master Showing Reduction Targets

Table 6-2. TAPING GUIDELINES

1.	Minimize handling of drafting film and tapes.
2.	Do not fold, crease or tightly roll the artwork
3.	Burnish tape when complete
4.	Run tapes to the center of the annular ring of the pad.
5.	Do not cut through tapes or pads (it may show up on the photograph).
6.	Do not pull the tape hard or it will creep or elongate.
7.	Place the center of pads on the grid coordinate or coincident with lines.
8.	Trim all corners as required.
9.	Be neat.

Making Copies

To make a copy of your tape-up, never run the tape-up through a blueprint machine as shown in Figure 6-31. The lamps and rollers inside the machine can damage the tape-up.

The correct method for running copies of the artmaster tape-up is to use a (plate maker) as shown in Figures 6-32, 6-33, and 6-34. This technique is used because the tape-up is flat at all times and under no stress.

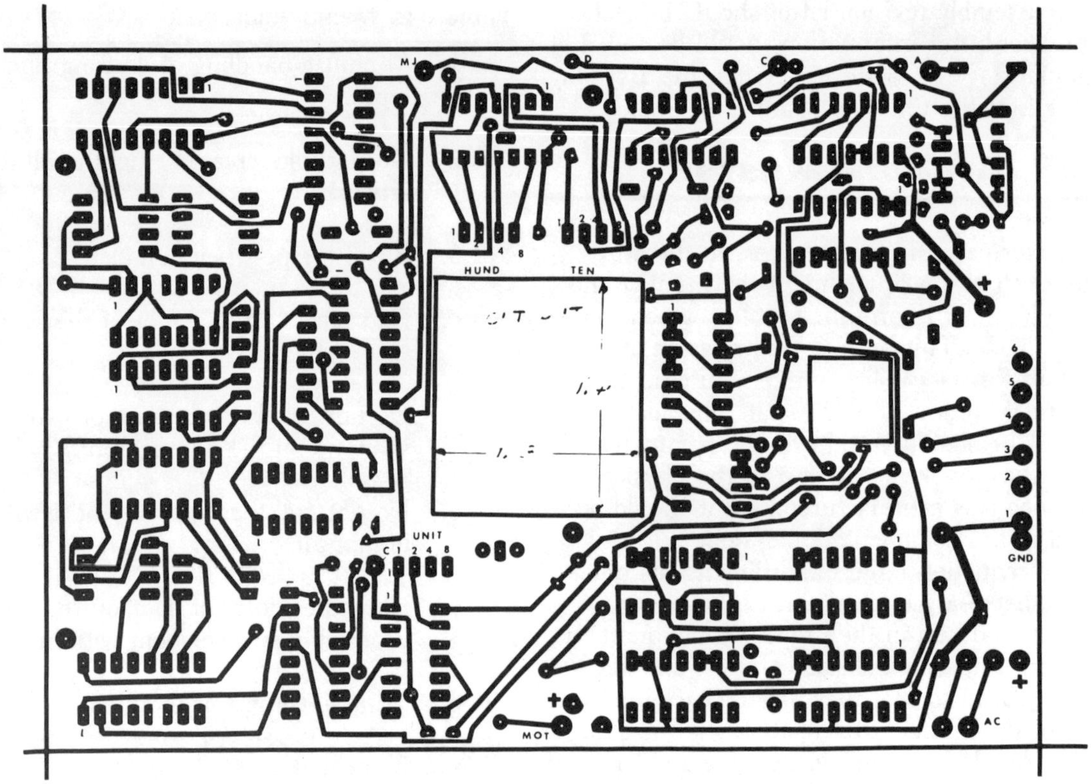

Figure 6-29. Example of Poor Taping

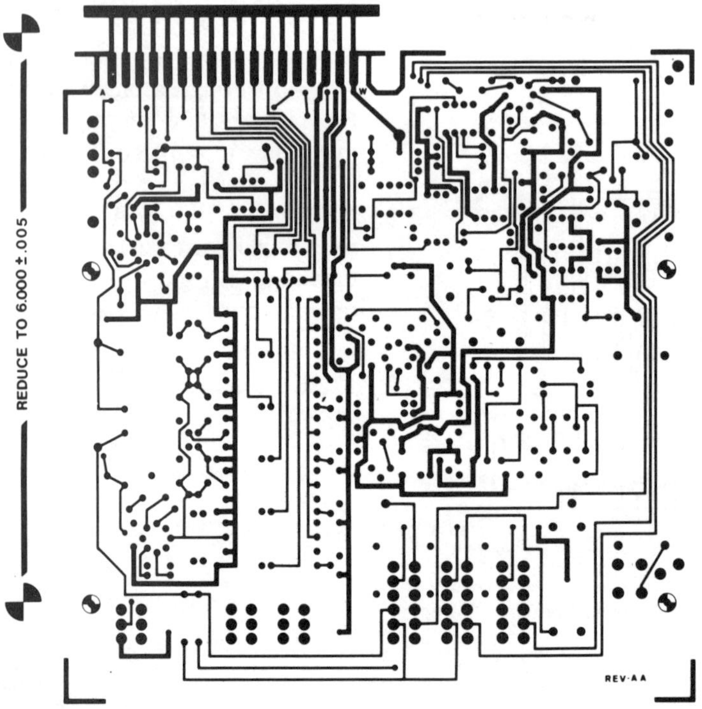

Figure 6-30. Example of Quality Taping

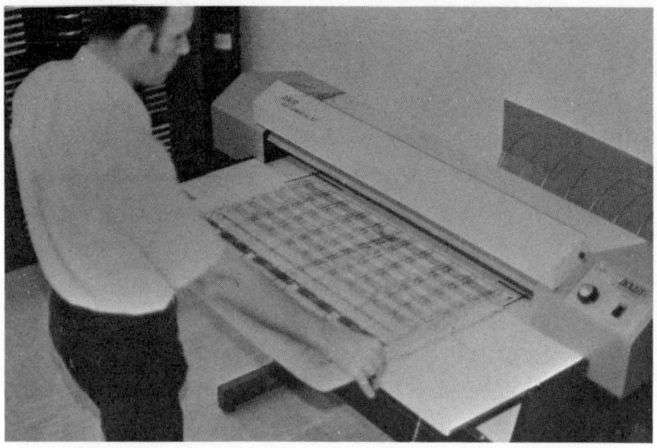

Figure 6-31. Thru a blueprint machine

Figure 6-32. The glass door is raised and the artwork is positioned.

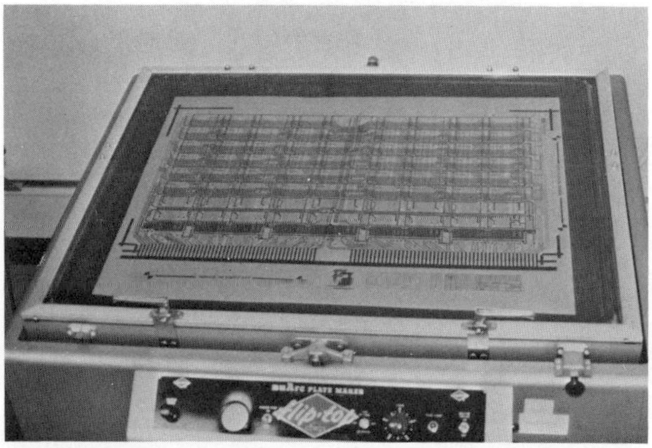

Figure 6-33. The glass door is lowered and the tape-up is placed under a vacuum.

Figure 6-34. The top is turned over and the tape-up is exposed.

Figure 6-35A shows the technique for producing the master pattern. This is called the film, which is a production tool used by the PC manufacturer to create the finished fabricated PC board. The photographer uses a high-precision, graphic-art process camera.

The graphic arts process of camera copies line or half-tone material. It must be capable of repeatedly producing the same photo reduction. The camera must be dimensionally stable and have distortion-free optics. Figure 6-35B shows the camera lens.

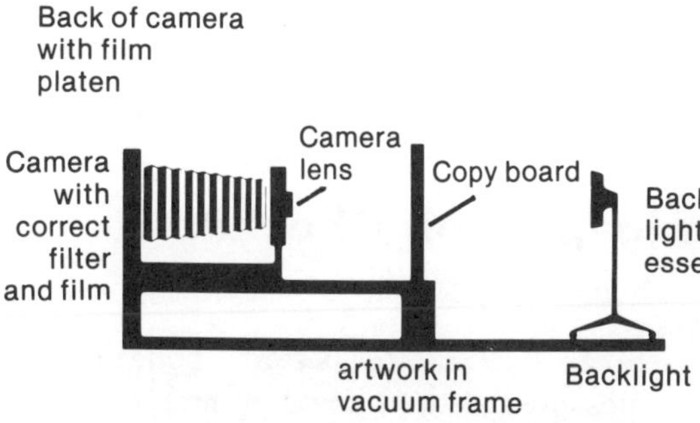

Figure 6-35A. Graphic Arts Process Camera

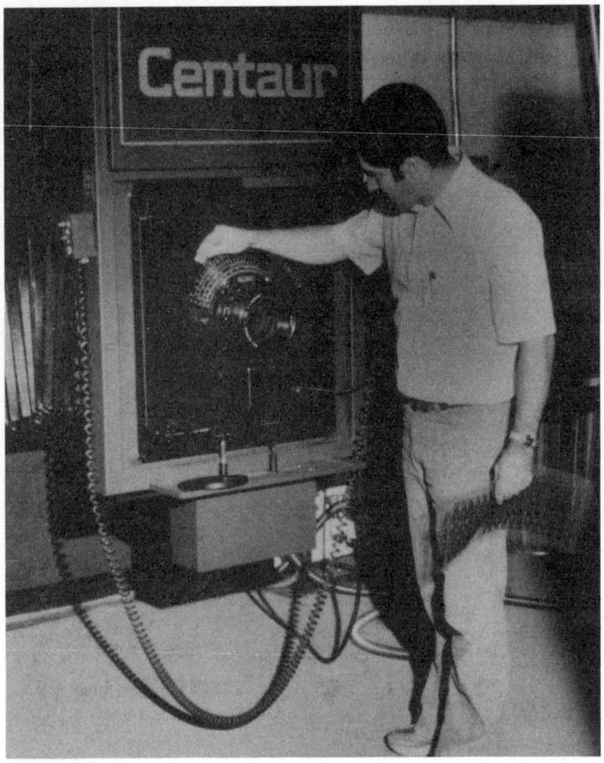

Figure 6-35B. Graphic Arts Process Camera

Final Inspection

The film is a production tool. It should be inspected and controlled in the same manner as other precision tooling.

Film Inspection Checklist

1. SI/1 - Each master pattern should be measured at the photographic reduction marks, and at least one other dimension at right angles to the first measurement. This insures proper reduction and size.

2. REGISTRATION - All patterns required to produce a printed circuit board must be inspected together to insure that registration has been achieved as stated on the master drawing.

3. SHARPNESS - Image sharpness and resolution are necessary to prevent ragged conductor edge definition on the etched printed circuit board. A sample conductor line width and spacing measurements must be made to verify proper exposure and development.

4. DENSITY - The image on the master pattern must be opaque enough to insure that proper exposure of the photosensitive coating can be made.

5. WORKMANSHIP - All master patterns should be inspected for pinholes, scratches, fingerprints and general cleanliness. All touch-up work should be made on the non-emulsion side of the film.

Ground and Voltage Planes

Ground planes are areas on the PC board that have not been consumed by traces or pads. They are left with the copper that is tied to the ground circuit (See Figure 6-36). The ground plane is generally on the component side especially for flow soldering or dip soldering. It is isolated from the other traces and pads. In addition to ground planes, which are tied to the ground circuit, voltage planes are tied to the voltage circuit.

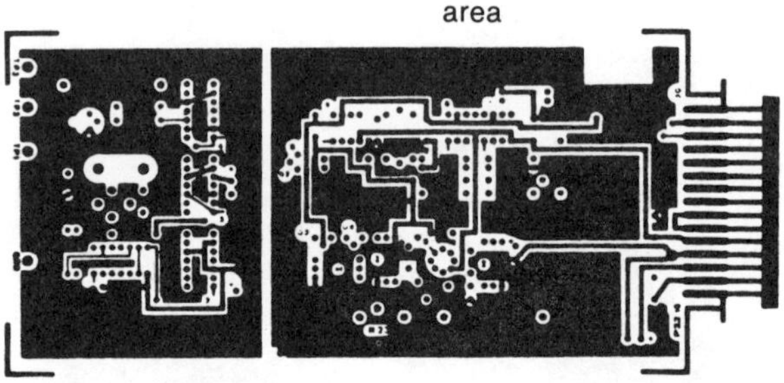

Figure 6-36. Ground Plane

Planes on the noncomponent side of the PC board, wider than .50 inches, should be separated into strips or checkered patterns (See Figure 6-37). This prevents blistering and warping during flow-or dip-soldering operations. On external board layers, adequate clearance should be provided around the non-functional terminated areas of ground and voltage planes.

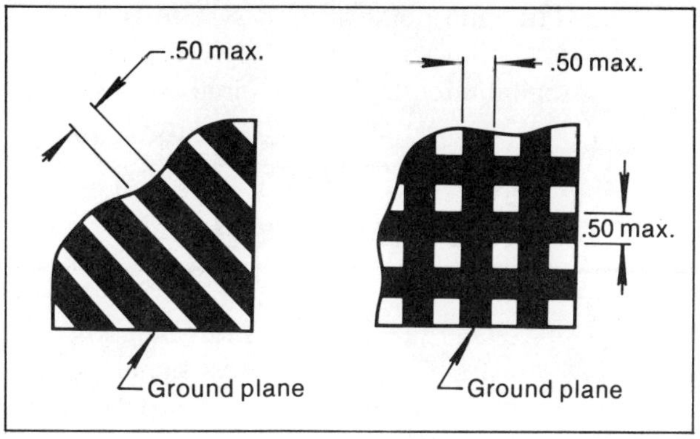

Figure 6-37. Non Component Side Ground Plan Patterns

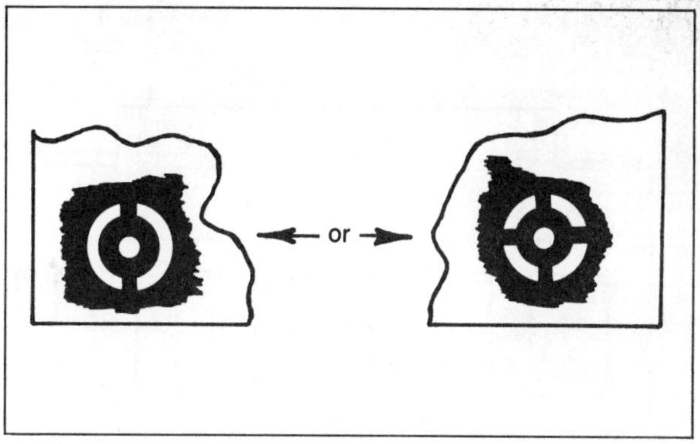

Figure 6-39. Clearance Between Terminal Pad and Ground Plane.

On internal layers (such as multilayer boards), the use of non-functional terminal area pads is not necessary. If the pad is unused, it should be removed (See Figure 6-38). The only requirement is to provide clearance around the hole to prevent electrical continuity from the plane to the hole. When a hole terminates in a plane and electrical continuity is required, terminal areas or pads should be used. Provide clearance between the terminal area or pad and ground plane for two to four connections. This preserves circuit continuity (See Figure 6-39), and prevents heat sinking of the terminal area or pad during dip-or wave-soldering operations. Otherwise, an inferior solder joint may occur if proper clearance is not provided around the terminal area of pad.

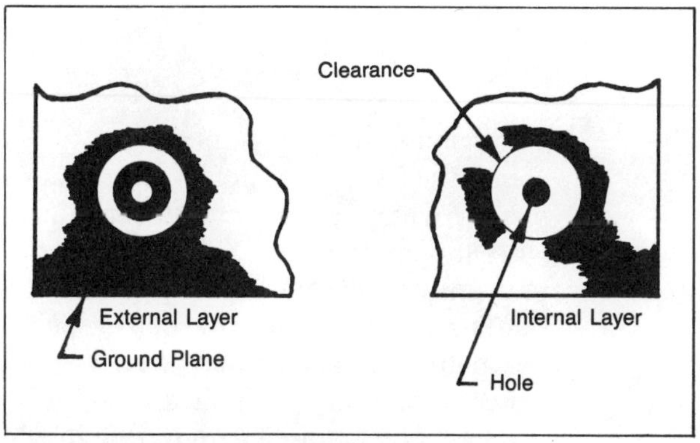

Figure 6-38. Nonfunctional Terminal Area Such As Mounting Holes, Tooling Holes

METHODS

There are three different methods to connect ground. For P.C. circuits, the parallel-connected ground system separates the main busses from the connector to the various circuits onto the PC board (Figure 6-40).

The series-connected ground system is used in digital, low-frequency design. A common buss links one IC (or circuit) to another (Figure 6-41).

The multipoint-ground system uses a ground plane on an external or internal ground plane. This is called a multilayer PC board.

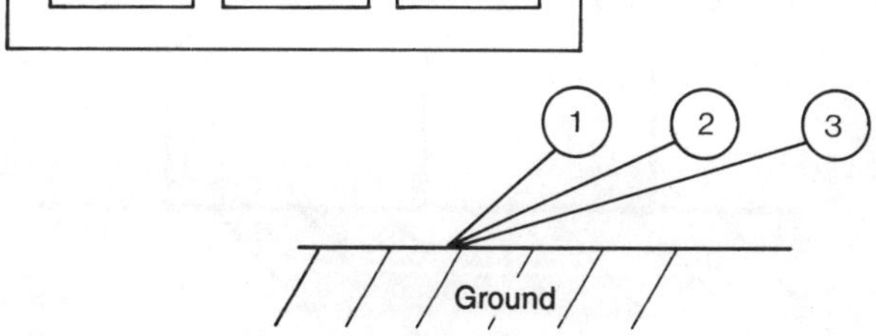

Figure 6-40. Parallel Connected (Separate Ground System)

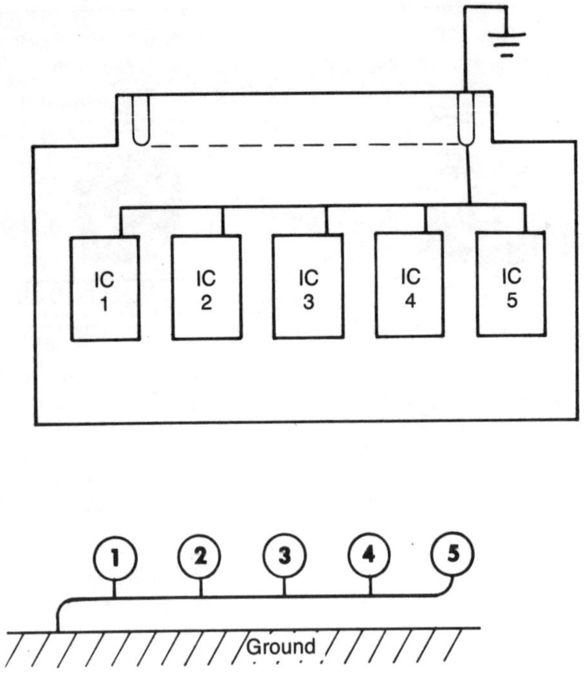

Figure 6-41. Series Connected (Common Ground System)

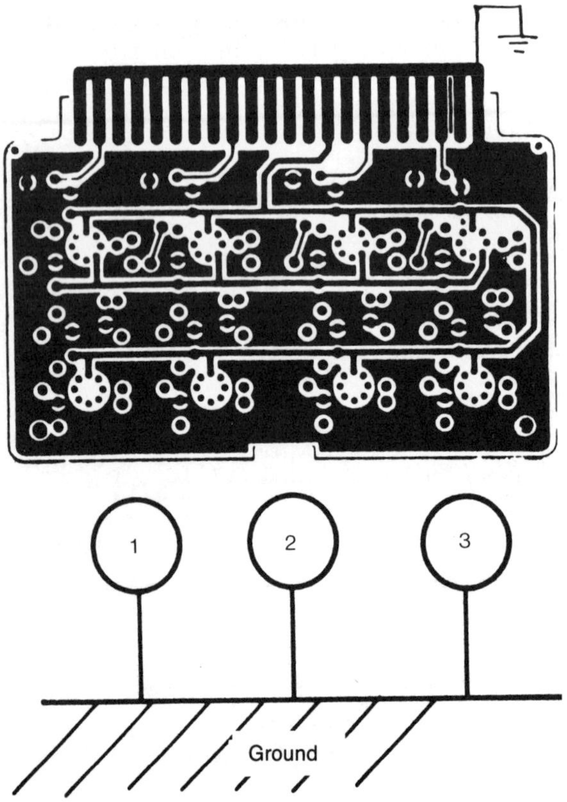

Figure 6-42. Multipoint Ground

The multipoint ground system is used for either shielding or high-frequency digital circuits. IC's or circuits terminate to a massive area of copper (referred to as a plane). See Figure 6-42.

ARTMASTERS

Ground or voltage planes should not be connected by means of tape-like traces, signal paths, or voltage ground busses. There are two methods used to construct a voltage ground plane. Besides CADD, one is manual cutting and the other is photographically produced. These two techniques are itemized step-by-step below.

Using Red and Blue Artwork

Ground and voltage planes can be created by using red Para-paque along with blue and red tape. Figure 6-48 shows the creation of large massive areas for ground or voltage.

There is one main difference between using the red and blue tape and cutting (taping) Para-paque over black crepe. Red and blue tape for traces and pads is applied last rather than first. After all of the plane is applied, the remainder is cut away. The red traces, black pads, or red pads are taped down.

Method No. 1

MANUAL CUTTING

The hand cutting of planes is accomplished with a material called Para-paque* which is a red translucent film with an adhesive back. This is applied to a finished PC board. All pads and traces are placed down in their respective patterns. Para-paque has an adhesive back and is applied over the surface. Para-paque is removed manually by cutting the areas that should not

* Para-paque can be obtained through Bishop graphics Drafting Supply Centers

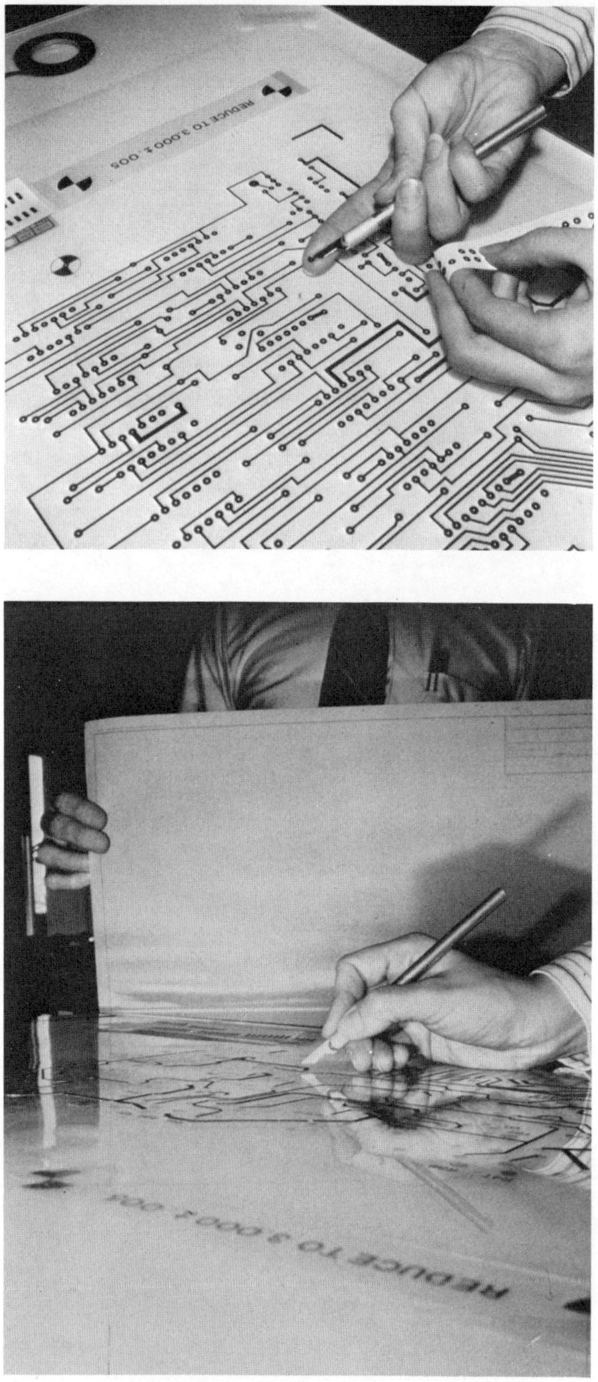

Figure 6-43. Applying Artmaster Patterns and Traces

be covered by voltage or ground planes. Following is the technique used to apply Para-paque to the artwork.

STEP 1

Apply crop marks, all pads, and all voltage ground and signal paths that are used on a double-sided PC board (Figure 6-43).

STEP 2

After completing the 2:1 artwork by applying all pads and traces, completely cover the component or non-component side with Para-paque. Apply this to the surface by brushing and burnishing over the traces and pads. Cut away the areas that should not have ground or voltage planes. Make a small circle using a tool called a drop-bow compass. This has a knife blade to score circles and to create a clearance around the pads so that they do not connect. The compass also creates clearances around the pads that strap to the plane. Once the clearance has been cut, lift out the Para-paque with an X-Acto knife. Using .100 tape, strap the pad to the plane as demonstrated in Figures 6-38 and 6-39.For this cutting sequence, see Figure 6-44A, B and C.

STEP 3

To remove large sections from the center of the artwork or to cut long straight edges for strip lines around the edge of the artwork, use a straight edge along with an X-Acto knife. The drop-bow compass is only used to cut circles around the pads or to clear out areas for tooling or mounting holes (See Figure 6-45).

STEP 4

Step 4 demonstrates the final product. This is the finished plane applied to the component side or the non-component side. Figure 6-46 shows the finished ground plane applied to the component side of the PC board.

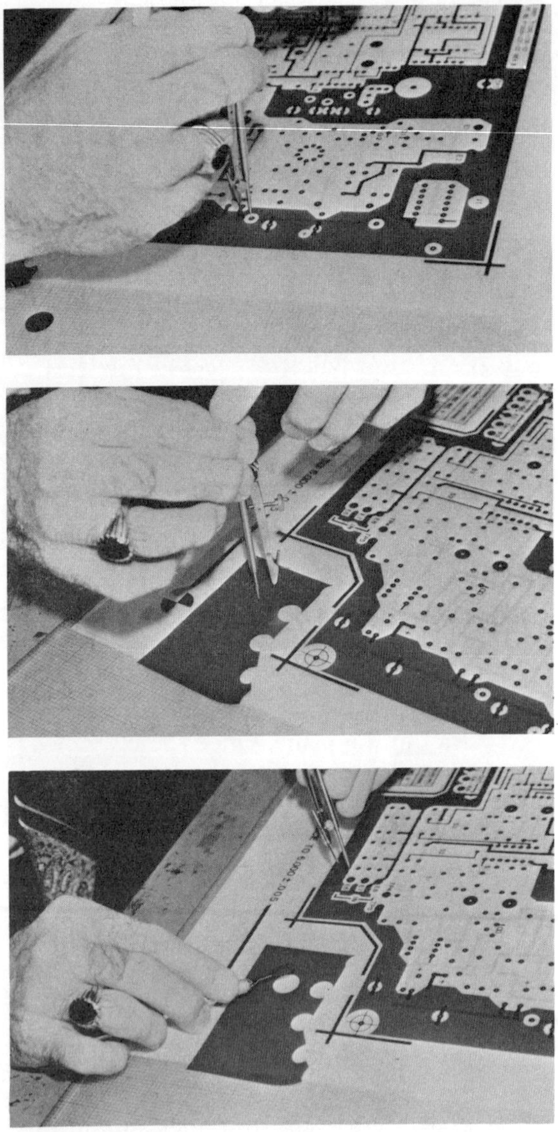

A

B

C

Figure 6-44. Cutting Sequence

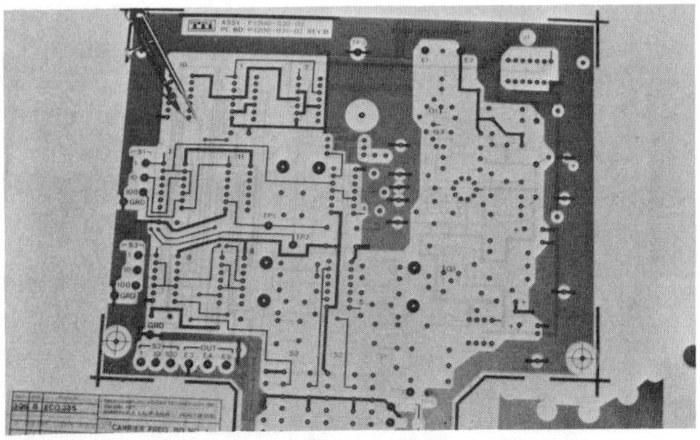

Figure 6-45. Cleared Areas For Tooling Or Mounting Holes

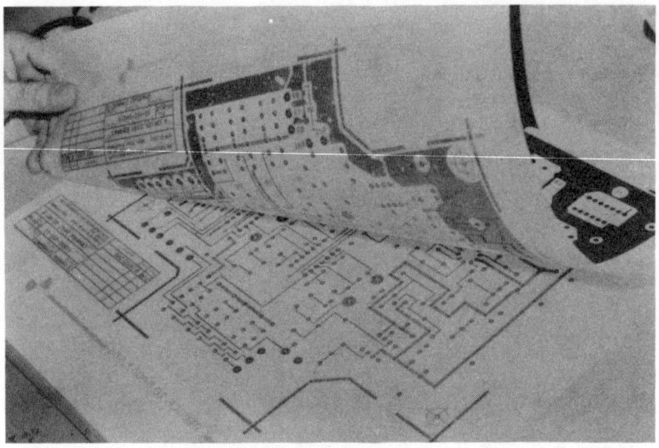

Figure 6-46. Finished Ground Plane

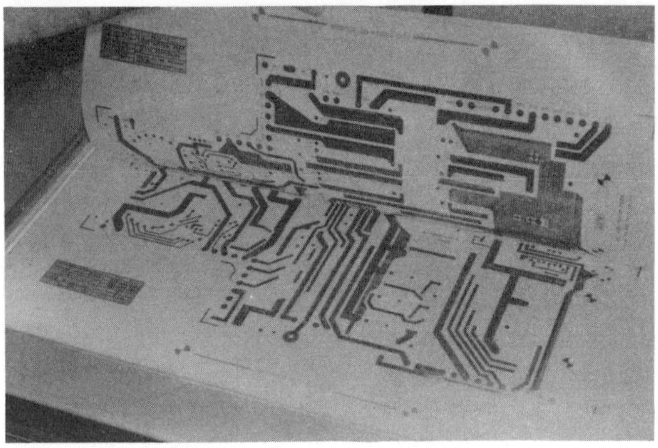

Figure 6-47. Hand Tapped Small Islands

There are two ways to create small islands for ground planes, voltage planes, or major bus lines. It can be cut using Para-paque or it can be taped by hand as shown in Figure 6-47.

Method No. 2

PHOTOGRAPHY AND VOLTAGE PLANES

Ground and voltage planes can be produced photographically. This is faster than the manual technique. The following steps explain the photographic production of the ground or voltage planes.

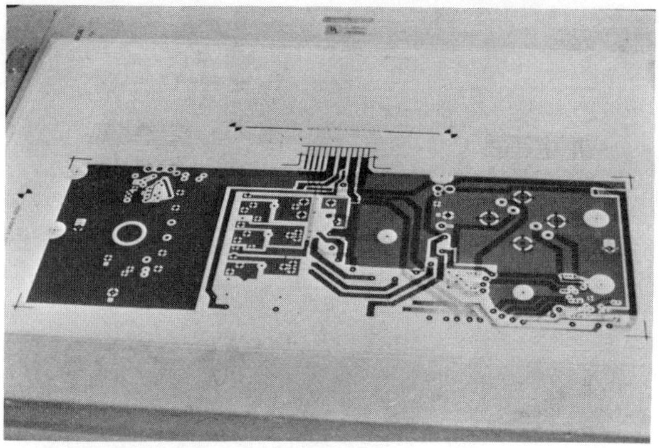

Figure 6-48. Use of Para-paque

STEP 1

The 2:1 artmaster is taped using red or black pads and traces. This is exactly the same as a hand-cut ground plane to this point. The ground plane is transposed onto film with a graphic process camera using the 1:1 positive. You must have one continuous flow of the voltage or ground plane. After the board is taped, the photographer is shown a blueprint of the side of the artmaster that will have the plane on it (typically the component side). The blueprint shows the ground plane limitations and all connections to be made on this plane. The blueprint is sent to the photographer along with the artmaster. Both are used to create the ground plane. Figure 6-49 shows the instructions to the photographer: the notes describing the air gap to be created around the pads; the traces and the pads to be connected to the plane, the isolated areas of copper to be removed and any special instructions.

STEP 2

The photographer receives a blueprint with its instructions along with the 2:1 artmaster. He photographs the 2:1 artmaster in its positive state. He reduces it to 1:1, creating a negative (Figure 6-50).

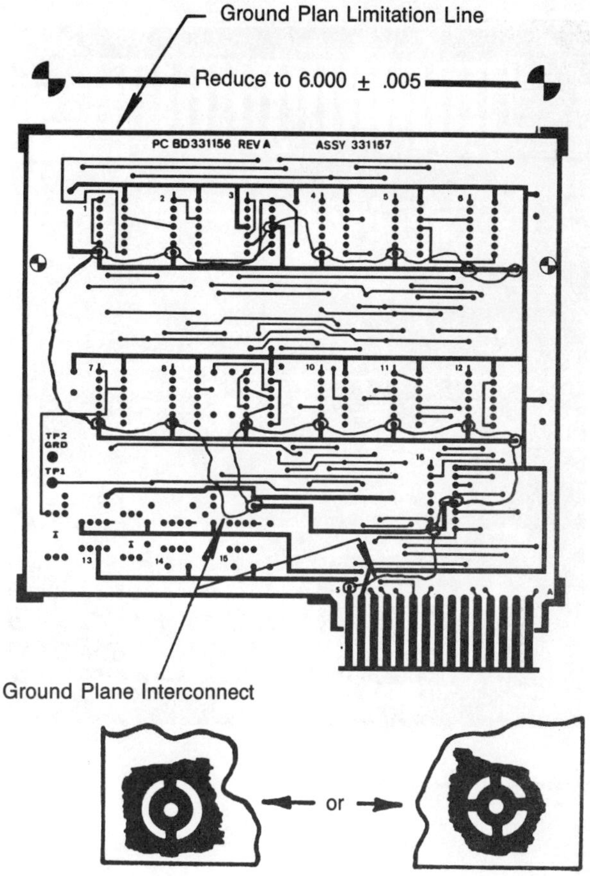

Notes:

1. Ground plane side shown only
2. Air gap to be .xxx at 1/1 scale
3. Connect all indicated pads to ground level
4. Remove all isolated grounds
5. Remove ground plane limitation line from fininshed neg. and positive.

• Notes for the photographer—(These notes are to be place on or accompany the blue print.)

Figure 6-49. Instructions To Photographer

STEP 3

Once the negative has been completed at 1:1, a contact sheet is made creating a 1:1 positive (Figure 6-51).

STEP 4

Using the positive and the negative that was just created by the photographer, place these into a special ground-plane, photographic camera.

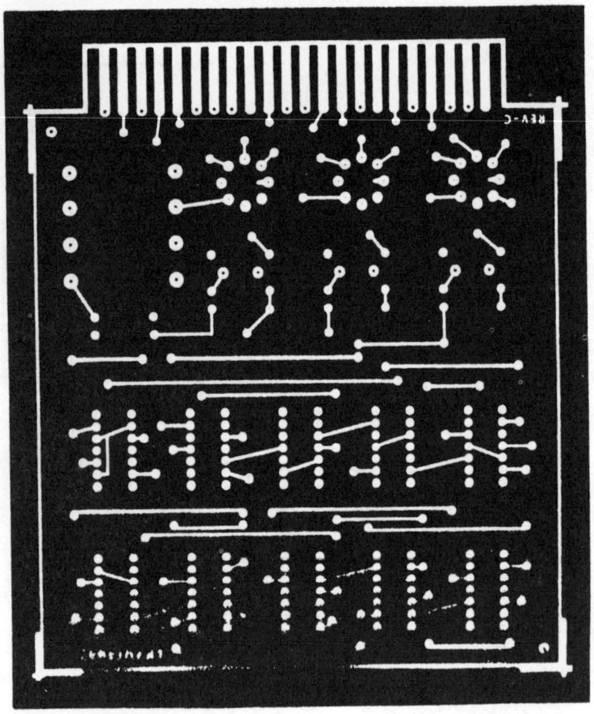

Figure 6-50. Negative

This holds the positive and the negative superimposed upon each other in exact registration. The positive will hold the image on the new film which will become the new ground plane. The positive has black pads and black traces blocking the light. The negative creates the air gap or the clearance around all pads and traces through a rotating process on this special camera. Because it is a negative, light penetrates through the pads and traces. Overexposing through the rotating negative, creating the air gap. The end result is a ground plane negative (See Figure 6-52).

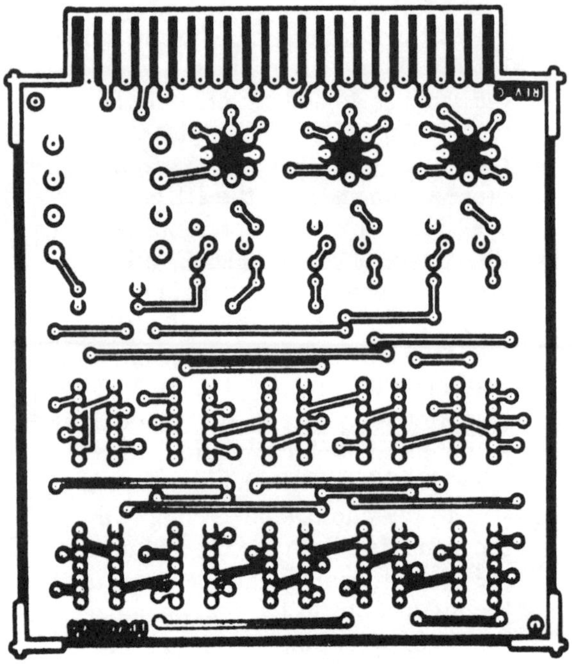

Figure 6-52. Ground Plane Negative

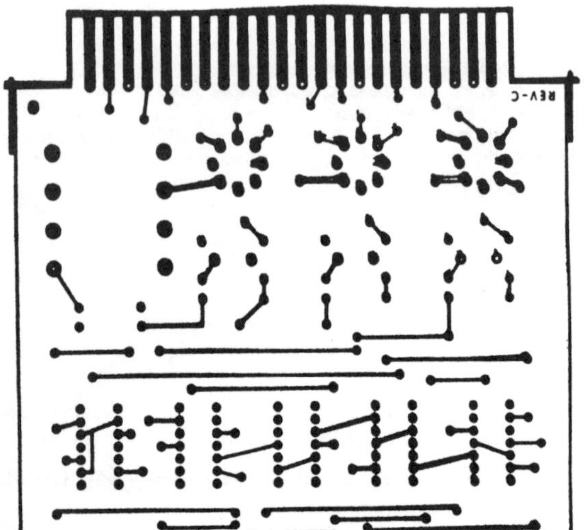

Figure 6-51. Positive

STEP 5

The negative should have no ties because the photographer does not strap off the pads that must connect the plane. The photographer removes all isolated ground islands along with tying off those unconnected pads (Figure 6-38). Once the isolated islands (that could not be 6-35 connected to the plane) and the pads have been tied off, one more contact would be made which creates the positive ground plane (Figure 6-53).

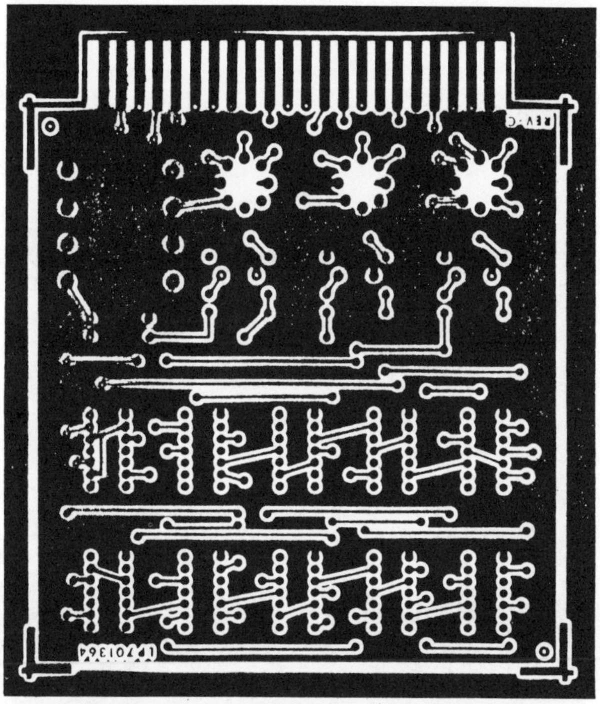

Figure 6-53. Positive Ground Plane

Note: A tiemaster can be used instead of tape to strap off the pads that tie to the ground plane (as discussed in Step 4). When the contact is made to create the positive ground plane, the photographer inserts the tiemaster to make the tie (Figure 6-54). This would be a separate sheet showing only the small pieces of tape tying the pads that connect to the plane.

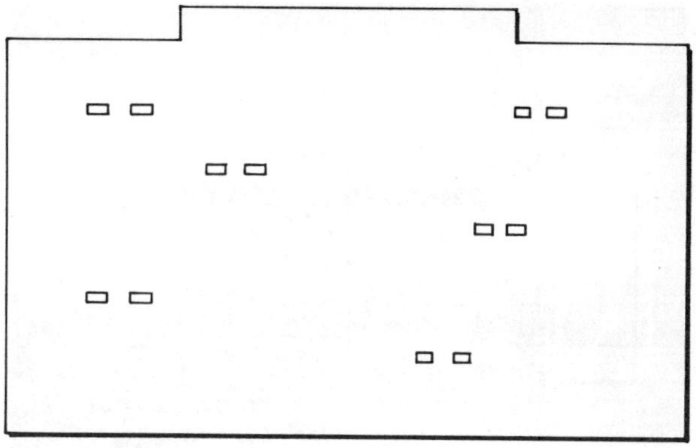

Figure 6-54. The Tie Master

Design Considerations

Separate grounds are used for signal sensitive circuits, high current power circuits, and chassis connected grounds.

Ideally, a layer is a complete surface (such as the component side), an internal plane on a multilayer board, or the noncomponent side covering the entire surface.

Isolated Areas

Sections of a modified ground plane should neither be isolated nor connected by long narrow conductors. If possible, route as many lines as necessary to those isolated areas. This provides a solid path to the isolated ground area. Figure 6-55 shows the non-preferred ground and voltage plane design. Note that the ground and voltage conductive areas are interconnected by long thin conductors. This is different from the preferred ground plane shown in Figure 6-56. By having additional conductors to the voltage and ground planes, more points are provided on the connector. This ensures a more reliable electrical contact to the plane areas that were isolated by the long thin conductor in the non-preferred example of Figure 6-55.

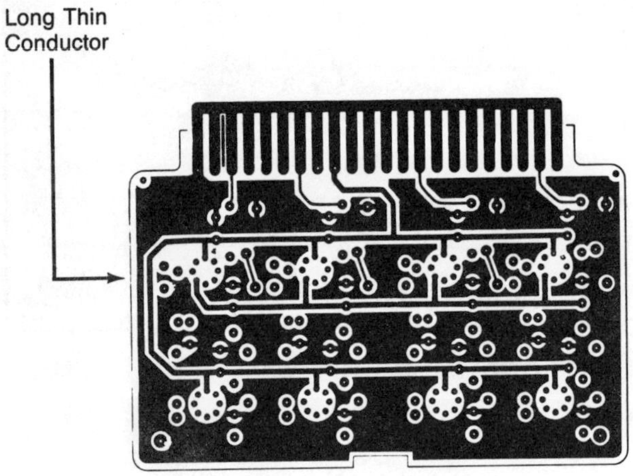

Figure 6-55. Non-Preferred

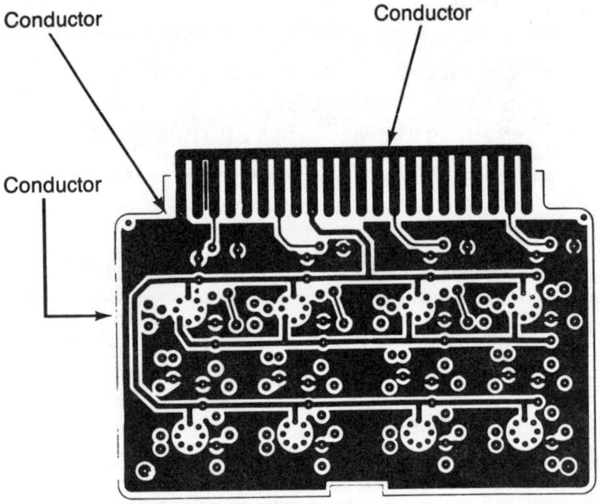

Figure 6-56. Preferred

Note:

A. Preferred ground and voltage design practices should have large areas of continuous conductive copper.

B. Ground and voltage planes should have several equally spaced printed contacts at the edge board connector.

C. When wide conductors must be transferred from one conductive layer to another, use more than one plated-through hole (as shown in Figure 6-57).

D. When ground and voltage planes become impractical because of component density or trace density, bus bars can substitute for voltage or ground planes. They provide greater component and wiring density, without requiring multilayer boards. They reduce power distribution inductance and resistance (Figures 6-58 and 6-59). Bus bars also can be used as a shielding technique, when substituting for a ground plane (Figure 6-60).

Figure 6-57. Wide Conductor Traces From Layer to Another.

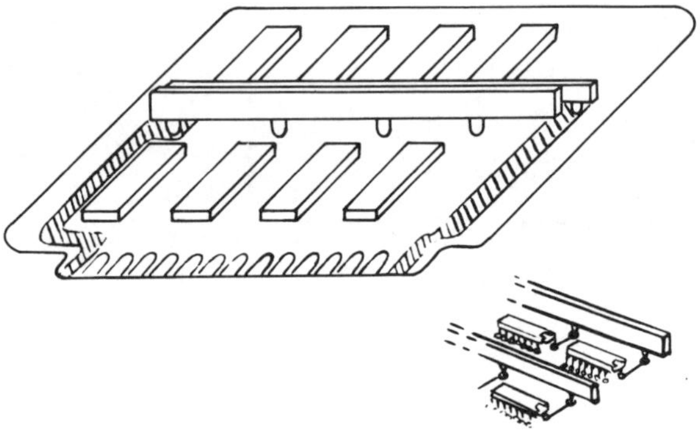

Figure 6-58. Bus Bars

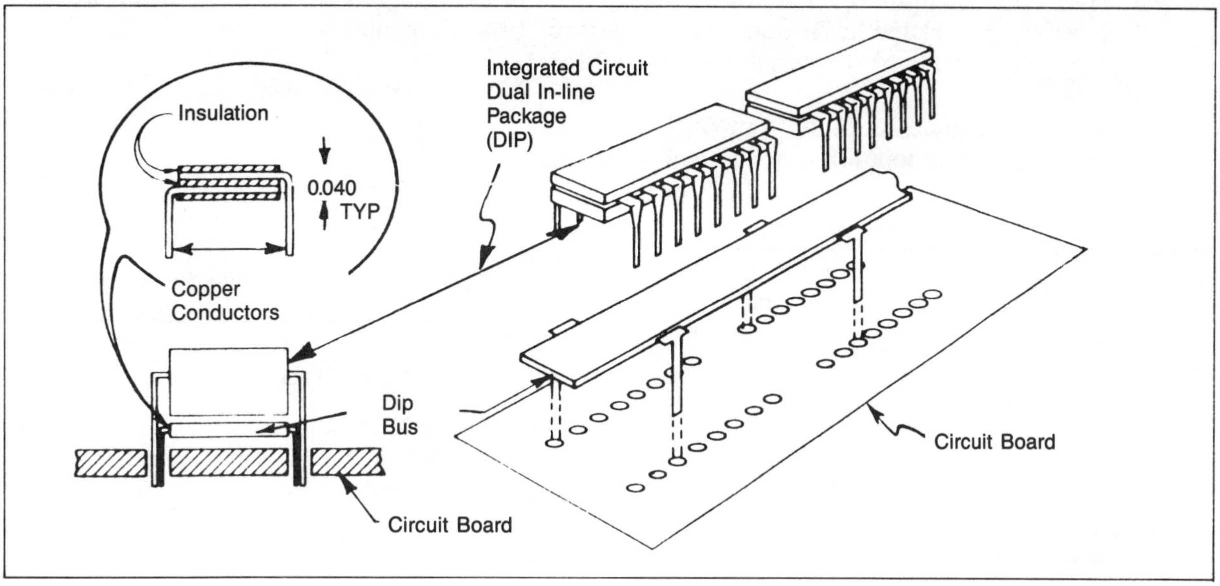

Figure 6-59. Bus Bars

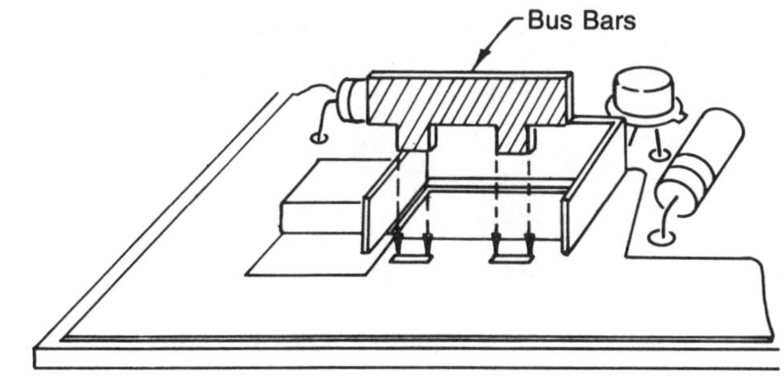

Figure 6-60. Shielding Considerations, Metal Vertical Shielding Plates Are An Effective But Sometimes Expensive Solution For Unwanted Coupling Problems

Layout Checklist

General

1. Have layout grid, scale and number of layers been indicated?

2. Check electrical continuity using copies of the layout and the schematic/logic diagram by marking out each connection with a colored marker.

3. Have all mechanical dimensions pertaining to board size, mounting locations and clearances been indicated?

Components

1. Have all components on the schematic been included and are they properly designated?

2. Are all component shapes in the proper scale?

3. Are maximum dimensions used on all components for eliminating body interference?

4. Have components been located in an orderly fashion?

5. Have standard lead spacings been used for like components?

6. Has orientation been indicated for multiple lead and polarized components?

7. Is there adequate clearance between components and board edges, mounting hardware, and other component?

8. Have special mounting requirements been accommodated (insulation, heat sinks, supports, hardware, etc.)?

9. Is there adequate access for components requiring adjustment after installation?

Holes

1. Are all holes located on a standard grid or dimensioned from a grid location?

2. Has a separate hole been provided for each component lead or terminal?

3. Have all hole sizes and types been indicated, and do they meet design requirements?

4. Does all hole-to-hole spacing meet requirements?

Conductors

1. Have all conductor widths and spacing been properly indicated, and do they meet design requirements?

2. Have all terminal area sizes been properly indicated, and do they meet design requirements?

3. Have conductors been routed in the most efficient manner (smooth flow, short as practical, minimum quantity of jumpers, etc.)?

4. Have critical circuit points been accommodated (conductor length, shielding, isolation, ground planes, voltage planes, heat sinks, etc.)?

5. Has the difference between conductors on separate layers been indicated clearly?

Registration Techniques

The relative position of one or more printed wiring patterns, or portions thereof, with respect to their desired location on the printed wiring board base material is called register.

Printed circuit board registration consists of three basic elements:

1. Overall pattern alignment from layer to layer

2. Exact alignment of individual terminal areas (pads) from layer to layer

3. Exact alignment of drilled holes to the pattern during board manufacturing.

Artwork Markings

Printed circuit board artwork markings relating to registration include datum points, register marks, indexing holes, and photo reduction points. (Other types of board markings are part numbers, reference designations, corner markers, etc.)

DATUM POINTS are points, assumed to be exact, from which the location of printed wiring board features may be established by dimensional computation. Three datum points establish the perpendicular Cartesian coordinate system of grid dimensioning. Datum points are normally indicated by distinctively shaped pads when they are located directly on the printed circuit board. All mechanical dimensioning of the master pattern should be made from datum points.

Bishop's Universal Target is designed to achieve optimum visual register. Lines sighted through the apertures in figure (A) above are in a tightly controlled position and are in proper alignment. The line sighted through the apertures in figure (B) are obviously out of alignment. REGISTER MARKS are bold geometric shapes (e.g., Bishop's Universal Targets) which are used to establish the registration. Register marks are affixed to the artwork on all layers so, when the layers are overlaid, the patterns will be superimposed. Register marks may be located inside or outside of the board outline as appropriate. When located inside the board area, they may be drilled for mounting holes after they have served their original purpose.

INDEXING HOLES are holes placed in a printed circuit board to accurately position the board during manufacturing processes. They may or may not appear within the finished board area, but they must appear on the master pattern.

A minimum of two indexing holes are required, and they should be diagonally located to encompass the test area. A third hole is sometimes used, and should form a right triangle with the first two. They should always be located on grid intersections.

Datum points, register marks, and indexing holes may be combined in the same locations, or they may be completely independent of each other. A registration target such as Bishop's Universal Target or Modified Crosshair may be used as a combination of datum point symbol, indexing hole symbol and register mark.

PHOTO REDUCTION POINTS are two points spaced an exact distance apart to visually indicate the extent to which printed circuit artwork is to be photographically reduced. Care should be used in selecting a mark which clearly indicates the precise measurement point.

Use reduction marks or simple quadrants since the exact point of measurement cannot be confusing—it must be taken from the edge, in addition, the points of intersection indicate proper photo exposure and development.

DO NOT USE Crosshairs for reduction points since the width of the line makes the exact measurement point ambiguous.

Pad Master

The Pad Master concept eliminates misregistration of individual terminal pads and holes. Many methods of executing the concept exist, but all employ the principle of using one original artwork to position all terminal area pads and holes for all layers.

The original artwork is prepared with only terminal areas, registration, reduction and datum marks, and contact prints are made for each layer of the printed circuit board. Even through the contact copies should be identical for each layer, and therefore, totally eliminate individual pad misregistration, minor dimensional variations can exist due to stability differences in individual pieces of film. Therefore, contact copies should be thoroughly checked before completing the individual layers.

Utilizing this same pad master to prepare a drill template will greatly assist in accurate hole-to-terminal area alignment.

Pin Registration

Various pin registration systems have been adapted from the graphic arts industry to simplify alignment of multiple layers of artwork. These systems are no more accurate than properly designed targets, but they do save time. They are particularly helpful when used for all artwork preparation steps, or as a total system from preparation of the master pattern through photo resist exposure during board fabrication.

Remember that proper registration still depends upon the accuracy and alignment of all holes in the artwork.

Registration with Bishop's Red and Blue System

Perfectly registered artwork for double-sided printed circuit boards can be created using a single sheet of film and Red and Blue transparent drafting aids.

When using this system, tape-ups should be done on a high-quality, clear polyester film over a black-line grid to insure professional results and economy. When taping directly to a grid, a non-repro black pattern may be used. When conductor density is high, spacing minimal, and many red-and-blue crossovers occur, use both sides of a thin sheet of film to minimize parallax problems.

All terminal areas common to both sides of the board must be represented by a black pad. Black cannot be filtered out, and will appear in both sides as the major feature with absolute registration. All crossover areas should be burnished to eliminate possible line deviations or ragged edges on the negative.

Common practice is to use Blue tape for the component side of the board (least amount of printed wiring), and Red tape for the opposite side (greater amount of printed wiring). On crossovers, either tape can overlay the other without affecting the filter efficiency.

The specially formulated hue and density factors of Bishop's materials were developed in cooperation with Eastman Kodak Bishop Red and Blue Transparent Tapes, PreKut Tape Shapes, and PreKut StikOns are completely compatible, and delineation such as unequal line widths do not occur. Bishop's exclusive control of hue and density in both Red and Blue transparent materials assures uniform reproduction year after year.

Avoiding Problems

1. Don't tape directly to a blue line grid. Some blue pattern grids cannot be photographically "dropped" when trying to hold blue conductor traces.

2. Don't place a red pad over a blue pad to achieve a common terminal area. While the result appears black to the camera, normal drafting tolerances make misregistration possible.

3. IMPORTANT: Don't overlay three layers of the same color. As buildup occurs with the same color tape, sharp edges distort, color densities change and line definition can suffer in photography.

Quiz—Chapter 6

1. The artmaster is a tool which is generally prepared at an enlarged scale of ____________
2. Artwork base material should be ________ to .007 thick stable-based polyester film.
3. Black crepe tape & pads provide a material which is easy to work with. T F
4. The 3-layer method is essentially the same as the 2 layer method except ____________
5. Red and blue method is essentially the same as the 3-layer method T F
6. Red and blue tape is very easy to work with. T F
7. Red & blue tape can not be bent without distortion. T F
8. Conductor (traces) should be taped at a preferred angle of ________
9. The minimum angle that any trace should be taped is at 90°. T F
10. Traces should not come any closer to the edge of the board than .100. T F
11. Do not overlap tape or cut your pads. T F
12. Terminating traces to pads should be very deliberate. T F
13. Routing traces on very dense digital P.C. boards can be a problem to conserve space you should ____________
14. Routing traces thru IC pads on digital P.C. boards is not a common practice. T F
15. The designer should avoid placing a feed-through hole under an IC pattern. T F
16. It is not uncommon for the engineer to require a space IC position, especially in the prototype stage. T F
17. Component marking: in cases where rigid identification is necessary, it is still important to identify components such as: connectors, test points, switches or ____________

18. Polarity/Indexing: all components require polarity indexing. **T F**
19. IC's should have indexing on Pin #1. **T F**
20. It is very helpful to index various pins on a connector. **T F**
21. IC's should be placed on the board so that all are oriented in the ____________ direction.
22. The revision letter of the board must be identified somewhere outside the board parameters. **T F**
23. The master pattern number refers to the artmaster itself. **T F**
24. Reduction targets are used by the photographer. **T F**
25. The correct method for making copies of your tape-up is to use a ____________________________
26. Each film produced on a large process camera should be inspected for:
    A. Proper reduction
    B. Sharpness
    C. Density
    D. Workmanship
    E. ________________________
27. Ground planes are areas on the P.C. board that have not been consumed by ______________ or pads.
28. The ground plane is generally on the non-component side of the board. **T F**
29. Ground planes are tied to the ground circuit. **T F**
30. There are three different methods to connect ground for P.C. boards. 1. Parallel connected ground system. 2. Series and __________ ground system.
31. There are two methods used to produce ground planes on artwork 1. Manual cutting 2. __________
32. When side conductors must be transferred from one layer to another you should use more than one plated through hole. **T F**
33. When ground and voltage planes become impractical __________ bars can be substituted.
34. Designers should avoid taping directly to a blue line grid. **T F**
35. The pad master concept for P.C. boards eliminate misregistration of pads & holes. **T F**

The answers for each question can be found on the pages listed below:

Question 1. Pg. 152
Question 2. Pg. 152
Question 3. Pg. 152
Question 4. Pg. 152
Question 5. Pg. 153
Question 6. Pg. 154
Question 7. Pg. 154
Question 8. Pg. 154
Question 9. Pg. 154
Question 10. Pg. 155
Question 11. Pg. 155
Question 12. Pg. 156
Question 13. Pg. 156
Question 14. Pg. 157
Question 15. Pg. 157
Question 16. Pg. 159
Question 17. Pg. 163
Question 18. Pg. 163
Question 19. Pg. 163
Question 20. Pg. 163
Question 21. Pg. 163
Question 22. Pg. 164
Question 23. Pg. 164
Question 24. Pg. 165
Question 25. Pg. 165
Question 26. Pg. 168
Question 27. Pg. 168
Question 28. Pg. 168
Question 29. Pg. 168
Question 30. Pg. 169
Question 31. Pg. 173
Question 32. Pg. 178
Question 33. Pg. 178
Question 34. Pg. 180
Question 35. Pg. 179

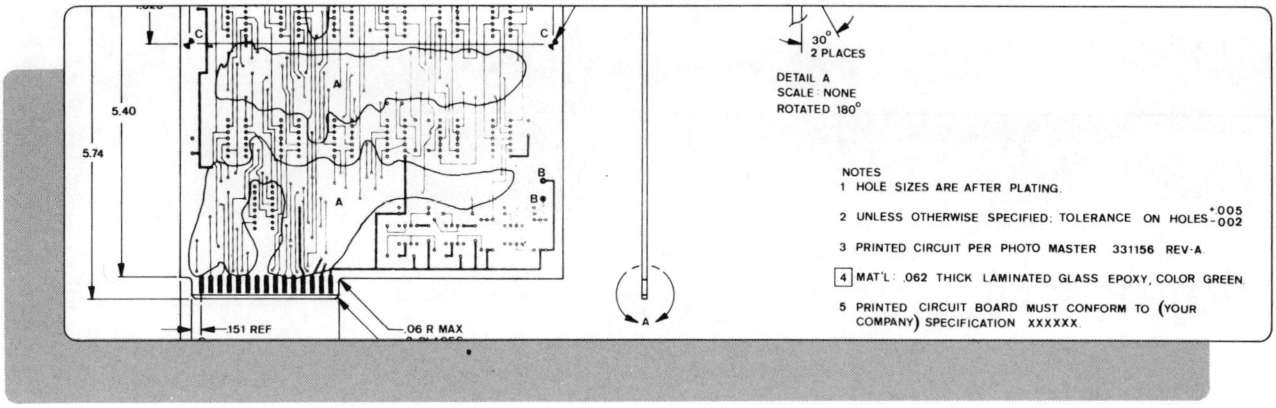

7

PC DRAWINGS AND DOCUMENTATION

This chapter contains general information about the most commonly used printed circuit drawings and documentation. It also includes preferred practices which are recommended when preparing each type of drawing or document.

Drawings to be covered in this chapter are as follows:

1. Schematics
2. Fabrication drawing
3. Assembly drawing and parts list
4. Silkscreen drawing
5. Solder Mask
6. Photomaster

Figure 7-1 shows an example of a complex schematic.

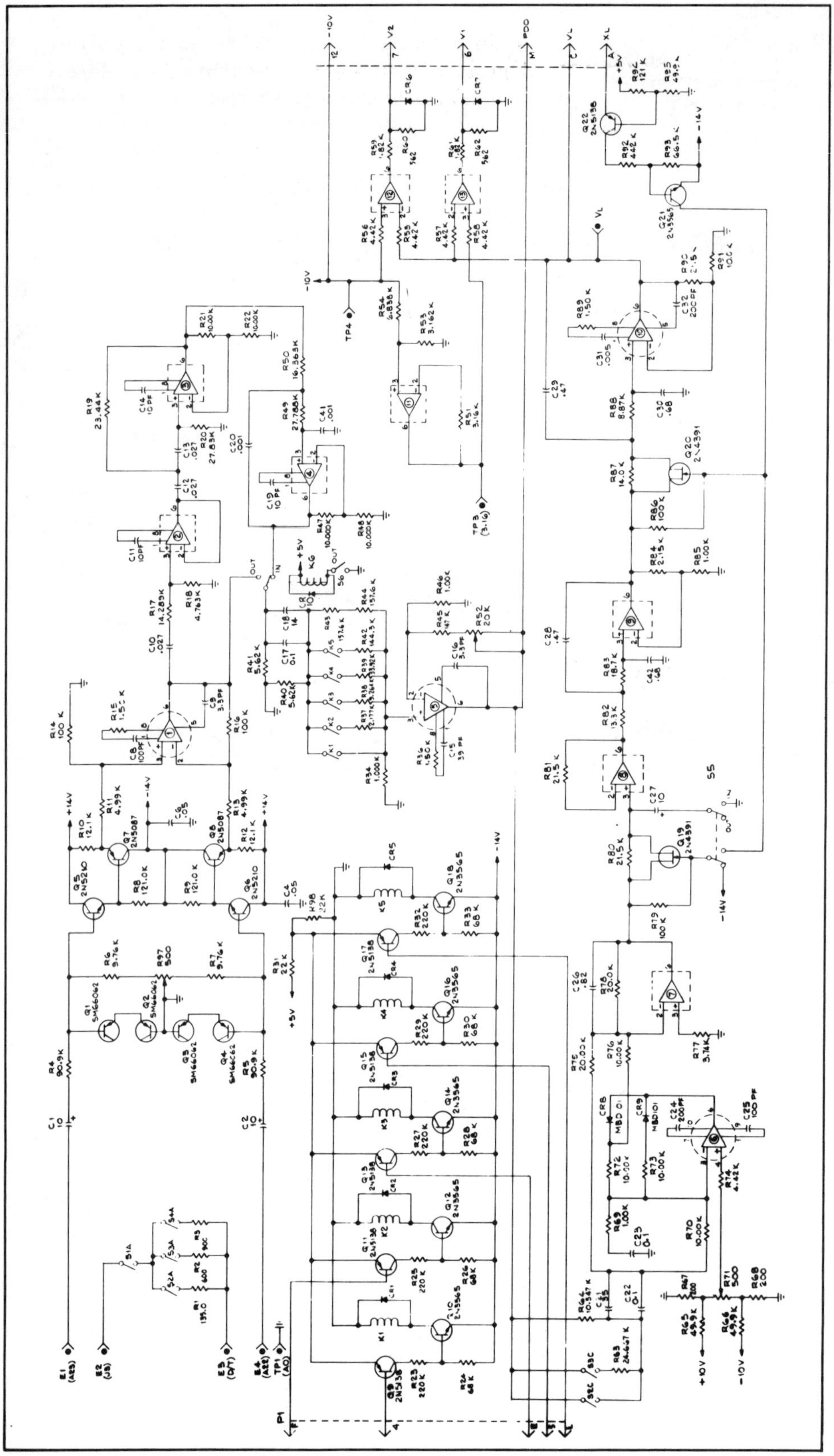

Figure 7-1. Complex Schematic

Schematic

Schematic drawings can be simple conveying only basic or complex information showing all fine details about the operation and function of a circuit.

The simple drawing, often called a block diagram, omits most of the detailed information shown on a typical schematic or logic diagram. A block diagram is a reference drawing which shows the various separate functional groups of circuits as block symbols. The lines interconnecting these blocks represent the intercommunication or flow from block to block or circuit to circuit. Figure 7-2 shows a block diagram. The basic rule is to use the type of diagram that will adequately define the circuit in the simplest possible way.

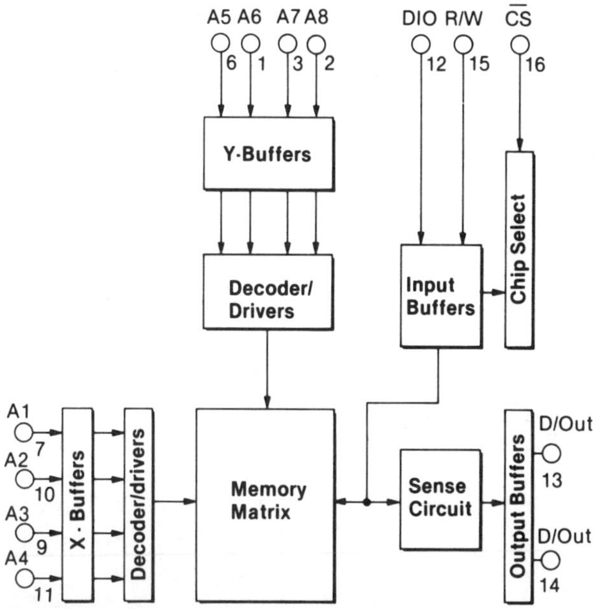

Figure 7-2. Block Diagram

The schematic diagram shows the electrical connections and functions of a specific circuit by means of graphic symbols. Since the fundamental purpose of this document is electronic, it does not show any of the mechanical details of the printed circuit board. In addition to graphic symbols and the interconnections between them, the schematic diagram contains reference designations to distinguish one identical symbol from another. Figures 7-3 (Analog) and 7-4 (Digital logic) show an example of a schematic diagram.

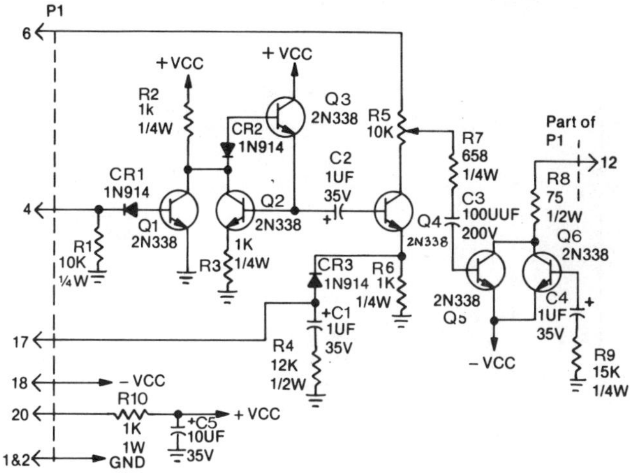

Figure 7-3. Six-transistor Schematic

Logic Diagram

The schematic or logic diagram shows the various electrical and electronic parts connections and functions of a specific circuit arrangement by means of graphic symbols. The diagram facilitates tracing the circuit and its functions without regard to the actual physical size, shape, or location of the components on the PC board. The electronic meaning of logic symbols is not as precise and clear as the more elementary schematic symbols. Figure 7-4 shows an example of a logic diagram.

The schematic or logic diagram shows the exact electrical values and types of components that are used in the design. These values or part identification are usually listed in a table on the drawing. For example in the case of resistors the resistance in ohm, its tolerance and wattages, in the case of capacitors the capactance value and type. In the case of transistor, its part number: --- etc. are shown.

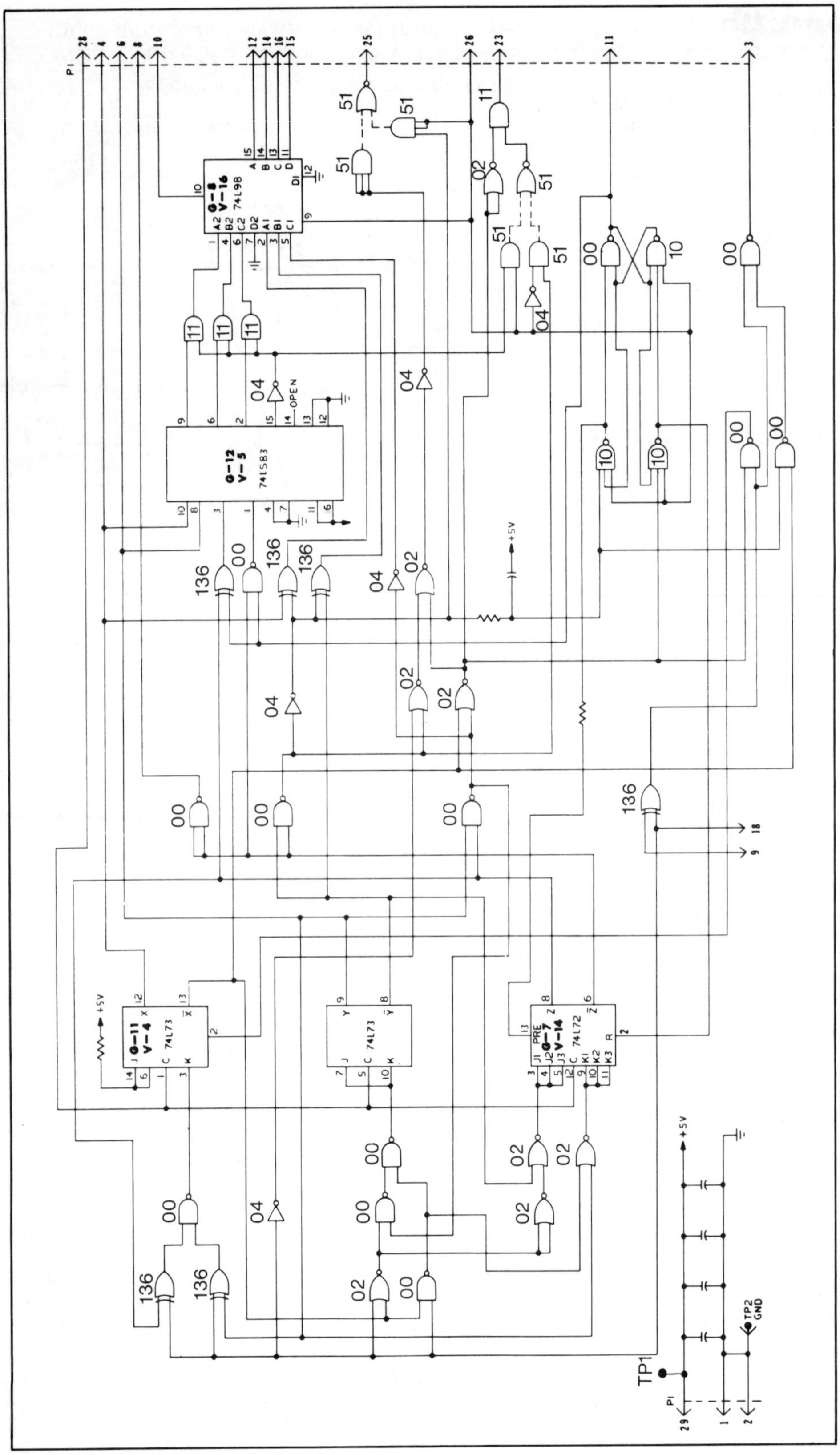

Figure 7-4. Digital Logic Diagram

Detail Fabrication Drawing

The detail fabrication drawing of the printed circuit board depicts the dimensional configuration of the board, size, location of holes, and specifies the material, process specifications and (if applicable) notes and other information necessary for the fabrication of printed circuit board. Figure 7-5 shows a typical PC board detail drawing.

The detail fabrication should be drawn in the same scale as the film or photomaster (normally). If more than one sale is used, each scale shall be clearly identified.

All dimensions needed to fabricate the PC board should be shown with the exception of those features located by the film or photomaster, such as pads, traces, connector fingers, etc. Dimensions for board configuration should be referenced from the tooling holes or register marks etched on the printed circuit board. All holes, cutouts, and the board outline should be dimensioned from these points. See Figure 7-6.

The detail fabrication drawing should show the noncomponent side of the PC board. One reason is because the manufacturer would like to view the side of the board that is drilled. Another reason is because of the burr left on the opposite side of drilling after the drill passes through the PC board.

The purpose of having the burr on the component side of the PC board is because of the widespread use of wave or flow soldering (see Chapter 4). This process requires that the side passing through the wave of solder be smooth, and that the noncomponent side must be the side that passes through or over the wave of solder.

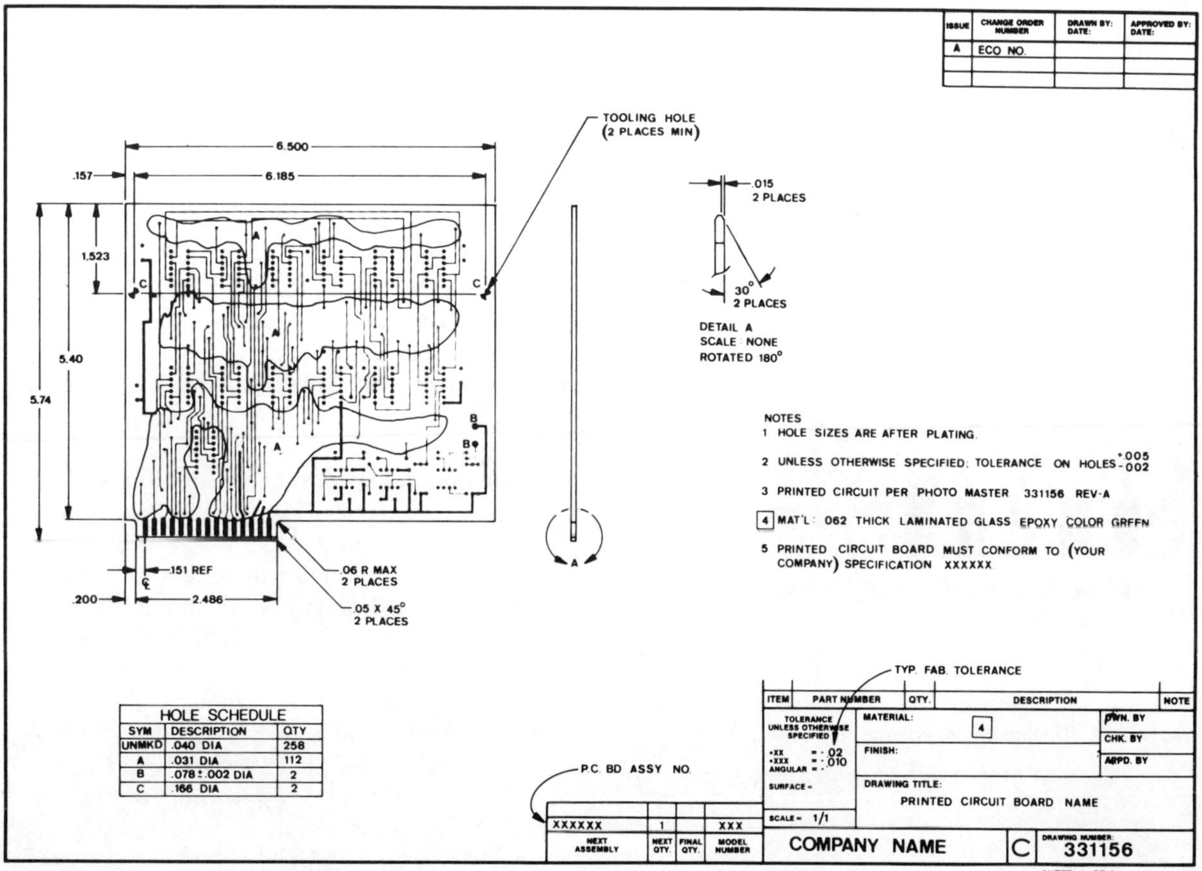

Figure 7-5. Fabrication Drawing

The fabrication drawing presents a complete engineering description (notes) of all design features essential to printed circuit board fabrication.

The fabrication drawing should always include.

A. Board outline and thickness.
B. Materials (board, conductors and plating).
C. Hole locations and size.
D. Film (master pattern).
E. Test and qualification requirements.
F. Workmanship and acceptance criteria.
G. Change control information.

Subsequently, the PC designer must collect mechanical considerations and PC design rule. He/she must combine them with the artwork, fabrication instruction, and workmanship criteria in the form of notes on the fabrication drawing (see notes example). Figure 7-5 shows a number of examples.

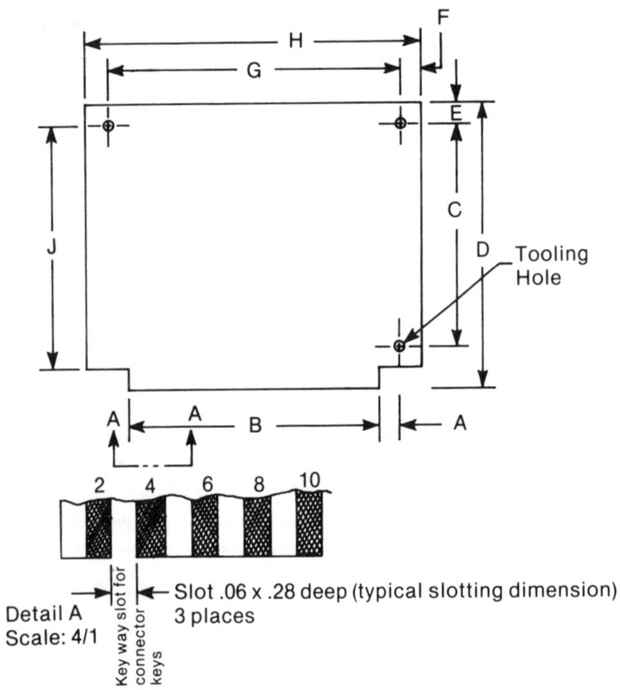

Figure 7-6. Dimensions needed to fabricate.

NOTES

1. Hole sizes are after plating.
2. Circuit side of card shown.
3. Printed circuit board must conform to XXXX specification (see the plating specifications below).
4. Printed circuit per photomaster XXXX, Revision X.
5. Hole diameter tolerance XXX +.005
6. Material -.002
7. Connetor Finish

The hole schedule chart appears on the face of the detail drawing and will give the symbol used to identify each hole called out.

It contains a description of each hole size and tolerance, and if necessary, the quantity of each type or size used in tabulated form. See Table 7-1.

Table 7-1. Hole Schedule Chart

Symbol	Description	Qty.
Unmarked	.038 Dia.	105
A	.048 Dia.	59
B	.187 Dia.	7

Listed below are most of the basic plating specification requirements that must be provided to the PC board manufacturer. These specifications will not conform to all companies' standards; therefore, use these specs where applicable.

1. Base laminate: Base material shall be laminated epoxy glass type G10, .06 ± .002 thick.

2. Conductor: Conductor material on double-sided (plated through) board should be 1 oz. or 2-oz. copper minimum before plating.

3. Solder plating: The tin lead alloy plating should contain 40% ± 5% lead and 60% ± 5% tin. All solder-plated areas should be subject to hot oil solder reflow process.

4. Minimum plating thickness on trace surface and through holes (inches):
   Copper .001
   Tin-lead .0003
   Gold .00005
   Nickel-gold .003

5. Minimum plating thickness for fingers (inches):
   Nickel-gold .0003
   Nickel-rhodium .0003
   Gold .00005

6. Conductor defects:
   a. Pin holes, nicks, and scratches: Conductor defects such as pin holes and nicks which do not reduce the conductor width by more than .002-in.
   b. Plating overhang: The plated resist material overhang remaining after the copper foil has been removed by etching shall not exceed .005-in.
   c. Under-cut: The reduction of conductor caused by etching under the edge of plating resist shall not exceed .005-in.
   d. Excess conductor material: Isolated spots of conductor material no greater than .015-in. in diameter are acceptable providing spacing requirements are not violated.

7. Warpage: Warpage or twist of a printed circuit board should not exceed .005-in. per inch.

8. Solderability: The solder coating should be adherent and nongranular in appearance. The base material should not show evidence of blistering or delamination, nor should conductors or terminal areas show evidence of separating from the base laminate after dip soldering.

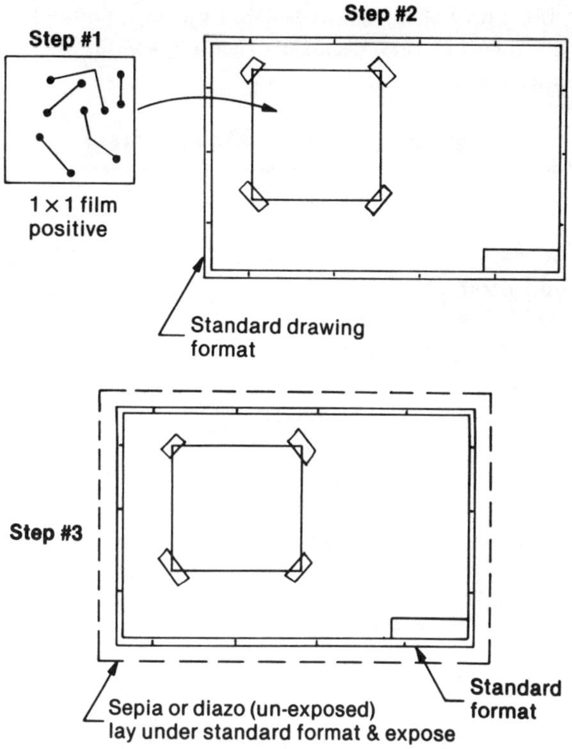

Figure 7-7. Print of the Noncomponent side

Techniques

There are several ways to delineate the circuit pattern on a PC board fabrication drawing (see Pads and Traces Figure 7-5). One of the simplest methods is to make a diazo or sepia print of the noncomponent side using the 1:1 film positive (see Figure 7-7). This print should be made on a flat light table, or a vacuum frame (see Figures 6-51, 6-52, and 6-53). Never place the original artwork through a print machine that has rollers, this can damage the original artwork. A "Screen Back" image should be used because, by fading the traces and pads slightly, it allows the board outline, dimensions and other necessary inforation to appear more prominently. The circuit images may be prepared first, then "stripped" into a standard or company drawing format.

Another method that has several benefits, is to make composite contact copies on matte finish diazo or photosensitive film. This is accomplished by contact printing a clear film master of a standard printed circuit fabrication drawing format with a properly positioned, screened back copy of the reduced circuit pattern. Information that is identical from board to board (ie: standard notes, board outlines, hole schedule, etc.) may be placed on the format master to reduce repetitive drafting effort. It may be desirable to include a clear finish grid pattern (accurately aligned with the circuit pattern) to aid in locating pads, or tooling holes or cutouts, etc. This is not normally practiced.

Assembly Drawing

Assembly data comes in many forms. The primary purpose of this document describes the end product: an assembled printed circuit board with all of the electrical components as well as all of the mechanical parts contained on a particular assembly. The printed circuit assembly drawing should meet the normal requirements of an assembly drawing. It should depict the PC board components, reference designation markings, assembly and test specs, and also include a list of material which contains all part information needed to order the parts. A typical assembly drawing is shown in Figure 7-8.

In some companies where large production requirements exist for PC board assemblies, assembly data may be separated into individual portions. This provides operators with only that information with which they are concerned. For example, an operator may be inserting only resistors into a PC board. He/she only needs to know the location of those resistors and not any other parts to be assembled. The information provided for this operator would therefore show only those components that are specifically inserted at this particular station. This technique coincides well with "binning" of components, or other semiautomated techniques for component assembly (see Figure 7-9).

The operator is delivered only the correct component and shown exactly the correct place to insert it on the board with the aid of a lighted arrow from an articulated overhead projector.

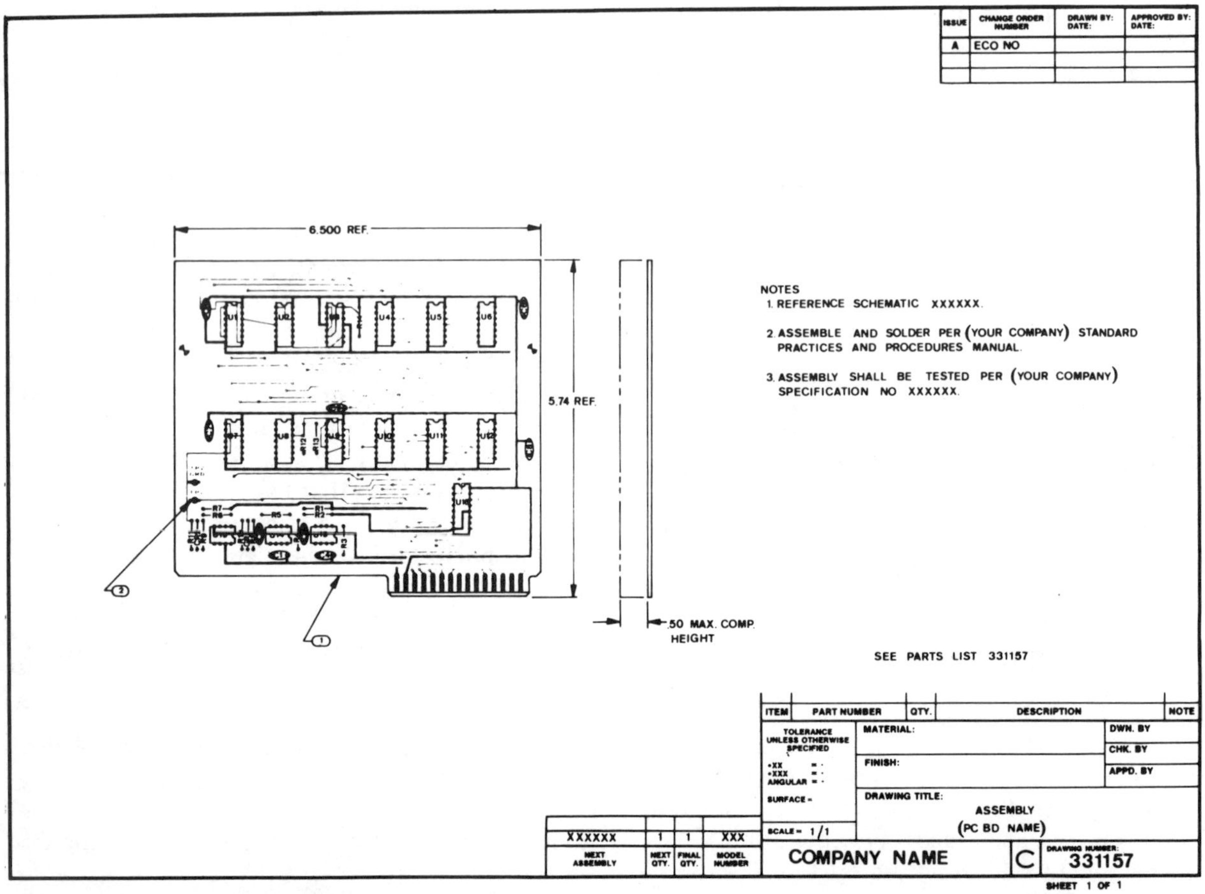

Figure 7-8. Typical PC Board Detail Assembly Drawing

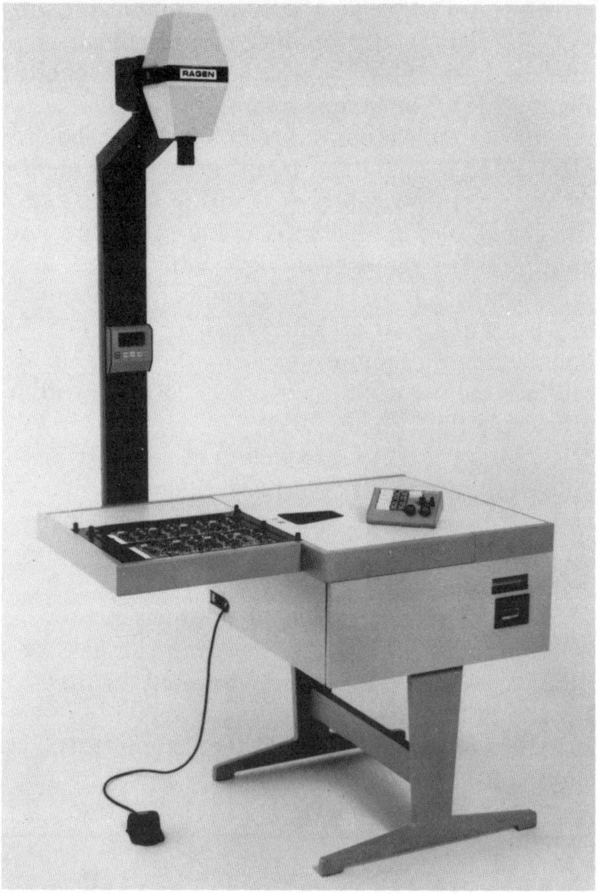

Figure 7-9. Automated Component Assembler

Changes in board design are accommodated by a keyboard command.

Identification

A PC board assembly is identified by a part number in the same way that detail parts are. Usually, the part number is the same as the assembly drawing number. The parts comprising an assembly are listed on a parts list. Each line of the parts list is identified by a description, such as reference designation (REF DES), item number, manufacturer's part number, description of the item, and total quantity (see Figure 7-10). The parts required to make up the assembly, as listed on the parts list, are identified on the assembly drawing by their item number enclosed in a circle (Figure 7-8). This practice should be confined to parts that have no reference designation (REF DES). Parts that have a component REF DES will be identified by the REF DES on the assembly drawing (Figure 7-8) and on the parts list (Figure 7-10).

ASSEMBLY NAME POWER SUPPLY
ASSEMBLY NUMBER 331578
PARTS LIST

REF DES	ITEM NO.	MANUFACTURERS PART NUMBER	DESCRIPTION	TOTAL QTY
C1		CKO5	CAPACITOR, CERAMIC 0.1 UF, 10V	3
C2		CKO5	CAPACITOR, CERAMIC 0.1 UF, 10V	-
C3		CKO5	CAPACITOR, CERAMIC 0.1 UF, 10V	-
R1		89PR5K	RESISTOR, VARIABLE 5K	1
R2		JANRCRO7	RESISTOR, CARBON 560, 1/4W, 5%	2
R3		JANRCRO7	RESISTOR, CARBON 4 7K, 1/4W, 5%	1
R4		89PR10K	RESISTOR, VARIABLE 10K	1
R5		JANRCRO7	RESISTOR, CARBON 560, 1/4W, 5%	-
Q1		2N3210	TRANSISTOR NPN	2
Q2		2N3210	TRANSISTOR NPN	-
U1		LM741	OP-AMP	2
U2		LM741	OP-AMP	-

Figure 7-10. Parts List

1. The assembly drawing is normally drawn in the same scale as the artwork or tape-up. If more than one scale is used, each scale is clearly identified.

2. The assembly drawing is drawn as viewed from the component side. Also a side view is shown to indicate the maximum component height relative to the board. Enlarged views or sections may be used, if needed for clarification.

3. Symbols and reference designations should be shown on the schematic diagram. Abbreviations, if used, shall be per MIL-STD-12C or company standards. Reference designations should also be used to identify all electrical components from the assembly drawing to the schematic and parts list, (Figure 7-10).

4. Methods for mounting the components, such as riveting, soldering and brackets, should be noted.

5. If wire terminations are used for circuit interconnections and are not etched or marked on the PC board, the wire terminations should

be identified as jumper wires and shown on the assembly drawing. Also soldering note for the jumper wire to be written.

6. All transistor tabs should be shown unless it is an epoxy-type transistor, which will be indicated by the flat side. TO18 and TO5 case pins 8 or 10 should be marked. All polarized capacitors (plus side) and all diodes (cathode end) should also be marked.

NOTES

Sample PC assembly drawing notes are listed below.

1. Reference electrical schematic diagram XXXXXXX.

2. Assemble and solder per XXXX spec (or XXXXX standard practice).

3. Observe polarity of capacitors, diodes, etc.

4. Mark assembly number and revision letter (per applicable company standard).

5. Board to be conformal coated in accordance with XXXX military specification. Test points and adjustable components shall be masked during coating.

6. Electrical reference designations are for reference only and need not appear on the parts or components unless otherwise specified, such as R1, C1, CR1, etc.

When a bulk material such as wire, solder etc., is included in the parts list, its quantity is marked "AR" (as required). When one or more assemblies are required (which differ, however, slightly), new and separate part numbers are required.

Techniques

Printed circuit assembly drawings are easily prepared by using a number of contact reproduction methods utilizing either wash-off or diazo photosensitive drafting films.

Using the artwork or film master pattern of the component side of the board and a half-tone screen (120 line, 40%) a screened copy of the artwork configuration is produced on a standard drawing format (see the example of a half-tone in Figure 7-11). The component outlines and reference designations are then added in solid black using a pencil and template, or preprinted, pressure-sensitive, adhesive decals (puppets). See the example of component outlines in Figure 7-12 and 7-13. The screened artwork pattern is a reference which allows components to be easily located during assembly and testing. Minimum documentation requires only that the component outlines are placed in their proper location within a board outline (Figure 7-12).

The quickest, easiest method for preparing an assembly drawing is to make a reproducible copy of the component outline of a layout made with Bishop puppets (see the example in Figure 7-14).

The silkscreen drawing and artmaster provide a tool from which silkscreens, stamps, stencils or other marking mediums are fabricated for marking the printed circuit board with reference markings for the assembly of the components.

Silkscreen Artmaster

The artmaster seen in Figure 7-15 should have sharp definition and high contrast because it is reproduced by a photographic process.

The silkscreen artmaster should be a transparent, stable material either with black self-sticking printed circuit tape, black matte characters having a heat resistant, pressure-sensitive adhesive backing, or black ink lettering with a "Leroy" or similar pen. The tape and preprinted characters are preferred because of their well-defined lines and uniform contrast.

Ragged edges, splits on the line work or characters, and pinholes should be avoided in order to reduce the amount of retouching. The artmaster should be made to the same scale as the printed circuit artmaster. The silkscreen artmaster is an undimensioned master. The artmaster should show the accurately scaled distance between two target points in the horizontal or vertical planes; both horizontal and vertical are not necessary, but may be used.

Tooling holes or fabrication lines for the artmaster should be located on the artmaster to correspond with those on the printed circuit art-

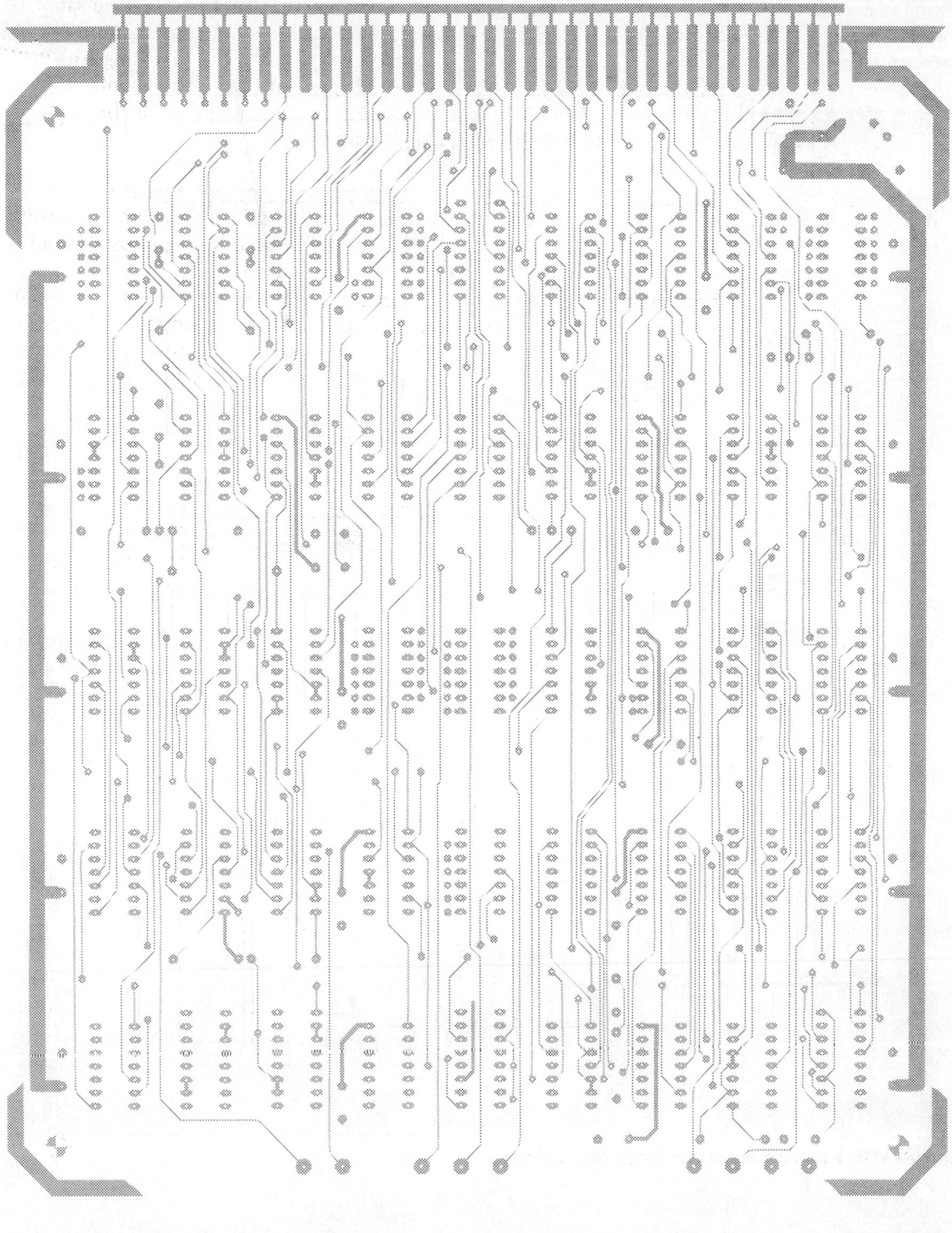

Figure 7-11. Half-tone Copy of the Artwork

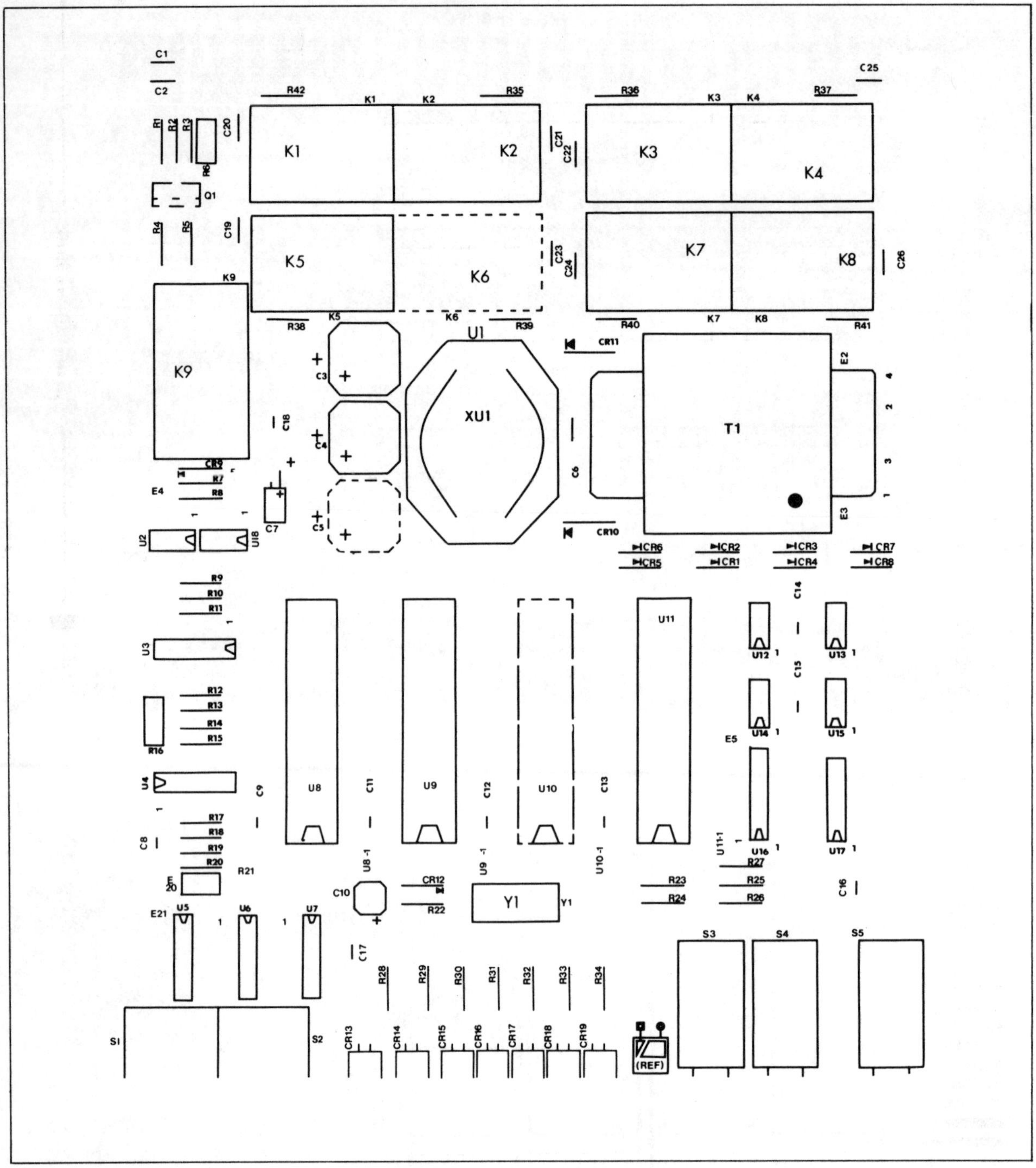

Figure 7-12. Silkscreen Artwork for Component Outlines

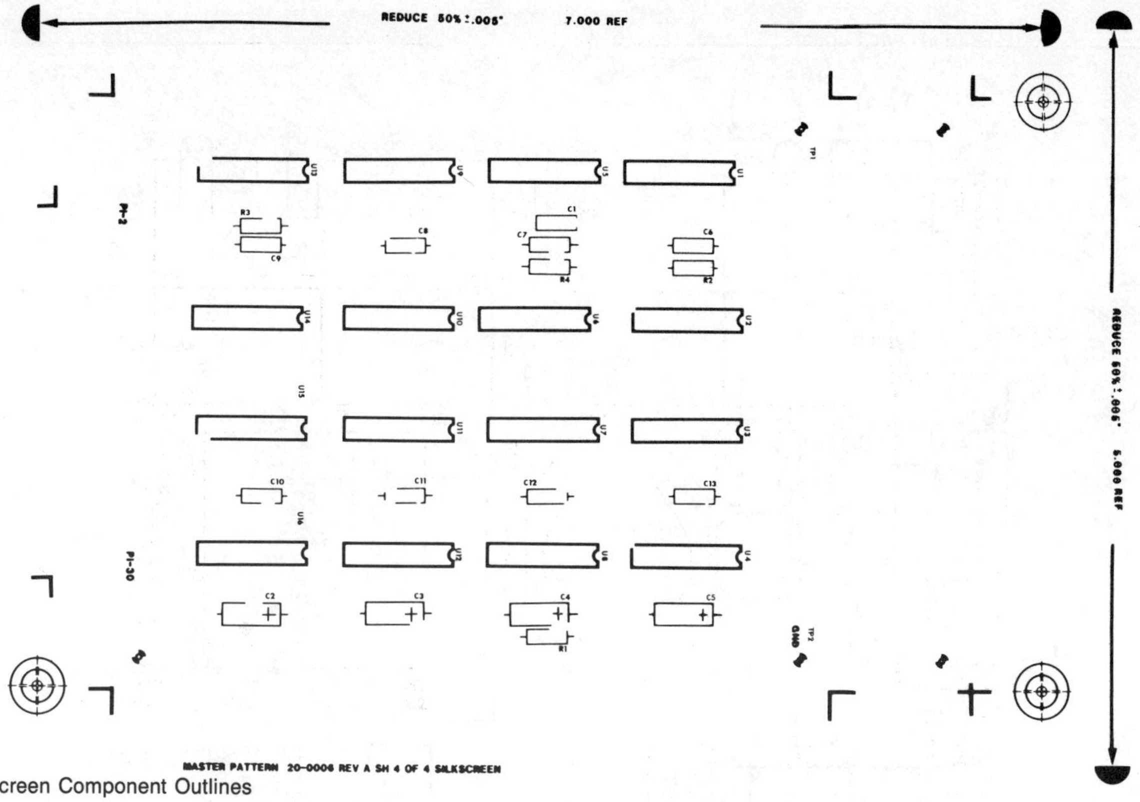

Silkscreen Component Outlines

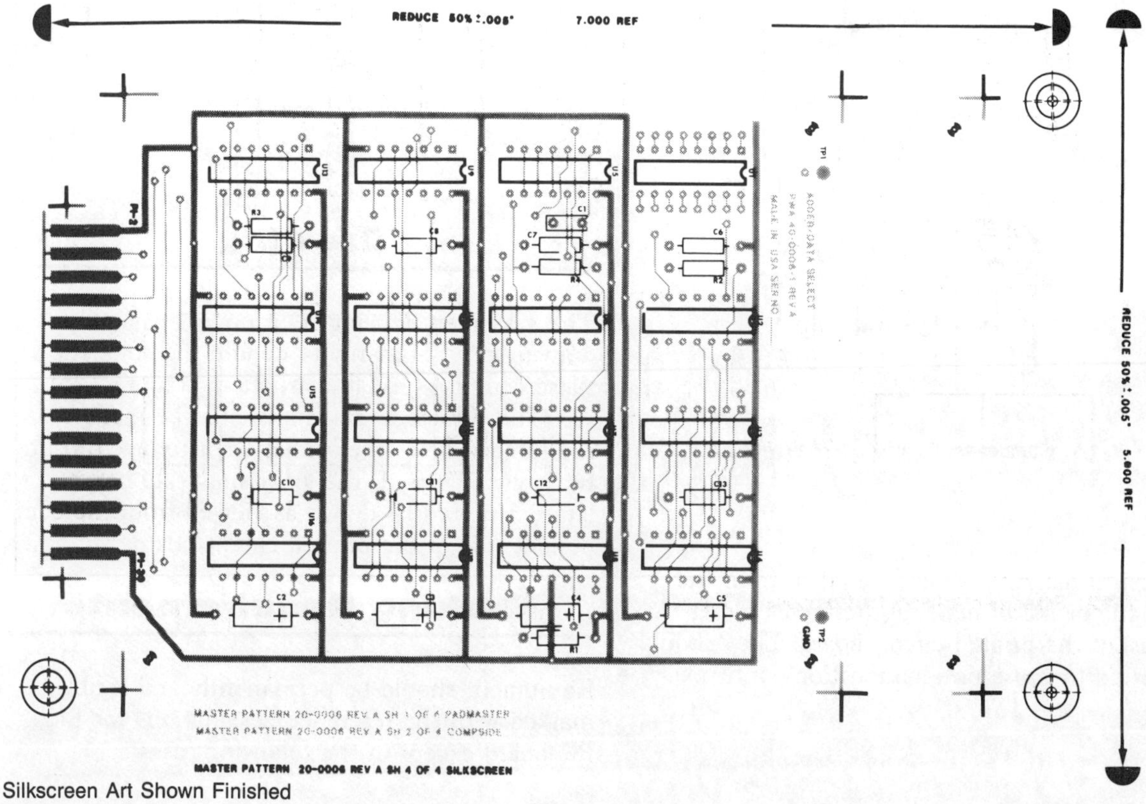

Silkscreen Art Shown Finished

Figure 7-13.

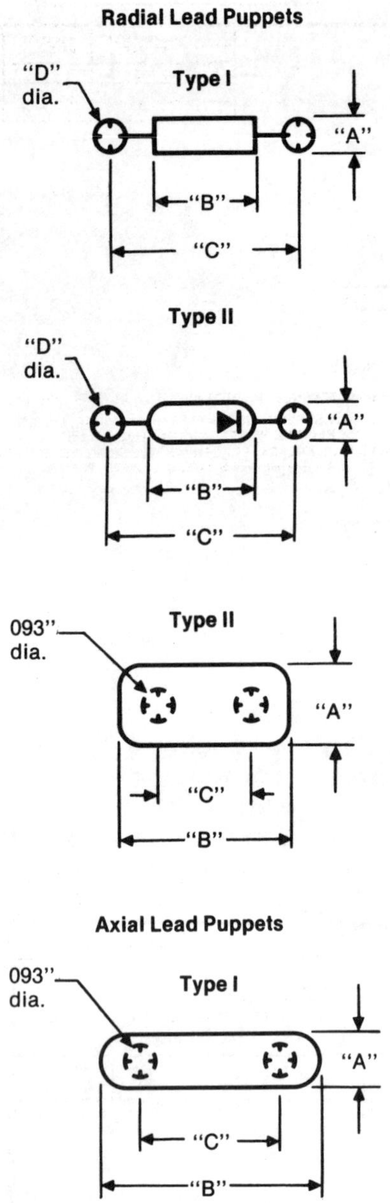

Figure 7-14. Component Outline

master. All markings should be taken from the schematic or bill of materials (reference designations) of the printed circuit board. They should be located to cause no identification confusion.

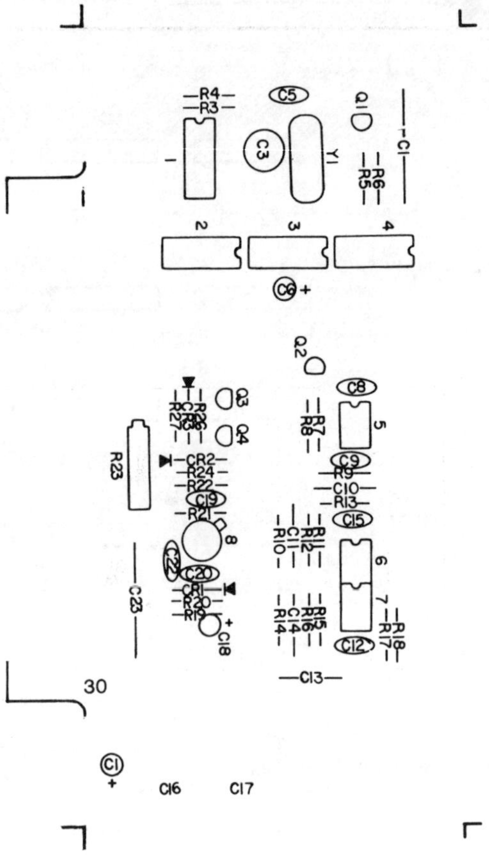

Figure 7-15. Silkscreen Artmaster

Drawing

The silkscreen drawing is an undimensioned drawing used primarily to denote the color the silkscreen and special instructions (Figure 7-16).

The silkscreen drawing should be made to the same scale as the artmaster or reduced to fit the drawing size. If the artmaster scale is too large, it should be drawn as viewed from the component side of the printed circuit board.

Marking Requirements

Equipment should be permanently and legibly marked with the reference designations for each PC board except in the following cases:

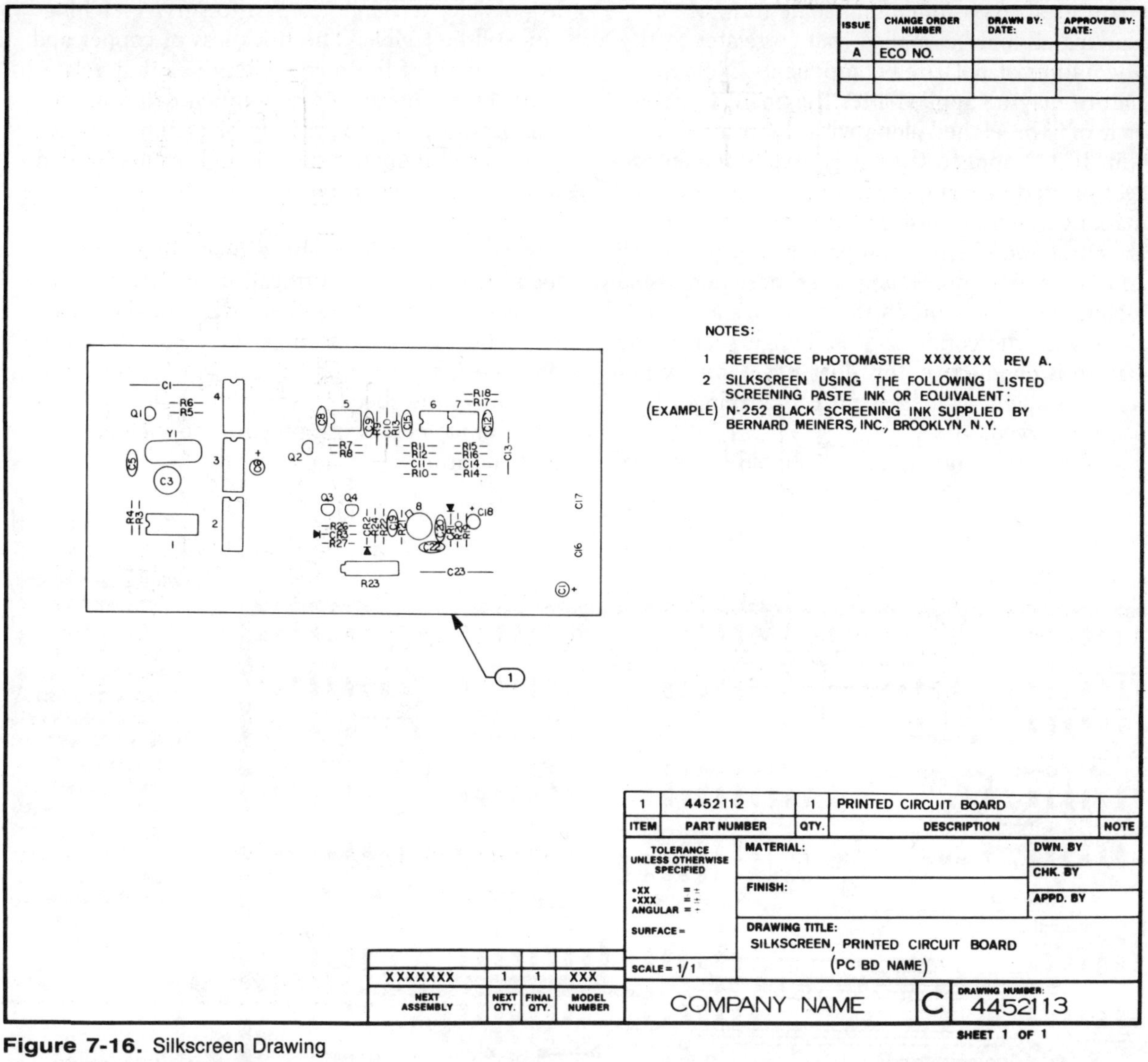

Figure 7-16. Silkscreen Drawing

1. Where space limitations prevent such marking.
2. When the customer specifically requires that markings be omitted.
3. Where it is customary in your company to omit some types of components.

The decision to provide board marking is based primarily on the maintenance of that board. ANSI Y32.16 also requires that the reference designation be marked in such a position as to physically locate the parts and yet be readily visible for purposes of maintenance without removal of the part of other parts. The primary intent of these requirements is that the removal of a part should not result in loss of the identification of the physical location of that part. There are several ways that identification of parts can be accomplished. It is not always possible to have the reference designator visible once the part has been assembled. Some fabricators use the reference designator as a guide during the assembly operation. They are not concerned if the symbols are covered after the component is inserted.

Almost all printed circuit boards require the use of nomenclature. If not the components, the nomenclature identifies the test points or the board itself. As a minimum, some form of marking should be applied that correlates to the orientation of polarized components. Nomenclature may be applied after the board is etched, or it may be etched along with the printed circuit. If it is applied later, it is usually screen process printed or stamped. In either case, the marking material must be carefully chosen for its electrical and chemical properties. Because of this and other extra processing steps, designers usually include nomenclature in their artwork and etch it along with the conductors. Although such nomenclature is conductive, this does not normally present a problem as long as proper clearances have been observed.

The size of nomenclature should be as small as possible, yet large enough to survive etching and still be legible. The thickness of copper and the location of the nomenclature as they relate to conductors determines how much deterioration will occur during the plating or etching process. (Very small nomenclature should not be located in open areas when two-or three-ounce copper is being used).

Some companies also display their company logo, federal code identification, or UL (Underwriters Laboratory) number. Many of these can be etched at the same time the conductors are processed, so all nomenclature information should appear on the specific artwork. Figure 7-17 shows an example of artwork prepared to include nomenclature.

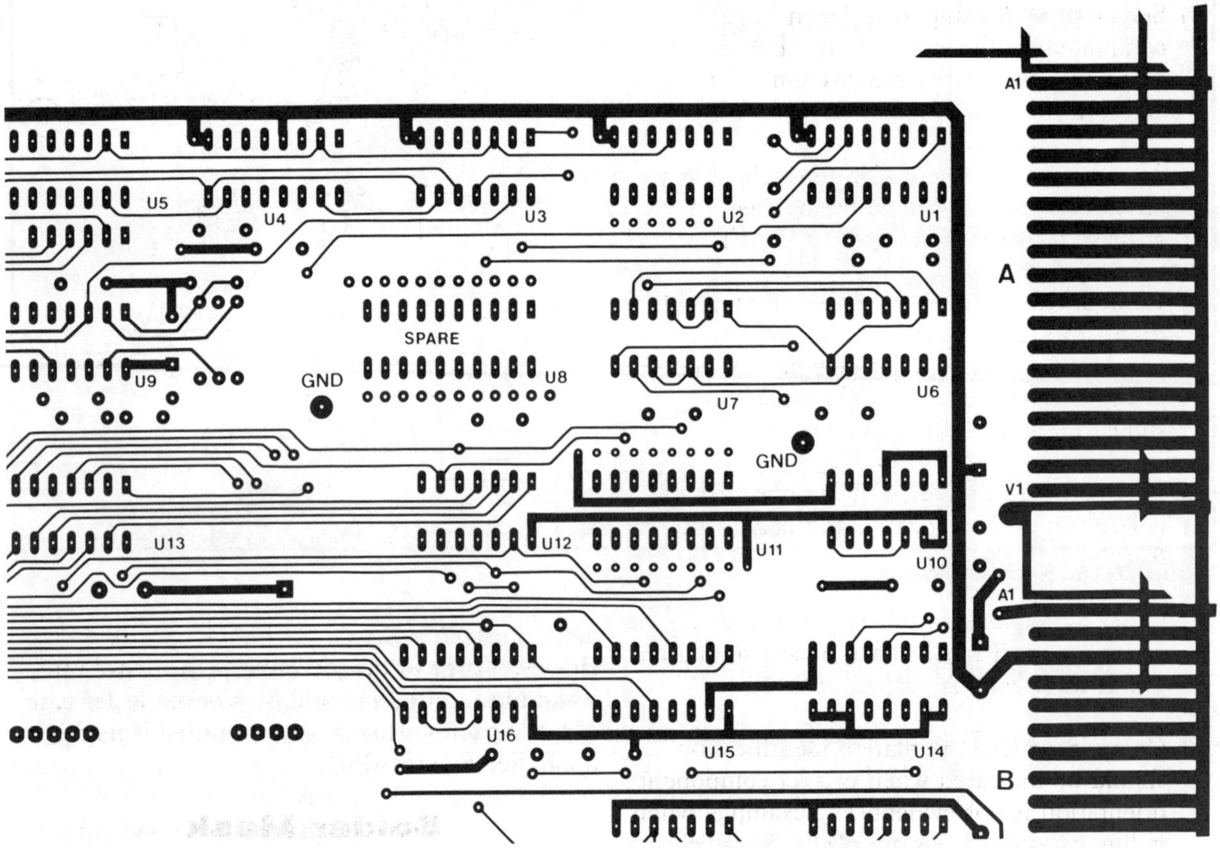

Figure 7-17. Artwork with Nomenclature

Design Guide

The following items must appear on the component side of a PC board: capital letters in parentheses refer to item in Figure 7-18.

1) The first and last printed-contact numbers/letters must appear if board edge contacts are used. (A)

2) The etched component legends should not reduce the conductor spacing below design requirements. (B,C) Nor they should create electrical short between two adjacent conductors.

3) Screen-printed component legends and component outlines: should not be located on any lands or other solder connection areas. (D)

4) The component orientation pin, tab, or mark on all components should be indicted on the artmaster as shown in Figure 7-18. (E)

5) Component numbers for active components (transistors, IC's, operational amplifiers, or power supplies) and adjustable passive components (trimpots, or variable capacitors) should be visible after assembly. (C)

6) Test point identification must appear (diode polarity, capacitor polarity, and board title). (B)

7) Other component numbers (for resistors, capacitors, and small signal diodes) should be visible after assembly when possible.

8) Component lead orientation identification should be indicated when correct component orientation is not evident. For example, when in-line transistor leads are placed in radial component holes or vice-versa, an E, B, or C designation should be included. (SEE FIGURE 7-19).

When preparing the component marking, whether etched or screen printed, it is not necessary to block (draw body outline) in small resistors, capacitors, and diodes (provided mounting-hole locations are unobstructed). However, large or unique components, hardware, transistors, and IC's should be blocked in for ease of loading (only when a screen-printed legend is used). See Figure 7-18D.

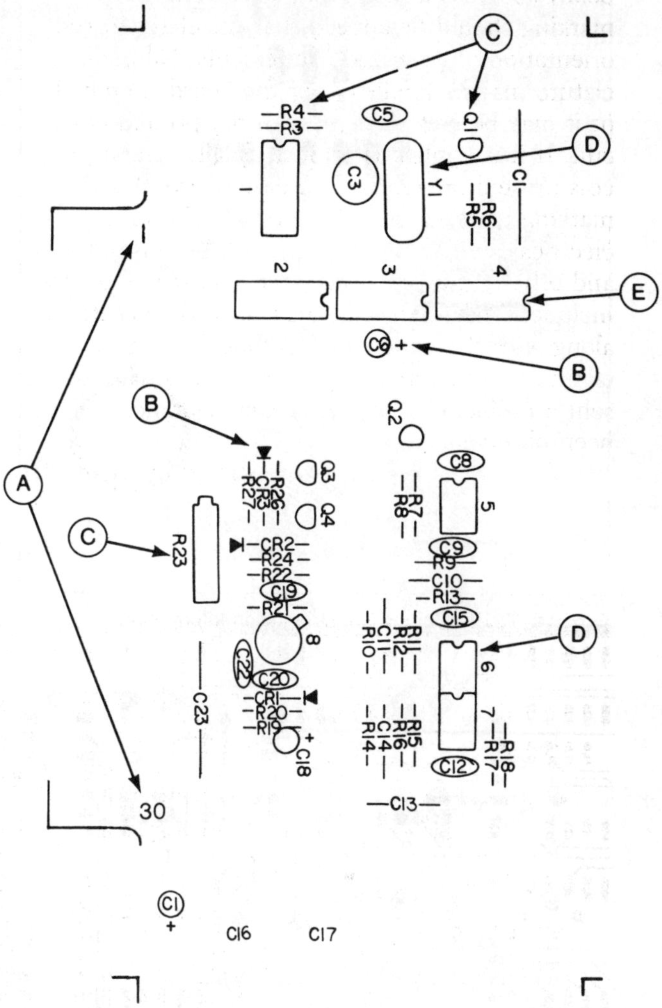

Figure 7-18. Artmaster, Component Side Silkscreen

Solder Mask

Solder masking is a coat of epoxy resin, covering all areas of the PC board, except pads which require soldering (see Figure 7-20).

With the current industry trend towards higher density boards with tighter line widths and spacing, solder masks are increasingly being used to eliminate bridging between adjacent conduc-

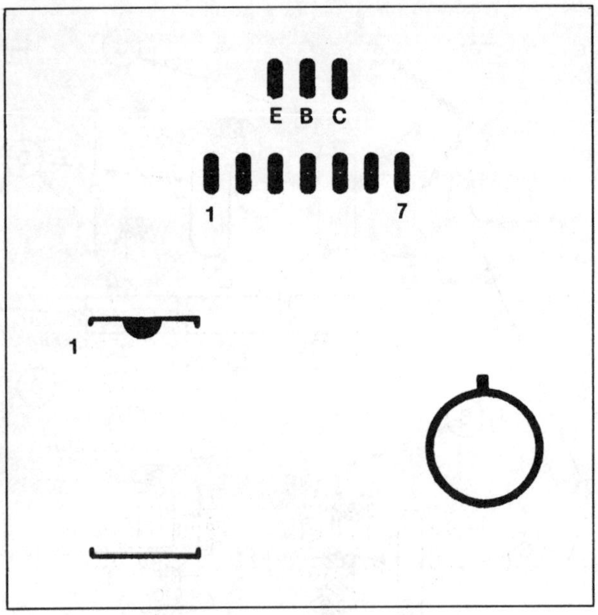

Figure 7-19. Component Orientation Information

tors during wave soldering.

Wave soldering is a process whereby component leads are soldered to a printed circuit board by traveling at a predetermined rate of speed through a wave of molten solder, created by pumping to form a head of solder (see Figure 7-21).

The movement of the solder as it wipes the noncomponent surface of the board can cause bridging between pads and traces. A solder mask on this side would eliminate the problem. However, when a solder mask is applied to a conventional circuit with the usual tin-lead overplate, the mask has a tendency to chip and peel when the overplate reflows under the mask (which is covering over the copper).

The solder mask will cover all areas except the pads. "Clearance" around these pads would be .005 – .015 inch. See Figure 7-22.

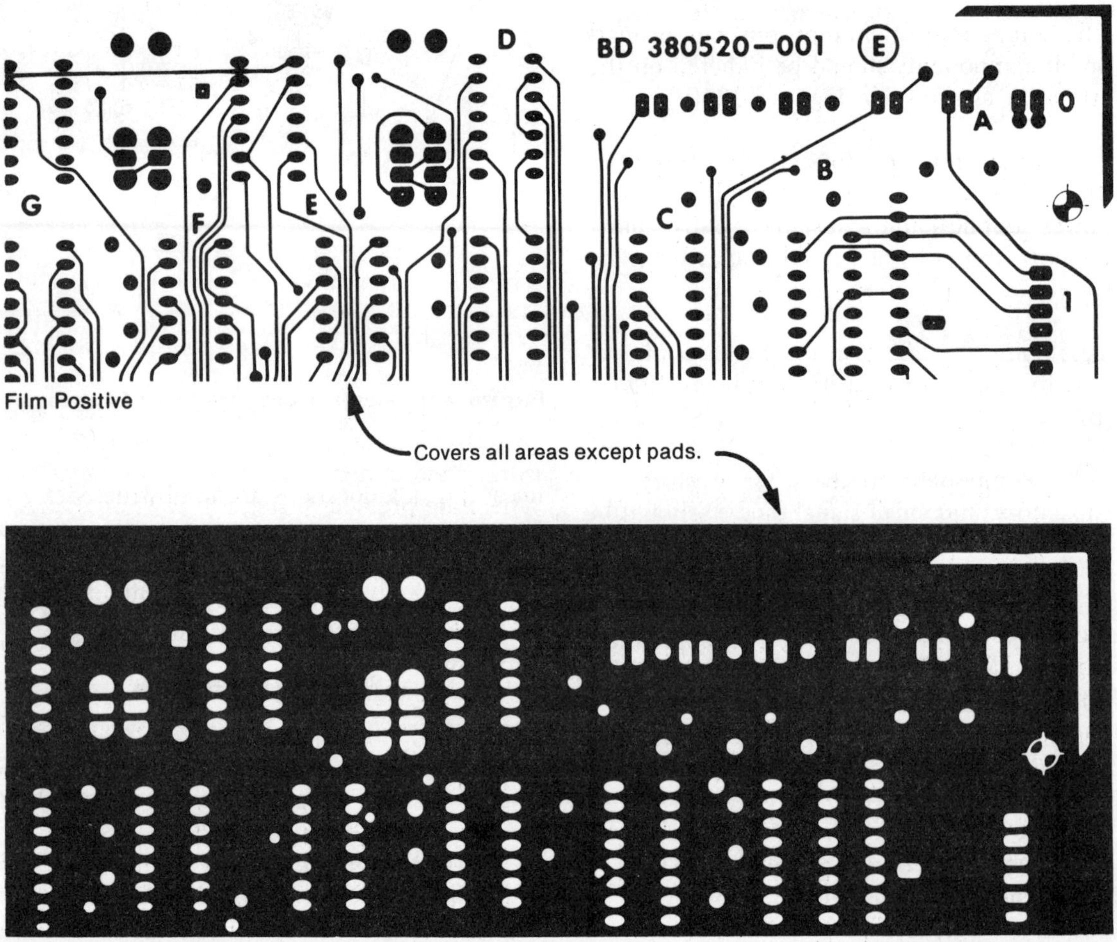

Figure 7-20. Solder Mask

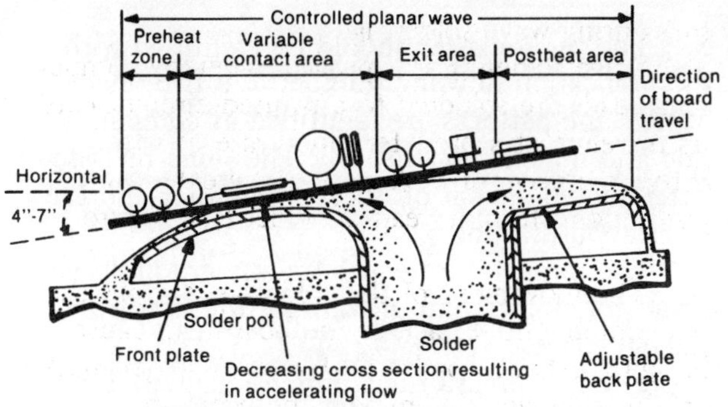

Figure 7-21. Wave Soldering

Solder masks are generally on the noncomponent or solder side of the board only. However, many companies place masks on both sides. Masks are created by a photographic process by using the pad master on a three-sheet taping technique (see Figure 7-23). The photographer creates the desired clearance around the pads photographically by using the pad master sheet and a special technique.

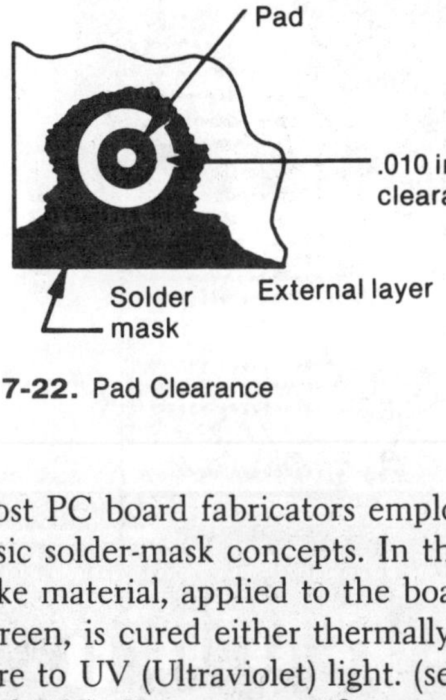

Figure 7-22. Pad Clearance

Most PC board fabricators employ one of two basic solder-mask concepts. In the first, a wet paint-like material, applied to the boards through a silkscreen, is cured either thermally or through exposure to UV (Ultraviolet) light. (see Figures 7-24 and 7-25). You can specify several different grades of these wet solder-mask materials. You should choose one of the higher grades when working with leakage-sensitive circuits.

The second solder-mask concept employs a dry photosensitive film consisting of a layer of polyethylene, a layer of photopolymer, and a layer of drafting film. A special machine applies this film by first removing the polyethylene layer,

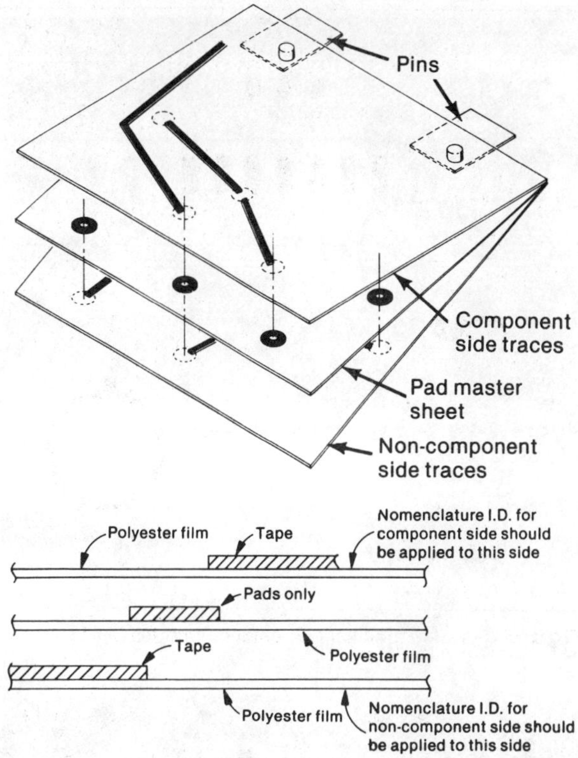

Figure 7-23. Three-sheet Taping Technique

then laminating the remaining two layers onto the PC board surface under heat and pressure. The board is then photographically exposed through a photographic mask, and the drafting film is removed. The developed photopolymer acts as the solder mask.

PC fabricators often use this dry-film approach in dense designs, where registration accuracy and dimensional stability are important. There are problems using the dry-film approach. If not properly applied, the film might not conform to the grooves between closely spaced conductors. The resulting cavities can trap impurities and moisture, resulting in serious leakage problems.

The solder mask is applied by screening or photo-printing. Figure 7-24 shows a typical screen. To create the images on the screen, the screen is painted with a red emulsion. The emulsion is then photographically exposed through the negative using ultraviolet rays. These rays make the exposed areas of the emulsion harden. The screen is processed, leaving only the hardened areas. This produces the image.

Figure 7-24. Applying the Solder Mask

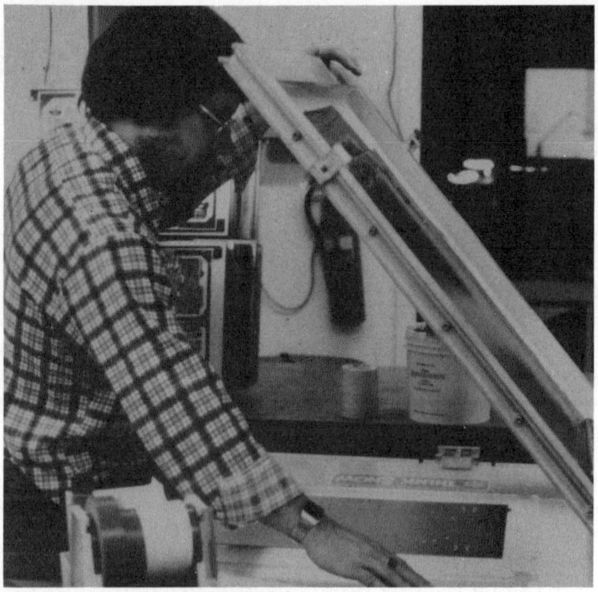

Figure 7-25. Board Positioning

Photomaster

A photomaster is a one-to-one scale PC board pattern used to produce a printed circuit board within the accuracy specified on the fabrication drawing. See Figure 7-26 for an example of a photomaster positive and negative. One of the most important steps in printed circuit board photo-fabrication is the artwork photography since the quality of the finished part depends directly on the quality of the photomaster used to produce the PC board.

The photomaster should be identified with the fabrication drawing number. On two-sided boards, the patterns are identified as component side and noncomponent side. The same revision letter should appear on the photomaster and the fabrication drawing.

The temperature and humidity are important either for storing film or transporting film from place-to-place.

The photomaster is a dimensionally stable base material (typically polyester film). The photosensitive coating should have high contrast, resolution and stability.

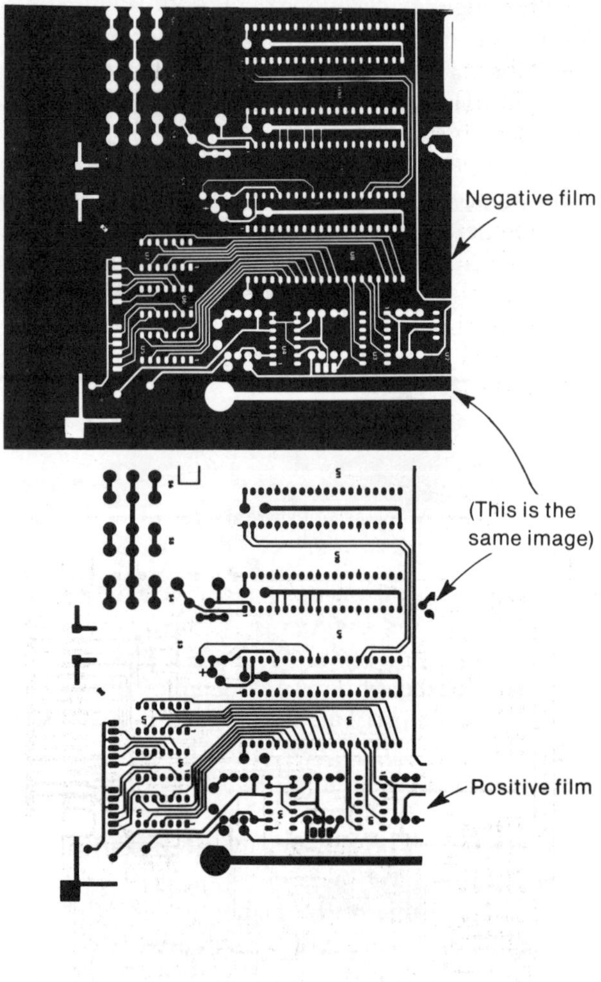

Figure 7-26. Photomaster Negative (Top) and Positive (Bottom)

Check List

1. The size of reduction: Each photomaster should be measured to the reduction marks to ensure proper reduction and size (see Figure 7-27).

2. All photomasters (films) required to produce PC boards must be inspected together to ensure registration from the component side to the noncomponent side.

3. Sharpness and density: The image on the film must be opaque enough to ensure that proper exposure will produce sharp conductor edges.

4. Check: Photomasters should be inspected for pin holes, scratches, fingerprints, and general cleanliness.

5. Storage: Store film in an area where the temperature and humidity remain relatively constant. Don't expose to temperatures in excess of 120°F (40°C) or below 33°F (1°C). Film should be transported in flat closed containers to eliminate scratches or damage to it surface.

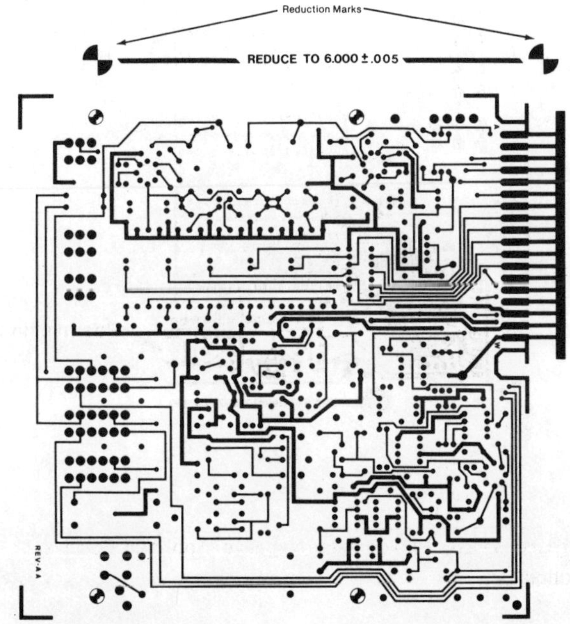

Figure 7-27. Photomaster Reduction Marks

Quiz—Chapter 7

1. Block diagrams are reference drawings which show functional groups of circuits as block symbols. T F
2. Block diagrams have lines interconnecting these blocks which do not represent electrical flow. T F
3. The schematic diagram shows the electrical connections and functions of a specific circuit. T F
4. The schematic: since the function purpose of this document is electronic, it shows mechanical details. T F
5. Schematic's use reference designations to distinguish the different interconnections. T F
6. Logic symbols are as precise and clear as the elementary schematic symbols. T F
7. The logic diagram shows the exact electrical value and type of components that are used in the design. T F
8. The detail fabrication drawing of the P.C. boards depicts the board size, holes & specifies the material. T F
9. Normally the detail fabrication drawing is drawn at the same scale as the photomaster. T F
10. Detail fabrication drawings should show all dimensions to fabricate the P.C. board including those features located by the photomaster. T F
11. The detail fabrication drawing should show the ____________ side of the board.
12. The detail fabrication drawing should always include workmanship criteria but not test qualifications. T F
13. A hole schedule describes the hole size, tolerance, and minimum plating thickness. T F
14. Assembly drawing: the primary purpose of this document describes the electrical flow. T F
15. Assembly drawing should depict the components, references markings, and test specs. T F

16. The parts comprising an assembly are listed on a ________________ ________________
17. The assembly drawing is normally drawn in the same scale as the ________________
18. The assembly drawing is drawn as viewed from the ____________ side.
19. Assembly drawings should show polarity for all polarized components. **T F**
20. Silkscreen artmasters provides a tool for marking the printed circuit board. **T F**
21. The silkscreen artmaster should be made to the same scale as the ____ ________________ artmaster.
22. The silkscreen artmaster must be dimensioned. **T F**
23. Silkscreen artmaster fabrication lines should correspond with those on the ________________ ________________ artmaster.
24. Board marking requires all components to be marked permanently and legibly with no exceptions. **T F**
25. The decision to provide board marking is based primarily on the ________________ of that board.
26. Marking can be accomplished by etching, screening or __________
27. Marking: some companies display their company logo's. **T F**
28. Silkscreen artmaster: when preparing artwork it is not necessary to draft the body for small components. **T F**
29. Solder masking is a coat of ________________, resin, covering all areas of the board except pads.
30. Solder masks are used to control a bridging problem created by the wave soldering process. **T F**
31. Solder masks are on the noncomponent side only. **T F**
32. Most fabricators employ only one method to apply solder mask, silkscreening. **T F**
33. The photomaster is the most important step in the process up to the beginning of the fabrication process. **T F**
34. When storing or transporting film (photomaster) the temperature and ________________ are important.
35. Photomasters should be inspected for scratches, pinholes & proper ________________ in size.

The answers for each question can be found on the pages listed below:

Question 1. Pg. 185	Question 19. Pg. 192
Question 2. Pg. 185	Question 20. Pg. 192
Question 3. Pg. 185	Question 21. Pg. 192
Question 4. Pg. 185	Question 22. Pg. 196
Question 5. Pg. 185	Question 23. Pg. 196
Question 6. Pg. 185	Question 24. Pg. 196
Question 7. Pg. 185	Question 25. Pg. 197
Question 8. Pg. 187	Question 26. Pg. 198
Question 9. Pg. 187	Question 27. Pg. 198
Question 10. Pg. 187	Question 28. Pg. 199
Question 11. Pg. 187	Question 29. Pg. 200
Question 12. Pg. 188	Question 30. Pg. 201
Question 13. Pg. 188	Question 31. Pg. 201
Question 14. Pg. 190	Question 32. Pg. 201
Question 15. Pg. 190	Question 33. Pg. 202
Question 16. Pg. 191	Question 34. Pg. 202
Question 17. Pg. 191	Question 35. Pg. 202
Question 18. Pg. 191	

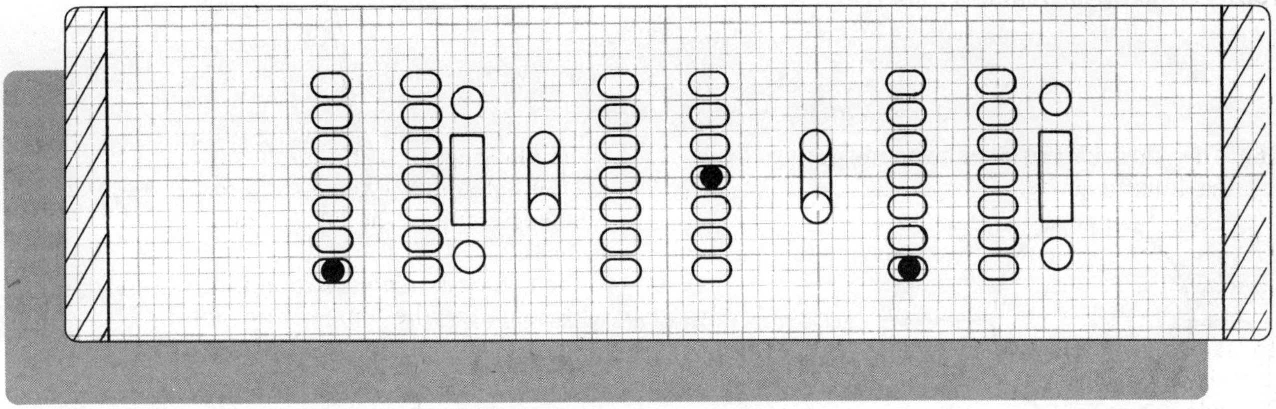

8

3-IC DIGITAL BOARD

This chapter illustrates the design of a 3-IC digital printed circuit board layout. You will learn to design from schematics, know what questions to ask the engineer, and how to use specification work sheets. You will determine component lead spacing. Component layout design rules are presented. Space requirements to fit the components of the schematic within the board outline will be discussed. We will show you several ways to group and plot a schematic, using printed circuit templates. Each is explained while you construct the formal digital printed circuit design.

Supply List Needed For Chapter Eight

1. Component Template: Bishop Graphics Catalog No. 3367 (one)
2. Black Drafting Film Lead Pencil: Mars 100-30 Duralar (one)
3. Colored Pencils (dark colors): Venus Col-erase or equivalent (red, green, yellow, blue), one each
4. Eraser, Drafting Film: Castell #1954 or equivalent (one)
5. Drafting Dots

This is a flowchart of what happens during the design concept phase.

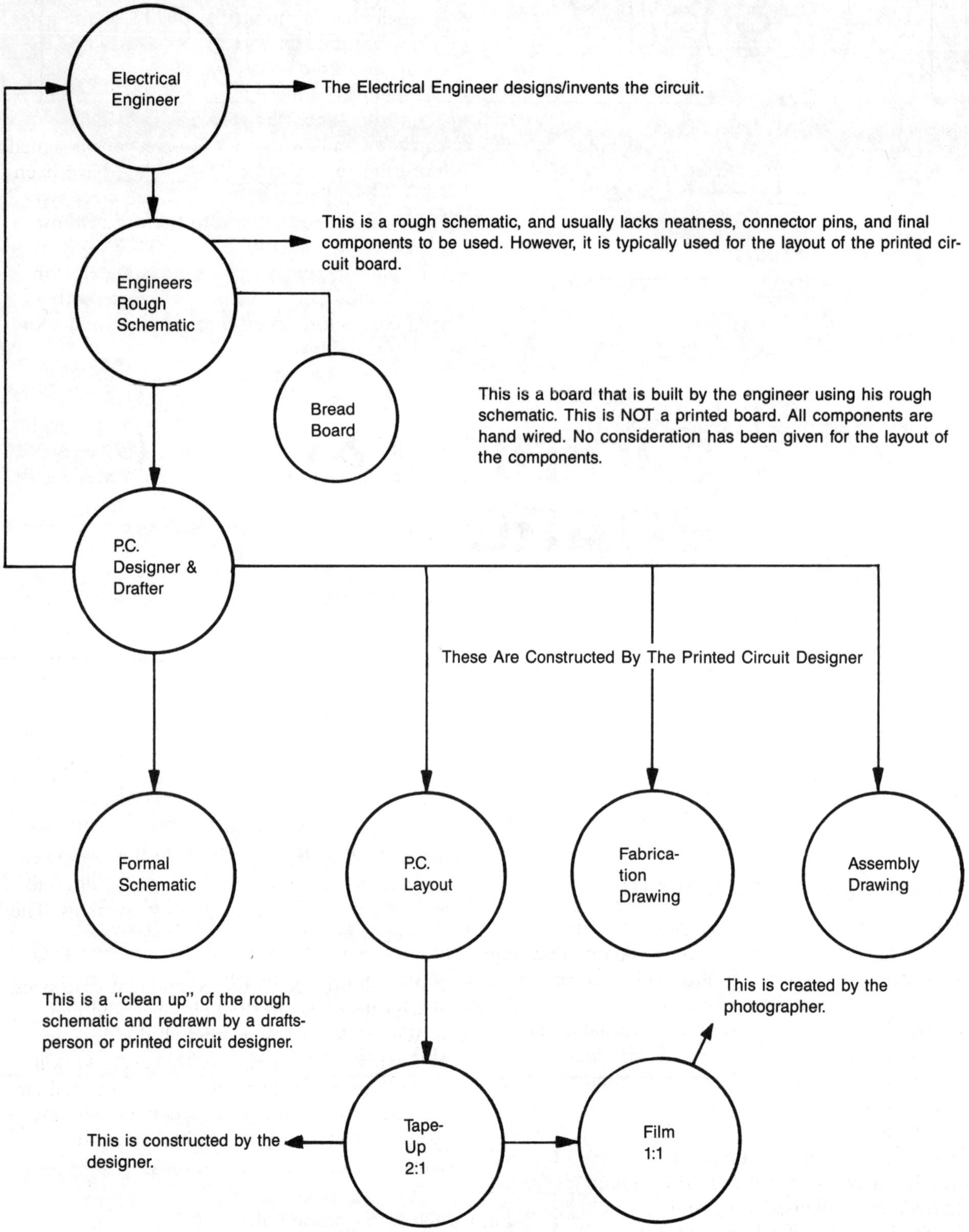

Figure 8-1. Design Concept Phase

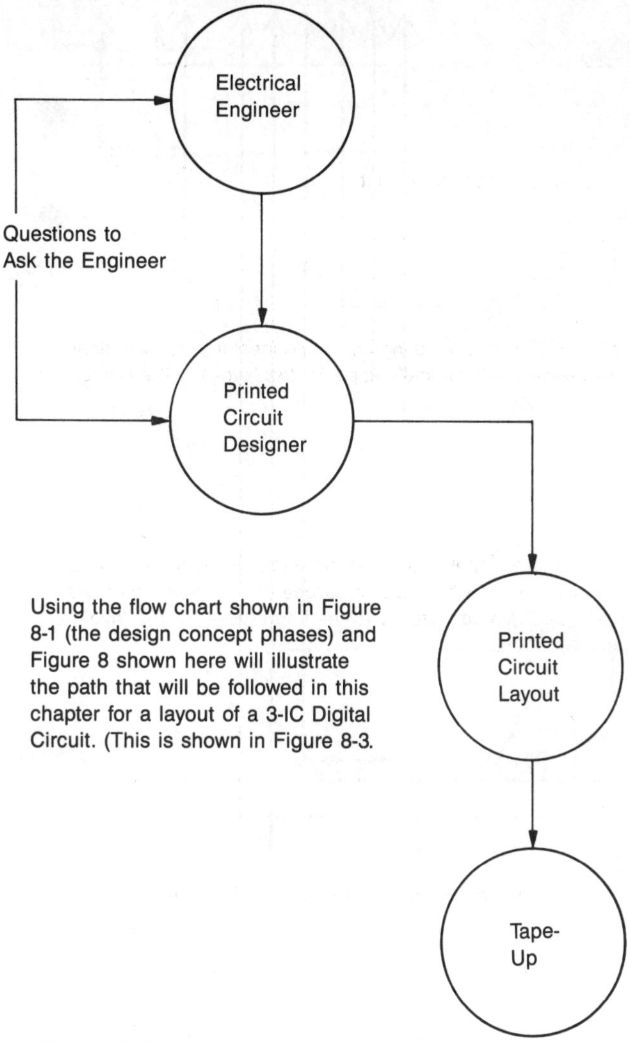

Figure 8-2. Layout

From Sketch to Layout

The flow chart in Figure 8-2 illustrates the process of laying out a printed circuit board. The steps listed below simplify the design process.

Step #1—Schematics should be readable: all symbols, reference designations, and identification of all logic gates used by the engineer. (See the schematic of Figure 8-3).

Examine the schematic in Figure 8-3. Identify all gates with the manufacturer's specification number. Note 1 identifies four different specification sheet numbers. As mentioned in chapter 2 (under logic identification), the last two digits becomes the main identifier. The first two digits belong to the IC series. Specification sheets are provided by the engineer. The schematic (Figure 8-3) usually has no numbers assigned to the gate inputs or outputs. If numbers were assigned to the gates inputs and outputs, we would recommend removal of these numbers from the schematic at this time. The reason is that the engineer's numbers that you would have assigned to the gates inputs and outputs would have been used in a breadboard design. The engineer typically would not restrict you to the use of those numbers.

P1 is on Figure 8-3 is a male connector on the PC board. This connector will mate with a female connector. An example of this connector is shown in Figure 4-16.

If we use the same connector on the PC board as shown in chapter 4, we can use the specification sheet to get the pin numbers of the connector as we assign them to the PC board. Figure 8-3 does not have pre-assigned pins for P1. Because the schematic could have pre-assigned pins, this is one of the questions we will ask the engineer.

The schematic is not a complete source of information. We do not have all the mechanical data required either for the component case sizes or for the board size. Many questions must be asked of the engineer before proceeding into the PC layout.

Notice the letters assigned on the schematic inside the flip-flop (called a 74-73). These letters will help assign the pin numbers to the inputs and outputs of the flip-flop.

Step #2—Specification sheets identify the case and body size, width and length of all discrete components, and all logic gates or symbols. The engineer provides the specification sheets.

We will provide all the component case sizes, lead sizes (to compute the hole size), and pad size. We will learn how to find all the discrete components on the template. However, we will provide the specification sheets for integrated circuits. Illustration 1 provides all component case sizes for discrete components and integrated circuits.

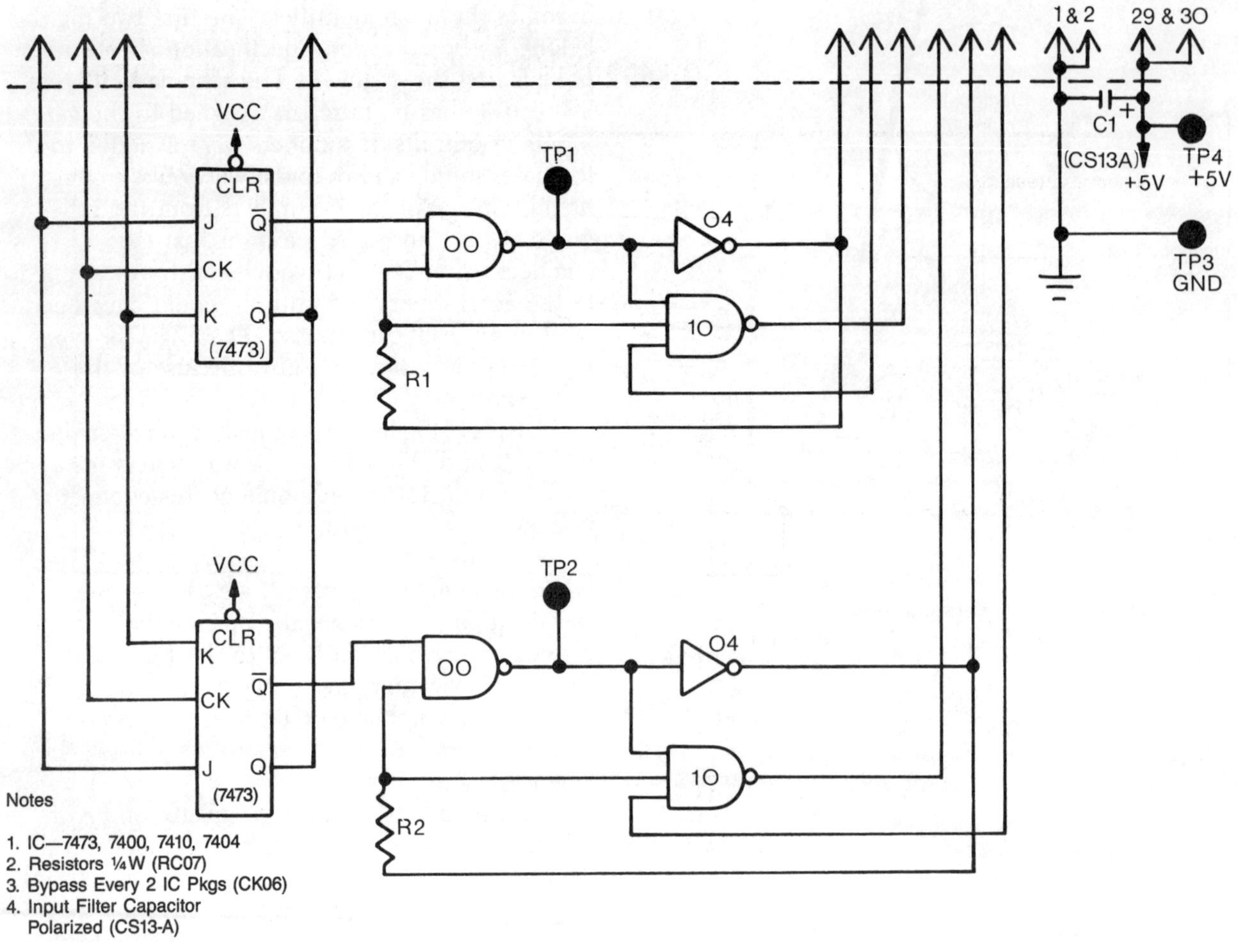

Figure 8-3.

Step #3—Determine the component lead spacing and body size. Space the discrete component pads where the leads insert into the board according to the specification sheets. (This information can be reviewed in chapter 3, of the Analog Printed Circuit Design & Drafting book). Determine the IC body size, such as 14-, 16-, 24-pin, etc. Identify each component on the schematic (Figure 8-3) using the specification sheets. See Illustration 1.

Illustration #1: Component Identification

Refer to the notes in Figure 8-3. Note 1 identifies the specification sheets that we must obtain from the engineer. See Figures 8-4 and 8-5 for these specifications. Figure 8-4 shows a complete specification sheet removed from a manufacturer's catalog. This is what you would usually receive from the engineer. Figure 8-5 gives the designer the minimum useful information from the engineer. Look at case style. Figure 8-4 indicates that there are two styles to choose from: the flat pack and the dual inline package. Cross out the flat pack style. This leaves the dual inline package shown in Figure 8-6 which shows two case sizes, 14- and 16-pin. The specification sheets (Figures 8-4 and 8-5) identify the number of pins on the package. Note that all IC's are 14-pin.

Review chapter 3 for the different style packages and case sizes. Figure 3-7 gives the mechanical data required to compute pad spacing, hole size, and pad size for the dual inline.

CIRCUIT TYPES SN5400, SN7400
QUADRUPLE 2-INPUT POSITIVE NAND GATES

schematic (each gate)

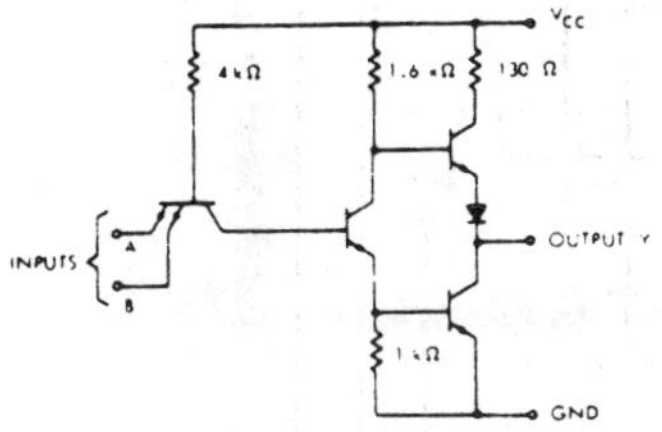

NOTE: Component values shown are nominal.

W FLAT PACKAGE (TOP VIEW)

J OR N DUAL-IN-LINE PACKAGE (TOP VIEW)

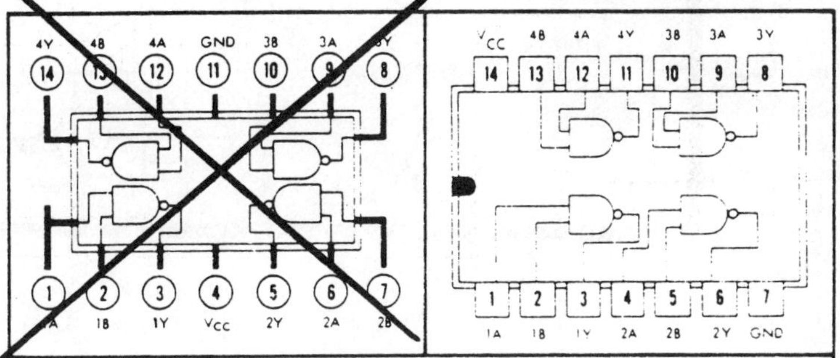

positive logic: $Y = \overline{AB}$

recommended operating conditions

	MIN	NOM	MAX	UNIT
Supply Voltage V_{CC}: SN5400 Circuits	4.5	5	5.5	V
SN7400 Circuits	4.75	5	5.25	V
Normalized Fan-Out From Each Output, N			10	
Operating Free-Air Temperature Range, T_A: SN5400 Circuits	−55	25	125	°C
SN7400 Circuits	0	25	70	°C

electrical characteristics over recommended operating free-air temperature (unless otherwise noted)

PARAMETER		TEST FIGURE	TEST CONDITIONS†			MIN	TYP‡	MAX	UNIT
$V_{in(1)}$	Logical 1 input voltage required at both input terminals to ensure logical 0 level at output	1				2			V
$V_{in(0)}$	Logical 0 input voltage required at either input terminal to ensure logical 1 level at output	2						0.8	V
$V_{out(1)}$	Logical 1 output voltage	2	V_{CC} = MIN, I_{load} = −400 μA	V_{in} = 0.8 V,		2.4	3.3		V
$V_{out(0)}$	Logical 0 output voltage	1	V_{CC} = MIN, I_{sink} = 16 mA	V_{in} = 2 V,			0.22	0.4	V
$I_{in(0)}$	Logical 0 level input current (each input)	3	V_{CC} = MAX,	V_{in} = 0.4 V				−1.6	mA
$I_{in(1)}$	Logical 1 level input current (each input)	4	V_{CC} = MAX,	V_{in} = 2.4 V				40	μA
			V_{CC} = MAX,	V_{in} = 5.5 V				1	mA
I_{OS}	Short-circuit output current§	5	V_{CC} = MAX		SN5400	−20		−55	mA
					SN7400	−18		−55	
$I_{CC(0)}$	Logical 0 level supply current	6	V_{CC} = MAX,	V_{in} = 5 V			12	22	mA
$I_{CC(1)}$	Logical 1 level supply current	6	V_{CC} = MAX,	V_{in} = 0			4	8	mA

switching characteristics, V_{CC} = 5 V, T_A = 25°C, N = 10

PARAMETER		TEST FIGURE	TEST CONDITIONS		MIN	TYP	MAX	UNIT
t_{pd0}	Propagation delay time to logical 0 level	65	C_L = 15 pF,	R_L = 400 Ω		7	15	ns
t_{pd1}	Propagation delay time to logical 1 level	65	C_L = 15 pF,	R_L = 400 Ω		11	22	ns

† For conditions shown as MIN or MAX, use the appropriate value specified under recommended operating conditions for the applicable device type.
‡ All typical values are at V_{CC} = 5 V, T_A = 25°C.
§ Not more than one output should be shorted at a time.

Figure 8-4. Typical Specification Sheet

7404

HEX INVERTERS

positive logic:

$Y = \bar{A}$

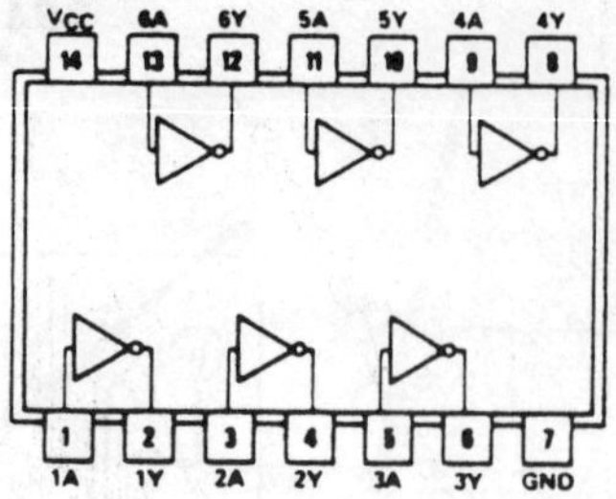

7410

TRIPLE 3-INPUT
POSITIVE-NAND GATES

positive logic:

$Y = \overline{ABC}$

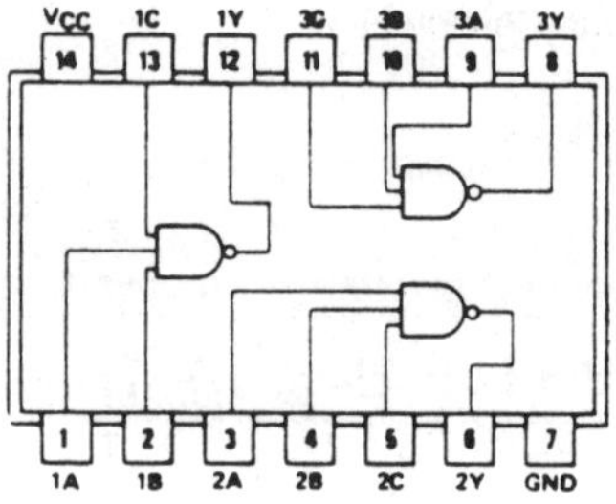

DUAL J-K FLIP-FLOPS WITH CLEAR

73

'73, 'H73, 'L73
FUNCTION TABLE

INPUTS				OUTPUTS	
CLEAR	CLOCK	J	K	Q	$\bar{Q}$
L	X	X	X	L	H
H	⎍	L	L	Q_0	$\bar{Q}_0$
H	⎍	H	L	H	L
H	⎍	L	H	L	H
H	⎍	H	H	TOGGLE	

'LS73
FUNCTION TABLE

INPUTS				OUTPUTS	
CLEAR	CLOCK	J	K	Q	$\bar{Q}$
L	X	X	X	L	H
H	↓	L	L	Q_0	$\bar{Q}_0$
H	↓	H	L	H	L
H	↓	L	H	L	H
H	↓	H	H	TOGGLE	
H	H	X	X	Q_0	$\bar{Q}_0$

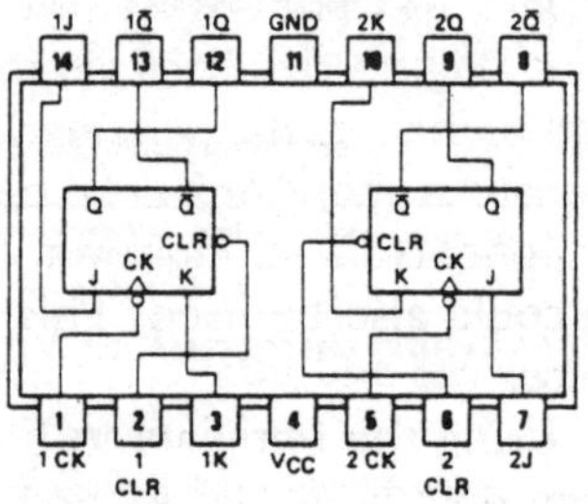

SN5473 (J, W) **SN7473 (J, N)**
SN54H73 (J, W) **SN74H73 (J, N)**
SN54L73 (J, T) **SN74L73 (J, N)**
SN54LS73 (J, W) **SN74LS73 (J, N)**

Figure 8-5. Specification Information

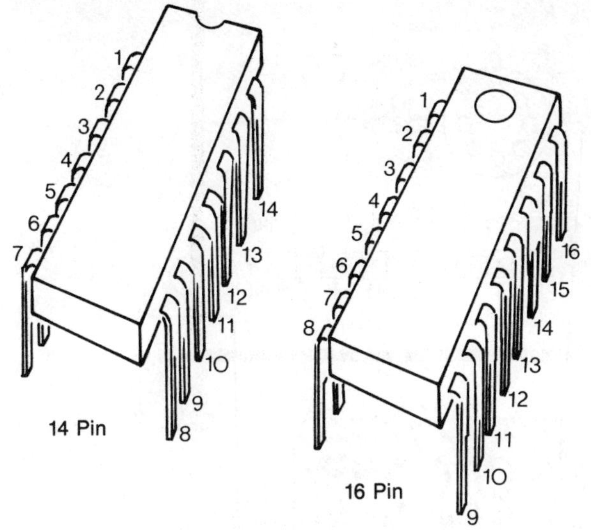

Standard Package Configuration

Figure 8-6. Dual Inline Package

In this case, the resistor is called out as a ¼W and given a case code of RC07.

Refer to the analog design book. To obtain the case size RC07, page 48 Figure 3-9, you need the following information.

Type:
Fixed or Variable

Composition:
What is the resistor made of? Carbon, metal film, wire wound, etc.

Rating:
Expressed in Wattage. ¼ Watt, ½ Watt, etc.

Tolerance:
Expressed in percentage. ±2%, ±5%, etc.

Value:
Expressed in ohms.

Review Figures 5-10 and 5-12 in chapter 5. They show the recommended sizes for low, medium, and high densities for traces.

Figure 8-3 is a low density layout (which you will calculate later on in this chapter). Look at the pad configuration for Figure 5-12 (called low trace density). Compare this to Figure 8-7, a dual inline template with each pattern identified. This template indicates a .070 by .125 pad size for low trace density. Figure 8-8 (PC Design Template I), has many discrete components, dual inline, and flat pack pad patterns. Use this template for the layout (Figure 8-8) because there are discrete components on the schematic in Figure 8-3. This template is more universal because it has more components on it. However, IC template (Figure 8-7) could also be used. This becomes a personal choice.

Using the template in Figure 8-8, for our 14-pin dual inline IC pattern, draft the 14-pin dual inline pad pattern onto a grid pattern (shown in Figure 8-9). Assign the lead diameter for each pin, which will be required to compute the hole size and pad size for the dual inline. Using the schematic in Figure 8-3 note 2, identify the case size of the resistor and obtain the lead diameter.

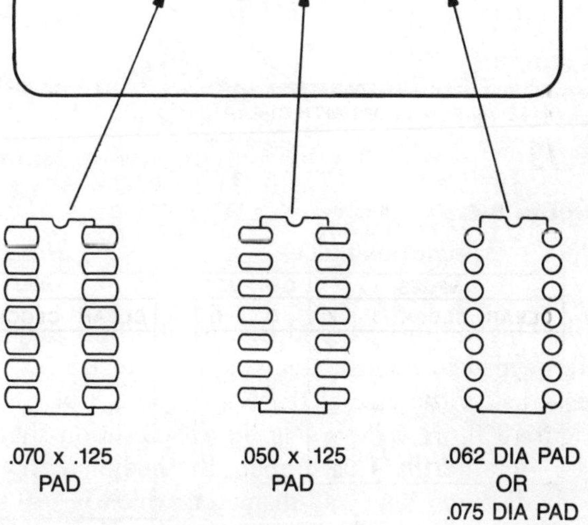

Figure 8-7. Dual Inline Template

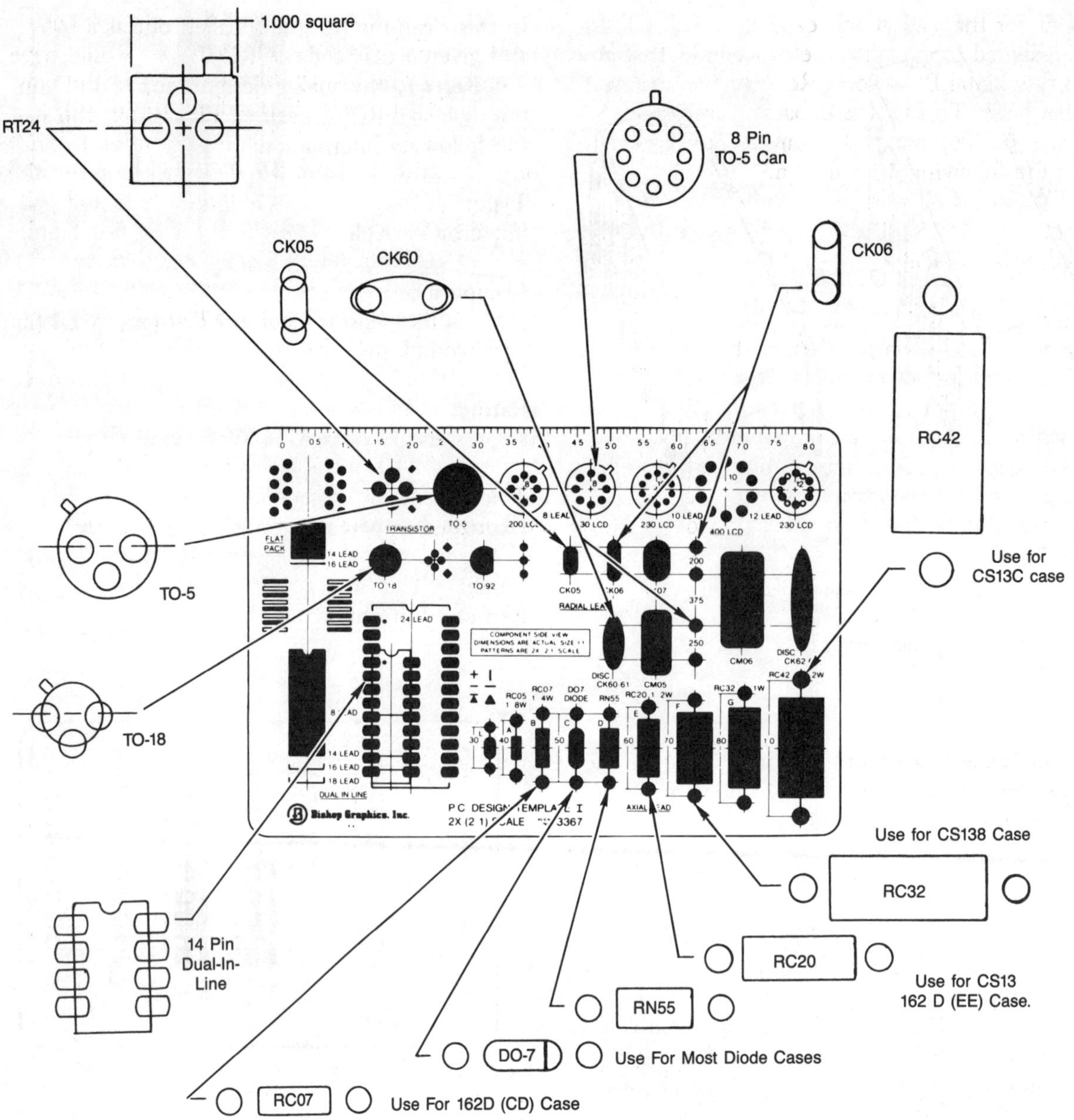

Figure 8-8. P.C. Template

This information and the specification sheets are needed to find the component case size for resistors. In the case of the RC07, (note 2 of Figure 8-3), we will provide the specification sheet (see Figure 8-10). This supplies the body length, body diameter, and lead diameter, which is necessary to compute the pad spacing (if we did not have the RC07 case code on the template). Refer to Figure 8-8. Notice that the RC07 outline is provided on the template. We have drafted a RC07 resistor outline onto the grid pattern in Figure 8-9 and have provided the lead diameter from the resistor specification sheet (Figure 8-10).

Refer to Figure 8-3, note 3. This note discusses bypass capacitors. Use one bypass capacitor for every two IC's on the digital PC board layout. This has been discussed in chapter 3 of this book. It makes reference to a case

(CK06) for the size of this capacitor. The CK06 was assigned to expedite the procedures that go into the digital PC layout. Refer to the analog design book. To find the capacitor case code number (CK06), page 60 Figure 3-28, we must have the following information.

Type:
Fixed or Variable.

Materials:
This is the body composition, such as ceramic, mylar, metal foil, or dip mica, etc.

Electrical description of Component:
Value - expressed in farads. Rating - expressed by voltage. Tolerance - expressed by percentage. This and the specification sheet provides a case code number.

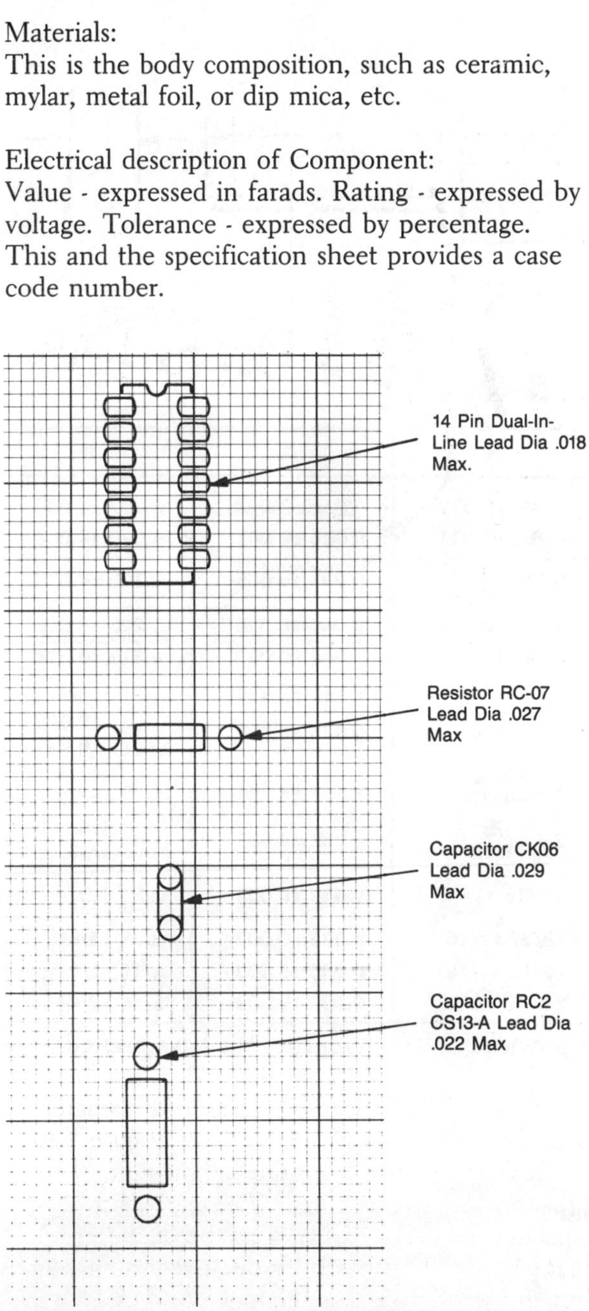

Figure 8-9. Component Outline Drafted on Grid

The template in Figure 8-8 shows a CK06 body outline. If it were not there, we would have to obtain a specification sheet similar to the one provided for the resistor (Figure 8-10). In this exercise, you will be given the body outline to draft onto the grid in Figure 8-9. You will be given the lead diameter required to compute hole and pad size. Refer to the schematic in Figure 8-3. Note 4 calls for an input filtering capacitor that is polarized. The engineer has called out a case code number of CS13A. Typically, this case code would not be provided by the engineer. But a value, rating and tolerance would be given on the schematic. As we've just discussed in the information in note 3, along with the composition (ceramic, mylar, metal foil, or dip mica) would have to be known. Examine the PC template (Figure 8-8). There is no case outline identified as a CS13A. Use the specification sheet and formula in the analog design book to compute pad spacing. See the specification sheet in Figure 8-11, the CS13A case outline, along with the formula to compute pad spacing, hole diameter, and pad size (see Figure 8-12).

Problem: Find component pad spacing for C1, the CS13A capacitor.

Procedure #1:

Using your specification sheet (Figure 8-11), find the dimension block that has the case size listed A, B, C, and D. We are using a CS13 family. Note that the specification sheet calls out CSR13, which is exactly the same as CS13.

Problem: Find case size A (CSR13-A).

Procedure #2:

Find the maximum body length listed as "J" under the dimension block on Figure 8-11. "J" calls out a maximum body length of .422.

Procedure #3:

Using the formula (Figure 8-12, note 1), notice that component pad spacing requires the addition of the maximum body length .422 to the .060 minimum on each side. (Each side means there are two ends to it). Add .060 twice to the max-

INSTITUTE OF PRINTED CIRCUITS
1717 HOWARD STREET EVANSTON, ILLINOIS 60202

IPC

PRINTED WIRING DESIGN GUIDE

NUMBER	
7.1.1.1	
SUBJECT	
Axial Lead Resistors	
DATE	REVISION
8/70	–

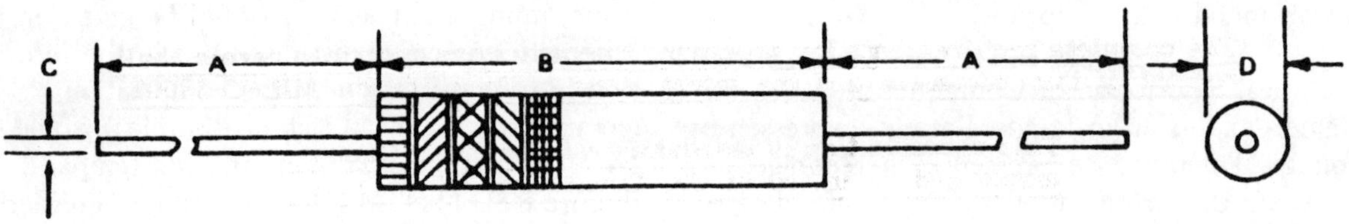

FIGURE 1. Axial Lead Resistor

LEAD DIA

SPECIFICATION	STYLE	A	B	C	D
MIL-R-11 1/10W	RC06	1 ±1/8	0.140 ±0.020	0.015 ±0.003	0.067 ±0.008
MIL-R-11 1/4W	RC07	1-1/2 ±1/8	0.250 ±0.031	0.025 ±0.002	0.090 ±0.008
MIL-R-11 1/2W	RC20	1-1/2 ±1/8	0.375 +0.041 −0.031	0.031 ±0.005	0.138 ±0.023
MIL-R-11 1W	RC32	1-1/2 ±1/8	0.562 ±0.031	0.040 ±0.005	0.225 ±0.015
MIL-R-11 2W	RC42	1-1/2 ±1/8	0.688 ±0.040	0.045 ±0.003	0.318 ±0.018
MIL-R-10509	RN55	1-1/2 ±1/8	1/4 ±1/32	0.025 ±0.002	7/64 ±1/32
MIL-R-10509	RN60	1 1/2 ±1/8	3/8 ±1/16	0.025 ±0.002	0.125 +0.040 −0.031
MIL-R-10509	RN65	1 1/2 ±1/8	5/8 +1/32 −3/32	0.025 ±0.002	3/16 +1/16 −1/32
MIL-R-10509	RN70	1-1/2 ±1/8	3/4 +1/8 −1/16	0.032 ±0.002	1/4 +5/64 −1/32
MIL-R-10509	RN75	1-1/2 ±1/8	1-1/16 ±1/16	0.032 ±0.002	3/8 +1/16 −1/32
MIL-R-11804	RD60	1-1/2 ±1/8	19/32 ±1/16	0.033 ±0.003	3/16 ±3/64
MIL-R-11804	RD65	1-1/2 ±1/8	15/16 ±1/16	0.033 ±0.003	5/16 ±3/64
MIL-R-11804	RD70	1-1/2 ±1/8	2-1/16 ±1/16	0.033 ±0.003	5/16 ±3/64

(ALL DIMENSIONS IN INCHES)

STYLE
RB - Wirewound
RC - Composition
RD - Film (power)
RE - Wirewound (power)
RL - Film
RN - Film (hi stability)

Figure 8-10. Spec Sheet

MIL-C-39003/1C
15 May 1970
SUPERSEDING
MIL-C-39003/1B
26 March 1970

MILITARY SPECIFICATION SHEET

CAPACITORS, FIXED, ELECTROLYTIC, TANTALUM, SOLID-ELECTROLYTE,

ESTABLISHED RELIABILITY, STYLE CSR13

The complete requirements for procuring the capacitors described herein shall consist of this document and the latest issue of Specification MIL-C-39003.

This specification is mandatory for use by all Departments and Agencies of the Department of Defense.

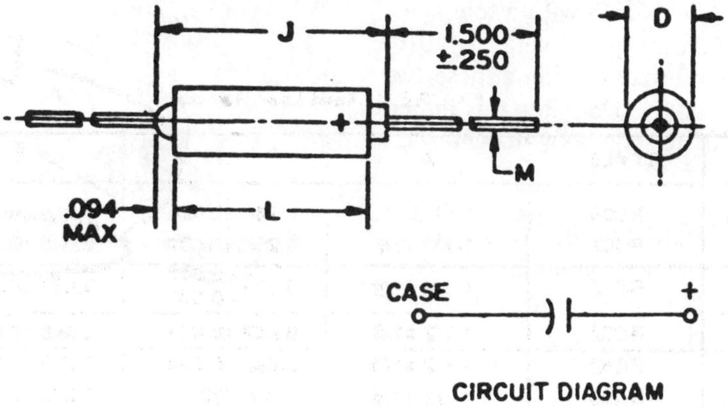

Case size	Dimensions			
	L ±.031	D +.016 −.015	M ±.002	J Max
A	.286	.135	.020	.422
B	.474	.185	.020	.610
C	.686	.289	.025	.822
D	.786	.351	.025	.922

INCHES	MM	INCHES	MM	INCHES	MM
.002	.05	.135	3.43	.474	12.04
.015	.38	.185	4.70	.610	15.49
.016	.41	.250	6.35	.686	17.42
.020	.51	.286	7.26	.786	19.96
.025	.64	.289	7.34	.822	20.88
.031	.79	.351	8.92	.922	23.42
.094	2.39	.422	10.72	1.500	38.10

NOTES:
1. Dimensions are in inches.
2. Metric equivalents (to the nearest .01 mm) are given for general information only and are based upon 1 inch = 25.4 mm.
3. The case insulation shall extend .015 (.38 mm) minimum beyond each end. However, when a shrink-fitted insulation is used, it shall lap over the ends of the capacitor body.

Figure 8-11. Spec Sheet

imum body length which equals .542 minimum from center line of one lead to the center line of another:

.422 max. body length
+.060 min. bend length
+.060 min. bend length
.542

Round up to the nearest grid at a scale of 2 x 1. .542 is at a scale of 1:1. Double .542 to equal 1.084. Rounding up to the nearest grid equals 11 grids or 1.100. 11 grids is the minimum pad spacing for a CS13A electrolytic capacitor.

Procedure #4:

Find the maximum body diameter on Figure 8-11. Labeled "D" in the dimensions block for a case size A. Notice that "D" is .135 with a tolerance of +.016, − .015. Add the plus tolerance, (.016) to the "D" dimension of .135. Adding these two together equals .151 at 1:1. Doubling this dimension (.151) equals .302 for the body diameter at a scale of 2:1.

Conclusion: The body diameter is .302 (or roughly 3 grids at 2:1 with a minimum pad spacing of 11 grids). Find a component outline on the template that is close to the pad spacing and body dimensions. Examine the template in Figure 8-8. Notice that the RC20 resistor outline accommodates the 2:1 dimensions that we have just calculated. The RC20 has 12 grids from center of pad to center of pad. 11 grids was minimum; it has a 3-grid diameter body at 2:1. Using the RC20 outline for the capacitor CS13A, draft it onto the grid pattern in Figure 8-9.

Indicate the lead diameter. The lead diameter is obtained from the specification sheet in Figure 8-11 labeled "M" in the Dimensions Block. The lead diameter was called out as .020 ± .002. Adding the tolerance to the dimension would be .020 + .002 = .022. The maximum lead diameter is required to compute the hole and the pad sizes.

The template is only a guide.

NOTE: It is not a precision instrument. It only shows the rough body outline and grid intersection lines for the lead pads, not the actual pad size.

All the components case sizes are determined. Find the lead size diameter of each component. Normally we would use a specification sheet to find this. However, we will give these sizes in this exercise. The whole formula is in Figure 8-12. This formula is adequate for good commercial practices. For the military hole diameter formula (Mil. standard practice), see Figure 8-13. These two formulas give an example of how to take a specification sheet, find the lead size, and compute the minimum and maximum hole sizes. In this chapter we will use this procedure. This is why we are giving you the lead sizes in Figure 8-9.

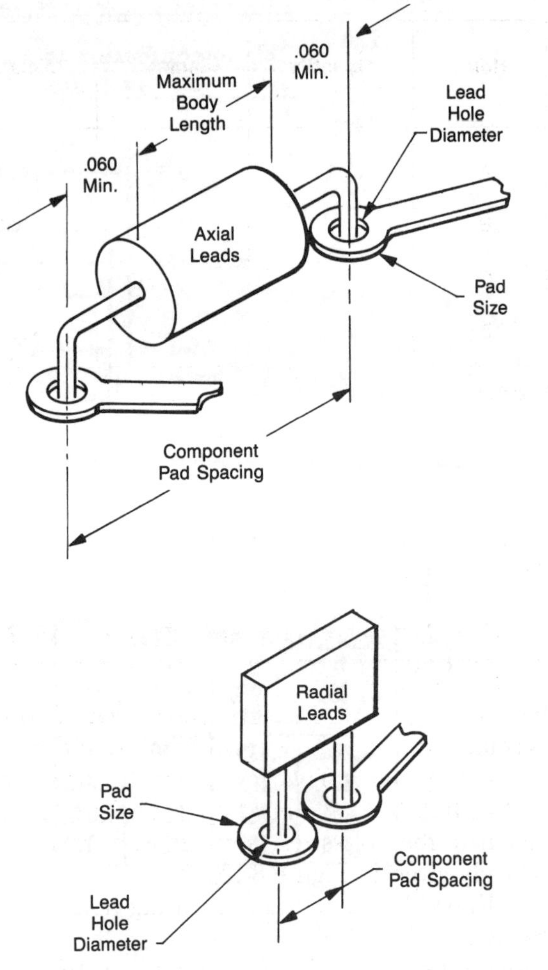

Figure 8-12. Component Mounting Information

Equation:

1. Max lead − Minimum clearance = Minimim Hole Size
2. Max lead + Minimum clearance = Maximum Hole Size

Definition of Equation:

Max Lead = Lead Diameter + Tolerance
Minimum Clearance = .004 inch over maximum lead diameter
Maximum Clearance = .028 inch over maximum lead diameter

Hole	Quantity	Diameter	Tolerance
A	2	.375	±005
B	13	.093	±002
C	7	.125	±003
D	All other	.040	±003

Figure 8-13. Hole Diameter Formula: Mil Standard Practice

With these lead sizes, compute the hole size. Table 8-1 gives the pad size.

Explanation of Table 8-1

This table gives an itemized breakdown of pad spacing, mounting hole requirements, and pad size requirements. Follow each step as described. Notice that the table has steps 1-13. The formulas calculate the hole sizes from the lead sizes (already given in Figure 8-12).

Know the board manufacturing hole tolerance. Every hole on a PC board that is going to be drilled by the board manufacturer must have both an upper and a lower limit to the size called out. This is referred to as the tolerance. In this case, the PC board manufacturing tolerance uses an upper tolerance (maximum tolerance) of .004 and a lower tolerance (minimum tolerance) of .002 for all the hole sizes.

Column 1, reference designations: This column lists all the components on the schematic (Figure 8-3). We know we will have one capacitor (Cl) at least two bypass capacitors, but we are not sure exactly how many. List them as bypass capacitors with a question mark for their reference designations.

List all the resistors. Check the schematic in Figure 8-3. There are two: Rl and R2.

List all of the integrated circuits. Their proper designation is U. Because we have not yet grouped this schematic, we do not know how many IC's there are. List them as U with a question mark for their quantity.

This includes all the components that we have for the schematic in Figure 8-3.

Column 2, component types: From the information obtained from the engineer and specification sheets, Cl is a CS13A type of capacitor. For the bypass capacitors without reference designations, list CK06. For the resistors, Rl and R2, list RC07. For the IC's list 14-pin dual inline.

Column 3, maximum body length: This column provides the maximum body length from the specification sheets (usually given by the engineer). Using the specification sheets given in this chapter, notice that the capacitor, Cl, has a maximum body length of .422. The bypass capacitors, CK06, have a maximum diameter of .300 (from the template). The specification sheet for RC07 resistors, Rl and R2 lists .281 to be the maximum body length. For the 14-pin dual inline of the IC's, list the overall length from the template.

Column 4, 1:1 pad spacing (inches): The information in this column is used to compute the pad spacing for the components that we have just listed. The spacing for these pads (according to good commercial practices) can be calculated by adding the maximum length to the minimum of .120. As shown in column 3, add column 3 plus .120 to column 4.

For example, the CS13A component, Cl, has a maximum body length of .422. If you add .120 to .422, the answer equals .542. Round up to the nearest grid. In this .542 rounds up to .600, which would be the nearest tenth grid. Write .600 in column 4 for CS13A. This is the minimum pad spacing. The template indicates that CS13A is a RC20, with a pad spacing of 12 grids at a 2:1 scale. Double the dimension, .600, since we are

Table 8-1. Component Mounting Table

Reference Designation	Component Type	Maximum Body Length	1:1 Pad Spacing (Inches)	Maximum Lead Diameter	Lead Mounting Hole (Min.)	Lead Mounting Hole (Min.)
1 From Schematic	2 From Specification Sheet	3 From Specification Sheet	4 Column 3+.120 (Round Up)	5 From Specification Sheet	6 Column 5 + .006	7 Column 5 + .020
C1	CS13-A	.422	.600	.022	.028	.042
Bypass Cap's	CK06	.300	Radial Leads (.200)	.029	.035	.049
R1 & R2	RC07	.281	.500	.027	.033	.047
All U? Dual-in-Line	14 pin template	per template	.100 x .300	.018	.024	.038

With a PCB MFG Tolerance of +.004 − .002 for the mounting holes, the above components require:

	Reference Designation	Mounting Hole Min W/Tolerance	Mounting Hole Max W/Tolerance	Hole Call-Out	Min Pad Size	Selected Pad Size
	8 Column 5 In Increasing Size	9 Column 6 Plus Tolerance (Add .002)	10 Column 7 Minus Tolerance (Sub .004)	11 From Column 9 and 10	12 Column 11 + .030	13 Standard Pad Size at 2:1
U ?	.018	.026	.34	.032	.062	.125
C1	.022	.030	.038	.032	.062	.125
R1 & R2	.027	.035	.043	.040	.070	.150
C ?	.029	.037	.045	.040	.070	.150

going from a spacing of 1:1 to a scale of 2:1. Use the minimum spread of 12 grids. Keep in mind that this was a minimum. The template we use has 12 grids. However, we could have gone to a larger spread: 13 or 14 grids.

Compute each component. The radial lead (CK06 bypass capacitors) has a pad spacing of .200 at 1:1 according to the template. This is 4 grids at 2:1.

The resistors, R1 and R2, have a maximum body length of .281. Adding the .120 spacing for the pads equals .501. This is so close to .500, round off to .500 at 1:1. This is 10 grids at 2:1. The IC's (14-pin dual inlines) have a 100 pad spacing from pad-to-pad on the length and .300 on the pad width. Use the template as our guideline (for the pad spacing formula used for military standards, see Figure 8-14).

Column 5, maximum lead diameter: List the maximum lead diameter given in Figure 8-9. C1 has a maximum lead diameter of .022 (which will list in column 5). The bypass capacitors (for which there are no C reference designations) have a maximum lead diameter of .029. R1 and R2 have a maximum lead diameter of .027. The IC 14-pin dual inline has a maximum lead diameter of .018.

Column 6, lead mounting hole minimum: This column lists the minimum clearance. Add .006 to the maximum lead diameter. Always have a minimum clearance of .006 over the maximum lead diameter (listed in column 5). Add the C1 (CS13A) maximum lead diameter of .022 to .006 to equal .028 in diameter.

Compute the following for the remaining components and list them in column 6. For bypass capacitors the maximum lead diameter is .029 plus .006 equaling .035 in diameter. For R1 and R2 the maximum lead diameter is .027 plus .006 equaling .033 in diameter. For the IC (listed as a 14-pin dual inline) the maximum lead diameter is .018 plus .006 equaling .024 in diameter.

Column 7, lead mounting hole maximum: The lead mounting hole in column 6 is the minimum. Compute the largest diameter that the hole could ever be and still allow the solder to properly flow through the hole.
Example: C1, CS13A, has a maximum lead diameter of .022. Add .022 to .020, equaling .042.
Bypass capacitors: The maximum lead diameter is .029 plus ,020 equaling .049 in diameter.
R1 and R2: The maximum lead diameter is .027 plus .020 equaling .047.

The pad spacing formula used for mil standards. (See Figure 8-14).

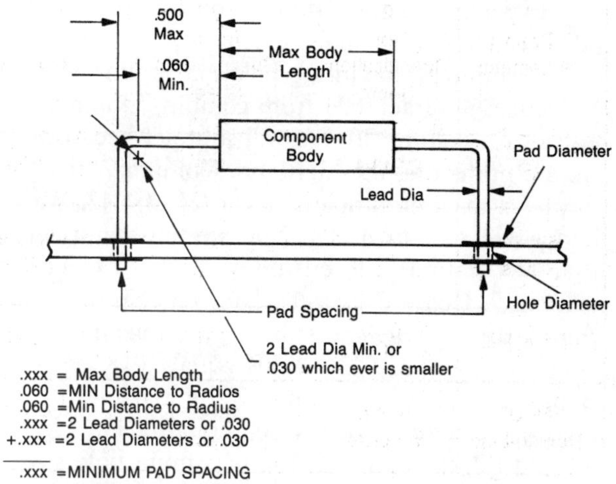

Figure 8-14. Pad Spacing Formula

IC (14-pin dual inline package): The maximum lead diameter is .018 plus .020 equaling .038 in diameter.

Column 8, reference designations: List all of the components on the schematic in numerical order, starting from the smallest lead to the largest lead. The smallest lead in column 5 is .018 which is the integrated circuits. The next largest lead is the capacitor, C1, at .022. Following C1, is R1 and R2 at .027. The last component is the bypass capacitors at .029 (with unknown C reference designations). These leads must be listed starting from the smallest and progressing to the largest.

Column 9, mounting hole minimum, with tolerance: We computed in column 6, the minimum lead mounting hole. However, we have not considered tolerance. The .022 lead (called C1 or CS13A) has a minimum lead mounting hole of .028. But this did not take into consideration the tolerance for the hole. The manufacturer cannot drill this exactly at .028. We must have a tolerance. In this case the tolerance will be .028, plus .004 minus .002. If .028 was the smallest dimension for the .022 (C1) then we must protect the smallest size called .028. In column 9 we will add the lower tolerance. Add .028 plus .002 (C1 equals .030). Compute the remaining components integrated circuits (14-pin dual inline): Add column 6 to column 9 (.024 + .002), this equals .026.

Rl and R2: Add column 6 to column 9 (.033 + .002), this equals .035.

Bypass capacitor: Add column 6 to column 9 (.035 + .002), this equals .037.

Column 10, mounting hole maximum, with tolerance: Subtract .004 from column 7 to create entries for column 10. We will subtract because we are protecting the high end. Column 7 indicates a maximum hole size for Cl of .042. We had a tolerance of +.004. Now we must subtract that .004 from .042 to ensure we do not exceed column 7. Column 10 will take from column 7, the maximum hole and subtract the tolerance which is .004. So, .042 – .004 equals .038. Bypass capacitors: In column 7 the maximum hole size is .049 – .004 which equals .045. Rl and R2, in column 7, the maximum hole size is .047 – .004 equals .043 added to column 10. Integrated circuits (14-pin dual inline): The maximum hole size is .038 (from column 7). Subtract the .004 from column 10 which equals .034.

Column 11, hole call out: This column lists the finished hole call out for all the components listed in column 8. Compare column 10 to column 9. We would like to have the same size hole for every component on the PC board, if possible. See Figures 8-12 and 8-13 (the hole formulas). The most economical board to build has component lead and every hole the same size. "U" is the integrated circuits not assigned numbers. Notice in column 10 that the maximum hole size is .034. This has a tolerance of +.004 - .002 built into it. Compare the maximum hole size to the next size down (in this case Cl). Can we use a hole, .034 for Cl? The minimum hole for Cl is .030. Its maximum is .038. Therefore, the answer is yes. .034 can be used for Cl.

Can .034 be used for Rl and R2? The minimum hole is .035; the maximum is .043. Therefore, we cannot use the same size hole for Rl, R2, Cl and "U". However, we do know that the IC's, "U?" and Cl, can be the same size hole from column 11. Split the difference between the minimum size for Cl and the maximum size for "U?" The difference between the two is .004. .030 to .034 is a difference of .004 divided by two. Add .002 to .030 or subtract it from .034. Place the answer, .032, into column 11. Return to column 10, find the next hole size. Rl and R2 have a maximum hole size, column 10, of .043. Compare that to "C?" (the bypass capacitor). This has a minimum of .037, column 9, to a maximum of .045, column 10. Could Rl, R2 and "C?" the same size R2, and C? (bypass capacitor), and the answer would be: Yes. We could use. 037 or we could use .043. We would probably want to use not the low and not the high, but the middle. Therefore, in column 11, I would choose, .040 as a hole size for Rl, R2 and "C?" (the bypass capacitors). However, we could use .037, .038, .039, .040, .041, .042, or .043 as a hole size for Rl, R2 or "C?" (the bypass capacitors).

Column 12, minimum pad size: See Figure 8-12 (component mounting information, note 3). The minimum annual ring (meaning "all the way around") of .015 over the maximum hole diameter will be used to compute the pad size. The pad size is the copper left around the hole after drilling. In this case, the plan is to leave .015 of copper all the way around the hole. .015 is not a standard annular ring; but it is the number we chose for our example. We could use a .010 annular ring. Doubling .015 equals .030. This is called out in column 12. The .030 is added to the hole size that we have computed for each component (in this case for "U?" and Cl). Adding .030 + .032 equals a pad size (at 1:1) of .062 for "U?" and Cl. 2:1 is the pad size we will use on our tape-ups. Therefore, double the pad size in column 12 to obtain a finished 2:1 pad size on column 13. .062 × 2 = .124. The closest standard pad to .124 is .125. We could increase the pad size to .150 if we wanted a larger annular ring. Look at column 11 for Rl, R2, and "C?". The hole call out in column 11 is added to column 12. This is added to the hole call out of .040. .030 = .070 at 1:1. Double it for the 2:1 scale (.070 × 2 = .140). The closest standard pad to .140 equals .150.

Column 13, selected pad size: Finding standard size pads was covered extensively in chapter 3 of the Analog Printed Circuit Design & Drafting book. Use a pad size of .150 for column 13 on Rl, R2, and "C?" (the bypass capacitors). You have all the case codes for the components; the spacing pads, the leads in the pads, the pad sizes, and the finished hole diameters with their tolerances (in this case, .032 + .004 − .002 and .040 + .004, − .002).

Step #4—Establish the board layout design rules (illustration 2). These rules are used for designing and taping the printed circuit boards in this book.

Illustration #2: Design Rules.

It includes the general requirements for formal layout and artmasters.

Establishing Design Rules

Each company may have different standards for pad sizes, trace sizes, and air gaps. There are many different standards: Military, ANSI, IPC, and good commercial practices. For the boards in chapters 8 through 12 we will establish our own standards. From good commercial practices see procedure 1.

Procedure #1

1. Board layouts shall be made to the same scale that the artwork (tape up) will be made, when practical. The scale will be 2:1.

2. The grid pattern shall be .100 inch spacing.

3. All component leads, holes/pads shall be located at the intersection of grid lines. See Figure 8-15.

4. When locating components on the grid, the mounting length shall be determined by the maximum body size. Component parts shall be mounted on only one side of the board. Parts shall be so placed and located, that any part may be removed without removing any other part, unless the assembly is not intended to be repaired. Only one component lead shall be placed in a lead hole. When jumpers are used, they shall be placed on the component side of the board, and constructed of solid, insulated wire designed to be as straight and as short as possible. They shall not cross over other component parts.

5. Polarity and indexing features shall be as indicated according to the format in Figure 8-16.

6. Components shall be indicated that require hold-downs because of shock and vibration considerations, specified: clamping, bolting or adhesive coating. See chapter 4 in the analog design book for further information.

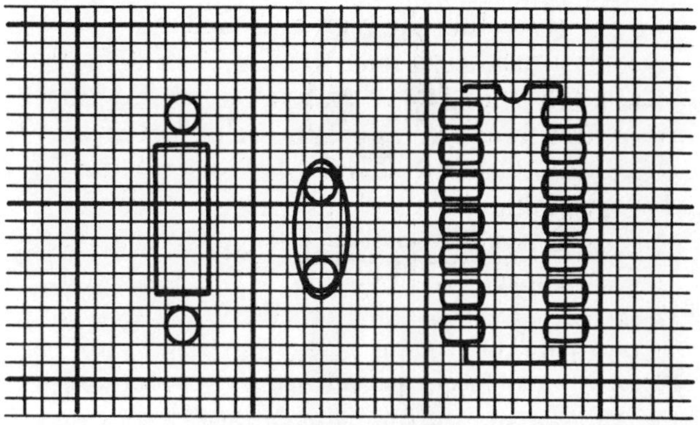

Figure 8-15. Pads on Grid Intersection

7. Test point location.
   A. When possible, all test points should be brought out to any three edges of the board or to the accessible edge of the board.
   B. All test points should be arranged in a straight line and fall on grid intersections with a minimum distance, center-to-center of .200 at 1:1. See Figure 8-17 for a further explanation of test points.

8. Conductor width on the finished board shall be determined as described below:
   A. The standard conductor width for signal lines at 1:1 scale shall be .025. At a 2:1 scale it shall be .050. The minimum conductor width for signal line at 1:1 shall be .015. At 2:1, it shall be .030.
   B. The standard conductor width for voltage and ground lines at 1:1 shall be .050. At a 2:1 scale it shall be .100.

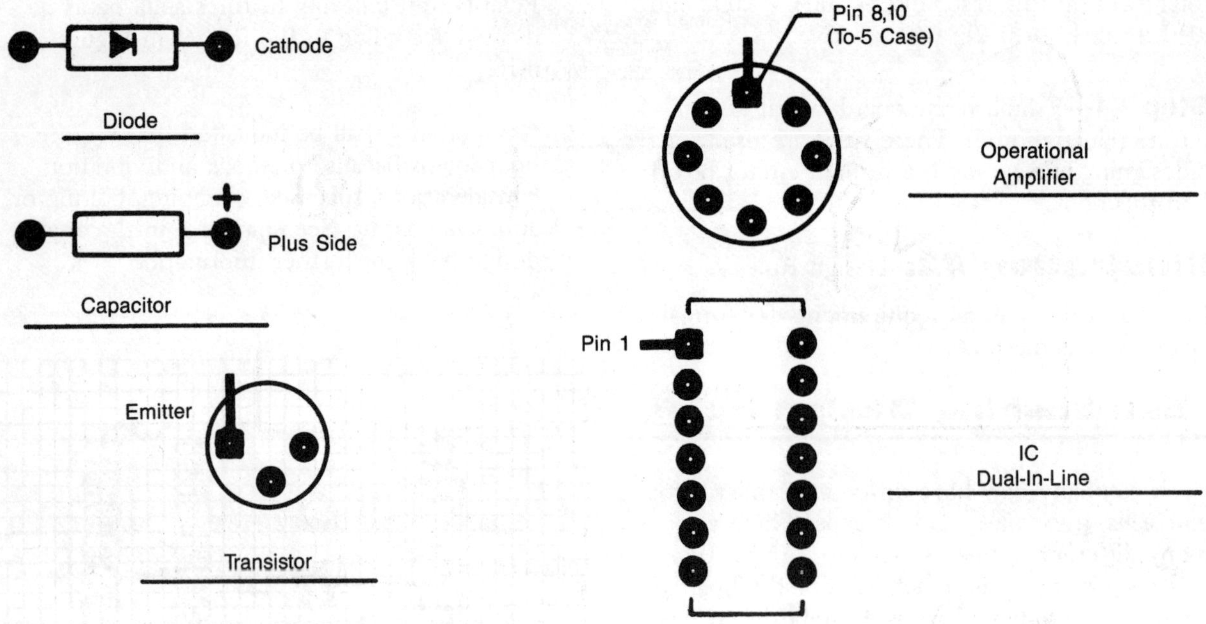

Figure 8-16. Polarity and Indexing

NOTE: The two sizes labeled A and B shall be the standard conductor/trace widths, unless otherwise specified.

9. Conductor spacing to board edge explanation: Traces shall not come closer to the edge of the board than .050 minimum. If at all possible, leave at least a .100 at a 1:1 scale air gap to the board edge. See Figure 8-18.

10. Spacing explanation: The standard spacing between pads and conductors shall be .025 at 1:1, or .050 at a 2:1 scale, unless the conductor density requires otherwise. The minimum spacing shall be .015 at 1×1 or .030 at 2×1: See Figure 8-19.

11. Hole sizes. The inside diameter after plating of a plated through hole shall be no more than .020 and no less than .006 over the maximum diameter of the lead (to be inserted into the hole). This is according to good commercial practices. See Figure 8-12 for component mounting information. For military practices, see Figure 8-13.

12. Feed through hole sizes shall be determined by using a hole diameter of .025 at a 1:1 scale. This allows the use of a pad of either .062 at 1:1 or .125 at 2:1 (using an annular ring of .020). If we choose to reduce our annular ring to a minimum of .015, we could use a pad size of either .050 at a 1:1 scale or .100 at 2:1.

13. Tooling hole sizes shall be either .100 at 1:1 scale or .125 at 1:1 scale. See Figure 4-11 for further information.

This is an explanation of test points used on printed circuit boards (Figure 8-17)

Example 1 shows the socket style for test points used on printed circuit boards. In some printed circuit board applications, frequent insertion and removal of test probes are required. You would use this style of socket test point on your printed circuit board.

Example 2 shows the simpler type of test point, called a standoff or terminal. This style can be used with test probes that the technician or

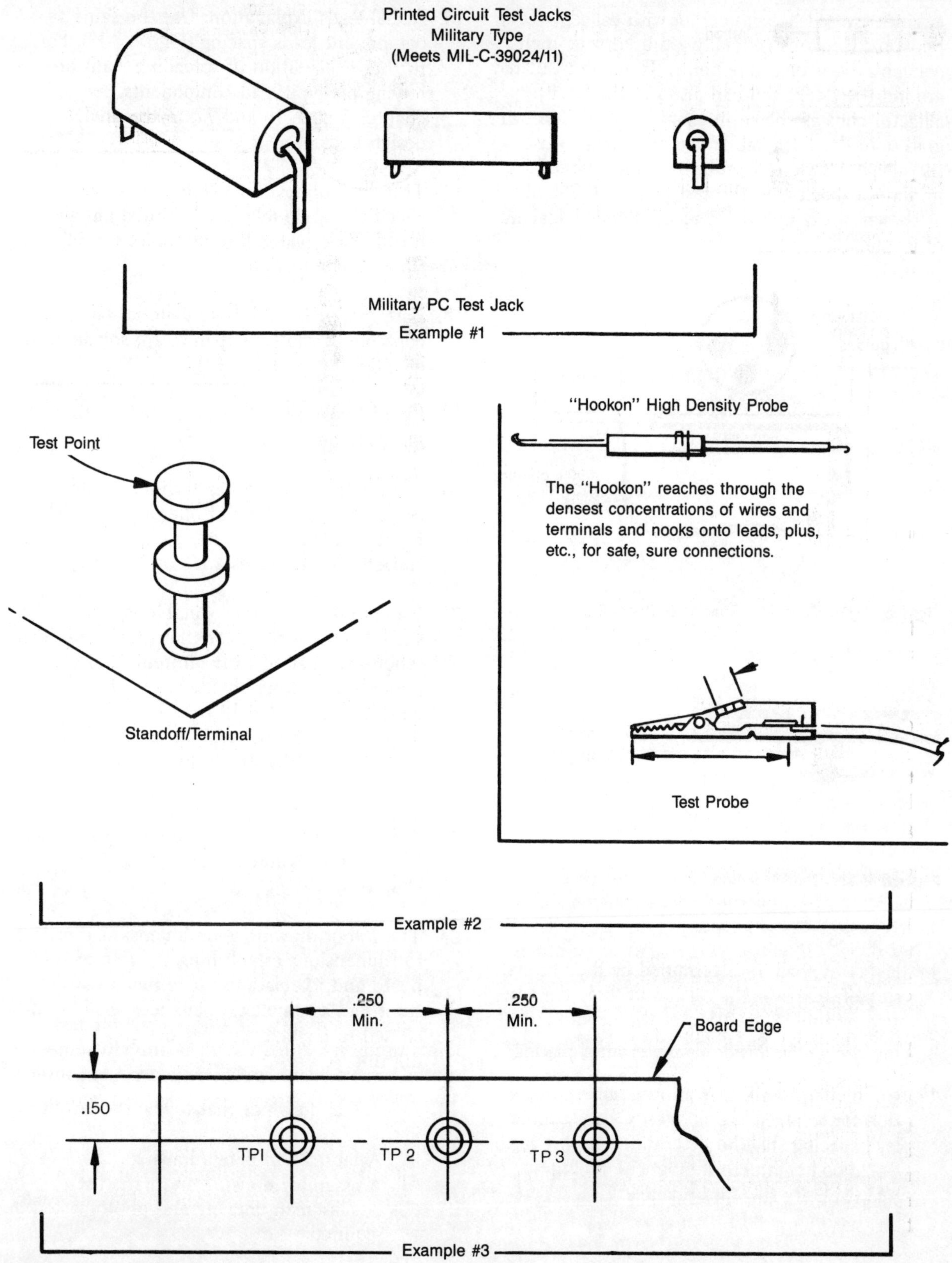

Figure 8-17. Test Points

engineer would be connecting with called hookons. Hookons generally are used with high concentrations of components. They are insulated around the hook. You can also use them with alligator clips as shown in example #2, which can grab onto the terminal. If you are going to use the alligator style test probes, we recommend a minimum spacing on the test points of .250 at 1:1 as shown in example 3 for recommended spacing of terminals.

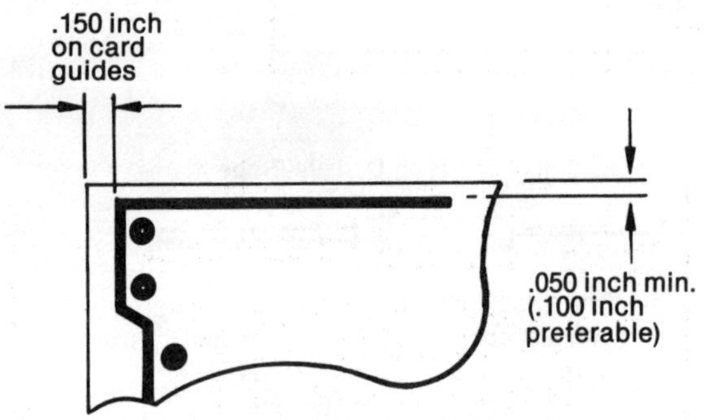

Figure 8-18. Conductor Spacing to Board Edge.

Figure 8-19. Conductor Spacing.

14. Pad sizes shall be determined by the annular ring of .020 over the maximum hole diameter or a minimum of .015 over the maximum hole diameter. See Figure 8-12.

15. Component mounting information explanation for axial lead components: Pad spacing/lead spacing shall be determined from Figure 8-12, good commercial practices, or Figure 8-13 (Military standard practices).

16. Radial leads explanation: Use the same as component leads spacing (Figure 8-12). For a further explanation of spreading leads or buying pre-bent lead components, see chapter 3 page 76 and 77, in the analog design book.

17. Through connections: Unless otherwise specified, all double sided printed circuit boards have plated-through holes for all through connections.

18. Base materials. The base material types acceptable by most companies are specified in chapter 4. (See Table 4-1).

19. Board thicknesses shall be stated on the fabrication drawing.

20. Conductors shall be based upon a copper conductor "thickness of one ounce". Use Figure 8-20 for determining current carrying capacity for all traces and conductors.

21. Part numbers or other suitable identification such as reference designations taken from the schematic shall identify all items and components appearing on the layout. Component legend lettering height on 2:1 master drawings should be (.125) inch. Lettering and other line widths should be 0.5 mm. See Figure 8-21.

22. Abbreviations, symbols, and coding shall conform to the requirements discussed in chapter 2.

23. The layout drawing is undimensioned except for dimensions establishing the size of the board and its relationship to mating structures and assemblies. Tolerances shall be included to facilitate further design and drafting.

24. Additional drawings that follow the layout are:
(A). A fabrication detail drawing.
(B). A assembly drawing and parts list.
(C). A silkscreen drawing and photomaster, if required.

CONDUCTOR THICKNESS AND WIDTH

(For use in determining current carrying capacity and sizes of etched copper conductors for various temperature rises above ambient)

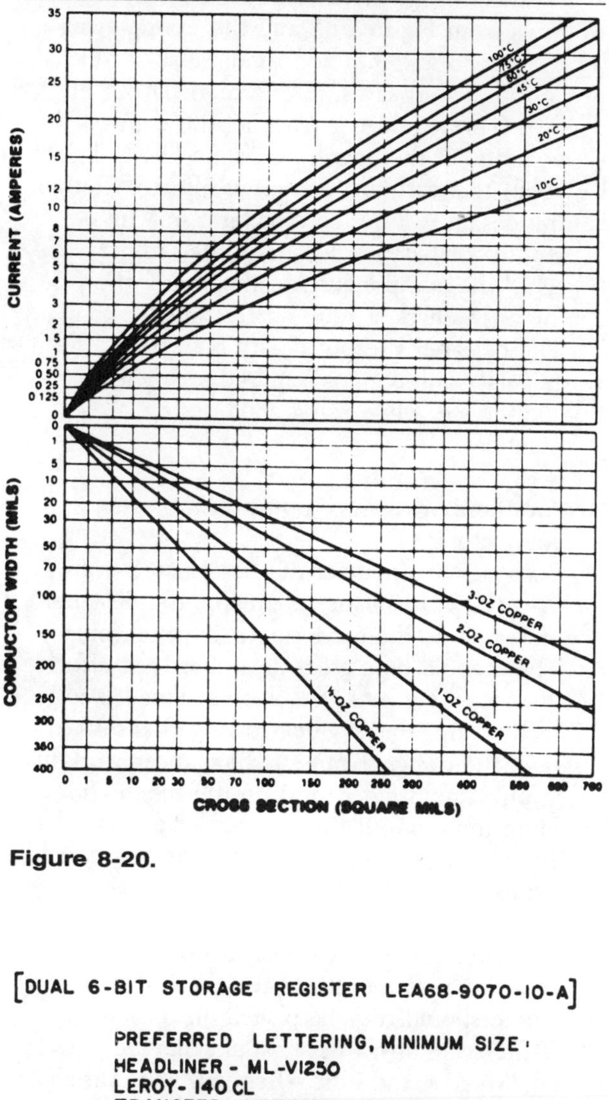

Figure 8-20.

[DUAL 6-BIT STORAGE REGISTER LEA68-9070-10-A]

PREFERRED LETTERING, MINIMUM SIZE:
HEADLINER - ML-V1250
LEROY- 140 CL
TRANSFER LETTERS- 707 OR 14 PT.

Figure 8-21. Sample Lettering Size And Style (Twice Sizes)

Step #5—Refer to the list of questions to ask the engineer about the 3-IC digital schematic. See Illustration 3.

Illustration 3: Questions To Ask The Engineer

1. Check to see if there are any special size components, or if all the components are standard. For example:
   (a) Resistor type and power rating.
   (b) Capacitor material and working voltage.
   (c) Transistor case type.
   (d) IC family and package type.
   (e) Diode case size.

   Are we using components that are standard to the company? Every company stores and stocks standard components purchased as standard items rather than buying special components. The engineer would use these standard stocked components. The company should have specifications sheets or a standard parts book necessary to design to that component. Company information has the component lead spacing, the component lead size, and all the information needed to mount the component on the PC board.

2. How many bypass capacitors per IC? The bypass capacitor is also called a decoupling capacitor. Chapter 3 has an explanation for a decoupling capacitor. See Figure 3-13. The engineer might answer "one per IC", "or one decoupling/bypass capacitor for every two or three IC's", or "sprinkle some around". These capacitors are placed between the voltage pin and the ground pin bus lines on the IC rows. Do not confuse these capacitors with input filtering capacitors, which would be placed very close to the input voltage pin on the PC board connector. For example of placement on the bus line, see Figures 3-14 through 3-18.

3. Are there any critical leads on the board? Critical leads pertain to signal paths or any traces on the PC board that is not a voltage or ground trace. Signal paths could be critical in length from component connection to component connection or gate to gate. A series of connections could be critical in length. If there are critical connections on the PC board, where should they be routed

or placed on the board? This question should be asked before designing the PC board. You would code these connections (or lines) on the schematic.

4. Are there any special size filtering capacitors required on the input voltage pins (decoupling network)?
   This question refers to an input filtering capacitor. (chapter 3 and Figure 3-19). There are several techniques used for input filtering capacitors. These are diagrammed in Figure 3-20. Find out which voltages require these capacitors. Obtain the size of the components in the input filtering circuits. Determine the value, voltage, composition, and the manufacturer's specification sheet. Keep in mind that the input filtering capacitors or circuits components should be as close as possible to the input point location (whether it be a male or female connector). Consult the responsible electrical engineer to the best placement on the digital PC board. Various capacitor types of materials are used as filters depending on the frequency coming in close contact with the circuit.

5. Should there be any spare IC gates?
   We are not asking whether every available gate on every IC be used. We know from Figure 2-63 that there would be some spare gates inside some of the IC's. Should there by any particular type of gates? For example, the engineer might ask for one 3-input NAND, or two 3-input NORS, etc. The engineer could be planning for future electrical modifications on his circuit design. If we are in a prototype stage, the engineer may require certain spare gates. Use as many gates as possible to reduce the total count of IC's on the PC board (see chapter 2). If the engineer requests spare gates, it would be like those gates belong to the schematic. But they must not be used. They are for future electrical design modifications. There may even be spare pads (Figure 6-18).

6. By using the maximum size of all the components, will all components fit on the board? (This is your responsibility)
   The designer calculates the maximum size of all the components (discrete and integrated circuits) on the schematic. He uses the formula from Figure 5-18 and the examples in chapters 8 and 9. The formula gives a 90% confidence. It saves an enormous layout effort. If we could not place all the components on the board, we would like to know about this early in the design effort. Whether the design will be loose or tight or whether you might have to do out of the ordinary to fit all the components on the board, the formulas in chapter 5 are a very valuable tool. The designer's responsibility is to precalculate component required area with some confidence, while using good design practices.

7. Are there pre-assigned pins on the connectors?
   Because the engineer did not place the preassigned pins on the schematic, the designer may assume there are no pre-assigned pins. There could be a great deal of wasted effort disconnecting pins that were assigned incorrectly. After the engineer discovers the situation, the designer must redraw and rework his layout. Pre-assigned pins on the schematic require more layout time. It can take from 10-15% longer to interconnect the board on complex schematics.

8. Are there any variable resistors or variable capacitors, or frequency switches? If so, where should they be placed on the board? If there are any, where should they be placed? Most of the time when you have variable components such as resistors or capacitors, their adjustment is required. Access is not always easy. These types of components may require mandatory placement on the PC board for easy accessing. The enclosure may have limited access points for a screw driver to adjust them while operating. Find out before designing where these components should be placed on the PC board. Determine if there are restrictions to design for.

9. Should there be any test points? (Try to place all test points on the edge.) Check for test points on voltages and grounds, if not so indicated on the schematic.
   Test points are terminals or pins that check the function of the PC board. These test points sometimes have limited access. Frequently the engineer will not draft them on the schematic, even though they should be there. Whenever there are one or several test points required by the engineer on the schematic, always place a ground test point with the required test points. This must be done even if the engineer has not called for a ground test point. The ground test point must be within a 6-inch radius of any other test point on the PC board. The oscilloscope used with the test point has a ground lead on it. Connect the ground lead in order to make a proper electrical test for each test point being used.

10. Do any components require heat sinks? If so, where should the components be placed in relationship to the other components within the board outline?
    Heat sinks are discussed in chapter 4 of the analog design book. Heat sinks extract the heat away from components that typically run hot. Transistors, resistors and power transistors sometimes need heat sinks. Integrated circuits usually do not require heat sinks. Heat sinks have fins that project away from the component body. It could look like a metal hat that sits on top or surrounds a transistor. Therefore, they require more surface area on the PC board. We must know before making the layout whether heat sinks are needed for any component that runs hot during operation. Also, if the heat generated by a given component it must be transmitted to a heat sink such as the chassis or cooling element.

11. Have all the package considerations been examined? (See chapter 4).
    The PC designer must have all the pertinent information about the board size, the female and male connector that interfaces with the PC board, mounting restrictions, how the board is secured into the enclosure, special trace widths, voltages, or grounds. Identify special air gaps required between traces, special pad sizes, structures mating with the board (with clearances), or areas that might have to stay clear of mating structures. Determine where PC board is mounted, how it is extracted, or how it's accessed. All of these questions must be answered before the design can begin.
    Before laying out a schematic, take a few moments to compile a list of questions similar to what we have discussed. Discuss all of these questions at one time rather than piecemeal. Otherwise you lose the continuity of understanding everything the engineer is trying to tell you. You will be more successful by taking a few moments to study the schematic. List all of the questions that you have from an itemized list that we've given you plus any other questions that you might come up with. Such a professional attitude will build rapport with the engineer, make the job smoother, and save you many headaches in the future.

At this point let's cover questions 1 through 11 for the schematic in Figure 8-3. The responsible electrical engineer would give the answers.

Question 1: Check to see if there are any special size components, or if all the components are standard. The schematic in Figure 8-3 identifies all the discrete components and IC's for which we have found the standard case size of each component.
Question 2: How many bypass capacitors per IC? The schematic in Figure 8-3 (note 2) specifies a bypass capacitor every two IC packages.
Question 3: Are there any critical leads on the board? In this case the engineer will not specify any leads critical in length.
Question 4: Are there any special size filtering capacitors required on the input voltage pins in the schematic? Figure 8-3 shows an input filtering capacitor, C1. Note 4 on Figure 8-3 specifies C1 as being a polarized

capacitor (CS13A case size). The template in Figure 8-8 shows that the CS13A case size can be substituted with an RC20 body outline.

Question 5: Spare IC gates, should there be any? Most likely you will have spare IC gates. The engineer on this particular schematic does not require any particular type of spare gate.

Question 6: By using the maximum size of all the components, will all the components fit on the PC board? This question will be answered extensively in step 8 once the board size has been determined and the grouping of the schematic has been accomplished.

Question 7: Are there pre-assigned pins on the connector? Schematic Figure 8-3 shows pre-assigned pins for voltage and ground. Voltage is pins 29 and 30, and ground is pins 1 and 2 on the connector Pl.

Question 8: Are there any variable resistors or variable capacitors, or frequency adjustment switches? If so, where should they be placed on the PC board? In this case there are no variable resistors or capacitors for Figure 8-3.

Question 9: Should there be any test points? Schematic Figure 8-3 does show test points, TP1 through TP4. TP1 and TP2 are for signal paths inside the circuit. TP3 is ground test point, and TP4 is the voltage for +5V. All of these test points should be accessed from the opposite edge of the connector on the PC board. This board will be inserted into a chassis on card tracks sliding into a female connector in the back of the enclosure. See Figure 8-17 for more information; location and size of test points.

Question 10: Do any components require heat sinks? For Figure 8-3, the answer is no. There are no components that require heat sinking.

Question 11: Have all the package considerations been examined? Question 11 will be answered extensively in Step 6 which covers board outline. We will look at the enclosures and the mounting required for the PC board.

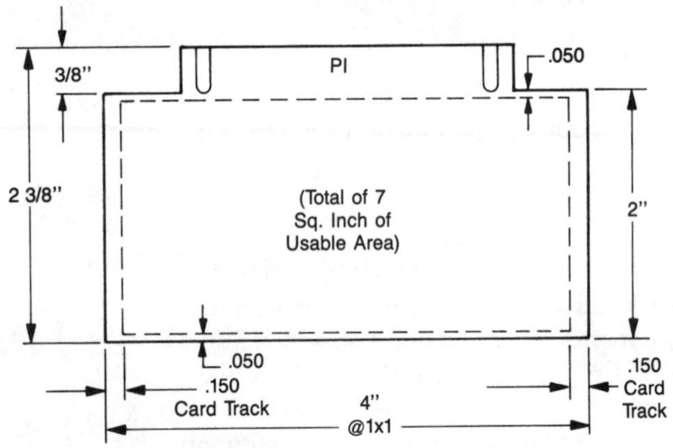

Figure 8-22. Board Outline

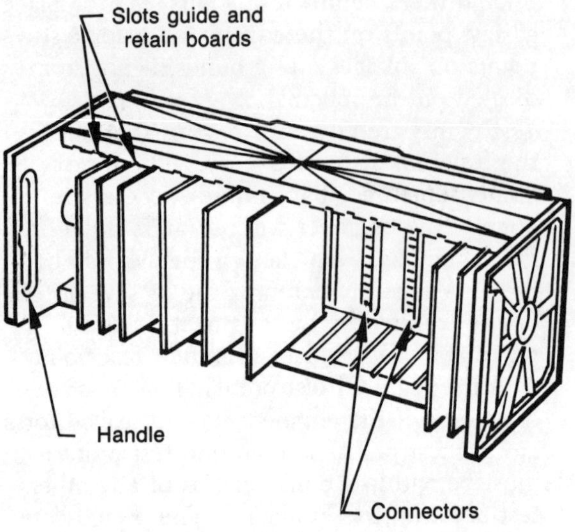

Figure 8-23. Housing or Enclosure

Step #6—Determine the board outline (illustration 4). The board size must be known in order to compute the total square inches available for the engineer's schematic (Figure 8-3). Besides the size of the board, find out what method of fastening or mounting is being used (card guides, screws, stand-off, etc.). What type of electrical interface is required (connector, cables, wires, etc.). For review of the board outlining, see Chapter 4.

Illustration 4: Board Size and Mounting

See procedures 1 through 3.

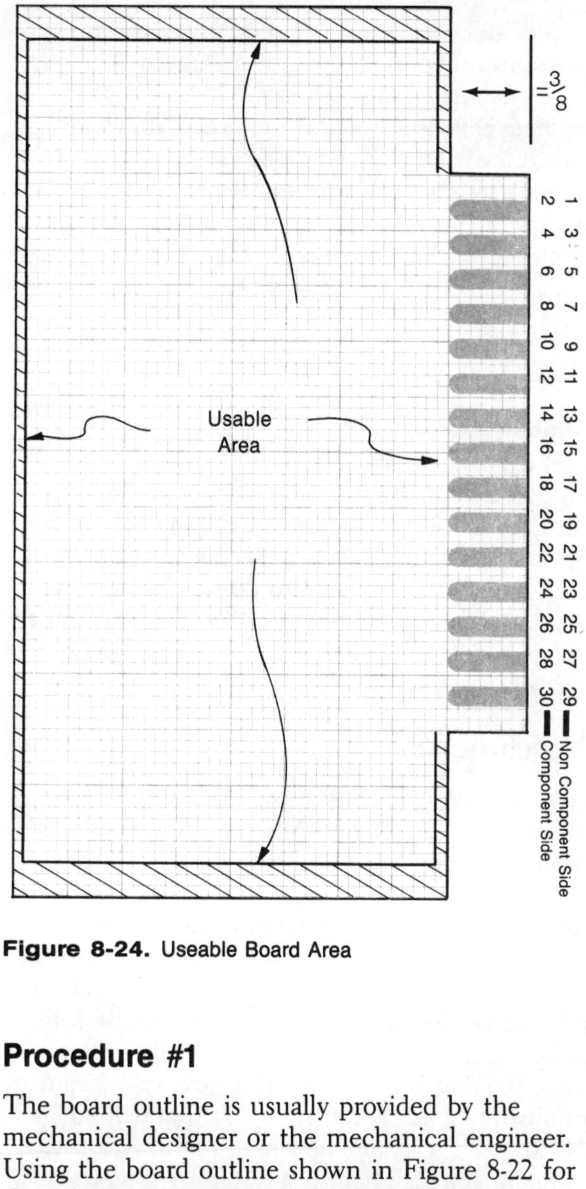

Figure 8-24. Useable Board Area

Procedure #1

The board outline is usually provided by the mechanical designer or the mechanical engineer. Using the board outline shown in Figure 8-22 for the schematic in Figure 8-3, we find a total of 8 square inches on the board (not counting the connector tongue, P1). In order to calculate the total number of useable square inches, define what interfaces to the PC board. How is it secured to an enclosure? See procedure 2.

Procedure #2

The board outline is inserted into an enclosure that uses a card track which holds the board in position (Figure 8-23). Therefore, the board has restrictions on its edges. See Figure 8-22 for card track clearance. Also, see Figure 8-18. Using the restrictions, re-compute the useable area by checking off the following steps:

1. 4" × 2" = 8 square inches.
2. 4 inches will be reduced by the two card tracks clearance.
   Example: 4.000"
   − .300 card tracks
   = 3.700
3. 2 inches will be reduced by the board edge clearance.
   Example: 2.000
   − .100 Board edge clearance
   = 1.900
4. Multiplying 3.700 × 1.900 equals 7.06. Round off to 7 square inches. This is the amount of useable area. See Figure 8-24 is an example of a 2:1 scale layout of useable area for the schematic in Figure 8-3.

Special Note

The card track clearance of .150 on each side of the board and the board edge clearance of .050 for the top and bottom edges of the board were obtained from design rules (Figure 8-18) used to compute the total useable square inches.

Procedure #3

Obtain a connector specification sheet from the engineer. This indicates the number of contacts and the size of the male tongue on the PC board (called P1). The specification sheet is similar to Figure 4-15. In this example we will provide all the necessary information without determining the size and the number of contacts, and the numbering of each contact. See Figure 8-24 for connector size and numbering. Use colors to represent the two different sides of the connector. You can only see the component side, which we have colored in red (1 through 29). The noncomponent side of the contacts (which you cannot see) are colored in blue (2 through 30). Red represents the component side; blue represents the noncomponent side (which is the side of the connector you are actually terminating to). Return to the schematic in Figure 8-3 to group the logic gates using the specification sheet (see step 7).

Step #7—Group the logic gates on the schematic (Figure 8-3) using the logic considerations in illustration 5. Grouping must be done at this time in order to compute the total number of IC's required to fit within the assigned board outline. A minimum of one square inch is required for each IC. For a review of grouping, see Chapter 2.

Illustration #5: Grouping

Procedure #1

Using the schematic in Figure 8-25, group and combine to find the lowest number of IC's required on this schematic. At the same time, combine gates for the shortest trace connection by following the steps below.

1. There are 2 reasons to identify all gates on the schematic. The first reason is that there are different logic families such as open collector and non-open collector. (Another way of saying non-open collector is totem pole). Do not mix these two logics together. The second reason is that you must have the number of the logic to obtain the specification sheet. For the schematic in Figure 8-25, all gates have been identified with the proper manufacturer's specification number, such as 04, 00, etc.

2. Specification sheets identify logic families and symbols. The specification sheets required for the schematic in Figure 8-25 have been provided by the electrical engineer. See Figures 8-4 and 8-5. These specification sheets identify each type of logic gate on the schematic. They indicate the total quantity of gates within each package. Provide the input and output assignments for each and provide the voltage and ground pin numbers for each type package. They also provide information about open collector versus non-open collector. If we were working with an open collector gate, it would clearly call out in bold letters: OPEN COLLECTOR. Figures 8-4 and 8-5 show no open collector. All IC packages are totem pole style.

Procedure #2

Before we proceed with grouping, review the questions asked of the engineer (Figure 8-25). Are there any spare gates required? In this case the engineer said that there were none. Are there any critical leads on the board? These questions were particularly important for grouping. This is because it directly influences which gates would be placed in the same IC package. On this particular board there are no critical leads and no spare IC gates required.

Procedure #3

Combine the gates. For the schematic in Figure 8-25, use the combining rules in chapter 2 (Figures 2-56 through 2-61). When combining gates using the technique called tying off unused inputs, tie off to another used input. Always seek the engineer's permission to do so. This is a more popular technique than tying the unused input to a high or a low point (meaning the voltage or ground).

Procedure #4

Combine the gates that most obviously communicate to each other. Make the traces as short as possible. See Figure 8-25. Look at the schematic in Figure 8-25. We combined the JK flip-flops 7473. Examine the specification sheet in Figure 8-5. Notice that there are two JK flip-flops in one package. We will call these two U1. After marking U1 inside the gate symbols on the schematic, reference the spare count on the schematic. U1 has zero spares. The next set of communicating gates is the 7400, 7404, and 7410 (at the top of the schematic where the resistor, R1, is interconnected). There is a duplicate circuit where R2 is located. All three gates should be located inside the same IC (the 00, 04 and 10). Can we place all three gates inside the same IC package? Review the specification sheet in Figure 8-5 and the design rules in Figure 2-57. Using Figure 8-5, determine how many gates there are in a 7410. The answer is three. There are three gates to put inside the same package: 00, 04, 10.

Using your combining rules (Figure 2-57), find out if a 3-input NAND can act like a 2-input NAND called a 00. The answer is yes. Tie one of the unused inputs of the 3 on the 10 to a used input. Also you could tie one of the unused in-

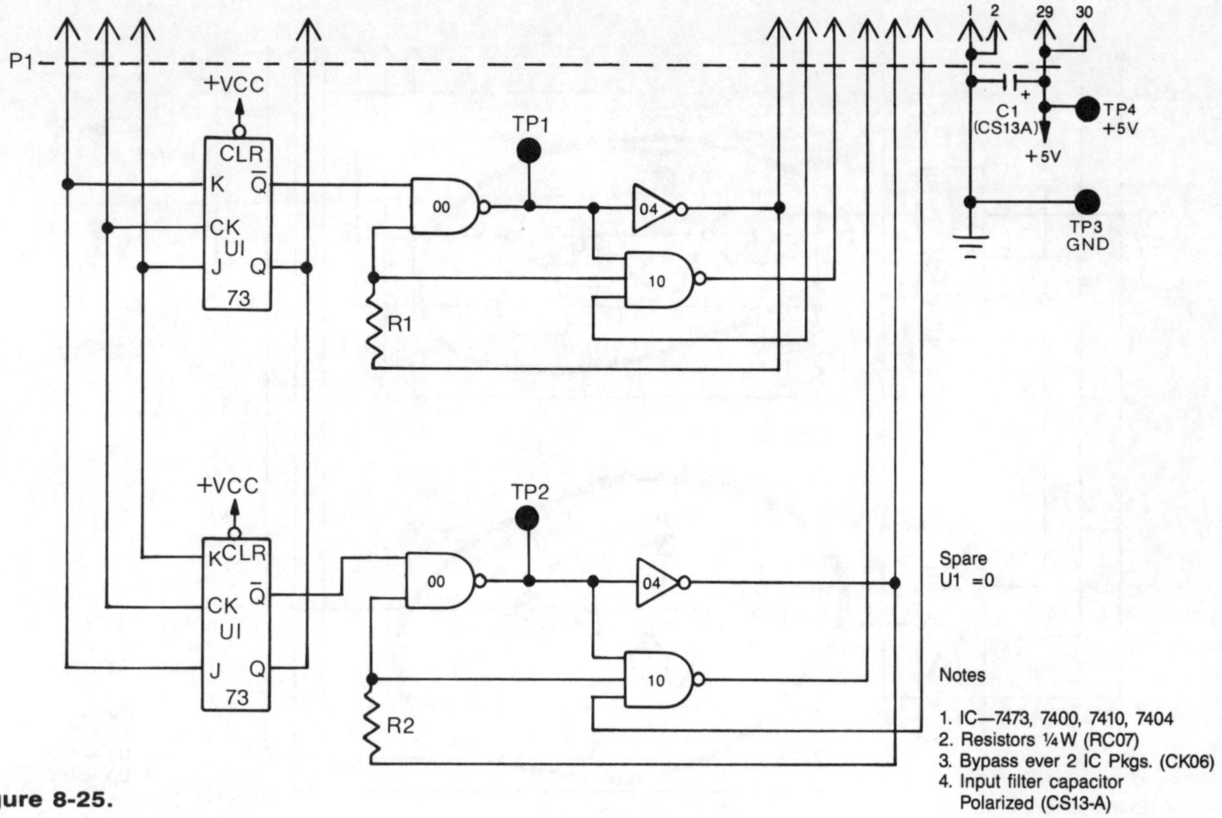

Figure 8-25.

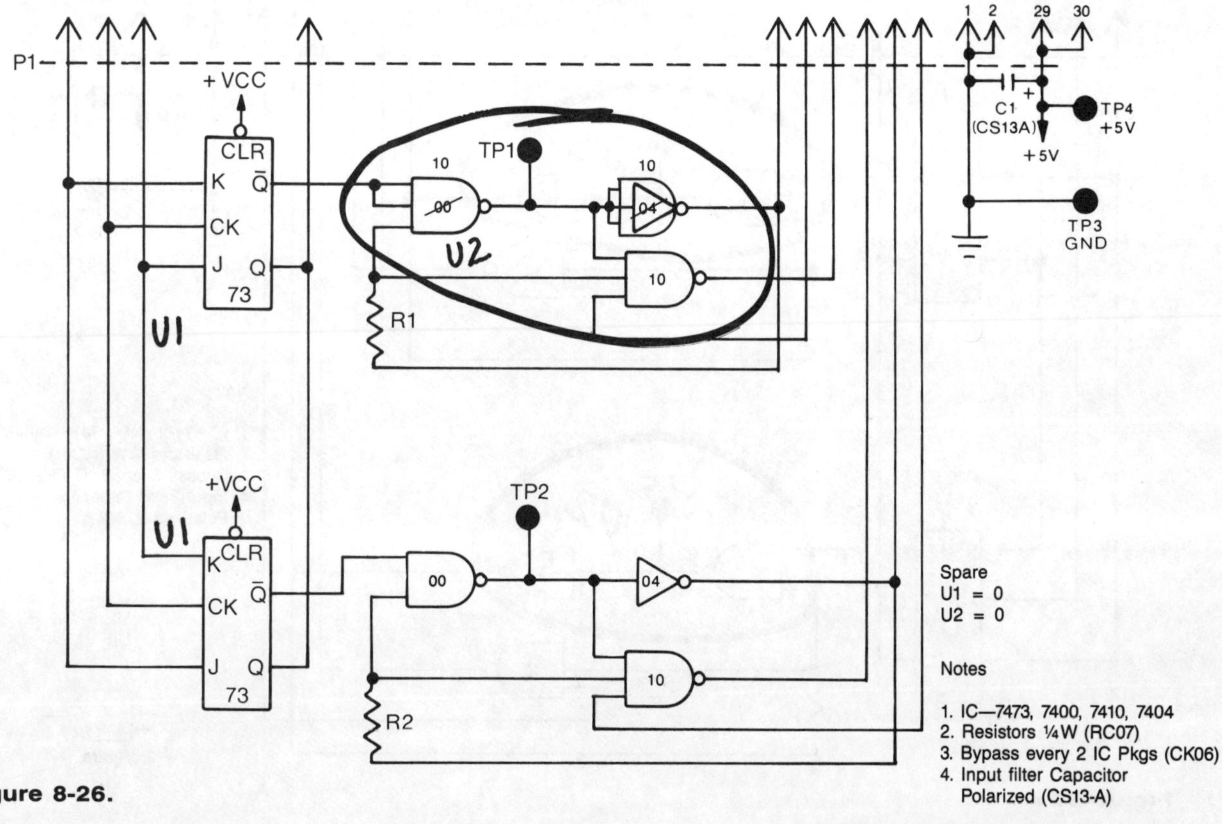

Figure 8-26.

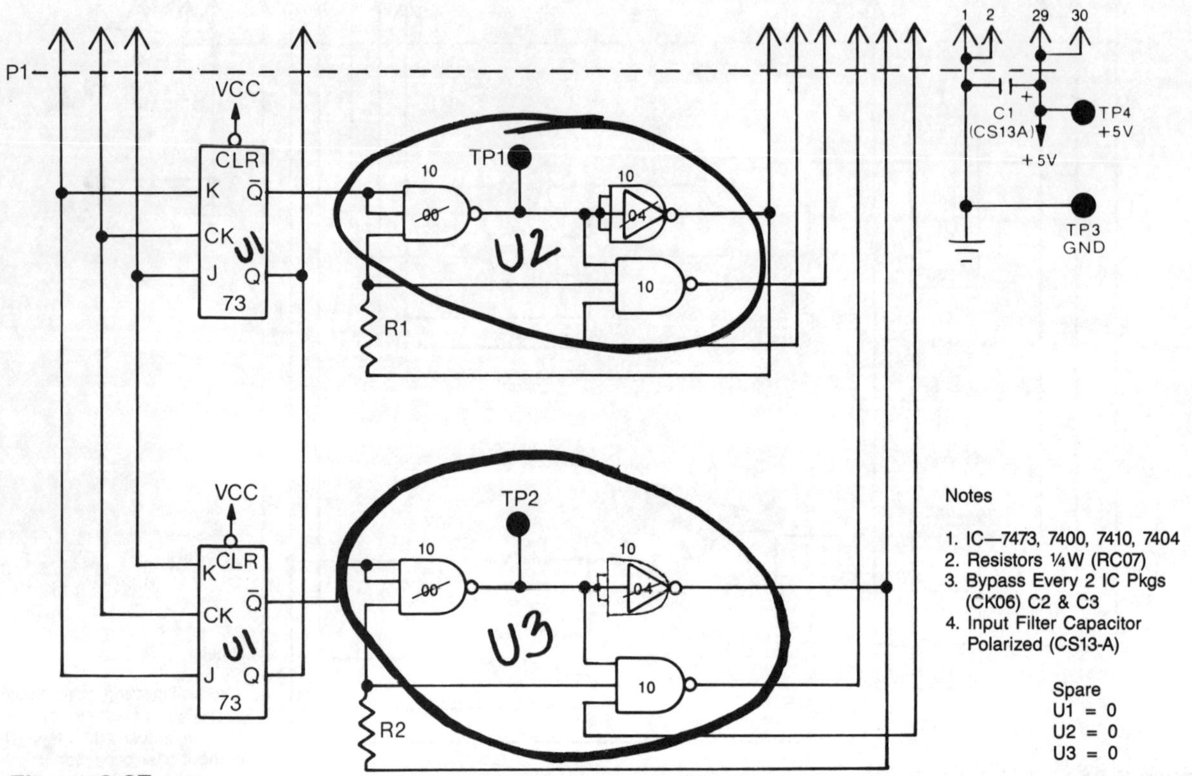

Figure 8-27.

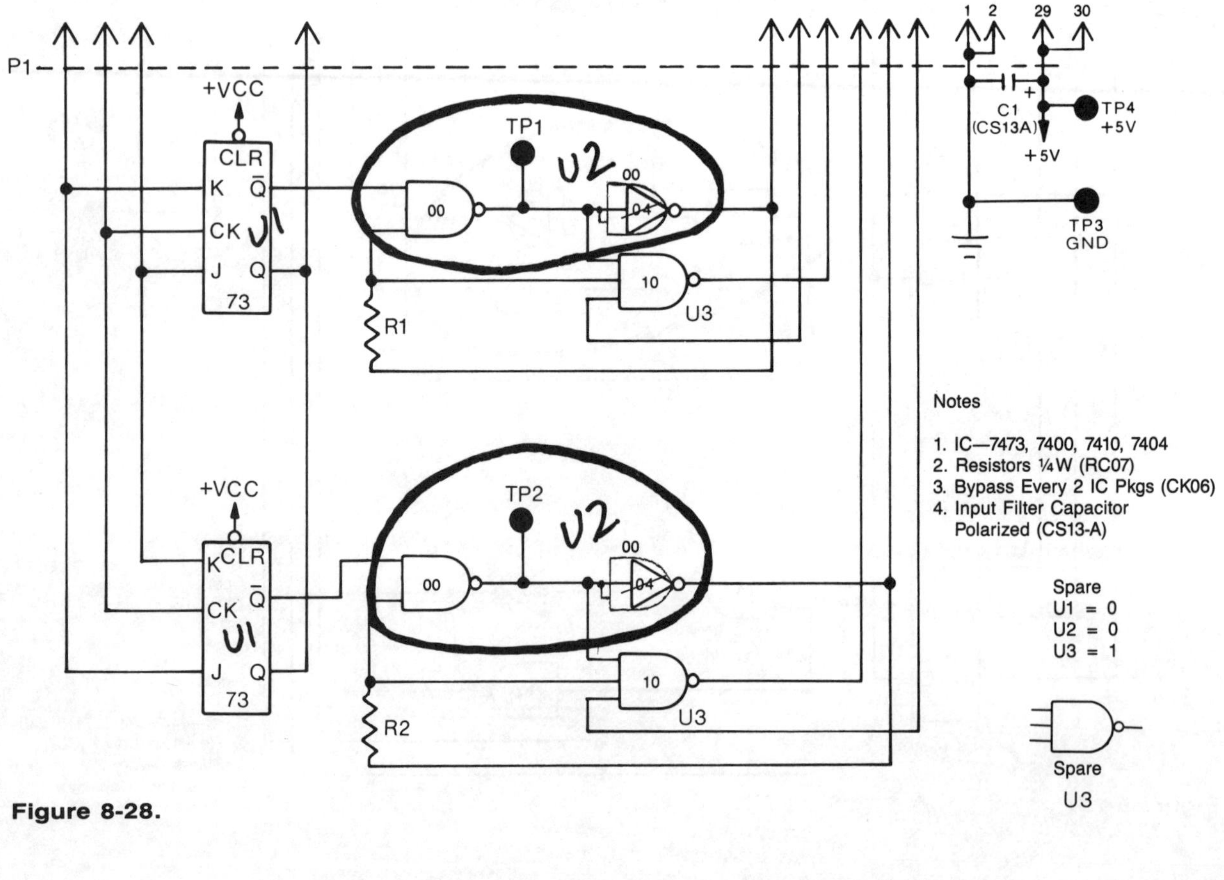

Figure 8-28.

puts to a high. The engineer has requested that, for this schematic, we tie all unused inputs to a used input.

Of the three gates on the 7410, (Figure 8-5) two are used. One of them is a 7410. The second one was to take the place of the 7400. The last gate in the package should be used before the 7404. Refer to the combining rules in Figure 2-57. A 3-input NAND can act electrically like a single input inverter by tying all three inputs together to create a single input with a single output. That is what we will do with the 04. We will turn it into a 7410 by tying all three inputs together. We will call these three gates (inside one IC package) 7410: U2. See the schematic in Figure 8-26 for this combining exercise. Note that U2 has zero spares. Make the traces as short as possible on the layout by keeping the gates (that intercommunicate) within the same IC package. Let us mirror image this combining technique with the remaining three gates on the schematic, by duplicating the grouping procedure. See Figure 8-27 for U3. Note that U3 would have zero spares. The grouping exercise has given a total of 3 IC's, with zero spares.

Procedure #5

Using the same rule for the schematics in Figures 8-25, 8-26, and 8-27, choose an alternate combining way of grouping and combining our gates. See the schematic in Figure 8-28.

Grouping and Combining for Figure 8-28

1. Using the specification sheet in Figures 8-4 and 8-5, group the 7473 JK flip flops. Call this U1. List the spares as we group. List U1 on the side of this schematic as having zero spares.

2. Use the specification sheet in Figure 8-4 and the combining rules in Figure 2-57. Can we combine a 7404 inverter into the same package as the 7400 shown in Figure 8-4? The answer is yes; there are four gates in the 7400. Using the two 7400's on the schematic for 7410's of those gates leaves a remainder of two spares. We changed the 7404 into a 7400. We called all four of them "U2". List U2 in the spare count as having zero spares.

3. Examine the schematic. Note that only two 7410's remain. Review the specification sheet shown in Figure 8-5 for a 7410. There are three 7410's in the package. Call the two remaining on the schematic U3, which would have one spare. Show the spare in the spare column (listed on the side of the schematic). All spare gates will be drafted onto the schematic and labeled "spare". Label the spare gate with its pin numbers (inputs and outputs). This alternate practice exercise for grouping and combining demonstrates the same total count of IC's. In this case, there are three IC's: U1, U2, and U3. However, there is a different spare count. The schematic in Figure 8-27 has no spares as compared to the alternate practice exercise in Figure 8-28, which has one spare 3-input NAND. It does not exceed the three IC's that we finished with on Figure 8-27. However, the grouping in Figure 8-27 (concerning intercommunication and short trace lengths) is superior. On the other hand, while sacrificing some trace length, Figure 8-28 may be a better approach. This would be true if we require a spare gate, such as a 3-input NAND or a 2-input NAND, or a single input inverter. The 3-input NAND spare on schematic Figure 8-28 could act as all three gates. It could be the 7410 which is a 3-input NAND. It could be the 7400 (a 2-input NAND), or it could be the 7404 inverter. Refer to Figure 2-57 for combining rules. This becomes valuable if the engineer needs spare gates. So, both schematics are correct, Figures 8-27 and 8-28. If we are looking for short trace lengths on combining and grouping, then use Figure 8-27. If we are looking for spare gates, use the schematic in Figure 8-28.

NOTE: We will use schematic in Figure 8-27 for the layout.

Step #8—Calculate the required area. Before plotting the IC's and the discrete components within the PC board outline, determine the space requirements (illustration 6). For a review of calculating the required area, see Chapter 5.

COMPONENT COUNT (Equivalent 14/16 pin IC calculations)
Convert all components on the board to an equivalent 14/16 pin IC count using the following conversion factors:

	Total No. on Board		Equiv. 14/16 IC
14- and 16 pin IC	3	× 1.00 =	3 Sq In.
20 pin IC		× 1.25 =	
24 pin IC		× 1.50 =	
28 pin IC		× 2.00 =	
40 pin IC		× 3.5 =	
64 pin IC		× 6.5 =	
8 pin IC		× .75 =	
test points	4	÷ 10 =	Approx. ½ Sq. In.
jumper pads		÷ 10 =	

	Total No. on Board		Equiv. 14/16 IC
1/4 -watt resistors	2	÷ 7 =	¼ Sq In.
1-watt resistors		÷ 4 =	
disc type caps	2	÷ 7 =	¼ Sq. In.
large caps	1	÷ 4 =	¼ Sq. In.
3-pin components (Transistors, variable resistor, etc.)		÷ 3 =	
pre assigned connector pins used	4	÷ 10 =	½ Sq. In.
Other components (list)			

Total Equivalent 14 – and 16-pin IC = 1 ¼ Sq In.
Total Sq. Inch Req'd on P.C. Board = 4 3/4

Figure 8-29. Component's Equivalent to 14/16 Pin IC

Illustration #6: Space Requirements

As a rule of thumb, one square inch for each IC is a minimum for the average PC board. IC's are not the only components on a digital PC board. Use the 14- or 16-pin IC as a reference to the area required for these discrete components (such as resistors, capacitors, transistors, etc.).
The similarity between the IC's and discrete components is the number of pads used. Using the component equivalent chart in Figure 8-29, calculate the minimum area required for the schematic in Figure 8-27. Use the following procedures:

1. On the schematic in Figure 8-27, count the total number of 14- or, 16-pin IC's. Use the specification sheets in Figures 8-4 and 8-5. All of the IC's are 14-pin dual inline. Enter this on the component equivalent chart. Figure 8-29 shows that a total of three 14- or 16-pin IC's times one square inch equals a minimum requirement of 3 square inches for the IC's.

2. Refer to the schematic in Figure 8-27. Find the total number of ¼W resistors. There are two, R1 and R2. List this on the component equivalent chart in Figure 8-29 under "¼Watt resistor". Two on the board divided by seven equals ¼ square inch.

3. Refer to the schematic in Figure 8-27. How many disc type of capacitors are we going to have on our schematic? Note 3 on Figure 8-27 instructs us to bypass every two IC's with a CK06 disc capacitor. Because there are three IC's, we may end up with two disc capacitors. List this on the component equivalent chart in Figure 8-29 under "disc type caps". Total number on the board equals two divided by seven approximately ¼ square inch.

4. There is one input filtering capacitor on Figure 8-27 (called C1). List it on the component equivalent chart in Figure 8-29 as a large capacitor ("large caps"). There is only one on the board. It is divided by four which equals ¼ square inch.

5. Review the component equivalent chart in Figure 8-29. There are entries for test points and jumper pads. We do have 4 test points. List these four under "test points". Ten test points equal one square inch. Four test points divided by ten is approximately ½ square inch.

6. Review the component equivalent chart in Figure 8-29. There are entries for pre-assigned connector pins. Refer to the schematic in Figure 8-27. P1 has only 4 pre-assigned pins as voltage (+5) and ground. Assign this on the component equivalent chart; dividing four into ten equals approximately ½ square inch.

7. Add the column for IC's. There are a total of three and a half square inches: 3 square inches for the 14-pin IC's and ½ square inch for the test points. Add the column for discrete components. The total equivalent IC's, equals 1¼. Add the resistors, capacitors, and pre-assigned pins. They equal: 1¼ square inches. Add the IC column and the discrete component columns together. This equals a total of 4-3/4 square inches on the PC board.

NOTE: Review the board outline assigned in Figure 8-22. We calculated a total of seven square inches of useable area on the PC board. Compare that to the total square inches as a minimum requirement. The chart in Figure 8-29 shows a minimum of 4-3/4 square inches. We can place all components (IC's, resistors, capacitors, test points, etc.) within the board outline assigned in Figure 8-22.

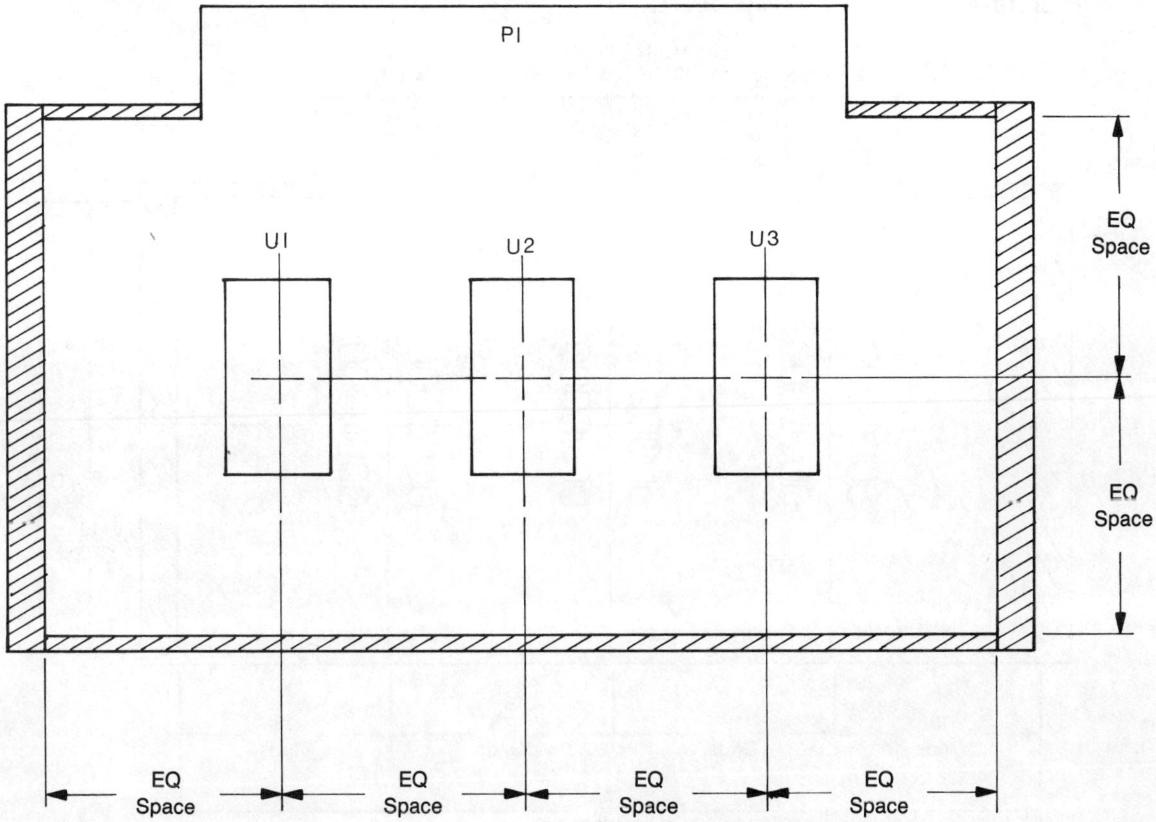

Figure 8-30. Equal Package Spacing

Step #9—Plot the grouped IC's, schematic (Figure 8-27) onto the PC board (Figure 8-24). Assign IC numbers to locations on the board. IC's should communicate to each other in line or in the same rows with each other (illustration 7). For a review on plotting, see Chapter 5.

Illustration 7: Plotting

Using Figure 8-27, plot the grouped IC's onto a board outline shown in Figure 8-30. Use the PC design template shown in Figure 8-8.

1. Notice that the gates in Figure 8-27 have been grouped to ensure signal flow. This flow determines how the PC board is to be layed out. For example, Figure 8-27 shows that IC, U1, communicates directly to IC U2 and U3. There are only 3 IC's. We know that there are no pre-assigned pins. Therefore, we are only interested in the intercommunication of IC's to each other. Line them in one common row. All three IC's are shown in Figure 8-30. Assign IC U1 to the center of the PC board so that it can communicate directly with IC U2 and IC U3. Equally space the IC's in one common row. The IC's are given equal consideration inside the useable board area. Label U1, U2 and U3. Note that all three IC's have equal access to connector P1.

2. Use the template in Figure 8-8. Draft in the pads by determining the equal spacing shown in Figure 8-30. Figure 8-31 shows the board outline with grids at a 2:1 scale. Counting the grids on the board gives a total useable area of 38 grids vertically (at 2:1) and 74 grids horizontally. Place the IC pads on the grid pattern. Center the IC pads on a grid cross-hair. Divide 74 grids by 4 to equal 18.5 grids for each space, giving you four equal spaces. Round this off to 18 or 19. In Figure 8-31, we placed the first IC pattern 18 grids (on the horizontal) to the center line of the IC. The next IC which made it fall on the center line of the board, at 19 grids. The third one is placed at 19 grids, which gives a balance of 18 grids for the last space. In the vertical

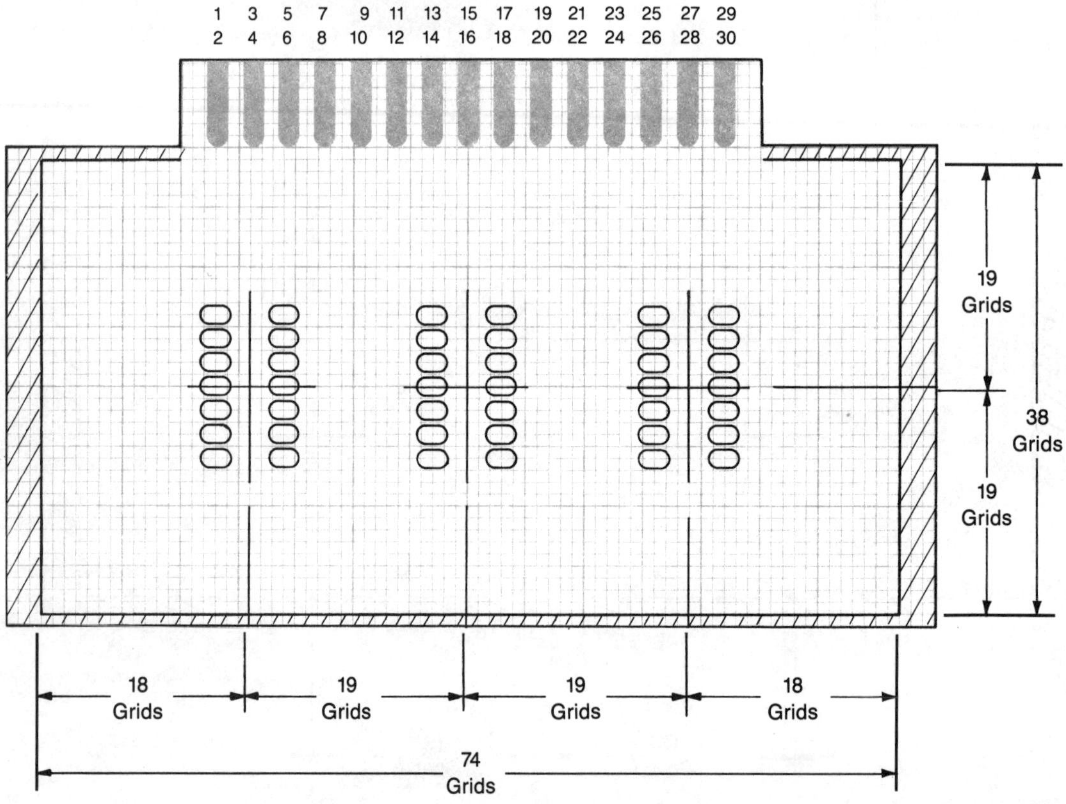

Figure 8-31. Scale 2 × 1

direction, 38 grids are divided by 2 equal spaces to equal 19 grids each. We found the center line of each IC. Using the template in Figure 8-8, we positioned the center of the IC pattern (14-pin) on the crosshairs and drafted in the pads.

Step #10—We will conduct a formal template layout at this time, establishing our conductor paths in an X-Y (horizontal-vertical) direction. View the layout from the component side of the printed circuit board (illustration 8).

Illustration #8: Formal Layout

The formal template layout is the next step. Using the IC plotted locations (from Figures 8-30 and 8-31) bypass the rough sketch normally done for an analog board. The following procedures itemize each design step.

1. Review the design rules for formal PC board layout (step 4). This establishes conductor/trace air gaps and widths along with the pad sizes for components and feedthroughs, tooling hole sizes, and card edge spacing minimums.

2. Identify the layout material. Use something similar to ACCUFILM®, an anti-static stable based matte film with a grid underlay (in this case, a .100 grid). See Figure 8-32.

3. We used the IC plot in Figure 8-30. We established the IC pad locations (Figure 8-31). We have itemized the beginning steps for our formal digital layout. See Figure 8-33.
   a. Place the U numbers inside each IC on the formal layout. Begin from the left (U2) then the center position (U1), and finally the extreme right (U3). Number each IC pad using the specification sheets in Figures 8-4 and 8-5. Start with pin 1 in the upper left hand corner. We could have rotated the IC position so that pin 1 would have been in the lower right hand corner. This is a personal choice.

Special Note

Remove the specification sheets from the book (Figures 8-4 and 8-5). You will

ACCUFILM® Anti-Static
Stable-Based MATTE
DRAFTING FILMS

- Anti-static...NO static electricity build-up from erasing, brushing, etc...ideal for pressure-sensitive PC artwork symbols.
- Stable-based...highly resistant to remperature and humidity changes. Thermal coefficient of linear expansion: 1.7 x 10 inches/inch/Centigrade; 9.5 x 10 inches/inch/Fahrenheit. Hygroscopic coefficient of linear expansion: 0.6 x 10 inches/inches/% Relative Humidity. Meets or exceeds all requirements of MIL SPEC: MIL-F-519C.
- Ultra high translucency insures superior reproduction.
- Excellent surface adhesion—accepts all commonly used drafting pens, pencils and inks.
- Sharness, clarity and quality of originals permits "scissors" drafting, or borrowing from earlier drawings.
- Reproducibility—resists "gain" during reproduction.
- Permanent.
- Erases easily without damage to film surface.
- Fast drying; maintians consistent line widths even after frequenty erasures.
- One-swipe cleaning...won't attract lint, dust and erasure residue.
- Over the long term, save 15% of the cost of drafting vellum, with improved sharpness and reproduction qualities.
- Waterproof, stainproof, curl-resistant.

Figure 8-32.

be using these extensively throughout the remaining part of the design.

b. Place the input capacitor, C1, using the template in Figure 8-8. Label RC20, close to the input pin called +5V, pins 29, and 30. Place the positive end of the capacitor close to those two pins. See Figure 8-33.

c. Use a color code to identify voltage and ground pads, and component side traces and noncomponent side traces. Red circled pads equal +5V, black circled pads equal ground. This means that the +5 voltage interconnects to each red color coded pad. It means that the ground bus line interconnects to each black pad and terminates to their respective pins on the P1 connector. Also we have identified red lines as component side and blue lines as the noncomponent side of the board. These codes should be placed on your layout, as shown on Figure 8-33.

Special Note

After identifying the component side with red traces and the noncomponent side with blue traces, number the connector pins as shown in Figure 8-33. All odd numbers (1, 3, 5, 7, through 29) are color coded red and labeled "component side for P1". All even numbers (2, 4, 6, 8, through 30) are color coded blue and labeled "noncomponent side for P1". Obtain this information from a connector specification sheet. We provide this information on the schematic in Figure 8-27.

Using the color coded dots, red equals +5V, and black equals the ground pads, examine the specification sheets in Figures 8-4 and 8-5. Identify the voltage and ground pins for U1, U2, and U3. For example, U1 (7473) has voltage called VCC (pin 4). VCC and +5V are the same. U2 (7410) in Figure 8-5 has VCC which is +5V (pin 14). U3 is the same. On U1 place a red dot on pin 4. On U2, place a red dot on pin 14. On U3, place a red dot on pin 14. This red dot indicates where to draft in the bus line for +5V. The line will come from pins 29 and 30 on the P1 connector. Color coding may seem unnecessary for a

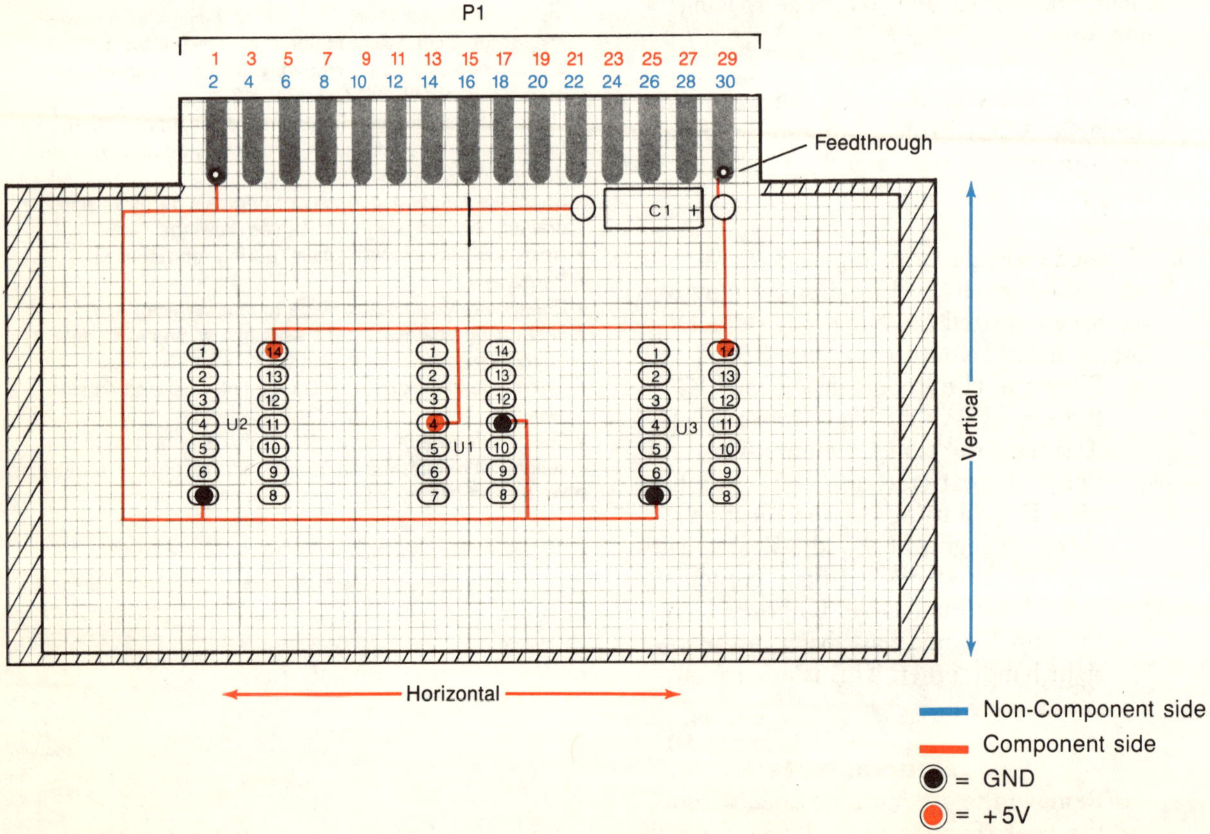

Figure 8-33.

3-IC layout. It is helpful on more complex PC boards. Color code the black pads called ground. Return to the specification sheet for U1 7473 (Figure 8-5). Identify ground as pin 11. Place a black dot on pin 11, which denotes ground. U2 7410 (Figure 8-5) denotes ground as pin 7. Place a black dot on pad 7 for U2 and the same for U3. This now identifies all the ground points for the ground bus to connect to. The ground bus starts with the P1 connector, pins 1 and 2.

4. Draw the bus line for voltage +5. Start from the connector pins 29 and 30. Refer to the schematic in Figure 8-27. Pins 29 and 30 are called out for the +5V line for P1. We must strap both sides of the PC board. Pin 29 is on the component side and pin 30 is on the noncomponent side. Following is the technique called placing a feedthrough on the connector finger. This conserves board space. We simply drill a hole which will be a feedthrough connecting pins 29 and 30. The size of this trace (called our +5 bus line) will be .100 at a scale of 2:1. We obtained this from the design rules, step 4. Figure 8-33 shows this to be .100 thick on the layout. However, this is not necessary. We could show a single line width to represent the wide trace (called our +5v). Figure 8-33 shows the wide and thin drafting of this bus line. Draw from the feedthrough on pins 29 and 30 (P1), connect to the capacitor (C1 positive end), continue to pin 14 on U3 (color coded red). Pin 4 on U1 is color coded red and pin 14 on U2 is color coded red.

5. Draw the ground bus connecting the black dots to connector pins 1 and 2 on P1. Refer to the schematic in Figure 8-27. The ground connects to pins 1 and 2 on P1. You have placed a feedthrough for +5V, pins 29 and 30. We will place a feedthrough on the connector finger for pins 1 and 2, strapping the component and the noncomponent sides together. Continue out of pins 1 and 2 (on P1) for ground. Connect to the negative end of C1, the input filtering capacitor. Continue down the side of the board. Cross and connect to the black pads for U2, pin 7. Connect

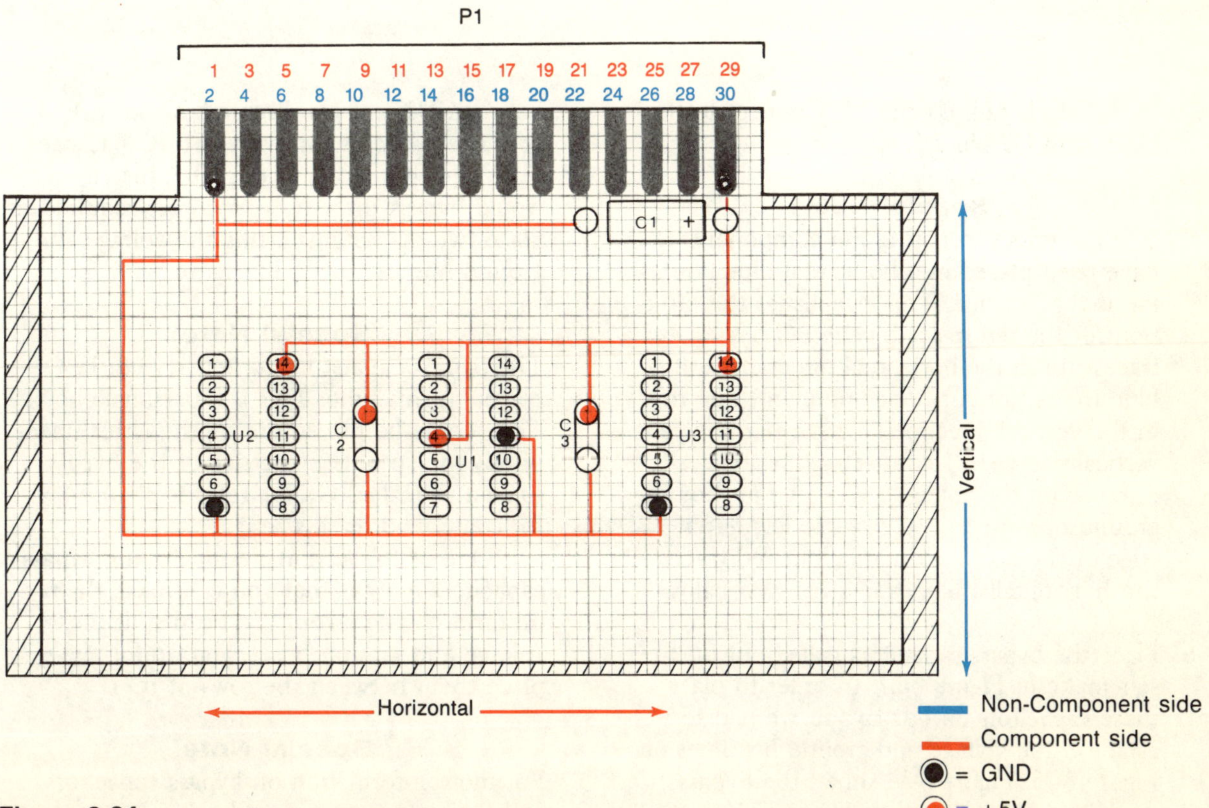

Figure 8-34.

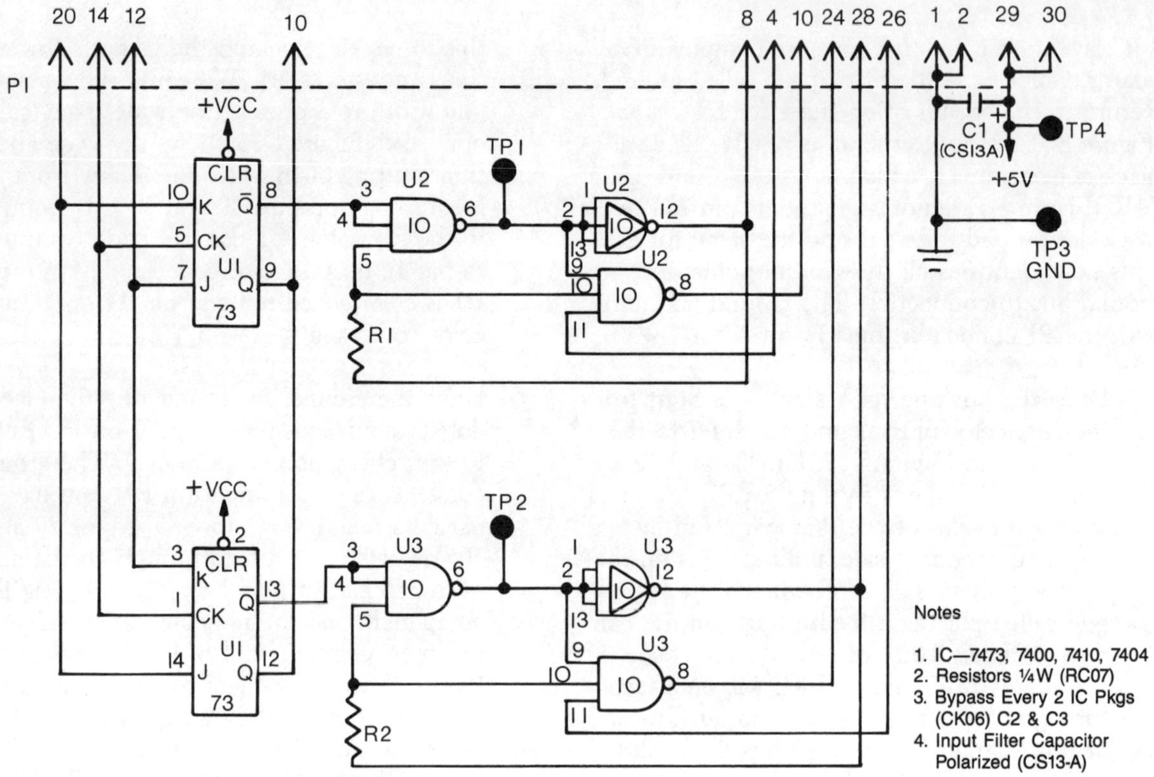

Figure 8-35.

to the black pad, U1 pin 11. Connect to the black pad, U3 pin 7.

Special Note

The red traces (denoting the component) have been placed in a horizontal direction as far as the bus line. This establishes the direction for the red traces on the PC board. Red traces run in the horizontal direction, and blue traces (for our noncomponent) side run in the vertical direction. We have connected vertically to the P1 connector (for a short distance on the voltage). We have pushed the ground bus line to the extreme left outer edge of the board. Primarily, our red traces run horizontally; blue traces run vertically.

6. Place the bypass capacitors called out on the schematic in Figure 8-27. In order to place these capacitors on our layout, we had to establish the voltage and ground bus lines on Figure 8-33. Figure 8-34 shows the bypass capacitors placed between U2 and U1 and between U1 and U3. Figure 8-27, note 4 specifies a bypass for every two IC packages. One capacitor is not enough because it would bypass three IC's. We were required to place two of them, splitting the voltage and ground bus (as shown in Figure 8-34).

Special Note

The bypass capacitors do not have to be placed at this time. This is a personal choice. They could be placed during the latter stages of design. To obtain the size of the bypass capacitors (called CK06), refer to Figure 8-9 for component case sizes.

One of the advantages of placing bypass capacitors early is that you can make them uniform on the layout. You can split the voltage and ground buses, and uniformly place them between the rows of IC's.

Special Note

For more information on bypass capacitors and their placement, see chapter 3.

Connect these capacitors to the voltage and ground bus (without feedthroughs, if possible). We will make an exception to the X-Y direction given for red and blue traces. Bus into the bypass capacitors with the color red off the voltage and ground bus (which is also colored red). They will carry the same trace widths as the voltage and ground bus (which will be .100 at a scale of 2:1). Identify these capacitors as C2 and C3. These will also be either noted on the schematic in note 4 or electrically diagrammed on the schematic in Figure 8-27.

7. Before interconnecting the signal lines and placing the discrete components, clean up the schematic in Figure 8-27. Refer to the schematic in Figure 8-35. Identify every trace line that connects from point A to point B. For each interconnection, label the pin numbers on the gates. Label the pin numbers on the connector (P1) while interconnecting each trace. This is done on the schematic in Figure 8-35 and the layout in Figure 8-36.

Connection Sequence

1. Start with discrete components that have both ends connecting to the same IC (such as R1). This is because both ends can connect to the same side of the IC. This saves the effort of routing around the IC. Place R1 (that connects to U2) on the high pin (numbered 8 through 14). Refer to Figure 8-9 for an RC07 configuration for R1. Place it down using the template. Place the pads onto the grid as shown in Figure 8-36. Label it R1. Refer to Figure 8-5. Use this specification sheet to identify the pin assignments for the inputs and outputs of the gates. R1 has two ends: one end is on the lower numbers of U2 (8, 9, or 10). The other end of R1 is on the higher numbers of U2: (11, 12, or 13). Choose number 10 which is the input of 7410 for the lower end of R1. Place the connection in the color red. Figure 8-35 shows the number 10 placed for the input of the gate. There is an additional input connecting to the same side of R1. It is easy to

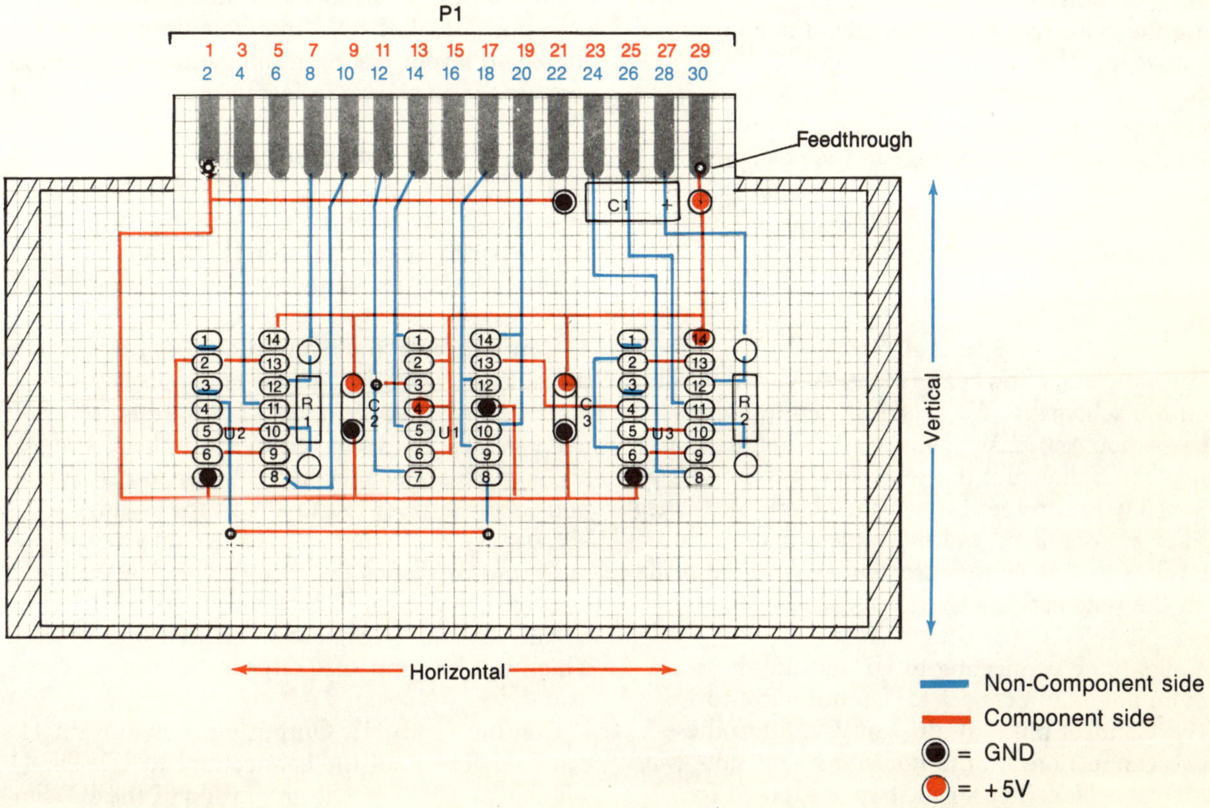

Figure 8-36.

continue the connection straight across to pin 5 on U2 (which is also an input of another gate).

Special Note

After assigning one input to a gate (as we did on U2 for input pins 5 and 10), assign all of the inputs and outputs for those respective gates. Figure 8-37 shows the inputs 5 and 10 on U2 (labeled on the schematic in Figure 8-35). Look at Figure 8-38. Assign the rest of the inputs and outputs to those two gates inside the U2 package. Notice the third gate on U2. Label the pins 1, 2, and 13 on the input and label pin 12 on the output. The reason is there were no other gates remaining. The numbers had to be assigned to that gate. You may change the inputs and outputs when you are assigning these numbers in advance. Later you can always change it for a design advantage. Use the specification sheet to obtain these numbers.

2. Continue with the R1 end that connects to U2, pin 12 (assigned previously). It was the last gate inside IC package 7410. It was the last assigned output. Connect R1 to pin 12 of U2 using the color red. From there, continue on to connector P1.

Change colors because a red trace crosses through a bus line. Change to the color blue. Draw the trace to easiest connector pin since there is not a pre-assigned one on the connector. Go to pin 8 on P1 and label that on the schematic in Figure 8-35.

Special Note

We are completing connections as we did in R1 on the schematic. We assigned these numbers to the various gates. We completed the entire connection. Yellow out the schematic using a yellow pencil (you can see through the color). This keeps track of completed and incomplete traces. This ensures that all of the interconnects will be made on the schematic.

3. Continue connecting to U2 because there are more points to connect to. Do not move to U3 yet. Connect pin 3 to pin 4 on U2. Strap these two connections on the noncomponent side using the color blue. The small strap is exposed for changes, if the engineer should decide to modify the board. Strapped on the component side, you may have to remove the entire IC to cut the connection. U2 (pins 3 and 4) also connects to U1. Consider the Q-Bar. Review the specification sheet for U1, 7473 (see Figure 8-5). There are two flip-flops in one package. Both have Q-Bars. Decide which flip-flop we will connect to. We have the choice of Q-Bar, pin 8, or Q-Bar, pin 13. Choose U1, pin 8 to connect to U2, pins 3 and 4. Draft a blue trace in the vertical direction down to the aisle closest to pin 8 of U1. Come out of pins 3 and 4 of U2 with a blue trace down to the aisle. Draw a feedthrough at the end of the blue trace. Draw another feedthrough just below U1, pin 8. Connect a blue trace to pin 8 on U1. Draft a red trace between the two feedthroughs. This completes the connection from U2, pins 3 and 4 to U1, pin 8. The feedthroughs may seem unnecessary. We could have made a complete connections just using blue (from U2 pins, 3 and 4 to U1, pin 8). Consider a more complicated board, with IC's below the row of U1, 2 and 3. Blue traces would run vertically through the IC's. You would be blocked by both red and blue in the vertical direction (because of your red voltage and ground buses). Draft in the feedthroughs; this does not disturb the X-Y red blue direction. After the board layout has been interconnected, we can remove the unnecessary feedthroughs.

Special Note

Assign all pin numbers to both JK flip-flops because we have identified one point on the flip-flop Q-Bar as being pin 8 for U1. Now we can identify all the remaining pins for both flip-flops.

4. Continue with the connections of U2. Examine the output in pin 6. It also is common with U2 inputs called 1, 2, 13, and 9. To make those connections, use the color red to strap pins 2 and 13, because we are connecting in a horizontal direction. Use the color red to strap pins 6 and 9 because we are connecting horizontally.

Now tie all five points together. Use a blue trace in the vertical direction to tie pins 1, 2, and 6. The remaining connection on that line is TP1 which will be connected later in the design.

5. Continue with U2. Output pin 8, input pin 11 can complete all of the connections to U2. Pin 11 will have a blue trace to the inside of the IC. Pin 4 is the easiest pin to connect on P1. Assign this

to the schematic, pin 4 for U2 and pin 11 on P1. Using the inside of the IC to route traces rather than the outside is a personal choice. To complete U2, connect pin 8 to the connector P1. Route in the easiest path. In this case, we took the outside of the IC and connected to P1, pin 10. Assign this to the schematic, P1. You have now completed all the connections to U2.

Consider U1 or U3. Since the circuits, U2 and U3, are identical, interconnect U3 before attempting U1. Transfer all of the numbers directly down (assignment to the input and output of the gate). Since the two circuits are identical, transfer the numbers down, duplicate the two circuits, and continue interconnecting.

6. Duplicate the numbers from the U2 circuit to the U3 circuits for inputs and outputs.

7. Continue interconnecting the inputs and outputs of U3 until completed. Draft R2 on the high side of U3 (similar to R1 in U2). Connect R2 to U3, pins 10 and 5, using a red trace. Red is primarily in the horizontal direction. To practice for a more complicated board, we would have blue traces coming up through U1, U2, and U3). Connect the opposite end of R2 to U3, pin 12, and the easiest connection to P1. Use blue because the outside of U3 will have no interference for the routing of blue traces in the vertical direction. Route the entire connection R2 to U3, pin 12 to P1, pin 28 in blue. Assign pin 28 to the schematic. Connect U3 pins 1 and 2, strap in blue. U3, pin 2 to U3, pin 12 draft in red. U3 pin 6 to U3 pin 9 draft in red. Now tie all five points together by connecting in the vertical direction in blue from U3 pin 2 to U3 pin 6. The only remaining connection is the test point TP2 which will be done later in the design sequence. To complete the connections for U3, connect U3, pin 8 to the easiest pin on P1. In this case use the inside of U3 to route the trace. Connect to P1, pin 24. Assign this to the schematic. Connect U3 pin 11 to the easiest pin on the P1 connector. Routing to the inside of U3, connect to pin 26. Assign this to the schematic. The last U3 connection is pins 3 and 4 (strapped in blue on the noncomponent side routing to U1, pin 13 in red). Route under and between the pads of C3 to cross over to U1, pin 13, from U3.

8. Complete the connections on U1. U1, pins 9 and 12 tie together with a blue trace in the vertical direction using the inside of U1. Continue in the vertical direction. Complete the connection to P1. Terminate to the closest pin (P1, pin 18). Assign this to the schematic. Connect U1, pin 1, to U1, pin 5, using the outside of U1. Use a blue trace in the vertical direction, terminating to P1, pin 14. Connect U1 to U3 on the outside of U1. Terminate to P1, pin 12. Route from U1, pin 7 to P1, pin 12, in blue. To tie into that trace with U1, pin 3, insert a feedthrough on that line. Connect with a red trace in a horizontal direction to complete the interconnect. Connect U1, pin 10, to U1, pin 14. Terminate P1, pin 20, in blue using the outside of U1. Use the outside in the vertical direction in the color blue because of the routed vertical blue trace connecting U1, pins 9, and 12 on the inside of U1. Assign pins 20 and 12 on P1 to the schematic.

9. All the signal traces for U1 have been completed. There are two connections left on the JK flip-flops, pins 2 and 6 (CLR). Both of these pins on U1 must tie to VCC (which is the +5V in this case). Tie pins 2 and 6 on U1 to the +5V bus line. The layout in Figure 8-3 shows the voltage bus line connecting to U1, pin 4, on the inside of the IC. This passes pin 2 to which we will tie off that bus line. Continue that line down. Connect to U1, pin 6, with the +5V bus line. This completes the connections on the schematic for U1.

10. At this point, we have completed all the interconnecting for the signal paths on the layout in Figure 8-36. We routed the voltage and ground buses. All inputs and outputs of gates should be numbered. All pins on P1 should be identified and every signal line and voltage line should be yellowed out as a completed connection. Test points will be covered in step 11 which will be the clean-up phase of the design. Also tooling holes, etc. will be placed.

Step #11—This step establishes the test point locations. It removes all unnecessary via holes (feedthroughs). In the formal layout (step 10) we placed feedthroughs, because the X-Y system was used for routing traces. Remove the vias that are not necessary (illustration 9).

Illustration #9: Layout Clean-up

Complete and clean up the layout in Figure 8-36. This step establishes test point and tooling hole locations, and cleans up all unnecessary feedthrough holes. It places all the IC's in numerical order from left-to-right, top-to-bottom (or any practical alphanumerical order on the PC board). This step matches Figure 8-35 with Figure 8-36.

When we group IC's, U1, U2, and U3 on the schematic, we did not know where their final placement would be on the board outline. There was no control between the numerical order and the intercommunication with each other and to the P1 connector. It would have been impossible to try to arrange the IC's in numerical order. After completing the layout, renumber the IC's. The schematic should correspond to those new numbers. The procedures below itemize the remaining steps that complete the formal digital layout. See Figure 8-39 for the final layout and Figure 8-40 for the final updated schematic.

1. Remove all unnecessary feedthroughs. Examine the layout in Figure 8-36. There are two feedthroughs used on the routed connection from U2, pins 3 and 4, to U1, pin 8. Can we complete the routed connection without the aid of the two feedthroughs? We can by routing from U2 pins 3 and 4 and U1 pin 8. Use the color blue to route the entire connection. For the removal of those feedthroughs, see Figure 8-39. Continuing to remove unwanted feedthroughs. Notice that U1, pin 7, connects to U1, pin 3, and also P1, pin 12, with the aid of a feedthrough off U1, pin 3. Can we eliminate that one feedthrough and still connect U1, pin 7, to U1, pin 3, and to P1, pin 12? The answer is no, with a qualification. We could connect U1, pin 3, to U1, pin 7; then connect all the way through U1, pin 7, to P1, pin 12, without the feedthrough. However, it would be a loopback. It is better to disconnect from P1, pin 12, since there is no pre-assigned pin connector. We can choose the pins. It would be much simpler to reroute by connecting U1, pin 7, to U1, pin 3, on the inside of U1 and by continuing up in the color blue, connecting to P1, pin 16. See the final layout in Figure 8-39 for an updated version. Update the schematic by changing P1, pin 12, to P1, pin 16. See the final schematic in Figure 8-40.

2. Renumber the IC U numbers on the layout in Figure 8-36. Use numerical order, starting from the left and proceeding to the right. U2 would now be U1. Ul would now be U2, and U3 would not change. See the final layout in Figure 8-39. On the final schematic (Figure 8-40), change U2 to U1, and U1 to U2.
The reason that we change the IC numbers and place them in numerical order is to help the engineer or technician working on or testing the PC board. If a complex high-density PC board with 50 or 100 IC's did not have them in numerical or alphanumerical order, it would be difficult to reference a particular IC.

3. One of the final clean-up steps includes color change. It is better to have one continuous

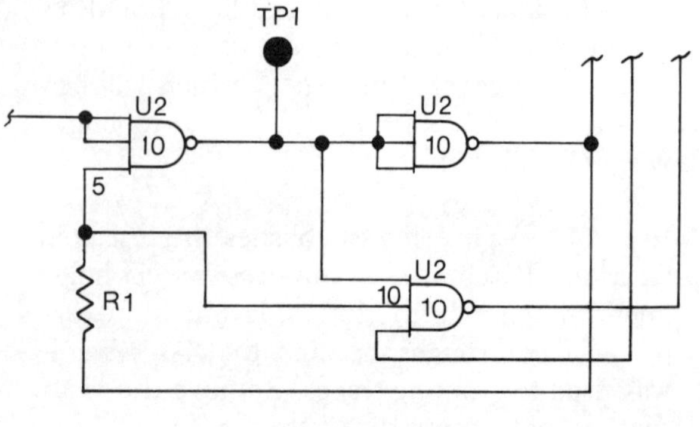

Figure 8-37. Pin Assignment

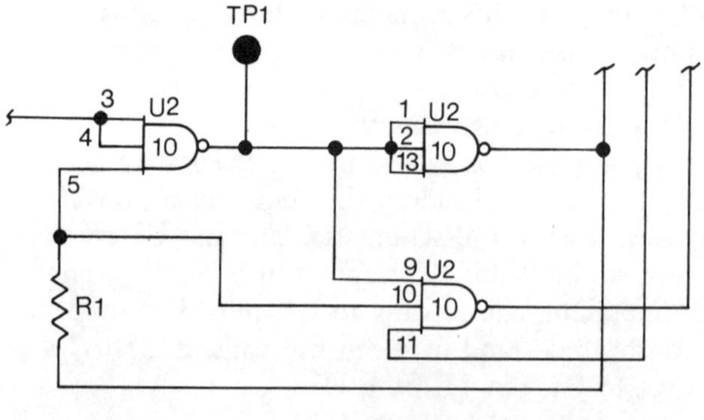

Figure 8-38. Pin Assignment to Respective Gates

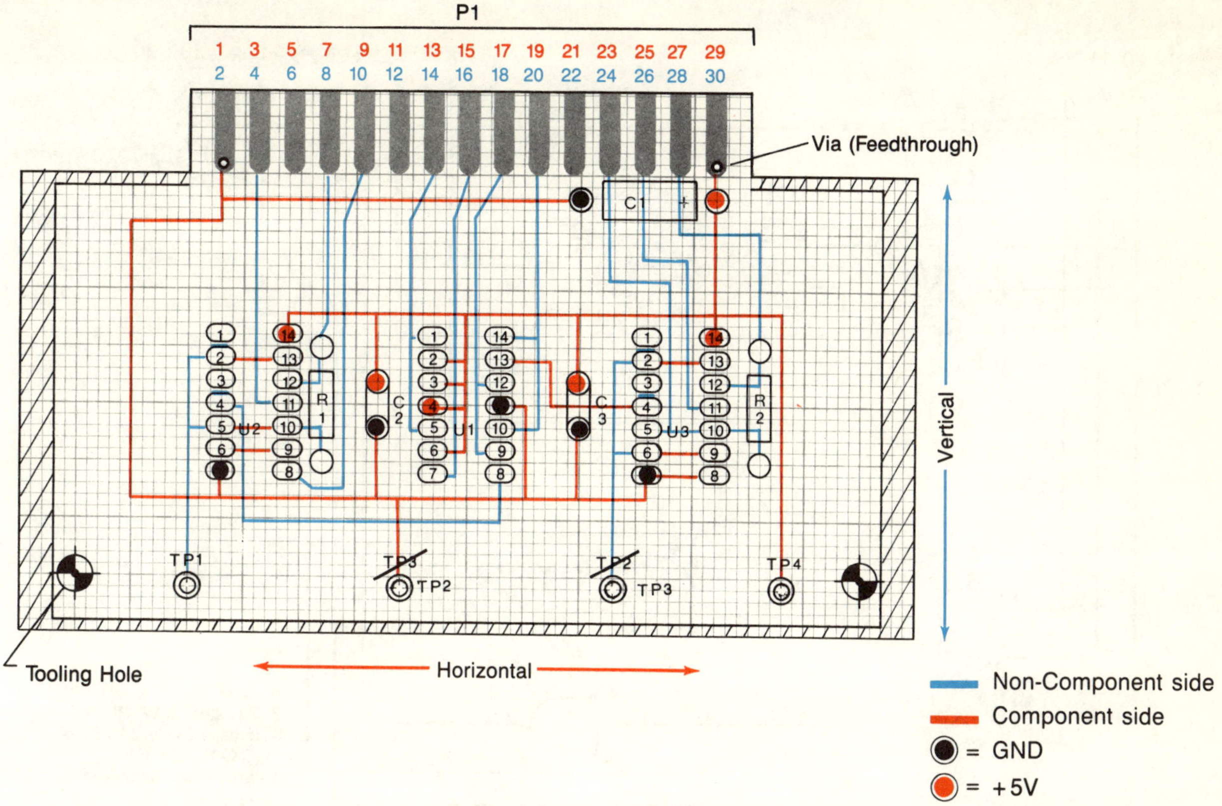

Figure 8-39. Final Layout

connection on the same side of the PC board (component side or noncomponent side). For example the layout in Figure 8-36 shows a connection from U1, pin 12 to R1 in red. It continues from R1 to P1, pin 8 in blue. It is better to have one continuous connection from P1, pin 8, R1 in blue. Then continue from R1 to U2, pin 12, in blue. We can only do this when it does not violate a crossover in colors. Do not change it arbitrarily. It may cause a direct, short crossing of two different colors.

4. Establish the location for the test points. When placing test points, consider the accessibility to the PC board itself. This board slides in the card tracks and connects to a female connector in the back of the unit. Our accessing point will be the opposite edge of the P1 connector. Place the test point on that edge. Using the schematic in Figure 8-35, start with TP1. Notice that TP1 is common to the blue vertical line connecting from U2, pins 1 and 2, to U2, pin 6. Continue that blue vertical line down to the accessing edge of the PC board. Place the test point close to the back edge of the board. Use a large pad (.250 diameter at a 2:1 scale). Test points typically require larger pads because of the size of the terminal or stake that is used. For a review of test points, see Figure 8-17. Connect TP2. Notice that TP2 is common with the blue vertical line connecting U3, pins 1 and 2, to U3, pin 6. Continue the blue vertical line down to the back edge of the PC board. Place a .250 diameter pad. Continue to TP3 which is a ground test point on the schematic in Figure 8-35. The ground test point will be used when testing any signal path such as TP1 or TP2. Since it would be used with those terminals, split the difference between the two by placing TP3 between TP1 and TP2. The final test point is TP4. The schematic in Figure 8-35 shows that TP4 is the +5V line test point. Route the +5V to the extreme right of the layout using the

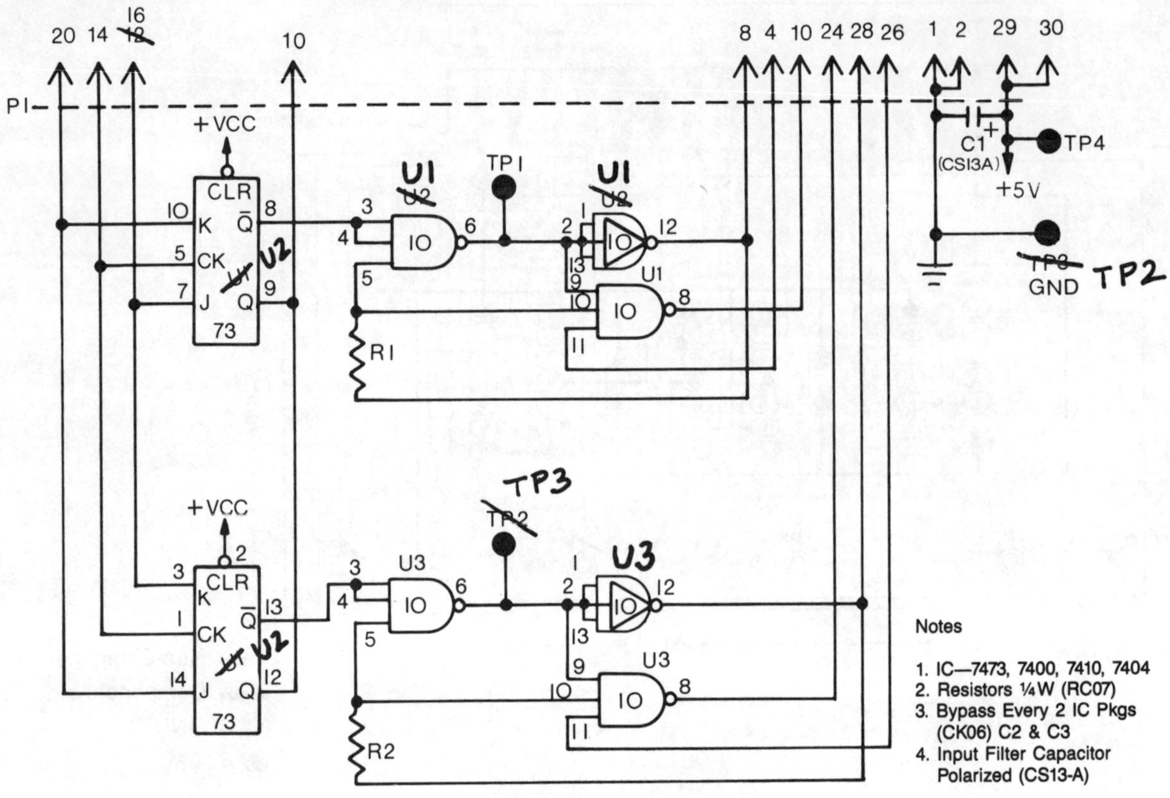

Figure 8-40. Final Schematic

color red. Place the .250 diameter pad at the back edge of the board. Like IC locations as far as placing them in numerical order, place the test points in numerical order by numbering left-to-right. We have placed TP1 on the extreme left, which will remain TP1. TP3 was between TP1 and TP2. Change TP3 to TP2. Change TP2 to TP3. TP4 remains the same. The test points now read in numerical order from left-to-right. The reason we do this is to make the engineer's and the technician's job easier. These changes must be done on the final schematic in Figure 8-40.

5. Adding tooling holes. Tooling holes must be added as part of the layout procedure, because they require a useable area of the PC board. Tooling holes are used as PC board reference points which other dimensions are based. For example, during the manufacturing process, the manufacturer will tool up to design his fixtures for drilling and routing the outer edges of the PC board. He uses these tooling holes as a reference point. The dimensions (in reference to the tooling holes) must be located accurately. Dense boards may not have sufficient room to locate the pads for tooling holes. Locate them outside the perimeter of the board outline. They will appear on the final positive and negative film which the manufacturer will use to produce the board. The same accurate dimensioning applies. After the board is completed, they would be trimmed away with the outer access material. This is a

disadvantage for an inspection department. They need these tooling holes to measure and to check the PC board dimensionally.

The typical hole size for tooling holes is .125 inch in diameter. A target material is usually left around the hole as a means of registering the board interior (traces and pads) to these tooling holes. For placement on the final layout in Figure 8-39, place the tooling holes in the lower left and right hand corners. There is a minimum of two. Large boards should have three. Place these tooling holes on the grid (in this case, both .250 in the horizontal direction, and in the vertical direction, in each corner, at a scale of 1:1 on the layout in Figure 8-39, five grids vertically and horizontally from each corner).

Step #12—After completing the layout drawing, an artmaster is taped. The artmaster is made to an enlarged scale (in this case 2:1). Then it is photoreduced to a 1:1 scale negative and positive (photomaster). This becomes a tool for manufacturing the PC board. The schematic in Figure 8-3 will not be taped in this exercise; taping will be done for the schematic in chapter 9.

Step #13—The information used for this PC board becomes the foundation for future boards.

Summary

We listed all the points that are unique to the schematic. We listed the procedures to find the information necessary to lay it out. We did not itemize the steps for the layout. We listed only what was important to this particular schematic (each schematic that we are doing in this book will have different point of interest).

This particular schematic (called board 1, 3-IC circuit), shows. What does a designer think about first when handed his digital schematic? We provided a list of questions to ask the engineer. We determined the board size. How many components will fit within the board? Once the board size has been determined, where do you place the IC's? How do you plot them onto the board outline? You were presented with three beginning steps.

To start the layout, we first had to identify the components and gates, using the specification sheets provided by the engineer. When we were using the specification sheets, we had to determine the component shape size, spacing, pad size, and hole size, using the component mounting information in Table 8-1. We explained each column in the table.

Second, the board outline was determined. Without knowing the board outline, we cannot calculate the useable area on the schematic. We considered the environment of the PC board. We discussed card tracks, which edge of the board is accessible, whether there are any special mounting instructions, etc. We practiced grouping logic gates. Even with a simple schematic, we found that there were different ways to group logic gates. We considered whether the engineer needed spare gates, or needed short lead lengths, or just ignore the spare gate count. We also calculated the area, once the schematic was grouped.

Third, we showed the use of color coding techniques to set up voltage and ground bus lines. We presented a sequence to make the formal layout. We plotted the voltage and ground bus lines which established our X-Y direction for the proper color. We interconnected the signal paths. We placed the discrete components. After the formal layout, we discovered there is a clean-up phase to remove unwanted feedthroughs and to add test points, tooling holes, etc.

There is a difference in board layout technique between digital and analog design. Digital boards require no free hand sketching for the layout. You go directly into a formal template layout once the schematic has been grouped. We introduced the concept of digital design. We provided the formulas to calculate spacing, size of holes, etc. We presented questions to ask the engineer. We discussed the beginning steps of the actual layout: grouping, plotting the IC's, and the routing of voltage, ground, and signal path traces.

Quiz—Chapter 8

1. Before starting your layout the schematic must be readable: all symbols, REF designations and component identification must be supplied by the______________.
2. The schematic provided by the engineer usually has no numbers assigned to the gate ______________ or outputs.
3. The connector specification sheet will help to identify or verify connector ________ numbers.
4. Pre-assigned pin's on the connector is one of the questions which should be asked by the P.C. Board ____________________.
5. The schematic is ____________ a complete source of information.
6. To determine the component lead spacing and body size the designer must consult the ____________ sheet.
7. IC specification sheets could give you different case styles to choose from such as flat packs and __________________ packages.
8. The style most often used today is the ______________________.
9. The IC specification sheet helps to identify case size & style and ____________ numbers.
10. The density of components will determine your pad size and ________________________.
11. The ______________ diameter will be used to compute the hole size and pad size.
12. If the designer does not have the case outline on his/her template they would be required to compute the pad spacing by the infomation provided by the specificaion sheet for a resistor it would require:
    1. Body: __________________
    2. Body: __________________
    3. Lead: __________________
13. In this chapter the input filtering capacitor is called out as a case code CS13-A, typically the engineer would provide only the component:
    1. Value
    2. ________________________
    3. ________________________
    4. ________________________
14. Using the formula for axial lead components to find the required pad spacing, the maximum body length was added to the minimum lead length of ____________ on each side of the component.
15. When using minimum dimensions to compute the pad spacing of an axial lead component the designer should round ____________ to the nearest grid.
16. In this chapter the input filter capacitor CS13-A is an electrolytic capactor. Question is it's composition.
    A. Ceramic
    B. Mylar
    C. Metal Foil
    D. Dipped Mica
17. When selecting component outline's on a template. The designer should select a size that is ______________ to the pad spacing and body dimensions that were calculated.
18. The template is only a __________ it is not a precision instrument.
19. The template show a rough body outline & ____________________.
20. In this chapter you are given a commerical & MIL standard formula to compute the minimum & maximum hole size, from the ________________________
21. The component mounting table provided in this chapter is used to find the __________ size and ______________ size of each component.

22. The component mounting table needs the minimum clearance around the lead of a component. For good commercial practices the designer should use ________ for MIL standard ________.
23. The component mounting table needs the maximum clearance around the lead of a component. For good commerical practices the designer should use .020 for MIL standard ________
24. The component mounting table column 11 would like to have the same size hole for every component on the P.C. board if possible. Why?

____

____

25. The component mounting table colmum 12 is used to find the minimum pad idze based from an annual ring of .015 over the maximum hole. What would be the smallest annular ring you could use in this book.

____

26. The component mounting table column #13 will select a pad size. This pad size in most companies is based on which sizes are ______.
27. Before the board can be layout out, the designer must establish the board layout design rules: List four different standards:
    1. ________________
    2. ________________
    3. ________________
    4. ________________
28. All component lead holes shall be located at grid intersection if ________________.
29. Components that require hold-downs because of shock or vibration can be secured by: List three methods:
    1. ________________
    2. ________________
    3. ________________
30. Test points: when possible all test points should be brought out to an ________________ edge of the board.
31. Traces on most boards should not come closer to the edge of the board than ____________ min. at 1 x 1.
32. Feed through hole sizes shall be determined by using a _____ __________ diameter rather than a component lead diameter.
33. Radial lead components could have their lead spacing
    A. Same as the components
    B. ________________
    C. ________________
34. Questions to ask the engineer: the question "Is there any special size components or are all components standard?" Means:

____

____

35. The question - "Are there critical leads on the board " __________

____

or ________________

36. Input filtering capacitors if required belong very close to the: ________________ pin on the connector.
37. The designer calculates the maximum size of all the components to insure that all components will fit on the: ________________
38. The P.C. designer must have all the pertinent packaging information about the board size & ________________ interface.
39. Cards tracks are used to secure the P.C. board to the chassis. Card tracks should have clearances of __________ for both top & bottom edges.
40. The designer should group and combine IC gates to find:

____

41. The designer should group and combine gates that most obviously ______________ to each other.

42. Space requirements—as a rule of thumb, ________ square inch for each IC is minimum for the average P.C. board.
43. For calculating area required—The similarity beween the IC's and discreet components is the number of ____________ used.
44. Plotting IC's - IC's should communicate to each other ________ or in the same __________ with each other.
45. A color code should be used to identify:
    1. ____________________
    2. ____________________
    3. ____________________
    4. ____________________
46. The first lines to be placed on the P.C. board layout are the:

    ____

    ____
47. On digital boards component side traces establish one direction vertical or horizonatal and the non-component traces establishes the ____________________ direction.
48. When completing connections on your schematic, the designer should yellow out the schematic. Why?

    ____

    ____
49. Layout clean-up-this step is used for test points, tolling holes and removal of unnecessary ____________________ holes.
50. Another final clean-up step includes color change? Why?

    ____

    ____

    ____

The answers for each question can be found on the pages listed below:

Question 1. Pg. 211
Question 2. Pg. 211
Question 3. Pg. 211
Question 4. Pg. 211
Question 5. Pg. 211
Question 6. Pg. 212
Question 7. Pg. 212
Question 8. Pg. 212
Question 9. Pg. 215
Question 10. Pg. 215
Question 11. Pg. 215
Question 12. Pg. 216
Question 13. Pg. 217
Question 14. Pg. 217
Question 15. Pg. 220
Question 16. Pg. 220
Question 17. Pg. 220
Question 18. Pg. 220
Question 19. Pg. 220
Question 20. Pg. 220
Question 21. Pg. 221
Question 22. Pg. 223
Question 23. Pg. 223
Question 24. Pg. 224
Question 25. Pg. 224
Question 26. Pg. 225
Question 27. Pg. 225
Question 28. Pg. 225
Question 29. Pg. 225
Question 30. Pg. 225
Question 31. Pg. 226
Question 32. Pg. 228
Question 33. Pg. 228
Question 34. Pg. 229
Question 35. Pg. 230
Question 36. Pg. 230
Question 37. Pg. 230
Question 38. Pg. 231
Question 39. Pg. 232
Question 40. Pg. 233
Question 41. Pg. 233
Question 42. Pg. 237
Question 43. Pg. 237
Question 44. Pg. 238
Question 45. Pg. 241
Question 46. Pg. 242
Question 47. Pg. 242
Question 48. Pg. 245
Question 49. Pg. 247
Question 50. Pg. 248

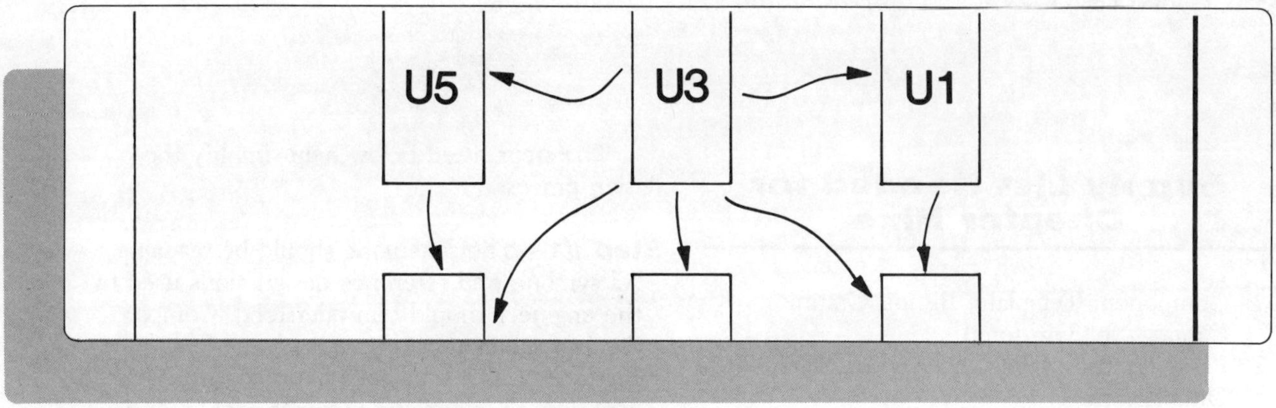

9 DESIGN PROCESS FOR 6-IC DIGITAL BOARD

The following chapter illustrates a step-by-step process to design a 6-IC digital printed circuit board using an X-Y system on a two-layer/double sided board. One other additional feature of this chapter will be the two-sheet procedure for taping. You will learn to design from an engineer's digital logic schematic and how to calculate the area for that schematic. You will also be given an itemized list of questions to ask the engineer about his schematic. This list will become the foundation for restrictions or special requirements.

Questions will be borrowed from Chapter 8 along with their explanation. You will learn about the design and placement of variable components such as a variable resistor (sometimes called a pot). There will be two different types of logic on the schematic. Open collector and non-open collector (totem pole logic).

We will discuss the running of traces between IC pins compared with a layout that does not have traces running through the pins. Special points of interest will be presented on the taping exercises.

Supply List Needed for Chapter Nine

1. Component Template: Bishop Graphics Catalog No. 3367 (one)
2. Black Drafting Film Lead Pencil: Mars 100-30 Duralar (one)
3. Colored Pencils (dark colors): Venus Col-erase, or equivalent (red, green, yellow, blue) one each
4. Eraser, Drafting Film: Castell #1954 or equivalent (one)
5. Drafting Dots
6. Knife: X-acto, Bishop Graphics Catalog No. 3141
7. Knife Blades: X-acto, Bishop Catalog No. 3116
8. Donut Pads: .150", 50, Bishop Graphics Catalog No. D144*
9. Donut Pads: .250", 4, Bishop Graphics Catalog No. D141*
10. Precision Slit Tape: Black Crepe .050", 2 rolls Bishop Graphics Catalog No. 20105011
11. Precision Slit Tape: Black Crepe .100", 1 roll, Bishop Graphics Catalog No. 20110011
12. Connector Pattern: .156" C-C, 30 Fingers, Bishop Graphics Catalog No. 6720*
13. Reduction Mark: 8, Bishop Graphics Catalog No. 4052*
14. Datum Target (tooling holes): 4, Bishop Graphics Catalog No. 4011*
15. Drafting ACCUFILM® : matte one side, .003", 8½" x 11", 2 sheets
16. Drafting ACCUFILM® : matte one side, .005", 8½" x 11", 2 sheets
17. Dual In Line Pattern: 14-lead, round pads, 24, Bishop Graphics Catalog No. 6237*

* Bishop catalog number shown for these products represent order numbers for Bishop's standard industrially packaged products. The package quantities for these catalog numbers may be greater than supplied with the "Digital" project kit, #10009C, available from Bishop.

All the steps listed in Chapter 8 apply to this 6-IC digital schematic. For example, we will borrow Step 4 which establishes the board layout design rules. There are minor changes for this chapter, which we will cover.

The steps listed below help simplify the design process.

Step #1—The schematic should be readable. All symbols and reference designations used by the engineer should be understood. Compare the two schematics (Figures 9-1 and 8-40). Figure 8-40 is a final schematic; Figure 9-1 is an early one to which the designer adds information as he proceeds into the layout. Notice that Figure 9-1 has no input or output pin number assignments on the gates, except for two IC's. One is numbered 7495 and the other is numbered 74141. The engineer had to pre-assign these types of IC's because of the way they intercommunicate to the rest of the logic gates. This electrical function is the responsibility of the engineer. He/she determines how they electrically connect to the gates within the logic circuit. You must use the pins assigned to the schematic on these types of IC's.

Comparing the two, we find that P1 is a male connector on the PC board. In both cases there are no pre-assigned pins for the signal paths. However, in both cases there are pre-assigned pins for voltage and ground on the PC board: pins 1 and 2 for ground, and pins 29 and 30 for +5V.

Figure 8-40 had its discrete component reference designations numbered on the schematic; the case code was given for each component. However, no values were given.

Figure 9-1 does have reference designations and values or part numbers. These values will be used along with the specification sheets to find the component case size. Examine the discrete components. We must ask the engineer about the variable resistor: its placement, case size, accessibility, etc. Figure 9-1 has more gates and IC's than Figure 8-40. Other differences concerning the types of gates being used (or IC's), component values, and types. Is the schematic readable? Proceed to Step 2.

Step #2—The engineer provides the specifications sheets. See Illustration 1.

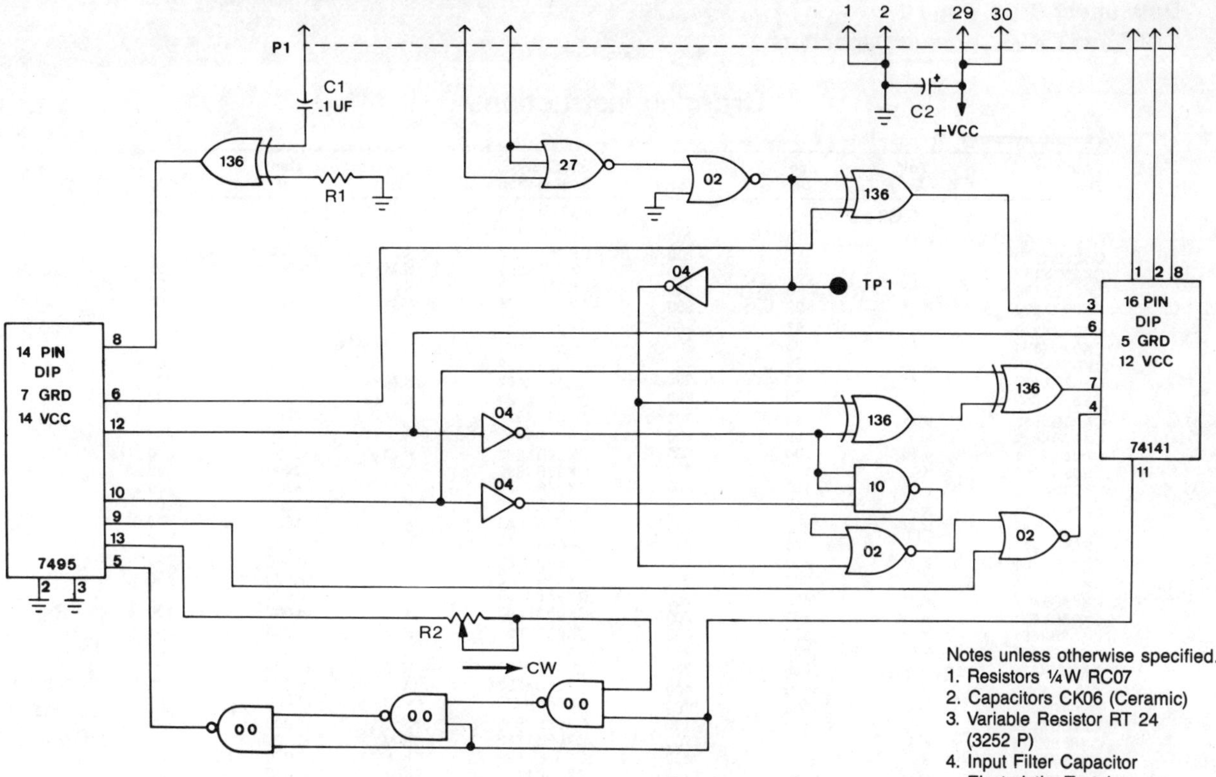

Notes unless otherwise specified.
1. Resistors ¼W RC07
2. Capacitors CK06 (Ceramic)
3. Variable Resistor RT 24 (3252 P)
4. Input Filter Capacitor Electrolytic, Tantalum
5. Bypass every 2 IC pkgs .1 UF, 100V ±10% (Ceramic)

Figure 9-1. Digital Schematic

Illustration #1: Component Identification

Specification sheets identify the case and body size of each component on the schematic (discrete resistors, capacitors, diodes, and integrated circuits). The engineer provides the specification sheets of all the components on the schematic in Figure 9-1. The following is an itemized breakdown from the specification sheets for each component.

Capacitors

C1: this capacitor is non-polarized. Note 2 in Figure 9-1 calls out a CK06 ceramic case style. For the specification sheet, see Figure 9-2.

C2: this is polarized and an input filtering capacitor. Note 4 of schematic Figure 9-1 calls out an electrolytic tantalum capacitor. For the specification sheet, see Figures 9-3A and 9-3B.

C3, C4 and ?: these are bypass capacitors. Note 4 of schematic Figure 9-1 calls out bypass every two IC packages. The quantity will be unknown until the total number of IC packages has been determined. However, we do know that they are ceramic capacitors, the same style as C1. Therefore, use the same specification sheet for C1, Figure 9-2.

Resistors

R1: this is a fixed resistor covered in Note 1 of Figure 9-1. It is a 1/4W RC style of resistor. For the specification sheet, see Figure 9-4.

Data Sheet C13F, Page 2
CK05 and CK06 Ceramic Capacitors

MIL-C-11015

Ordering Instructions

Military Designation	Tol* Avl.	Cap. pF	Vdc
CK05			
CK05BX100	K,M	10	200
CK05BX120	K	12	200
CK05BX150	K,M	15	200
CK05BX180	K	18	200
CK05BX220	K,M	22	200
CK05BX270	K	27	200
CK05BX330	K,M	33	200
CK05BX390	K	39	200
CK05BX470	K,M	47	200
CK05BX560	K	56	200
CK05BX680	K,M	68	200
CK05BX820	K	82	200
CK05BX101	K,M	100	200
CK05BX121	K	120	200
CK05BX151	K,M	150	200
CK05BX181	K	180	200
CK05BX221	K,M	220	200
CK05BX271	K	270	200
CK05BX331	K,M	330	200
CK05BX391	K	390	200
CK05BX471	K,M	470	200
CK05BX561	K	560	200
CK05BX681	K,M	680	200
CK05BX821	K	820	200
CK05BX102	K,M	1000	200
CK05BX122	K	1200	100
CK05BX152	K,M	1500	100
CK05BX182	K	1800	100
CK05BX222	K,M	2200	100
CK05BX272	K	2700	100
CK05BX332	K,M	3300	100
CK05BX392	K	3900	100
CK05BX472	K,M	4700	100
CK05BX562	K	5600	100
CK05BX682	K,M	6800	100
CK05BX822	K	8200	100
CK05BX103	K,M	10,000	50
CK05BX123	K	12,000	50
CK05BX153	K,M	15,000	50
CK05BX183	K	18,500	50
CK05BX223	K,M	22,000	50
CK05BX273	K	27,000	50
CK05BX333	K,M	33,000	50

Military Designation	Tol* Avl.	Cap. pF	Vdc
CK05BX393	K	39,000	50
CK05BX473	K,M	47,000	50
CK05BX563	K	56,000	50
CK05BX683	K,M	68,000	50
CK05BX823	K	82,000	50
CK05BX104	K,M	100,000	50
CK06			
CK06BX122	K	1200	200
CK06BX152	K,M	1500	200
CK06BX182	K	1800	200
CK06BX222	K,M	2200	200
CK06BX272	K	2700	200
CK06BX332	K,M	3300	200
CK06BX392	K	3900	200
CK06BX472	K,M	4700	200
CK06BX562	K	5600	200
CK06BX682	K,M	6800	200
CK06BX822	K	8200	200
CK06BX103	K,M	10,000	200
CK06BX123	K	12,000	100
CK06BX153	K,M	15,000	100
CK06BX183	K	18,000	100
CK06BX223	K,M	22,000	100
CK06BX273	K	27,000	100
CK06BX333	K,M	33,000	100
CK06BX393	K	39,000	100
CK06BX473	K,M	47,000	100
CK06BX563	K	56,000	100
CK06BX683	K,M	68,000	100
CK06BX823	K	82,000	100
CK06BX104	K,M	100,000	100
CK06BX124	K	120,000	50
CK06BX154	K,M	150,000	50
CK06BX184	K	180,000	50
CK06BX224	K,M	220,000	50
CK06BX274	K	270,000	50
CK06BX334	K,M	330,000	50
CK06BX394	K	390,000	50
CK06BX474	K,M	470,000	50
CK06BX564	K	560,000	50
CK06BX684	K,M	680,000	50
CK06BX824	K	820,000	50
CK06BX105	K,M	1,000,000	50

Same As .1mp

*K = ±10%, M = ±20%

Dimensions are in inches

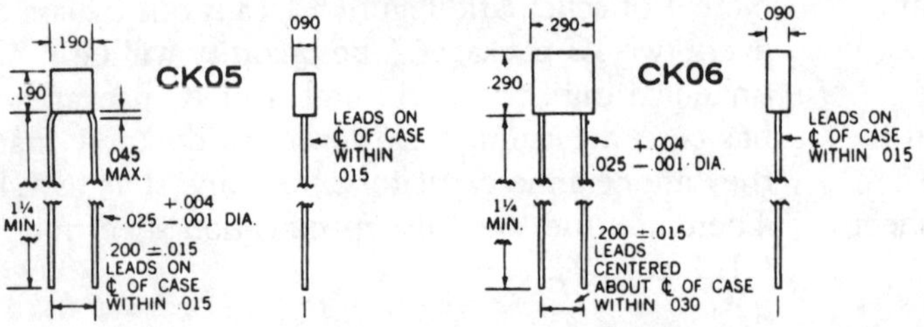

Dimensions are ±.010 except where noted

Metric Conversion Table

Inches	MM	Inches	MM
.001	.03	.090	2.29
.004	.10	.100	2.54
.010	.25	.190	4.83
.015	.38	.200	5.08
.025	.64	.290	7.37
.030	.76	1.250	31.75
.045	1.14		

Vitramon®

Vitramon North America
Box 544
Bridgeport, Connecticut 06601
TEL. (203) 268-6261
TWX 710-450-3640
Division of Vitramon, Incorporated
Bridgeport, Connecticut 06601, U.S.A.

Subsidiaries & Divisions
Vitramon North America
Vitramon Limited (United Kingdom)
Vitramon Pty. Limited (Australia)
Vitramon Japan Limited (Japan)
Vitramon GmbH (West Germany)
Vitramon France S.A.R.L. (France)

Figure 9-2. Spec Sheet

MIL-C-39003/1C
15 May 1970
SUPERSEDING
MIL-C-39003/1B
26 March 1970

MILITARY SPECIFICATION SHEET

CAPACITORS, FIXED, ELECTROLYTIC, TANTALUM, SOLID-ELECTROLYTE,

ESTABLISHED RELIABILITY, STYLE CSR13

The complete requirements for procuring the capacitors described herein shall consist of this document and the latest issue of Specification MIL-C-39003.

This specification is mandatory for use by all Departments and Agencies of the Department of Defense.

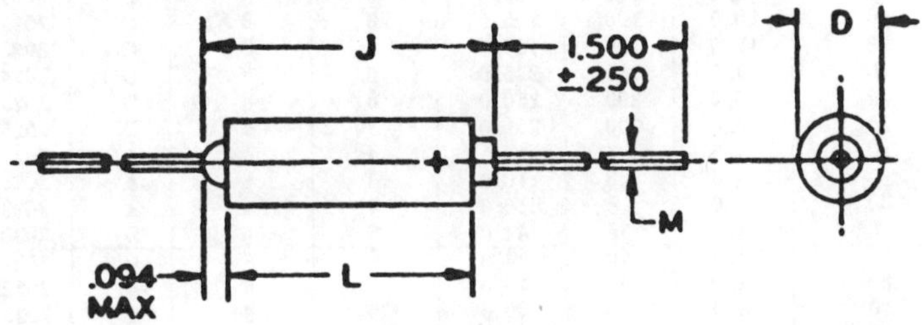

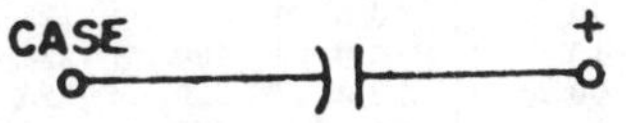

CIRCUIT DIAGRAM

Case size	Dimensions L ±.031	D +.016 −.015	M ±.002	J Max
A	.286	.135	.020	.422
B	.474	.185	.020	.610
C	.686	.289	.025	.822
D	.786	.351	.025	.922

INCHES	MM	INCHES	MM	INCHES	MM
.002	.05	.135	3.43	.474	12.04
.015	.38	.185	4.70	.610	15.49
.016	.41	.250	6.35	.686	17.42
.020	.51	.286	7.26	.786	19.96
.025	.64	.289	7.34	.822	20.88
.031	.79	.351	8.92	.922	23.42
.094	2.39	.422	10.72	1.500	38.10

NOTES:
1. Dimensions are in inches.
2. Metric equivalents (to the nearest .01 mm) are given for general information only and are based upon 1 inch = 25.4 mm.
3. The case insulation shall extend .015 (.38 mm) minimum beyond each end. However, when a shrink-fitted insulation is used, it shall lap over the ends of the capacitor body.

Figure 9-3A. Spec Sheet

MIL-C-39003/1C

TABLE I. Electrical characteristics and part numbers.

DC rated voltage	Capacitance	Capacitance tolerance	DC leakage at			Dissipation factor at		Case size	Part No. M39003/01- Failure rate level for (%/1,000 hr)				
			+25° C	+85° C	+125° C	-55° C +25° C	+85° C +125° C		L	M	P	R	S
volts	μF	percent	μa	μa	μa	percent	percent		2.0	1.0	0.1	0.01	0.001
6	5.6	10	.5	6	7.5	4	4	A	2001	2241	2481	2721	2961
6	6.8	10	.5	6	7.5	6	6	A	2002	2242	2482	2722	2962
6	6.8	20	.5	6	7.5	6	6	A	2003	2243	2483	2723	2963
6	47.0	10	3.0	24	30.0	6	6	B	2004	2244	2484	2724	2964
6	47.0	20	3.0	24	30.0	6	6	B	2005	2245	2485	2725	2965
6	56.0	10	3.0	24	30.0	6	6	B	2006	2246	2486	2726	2966
6	150.0	10	9.0	90	113.0	8	8	C	2007	2247	2487	2727	2967
6	150.0	20	9.0	90	113.0	8	8	C	2008	2248	2488	2728	2968
6	180.0	10	11.0	110	138.0	8	8	C	2009	2249	2489	2729	2969
6	270.0	10	13.0	130	163.0	8	8	D	2010	2250	2490	2730	2970
6	330.0	10	15.0	150	188.0	8	8	D	2011	2251	2491	2731	2971
6	330.0	20	15.0	150	188.0	8	8	D	2012	2252	2492	2732	2972
10	3.9	10	.6	6	7.5	4	4	A	2013	2253	2493	2733	2973
10	4.7	10	.7	7	8.8	4	4	A	2014	2254	2494	2734	2974
10	4.7	20	.7	7	8.8	4	4	A	2015	2255	2495	2735	2975
10	27.0	10	4.0	40	50.0	6	6	B	2016	2256	2496	2736	2976
10	33.0	10	5.0	50	63.0	6	6	B	2017	2257	2497	2737	2977
10	33.0	20	5.0	50	63.0	6	6	B	2018	2258	2498	2738	2978
10	39.0	10	5.0	50	63.0	6	6	B	2019	2259	2499	2739	2979
10	82.0	10	8.0	80	100.0	6	6	C	2020	2260	2500	2740	2980
10	100.0	10	10.0	100	125.0	8	8	C	2021	2261	2501	2741	2981
10	100.0	20	10.0	100	125.0	8	8	C	2022	2262	2502	2742	2982
10	120.0	10	12.0	120	150.0	8	8	C	2023	2263	2503	2743	2983
10	180.0	10	18.0	180	226.0	8	8	D	2024	2264	2504	2744	2984
10	220.0	10	20.0	200	250.0	8	8	D	2025	2265	2505	2745	2985
10	220.0	20	20.0	200	250.0	8	8	D	2026	2266	2506	2746	2986
15	2.7	10	.6	6	7.5	4	4	A	2027	2267	2507	2747	2987
15	3.3	10	.8	8	10.0	4	4	A	2028	2268	2508	2748	2988
15	3.3	20	.8	8	10.0	4	4	A	2029	2269	2509	2749	2989
15	18.0	10	3.5	35	44.0	6	6	B	2030	2270	2510	2750	2990
15	22.0	10	4.0	40	50.0	6	6	B	2031	2271	2511	2751	2991
15	22.0	20	4.0	40	50.0	6	6	B	2032	2272	2512	2752	2992
15	56.0	10	8.0	80	100.0	6	6	C	2033	2273	2513	2753	2993
15	68.0	10	10.0	100	125.0	6	6	C	2034	2274	2514	2754	2994
15	68.0	20	10.0	100	125.0	6	6	C	2035	2275	2515	2755	2995
15	120.0	10	18.0	180	226.0	8	8	D	2036	2276	2516	2756	2996
15	150.0	10	20.0	200	250.0	8	8	D	2037	2277	2517	2757	2997
15	150.0	20	20.0	200	250.0	8	8	D	2038	2278	2518	2758	2998
20	1.2	10	.6	6	7.5	4	4	A	2039	2279	2519	2759	2999
20	1.5	10	.6	6	7.5	4	4	A	2040	2280	2520	2760	3000
20	1.5	20	.6	6	7.5	4	4	A	2041	2281	2521	2761	3001
20	1.8	10	.6	6	7.5	4	4	A	2042	2282	2522	2762	3002
20	2.2	10	.8	8	10.0	4	4	A	2043	2283	2523	2763	3003
20	2.2	20	.8	8	10.0	4	4	A	2044	2284	2524	2764	3004
20	8.2	10	2.0	20	25.0	6	6	B	2045	2285	2525	2765	3005
20	10.0	10	3.0	30	38.0	6	6	B	2046	2286	2526	2766	3006
20	10.0	20	3.0	30	38.0	6	6	B	2047	2287	2527	2767	3007
20	12.0	10	3.5	35	44.0	6	6	B	2048	2288	2528	2768	3008
20	15.0	10	4.0	40	50.0	6	6	B	2049	2289	2529	2769	3009
20	15.0	20	4.0	40	50.0	6	6	B	2050	2290	2530	2770	3010
20	27.0	10	5.0	50	63.0	6	6	C	2051	2291	2531	2771	3011
20	33.0	10	7.0	70	88.0	6	6	C	2052	2292	2532	2772	3012
20	33.0	20	7.0	70	88.0	6	6	C	2053	2293	2533	2773	3013
20	39.0	10	8.0	80	100.0	6	6	C	2054	2294	2534	2774	3014
20	47.0	10	9.0	90	113.0	6	6	C	2055	2295	2535	2775	3015
20	47.0	20	9.0	90	113.0	6	6	C	2056	2296	2536	2776	3016
20	56.0	10	11.0	110	138.0	6	6	D	2057	2297	2537	2777	3017
20	68.0	10	14.0	140	175.0	6	6	D	2058	2298	2538	2778	3018
20	68.0	20	14.0	140	175.0	6	6	D	2059	2299	2539	2779	3019
20	82.0	10	16.0	160	200.0	6	6	D	2060	2300	2540	2780	3020
20	100.0	10	20.0	200	250.0	8	8	D	2061	2301	2541	2781	3021
20	100.0	20	20.0	200	250.0	8	8	D	2062	2302	2542	2782	3022
35	5.6	10	2.5	25	32.0	4	4	B	2063	2303	2543	2783	3023
35	6.8	10	3.0	30	38.0	6	6	B	2064	2304	2544	2784	3024
35	6.8	20	3.0	30	38.0	6	6	B	2065	2305	2545	2785	3025
35	22.0	10	8.0	80	100.0	6	6	C	2066	2306	2546	2786	3026
35	22.0	20	8.0	80	100.0	6	6	C	2067	2307	2547	2787	3027
35	27.0	10	9.0	90	113.0	6	6	D	2068	2308	2548	2788	3028
35	33.0	10	11.0	110	138.0	6	6	D	2069	2309	2549	2789	3029
35	33.0	20	11.0	110	138.0	6	6	D	2070	2310	2550	2790	3030

Figure 9-3B.

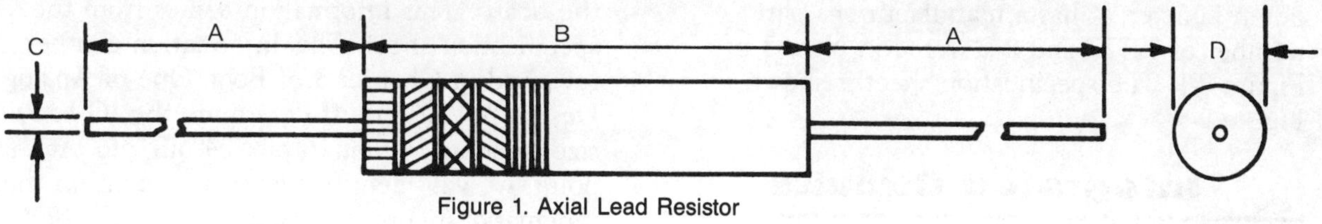

Figure 1. Axial Lead Resistor

Specification	Style	A	B	C	D
MIL-R-11	RC06	1 ±1/8	0.140 ±0.020	0.015 ±0.003	0.067 ±0.008
MIL-R-11 1/4W	RC07	1-1/2 ±1/8	0.250 ±0.031	0.025 ±0.002	0.090 ±0.008
MIL-R-11 1/2W	RC20	1-1/2 ±1/8	0.375 +0.041 −0.031	0.031 ±0.005	0.38 ±0.023
MIL-R-11 1W	RC32	1-1/2 ±1/8	0.562 ±0.031	0.040 ±0.005	0.225 ±0.015
MIL-R-11 2W	RC42	1-1/2 ±1/8	0.688 ±0.040	0.045 ±0.003	0.318 ±0.018
MIL-R-10509	RN55	1-1/2 ±1/8	1/4 ±1/32	0.025 ±0.002	7/64 ±1/32
MIL-R-10509	RN60	1-1/2 ±1/8	3/8 ±1/16	0.025 ±0.002	0.125 +0.049 −0.031
MIL-R-10509	RN65	1-1/2 ±1/8	5/8 +1/32 −3/32	0.025 ±0.002	3/16 +1/16 −1/32
MIL-R-10509	RN70	1-1/2 ±1/8	3/4 +1/8 −1/16	0.032 ±0.002	1/4 +5/64 −1/32
MIL-R-10509	RN75	1-1/2 ± 1/8	1-1/16 ±1/16	0.032 ±0.002	3/8 +1/16 −1/32
MIL-R-11804	RD60	1-1/2 ±1/8	19/32 ±1/16	0.033 ±0.003	3/16 ±3/64
MIL-R-11804	RD65	1-1/2 ±1/8	15/16 ±1/16	0.033 ±0.003	5/16 ±3/64
MIL-R-11804	RD70	1-1/2 ±1/8	2-1/16 ±1/16	0.033 ±0.003	5/16 ±3/64

(All dimensions in inches)

Style
- RB - Wirewound
- RC - Composition
- RD - Film (power)
- RE - Wirewound (power)
- RL - Film
- RN - Film (hi stability)

Figure 9-4. Spec Sheet

R2: this is a variable resistor also called a potentiometer. It has a manufacturer's part number of 3252P and is covered in Note 3 of Figure 9-1. The specification sheet is shown in Figure 9-5.

Integrated Circuits

Figure 9-1 has eight different types of IC specification numbers called out in the various gates and IC symbols. This does not mean that there will be eight IC's on the final design. We can combine some of these gates into other IC's to reduce the IC count.

The following identifies each symbol and provides the specification sheet. The gate identified on Figure 9-1 as 00, is called a 2-input NAND. It has four gates to a package. The specification sheet in Figure 9-6 calls out a 7400. The gate identified on the schematic in Figure 9-1 is a 02, a 2-input NOR gate. It has four gates to a package. The specification sheet in Figure 9-6 called out a 7402. The gate (identified on Figure 9-1 as 04) is an inverter gate. It has six gates to a package. The specification sheet in Figure 9-6 called out a 7404. The gate identified on Figure 9-1 is a 10, a 3-input NAND. It has three gates to a package. The specification sheet in Figure 9-6 calls out a 7410. The gate identified on Figure 9-1 is a 27, a 3-input NOR. It has three gates to a package. The specification sheet in Figure 9-6 calls out a 7427. The gate identified on schematic Figure 9-1 as 7495 is called a 4-6it shift register. There is one to a package. Do not change the pin numbers assigned on the schematic for this IC. The specification sheet in Figure 9-7 calls out a 7495.

The gate identified on Figure 9-1 as a 136 is called a 2-input EXCLUSIVE OR gate. It has four gates to a package. The specification sheet in Figure 9-7 calls out a 74136. The gate identified on Figure 9-1 as a 74141 is called a BCD-to-decimal decoder/driver. There is one to a package. Do not change the engineer's pin numbers (assigned to it on the schematic). The specification sheet in Figure 9-7 calls out a 74141. At this point, all the specification sheets have been provided. Proceed to Step 3.

Step #3—Determine the component lead spacing and body size. Spacing of all discrete component pads is where the leads insert into the board. This information comes from the specification sheet. This information can be reviewed in Chapter 3 of Book One of Analog Design. We also must determine the IC body size, such as 14 pin, 16 pin, 24 pin, etc. At this point, we will identify each component on the schematic, Figure 9-1, using the specification sheets, see Illustration 2.

Illustration #2: Component Mounting Table

Using the specification sheets provided in Step 2, determine the component lead spacing, body size, lead size, hole diameter, and pad size for both the discrete and IC packages of the schematic in Figure 9-1. Use the component mounting table (Table 9-1) along with PC design template, shown in Figure 8-8. Table 9-1 gives you an itemized breakdown of pad spacing, mounting hole requirements, and pad size requirements.

The formulas that calculate the hole and pad sizes, are taken from good commercial practices. For military formulas, see Figure 8-13. Use an upper tolerance of + .004 and a lower tolerance of –.002 for all holes drilled in the PC board. This is not a universal standard; it is one good commercial practice we chose to use. The table has steps 1-13. We will take you through each step for each component. At the end of each step refer to the PC design template. Draft on a practice grid (Figure 9-9) to determine each component body outline, pad spacing and pad size.

Explanation of Table 9-1

Information About C1

Column 1 lists the first component, C1.

Column 2 lists the component type which is the case size. This is found on the specification sheet in Figure 9-2 (called out under military designations as a CK06). Column 3 lists the maximum body length. According to the specification sheet in Figure 9-2, the body is .290 in length X .090 in width.

Model 3252 TRIMPOT® Potentiometer

SPECIFICATIONS

ELECTRICAL CHARACTERISTICS

Standard Resistance Range	10 ohms to 2 megohms
Resistance Tolerance*	±10% standard, closer tolerances available
Absolute Minimum Resistance	1 ohm, all resistances
Continuity*	Maintained for full mechanical range
Insulation Resistance, 500 volts DC	1000 megohms minimum
Contact Resistance Variation, Maximum*	1.6 ohms or 1.6%, whichever is greater

ENVIRONMENTAL CHARACTERISTICS

Power Ratings:	
70°C	1.0 watt or 400 volts maximum
150°C	0 watt
Operating Temperature Range	—65 to +150°C
Temperature Coefficient per MIL-R-22097	±100 ppm/°C maximum all resistances
Humidity	MIL-R-22097 100 megohms minimum insulation resistance after removal from chamber
Vibration①	MIL-R-22097, 30G
Contact Bounce	0.1 millisecond maximum
Wiper Shift, Maximum	1.0% voltage ratio
Shock	MIL-R-22097, 100G
Contact Bounce and Wiper Shift	Same as Vibration
Sand and Dust	Materials meet MIL-E-5272, procedure 1
Salt Spray	Meets MIL-STD-202, method 101A
Fungus	Materials meet MIL-E-5272

Load Life	1000 hours per MIL-R-22097
Resistance Shift, Maximum	±2.0%
Mechanical Life	200 cycles without discontinuity
Dielectric Strength	MIL-R-22097
Room Conditions	1000 volts AC
80,000 Feet (0.8" Hg)	400 volts AC

PHYSICAL CHARACTERISTICS

Immersion Leak Test*	No leaks
Shaft Torque*	5.0 oz.-in. maximum
Markings*	Manufacturer's name, wiring diagram, date code resistance, and manufacturer's part number (customer's part number optional)
Appearance*	Legible markings, no physical defects
Mechanical Adjustment	22 turns nominal
Mechanical Stops*	Wiper assembly idles
Weight	Approximately 0.1 oz.
Terminals:	
(L)	Teflon insulated stranded leads, 30 AWG (.024 O.D.) 7 strands/38 AWG
(P) (W) (X)	Gold plated grade A nickel printed circuit pins
Mounting	Mount by printed circuit pin or corner slotted mounting hole
*Inspection Note	Applicable 100% or statistical sampling inspection performed to insure highest quality.

① When mounting the "W" & "X" version by the printed circuit pins, the vibration specification is 20G.

Specification Note: Closer performance tolerances can be supplied upon request

Specifications are subject to change without notice.

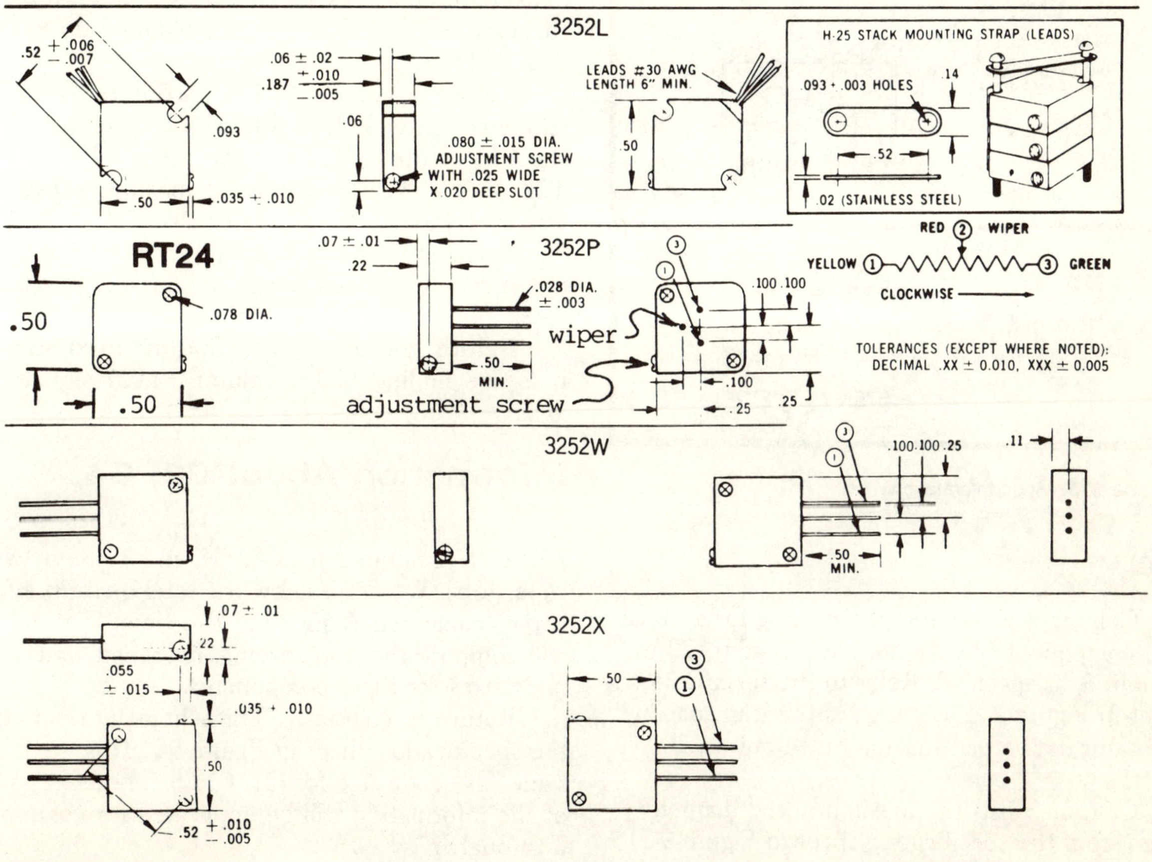

Figure 9-5. Spec Sheet

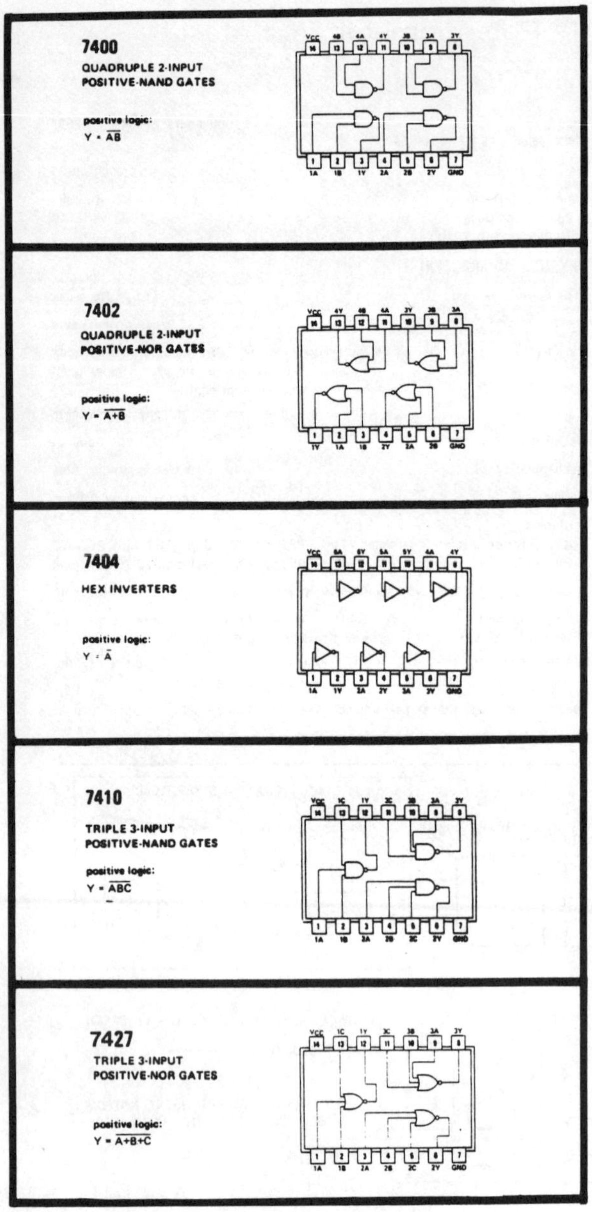

Figure 9-6. Spec Information

Column 4 shows the pad spacing for a radial lead component. We do not have to add .120 to column 3, as specified. Refer to the specification sheet in Figure 9-2, which gives the lead spacing (the same as the pad spacing called out as .200 ±.015).

Column 5 lists the maximum lead diameter taken from the specification sheet in Figure 9-2. It calls out .025 + .004 - .001. To find the maximum lead diameter, add .025 + .004 to equal .029.

Column 6 lists the minimum size hole that can be used for the maximum lead of .029. Using good commercial practices, add .006 to .029 to equal .035.

Column 7 lists the maximum size hole. The hole can be too large for a lead if it is allowed to exceed .020, in this case. For the maximum size hole that can be used for the lead in column 5 of .029, add .020 + .029 to equal .049.

Information About C2

Return to column 1 (C2).

Return to column 2 (Component type). Using the specification sheet in Figures 9-3A and B, we find a CS13A case type for C2.

Return to column 3. See the component specification sheet in Figure 9-3A. The case size A has a maximum body length of .422.

Return to column 4. To compute the pad spacing for an axial lead component, add .422 (maximum body length) + .120 and round up. In this case .422 + .120 = .542. Rounding up to the nearest tenth grid equals .600.

Return to column 5. The maximum lead diameter can be obtained from the specification sheet in Figure 9-3A. Case size A has a lead of .020 + or - .002. To obtain the maximum lead size, add .002 to .020 which equals .022.

Return to column 6. To find the minimum hole size, adding .006 to column 5 (.022) equals .028.

Return to column 7. To find the maximum hole size, adding .020 to column 5 (.022) equals .042.

Information About C3, C4, and ?

Return to column 1 for C3, C4, and ? (the bypass capacitors). We don't know the total quantity of bypass capacitors required at this time. We can still compute the lead spacing, hole, pad, and body sizes for these components.

Return to column 2. The schematic refers to the specification sheet in Figure 9-2. It is the same case code size for C1, CK06. Therefore, all of the information will be duplicated for columns 2 through 7.

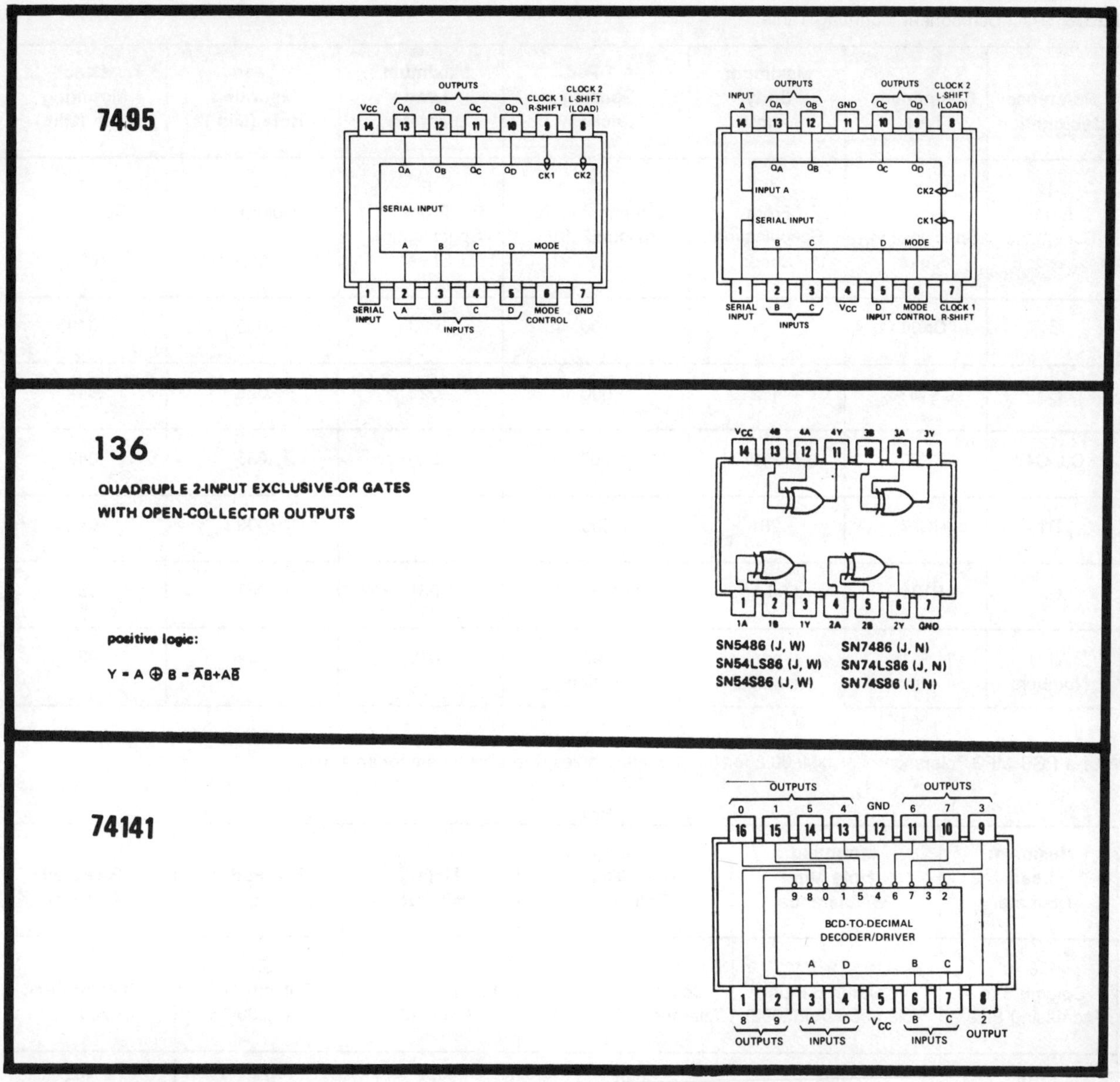

Figure 9-7. Spec Information

Information About R1

Return to column 1 for R1. Return to column 2 for the component type. R1 is specified as an RC07. See specification sheet Figure 9-4.

Return to column 3. The maximum body length for an RC07 (on the specification sheet in Figure 9-4) is .250 ± .031 which equals .281.

Return to column 4. The pad spacing is determined by adding the maximum body length to .120; .281 + .120 = .401. Rounding up to the nearest tenth equals .500.

Return to column 5. Find the maximum lead diameter. The specification sheet in Figure 9-4 calls out the lead diameter of .025 ± .002. Adding the tolerance to the lead diameter equals .027.

Return to column 6. Find the minimum hole size. Adding .006 to column 5 (.027) equals .033.

Return to column 7. Find the maximum hole size adding .020 to column 5 (.027) equals .047.

Table 9-1. Component Mounting Table

Reference Designation	Component Type	Maximum Body Length	1:1 Pad Spacing (Inches)	Maximum Lead Diameter	Lead Mounting Hole (Min.)	Lead Mounting Hole (Min.)
1 From Schematic	2 From Specification Sheet	3 From Specification Sheet	4 Column 3+.120 (Round Up)	5 From Specification Sheet	6 Column 5 + .006	7 Column 5 + .020
C1	CK06	.029	.200	.029	.035	.049
C2	CS13-A	.422	.600	.022	.028	.042
C3, C4?	CK06	.290	.200	.029	.035	.049
R1	RC07	.281	.500	.027	.033	.047
R2	RT24	.500 sq.	per template	.031	.037	.051
All U Numbers	14 & 16 pin	per template	per template	.018	.024	.038

With a PCB MFG Tolerance of +.004/-00.2 for the mounting holes, the above components require:

Maximum Lead Diameter	Mounting Hole Min W/Tolerance	Mounting Hole Max W/Tolerance	Hole Call-Out	Min Pad Size	Selected Pad Size
8 Column 5 In Increasing Size	9 Column 6 Plus Tolerance (Add .002)	10 Column 7 Minus Tolerance (Sub .004)	11 From Column 9 and 10	12 Column 11 + .030	13 Standard Pad Size at 2:1
.018	.026	.030	.032	.061	.125
.022	.034	.038	.032	.062	.125
.027	.035	.039	.041	.071	.150
.029	.037	.045	.041	.071	.150
.031	.043	.047	.041	.071	.150

Information About R2

Return to column 1 for R2 (component type). Use the specification sheet in Figure 9-5. The component type is an RT24, variable resistor. The maximum body length is .50 square. The template shown in Figure 9-9 (Bishop Graphics #336B) shows the body outline.

Return to column 4 (pad spacing). This is not an axial lead component. Do not use the formula given in this column. Use the template outline shown in Figure 9-8 and 9-9 for the pad spacing.

Return to column 5 (maximum lead diameter). The specification sheet in Figure 9-5 specifies the lead is .028 $\pm$ .003. Adding the tolerance to the lead diameter equals .031.

Return to column 6. Find the minimum hole size. Adding .006 to column 5 (.031) equals .037.

Return to column 7. Find the maximum hole size. Adding .020 to column 5 (.031) equals .051.

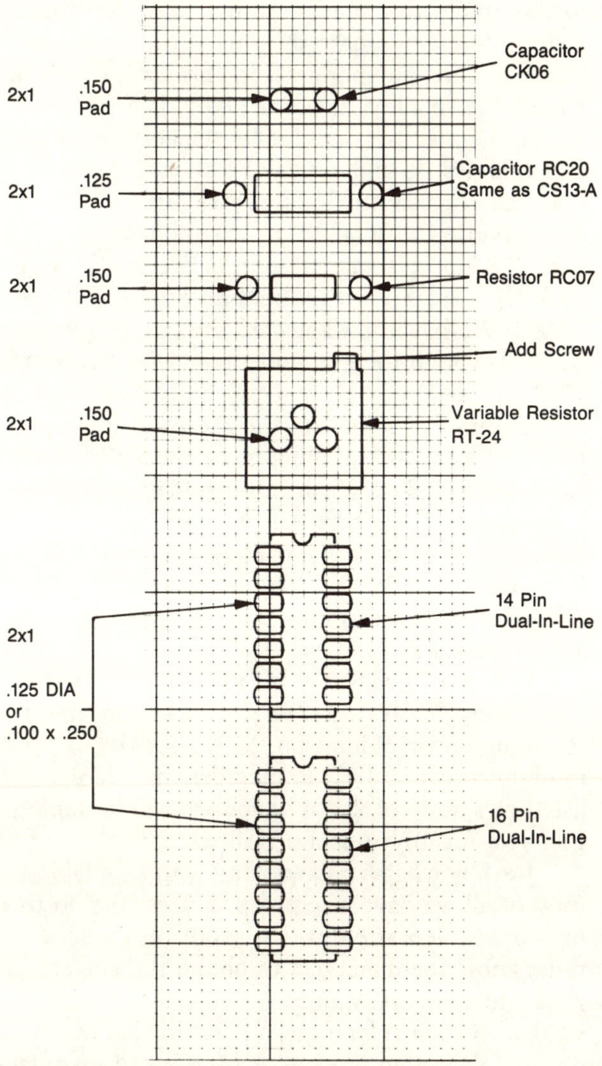

Figure 9-8. View From Component Side

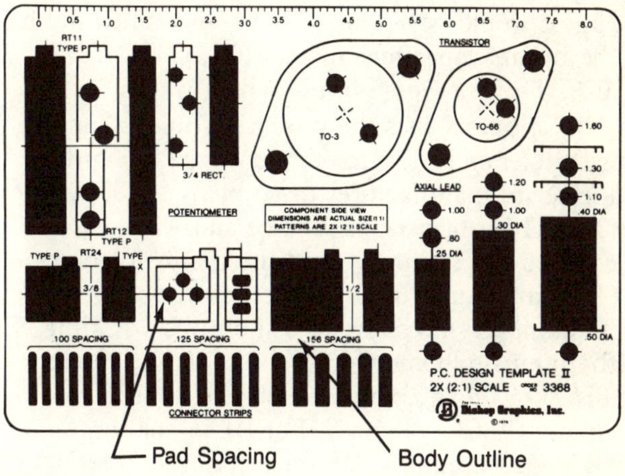

Figure 9-9. RT24

Information About U Numbers

Return to column 1. All U numbers are the IC's, 14 and 16 pin.

Return to column 2 (component type). All IC's are in the dual-in-line configuration as shown in Figure 8-6.

Return to column 3. The body lengths are determined by the template, Figure 8-8.

Return to column 4. The pad spacing is determined by the template shown in Figure 8-8. The maximum lead diameter is found on the specification sheet in Figure 5-11. it specifies .018 max.

Return to column 6. Find the minimum hole size. Adding .006 to .018 equals .024.

Return to column 7. Find the maximum hole size. Adding .020 to column 5 (.018) equals .038.

We are ready to compute columns 8 through 13. This is the most important part of the table. We will select the common hole and pad size. We would like only one size hole for all the components listed in column 1. Usually this is not possible. But the objective is to have as few different size holes as possible for the board manufacturer to drill.

Column 8 lists reference designations and maximum lead diameters. List from column 5 the maximum lead diameter. List the sizes in numerical order, starting from the smallest lead diameter to the largest. The IC's (U numbers) are the smallest lead. Therefore list .018 first. Next largest in size would be C2 at .022. The third largest would be R1 at .027. The fourth would be the bypass capacitors and C1 (C1, C3, C4, ?) at .029. The last would be R2 which is .031.

Column 9 lists the mounting hole minimum with tolerance. The tolerance we are using (for the board manufacturer to go by) is + .004 – .002. This is a personal choice and will vary from company to company. Add the minus tolerance (.002) to column 6 for each component. The IC's (U numbers) in column 6 call out .024. Adding the minus tolerance for column 9 (.002) equals .026. For C2 in column 6 add .002 to .028 to equal .030 for column 9. For R1 in column 6, .033 + .002 equals .035 for column 9. For C1, C3, C4, and ? in column 6, .035 + .002 equals .037 for column 9. For R2 in column 6, .037 + .002 equals .039 for column 9.

Column 10 lists the mounting hole maximum with tolerance. Find the largest the hole can be, with its own tolerance. In column 9, we are trying to find the smallest size the hole can be with its own tolerance. For column 10, subtract .004 from each size listed in column 7 (for each component). For example, consider the IC's (U numbers). Column 7 specifies .038; subtract .004 to equal .034. C2 in column 7 specifies .042 - .004 = .038, and so on through R2 which is in column 7 (.051 - .004 equals .047).

We are comparing column 9 to column 10 to find the hole call out for column 11. If we look at column 10 for the smallest lead size (.018 for the IC's) and for the largest that the hole can ever be with its own tolerance, we find that it is .034. Comparing .034 in column 11 to the smallest that the hole size can be in column 9. See how many components with the smallest hole will fit under or at least at the size of .034. For C2 in column 9, .030 fits under .034. R1 however has the smallest hole listed as .035, for column 9. Therefore, R1 cannot have the same hole size as both C2 and the IC's. Compare C2 and the IC's (U numbers); choose a hole size between .030 and .034. Choose .032 because it is between the two numbers .030 and .034. List .032 in column 11. This is the first hole size; it has a tolerance of +.004, -.002. Choose the next hole size. Keep in mind we want to find as few different hole sizes as possible. The next smallest lead size would be R1. Its maximum hole size is .043. We are trying to stay under that maximum hole size. Return to column 9 and compare the smallest hole size of the remaining components. In this case, C1, 3, 4, ? have the smallest hole size (.037). It easily fits under .043. R2 has .039 for its smallest hole size and will also fit under .043. To choose the next hole size for column 11, compare .039 to .043. Why choose the smallest, or why choose the largest? Choose somewhere in between .039 and .043. Choose .041 with a hole tolerance of +.004, –.002. We have now selected the two hole sizes which cover all the components on the PC board (.032 and .041).

Column 12 selects a pad size with an annular ring of .015 or a total amount of copper on both sides of the hole of .030. Adding .030 to the first hole size (.032) we find that we come up with .062. Continue to the next hole size. Add .030 (for column 12) .041 to column 11 which equals .071. There are two minimum pad sizes: .061 and .071 for column 12.

Column 13 selects standard size pads from the catalog. In this case we will round up, never rounding down. Round up to .125 for the minimum size of .061. Remember that .125 is listed at a scale of 2:1 at which we will be taping. Likewise for .071, select a .150 size pad.

Review what we have done with this table. First of all we have selected pad sizes. We do this before we even attempt the layout because we must know the diameters to obtain the correct air gaps when we are taping. We established hole sizes with their tolerance. We do this before we do the fabrication drawing for the board manufacturer. We established all the standard case sizes that we will be using on the template in Figure 9-8 and 9-9. We have accomplished many objectives with this chart. We understand the component configurations and the requirements for each component, pad, and hole.

Step #4—You should establish your board layout design rules. These rules will be used for designing and taping your printed circuit board in this chapter. See design rules, Illustration 3.

Illustration #3: Design Rules

The design rules from Chapter 8 will be used in Chapter 9. This section establishes the electronic engineering PC board design conventions for this book. It includes the general requirements for the formal layout and the artmasters. Before starting the design, review Chapter 8, Illustration 2 for the design rules.

Step #5—Refer to the list of questions to ask the engineer about the 6-IC digital schematic. See Illustration 4.

Illustration #4: Questions to Ask the Engineer

The questions were explained in detail in Chapter 8.

Discuss each point with the engineer to make sure you understand all the problems and all the engineer's requirements before beginning the layout.

1. Are there any special size components, or are all of the components standard? For example:
   a) Resistor type and power rating
   b) Capacitor material and working voltage
   c) Transistor case type
   d) IC family and package type
   e) Diode case size

   Standard refers to the components that are standard to a particular company. See the explanation to this question in Chapter 8. When we obtained the specification sheets from the engineer in Step 2, we found that all the resistors were in the case code RC family. The exception was the variable resistor, R2, which had a standard outline in the RT family. We could use the template in Figure 9-9. The capacitors all have a standard case code type. See Figure 9-1.

2. How many bypass capacitors are there for each IC?

   That question has been answered on the schematic in Figure 9-1; note 5 specifies to bypass every two IC packages.

3. Are there any critical leads on the board?

   In this case the engineer will not specify that any leads are critical in length.

4. Are there any special size filtering capacitors required on the input voltage pins (decoupling network)?

   Figure 9-1 shows an input filtering capacitor called C2 which is tied between +5V and ground. Note 4 on the schematic specifies that the input filtering capacitor is an electrolytic tantalum polarized capacitor. We found this to be a CS13A on the specification sheet.

   SPECIAL NOTE: This capacitor should be placed close to voltage pins 29 and 30 for +5V on the connector.

5. Should there be any spare IC gates?

   Most likely you will have spare IC gates. The engineer in this particular schematic (Figure 9-1) does not require any spare gates.

6. By using the maximum size of all the components, will all components fit on the board? (This is your responsibility)

   This question will be explained fully in Step 8 once the board size has been determined and the grouping of the schematic has been accomplished.

7. Are there pre-assigned pins on the connectors?

   The schematic in Figure 9-1 showed pre-assigned pins both for ground (pins 1 and 2) and for +5V (pins 29 and 30). All other signal inputs and outputs are not pre-assigned.

8. Are there any variable resistors or variable capacitors? If so, where should they be placed on the board?

   The schematic in Figure 9-1 shows a variable resistor R2. We request that the access for R2 be at the opposite edge of the P1 male connector. The adjusting screw must be at the back edge of the board so a technician or engineer can adjust it from that edge.

9. Should there be any test points? (Try to place all test points on the edge). Check for test points on voltages and grounds, if they are not indicated on the schematic.

   The schematic in Figure 9-1 shows test points TP1 for a signal path in the logic circuit. Check with the engineer to see if we should have test points for the +5V and ground. The engineer usually does not draft them onto the schematic even though they should be there. In this case, the engineer requests that you add test points for the +5V and ground. Update the schematic to add these test points at this time. See Figure 9-10. All test points should be accessed from the opposite edge of the connector on the PC board (similar to the variable resistor R2). See Figure 8-17 for location and size of these test points.

10. Do any components require heat sinks? If so, where should the components be placed in relationship to the other components within the board outline?

    For the schematic in Figure 9-10, the answer is no. There are no components that run hot and require heat sinks.

11. Have all the package considerations been examined? See Chapter 4.

    Question 11 will be answered extensively in Step 6, which covers the board outline. We will look at the enclosure and the mounting required for the PC board.

Step #6—Determine the board outline. Besides the size of the board, find out the method of fastening or mounting. (card guides, screws, stand offs, etc.). What type of electrical interface is required (connector, cables, wires, etc.)? The board size must be known in order to compute the total square inches available for the schematic given by the engineer (Figure 9-1). See Illustration 5 for more information and a review of board outlines. See Chapter 4 to review board size and mounting.

Illustration #5: Board Size and Mounting

Board outline must be determined at this time. Besides the size of the board, it is also important to find what method of fastening or mounting is being used: card guides, screws, stand-offs, etc.? To find this information, see Procedures 1 and 2 below.

SPECIAL NOTE: Board size and mounting information can be reviewed in Chapter 4.

Procedure #1

Usually the board outline is given by the mechanical designer or the mechanical engineer. Using the board outline shown in Figure 9-11 for the schematic in Figure 9-10, we find a total of 12 square inches, not counting the P1 connector. To calculate the total number of square inches that are useable, we define what interfaces to the PC board. How is it secured to an enclosure?

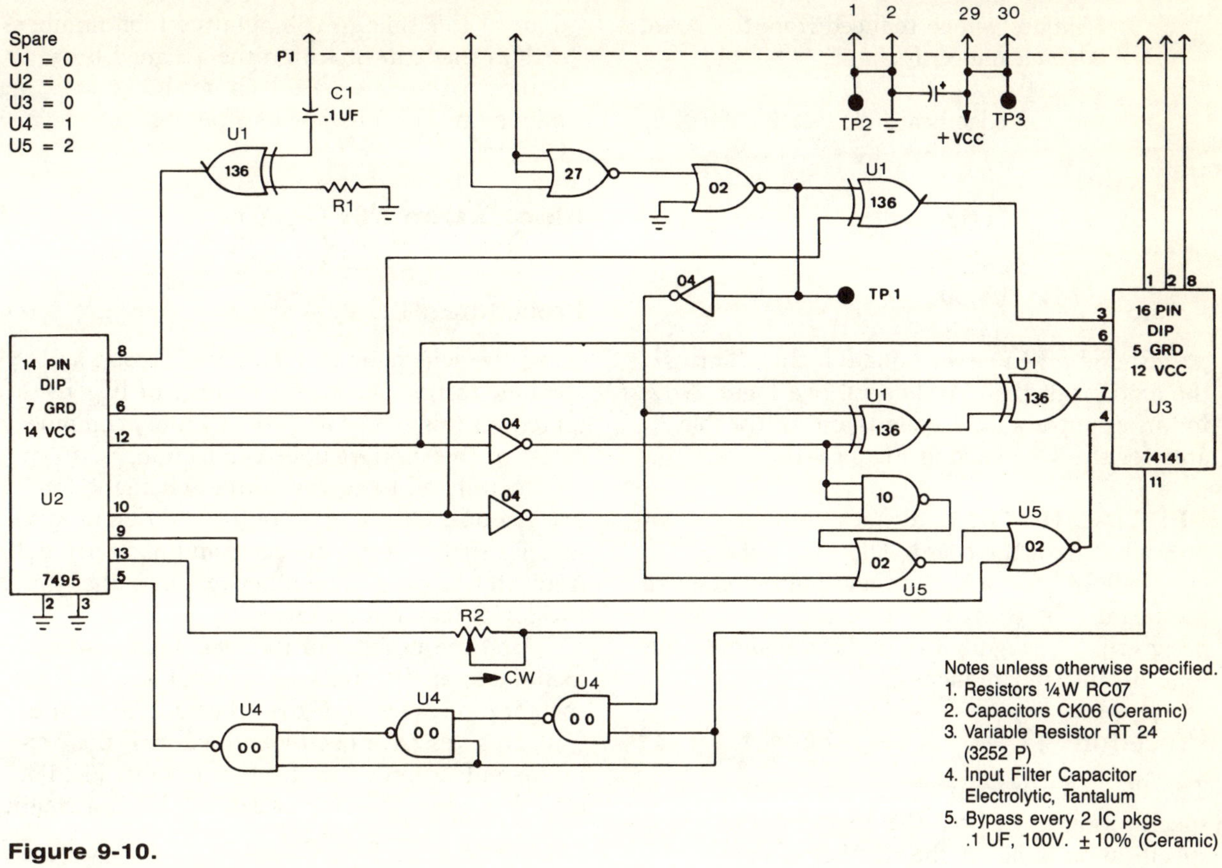

Figure 9-10.

Procedure #2

The board outline will be inserted into an enclosure that uses a card track as a means of holding the board into position. For an example of the enclosure, see Figure 8-23. The board has restrictions on its edges. The board outline in Figure 9-11 indicates a card track clearance of .150 on each side of the board. See Figure 8-18 for the design rules. Using the restrictions, recompute the useable area with the following:

1) 4 inches times 3 inches equal 12 square inches.

2) 4 inches will be reduced by the card track clearance. Example:

  4.000
– .300 card track (.150 x 2)
= 3.700

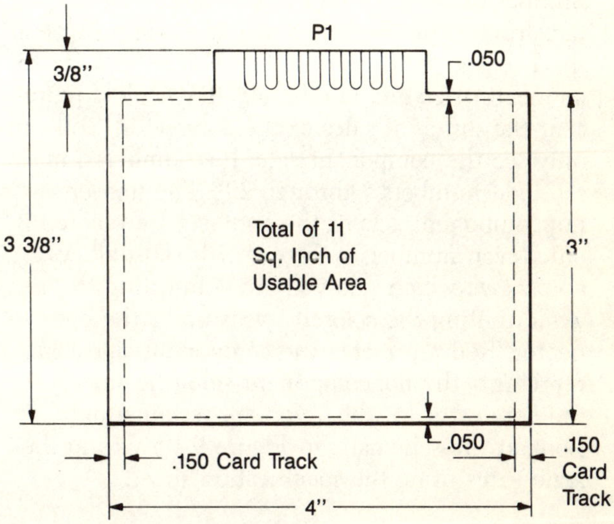

Figure 9-11. Board Outline

3) 3 inches will be reduced from the board edge clearance. Example:

$$\begin{array}{r} 3.000 \\ -\ .100 \\ \hline = 2.900 \end{array}$$ edge board clearance (.050 X 2)

4) Multiply

$$\begin{array}{r} 3.700 \\ \times\ 2.900 \\ \hline 10.730 \end{array}$$

Round off to eleven square inches. This is the amount of the useable area. See Figure 9-12 for an example a a 2:1 scale layout of useable area for the schematic in Figure 9-10.

SPECIAL NOTE: Card track clearance is .150 on each side of the board. The edge board clearance is .050 for the top and bottom edge of the board. The clearances were obtained from the design rules in Figure 8-18 (used to compute the total useable square inches).

Procedure #3

The PC designer must obtain a connector specification sheet from the engineer. The specification indicates the number of contacts and the size of the male tongue called P1. The specification sheet is similar to Figure 4-15. For this chapter, we will provide all the necessary information. We will not determine the size, number of contacts, and numbering of each contact from a specification sheet. We will provide all the necessary information. See Figure 9-12 for the connector size and numbering. Two colors represent the different sides of the connector. You can only see the component side. It's numbered in red (odd numbers 1 through 29). The unseen noncomponent side of the contacts are colored in blue (even numbers 2 through 30). This allows you to see which side you are connecting to while routing the colored traces up to the connector. Red represents the component side; blue represents the noncomponent side (the side of the connector to which you are terminating). Look at the schematic in Figure 9-10. Group the logic gates using the specification sheet.

Step #7—Group the logic gates on the schematic in Figure 9-1. Use the logic considerations in Illustration 6. Grouping must be done at this time to compute the total number of IC's that can fit within the assigned board outline. Use one square inch per IC as a minimum. For a review of grouping, see Chapter 2.

Illustration #6: Grouping

Procedure #1

Using the schematic in Figure 9-10. group and combine to find the lowest number of IC's required on this schematic. At this time, combine gates for the shortest trace connection.

Identify all the gates on the schematic for two reasons. One is that you have different logic families such as open collector and non-open collector (totem pole). The other reason is that these two logics may not be mixed.

Find the number of the logic family (which manufacturer the engineer is using) to obtain the specification sheet. For the schematic in Figure 9-10, all gates have been identified with the proper manufacturer's specification number (04, 00, etc.). For open collector or gates 74136, schematic

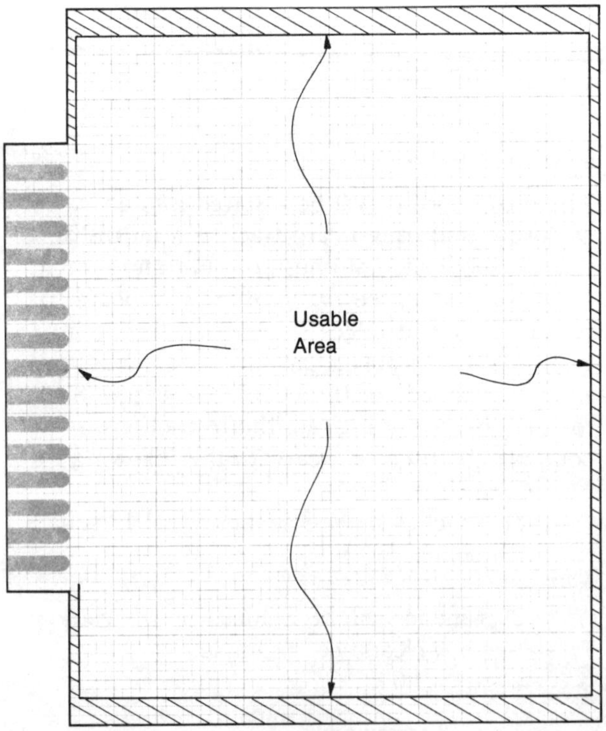

Figure 9-12. Usable Board Area

Figure 9-10, we should identify those gates by putting a small flag on the output side of the gate and reference to an example, Figure 9-13, as to this flag on the output side of the gate for open collector.

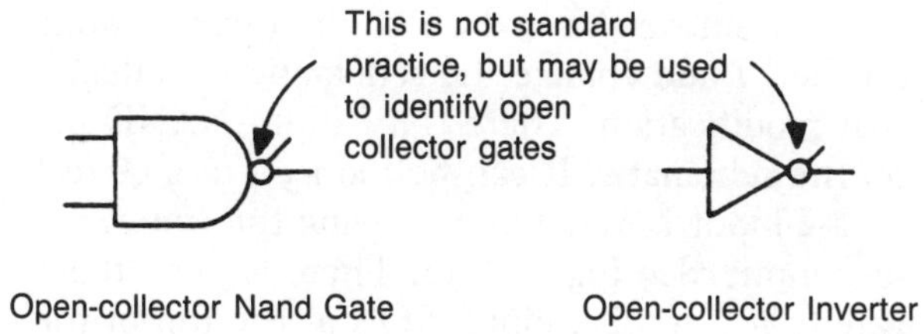

Figure 9-13. Open Collector Gates

SPECIAL NOTE: This is not a standard ANSI, IPC or military practice, but it is a good commercial practice for identifying open collector gates.

Specification sheets identify logic families and symbols. The electrical engineer provides the specification sheets required for the schematic in Figure 9-10. See Figures 9-6 and 9-7. These specifications identify each type of logic gate on the schematic. They provide the total quantity of gates within each package, input and the output assignments for each gate, and the voltage and ground pin numbers for each type of package.

If we were dealing with an open collector gate, the specifications call out in bold letters: OPEN COLLECTOR. Review the specification sheets in Figures 9-6 and 9-7. There is one open collector package called 74136, a 2-input exclusive OR with open collector outputs. All of the other IC packages are totem pole style.

Procedure #2

Before grouping the schematic in Figure 9-10, review the questions for the engineer. Are there any spare IC gates required? In this case the engineer said there are none.

Are there any critical leads on the board? This is particularly important for grouping because it determines which gates would be placed in the same IC package. In this particular board there were no critical leads.

Procedure #3

Combine the gates. For Figure 9-10, use the combining rules in Figures 2-56 - 2-61. Combine gates using the technique of tying off unused inputs; tie off to another used input. Obtain the engineer's permission to do so. This is a more popular technique rather than tying the unused input to a high or a low (meaning a voltage or ground point).

Procedure #4

Combine the gates that seem to be the simplest or the most obvious (intercommunication to each other). Make the traces as short as possible. Look at the schematic in Figure 9-10. The first thing is to combine the open collector gates, 74136. Examining the specification sheet in Figure 9-7, notice that there are four 74136 gates in one package. Label these four gates along the gate symbols on the schematic. We will also reference the spare count on the schematic. U1 has zero spares. 74136 is first because you can not intermix or combine open collectors with non-open collector gates. There are only four open collector gates on the entire schematic. They must be placed inside the same IC package called U1.

SPECIAL NOTE: Combine the easiest IC's first. The open collectors were first because they had to be within the same package. The next IC could be the 7495 or the 74141. Each of these two IC's must be made an individual IC within themselves. You can not change them or combine them with any other gates. Therefore, 7495 will now be called U2 with zero spares. The 74141 will be called U3 with zero spares. List the spare count on the schematic.

Continue with the grouping. Use the specification sheets for the gates. Find the easiest and most obvious gates (intercommunicating with each other). They are the 00's at the bottom of the schematic. There are three intercommunicating together and connecting to U2 and U3. These three 00's should be placed inside one IC package. Can this be done? Using the specification sheet in Figure 9-6, find the 7400. There are four 2-input NAND gates inside one package. There are three that should be placed inside that one package. Label the three 00's (2-input NAND) U4 with one spare.

Combining is now starting to become more difficult. The next most obvious combination would be the 02, 2-input NOR gates. Review the specification sheet in Figure 9-6. There are four 02 gates inside one package. Therefore do not place anymore than two (that connect to U3). Label these two 02's (2-input NOR) U5 with two spares.

Because it is getting more difficult to decide what to do with the remaining gates, move to another schematic (Figure 9-14) to represent the combining techniques more clearly. So far we have five IC's. Two of those five have spares, U4 and U5.

We would like to understand more completely what we need as a minimum for the remaining gates. For example, there is one gate called a 27 and shown as a 3-input NOR. This gate does not have to be a 3-input NOR; it can be a 2-input NOR. Figure 9-15 shows a 3-input NOR called a 27 that can be reduced to a 2-input NOR called an 02. Remove the tied off or the unused input on the 27 and change the schematic to reflect that modification. There is a 3-input NAND pin on the schematic. It can be reduced to a 00 (called a 2-input NAND) by removing the unnecessary input. See Figure 9-15. There is one other gate called an 02 2-input NOR at the top of the schematic. It can also be reduced down to an 04 (called an inverter) by removing the input tied to ground. Because it is an unnecessary input,

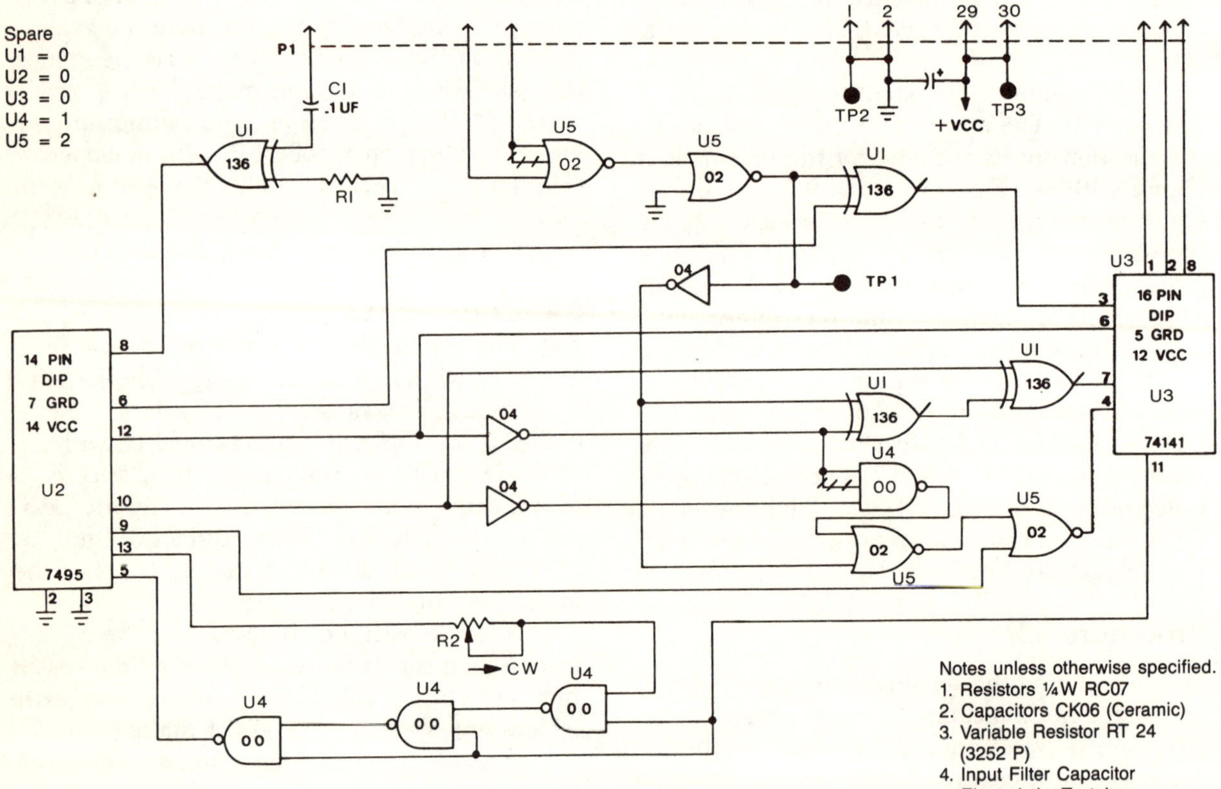

Figure 9-14.

Notes unless otherwise specified.
1. Resistors ¼W RC07
2. Capacitors CK06 (Ceramic)
3. Variable Resistor RT 24 (3252 P)
4. Input Filter Capacitor Electrolytic, Tantalum
5. Bypass every 2 IC pkgs .1 UF, 100V, ±10% (Ceramic)

eliminate it (see Figure 9-15). Notice that the spare count, U5 (which is also an 02 2-input NOR) has two spares. You might place the two 02's (called 2-input NORs) at the top of the schematic inside IC 5 (called U5). This reduces the spare count on U5 to zero spares. Why would you place the two 02's (2-input NOR gates) at the top of the schematic with the two 02 (2-input NORS) at the bottom, since they are not directly communicating? Sometimes you do not necessarily need direct communication of gates; but you would need a reduction of IC count. It would be better to place these four gates in the same package to reduce the IC count. This is because there are spares in U5.

Look at U4 with its one spare. Are there any gates on the schematic that could be placed inside U4 even though they are not directly communicating? The answer is yes. Change the 3-input NAND (called a 10) to a 2-input NAND (called a 00), which is the same as U4. Label that one last gate in the NAND family on the schematic, U4. Updating our spare count, we now see that U4 has zero spares.

Move to the schematic in Figure 9-16 for the final grouping. Review (the spare count. There are zero spares for all five IC's that we have listed. Therefore, we know we must add the sixth IC.

Examine the schematic in Figure 9-16. There are only three gates left that have not been assigned a U number. All three are 04 inverter gates. Use the specification sheet in Figure 9-6. There are six 04 inverter gates inside one package. The schematic has three gates to be assigned. Label all three U6 with three spares. Label these three as spare gates (called 04) at the bottom of the schematic. And when we assign the pin numbers to the gate inputs and outputs, assign to the spares their input and output numbers. If he wants to use these spares, the engineer would not have to search for which pins are available as spare gates.

We have completed the grouping exercise. This gives us a total of 6 IC's, with three 04 inverter spare gates.

SPECIAL NOTE: If you are satisfied with the grouping and combining shown in the schematic in Figure 9-16, plot on the PC board outline the six IC's that we assigned to the schematic. If you are not satisfied, obtain a fresh schematic and try another grouping idea. It is not uncommon to try two or three grouping ideas. This is the most critical stage in the PC board design. If we do a poor job at this point, it would be difficult to obtain a clean layout with short lead lengths.

SPECIAL NOTE: Use the schematic in Figure 9-16 for the layout.

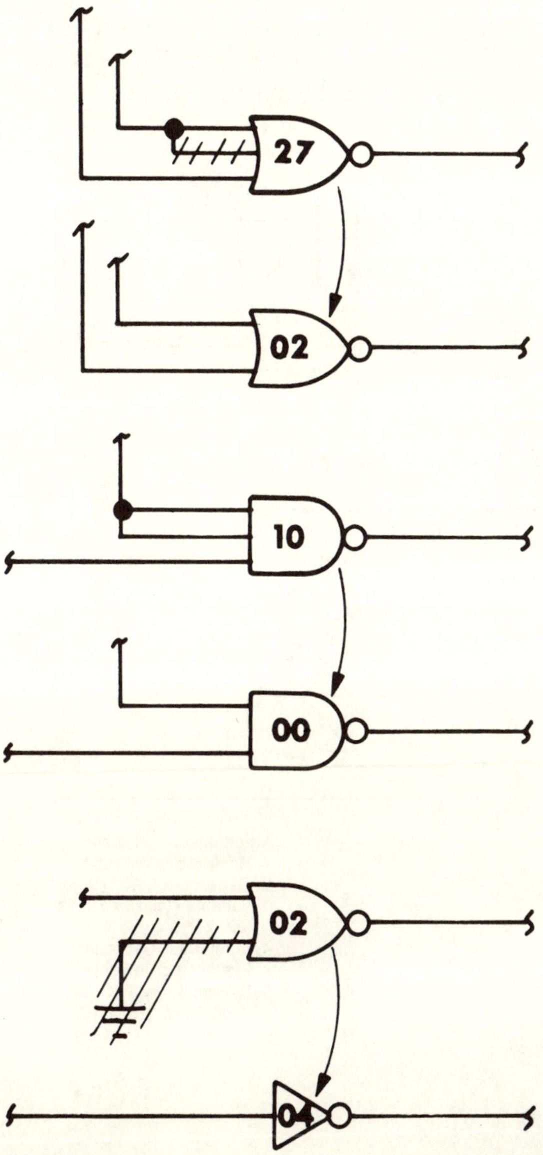

Figure 9-15. Reducing Gates

Step #8—For a review of design layout, see Chapter 5.

Illustration #7: Space Requirements

Calculate the area required before plotting the IC's and discrete components on the board outline. As a rule of thumb, one square inch for each IC is recommended for the average PC board. IC's are not the only components on the digital PC board. Use the typical 14- or 16-pin IC as a reference of the area required for discrete components such as resistors, capacitors, diodes, transistors, etc. The similarity between the IC's and the discrete components is the number of pads used. Using the component equivalency chart in Figure 9-17, we calculate the minimum area required for the schematic in figure 9-16. Use the following procedures.

Procedure #1

On the schematic in Figure 9-16, count the total number of IC's that have 14- or 16-pins. Use the specification sheets in Figures 9-6 and 9-7. Notice that all of the IC's are 14-pin dual in line, except for one. U3 is a 16-pin dual in line. Enter this on the component equivalency chart, Figure 9-17. Show a total of five 14-pin and one 16-pin IC. This is a total of six IC's (multiplied by 1 square inch) equaling six square inches.

Refer to the schematic in Figure 9-16. Find the total number of ¼W resistors. There is only one (called R1). List this on the component

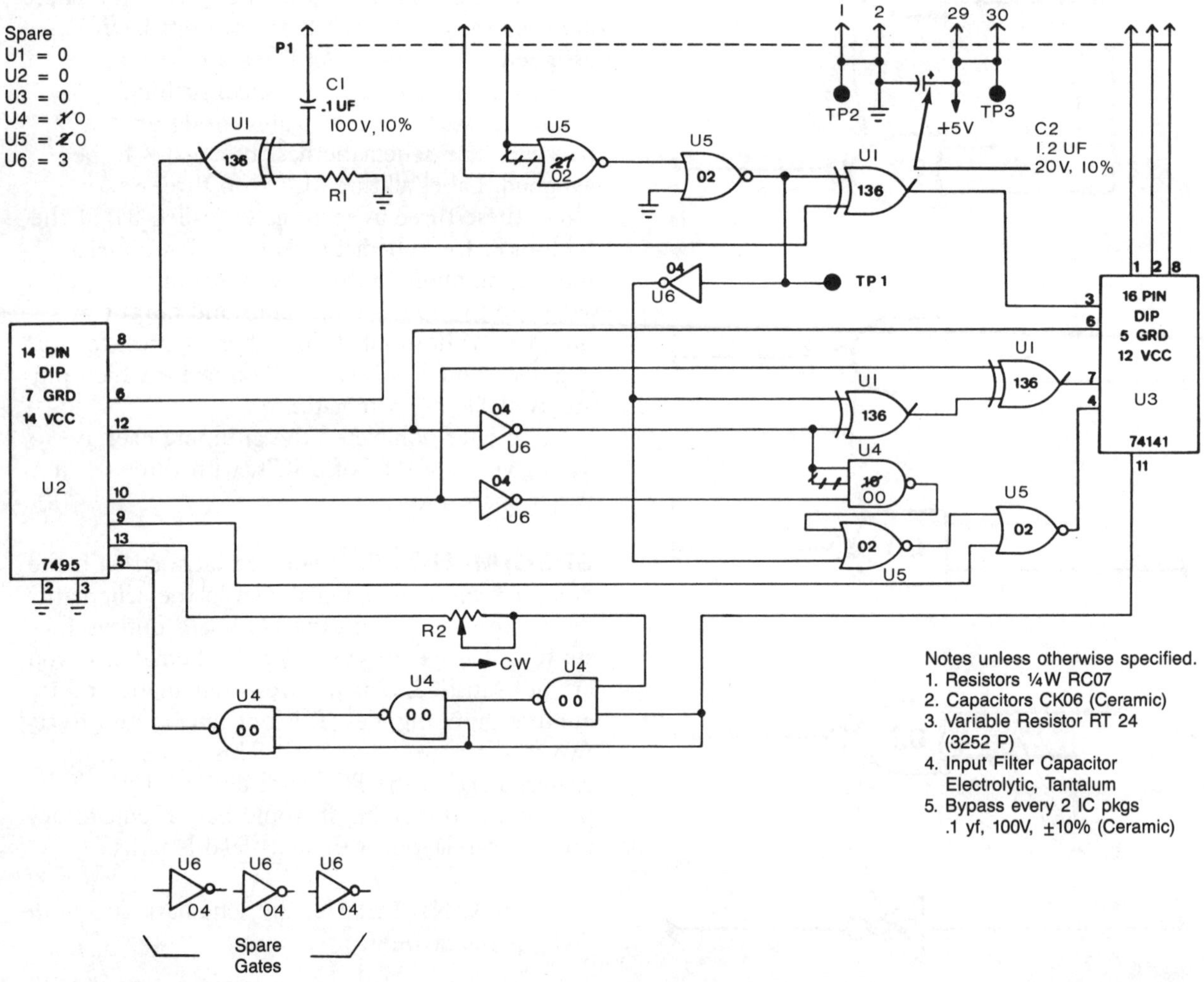

Figure 9-16.

COMPONENT COUNT (Equivalent 14/16 pin IC calculations)
Convert all components on the board to an equivalent 14/16 pin IC count using the following conversion factors:

	Total No. on Board		Equiv. 14/16 IC
14- and 16 pin IC	6	× 1.00 =	6
20 pin IC		× 1.25 =	
24 pin IC		× 1.50 =	
28 pin IC		× 2.00 =	
40 pin IC		× 3.5 =	
64 pin IC		× 6.5 =	
8 pin IC		× .75 =	
test points	3	÷ 10 =	¼
jumper pads		÷ 10 =	
Total =			6¼

	Total No. on Board		Equiv. 14/16 IC
1/4 -watt resistors	1	÷ 7 =	
1-watt resistors		÷ 4 =	
disc type caps	6	÷ 7 =	1
large caps	1	÷ 4 =	¼
3-pin components (Transistors, variable resistor, etc.)		÷ 3 =	¼
pre assigned I/O connector pins used	4	÷ 10 =	½
Other components (list)			
Total Equivalent 14/16-pin IC =			2
Total Sq. Inch Req'd on P.C. Board =			8¼

Figure 9-17. Component's Equivalent To A 14/16 Pin IC

equivalency chart (Figure 9-17) under ¼W resistor. Because there is only one of them, it is impractical to count this as a portion of one square inch. Rather than counting it as 1/7th of a square inch, include the ¼W resistor with the disc type capacitors (which will be covered in Procedure 3).

Refer to the schematic in Figure 9-16. How many disc type capacitors will there be on the schematic? Note 5 on the schematic requires a bypass for every two IC's with a CK06 disc capacitor. There are six IC's. We probably will finish with three or four disc type capacitors. Review the schematic in Figure 9-16. There is one more CK06 disc type capacitor (C1). Counting C1 and R1 (which we are going to include with our disc type capacitors), there is a total of six components to list on the component equivalency chart (Figure 9-17 under disc type capacitors). Dividing by 7 equals approximately one square inch; round it off.

There is one input filtering capacitor on the schematic in Figure 9-16 (called C1). List this on the component equivalency chart (Figure 9-17) as a large capacitor. There is only one on the board; dividing by four equals square inch.

The remaining component on the schematic in Figure 9-16 is the variable called R2. It has a similar number of pins and body size to a transistor. Place this under a 3-pin component; dividing one by three equals between 1/3 to 1/4 square inch. Round off to 1/4 square inch.

Review the component equivalency chart in Figure 9-17. Are there any test points or jumper pads? There are three test points. List these three under test points. Therefore, three divided into ten is approximately 1/4 square inch.

Review the component equivalency chart in Figure 9-17. It requires pre-assigned connector pins. Figure 9-16 indicates that the P1 pre-assigned pins are voltage +5 and ground (which has a total of four pins). This will be assigned on the component equivalency chart. Dividing ten by four equals approximately 1/2 square inch.

Adding the column for IC's, we have a total of 6 1/4 square inches: 6 square inches for the 14- and 16-pin IC's, and 1/4 square inch for the test points. Adding the column for discrete components (the total equivalent IC's) equals a total of two square inches. Adding the IC column and the discrete components together equals a total of 8 1/4 square inches required on the PC board.

SPECIAL NOTE: Review the board outline assigned in Figure 9-11. We calculated a total of eleven square inches of useable area on the PC board. Compare that to the total square inches as a minimum requirement found on the chart in Figure 9-17. We had a minimum of 8 1/4 square inches. We can easily place all components (IC's, resistors, capacitors, test points, etc.) within the board outline.

Step #9—IC's should communicate to one another in line or in the same rows. For a review on plotting, see Chapter 5.

Illustration #8: Plotting

Plotting the grouped IC's on the schematic in Figure 9-16, on the PC board outline (Figure 9-12) is the next step. Assign the IC location on the board according to how they communicate to each other. For a review on plotting, see Chapter 5. Using the schematic in Figure 9-16, plot the grouped IC's on the board outline shown in Figure 9-18. Use the PC design template shown in Figure 8-8.

Plotting Steps

Look at the schematic in Figure 9-16. Notice that the majority of gates have been grouped to facilitate signal flow. On this schematic, it is much harder to see the signal flow than the

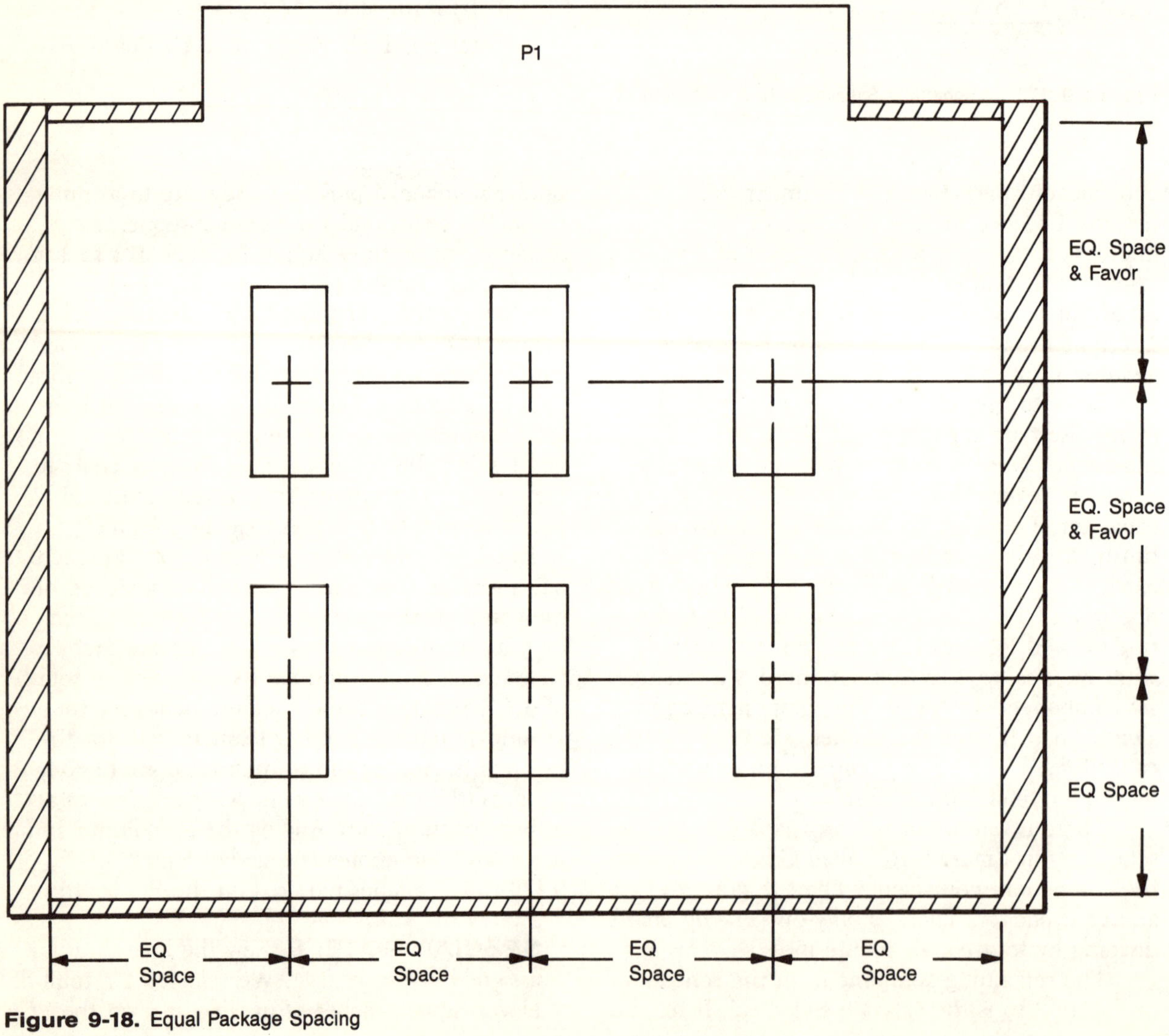

Figure 9-18. Equal Package Spacing

schematic we had in Chapter 8. This flow determines how the PC board is to be laid out. For example, the schematic in Figure 9-16 shows that IC U6 should be close to IC U2 because of the intercommunication between the two. IC U6 should be close to U1; perhaps U1 should be close to U3. There are no pre-assigned pins for the signal paths of the P1 connector. U3 should be close to the connector because of the number of pins that intercommunicate to the connector. Position the six IC's within the board outline shown in Figure 9-12. For the positioning of these IC's, see Figure 9-18. Equally space the IC's in two common rows. Do not favor one IC over the other by giving more space to two common rows. Do not favor one IC over the other by giving more space to any one individual IC. The IC's are given equal consideration inside the useable board area. It is common to favor a row of IC's. In Figure 9-19, look at the first row of IC's closest to P1. We may favor that row because of the high concentration of traces that travel down that aisle to intercommunicate or to interconnect with P1. Also, the second row or the space between the two rows of IC's might be favored because of the higher concentration of traces intercommunicating from the top row of IC's to the bottom row of IC's. Position the IC's into two rows of three. Placing four in one row and two in another imbalance the intercommunication from one IC to another.

Assigning U Numbers to Plotted IC Positions

Use the schematic in Figure 9-16 and the plotted IC positions in Figure 9-19. Assign IC U numbers to these positions. The first IC that we will place down is U3 because it intercommunicates to the connector. Place it on the front row closest to the P1 connector. Because P1 does not have pre-assigned pins, place U3 in the middle of the board (for the moment).

This does not mean that we cannot change these U numbers. Assign them to these positions temporarily. U3 communicates to P1, which is now close. What else does it communicate to and what else should be close to U3? U1 should be close because it directly communicates to U3. Also U5 should be close. U1 communicates with the connector through C1; U5 also communicates with the connector according to the schematic in Figure 9-16. Place U5 on the left side of U3 on the front row. Place U1 on the right side of U3. Because both U5 and U1 communicate to P1, it is a good choice to place them on the front row. U3 communicates with U5 and U3 communicates with U1. What else communicates with U3? U2 also intercommunicates with U3. Place U2 below U3. What IC communicates to both U2 and U5 which would fill the corner. U4 communicates with U5, U2 and U3. Therefore, place U4 in the bottom left-hand corner. U6 intercommunicates with U1, U2, and U3. This is not the only way to plot the schematic, but this has good intercommunication from one IC to another. We also have good intercommunication from the front row of IC's to P1. Try another plotting. This would be good practice for you to try at this time. See how you could shuffle the IC's around and still end up with close communication for the majority of IC's.

Using the template in Figure 8-8, draft in the pads. Determine the spacing shown in Figure 9-18. Figure 9-20 shows the board outline with the grids at a 2:1 scale. Count the grids on the board to give us the total useable area. There are 58 grids vertically at 2:1 and we have 74 grids horizontally. Placing the IC pads down onto the grid pattern, center the IC pads on the grid crosshair. If we divide 74 grids by four, we have

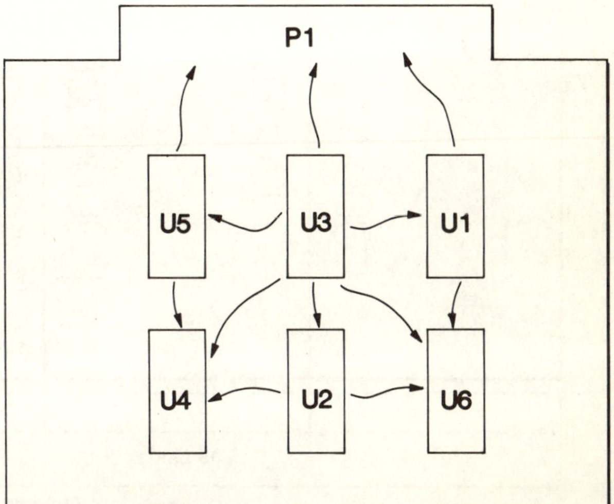

Figure 9-19. Plotted IC Positions

18.5 grids for each space (given you for equal spaces). Round this off to 18 or 19. In Figure 9-20, place the first IC row 18 grids to the center line. Place the second row of IC's 19 grids to the center line. Place the second row of IC's 19 grids to the center line. This leaves the last space at 18 grids.

This became a symmetrical pattern, giving equal space to all IC's in the horizontal direction. In the vertical direction, the total of 58 grids divided by 3 becomes equal spaces of 19 grids. However, we did favor the first row by giving it 20 grids because of the intercommunication with P1 connector. If there were many connections going to the P1 connector, we would favor it even more. Notice the second space between the rows of IC's. We gave it 22 grids because of the intercommunication from the top row to the bottom row. Again, we could favor it even more if we felt that we would have a higher concentration of traces in this row. Finally, the last space ended up at 16 grids, for a total of 58 grids.

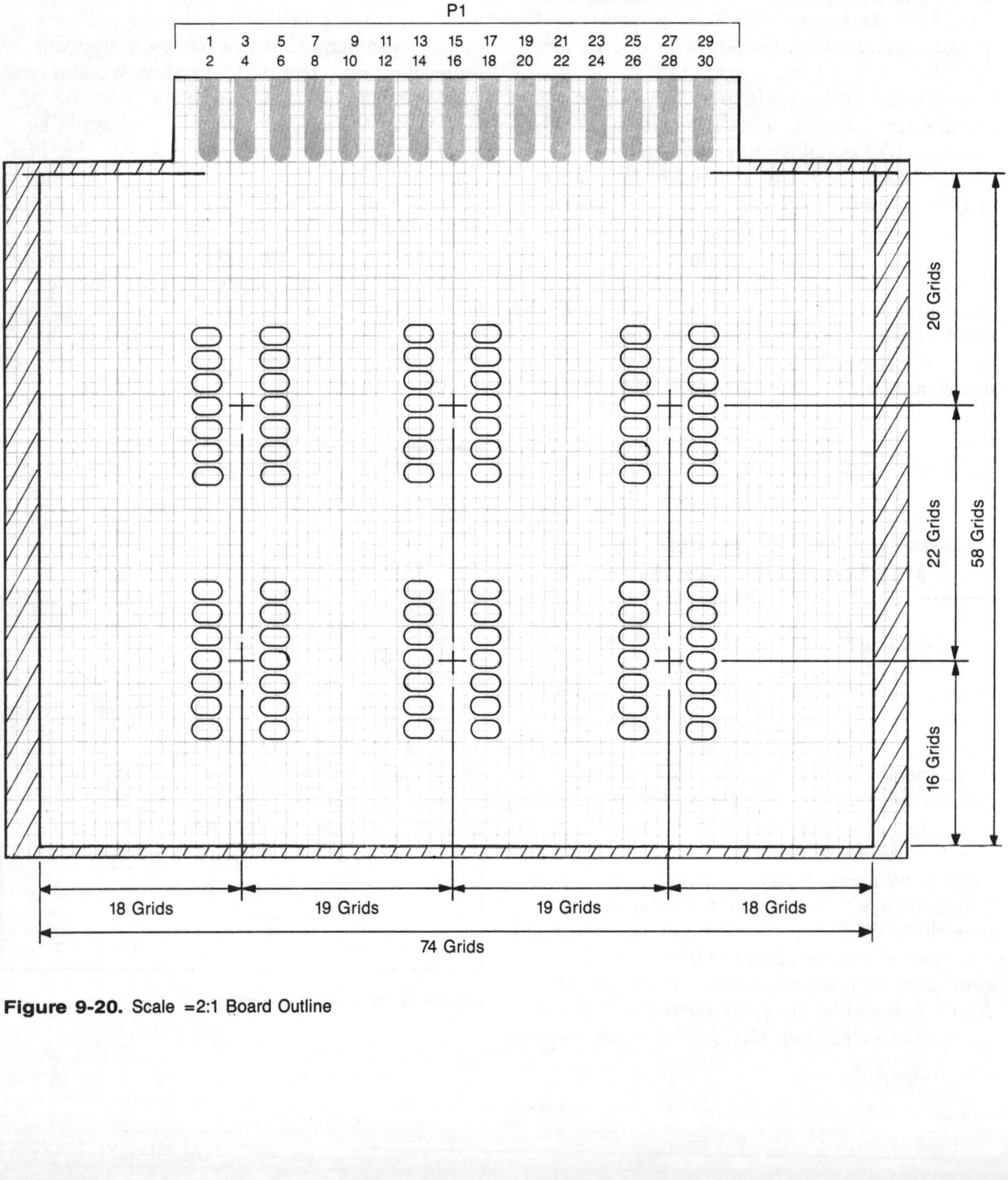

Figure 9-20. Scale =2:1 Board Outline

We did find the center line of each IC. Using the template in Figure 8-8, we positioned the center of the IC pattern (14-pin) on the crosshairs and drafted in the pads. Notice the 16-pin IC in the top row, middle position. We pushed the odd row of pads close to P1 to maintain the space between the top and bottom rows (for routing the traces from one IC to another). The positioning, of all IC's on the board outline is complete. Step 10 is the formal layout.

Step #10—The formal template layout is the next step. We will conduct a formal template layout establishing our conductor paths in an X-Y direction. The layout should be viewed from the component side of the printed circuit board. See Illustration 9.

Illustration #9: Formal Layout

Use the IC plotted locations obtained from Figures 9-19 and 9-20. Bypass the rough sketch (normally done for an analog type of PC board). Conduct a formal template layout by establishing conductor paths in an X-Y direction. The layout should be viewed from the component side of the printed circuit board. The following procedures itemize each design step.

Procedure #1

Review the design rules for formal PC board layout, Step 4. This establishes the conductor/trace, air gaps and widths, pad sizes for our components and feedthroughs, tooling hole sizes, and card edge spacing minimums.

Procedure #2

Identify the layout material. Use something similar to ACCUFILM® which is an antistatic, stable-based matte film with a grid underlay. Use a .100 grid. See Figure 8-32.

Procedure #3

Use the information from the IC plot in Figure 9-19, and from the established IC pad locations in Figure 9-20. Itemize the beginning steps for the formal digital layout. See Figure 9-21.

a) Place the U numbers inside each IC on the formal layout (established on Figure 9-19). Number each IC pad using the specification sheets in Figures 9-6 and 9-7. Start with Pin 1 in the upper left-hand corner, with the P1 connector up. Rotate the IC position so that Pin 1 would have been in the lower right hand corner (This is a personal choice).

SPECIAL NOTE: We recommend that you remove Figure 9-22 from the book; it is a duplicate of the board in Figure 9-21 without the traces. Duplicate each trace that we put on the formal layout. Use the template and a straight edge with the proper colors. Remove the specification sheets from the book (Figure 9-6 and 9-7). You will be using them during the design. Drafting will be done on Figure 9-22.

b) Place the input capacitor, C2, using the template in Figure 8-8. It is labeled RC20 close to the input pin called +5V, pins 29 and 30. Place the positive end of the capacitor close to those two pins. See Figure 9-21. Notice that we placed the capacitor off to the side of the board so that the capacitor neither takes up area in front of the P1 connector nor blocks this area of P1. (This is optional).

c) Set the color code to identify voltage and ground pads, the component and noncomponent side traces. The red circled pads equal +5V and the black circled pads equal ground. This means that the voltage (called +5) must interconnect to each red color coded pad; the ground bus line must interconnect to each black pad and terminate to their respective pins on the P1 connector. The red lines are the component side, and blue lines are the noncomponent side of the board. These codes should be placed on the layout as shown in Figure 9-21.

SPECIAL NOTE: Once the component side (red traces) and the noncomponent side (blue traces) have been identified, number the connector pins as shown in Figure 9-21. All odd numbers (1, 3, 5, 7, through 29) are color coded

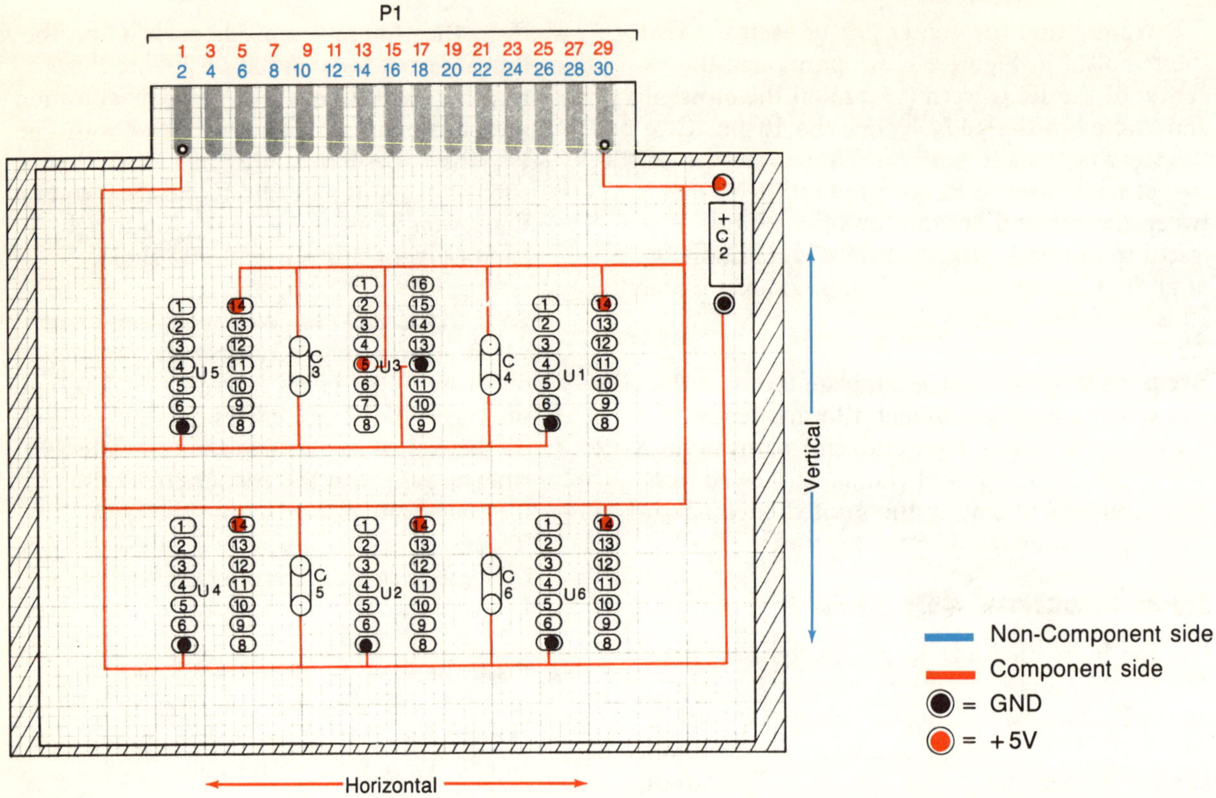

Figure 9-21.

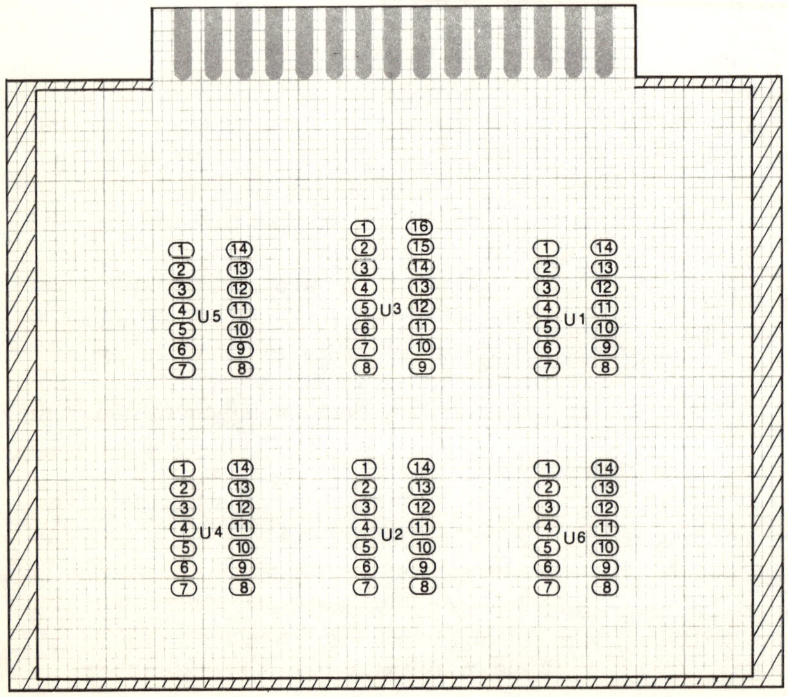

Figure 9-22.

red and labeled: component side for P1. All even numbers (2, 4, 6, 8, through 30) are color coded blue and labeled: noncomponent side for P1. This information is obtained from the connector specification sheet.

However, to expedite the layout, we provided this information for the schematic in Figure 9-21. This can be reviewed in Chapter 4 of this book, or it can be reviewed in the Analog Printed Circuit Design & Drafting Book.

Use the color code dots (red = +5V, black = ground pads). Figures 9-6 and 9-7 identify the voltage and ground pins for U1 through U6. U1 on Figure 9-16 is identified as a 74136. Figure 9-7 identifies VCC as +5V on Pin 14. Examine the other specification sheets. U2, U4, U5, and U6 also have pin 14 for +5V. However, U3 is a 74141. Shown also in Figure 9-7 identifies VCC as a +5V on Pin 5. U1, U2, U4, U5 and U6 have Pin 14 identified with a red dot; U3 will have pin 5 identified with a red dot.

Place the black dots on ground pads. Refer back to the specification sheet. U1, 74136, shown in Figure 9-7 calls ground, Pin 7. U3 (called a 74141) in Figure 9-7 had ground Pin 12. Place a black dot on Pin 7 for all IC's, except for U3 which will have a black dot for ground on Pin 12. This identifies all the voltage +5 and ground points for the bus lines. They connect to the IC's U1 through U6. They come from the connector pins 1 and 2 for ground and 29 and 30 for +5V.

Procedure 4

Draw the bus lines for voltage +5. Start from the connector pins 29 and 30. Refer to the schematic in Figure 9-16. Pins 29 and 30 are called out for the +5V line for P1. Place a feedthrough on the connector contact, tying pins 29 and 30 (29 on the component side and 30 on the noncomponent side).

The design rules require that for voltage and ground lines there be a .100 thick trace at a scale of 2:1. Route this trace starting from connector pins 29 and 30, down the side of the board in the vertical direction. Stop where the trace turns in the horizontal direction for the second row of IC's.

On that vertical run, connect the input filtering capacitor called C2 on the positive end. Make sure that C2 is connected close to the input connector pins 29 and 30 on P1.

Continue with the connection of the +5V line. We will now establish the horizontal traces. Tie to the vertical bus with a horizontal trace connecting off the first row of IC's on the red pads (U5, U3 and U1). To complete the +5V bus line, tie the second row of red pads with a horizontal bus (U4, U2 and U6). Connect to the vertical bus line. This completes the +5V bus line to all of the IC's into the input filtering capacitor, C2, and the P1 connector.

Procedure 5

Draw the ground bus connecting the black dots to the connector pins 1 and 2 on P1. Refer to the schematic in Figure 9-16. The ground does connect to pins 1 and 2 using a feedthrough on the contact. Starting with pins 1 and 2, put a .100 trace in the vertical direction down the side of the board. Stop where the horizontal trace meets in the last row. A horizontal trace connects the black dots for the first row (U5, U3, and U1). Completing the IC connections, pull the horizontal trace connecting the black dots on U4, U2 and U6. To connect the ground for our input filtering capacitor, continue the horizontal trace on the last row over until it lines up under C2. Then continue a vertical trace to make that final connection. This now completes the entire bus structure for the IC's for both +5V and ground.

SPECIAL NOTE: The red traces in the horizontal direction have established a direction for the red traces on the PC board. Red traces for the component side go in the horizontal direction; blue traces for the noncomponent side go in the vertical direction. This is the X-Y system.

Procedure 6

Place the bypass capacitors called out on the schematic in Figure 9-16, Note 5: bypass every two IC packages using a CK06 capacitor. As we mentioned in Chapter 8, tie the bypass capacitor between the voltage +5 bus line in the horizontal

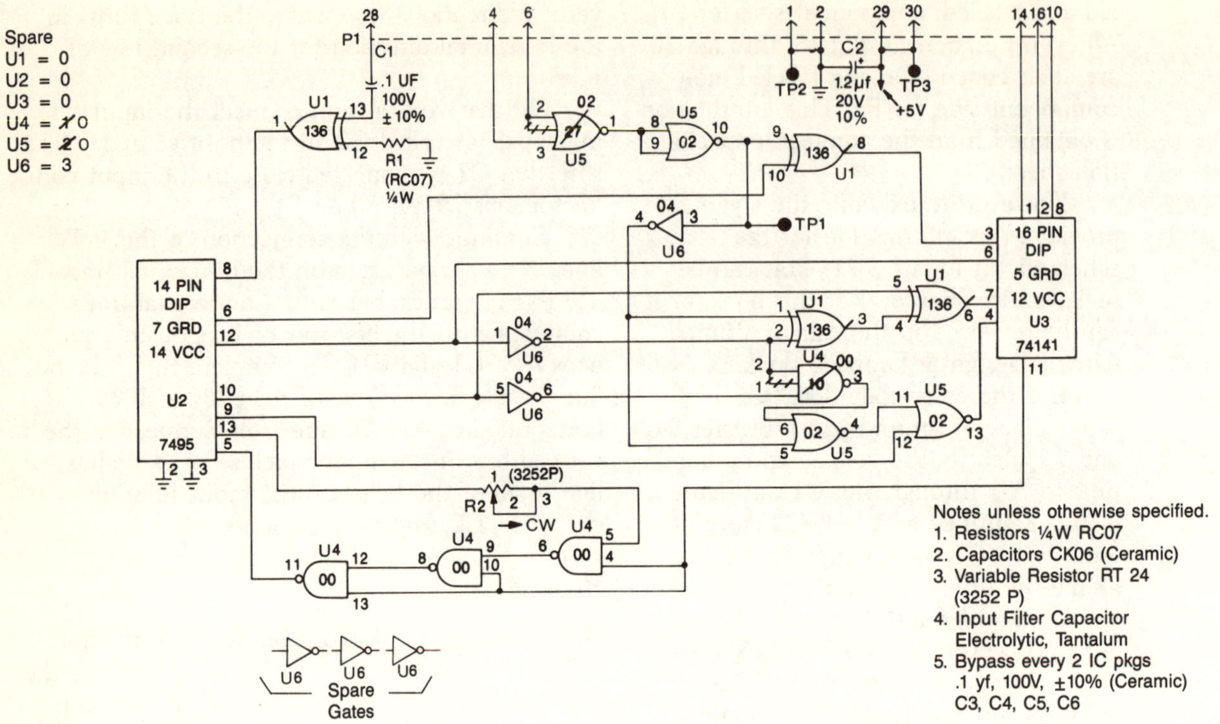

Figure 9-23.

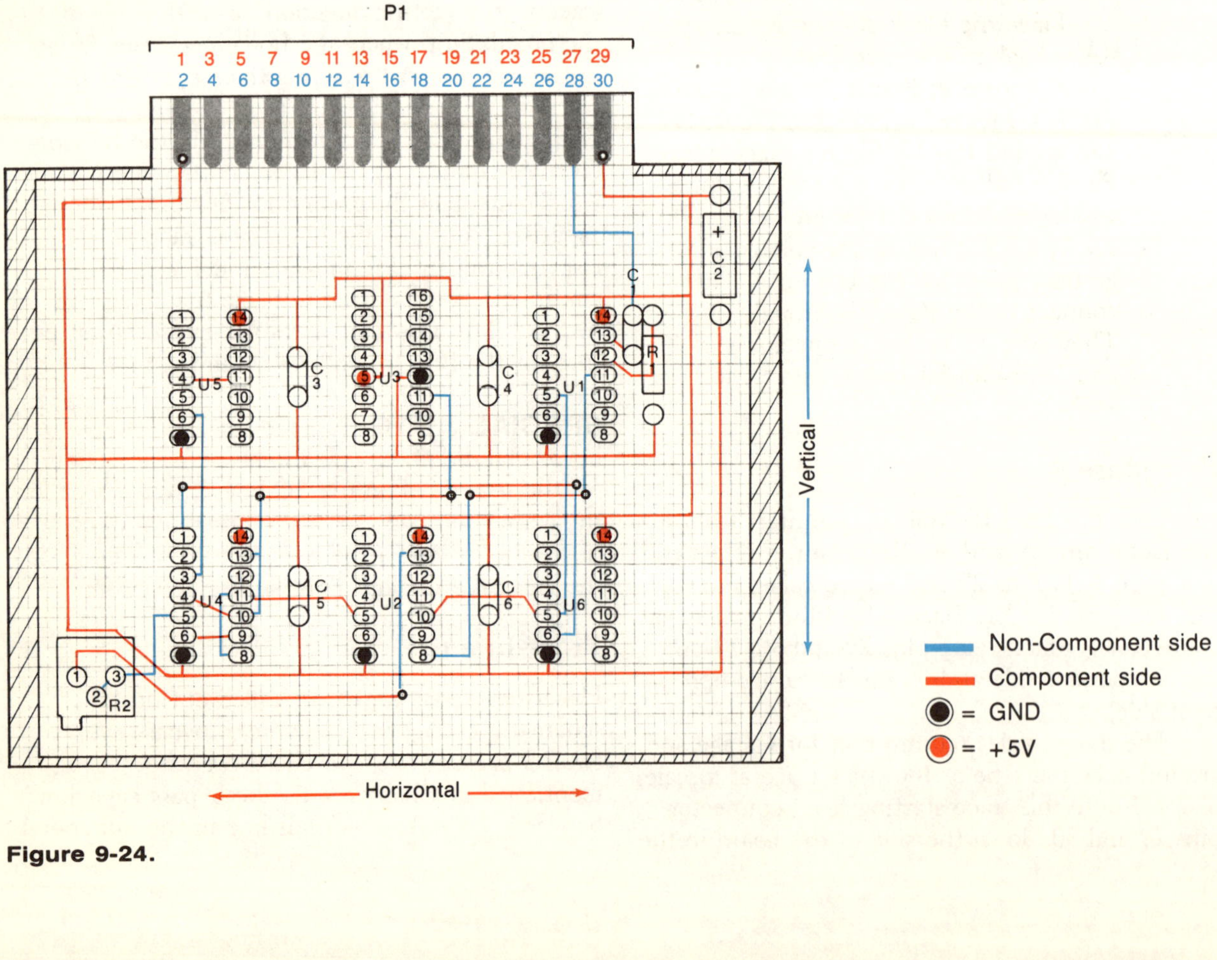

Figure 9-24.

direction to the ground bus line in the horizontal direction. Since there are two rows of three, place one capacitor between U5 and U3, and another between U3 and U1. Repeat that for U4 to U2, and U2 to U6. We will finish with a total of four IC bypass capacitors. The bypass capacitors do not have to be placed at this time. This is a personal choice. Their placement can be postponed. These capacitors should be connected to the voltage and ground bus without feedthroughs, if possible.

Ignore the X-Y direction for red and blue traces. Bus into the bypass capacitors with the color red off the voltage and ground bus (which is also colored red). They will carry the same trace widths as the voltage and ground bus. This is .100 at a scale of 2:1. Identify these capacitors as C3, C4, C5 and C6. Refer to the schematic in Figure 9-16. C1 and C2 are already being used. For placement and connection, see Figure 9-21. For reference to the size, CK06, see Figure 9-8.

SPECIAL NOTE: For more information on bypass capacitors and placement of bypass capacitors, see Chapter 3. The procedures that we have covered (1 to 6) should be duplicated on your board outline, Figure 9-22 at this time.

Procedure 7

Interconnect the signal lines and place the discrete components. Use the schematic in Figure 9-23 and the layout in Figure 9-24. Every trace that connects one point to an IC or one discrete component to another must be identified on the schematic. Figure 9-16 shows none of the input or the output pins assigned on the gates. The PC designer must choose these pins as he interconnects the board layout. The assignment for these pins will be shown on Figure 9-23. Each time we interconnect, we will label the pin numbers on the gates We will label the pin numbers on the P1 connector as we interconnect each trace. This will be reflected on the schematic in Figure 9-23.

a) When interconnecting signal paths from Figure 9-23, there is no correct starting point. To pick a point, start with R2, which is a variable resistor. The engineer made a special point that the variable resistor, R2, should be accessed from the back edge of the board (which is on the opposite edge to the connector P1). Knowing that R2 must connect to U4 and U2, place it close to those two IC's. The only logical place would be to place it to the left bottom corner of the board close to U4. In order to do this, we had to re-route the ground bus line, slightly. To find the case size for R2, see figure 9-8. Use the template in Figure 9-9.

The pads for R2 are numbered 1, 2, and 3. The specification sheet in Figure 9-5 (for R2) identifies pins 1, 2, and 3 in the clockwise direction. Figure 9-23 shows the arrow pointing to the CW which stands for clockwise. This clockwise direction should start with pin 3, pin 2 (the center wiper) and the remaining pin 1. Label this on the schematic in Figure 9-23. The pre-assigned pin for U2 calls out U2, pin 13 (now connecting to R2, pin 1). Make that connection by routing a red trace in the horizontal direction over to U2. Place a feedthrough and change the color to blue. Route up to pin 13 for U2.

To assign the pin numbers for U4, connect to R2, pins 2 and 3. Use the specification sheet for U4 called a 00 shown in Figure 9-6. Look for the closest available input pin for a 2-input NAND. Our specification sheet shows pins 4 and 5 as the closest input pins for R2 (pins 2 and 3). Select pin 5, connecting to R2 (pins 2 and 3).

SPECIAL NOTE: Once we have assigned one input to a gate (as we did with U4, pin 5), we can assign all of the remaining inputs and outputs for that particular gate. We did this for U4 on the schematic in Figure 9-23. Look at the specification sheet. The output was pin 6. The remaining input was pin 4. Once these connections have been completed (as shown in the layout in Figure 9-24), yellow them out as a completed connection on the schematic in Figure 9-23.

b) Which IC's should we continue interconnecting? Since we just connected R2, pins 2 and 3, to U4, pin 5, why not just continue connecting U4? For example, output pin 6 assigned to the first gate must connect to an input. Look for the easiest connection. Use the specification sheet in Figure 9-6 for the 2-input NAND 7400. The closest available input would be pin 9. Connect U4, pin 6, to U4, pin 9. Assign this to the schematic. Draft it onto the layout in Figure 9-24 using a red trace in the horizontal direction. Once we have assigned one input on the gate, we can now assign the rest of the inputs and outputs. The U4, pin 9 input would be pin 10; the output will be pin 8. Yellow out the connection from U4, pin 6, to U4, pin 9, on the schematic. Connect U4, pin 8, to the closest available input on one of the remaining two gates in U4. On Figure 9-23, the output on U4 (connecting to U2, pin 5) should be on the side facing U2, pin 5. Select U4, pin 8, to U4, pin 12, connecting the color blue in the vertical direction inside U4. Assign on the schematic, U4, pin 8 connecting to U4, pin 12. Use the specification sheet. Once pin 12 has been selected, the other input of that gate is pin 13 and the output is pin 11. It would be easy to use a red trace coming out of pin 11 on U4 connecting to U2, pin 5. Route between the two pads on the bypass capacitor, C5, and connect to U2, pin 5. Yellow out these connections on the schematic: U4, pin 8, to U4, pin 12, and U4, pin 11, to U2, pin 5. U4, pin 13 is connected to U4, pin 10, which is connected to U4, pin 4, and also to U3, pin 11. To make this connection, route out of U4 with these three connections, up into the center aisle, and move over to U3, pin 11. To do this, tie U4, pin 4, to U4, pin 10 in red in the horizontal direction. We will come up to the center aisle using blue out of pin 10. Tie to pin 13 on U4 while going up the aisle. Place a feedthrough in the center aisle. Connect a red trace going in the horizontal direction and route it under U3, pin 11. Place another feedthrough so that we can turn in the vertical direction with blue and route up to U3 pin 11. Yellow out these connections on the schematic: U4, pin 13, to pin 10 to pin 4 to U3, pin 11.

SPECIAL NOTE: Each time we change direction for any great length at all, we use feedthroughs to change from red to blue traces. These feedthroughs carry a pad size of .125 at a scale of 2:1 that we have obtained from our design rules in Chapter 8.

c) Consider the discrete components for placement. Simplify the interconnects (such as C1 and R1). Place them for easy connection (in this case to U1). Place C1 and R1 beside U1, in the area that has the most available space (on the right-hand side or the high pin numbered side of U1). C1 must also connect to connector P1. Place it close to P1. Choose the inputs available on that side of U1. Use the specification shown in Figure 9-7 for 74136. Pins 12 and 13 are inputs and pin 11 is an output. Connect pin 13 to C1 (using the color red). Continue through C1 to the opposite side. With the color blue, connect C1 to the P1 connector. The closest available pin, 28. Continue with the U1 input, pin 12. Connect with the color red to R1. Connect so that, when passing through R1 to its opposite side, there is a ground bus at the bottom of U1 to tie to (to complete the connection for R1). Yellow these connections out on the schematic in Figure 9-23 (once they have been completed on the layout Figure 9-24). Continue with U1, output pin 11.

Figure 9-23 shows that it connects to U2, pin 8. Using the inside of U1, route down with the color blue into the center aisle, Connect to a feedthrough which allows movement in the horizontal direction with the color red. Notice that we have stayed in line with the red trace which was put into that aisle to prevent wasting area. Move over until you line up with U2, pin 8. Then you can change the color to blue, routing down in the vertical direction and connecting to U2, pin 8. Yellow this connection out on the schematic.

d) Look at the yellowed schematic (Figure 9-23). All of the discrete and variable components have been placed. You may wish to go to U2 or U3, which already has pin assignments from the engineer. Perhaps you would like to continue with U1, since you have already selected one of the gates out of four. We would like to choose U2, pin 10, for no particular reason. U2, pin 10 connects both to an input on U2 (called an inverter gate) and to an input on U1 (called an exclusive OR). The specification sheets indicate that the U6 inverter gate is a 7404. There are six of them in the package. U1, 74136, is an exclusive OR gate. There are four of them in the package. We have used one out of four on U1.

Start with U2, pin 10. Find the easiest connection to U6. Looking at your specification sheet on Figure 9-6. Consider that your choices (for inputs on inverter gates to the low pin number side of U6) are pins 1, 3 and 5. Connect from U2, pin 10, to U6, pin 5. use a red trace in the horizontal direction (routing through C6 between the pads). To finish the interconnection, connect to an input on U1. The closest available input on U1 is pin 4 or 5. Connect (with a blue trace in the vertical direction) U6, pin 6, to U1, pin 5. Once you have assigned these pin numbers, return to the schematic in Figure 9-23. Assign pin 5 to the inverter gate 7404. Once you know the input number, you can assign the output number (in this case being pin 6). Once pin 5 on U1 has been assigned (using your specification sheet), you can see that pin 4 is the other input and pin 6 is the output. Yellow out on the schematic these completed connections: U2, pin 10 to U6, pin 5, to U1, pin 5.

e) The schematic now has more gates assigned (both inputs and outputs). It is now becoming easier to choose which points to pick up. Continue with U6, pin 6 (output) or we could go to U1, pins 4 or 6. Select U6, pin 6 (output) which connects to an input on U4. This is an easy selection because there is only one remaining gate on U4. According to Figure 9-6, that gate is input pin 1 and 2, and output pin 3. Your choice is U6, pin 6 to pins 1 or 2 on U4. Route to pin 2 on U4. Connect from the feedthrough to pin 1 on U4 using blue. You might question the shortness of the trace length for the feedthrough over U4, pin 1. We will return to it if we do not need that feedthrough. We might remove it and make the connection in red in the clean-up stage of the layout. Yellow out these connections on Figure 9-23: U6, pins 6, to U4, pins 1.

Continuing with U4, pin 3, look for an input on U5. The specification sheet for U5 is shown in Figure 9-6; it is called a 7402 2-input NOR gate. The closest available input are pins 5 and 6. Choose pin 6. Connect U4, pin 3, using blue in the vertical direction to U5, pin 6. Once we have assigned U5, pin 6, as input, we can assign the remaining input which would be pin 5 and the output pin 4. Yellow out the connection U4, pin 3 to U5, pin 6.

Continue with U5, pin 4. Look for input on the same IC, U5. The easiest possible connection would be a straight horizontal line to pin 11. Can we do this? Is pin 11 an input? Review Figure 9-6 (7402). U5 does have an input pin 11. Connect to pin 11 using a horizontal red trace. Assign pin 11 to the input. The remaining input can also be assigned (which would be pin 12 with an output pin 13). Yellow out the connection U5, pin 4, to U5, pin 11.

f) Assigning one input of a gate automatically picks up the other inputs and outputs of that same gate. This process indicates the points of connection. It is a self perpetuating function from this point on. We have completed almost half of the layout. It becomes relatively easy to assign the remaining gate inputs and outputs that have not been picked up by following the X-Y systems. It is easy to use aisles and feedthroughs to intercommunicate from one IC to the other. The connections to the P1 connector are untouched. It is not incorrect to put these points in first. It is easier to communicate with the rows of IC's before placing the inputs and outputs from the IC's to the connector. At this point, connect the remaining inputs and outputs. Assign the inputs and outputs to the gates that have not been picked up. Make the connections, using the schematic in Figure 9-25 and the layout in Figure 9-26.

g) This is an explanation of the remaining connections on the schematic in Figure 9-25 using the layout in Figure 9-26.

Consider U5, pin 12, and U2, pin 9. Use the center aisle as a path to interconnect these two points.

Consider U5, pin 13, and U3, pin 4. Use a straight horizontal line in red to connect these two points.

U1, pin 4, requires an output on the same IC. Choose pin 3. Take advantage of the closeness of this output pin. Strap on the noncomponent side using the color blue.

Consider U5, pin 5, to U1, pin 1, to U6, pin 4 (which we had to assign at this time). Once we have assigned 4 on the output of U6, we will assign 3 to the input. Route along the top aisle in red from U5, pin 5, to U1, pin 1 and in blue from U1 to pin 1 to U6, pin 4.

Consider U1, pin 2, to U4, pin 1, to U6, pin 2. This is a new assignment on U6. Once pin 2 is selected, assign pin 1 to the input side. Route from U1 to pin 2 down in blue to U6 pin 2. Intersect that line with a feedthrough in the center aisle. Cross over in red to U4 pin 1. Perhaps we are creating an unnecessary feedthrough on the U4 pin 1 connection (changing from red to blue). We will remove this in the clean up stage, if it is not necessary.

Consider U6, pin 1, to U2, pin 12, to U3, pin 6. Route from U6, pin 1 with a red trace directly to U2, pin 12, using a feedthrough (to connect U2, pin 12, in the vertical direction up to U3, pin 6).

Consider U2 pin 6 to U1 (new pin 10, new assignment). Once we have chosen pin 10 as the input for U1, we can assign the other input pin 9 and the output pin 8. This was the last gate available to us in the 74136 package. Route this connection by starting with U2, pin 6. Move into the bottom aisle. Cross with a red trace in a horizontal direction until it is on the extreme outside aisle. Move up in the vertical direction to U1, pin 10, with a blue trace. Notice how we have moved the blue trace going to the vertical direction out away from U6 and U1. This allows room if we wish to have additional blue traces in that area.

Consider U1, pin 8, to U3, pin 3. Make this connection starting from U1, pin 8. Use the center aisle in the horizontal direction with a red trace over to U3 up in the vertical direction. Use blue connecting to U3, pin 3. Notice how we moved away from pin 3 with the blue trace. By doing that we get one more connection from the U3 side.

Consider U3, pin 7, to U1, pin 6. Notice how short this connection looks on the schematic in Figure 9-25. However, to make this connection on the layout in Figure 9-26, it is more complicated. Notice the number of feedthroughs we must use, starting from U3, pin 7. Use a feedthrough to get to the top aisle in blue. Route in the horizontal direction in red over to another feedthrough by U1. Go down using Blue in the vertical direc-

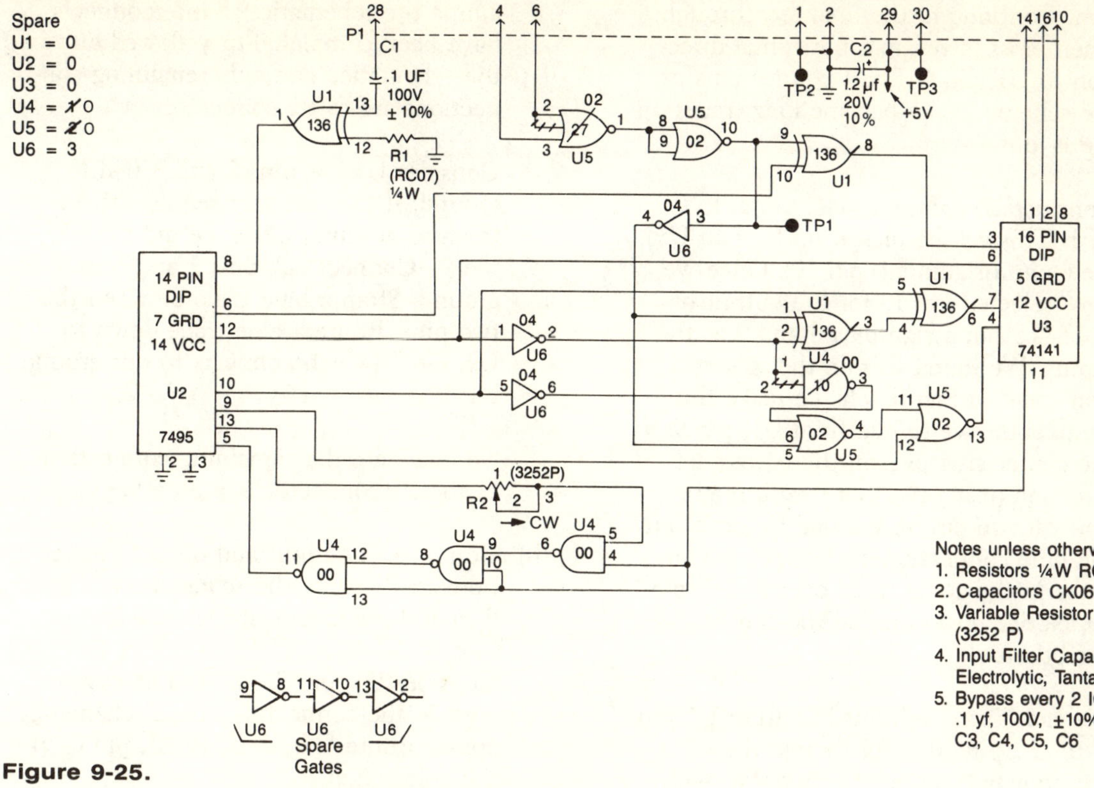

Figure 9-25.

Notes unless otherwise specified.
1. Resistors ¼W RC07
2. Capacitors CK06 (Ceramic)
3. Variable Resistor RT 24 (3252 P)
4. Input Filter Capacitor Electrolytic, Tantalum
5. Bypass every 2 IC pkgs .1 yf, 100V, ±10% (Ceramic) C3, C4, C5, C6

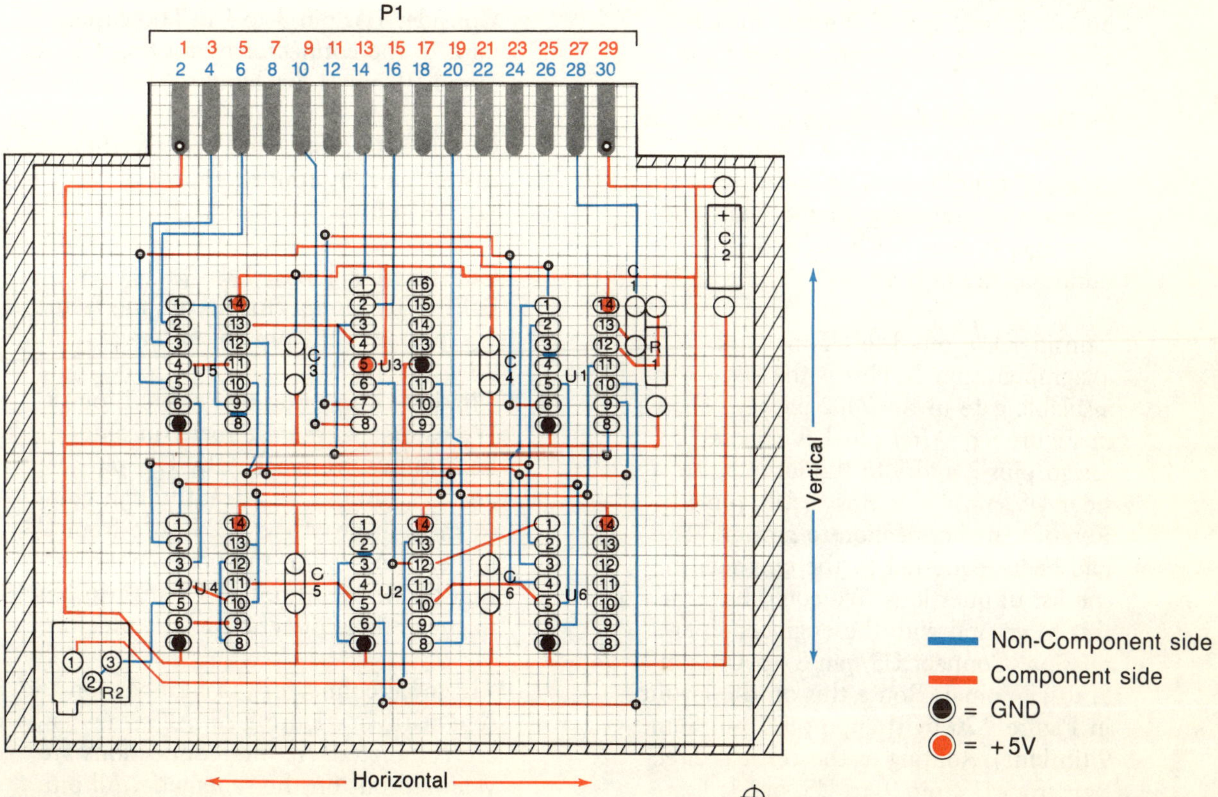

Figure 9-26.

tion. Continue to the final feedthrough. Then cross in red in a horizontal direction to U1, pin 6. Short connections on the schematic can become long traces on the layout.

Consider U1, pin 9, to U6, pin 3, to TP1. This will be picked up later to U5 (new pin assignment, pin 10). Once we have selected pin 10 for the output of 7402, we can assign pins 8 and 9 as the inputs. We found this on the specification sheet in Figure 9-6. To make this connection, we routed from U1, pin 9, to the center aisle over in the horizontal direction (using red to a feedthrough down to U6 pin 3). We had to move out the blue connection on U6, pin 4. Continue in the center aisle over to U5, pin 10. Notice how crowded the center aisle has become.

We can "fudge in" this line from U6 pin 3, to U5, pin 10, even though there is not enough room in the aisle. We will still draw in the line. When we tape the board, we will spread the two rows of IC's apart, creating more room in the center aisle. We will have enough room on the top and the bottom to spread the IC's. Be prepared to draw lines under crowded conditions. The layout is simply a means of preparing for the final tape-up stage. It does not have to be as accurate as the tape-up.

Consider U5, pin 8 to U5, new pin assignment, pin 1. This is the last available gate in the 7402 package shown in Figure 9-6. After pin 1 is selected, assign pins 2 and 3 to the input. Consider U5, pin 9. Tie this to U5, pin 8. Remove the connection to ground. This had been requested by the engineer in the list of questions. We could have tied this to ground with the engineer's permission. Connect U5, pin 8, to U5, pin 9, to U5, pin 1. Route this on the layout in Figure 9-26 by strapping U5 pin 8 and 9 (in blue). Routing in the vertical direction from U5, pin 9, to U5, pin 1. Examine the schematic. all interconnects have been completed or yellowed out on the schematic. The only remaining connections are to P1 connector.

Consider U5 has pins 2 and 3 tied to ground. If the engineer requires it, tie them to ground (unlike we did for U5, pin 9). Connect U2, pins 2 and 3 to ground. Strap a blue trace between the two pins. Route a blue trace down to U2, pin 7 (which connects to our ground bus).

To complete the remaining connections to the P1 connector, see Step 9.

h) Complete the input and output connections to P1 using the schematic in Figure 9-25 and the layout in Figure 9-26.

Consider U5, pin 2, to P1. The easiest path to the connector without changing colors. Route U5, pin 2, to P1, pin 6, in the color blue.

Consider U5, pin 3 to P1. The easiest path to the connector by routing U5 pin 3 and P1, pin 4, (in blue).

Consider U3, pin 1 to P1. Take the easiest path by routing from U3, pin 1 to pin 14 in blue.

Consider U3, pin 2. Take the easiest path to the connector by routing from U3, pin 2, to P1, pin 16, in blue.

The last connection to P1 is U3, pin 8. We cannot go there directly in blue because we are blocked by U3, pin 3, and U3, pin 6. Use a feedthrough going to the outside of the IC. Route from U3, pin 8, using red in the horizontal direction to the feedthrough. Change to blue in the vertical direction to P1, pin 10.

Assign all connector pin numbers on the schematic in Figure 9-25. All signal paths are completed. All interconnections are yellowed out on the schematic. All pin numbers are assigned to gates and to the P1 connector.

We need to assign the spare gate pin numbers (Step i). Illustration 10 is the clean-up stage of the layout. This includes test points, tooling holes, removing unnecessary feedthroughs, etc.

i) Assignment of the spare gate pin numbers. There are three spare gates on the schematic in Figure 9-25 that must have pin numbers assigned to the inputs and outputs. See the layout in Figure 9-26. U6 shows that pins 8 through 13 have not been used. Look at the specification sheet in Figure 9-6 for 7404. Because pin 8 is an output, assign pin 8 on one of the outputs of the three spare gates. Pin 9 is an input; pin 10 is an output; and pin 11 is an input. Assign that to the second spare gate. The third gate has pin 12 as the output and pin 13 as the input. This is helpful to the technician or the engineer for the future use of spare gates. They will not need to search for the pin numbers available as spare gates.

Step #11—Complete and clean up the layout in FIgure 9-26. This step establishes test point locations and cleans up all unnecessary via holes (called feedthroughs). In the formal layout we would have a tendency to place feedthroughs, because the X-Y system is used for routing traces. This step removes the unnecessary vias. Place all IC's in numerical order, left to right, top to bottom or any practical alpha/numeric order. This requires an update of the layout (Figure 9-26) to match the schematic (Figure 9-25).

Illustration #10: Layout Clean-up

After grouping IC's, U1 to U6 on the schematic, we did not know their placement. When plotting the IC's, we gave more consideration to intercommunication with each other and with the connector. Update both the layout and the schematic to bring these IC's into some form of alpha/numeric arrangement. This helps the technician or electrical engineer to troubleshoot these IC's. They can quickly find the IC's on the board. We have been working with small boards having few IC's. If there were 50 to 100 IC's, it would be difficult to find any particular one if there were no numerical arrangement. Renumber the IC's both on the board layout and on the schematic before proceeding to the final tape-up.

The following are remaining design steps to complete the formal digital layout. See Figure 9-28 for the final updated schematic and Figure 9-29 for the final layout.

Procedure #1

Change IC U numbers into numerical order as shown in Figure 9-27. Match this change on the schematic in Figure 9-28. For example, U1 starting in the upper left hand corner was U5. U2 in the lower left hand corner was U4. U3 in the upper middle remains the same. U4 in the lower middle was U2. U5 in the upper right hand side was U1. U6 remains the same. This completes the changeover of the U numbers. Transfer these numbers to the final layout shown in Figure 9-29. Match this with the schematic in Figure 9-28.

Procedure #2

Remove all unnecessary feedthroughs. The layout in Figure 9-26 shows the crossed out unnecessary feedthroughs in the Figure 9-27 layout. Examine each unnecessary feedthrough shown in Figure 9-27. We will show you the rerouted connection in the final layout in Figure 9-29.

For example, the U1, pin 5 feedthrough now can be routed to the U5, pin 1, feedthrough.

SPECIAL NOTE: All U numbers are now the new U numbers (in numerical order).

U3, pin 8, has been rerouted to the inside of U3 (using the color blue) to P1, pin 18. Change the schematic to reflect this change. U3, pin 8, was connected to P1, pin 10, with a feedthrough. Now connect to P1, pin 18. In order to make this change, we had to reroute U3, pin 6 (from the inside of U3 to the outside of U3).

U5, pin 9, has been rerouted (using red) around R1 to eliminate the feedthrough. U5 pin 8 has also been rerouted using red around R1 to eliminate the feedthrough. U5, pin 11 (connecting to U4, pin 8) has been rerouted entirely in blue

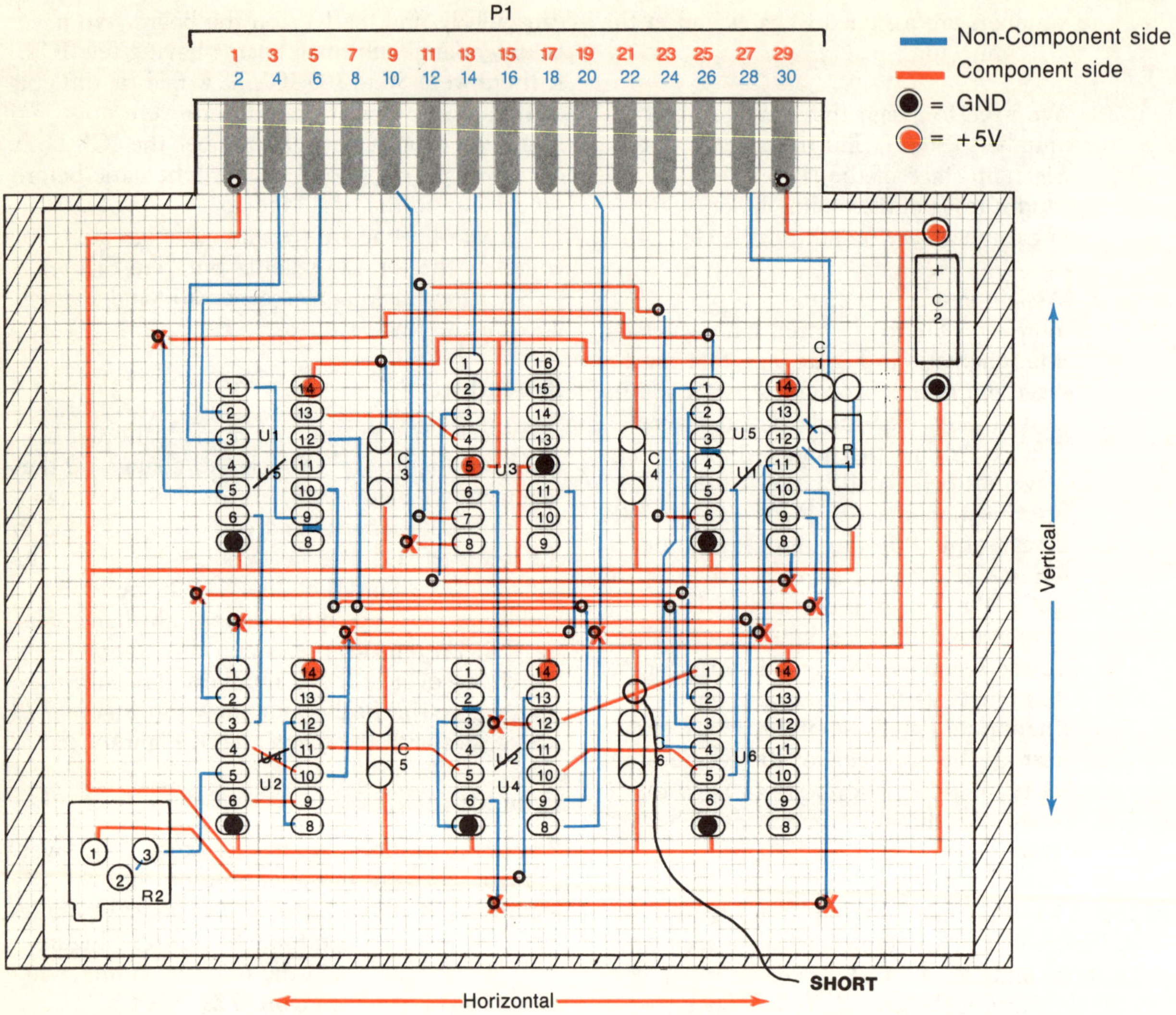

Figure 9-27.

through U6 to get to U4, pin 8. This eliminates two feedthroughs.

U4, pin 12, has a feedthrough under the IC. Remove this feedthrough, if possible. This does not mean there are never any feedthroughs under an IC. We prefer not to have feedthroughs under the IC. They are difficult to repair if there is poor, plated-through connection. Reroute (using blue) U4, pin 12, to U3, pin 6.

Consider U4, pin 6, to U5, pin 10. Route the entire path in blue, eliminating two feedthroughs.

U2, pins 13, 1 and 2, can eliminate their feedthroughs by turning the corner (in red).

Count the feedthroughs we have removed. We have eliminated twelve feedthroughs without needing to rework on the PC board layout. There may be more feedthroughs we could remove, but it is not worth the effort.

Procedure #3

Find short or obvious disconnections by using the schematic in Figure 9-28 and the layout in Figure 9-27. Scan the layout for obvious shorts or missed connections. There is one short. We had crossed two red traces that should not be connected. U6, pin 1, crosses over a +5V bus line connecting to U4, pin 12. This is an obvious short. There are two choices: reroute the red trace connecting U6, pin 1, to U4, pin 12. Otherwise change the +5V bus line (connecting to C6) to the color blue by using a feedthrough. This is what we did and is shown in the final layout in Figure 9-29. The red trace must avoid U6, pin 1, to U4, pin 12.

Procedure #4

Establish the locations of the test points. Consider the point of access for the PC board itself. Locate the test points on the same edge that we placed the variable resistor, R2. Refer to the schematic in Figure 9-28. There is a connection from U1, pin 10 that connects to U5 pin 9, and connects to U6, pin 3. The test point should be on that line, running out to the back edge of the board. Refer to the final layout in Figure 9-29. We have routed that connection for TP1 from the feedthrough off U1, pin 10. We used the color blue down through U2 to the test point, TP1.

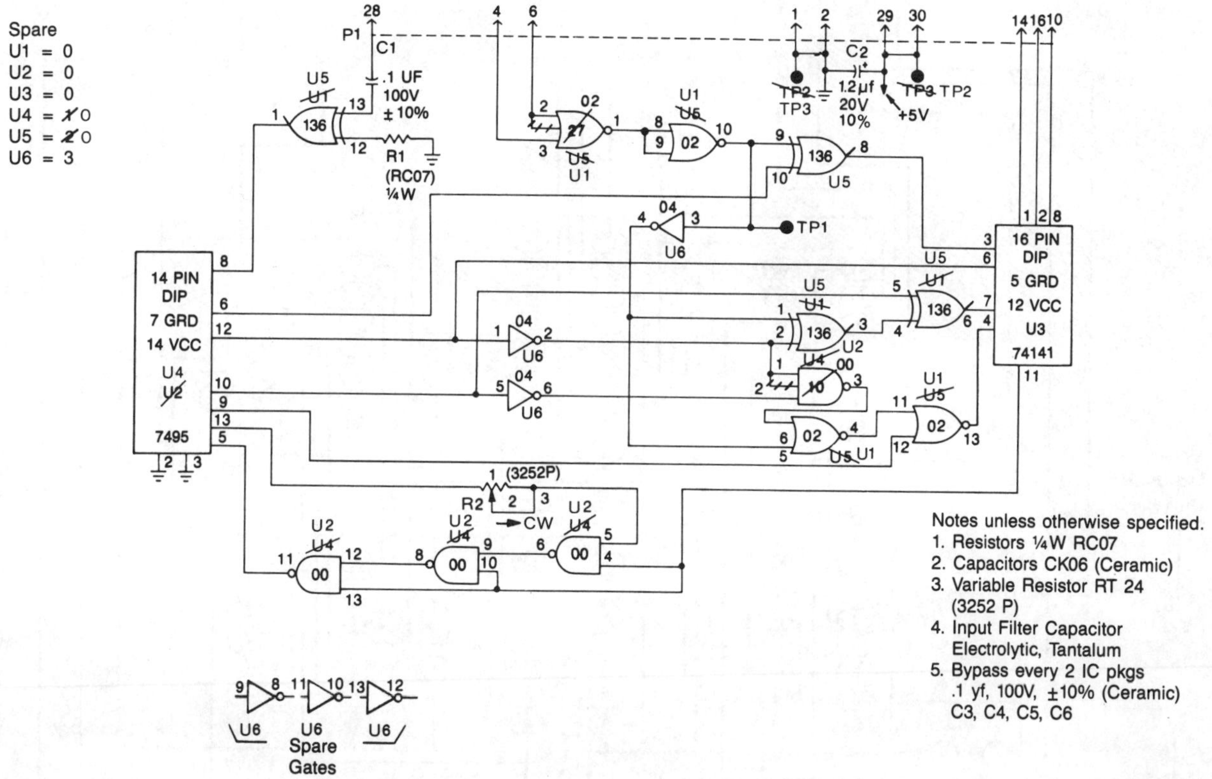

Figure 9-28.

Route the +5V the second test point. Refer to the schematic in Figure 9-28. It comes off either pins 29 and 30 or any red color-coded pad. U2, pin 14, is a red color-coded pad. Route using a blue trace off that 9-80 pad down to TP3 on the back edge of the board. Label it +5V.

According to the schematic in Figure 9-28, the third test point is TP2 (which is the ground test point). Ground pads are color-coded with black pads. The final layout in Figure 9-29 shows a ground bus connecting to black pads for U2, U4, and U6. Connect to that bus line with a red trace (connecting to TP2). To complete the test points, put the test point numbers in numerical order from left to right. Change TP3 to TP2 and TP2 to TP3. Reflect this on the schematic in Figure 9-28. For more information on test points, their location, and spacing, see Figure 8-17.

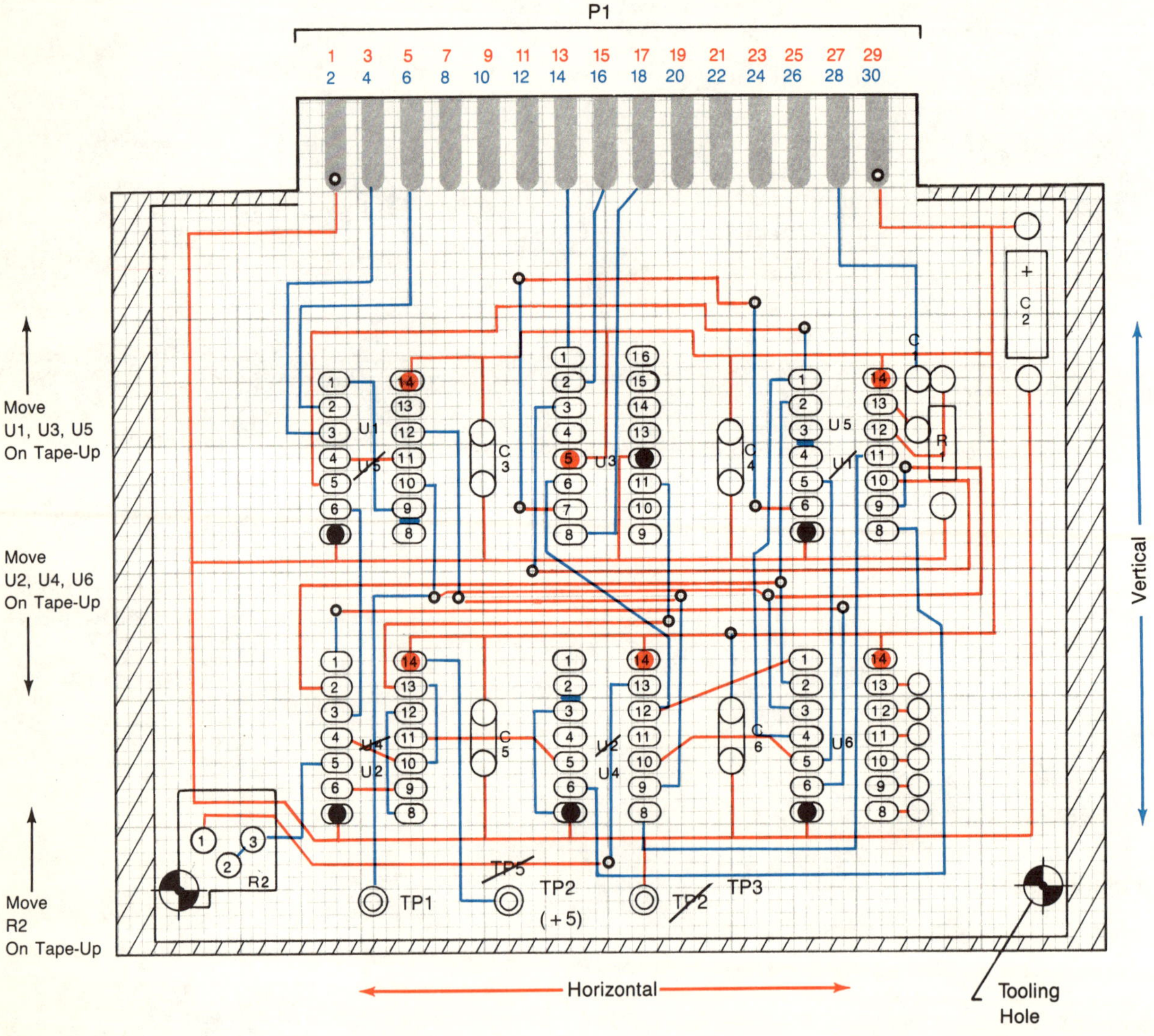

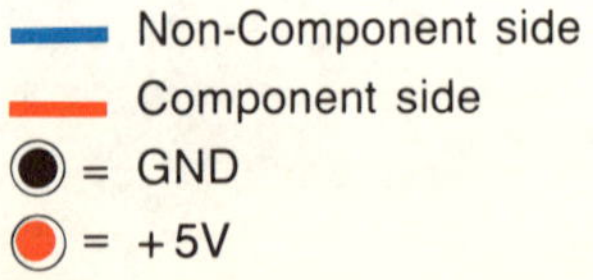

Figure 9-29.

Procedure #5

Add tooling holes. Tooling holes must be added as part of the layout procedure because they require the useable area of the PC board. Tooling holes are used as PC board reference points on which other dimensions are based. For example, during the manufacturing process, the manufacturer tools up or designs his/her fixtures for drilling and routing the outer edges of the PC board. The manufacturer uses these tooling holes as a reference point. The tooling holes must be located accurately. The typical tooling hole size is .125" in diameter. Usually some type of target material is left around the hole as a means of registering the board interior (traces and pads). For placement on the final layout, see Figure 9-29. Place the tooling holes in the lower left hand and right hand corners. There is a minimum of two. On large boards we would prefer to have three. Place these tooling holes on grids, if possible. The tooling holes are .250 in the horizontal direction and .250 in the vertical direction in each corner (at a scale of 1:1). On the layout in Figure 9-29, the tooling holes are five grids vertically and horizontally.

For a review on tooling hole locations and size, see Figure 4-11.

Procedure #6

Provide solder pads for the spare gates. This has not been discussed on previous boards. We did discuss spare IC positions (an example of a spare IC position is in Figure 6-18 which shows solder pads provided on U6, pins 8 through 13. This is a good feature for the future of these spare gates. Ask the engineer if he would like to have these pads. We will provide them on the final tape-up shown in Step 12.

Procedure #7

After completion of the layout drawing shown in Figure 9-29, an artmaster is taped. The artmaster is made to an enlarged scale (in this case, 2:1). Tape the layout in Figure 9-29 using a two-sheet technique shown in Step 12. However, before doing the tape-up, examine the layout for movement of IC rows or movement of discrete components.

For example, R2 (the variable resistor shown in the Figure 9-29 layout) should be moved closer to U2 by at least two or three grids. It is covering the tooling hole. The center aisle is overloaded with traces and feedthroughs. Therefore, move the top row of IC's (U1, U3, and U5) closer to P1 by at least three or four grids. Move the bottom row of IC's (U2, U4, and U6) down closer to the back edge of the board by at least two grids. This gives more space by routing the traces and feedthroughs in the center aisle. Make this movement only on the 2:1 tape-up, not on the layout. The layout is a guide for understanding where to connect on the final artmaster (called the tape-up).

Procedure #8

Route the traces through the IC patterns. On the last two layouts in Chapters 8 and 9, the designer did not run traces through the IC pads (as demonstrated in Figure 6-13). Allowing the designer to run traces through the IC pad patterns eliminates many feedthroughs and increases the IC packing density on most boards. This technique is shown on a completed board (Figure 9-30) using the same IC U numbers and pin numbers for all gates (as shown in Figure 9-28). This is the same layout as Figure 9-29, with the one exception. Routing is done through IC pads rather than through the aisles to cross communicate in rows of IC's. Figure 9-30 shows the benefits of routing through the IC pins. Not all companies approve of this technique. Check with the responsible engineer or supervisor. This technique is becoming very commonplace in the industry.

Compare the two layouts, Figures 9-29 and 9-30. Figure 9-30 shows routing through the IC patterns. Notice the center aisle. Figure 9-29 shows six red traces in the horizontal direction. Figure 9-30 shows one red trace in the horizontal direction. This eliminates the congestion in that aisle. Notice, also the top aisle. There are no traces in Figure 9-30; there are two traces in Figure 9-29. Figure 9-30 shows feedthroughs on the +5V line and the ground bus lines both for the bypass capacitors (C3, C4, C5, and C6) and for U3. One other modification we had to make was the U3, pin 8, to the P1 connector. We had to change the assignment of this pin to P1, pin 10. The layout in Figure 9-29, U3, pin 8, connects to P1, pin 18. Figure 9-30 is better: reduction of

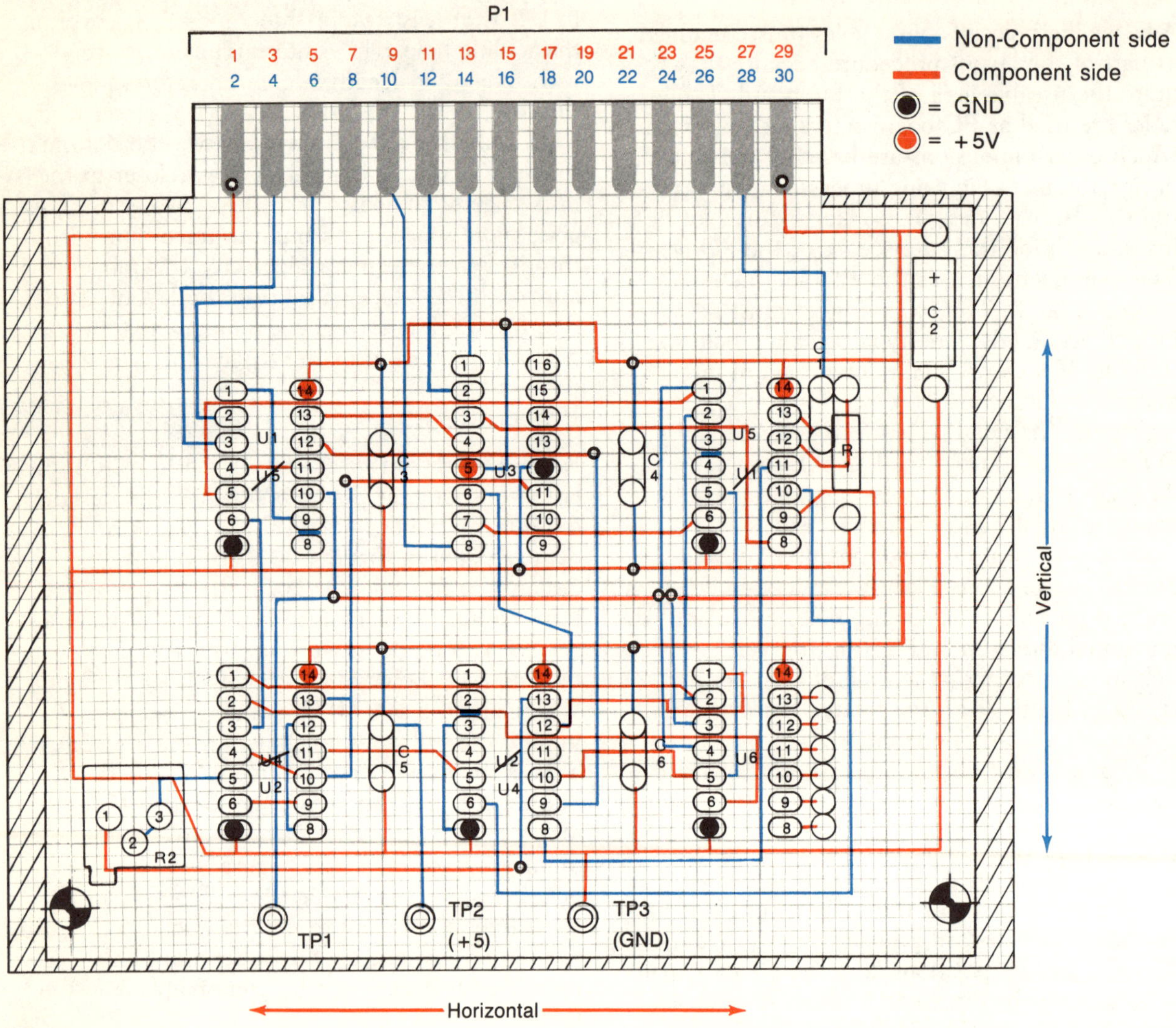

Figure 9-30.

area, routing of traces, reduction of the feed-throughs, and ease of taping (movement of the IC rows or discrete components). This state-of-the-art technique is used in Chapters 10 and 11. If your company does not allow this technique, refer to the routing technique used in Figure 9-29.

Step #12—After completion of the layout drawing an artmaster shall be taped. The artmaster is made to an enlarged scale, in this case: 2:1. And then photo reduced to provide a 1:1 scale negative and positive (photomaster, which is the tool for manufacturing the PC board). Our schematic, 6-IC digital board, Figure 9-1, will be taped using a 2-sheet method. For a step by step procedure, see Illustration 11.

Illustration #11: Artmaster

An artmaster tape-up is the next step in the board layout sequence. We will illustrate the tape-up technique called the two-sheet or the two-layer method.

This method is shown in Figure 6-2. The tape-up technique in Figure 9-29 does not allow the designer to route through the IC pins; Figure 9-30 allows routing through the IC pins. The 2-layer technique was the first taping system used in the PC board industry for 2-sided boards. This system has since been replaced by two other systems called the red and blue and the 3-sheet or 3-layer technique. It is important to understand the evolution of today's techniques. You should use the 2-sheet or 2-layer technique to appreciate the advantages of the red and blue and 3-sheet system. We will show you these other two techniques (red and blue, and 3-sheet) in future boards in this book. Figure 9-31 shows an example of the artmaster technique called the 2-layer or 2-sheet method. This technique requires a piece of polyester film for each side of the PC board. One represents the component side and the other represents the noncomponent side. Figures 9-32 and 9-33 show an example of a 2-sided PC board. It has two pieces of polyester film with taped pads and board outline. Each piece references both sides, one for each side. The first example of a tape up uses the layout of Figure 9-29, and is shown in Figure 9-34. Because the layout is done at a scale of 2:1, the artmaster will also be taped at the same scale (2:1). This is the most popular scale in the industry. Use a fixed stable based polyester film with a minimum of .005 to a maximum of .007.

Pads and Traces

Use black crepe tape for the tape-up and the pads. This material is easy to work with. The traces can be routed either by sweeping (called free-flow style) or by a cut and bend method. For minimum tape widths and air gaps, see Chapter 8, Design Rules. Before we can tape this board, we must know the correct pad size for all the components used on the layout in Figure 9-29. This has been predetermined on Table 9-1. Refer to Table 9-1, column 13. U1 through U6, and C2 have a pad size of .125, at a scale of 2:1. R1, C1, C3 through C6, and R2 have a pad size of .150, at a scale of 2:1. Test points have a pad size of .250 at a scale of 2:1. All feedthroughs (according to our design rules in Chapter 8) have a pad size of .125, at a scale of 2:1.

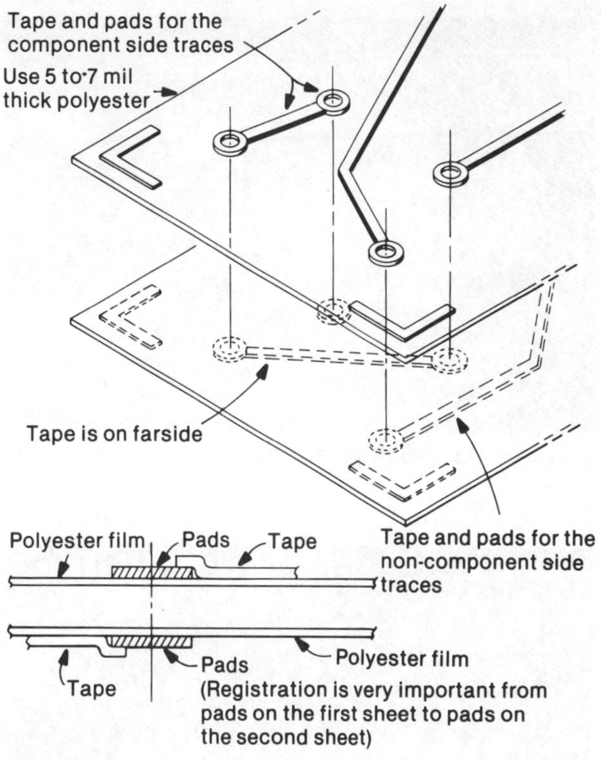

Figure 9-31. Two Layer Method

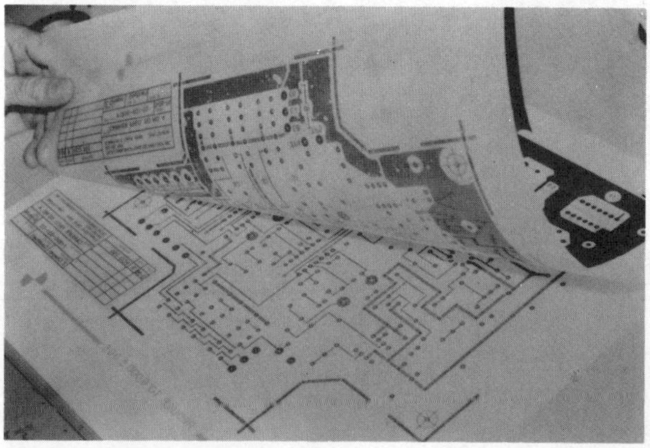

Figure 9-32. Two Layer Taping

SPECIAL NOTE: IC's U1 through U6 have a specified pad diameter of .125 at a scale of 2:1. Use the preprinted pattern shown in Figure 6-19, pattern A.

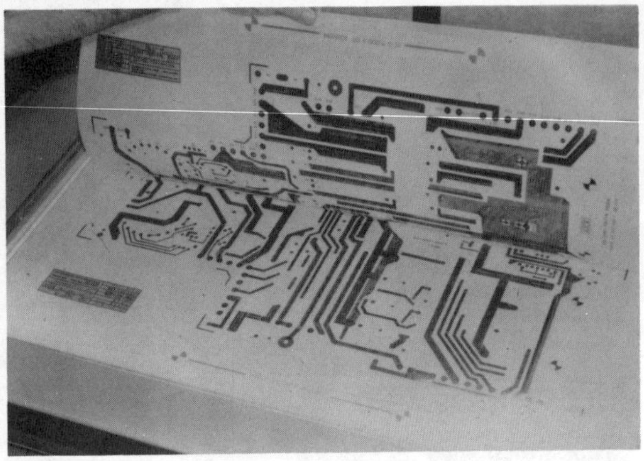

Figure 9-33. Two Layer Artwork

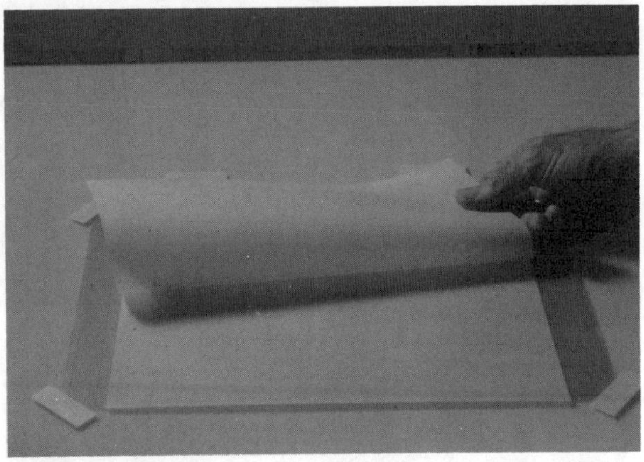

Figure 9-34. Two Sheet Tape-Up Of Layout Figure 9-29

Conductor Widths

For the layout in Figure 9-29 and the tape-up, use a standard conductor width for signal lines of .050, at a scale of 2:1. The standard conductor width for voltage and ground bus lines is .100, at a scale of 2:1.

Air Gap

The standard spacing between pads and conductors is .050 at a scale of 2:1 (unless the conductor density dictates otherwise). The minimum spacing is .030 at a scale of 2:1.

Before itemizing the steps on the two-sheet tape-up (Figure 9-34), use the following standards for this tape up procedure.

Consider conductor routing. The traces should have all turns either at 90 degrees or at 45 degrees. Do not bend the tape. With the free-flow style, this avoids relaxation of the tape and the loss of air gaps from one trace to another as the tape-up ages. Avoid overlapping the tape. Do not cut into the pads to prevent disconnection of the pad from the hole. Review Chapter 6 before taping the first PC board.

SPECIAL NOTE: This book does not present handling or cutting the tape. Basic taping practices and manual taping techniques are covered extensively in Chapter 6 of the Analog Design Book. The following taping exercises demonstrate red and blue, and the 3-sheet systems.

The following is a step-by-step procedure for the tape up of the layout in Figure 9-29.

Procedures for Tape-Up of the Layout

Procedure #1

The polyester film for the tape up can either be matte one-side or it can be clear. Because this technique uses two sheets, use the matte one side. The light table can reflect off clear film. After selecting the matte or the clear film, cut two sheets to the size that overlaps the layout itself. Leave enough room to put title blocks, reduction targets, identification, etc.

Procedure #2

Select the grid for the tape-up. Choose either .100 (Figure 5-33) or .050. The grid is a personal choice. The layout is registered with the grid. First, the grid is taped down over the light table (or whatever working surface).

Procedure #3

The layout is registered and taped down over the grid. Overlay one of the two sheets cut from the new polyester film. If you are using the matte polyester film, then the matte side must be taped up. This sheet will be the component side of the PC board. See Figure 9-35.

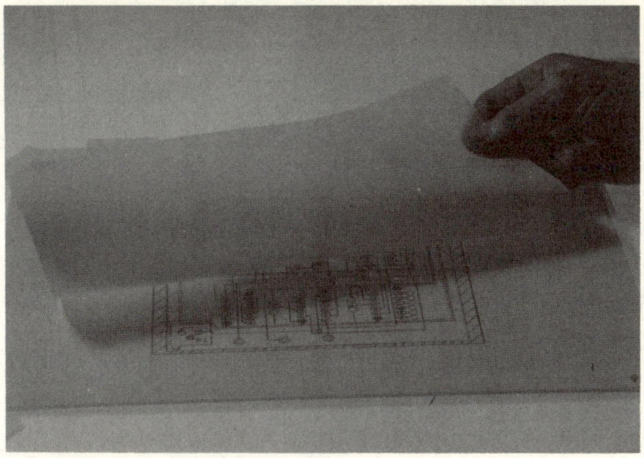

Figure 9-35. Procedure 3

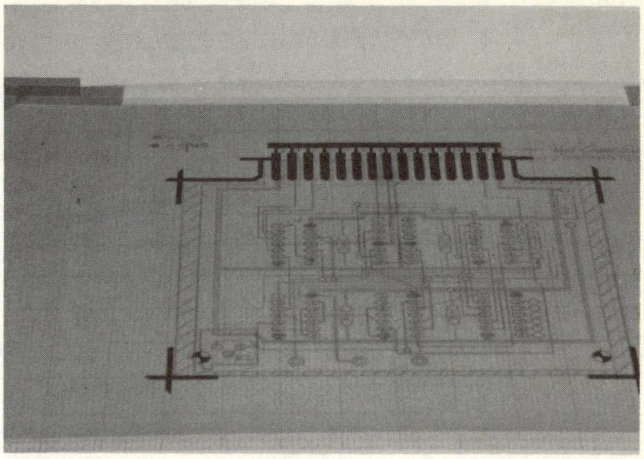

Figure 9-36. Procedure 4

Procedure #4

The first taping is the board outline. Tape the crop marks using .100 tape. The crop marks shown in Figure 9-36 will only be placed in the corners of the PC board along with the outline for the connector tongue (called P1). Place the connector tongue (Figure 9-37). Place the tooling holes, one in each corner. See Figure 9-36. The tooling holes register the component side with the noncomponent side. The first sheet of polyester film registers with the second sheet (representing the noncomponent side). Consider the first sheet (called the component side) to be complete. Remove the component side polyester sheet from the working surface. Remove the layout from the grid. Turn the component side polyester sheet over, facing down on the grid. Register it with the grid. Place the second polyester sheet on top of it, with the matte side up. Secure it firmly on top of the first sheet. Duplicate everything that you put on the first sheet: tooling holes, crop marks, and connector tongue. Registration ensures that one sheet always aligns with the next sheet, through the two tooling holes. You must register or align these two points very accurately.

Procedure #5

Both sheets look identical: the board outline and the P1 connector tongue on both sheets. Remove the second sheet (called the noncomponent side) from the working surface. Remove the component side from the grid (called the first sheet).

Figure 9-37. Connector Placement

Again place the layout on top of the grid to register it. Place the component side sheet on top of the layout, matte side up. We can place all the component pads computed in Table 9-1, column 13. Place the inside hole on the pad as close as possible to the crosshairs on the grid. See Figure 9-38 for an example of all pads placed on the component side.

SPECIAL NOTE: One IC row (U1, U3, and U5) is moved up. The other IC row (U2, U4, and U6) is moved down. This opens up the area in the center aisle. These adjustments must be made at this time.

Procedure #6

All pads are placed down for the component side of the PC board. Every component lead has a

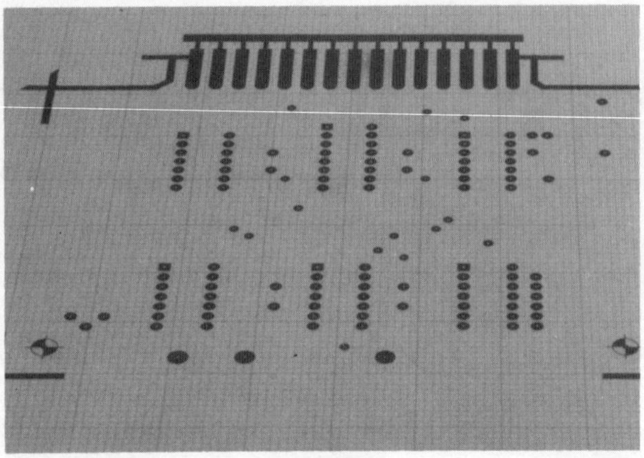

Figure 9-38. Pads, Feedthroughs

pad; all feedthroughs are spotted. Feedthrough locations can be readjusted. Interconnect or route the traces on the component side. Figure 9-29 shows that all red traces represent the component side of the board. Use the proper width of tape for signal paths. According to the design rules, .050 tape is used for signal paths; the color-coded voltage and ground traces have .100 tape. Complete all red traces. See Figure 9-39.

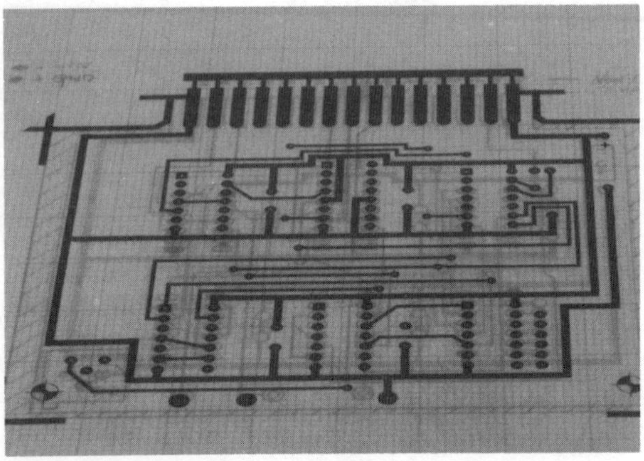

Figure 9-39. Red Traces (Component Side)

Procedure #7

After completing all component side traces and pads, remove sheet one from the top of the layout. Remove the layout from the grid. Turn over the layout so that the component side view is facing down on the grid. Turn over the component side tape-up. Place it down (over the layout) to view the noncomponent side of the board. Place the noncomponent side board outline on top of the component side, with the matte up. Register it to the tooling holes. Secure all three down to the grid.

Register the component side pads to the noncomponent side sheet pads. This is one of the more critical points or stages of the tape up. The registration of these two pads is demonstrated in Figure 9-31. Align as close as possible the one pad on the first sheet to the second pad on the noncomponent side sheet (they must be above and below each other as close as possible). Complete all of the blue traces representing the noncomponent side of the board. Signal traces are .050, at a scale of 2:1. Voltage and ground are .100, at a scale of 2:1. See Figure 9-40 for an example of a finished noncomponent side tape up of the layout in Figure 9-29.

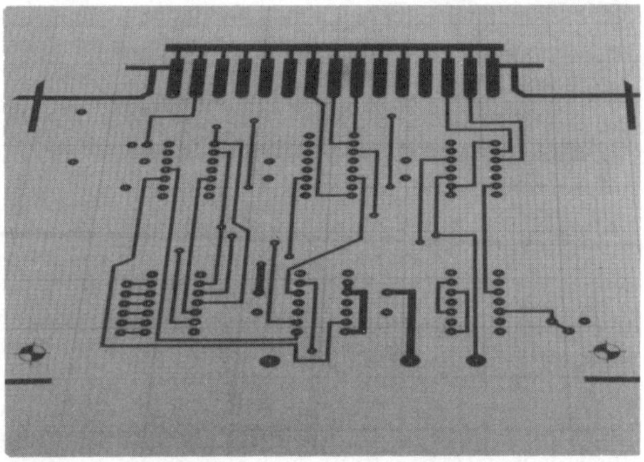

Figure 9-40. Blue Traces (Non Component Side)

Procedure #8

Consider component designation and marking for the PC board. Identify the location of all components on the schematic for the layout. Mark the identifying points and reference designations on the component side of the PC board. One way is to etch the designations and markings on the surface (like a trace/pad). One other technique is to silkscreen these marks for component identifications on the surface of the PC board (as shown in Figure 7-12). Silkscreening paints on the surface. The epoxy paint used is very strong and adheres to the board. Consider some of the components to be marked on the layout in Figure 9-29: U1 through U6, R2, TP1, TP2, TP3, the positive end of C2, and at least the first and last pins of the P1 connectors. These are the minimum designations that should appear on the component side of the board. You could mark all of the components: C1, C3, C4, C5, and C6, and R1. For proper placement of these designations, see Figure 9-41.

Procedure #9

Reduction targets and title block are the last steps on the tape-up procedure. We have placed all the component pads, traces (for the voltage, ground, and signal paths), connector tongue fingers, crop marks, tooling holes, component outlines, and designations to identify the components. Reduction targets and title block are used primarily by the photographer. The 2:1 tape up is reduced to a 1:1 film positive and negative. Tell the photographer how much to reduce the artwork. Chapter 6 discussed reduction targets. The larger the artwork, the larger the reduction target. The six inches reduce to three inches at 1:1 (Figure 9-42). There are reduction targets in both the X and the Y direction. These are used to measure the film from the photographer to see that the artwork has been reduced properly in both the horizontal and vertical directions. The reduction target instructs the photographer to reduce to 6 inches or reduce to 3 inches. In this case, reduce to 3 inches. The targets from center target to center target (at a 2:1 scale) are equal to 6 inches. The title block varies from company to company. We will show you the area where we choose to place the block on each side of the tape-up on each sheet. Usually the company's name and address, the part number and revision letter of the PC board are placed in the title block. You are ready to send the 2:1 artwork to the photographer, to request a reduction, and to receive one set of positives and one set of negatives to fabricate the printed circuit board.

WARNING: On the layout in Figure 9-29, we recommended that the IC rows be moved to relieve congestion in the aisles. This recommendation was only meant for the designer who is taping his own layout. If you have a taper working with you on the layout, or if your layout is to be digitized (using a Computer-Aided Design System) update the layout by drafting in the precise location of each trace. If necessary completely re-layout the entire board for the taper or the digitizer.

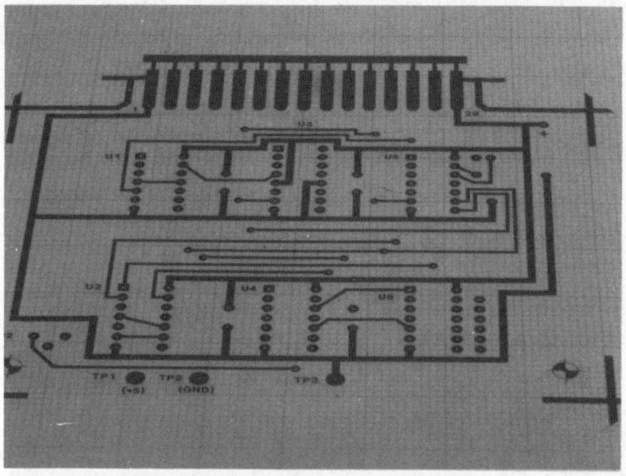

Figure 9-41. Reference Designation

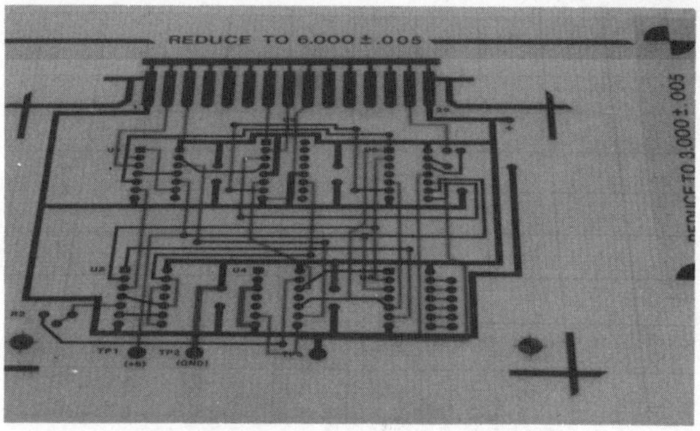

Figure 9-42. Reduction Targets

Procedure #10

Figure 9-43 and 9-44 shows the tape up for the layout in Figure 9-30. We routed the signal paths through the IC pins. The tape up for this layout illustrates the major differences between the two styles of layout required for taping. We are using the same two sheet taping method. Major differences between the two tape ups is the IC pad pattern and the trace widths. The IC pad pattern shown in Figure 6-19, pattern D, is an oval pad. The trace widths were reduced to .030, at a scale of 2:1. You can not run .050 tape at a scale of 2:1 through IC pad patterns. All other procedures (1 through 9) that were used for the layout in Figure 9-29, remain the same. This taping exercise shows the differences required for tape-ups running through the IC pins.

SPECIAL NOTE: Both taping techniques use the guide lines (1 through 9) in Figure 9-45 and the check list in Figure 9-45. Use these guide lines to check your own work.

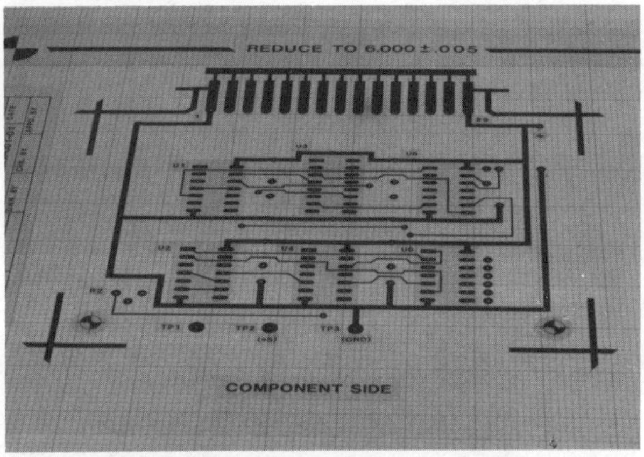

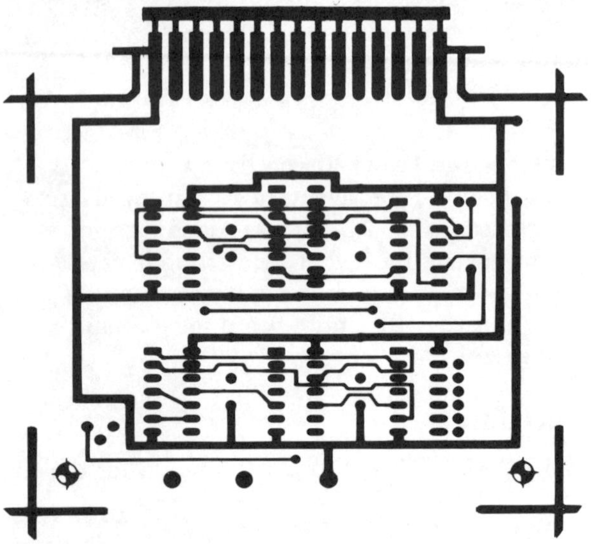

Figure 9-43. Component Side Artwork

Step #13—The material that we have used for our board 1, 3 IC digital board, and board 2, our 6 IC digital board, is what we will build our foundation from to gain knowledge for our remaining boards. At the end of this chapter, we will give a break down of all the material that has been added to this foundation.

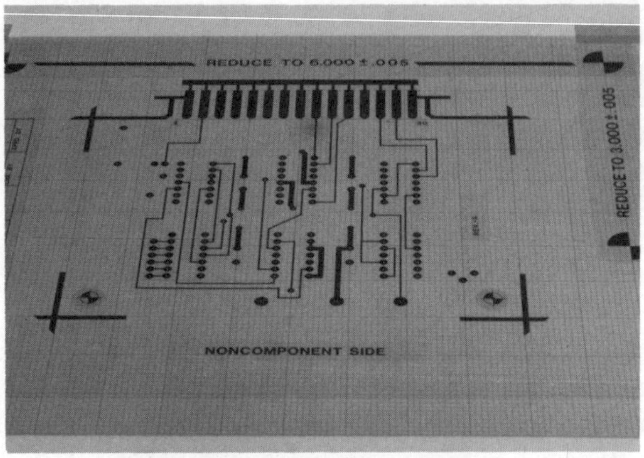

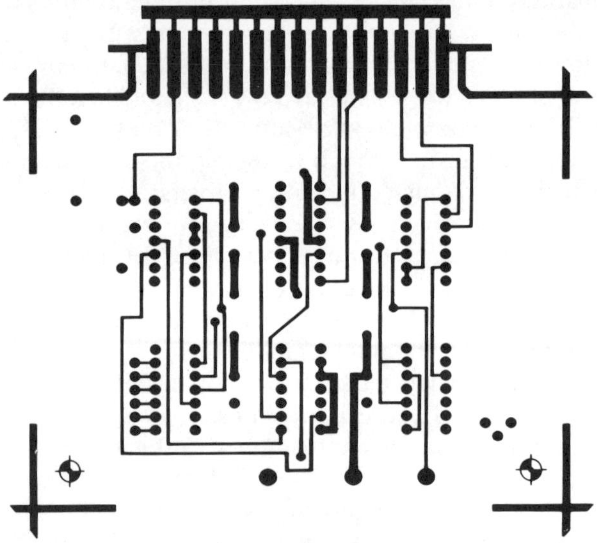

Figure 9-44. Non Component Artwork

Taping Guidelines

1. Minimize handling of drafting film.
2. DO NOT fold, crease or tightly roll the artwork.
3. Burnish the tape when complete.
4. Run the tapes to the center of the annular ring of the pad.
5. DO NOT cut through tapes or pads (it will show up on the photographs).
6. DO NOT pull the tape too HARD or it will creep.
7. Tape and place the pads either on the grid coordinate or coincident with the lines.
8. Trim all corners as required.
9. Be neat.

Artmaster Checklist

________ Tape/Pad/Drafting Film Cut
________ Pad Off Grid
________ Corner Trim
________ Air Gap/Tape Width
________ Run Tape to Center of Pad
________ Pad Center Covered
________ Not Tapes at 45/90 Degrees
________ Angle Less Than 90 Degrees
________ Electrical Short, Bridging
________ Ragged Edge, Ragged Cut, Crooked Trace
________ Trace Too Short/Open Circuit

Figure 9-45. Taping Guidelines, Artmaster Checklist

Summary

We listed all of the points unique to the schematic and board layout. We listed the procedures necessary to lay out the schematic. We itemized the steps important to this particular schematic. Each schematic in this book has different points of interest.

Compare the schematics in Chapters 8 and 9. The quantity of gates increased. There is a mixture of open and non-open collectors on the new schematic in Chapter 9 of spare gates. We drafted the spare gates on the schematic and assigned their pin numbers. The grouping of gates was more difficult than Chapter 8. We used specification sheets and Table 9-1 to identify the hole diameter and pad size for all of our components.

We discussed the variable resistor. We assigned pin numbers on both the schematic and the PC board layout for the variable resistor. We discussed the placement within the board outline. We introduced another area to place the input filtering capacitor, C2. We also provided a board outline to construct the layout in Figure 9-22. We presented all of the traces using the red and blue colors, and the components using the PC design template.

Plotting the IC's on the board outline, we found that the six IC's were more difficult than the three IC's in Chapter 8. We learned how to remove feedthroughs and how to find mistakes in the clean-up stage of the board layout.

We routed traces to the IC pads. We drafted the entire board using this technique. This technique will be used in Chapters 10 and 11 on double sided boards.

We introduced the two-sheet method (or two layer method) of taping. We gave examples of both styles of layout. The three major techniques are covered in this chapter and in chapters 10 and 11.

We combined what we have learned in Chapter 8 along with new procedures. Chapters 10 and 11 build on the knowledge required to design a layout for a digital type of PC board.

Quiz—Chapter 9

1. "The Schematic should be readable." The meaning of this statement is:
   A) All notes drafted by the Engineer should be legible
   B) All reference designations used by the engineer should be understood.
   C) All symbols and reference desginations used by the engineer should be understood.
2. The varaiable resistor in this chapter
   A) Has no important placement.
   B) Is placed off of the board.
   C) Must be checked for location with the engineer.
3. To obtain the specificaton sheets for all components
   A) Designer must find his/her own.
   B) The engineer should provide all spec sheets.
   C) The designer is typically not required to ask.
4. The manufacturer's part number called out in the various gates & IC symbols.
   A) Will help the designer combine gates to reduce the IC count.
   B) Will not help the designer combine gates
   C) Will only help provide the specification sheet.
5. Spacing of the component pads determine where:
   A) The component body should be placed on the P.C. board.
   B) The leads insert into the P.C. board
   C) The component identification should be placed on the P.C. board.
6. One of the important objectives on the component mounting table is:
   A) To find the smalled pad size.
   B) To find the largest component size.
   C) To have as few different hole sizes as possible.
7. The Board Layout Design Rules include the general requirements for:
   A) Trace widths & air gaps
   B) Layout Rules
   C) Artmasters Procedures
   D) All of the above.
8. The question: Are there any special size components, or are all of the components standard: means?
   A) The designer needs to obtain spec sheets on each component.
   B) The designer should find the correct mounting spacing for each component
   C) Standard refers to the components that are standard to a particular component
   D) All of the above.
9. Input filtering capacitors are placed:
   A) Next to the I.C.
   B) Very close to the voltage pin on the connector.
   C) Very close to the ground pin on the connector.
10. If the board has variable resistor, the designer must be concerned about?
    A) Where should the variable resistor be placed.
    B) Are there pre-assigned pins
    C) The adjusting screw must be at the edge of the board.
11. If the board requires test points, the designer must be concerned about?
    A) Test points on all signal lines
    B) Test points on voltage & ground
    C) Test points should be very close to connector.
12. Questions for the engineer. Explanation: The time to ask these basic questions is?
    A) When the engineer hands the schematic to you, and you have had time for review.
    B) At the end of the project as a final check.
    C) Just before the schematic is given to you by the engineer.

13. When determining the board outline, the designer should also:
A) Check for spare gates
B) Find out which components require heat sinks
C) Find out the method of fastening or mounting
14. The board size must be known in order to compute?
A) Total square inches available
B) Total number of components
C) Total number of holes on the board
15. Using the maximum size of all the components will all the components fit onto the board?
A) The engineeer
B) The drafting supervisor
C) The P.C. designer
16. In order to draft the P.C. board outline, the designer must know the following information:
A) The card track clearance
B) The connector size
C) How the board is secured into the enclosure
D) Both B & C
E) All of the above
17. Grouping and combining gates is required on a schematic for the following reason:
A) Helps the designer achieve short trace connections
B) Helps the designer find the lowest number of IC's required on his/her schematic
C) Helps the designer identify all the gates on the schematic
D) Both A & B
E) All of the above
18. The designer must find the logic family for each gate for the following reason:
A) To obtain the specification sheet
B) To identify open collector & totem pole
C) To find the shortest connection
19. IC specification sheets identify
A) Logic families
B) Logic Pin numbers
C) Total quanity of gates
D) Both A & B
E) All of the above
20. To find the voltage and ground pins on an I.C. the designer would check:
A) With the engineer
B) The specification sheet
C) The design rules
21. Critical leads on the board is important for grouping because:
A) It determines which gates would be placed in the same IC package.
B) It determines which IC packages would be placed close together
C) It determines which part of the circuit would be layed-out first.
D) All of the above.
22. Which gates should the designer combine first?
A) The gates that seem to be the simplest
B) The gates that have short traces
C) The totem pole gates
23. Once you have completed the grouping exercise, the next step would be:
A) Assign input & output pins
B) Find your spares
C) Plot on the P.C. board
24. The formula to calculate the square inches required on the P.C. board.
A) Is an exact formula to be followed
B) Is a simple rule of thumb formula
C) Is a formula that should be used only by the engineer.
25. When plotting the grouped IC's on the schematic on the board the next step is?
A) Calculate the total area required
B) Assign the IC location on the board
C) Connect the signal paths

26. When plotting IC's on to the board the:
A) IC's should be equally spaced
B) IC's should be given equal consideration
C) First row could be favored because of high concentration of traces
D) All of the above
27. The formal template layout:
A) Must start from a sketch
B) Must be viewed from the non-component side
C) Must establish conductor paths in an X-Y direction
28. Once the IC's are placed on the P.C. board and assigned U-numbers the Designer color codes volatge and ground for the following reason:
A) Helps in completing the bus lines for voltage and ground
B) Helps locate the placement of feed-thrus
C) Helps select color of interconnecting traces
29. Voltage and ground traces are:
A) Larger than signal traces
B) Placed on the component side
C) Establish the horizontal & vertical direction of red & blue traces
D) Both A & B
E) All of the above
30. The digital X-Y system would be defined as:
A) Red traces in the horizontal direction & blue traces in the vertical
B) Red & blue trace in the same direction
C) Red on the component side & blue on the non-component side
31. By-pass capacitors should connect?
A) To the input connector pin
B) Between voltage & ground of the same row of IC's
C) To the voltage bus only
32. By-pass capacitors are typically:
A) Large capacitors metal foil
B) Medium size mylar capacitors
C) Small ceramic disc capacitors
33. When interconnecting signal lines the designer should:
A) Assign input & output pin's on gates as he/she interconnects the layout
B) Assign all input & output in on gates first and then start interconnecting
C) Have the engineer assign all pin numbers
34. Each time a trace is interconnected on the layout, the designer should:
A) Label the pin number on the schematic
B) Notify the engineer
C) Yellow out the connection on the schematic
D) Both A & D
35. When starting to interconnect signal paths: the designer should
A) Pick a point starting with the connector
B) Start with resistors
C) There is no correct starting point, any point will work.
36. Once you have assigned one input to a gate the designer has the option of:
A) Not connecting the others
B) Assigning all of the remaining inputs & outputs
C) Asking the engineer for the remaining inputs & outputs
37. When interconnecting traces and a change of direction X to Y is required, the designer should:
A) Use a feed through to change from red to blue
B) Use the connectors to change from red to blue
C) Only use a component lead to change from red to blue
38. When interconnecting signal lines the designer at times's will:
A) Fudge in a line, even though there is not enough room
B) Spread IC rows on the tape-up to create room
C) Be prepared to draw lines under crowded conditions
D) All of the above

39. The clean-up stage of the layout includes:
A) Removal of unnecessary feedthroughs
B) Assign spare gates & pin numbers
C) Connector Pin assignment
D) Both A & B
40. Packing feedthrough under an IC is:
A) Never acceptable
B) Acceptable
C) Acceptable only if necessary.
41. The layout before the tape-up should renumber the IC's into some form of alpha/numeric arrangement, because:
A) It helps the designer
B) It helps the technician troubleshoot
C) It helps the engineer doing the schematic
42. To establish the location for test points, the designer should:
A) Consider the tooling holes
B) Consider the length of trace's
C) Consider the point of access for the P.C. board
43. Tooling hole are part of the layout procdure because:
A) Tooling holes are used as reference points
B) Tooling holes require the usable area of the P.C. board
C) The manufacture may use these tooling holes
D) All of the above
44. Solder pad on spare gates are used for:
A) Future expansion
B) Disconnecting
C) Adding componets
45. It is acceptable to change the tape-up from the layout because:
A) The designer makes mistakes.
B) The layout is a guide for understanding where to connect on the final tape-up
C) The layout has changes made by the engineer
46. The 2-sheet tape-up technique has been replaced with:
A) The red and blue system
B) 3-sheet technique
C) Both A & B
47. Black crepe tape is easier to use then red and blue because:
A) The material is easy to work
B) The tape can be turned or bent
C) Both A & B
48. The layout is registered to the grid by using this technique:
A) The grid is not taped down to the light table only the layout
B) First the lay-out is taped down over the light table
C) The layout is layed out on top of the grid and registered with it.
49. Reduction targets and title clock are the last step on the tape-up and are used by:?
A) The board manufacturer
B) The photgrapher
C) The P.C. designer
50. If your layout is to be digitized the designer should
A) Draw the traces very precise
B) Draw the feedthroughs more precise
C) Spend more time in the clean-up stage
D) All of the above.

The answers for each question can be found on the pages listed below:

Question 1. Pg. 256
Question 2. Pg. 256
Question 3. Pg. 257
Question 4. Pg. 262
Question 5. Pg. 262
Question 6. Pg. 262
Question 7. Pg. 268
Question 8. Pg. 269
Question 9. Pg. 270
Question 10. Pg. 270
Question 11. Pg. 270
Question 12. Pg. 270
Question 13. Pg. 270
Question 14. Pg. 271
Question 15. Pg. 271
Question 16. Pg. 271
Question 17. Pg. 273
Question 18. Pg. 273
Question 19. Pg. 273
Question 20. Pg. 273
Question 21. Pg. 273
Question 22. Pg. 273
Question 23. Pg. 275
Question 24. Pg. 276
Question 25. Pg. 279
Question 26. Pg. 279

Question 27. Pg. 281
Question 28. Pg. 283
Question 29. Pg. 283
Question 30. Pg. 283
Question 31. Pg. 284
Question 32. Pg. 284
Question 33. Pg. 284
Question 34. Pg. 284
Question 35. Pg. 284
Question 36. Pg. 285
Question 37. Pg. 285
Question 38. Pg. 290
Question 39. Pg. 291
Question 40. Pg. 292
Question 41. Pg. 292
Question 42. Pg. 293
Question 43. Pg. 295
Question 44. Pg. 295
Question 45. Pg. 295
Question 46. Pg. 297
Question 47. Pg. 297
Question 48. Pg. 298
Question 49. Pg. 301
Question 50. Pg. 303

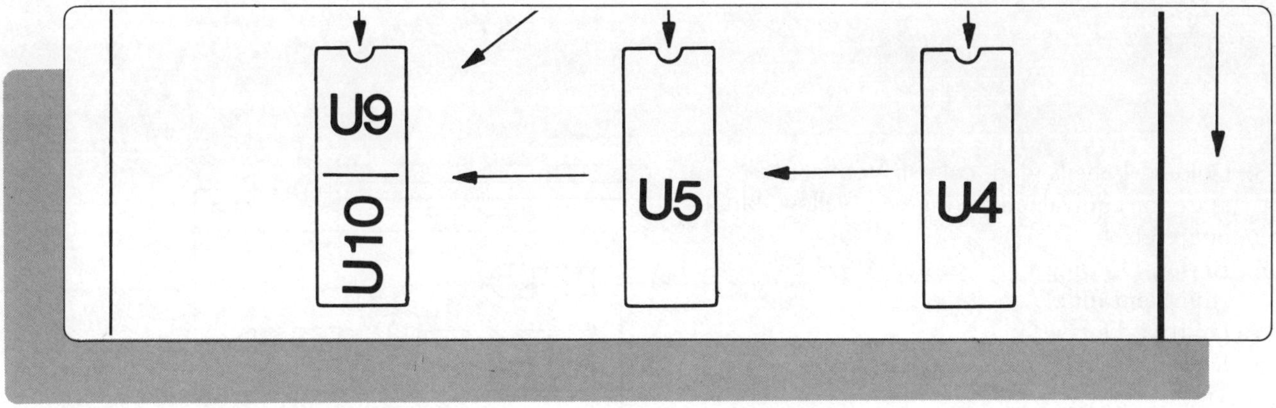

10-IC DIGITAL/ANALOG BOARD

10

This chapter illustrates the design of the ten IC digital printed circuit board using the X-Y system on a two-layer/double sided board. The student will make the digital layout. We have already plotted, and drafted one version of this layout. We have assigned the board outline, and placed the IC patterns within that board outline. You will plot the IC's and formally lay out your own version of the schematic.

We will show you each major step of the board layout. Refer to that layout as you design. Follow the same itemized steps and restrictions. You will ask the same questions of the engineer. The design rules are the same.

You will be presented with the feature of an operational amplifier. You will run traces between IC pins, and will work with pre-assigned pin connectors. You will use a red and blue taping technique.

Supply List Needed for Chapter Ten

1. Component Template: Bishop Graphics Catalog No. 3367 (one)
2. Black Drafting Film Lead Pencil: Mars 100-30 Duralar (one)

3. Colored Pencils (dark colors): Venus Col-erase, or equivalent: (red, green, yellow, blue) one each
4. Eraser, Drafting Film: Castell #1954 or equivalent (one)
5. Drafting Dots
6. Knife: X-acto, Bishop Graphics Catalog No. 3141
7. Knife Blades: X-acto, Bishop Catalog No. 3116
8. Donut Pads: .125", 50, Bishop Graphics Catalog No. D102*
9. Donut Pads: .250", 6, Bishop Graphics Catalog No. D141*
10. Donut Pads: .150", 50, Bishop Graphics Catalog No. D144*
11. Precision Slit Tape: Red Polyester .031", 1 roll Bishop Graphics Catalog No. 20503111
12. Precision Slit Tape: Blue Polyester, .031", 1 roll, Bishop Graphics Catalog No. 20510011
13. Precision Slit Tape: Blue Polyester, .100", 1 roll, Bishop Graphics Catalog No. 20703111
14. Precision Slit Tape: Blue Polyester, .100", 1 roll, Bishop Graphics Catalog No. 20710011
15. Connector Pattern: .156" C-C, 15 Fingers, Bishop Graphics Catalog No. 6720*
16. Reduction Mark: 4, Bishop Graphics Catalog No. 4052*
17. Datum Target (tooling holes): 2, Bishop Graphics Catalog No. 4011*
18. Drafting ACCUFILM® : matte one side, .003", 8½" x 11", 2 sheets
19. Drafting ACCUFILM® : clear taping film, .005", 8½" x 11", 1 sheet
20. Dual In Line Pattern: 14-lead, oval pads, 20, Bishop Graphics Catalog No. 6242*
21. Dual In Line Pattern: 16-lead, oval pads, 4, Bishop Graphics Catalog No. 6215*

* Bishop catalog number shown for these products represent order numbers for Bishop's standard industrially packaged products. The package quantities for these catalog numbers may be greater than supplied with the "Digital" project kit, #10009C, available from Bishop.

All the steps listed in chapters 8 and 9 apply to this schematic (the 10-IC digital/analog schematic). We will use the knowledge from the previous two chapters. We will borrow step 4, which establishes the board layout design rules (with minor modifications). As we have told you before in chapters 8 and 9, the steps listed simplify the design process.

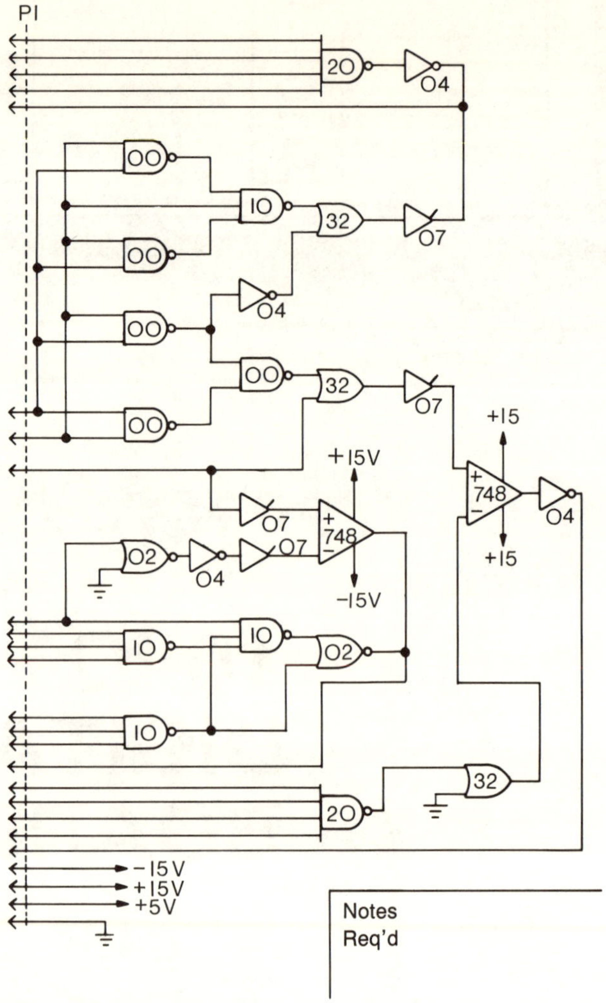

Figure 10-1. Digital Schematic

Step #1—Schematics should be readable and understood: symbols, reference designations and identification of all logic gates used by the engineer. See the schematic in Figure 10-1.

Illustration #1: Readable Schematic

Examine the two schematics: Figures 10-1 and 9-1. Figure 10-1 has three voltages with one ground (-15V, +15V, and +5V). Figure 9-1 has one voltage (+5V) and a ground. The additional two voltages (+15V and -15V) are for a component called an operational amplifier. There are two on the schematic in Figure 10-1. This is the first time we have used one in digital design. We did introduce the operational amplifier in the analog design book. The operational amplifier shares characteristics of both the digital and analog worlds. This is why we call this board a 10-IC digital/analog board. The operational

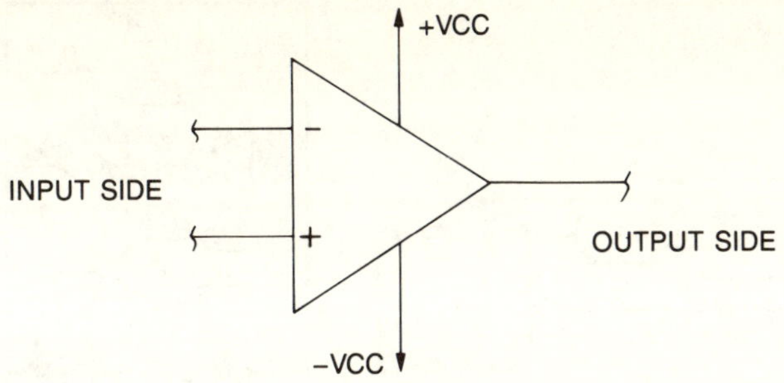

Figure 10-2. Operational Amplifier

amplifier in Figure 10-2 has both an input and an output side, just like the typical gate. However, there are two inputs with one output: a positive input and a negative input. The pin numbers must be identified using a specification sheet. The operational amplifier is briefly explained in Figure 10-2. We will completely layout one version of the schematic in Figure 10-1 (which is shown in Figure 10-3, along with the completed layout shown in Figure 10-4). You will be assigned to do a different version in the schematic in Figure 10-5.

On the schematic in Figure 10-3 (completed version), we will cover the following points. We will explain them in steps 2 through 13.

Points

A. Group all IC's.
B. Plot all IC's.
C. Assign all gate pin numbers and operational amplifier pin numbers.
D. Show the input filtering capacitor for +5V.
E. Show the bypass capacitors for the IC's and the operational amplifier.
F. Show the spare gates and assign the pin numbers.

Operational Amplifier (Abbreviation is op amp). An extremely stable (usually direct-coupled) linear amplifier originally intended for performing mathematical operations. Op amps's, especially the integrated-circuit variety, have a wide range of applications outside of computation.

G. Introduce the operational amplifier
H. Introduce three separate voltages.
I. Cover all questions to ask the engineer.

All of these points will be covered extensively in steps 2 through 13. After studying each step for the layout of the schematic in Figure 10-1, you will construct your own layout for the schematic in Figure 10-5. You will use the same steps and information supplied by the engineer. Use the grouping that we have completed for them. Figure 10-1 is a more complex schematic; it has more gates.

Step #2—Specification sheets identify the case, body size, width, and length of all discrete components. They identify all logic gates or symbols. The engineer would provide the specifications sheets for this schematic (Figure 10-3).

Illustration #2: Component Identification

The following gives a breakdown of each specification sheet for each component.

Capacitors

C1 is an input filtering capacitor (polarized), see note 2 on the schematic in Figure 10-3. It calls out 3.9 UF, 10V, ±10%. For the specification sheet, see Figure 10-6, sheets 1 and 2. C2 and C3 are operational amplifier bypass capacitors (non-polarized, ceramic disc), see note 3 on the schematic in Figure 10-3. This specifies op amps: bypass 1 .01 UF, 100V, 20% for two op-amps. For the specification sheet, see Figure 10-7. C4 through C8 are non-polarized ceramic disc capacitors. See note 1 on the schematic in Figure 10-3. This specifies to bypass every 2 IC packages: .01 UF, 100V, ±20%. For a specification sheet, see Figure 10-7.

SPECIAL NOTE: The number of bypass capacitors on the IC's (C4 through C8) were determined by the layout itself. The total number of capacitors were not known until the layout configuration was established. See Figure 10-4, formal layout.

Operational Amplifiers

The operational amplifiers are identified on the schematic in Figure 10-3. Note 4 specifies op amp's 748C 8-pin plastic dual inline. For the specification sheet and body configuration, see Figures 10-8 and 10-9.

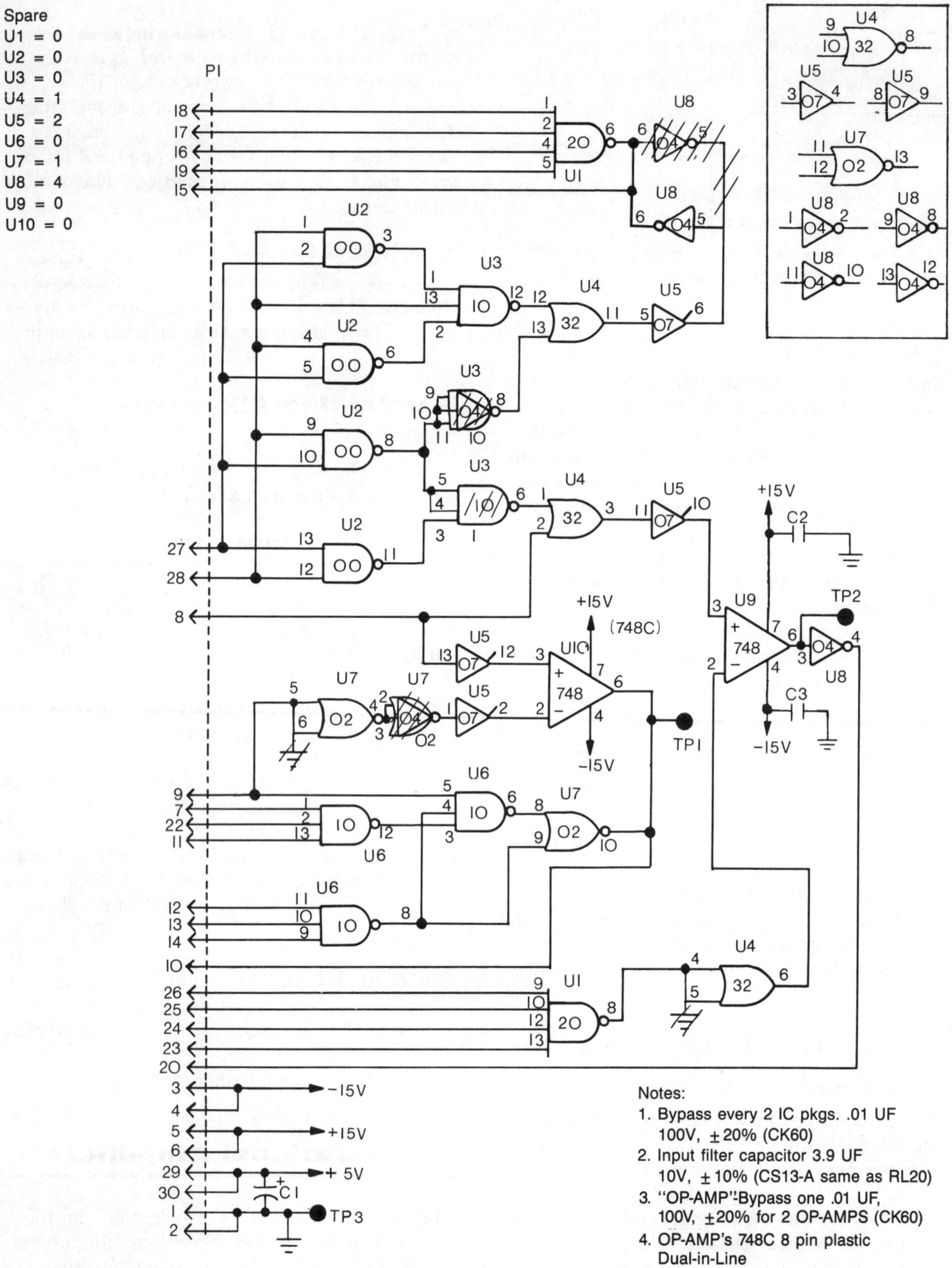

Figure 10-3. Schematic

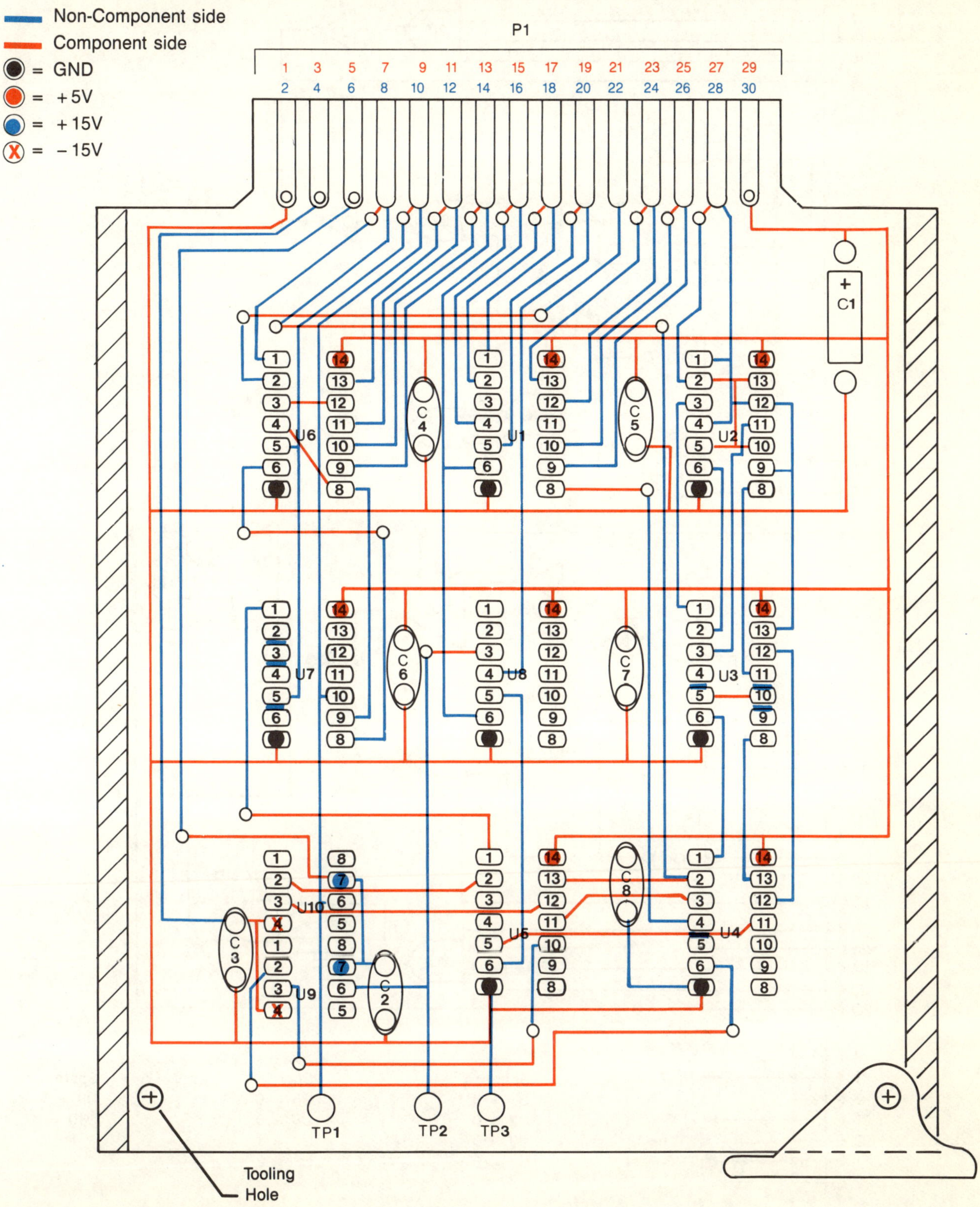

Figure 10-4. Completed Layout

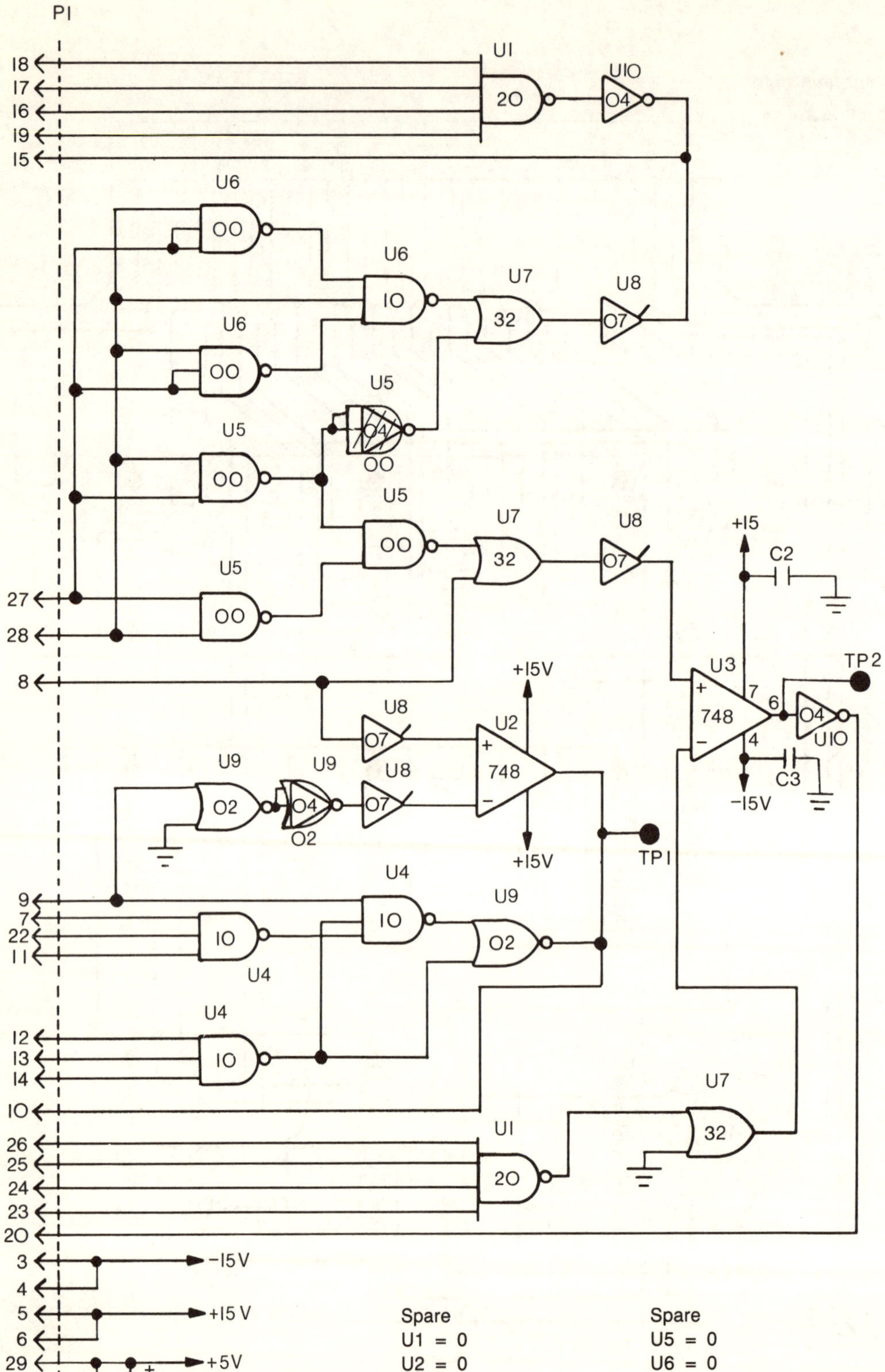

Spare	Spare	Spare
U1 = 0	U5 = 0	U9 = 1
U2 = 0	U6 = 0	U10 = 4
U3 = 0	U7 = 1	
U4 = 0	U8 = 2	

Figure 10-5. Student Layout

MIL-C-39003/1C
15 May 1970
SUPERSEDING
MIL-C-39003/1B
26 March 1970

MILITARY SPECIFICATION SHEET

CAPACITORS, FIXED, ELECTROLYTIC, TANTALUM, SOLID-ELECTROLYTE, ESTABLISHED RELIABILITY, STYLE CSR13

The complete requirements for procuring the capacitors described herein shall consist of this document and the latest issue of Specification MIL-C-39003.

This specification is mandatory for use by all Departments and Agencies of the Department of Defense.

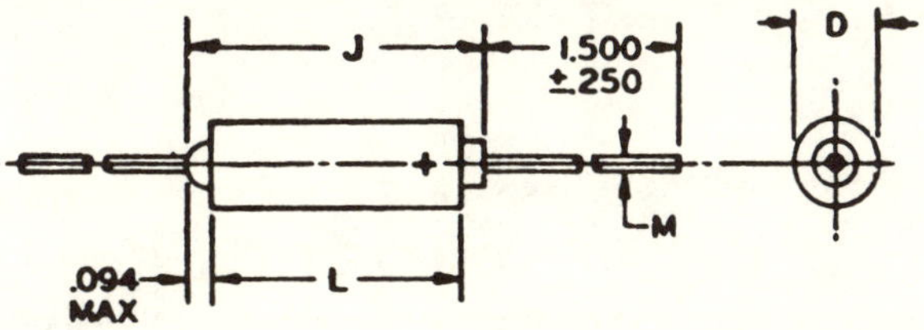

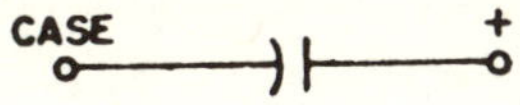

CIRCUIT DIAGRAM

Case size	Dimensions L ±.031	D +.016 -.015	M ±.002	J Max
RL20 ➤ A	.286	.135	.020	.422
B	.474	.185	.020	.610
C	.686	.289	.025	.822
D	.786	.351	.025	.922

INCHES	MM	INCHES	MM	INCHES	MM
.002	.05	.135	3.43	.474	12.04
.015	.38	.185	4.70	.610	15.49
.016	.41	.250	6.35	.686	17.42
.020	.51	.286	7.26	.786	19.96
.025	.64	.289	7.34	.822	20.88
.031	.79	.351	8.92	.922	23.42
.094	2.39	.422	10.72	1.500	38.10

NOTES:
1. Dimensions are in inches.
2. Metric equivalents (to the nearest .01 mm) are given for general information only and are based upon 1 inch - 25.4 mm.
3. The case insulation shall extend .015 (.38 mm) minimum beyond each end. However, when a shrink-fitted insulation is used, it shall lap over the ends of the capacitor body.

Figure 10-6. Sheet 1

MIL-C-39003/1C

TABLE I. Electrical characteristics and part numbers.

DC rated voltage	Capaci-tance	Capacitance tolerance	DC leakage at			Dissipation factor at		Case size	Part No. M39003/01- Failure rate level for (%/1,000 hr)				
			+25° C	+85° C	+125° C	-55° C +25° C	+85° C +125° C		L	M	P	R	S
volts	µF	percent	µa	µa	µa	percent	percent		2.0	1.0	0.1	0.01	0.001
6	5.6	10	.5	6	7.5	4	4	A	2001	2241	2481	2721	2961
6	6.8	10	.5	6	7.5	6	6	A	2002	2242	2482	2722	2962
6	6.8	20	.5	6	7.5	6	6	A	2003	2243	2483	2723	2963
6	47.0	10	3.0	24	30.0	6	6	B	2004	2244	2484	2724	2964
6	47.0	20	3.0	24	30.0	6	6	B	2005	2245	2485	2725	2965
6	56.0	10	3.0	24	30.0	6	6	B	2006	2246	2486	2726	2966
6	150.0	10	9.0	90	113.0	8	8	C	2007	2247	2487	2727	2967
6	150.0	20	9.0	90	113.0	8	8	C	2008	2248	2488	2728	2968
6	180.0	10	11.0	110	138.0	8	8	C	2009	2249	2489	2729	2969
6	270.0	10	13.0	130	163.0	8	8	D	2010	2250	2490	2730	2970
6	330.0	10	15.0	150	188.0	8	8	D	2011	2251	2491	2731	2971
6	330.0	20	15.0	150	188.0	8	8	D	2012	2252	2492	2732	2972
10	3.9	10	.6	6	7.5	4	4	A	2013	2253	2493	2733	2973
10	4.7	10	.7	7	8.8	4	4	A	2014	2254	2494	2734	2974
10	4.7	20	.7	7	8.8	4	4	A	2015	2255	2495	2735	2975
10	27.0	10	4.0	40	50.0	6	6	B	2016	2256	2496	2736	2976
10	33.0	10	5.0	50	63.0	6	6	B	2017	2257	2497	2737	2977
10	33.0	20	5.0	50	63.0	6	6	B	2018	2258	2498	2738	2978
10	39.0	10	5.0	50	63.0	6	6	B	2019	2259	2499	2739	2979
10	82.0	10	8.0	80	100.0	6	6	C	2020	2260	2500	2740	2980
10	100.0	10	10.0	100	125.0	8	8	C	2021	2261	2501	2741	2981
10	100.0	20	10.0	100	125.0	8	8	C	2022	2262	2502	2742	2982
10	120.0	10	12.0	120	150.0	8	8	C	2023	2263	2503	2743	2983
10	180.0	10	18.0	180	226.0	8	8	D	2024	2264	2504	2744	2984
10	220.0	10	20.0	200	250.0	8	8	D	2025	2265	2505	2745	2985
10	220.0	20	20.0	200	250.0	8	8	D	2026	2266	2506	2746	2986
15	2.7	10	.6	6	7.5	4	4	A	2027	2267	2507	2747	2987
15	3.3	10	.8	8	10.0	4	4	A	2028	2268	2508	2748	2988
15	3.3	20	.8	8	10.0	4	4	A	2029	2269	2509	2749	2989
15	18.0	10	3.5	35	44.0	6	6	B	2030	2270	2510	2750	2990
15	22.0	10	4.0	40	50.0	6	6	B	2031	2271	2511	2751	2991
15	22.0	20	4.0	40	50.0	6	6	B	2032	2272	2512	2752	2992
15	56.0	10	8.0	80	100.0	6	6	C	2033	2273	2513	2753	2993
15	68.0	10	10.0	100	125.0	6	6	C	2034	2274	2514	2754	2994
15	68.0	20	10.0	100	125.0	6	6	C	2035	2275	2515	2755	2995
15	120.0	10	18.0	180	226.0	8	8	D	2036	2276	2516	2756	2996
15	150.0	10	20.0	200	250.0	8	8	D	2037	2277	2517	2757	2997
15	150.0	20	20.0	200	250.0	8	8	D	2038	2278	2518	2758	2998
20	1.2	10	.6	6	7.5	4	4	A	2039	2279	2519	2759	2999
20	1.5	10	.6	6	7.5	4	4	A	2040	2280	2520	2760	3000
20	1.5	20	.6	6	7.5	4	4	A	2041	2281	2521	2761	3001
20	1.8	10	.6	6	7.5	4	4	A	2042	2282	2522	2762	3002
20	2.2	10	.8	8	10.0	4	4	A	2043	2283	2523	2763	3003
20	2.2	20	.8	8	10.0	4	4	A	2044	2284	2524	2764	3004
20	8.2	10	2.0	20	25.0	6	6	B	2045	2285	2525	2765	3005
20	10.0	10	3.0	30	38.0	6	6	B	2046	2286	2526	2766	3006
20	10.0	20	3.0	30	38.0	6	6	B	2047	2287	2527	2767	3007
20	12.0	10	3.5	35	44.0	6	6	B	2048	2288	2528	2768	3008
20	15.0	10	4.0	40	50.0	6	6	B	2049	2289	2529	2769	3009
20	15.0	20	4.0	40	50.0	6	6	B	2050	2290	2530	2770	3010
20	27.0	10	5.0	50	63.0	6	6	C	2051	2291	2531	2771	3011
20	33.0	10	7.0	70	88.0	6	6	C	2052	2292	2532	2772	3012
20	33.0	20	7.0	70	88.0	6	6	C	2053	2293	2533	2773	3013
20	39.0	10	8.0	80	100.0	6	6	C	2054	2294	2534	2774	3014
20	47.0	10	9.0	90	113.0	6	6	C	2055	2295	2535	2775	3015
20	47.0	20	9.0	90	113.0	6	6	C	2056	2296	2536	2776	3016
20	56.0	10	11.0	110	138.0	6	6	D	2057	2297	2537	2777	3017
20	68.0	10	14.0	140	175.0	6	6	D	2058	2298	2538	2778	3018
20	68.0	20	14.0	140	175.0	6	6	D	2059	2299	2539	2779	3019
20	82.0	10	16.0	160	200.0	6	6	D	2060	2300	2540	2780	3020
20	100.0	10	20.0	200	250.0	8	8	D	2061	2301	2541	2781	3021
20	100.0	20	20.0	200	250.0	8	8	D	2062	2302	2542	2782	3022
35	5.6	10	2.5	25	32.0	4	4	B	2063	2303	2543	2783	3023
35	6.8	10	3.0	30	38.0	6	6	B	2064	2304	2544	2784	3024
35	6.8	20	3.0	30	38.0	6	6	B	2065	2305	2545	2785	3025
35	22.0	10	8.0	80	100.0	6	6	C	2066	2306	2546	2786	3026
35	22.0	20	8.0	80	100.0	6	6	C	2067	2307	2547	2787	3027
35	27.0	10	9.0	90	113.0	6	6	D	2068	2308	2548	2788	3028
35	33.0	10	11.0	110	138.0	6	6	D	2069	2309	2549	2789	3029
35	33.0	20	11.0	110	138.0	6	6	D	2070	2310	2550	2790	3030

Figure 10-6. Sheet 2

CORNELL-DUBILIER

Ceramic Dielectric Capacitors

DISC STYLE MINIATURE GENERAL PURPOSE

The type MGP, 100 WVDC ceramic disc is designed to provide maximum capacitance in a minimum disc size for limited-space applications. These capacitors are designed for by-pass and coupling applications in low-voltage transistor circuits. Operating temperature range is −55° C to +85° C. Tolerance: ± 20%. Temperature characteristic is Z5V or better. Leads are #22 gauge, 1.5" long.

CAPACITY MF	CATALOG NUMBER	CAPACITY TOLERANCE	MAXIMUM DIAMETER (In.)	MAXIMUM THICKNESS (In.)	LEAD SPACING (In.)
.005	MGP005	± 20%	.390	.156	.250
.01	MGP01	± 20%	.390	.156	.250
.02	MGP02	± 20%	.590	.156	.375
.025	MGP025	± 20%	.590	.156	.375
.03	MGP03	± 20%	.590	.156	.375
.05	MGP05	± 20%	.690	.156	.375
.1	MGP1	± 20%	.718	.156	.375

GENERAL PURPOSE DISC

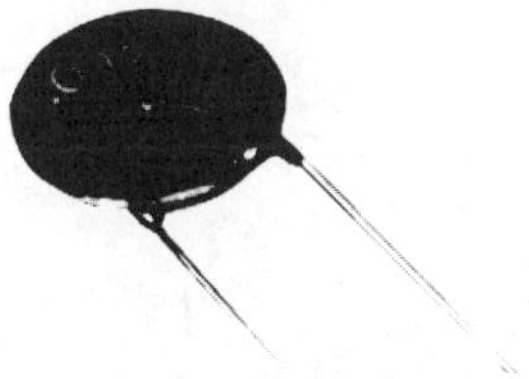

The type GP, 1000 WVDC is a general purpose ceramic disc capacitor. It is primarily designed for by-pass and non-critical coupling applications. Excellent moisture and humidity protection is provided. Operating temperature range is −55° C to +85° C. Leads are #22 gauge, 1.5" long.

CAPACITY PF	CDE TYPE	TOLERANCE	MAXIMUM DIAMETER (IN.)	MAXIMUM THICKNESS (IN.)
		1000 WVDC		
5	GP5	± 10%	.290	.156
10	GP10	± 10%	.290	.156
22	GP22	± 10%	.290	.156
33	GP33	± 10%	.290	.156
47	GP47	± 10%	.290	.156
50	GP50	± 10%	.290	.156
68	GP68	± 10%	.290	.156
100	GP100	± 10%	.290	.156
200	GP200	± 10%	.290	.156
270	GP270	± 10%	.290	.156
300	GP300	± 10%	.290	.156
330	GP330	± 10%	.290	.156
390	GP390	± 10%	.290	.156
470	GP470	± 10%	.290	.156
500	GP500	± 10%	.290	.156
680	GP680	± 10%	.290	.156
820	GP820	± 10%	.290	.156

CAPACITY MFD.	CDE TYPE	TOLERANCE	MAXIMUM DIAMETER (IN.)	MAXIMUM THICKNESS (IN.)
		1000 WVDC		
.001	GP1000	GMV	.290	.156
.0018	GP1800	± 10%	.385	.156
.002	GP2000	GMV	.385	.156
.0033	GP3300	GMV	.590	.156
.0047	GP4700	GMV	.590	.156
.005	GP5000	GMV	590	.156
.0068	GP6800	GMV	.590	.156
.01	GP10000	GMV	.690	.156
		600 WVDC		
.02	GP20000	+80% −20%	.690	.156
.05	GP50000	+80% −20%	.875	.250

TEMPERATURE COMPENSATING DISC

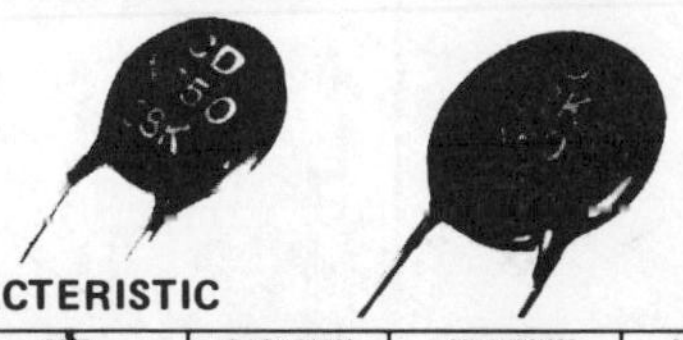

Available in NPO and N750 characteristics, these economically priced temperature compensating ceramic discs provide pre-determined linear capacitance change with changes in temperature. Many applications are in frequency determining and resonant circuits. Operating temperature range is −55° C to +125° C. Leads are #22 gauge, 1.5" long.

NPO CHARACTERISTIC

CAPACITY PF	CDE TYPE	CAPACITY TOLERANCE	MAXIMUM DIAMETER (IN.)	MAXIMUM THICKNESS (IN.)
1.5	NP01P5	± .5Pf	.320	.180
2.2	NP02P2	± .5Pf	.290	.156
3.3	NP03P3	± .5Pf	.290	.156
4.7	NP04P7	± .5Pf	.290	.156
6.8	NP06P8	± .5Pf	.290	.156
10	NP010	± 5%	.290	.156
15	NP015	± 5%	.290	.156
20	NP020	± 5%	.290	.156
22	NP022	± 5%	.390	.156
25	NP025	± 5%	.390	.156

N750 CHARACTERISTIC

CAPACITY PF	CDE TYPE	CAPACITY TOLERANCE	MAXIMUM DIAMETER (IN.)	MAXIMUM THICKNESS (IN.)
5	N5	± .5Pf	.320	.180
10	N10	± .5Pf	.290	.156
15	N15	± 5%	.290	.156
20	N20	± 5%	.290	.156
22	N22	± 5%	.290	.156
33	N33	± 5%	.290	.156
47	N47	± 5%	.290	.156
68	N68	± 5%	.390	.156
100	N100	± 5%	.390	.156
150	N150	± 5%	.590	.156
220	N220	± 5%	.590	.156
330	N330	± 5%	.690	.156

Figure 10-7. Ceramic Dielectric Capacitors

LINEAR INTEGRATED CIRCUITS

TYPES uA748M, uA748C GENERAL-PURPOSE OPERATIONAL AMPLIFIERS

FORMERLY SN52748, SN72748

- Frequency and Transient Response Characteristics Adjustable
- Short-Circuit Protection
- Offset-Voltage Null Capability
- Wide Common-Mode and Differential Voltage Ranges
- Low Power Consumption
- No Latch-up
- Same Pin Assignments as uA709

description

The uA748 is a general-purpose operational amplifier. It offers the same advantages and desirable features as the uA741 with the exception of internal compensation. The external compensation of the uA748 allows the changing of the frequency response (when the closed-loop gain is greater than unity) for applications requiring wider bandwidth or higher slew rate. This circuit features high gain, large differential and common-mode input voltage range, output short-circuit protection, and may be compensated under unity-gain conditions with a single 30-pF capacitor. A potentiometer may be connected between the offset null inputs, as shown in Figure 12, to null out the offset voltage.

The uA748M is characterized for operation over the full military temperature range of −55°C to 125°C; the uA748C is characterized for operation from 0°C to 70°C.

schematic

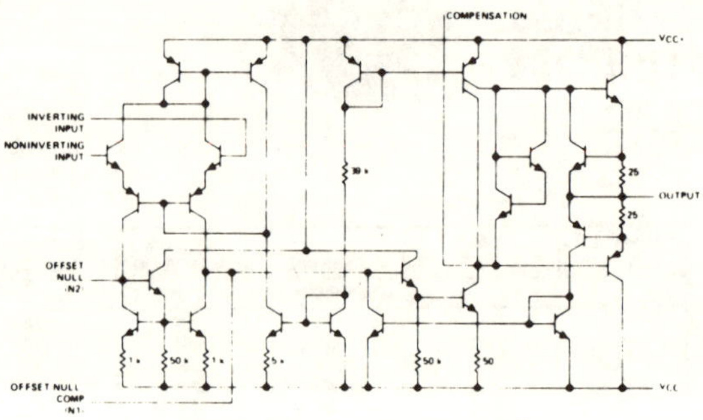

Resistor values shown are nominal and in ohms.

terminal assignments

√

J OR N DUAL-IN-LINE PACKAGE (TOP VIEW) | **JG OR P DUAL-IN-LINE PACKAGE (TOP VIEW)** | **L PLUG-IN PACKAGE (TOP VIEW)** | **U FLAT PACKAGE (TOP VIEW)**

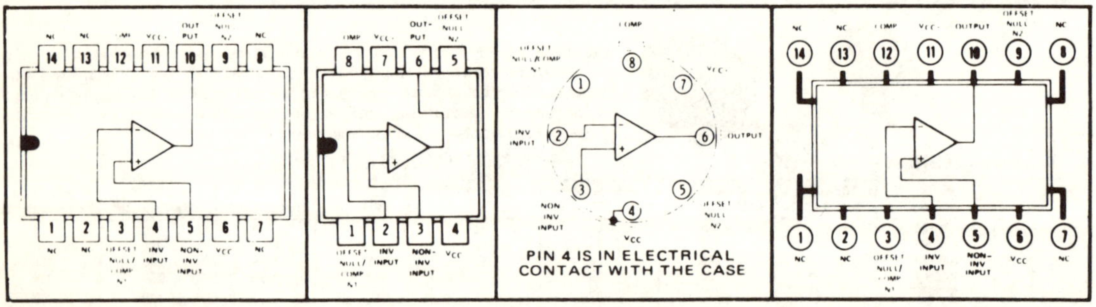

NC—No internal connection

+VCC = +15VDC

−VCC = −15VDC

Figure 10-8. Op Amps

P dual-in-line plastic package

This dual-in-line package consists of a circuit mounted on an 8-lead frame and encapsulated in an electrically nonconductive plastic compound. The compound will withstand soldering temperature with no deformation and circuit performance characteristics remain stable when operated under high-humidity conditions. This package is intended for insertion in mounting hole rows on 0.300 (7,62) centers (see Note a). Once the leads are compressed and inserted, sufficient tension is provided to secure the package in the board during soldering. Leads require no additional cleaning or processing when used in soldered assembly.

8-PIN P PLASTIC

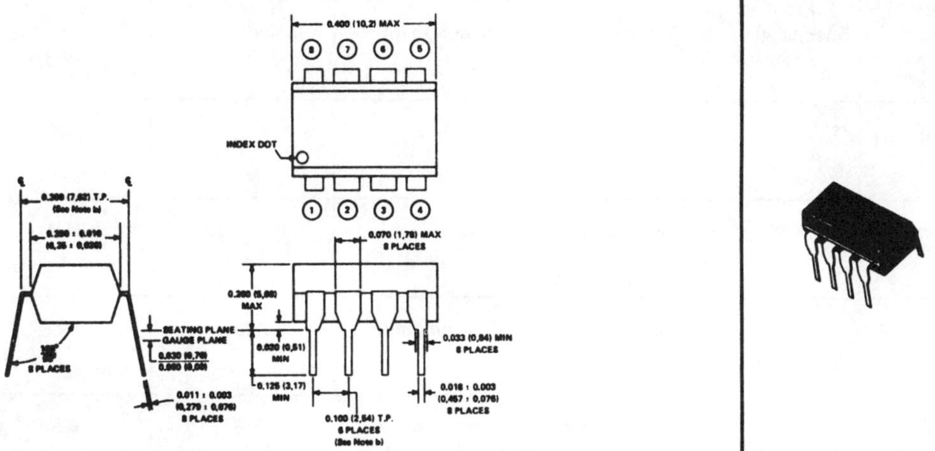

NOTES: a. All dimensions are in inches and parenthetically in millimeters. Inch dimensions govern.
b. Each pin centerline is within 0.005 (0,127) radius of true position at the gauge plane with maximum material condition and unit installed.

Figure 10-9.

Integrated Circuits

Figure 10-1 has seven different types of IC specification numbers called out on the various gates and IC's. This doesn't mean that we have to finish with seven IC's. You can use the same IC package type several times on the same schematic. The gate identified on our schematic Figure 10-1 as a 00 is called a 2-input NAND. It is shown on the specification sheet in Figure 10-10 (7400). The gate identified as 02 is a 2-input NOR. It is shown on the specification sheet in Figure 10-10 (7402). The gate identified as 04 is an inverter. It is shown on the specification sheet in Figure 10-10 (7404). The gate identified as a 10 is called a 3-input NAND. It is shown on the specification sheet in Figure 10-10 (7410). The gate identified as a 20 is called a 4-input NAND. It is shown on the specification sheet in Figure 10-10 (7420). The gate identified as a 32 is called a 2-input OR. It is shown on the specification sheet in Figure 10-10 (7432). All gates listed to this point are non-open collector, totem pole gates. The gate identified on the schematic (Figure 10-1) as an 07 is called a buffer with open collectors. It is shown on the specification sheet in Figure 10-10 (07).

Step #3—See Figure 10-3. Determine each lead size, hole diameter, pad size, component's lead spacing, and body size, using the specification sheets from step 2. Review chapter 3 of the analog design book. Determine the IC body size such as 14-pin, 16-pin, 24-pin, etc. At this point we will identify each component on our schematic, Figure 10-1. Using the specification sheets, see Illustration 3.

Illustration #3: Component Mounting Table.

Explanation of Table 10-1

This table gives a breakdown of pad spacing, mounting hole requirements, and pad size requirements. Each component will be discussed on the schematic in Figure 10-3 to compute this information. The formulas to calculate the hole and

Table 10-1

Reference Designation	Component Type	Maximum Body Length	1:1 Pad Spacing (Inches)	Maximum Lead Diameter	Lead Mounting Hole (Min.)	Lead Mounting Hole (Min.)
1 From Schematic	2 From Specification Sheet	3 From Specification Sheet	4 Column 3+.120 (Round Up)	5 From Specification Sheet	6 Column 5 + .006	7 Column 5 + .020
C1	CS13-A (RL20)	.422	.600	.022	.028	.042
C2 thru C8	CK60	.390 DIA	Radial Lead .250	22 Gauge (0.25)	.031	.045
U1 thru U8	14 Pin Dual-in-line	Per template	Per template	.018	.024	.038
U9 & U10	8 Pin Dual-in-line	Per template	Per template	.018	.024	.038

With a PCB MFG Tolerance of +.004/-00.2 for the mounting holes, the above components require:

	Maximum Lead Diameter	Mounting Hole Min W/Tolerance	Mounting Hole Max W/Tolerance	Hole Call-Out	Min Pad Size	Selected Pad Size
	8 Column 5 In Increasing Size	9 Column 6 Plus Tolerance (Add .002)	10 Column 7 Minus Tolerance (Sub .004)	11 From Column 9 and 10	12 Column 11 + .030	13 Standard Pad Size at 2:1
U1 thru U8	.018	.026	.034	.034 +.004 −.002	.064	Pattern "D" (Fig. 6-21)
C1	.022	.030	.038	.034 +.004 −.002	.064	.150
C2 thru C8	.025	.033	.041	.034 +.004 −.002	.064	.150

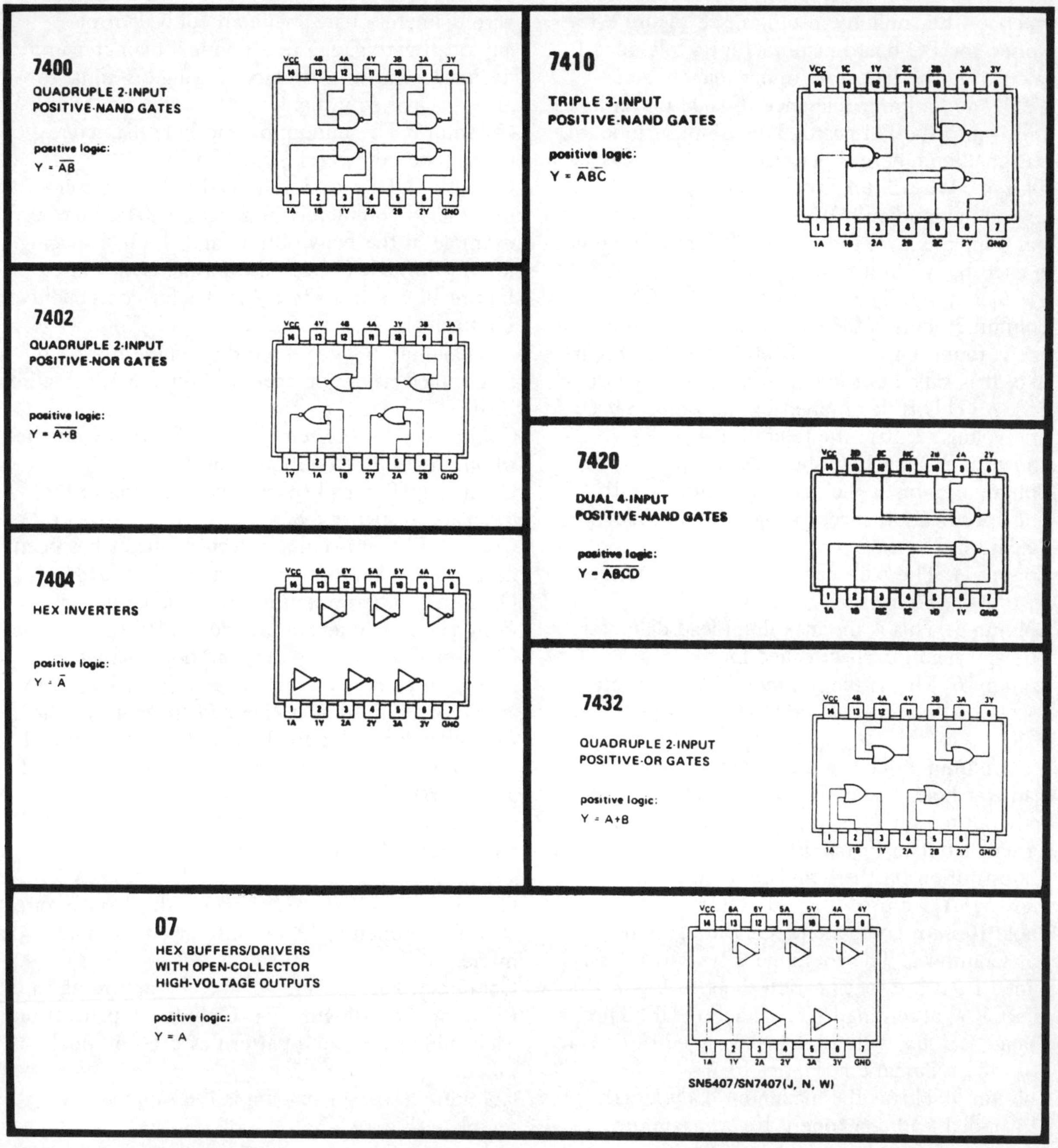
7400
QUADRUPLE 2-INPUT
POSITIVE-NAND GATES
positive logic:
Y = $\overline{AB}$
7410
TRIPLE 3-INPUT
POSITIVE-NAND GATES
positive logic:
Y = $\overline{ABC}$
7402
QUADRUPLE 2-INPUT
POSITIVE-NOR GATES
positive logic:
Y = $\overline{A+B}$
7420
DUAL 4-INPUT
POSITIVE-NAND GATES
positive logic:
Y = $\overline{ABCD}$
7404
HEX INVERTERS
positive logic:
Y = $\overline{A}$
7432
QUADRUPLE 2-INPUT
POSITIVE-OR GATES
positive logic:
Y = A+B
07
HEX BUFFERS/DRIVERS
WITH OPEN-COLLECTOR
HIGH-VOLTAGE OUTPUTS
positive logic:
Y = A
SN5407/SN7407(J, N, W)

Figure 10-10

the pad sizes are taken from good commercial practices. For military formulas, see Figure 8-13. Notice the PC board manufacturing tolerance for calculating the hole size: upper tolerance of +.004, and a lower tolerance of -.002 for all holes drilled into the PC board. This is not a standard; it is good commercial practice. The table has 13 columns. We will take you through each column for each component. We will refer you to the formal template layout in Figure 10-4, as a reference to each body outline and pad spacing.

Column 1: The first component is C1.
Column 2: This is the component type. The case size is found on the specification sheet in Figure 10-6. It is called out under military designation as CS13A. This is determined by the value: 3.9 UF. The voltage is 10V; the tolerance is $\pm 10\%$ (shown in Figure 10-6, sheet 2).
Column 3: This is the maximum body length. The specification sheet in Figure 10-6, sheet 1, specifies .422.
Column 4: The pad spacing at 1:1 was rounded up to .600. Figure 10-8
Column 5: This is the maximum lead diameter. The specification sheet called for .022.
Column 6: This is the minimum lead diameter hole. Good commercial practices allow us to use .028.

Column 7: This is the maximum lead diameter hole. Use .042. For an example of the body outline for C1 (with its pad spacing), see the formal layout in Figure 10-4. In this case we used a substitution on the template (Figure 8-8). For the CS13A, we used a RL20. Return to column 1, Table 10-1 for components C2 through C8.

Column 2: The component type was determined from the specification sheet in Figure 10-7 as a CK60 according to its value: .01 UF. The voltage is 100V with a tolerance of $\pm 20\%$ (taken from the schematic in Figure 10-3).
Column 3: This is the maximum body length. This radial lead component has a maximum diameter of .390. It was obtained from the specification sheet in Figure 10-7.
Column 4: The pad spacing at 1:1 was specified at .250 on the specification sheet in Figure 10-7.
Column 5: The maximum lead diameter was specified in Figure 10-7 as 22 gauge. Because we cannot use "gauge", we must convert it to a decimal equivalent. Figure 10-11 is a solid bare wire diameter chart giving an equivalent of american wire gauge to a nominal diameter in a decimal form. In this case, 22 gauge is equal to .025 decimal equivalent.
Column 6: The minimum lead hole diameter using good commercial practices is .031.
Column 7: The maximum lead hole diameter using good commercial practices is .045. For an example of the body outline and the pad spacing for C2 through C8, see the formal layout in Figure 10-4. Return to column 1 for components U1 through U8.

Column 2: The 14-pin dual inline has been determined using the specification sheet in Figure 10-10.
Column 3: Use the template in Figure 8-8 for the 14-pin dual inline configuration.
Column 4: The pad spacing is according to the template in Figure 8-8.
Column 5: The maximum lead diameter has been determined in the previous chapters as .018.
Column 6: The minimum lead hole diameter using good commercial practices is .024.
Column 7: The maximum lead hole diameter using good commercial practices is .038. For an example of 14-pin dual-in-line pad spacing, see the formal layout in Figure 10-4. Return to column 1 for components U9 through U10 (the operational amplifiers).

Column 2: Use the specification sheets for components U9 and U10 (the operational amplifiers).
Column 2: Use the specifications sheets in Figure 10-9. Components U9 and U10 are 8-pin dual inline.
Column 3: Pad spacing will also be according to the template in Figure 8-8. Only use four pads on each side of the same pattern as a 14-pin dual inline.
Column 4: 1:1 pad spacing is the same as the template (Figure 8-8).
Column 5: The maximum lead diameter is the same as the 14-pin dual inline (.018).
Column 6: The minimum lead hole diameter is the same as the 14-pin dual inline (.024).
Column 7: The maximum lead hole diameter is the same as the 14-pin dual inline (.038). For an

SOLID BARE WIRE GAGE DIAMETERS AND PRINTED CIRCUIT DRILL SIZES

AMER WIRE GAGE (AWG)	NOM DIA	MM DIA	AREA CIRCULAR MILLS (sq. in.)	OHMS PER 1000 FT at 20°C	FEET PER POUND	PC BOARD HOLE DIA
36	.005	.127	25.0	415.	13,210.	.016 (#78)
35	.0056	.142	31.5	331.	10,481.	.016 (#78)
34	.006	.152	39.8	261.	8,310.	.016 (#78)
33	.007	.178	50.1	206.	6.591.	.016 (#78)
32	.008	.208	63.2	162.	5,227.	.016 (#78)
31	.009	.228	79.7	131.	4,145.	.016 (#78)
30	.010	.254	101.0	104.	3,287.	.016 (#78)
29	.011	.280	127.0	81.2	2,607.	.016 (#78)
28	.013	.330	160.0	65.3	2,067.	.016 (#78)
27	.014	.356	202.0	51.4	1,639.	.026 (#71)
26	.016	.406	254.0	41.0	1,300.	.026 (#71)
25	.018	.457	320.0	32.4	1,031.	.026 (#71)
24	.020	.508	404.0	25.7	817.7	.031 (#68)
23	.023	.584	509.0	20.3	648.4	.031 (#68)
22	.025	.635	642.0	16.2	514.2	.031 (#68)
21	.028	.711	810.0	12.8	407.8	.040 (#60)
20	.032	.813	1,020.0	10.1	323.4	.040 (#60)
19	.036	.914	1,290.0	8,04	256.5	.052 (#55)
18	.040	1.016	1,620.0	6.38	203.4	.052 (#55)
17	.045	1.143	2,050.0	5.04	161.3	.052 (#55)
16	.050	1.270	2,580.0	4.01	127.9	
15	.057	1.448	3,260.0	3.18	101.4	
14	.064	1.626	4,110.0	2.52	80.44	
13	.072	1.829	5,180.0	2.00	63.80	
12	.081	2.057	6,530.0	1.59	50.59	
11	.090	2.286	8,230.0	1.26	40.12	
10	.102	2.590	10,400.0	.999	31.82	
9	.114	2.896	13,100.0	.792	25.23	
8	.128	3.251	16,500.0	.628	20.01	
7	.144	3.658	20,800.0			
6	.162	4.115	26,300.0			
5	.182	4.623	33,100.0			
4	.204	5.182	41,700.0			
3	.229	5.817	52,600.0			
2	.258	6.553	66,400.0			
1	.289	7.340	83,700.0			
0	.325	8.255	106,000.0			

NOTE: PC Board Hole Diameters are for manual insertion and NOT plated through holes.

Figure 10-11. Design Reference Materials

example of the component body outline and pad spacing, see Figure 10-4 for the 8-pin dual inline.

SPECIAL NOTE: We stacked the two 8-pin dual inlines together, body-to-body, to form the same configuration as a 16-pin dual inline. The two eights will fit within the same pattern as a 16-pin dual inline. Return to Table 10-1, column 8. This column is very important. The leads must be listed in increasing size, starting from the smallest listing to the largest (for the components listed in columns 1-7). For column 8, the first component is U1 to U10, with a lead diameter of .018. The second component in increasing size is C1 with a lead diameter of .022. The last component (C2 to C8) is .025.

Column 9 determines the minimum hole diameter with its tolerance (a tolerance of +.004, -.002 for all holes). Add the -.002 tolerance to column 6. Column 6 (U1 to U10): .024 + .002 = .026. Column 9 (C1): .028 + .002 = .030. Column 9 (C2 to C8): .031 + .002 = .033.
Column 10: This is the maximum hole size with a tolerance of +.004, -.002. Subtract the .004 from column 7. Column 7 (U1 to U10): .038 - .004 = .034. Column 10 (C1): .042 - .004 = .038. Column 10 (C2 to C8): .045 - .004 = .041.
Column 11: The final hole call out is selected. Compare the largest minimum lead diameter hole size to the smallest minimum hole diameter of the largest lead. We chose .034 as a final hole size for column 11, with a tolerance of +.004, -.002.
Column 12 is a selection of minimum pad size. The annular ring is .015; a total of .030 added to column 11 = .064 (for all components).
Column 13 is the final selection of the pad size at a scale of 2:1. For U1 to U10, column 13, select an oval pad shown in Figure 6-19 Chapter 6, pattern D. The reason for this IC pattern is that we will run traces through the IC pins on the component side of the printed circuit board. For an example of routing traces through the IC pins, see Figure 10-4.

Column 13 for C1 and C2 through C8 will be a .150 diameter pad at a scale of 2:1. There is one other pad that should be listed at this time. List the test point pads (TP1, TP2, and TP3). Use a .250 diameter pad at a scale of 2:1.

We have summarized the steps on Table 10-1, columns 1-13 for the new layout. For a review on each step, refer to chapters 8 and 9.

Step #4

Illustration #4: Design Rules

Establish the board layout rules. The design rules established in chapter 8 will once again be used in this chapter. This section establishes the electronic engineering PC board design conventions for this book. It includes the general requirements for the formal layout and artmasters. So, before starting the design, review chapter 8, Illustration 2 for the design rules.

Step #5

Illustration #5: Questions To Ask The Engineer

Questions to ask the engineer about his schematic were explained in detail in chapter 8.

While asking these questions about the schematic in Figure 10-3, we define the rules for the layout on the schematic in Figure 10-5. At the same time, we are answering and defining restrictions that were placed on the version of the layout shown in Figure 10-4.

Questions

1. Are there any special size components? Are all the components standard? The specification sheets indicated that C1 was a CS13A case size. After working with Table 10-1, we found that the component CS13A could be substituted with an RL20 outline. It has the same pad spacing and body size as the CS13A. We also found that C2 through C8 had a CK60 standard case outline. The IC's were all 14-pin dual inline. There were two operational amplifiers that were also in an 8-pin dual inline form. All of our components have a standard case code style. We can find the correct mounting spacing for each component on the schematic in Figure 10-3.

2. How many bypass capacitors per IC? That question has been answered on the schematic in Figure 10-3. Note 1, specifies to bypass every two IC packages. Refer to the layout in Figure 10-4.

We did bypass every two IC packages with C4 through C8. We used a CK60 ceramic disc radial lead capacitor. This question also applies to the operational amplifiers which also have bypass capacitors (C2 and C3) shown on the layout in Figure 10-4. This was covered on the schematic in Figure 10-3. Note 3 specifies to bypass one CK06 ceramic disc radial lead capacitor for 2 op amps. This would only apply if the op amps were in the same area. On the layout in Figure 10-4, note that the two operational amplifiers U9 and U10 are placed over and under each other. If we were to divide these two operational amplifiers and space them far apart on the board layout, then you would be required to provide bypass capacitors for the +15V and the -15V for each one. This would equal a total of four capacitors.

3. Are there any critical leads on the board? In this case the engineer will not specify any critical lead lengths for the integrated circuits. Consider the operational amplifiers. The Analog Printed Circuit Design & Drafting book mentioned that the inputs and outputs usually should be close; they should have short lead lengths. However, on a typical digital board similar to the layout in Figure 10-4, it is difficult to keep them as close as on an analog board layout. Arrange the IC's so that the inputs and outputs of the amplifiers are relatively close. Do not let them run for great lengths on the digital layout.

4. Are there any special size filtering capacitors required on the input voltage pin? The schematic in Figure 10-3 shows an input filtering capacitor (called C1). This is tied between the +5V and ground. Note 2 on the schematic in Figure 10-3 specifies the value and the case size for this capacitor. The engineer must also specify the fact that it was an electrolytic tantalum polarized capacitor. See the specification sheet in Figure 10-6. This capacitor should be very close either to voltage pins 29 and 30 or to +5V on the connector.

5. Should there be any spare IC gates? Most likely there may be spare IC gates. However, the engineer in this particular schematic (Figure 10-3) does not require any particular type of spares.

6. By using the maximum size of all the components, will all of the components fit onto the PC board? This question will be answered in detail in step 8 (once the board size has been determined and the grouping of schematic has been accomplished).

7. Are there any pre-assigned pins on the connector? The schematic in Figure 10-3 shows pre-assigned pins for P1, and for all voltages and ground. You cannot change the assignment of these pins.

8. Are there any variable resistors or capacitors? If so, where should they be placed on the board? The schematic in Figure 10-3 has no variable components.

9. Should there be any test points? The schematic in Figure 10-3 does show test points (TP1 and TP2) on the output lines for the operational amplifiers. The engineer placed TP3 on the ground line of P1. Check with the engineer to see if we should have test points for the +15V and the -15V. The engineer usually does not draft them on the schematic, even though they should be there. In this case, the engineer does not request that the added test points for +15V and -15V be added to the schematic. All test points should be accessed from the opposite edge of the P1 connector. See Figure 8-17 for the location and size of these test points.

10. Do any components require heat sinks? For our schematic in Figure 10-3, the answer is no. There are no components that run hot and need a heat sink.

11. Have all the package considerations been examined? We do have a new feature on the board layout in Figure 10-4: an extractor for boards that insert on card tracks. Question 11 will be answered extensively in step 6 which covers the board outline. Notice the enclosure and mounting techniques required for our layout in Figure 10-4.

Step #6

Illustration #6: Board Size and Mounting

Determine the board outline. Besides the size of the board, it is also important to determine the method of fastening or mounting: card guides, screws, stand-offs, etc. What type of electrical interface is required (connector, cables, wires, etc.)? Determine the board size area.

SPECIAL NOTE: Review board size and mounting in chapter 4.

Procedures

1. The board outline is usually provided by the mechanical designer or mechanical engineer. Use the board outline shown in Figure 10-12, for the schematic in Figure 10-3. There is a total of 17.4 square inches on the board, not counting the P1 connector. In order to calculate the total number of useable square inches, define what interfaces to this PC board. How is it secured to an enclosure? Figure 10-12 gives the board outline (its length, width, connector size, connector fingers, spacing, tooling holes, location, size, and card ejector location). It has all the necessary dimensions to draft the board outline in preparation to the layout. The basic board outline is shown in Figure 10-12. It is the first piece of information needed to answer question 4. Figure 10-13 shows the PC board mounted within its enclosure (called a chassis). It gives us some idea about mechanical restrictions. The board is secured by card guides, and plugs into a female connector (called J1). Figure 10-13, sheet 2, shows a card ejector, its physical dimensions, and where it mounts on the PC board. All component traces must remain clear of the area required for the ejector. Now we can calculate the useable square inches on the PC board. We know that we must design for a reasonable distance for the card guides. Chapter 8 design rules indicate that we must maintain at least .150 at a 1:1 scale clearance (from the edge of the board to the first trace or component, clearing back for the card guides). Procedure 2 computes the useable square inches on the PC board.

2. The board outline is inserted into an enclosure that uses a card guide track as a means of holding the board into position. For an example of the enclosure, see Figure 10-13 (sheet 1). The board has restrictions on its edges. Also, the board has restrictions about the edges that interface to the card track (from the design rules in chapter 8). Using the restrictions, re-compute the useable are:

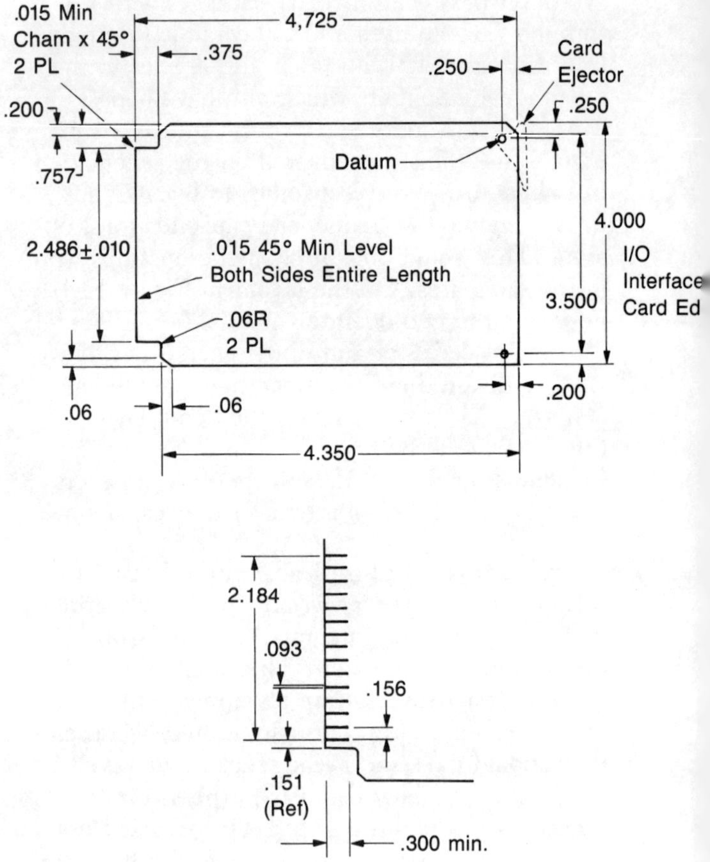

Figure 10-12 Basic Board Outline.

A. 4" (width of board) x 4.350 (length of the board, not counting connector) equals 17.400 square inches.

B. 4" will be reduced by the card track clearance.
Example: 4.00
- .300 Card track clearance
= 3.700

C. 4.350 will be reduced from the board edge clearance.
Example: 4.350
- .100 (Card board edge—clearance) (.050 x 2)
= 4.250

D. 3.700
x 4.250
= 15.725—Useable board area

The 15.725 can be rounded off to approximately 16 square inches. This is the actual amount of useable area that you can lay out the components and traces.

3. The PC designer must obtain a connector specification sheet from the engineer. The specification sheet provides the information about the number of contacts and the size of the male tongue (P1) on the PC board. The specification sheet is similar to Figure 4-15. For this chapter we will provide all necessary information. We will not go through the procedures for determining the size and the number of contacts, and the numbering of each contact (taken off the typical specification sheet). See the layout in Figure 10-4 and the board outline in Figure 10-12 for connector size and numbering. The component side is colored in red: pins 1 to 29. The noncomponent side of the contacts (which we cannot see) are colored in blue: pins 2 to 30. This lets you see the routing of the colored traces up to the connector. Red represents the component side; blue represents the noncomponent side (the side of the connector to which you are actually terminating). Return to the schematic in Figure 10-3. Group the logic gates, using the specification sheets (see step 7). For a review of board outlines, see chapter 4.

Step #7

Illustration #7: Grouping

Group the logic gates on the schematic in Figure 10-3. We will explain each decision. Grouping must be done at this time in order to compute the total number of IC's required for the board layout. We have also grouped the schematic in Figure 10-5 (which is very similar to the grouping in Figure 10-3). A few adjustments encourage the student to use his/her own plotting and version of the layout. The following explains the grouping in the schematic of Figure 10-3.

Procedures

1. Using the schematic in Figure 10-3, group and combine to find the lowest number of IC's required on this schematic. Combine the gates for the shortest trace connection using the following instructions. Follow the same instructions for his/her schematic (Figure 10-5).

A. Identify all the gates on the schematic. Use the IC specification sheet (Figure 10-10) for two reasons: one is that you have different logic families such as open collector and totem pole, the other is that you cannot mix these two logic families.

B. Determine the number of the logic family to know which manufacturer the engineer is using to obtain the specification sheet. For the schematic in Figure 10-3 and the student's schematic in Figure 10-5, all gates are identified with the proper manufacturing specification number: 00, 04, etc. Review the schematic in Figure 10-3 and the student's schematic in Figure 10-5. There are open collector gates labeled: 07. Flag these open collector gates with a tail (on the output side of the gates) as shown on both schematics. This prevents confusion when grouping and arbitrarily placing open and non-open collectors within the same package.

2. Before grouping, review the questions that we asked the engineer. Some affect the grouping on the schematics in Figures 10-3 and 10-5. Are there any spare gates required? In this case, the engineer said that there were no spare gates required. Are there any critical leads on the board? This question in particular is important to the operational amplifiers. On the digital board, we do not have to keep the leads extremely short (as we would on analog boards). However, keep the inputs and the output leads on the amplifiers reasonably short. The grouping and the plotting plays a most crucial role. We must be sure that the lead length communicating from one IC to the other (or from the operational amplifiers to our IC's) do not grow excessively in length.

3. Combine the gates for the schematics in Figures 10-3 and 10-5. Use the combining rules in Figures 2-56 through 2-61. Combine gates using the technique called tying off unused inputs. Tie off to another used input. Of course, you would always seek the engineer's permission to do so. This is a more popular technique than tying the unused input to a high or to a low (meaning voltage or ground point). The explanation for tying off to a ground or to a voltage is in chapter 2 of this book.

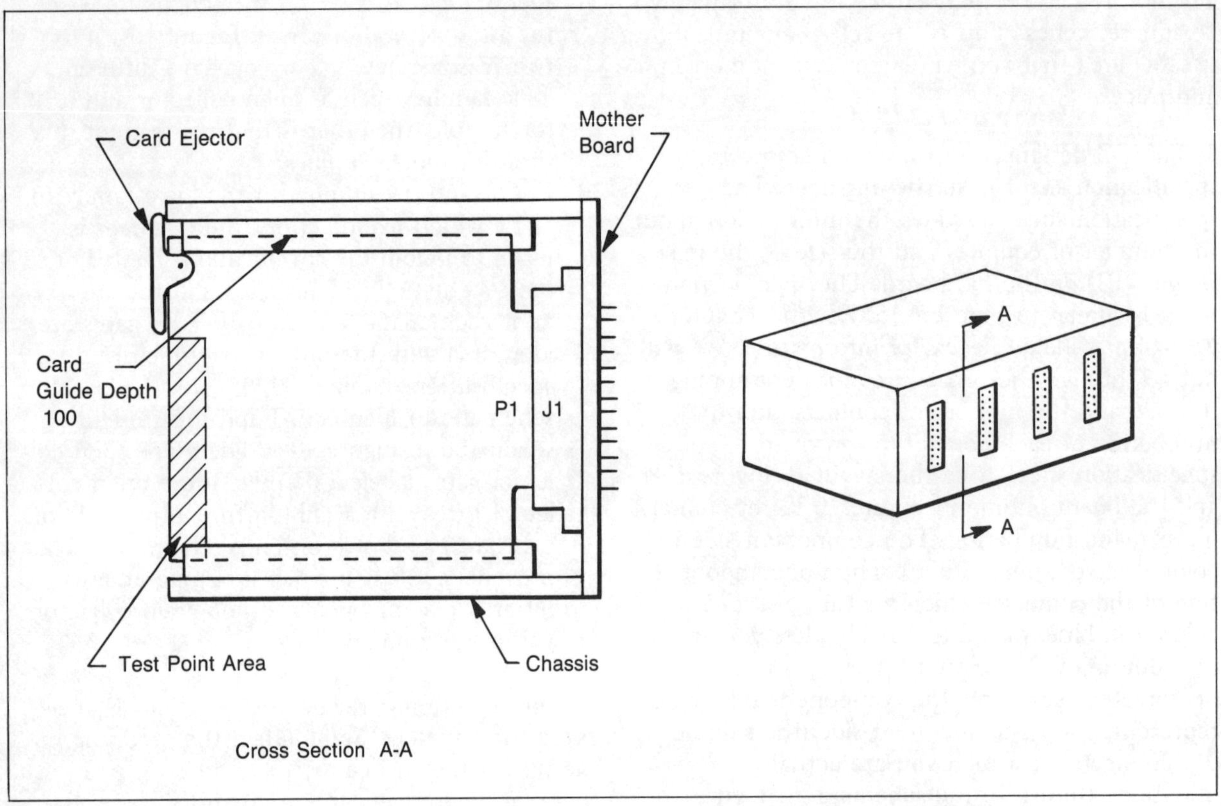

Figure 10-13. Sheet 1 of 2

4. See the schematic in Figure 10-3. Combine the gates that seem to be intercommunicating to each other. Make the traces as short as possible. Look at the schematic in Figure 10-3. While grouping, refer to the schematic in Figure 10-1.

Grouping and Combining the Schematic

Figure 10-3, U1, 7420, 4-input NAND: Review the schematic in Figure 10-3. There are only two 7420's. The specification sheet (Figure 10-10) shows there are two 7420 4-input NAND gates in one package. Select the 7420 as U1. Place both of these inside one package, leaving zero spares; both 4-input NANDS communicate directly to our connector, P1.

U2, 7400, 2-input NAND: The schematic shows four 2-input NANDS. Review the specification sheet (Figure 10-10), shows a total of four in the package. Select U2 as the four 2-input NANDS. U2 has zero spares. Look at the flow of the schematic. U2 would be talking to U3 which would be talking to U4 which would be talking to U5. This part of the schematic is the real key to the entire layout. The heart of the schematic involves flow and communication in one straight line.

U3, 7410, 3-input NAND: The specification sheet in Figure 10-10 shows that for a 7410, 3-input NAND gate, there are three in a package. Select U3 because U2 gets the flow of communication from one IC to another. With the engineer's permission, change the 04 (shown in the schematic in Figure 10-1) to a 3-input NAND by tying all three inputs and the spare (7410) together. The third gate has two inputs tied together for the 7410. Review the schematic in Figure 10-1 (the engineer's original input). This gate is shown as a 7400, 2-input NAND. By tying the third input on our spare 7410 together with one of the other used inputs, we can create the

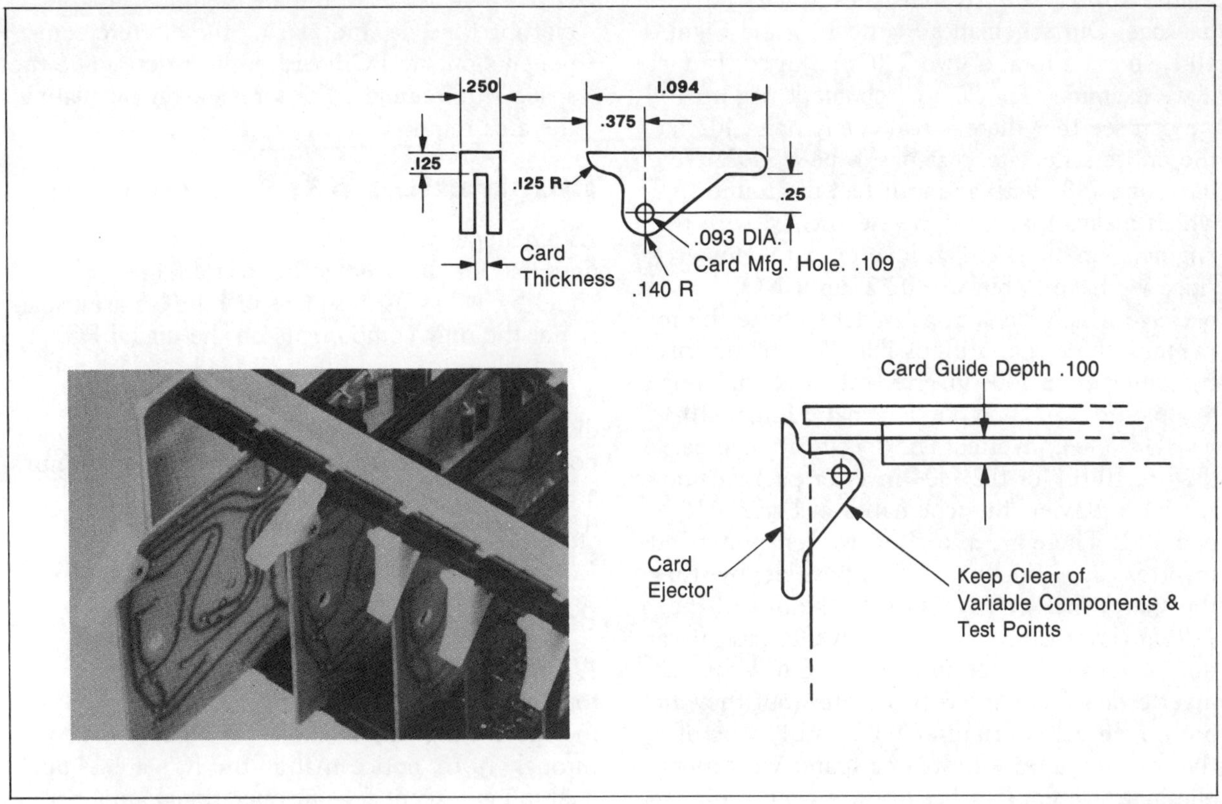

Figure 10-13. Sheet 2 of 2

same electrical function as the 2-input NAND. Therefore, U3 will have all three gates used, resulting in zero spares.

U4, 7432, 2-input OR gate: The specification sheet (Figure 10-10) shows a total of four gates in a 7432 package. Review Figures 10-3, and Figure 10-1. The engineer has three 2-input ORS shown. However, one of them has an input tied to ground on the schematic in Figure 10-1. This means that this does not have to be a 2-input OR. It can be a butter gate, non-open collector (Totem Pole). The assignment of U4 on the two 7432, 2-input OR gates (communicating to U3 is self explanatory. But notice the third OR gate. There are no totem pole buffer gates. Instead of adding another type of IC to the package, combine the OR gate shown with one input tied to ground into U4. Follow the engineer's instruction by tying both inputs together (as shown on the schematic in Figure 10-3). This gives U4 a total spare count of one.

U5, 7407, buffer gate with open collector: Use the specification sheets (Figure 10-10) for 7407. There are six buffers inside one package. Review the schematics in Figures 10-3 and 10-1. There are a total of four 7407 buffers with open collector gates. Assign U5 in the flow of U2, U3, and U4. Assign the remaining two buffers as U5. Because there is no other open collector on the schematic, this gives U5 a total spare count of two.

U6, 7410, 3-input NAND. Reviewing your specification sheet, Figure 10-10, we find that we have three 3 input NAND gates inside one package. Our schematics, Figure 10-3 and 10-1, shows three 3 input NANDS clustered together intercommunicating to themselves and to our connector, P1. It would only be logical to put all three of these inside one package called U6 with zero spaces.

U7: 7402, 2 input NOR. Explanation: Reviewing your specification sheets shown in

Figure 10-10, 7402 has four gates inside one package. Our schematics Figure 10-3 and Figure 10-1, shows a total of two 7402s required. In fact, if we examine very closely, schematic Figure 10-1, we can see that there is really only one 7402 on the entire schematic that has to be a 7402. We have one 7402 with an input tied to ground which means that it can be electrically, with the engineer's permission, an inverter gate. However, since we have to buy a 7402 2 input NOR package which has four gates, let's utilize the remaining three spares if possible. We will do this by calling U7 a 7404 inverter gate and changing our picture to match our schematic Fiugre 10-3.

U8, 7404, inverter: The specification sheet (Figure 10-10) for the 7404 inverter package shows six gates. Review the schematics in Figures 10-3 and 10-1. There are a total of two remaining 7404 inverter gates. Eliminate U8, if possible, by using the spares that we have. Can U4 be one of the 7404 inverter gates? No, U4 is an OR gate. It cannot be substituted for an inverter. Can U5 be an inverter gate? No, it has two spares (but they are both open collector buffers). U7 can be one of the inverter gates with its one spare. We cannot eliminate U8 entirely, because we would still need one inverter gate. Instead of taking the U7, label U8 a 7404 inverter package, using two out of the six. This results in four remaining spares.

U9 and U10, 748 operational amplifier: Use the specification sheet in Figure 10-8. Look at the dual inline P package. There is one operational amplifier in that package (an 8-pin package). Notice there is a positive and negative input with one output. Classify these as two separate IC's (U9 and U10).

Group and combine the schematic in Figure 10-5. We will not explain each grouping assignment for this schematic. A few points should be discussed. Maintain the flow of the schematic. U5 through U8 should all intercommunicate together. U4 and U9 should communicate to U8. The changes were made because you need to make your own decision about the plotting and flow of the layout.

SPECIAL NOTE: If you are satisfied with the grouping and combining completed on Figures 10-3 and 10-5, plot on the board outline the ten IC's that we assigned to the schematics.

Step #8—Calculate the area required. Before plotting the IC's and placing the discrete components on the PC board outline, determine the space requirements. For a review on calculating the area required, see chapter 5.

Illustration #8: Space Requirements

As a general rule, one square inch for each IC is desirable for the average PC board (However could be as low as .6 square inch.). IC's are usually not the only components on the digital PC board. However, use the typical 14- or 16-pin IC as a reference to the area required for these discrete components. Use Figure 10-14, the component equivalent chart. Calculate the minimum area required for the schematic in Figure 10-3. (Use the following procedure also for the schematic in Figure 10-5).

Procedures

1. On the schematic in Figure 10-3, count the total number of IC's that are 14 or 16-pins. Using the specification sheet shown in Figures 10-8 through 10-10, notice that all the IC's are 14-pin dual inline (except for our operational amplifiers, U9 and U10). The specification sheets in Figures 10-8 and 10-9 show that the operational amplifiers are 8-pin dual inline. Therefore, two 8-pin dual inline packages are equal to one 16-pin. On the component equivalent chart in Figure 10-14, list a total of 9 IC's in the IC count for 14- or 16-pin.

2. Refer to the schematic in Figure 10-3. Find the total number of disc type capacitors. There is a total count of seven bypass capacitors. Verify this on the formal layout in Figure 10-4. Place seven as the total count of disc type capacitors; this equals one square inch.

3. Refer to the schematic in Figure 10-3. Are there any large capacitors. There is one large filtering capacitor (C1). The component equivalent chart in Figure 10-14 should list one large capacitor equivalent to one quarter (¼) square inch.

COMPONENT COUNT (Equivalent 14/16 pin IC calculations)
Convert all components on the board to an equivalent 14/16 pin IC count using the following conversion factors:

	Total No. on Board		Equiv. 14/16 IC
14/16 pin IC	9	× 1.00 =	9
20 pin IC		× 1.25 =	
24 pin IC		× 1.50 =	
28 pin IC		× 2.00 =	
40 pin IC		× 3.5 =	
64 pin IC		× 6.5 =	
8 pin IC		× .75 =	
test points	3	÷ 10	
jumper pads		÷ 10 =	¼
Total IC's = 9¼			

	Total No. on Board		Equiv. 14/16 IC
1/4 -watt resistors		÷ 7 =	
1-watt resistors		÷ 4 =	
disc type caps	7	÷ 7 =	1
large caps	1	÷ 4 =	¼
3-pin components (Transistors, variable resistor, etc.)		÷ 3 =	
I/O connector pins used	29	÷ 10 =	3
Other components (list)			
Total Equivalent 14/16-pin IC =			4¼
Total Sq. Inch Req'd on P.C. Board =			13½

Figure 10-14. (For The Schematic In Figure 10-3)

4. Review the component equivalent chart in Figure 10-14. Are there any test points or jumper pads? There are three test points not only on the schematic in Figure 10-3, but also on the formal layout in Figure 10-4. List these three under test points. Ten equal one square inch. Therefore, three divided into ten is approximately ¼ square inch.

5. Review the component equivalent chart in Figure 10-14. There is an entry for pre-assigned connector pins. Figure 10-3, shows that P1 has a total of 29 pre-assigned pins (including all voltages and grounds). Divide 29 by ten to equal approximately three square inches.

6. Add the columns for the IC's. The total is 9¼ square inches. Add the column for the discrete components (the total equivalent IC's). The total is 4¼ square inches. Add the IC and discrete component columns together. The total required area for the PC board is 13½ square inches.

NOTE: Review the board outline (Figure 10-12). We calculated a total of sixteen square inches of useable area on the PC board. Compare that to the total square inches as a minimum requirement. The chart in Figure 10-14 indicates a minimum of 13½ square inches. We can easily place all the components, IC's capacitors, test points, etc. within the board outline, assigned in Figure 10-12. We can easily fit three more IC's (or equivalent IC's) on this board.

Illustration #9: Plotting

Plotting our grouped IC's (Figure 10-3) on the board outline in Figure 10-15 is the next step. Arrange the IC's in their proper location on the board according to the communication to each other, or in the same rows with each other. For a review on plotting, see chapter 5.

Plotting Procedure

Look at the schematic in Figure 10-3. Notice that the majority of gates have been grouped to make signal flow easier. The primary IC flow is with U2 through U5. The secondary signal flow is with U6 and U7. This flow determines the shape of the layout. The schematic in Figure 10-3 shows that U1, U6, and U2 have the majority of interconnections to the P1 connector. These three IC's should be close to the P1 connector. The flow among U2, U3, U4 and U5 should be unbroken. These IC's should be above, below, and beside each other. U6 and U7 should be close to each other to facilitate both signal flow and intercommunication between the two. U5 and U7 should be close to the operational amplifiers U9 and U10. This ensures relatively short interconnections on the inputs and outputs of these amplifiers.

How do we place the IC's in rows? The board is relatively square; it is four inches in width by 4.35 in length (not counting the connector). This means that we can place four IC's in the horizontal direction and four IC's in the vertical direction (for a total of 16 IC positions). There are ten IC's on the schematic in Figure 10-3. We would never use all sixteen positions for this schematic. We could use 12 positions by choosing a 4x3 combination as shown in Figures 10-15 and 10-16. If we chose Figure 10-15 with four IC's in the horizontal direction and three in the vertical, then we lose area in the vertical direction. The IC's would consume too much area in those aisles. If we chose Figure 10-16 (three IC's in the horizontal direction and four IC's in the vertical), then the aisles (communicating in the horizontal direction) become tighter. The ideal combination (for the schematic in Figure 10-3) is nine IC positions (see Figure 10-17). This is accomplished by using the operational amplifiers (with their 8 pins) in the area of one 16-pin pattern for both U numbers (U9 and U10). This becomes the ideal combination because it opens up the area in both the horizontal and the vertical directions. Use the IC pattern shown in Figure 10-17 on the layout.

Assigning U Numbers to Plotted IC Positions

We will use the schematic in Figure 10-3 and the plotted IC positions in Figure 10-17. Assign the U numbers for these positions, and explain the reasoning for the assignment. The first IC to place is U2. It intercommunicates to the connector and to the high pin numbers on the P1 connector. The flow is best directed downward through the board (U2 communicates to U3 communicates to U4 communicates to U5). See Figure 10-17. The flow is down through the board in the vertical direction and also across from U4 to U5 in the horizontal direction. Place U9 and U10 next to U5. Shorten the lead lengths for the input pins on the operational amplifiers. Place U6 and U7 down simultaneously. Keep U7 close to U10 because of the output pin on the operational amplifier, U10. Place U6 on the top row because it intercommunicates to the P1 connector, with low pin numbers. Place U7 below U6 because of the high intercommunication between the two shown on the schematic in Figure 10-3. The last two positions are filled with U1 and U8. U1 has many inputs and outputs to the P1 connector. U8 intercommunicates not only to U1 but to U5 and U9. This is our reasoning for the plotting of the schematic in Figure 10-3. We have given the same board outline in the same IC arrangement (shown in Figure 10-18). You can plot your own version of the schematic in Figure 10-5. Use the plotted IC locations shown in Figure 10-17. Proceed with the formal template layout (step 10).

Step #10

Illustration #10: Formal Layout

Use the IC plotted locations from Figure 10-17. Make a formal template layout. Establish the conductor paths in an X-Y direction. The layout should be viewed from the component side of the printed circuit board. The following procedures itemize the major design steps.

Procedures for the Formal Layout in Figure 10-4

1. Review the design rules in chapter 8, Table 10-1. Establish the conductor/trace, air gaps and widths, pad sizes, feedthroughs, tooling hole sizes, and card edge spacing minimums.

2. Choose an appropriate material to draw the layout. Use something similar to Accufilm (an anti-static stable base matte film with a grid underlay). Place the Accufilm on top of the grid. Use a grid pattern of .100. See the layout in Figure 10-4.

3. Use the information from the IC plot in Figure 10-17. Establish the IC pad locations on the formal template layout in Figure 10-4. Count the

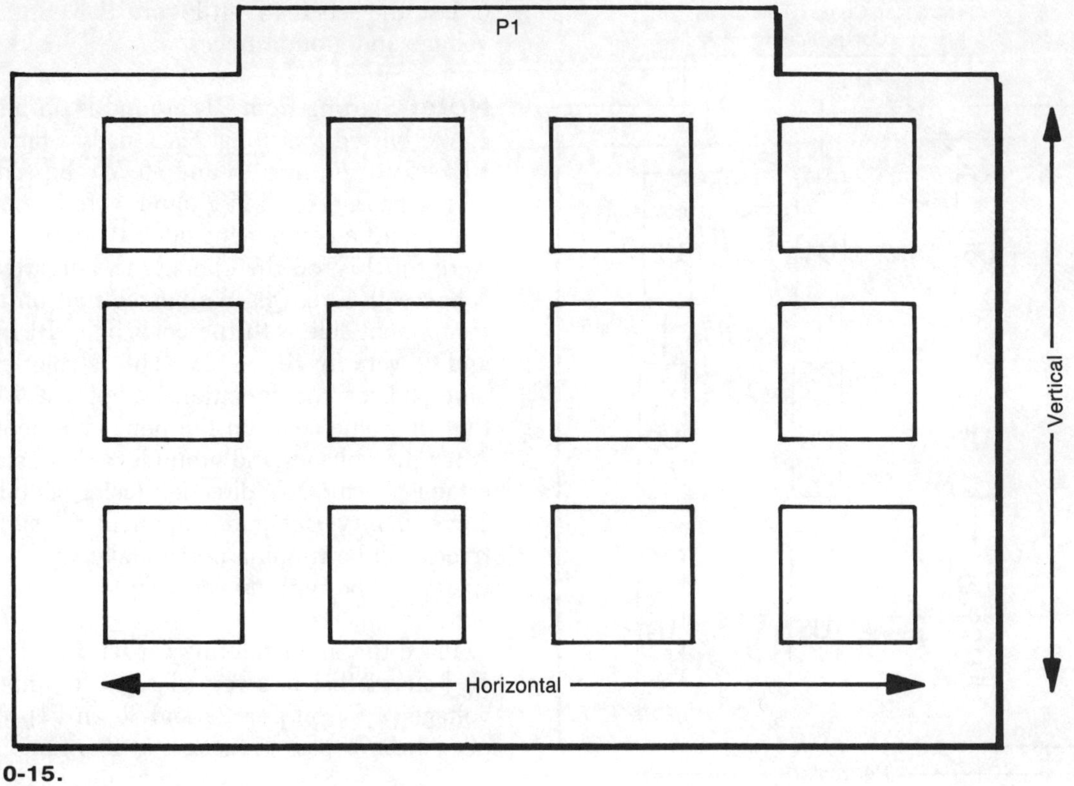

Figure 10-15.

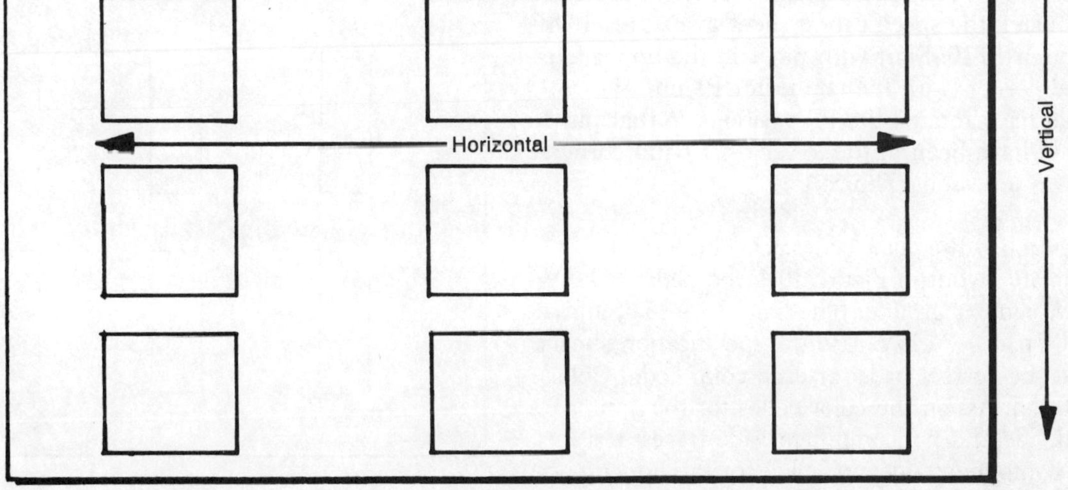

Figure 10-16.

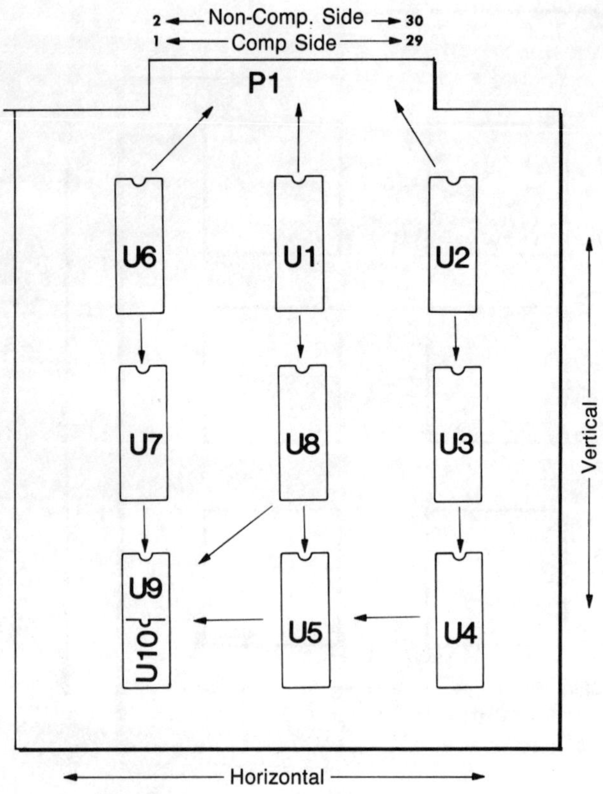

Figure 10-17.

grids. The IC's are spaced out in the horizontal and vertical directions.

4. Place the U numbers inside each IC pattern on the formal template layout. We obtained this from the plot in Figure 10-17. Number each IC pad using the specification sheet in Figures 10-8 through 10-10. Start with pin 1 in the upper left hand corner (with the connector P1 up). We could have rotated the IC positions so that pin 1 would have been in the lower right hand corner. This is a personal choice.

5. Establish the color codes. On the formal template layout in Figure 10-4, red pads = +5V; black pads = ground; blue pads = +15V, and a red ⊗ pad = −15V. Use the specification sheets. Find the correct pads for each color code. Color that pad. Assign the color codes for the component side and noncomponent side traces: red for the component side, and blue for the noncomponent side.

6. Use the schematic in Figure 10-3. Buss in the voltage and ground traces.

Note: Starting from P1, ground is on pins 1 and 2. We bussed that to all black pads. Starting from P1, +5V is on pins 29 and 30. We bussed that to all red pads. +5V and ground were bussed on the red side (the component side). P1 (pins 3 and 4) were for -15V on the operational amplifier (color coded with a red ⊗). We bussed that on the noncomponent side with the color blue. P1 (pins 5 and 6) were for the +15V. This is color coded in blue pads on the operational amplifier. We bussed that in a blue color on the noncomponent side. After the voltages and grounds were bussed, we established the X-Y direction (color code for the component versus noncomponent sides). Red traces will be running horizontally, and blue traces will be running vertically.

7. Place the input filtering capacitor, C1. Figure 10-4 shows that it is very close to the input voltage (+5V on pins 29 and 30 on P1). We also placed the bypass capacitors at this time. Refer to

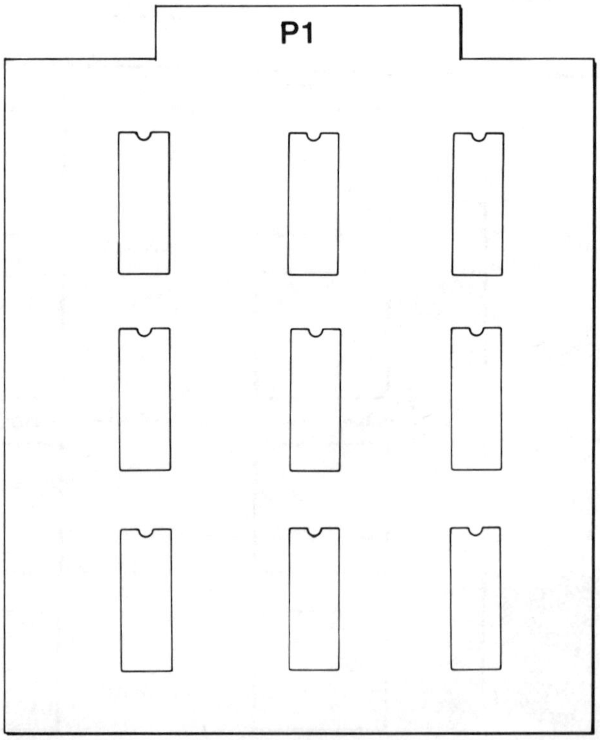

Figure 10-18. Student Layout Figure

the schematic in Figure 10-3, note 1. Place a bypass capacitor for every two IC packages. Notice on your formal template layout that we have placed a total of five IC bypass capacitors, which is one for every two. Connect them between the +5V bus line and the ground in the rows of IC's. We also placed the bypass capacitors for the operational amplifier for +15V and -15V. We connected them between the voltages (+15V to ground and -15V to ground).

8. Interconnect the signal lines. Route these traces in an X-Y direction. If routing in a vertical direction, use blue. If routing in a horizontal direction, use red. Choose U2 as the beginning point. Looking primarily at the major flow of the schematic: U2 to U3 to U4 to U5. Concentrate on the connections for the operational amplifier: U9 and U10. The schematic started with no input or output pin numbers assigned to the gates (as shown in Figure 10-1). Every line that was established as a signal trace had to be chosen by the designer. He/she used the specification sheets for each IC, and assigned the inputs and outputs pins to the schematic. As the designer chooses each pin, he assigns it to the schematic and yellows out the completed connection.

SPECIAL NOTE: The engineer allows us to route traces through the IC pattern for this board. When you are doing your own layout, you can route traces through the IC pattern. Also, another critical point on this board was the interconnection from P1 (which had pre-assigned pins) to its respective IC's. If you are not careful, you could create enormous "rat's nests" of interconnects in the first aisle close to P1 on the PC board layout. On the first two boards of this book, we analyzed each connection on the schematic. Trace back using the schematic in Figure 10-3. Compare it against the layout in Figure 10-4 (for a reference to each connection made on the board layout).

9. Review the schematic in Figure 10-3 for possible misconnections. Finish routing the test points: TP1, TP2, and TP3. Add the finishing touches to our board layout by placing tooling holes and any special mounting configurations such as the card ejector.

10. This is the final review of the schematic in Figure 10-3. Draft in all the spare gates. Assign the input and output pin numbers and their U numbers. The schematic in Figure 10-3 should have every signal line and every voltage line yellowed out (signifying a completed connection). Every input and output pin number on the gates should be assigned.

The clean up phase is covered in Step 11.

Step #11—Complete and clean up the layout in Figure 10-4. This step is used to clean up all unnecessary via holes (called feedthroughs). In the formal layout we place feedthroughs because the X-Y system is used to route traces. Remove unnecessary vias. Place all IC's in order, left-to-right, top-to-bottom or any practical alphanumeric order. Update the layout (Figure 10-4) and schematic (Figure 10-3). See illustration 11.

Illustration #11: Layout Clean-up

Figure 10-19 shows the formal layout in Figure 10-4 with the unnecessary feedthroughs crossed out in red. This alternate routing removes the feedthrough. Place the IC's in numerical order as shown in Figure 10-20. Instead of going left-to-right, top-to-bottom, place the U numbers in a top-to-bottom, left-to-right order. IC U numbers have now been assigned: U1 replaces U6, U2 replaces U7, U3 replaces U10, etc. You would transpose this onto a new schematic (which we did not do on this board). You now understand what to do. It is now time to complete your own layout using the schematic in Figure 10-5. Use the schematic in Figure 10-3 and the formal layout in Figure 10-4 as a reference ONLY.

Step #12—After completion of the layout drawing, an artmaster is taped. The artmaster is made to an enlarged scale (2:1). It is photo reduced to provide a 1:1 scale negative and positive. (The photomaster is a tool for manufacturing the PC board). The 10-IC digital/analog board in Figure 10-1 is taped using a red and blue method. For a step by step procedure, see illustration 12.

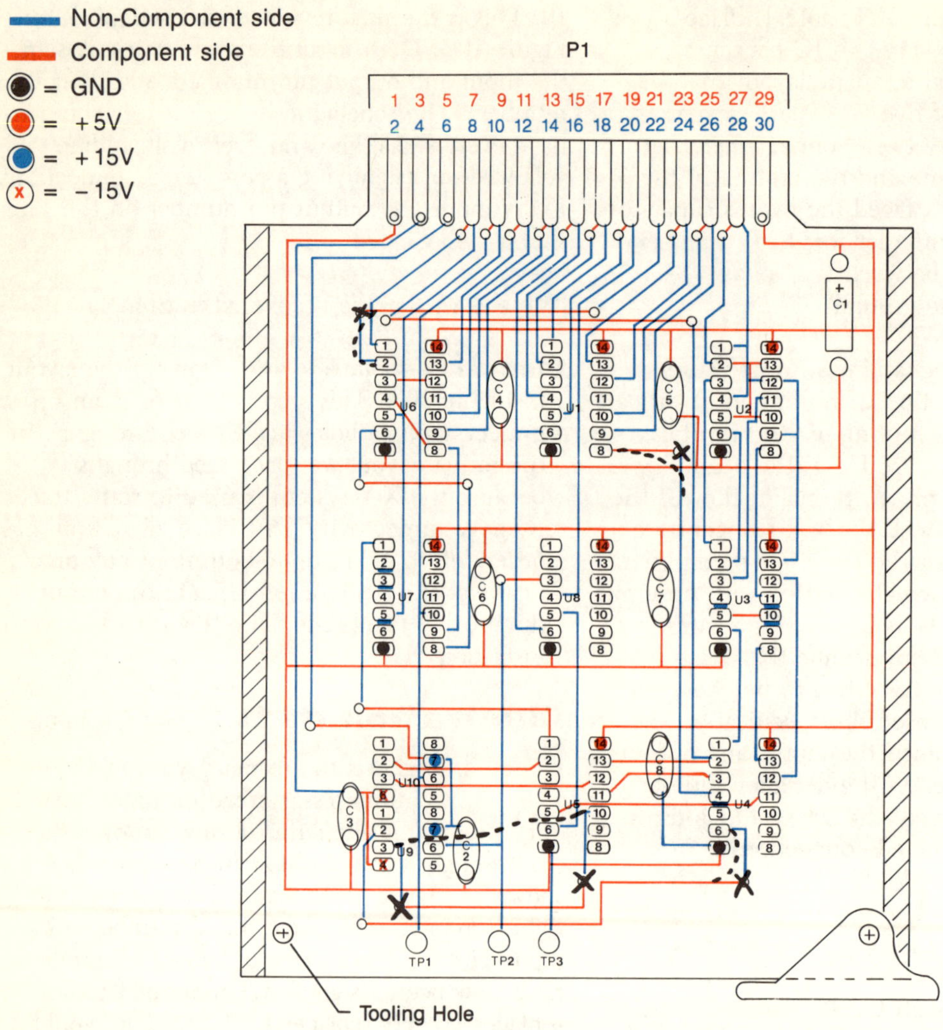

Figure 10-19.

Illustration #12: Artmaster

An artmaster tape up is the next step in the board layout sequence. Use the tape up technique of the one sheet, red and blue method (for the formal layout in Figure 10-4). This method is also shown in chapter 6. The red and blue taping technique provides tapes in widths, the same as the black crepe demonstrated in chapter 9, except that they can be taped on one sheet of polyester film. Chapter 9 showed two sheets, using a black crepe tape. Red is on one side, and blue is on the other side. The pads would be black crepe placed only on one side. The 2-sheet method places the black pads on both sheets. The photographer can use red and blue filters in his camera. He/she filters out the blue tape for photographing one side of the board, and filters out the red tape when photographing the opposite side of the board. The black pads show up for the blue side when photographing that side. The black pads show up for the red side when photographing the red side. The end result is an accurate registration, from the component side of the board pads to the noncomponent side of the board pads. One pad is used for both sides. This system has been in use for a number of years. This system was invented to replace the two-sheet technique shown in chapter 9. Registering the black pads for both the component side and the noncomponent side provide nearly perfect registration.

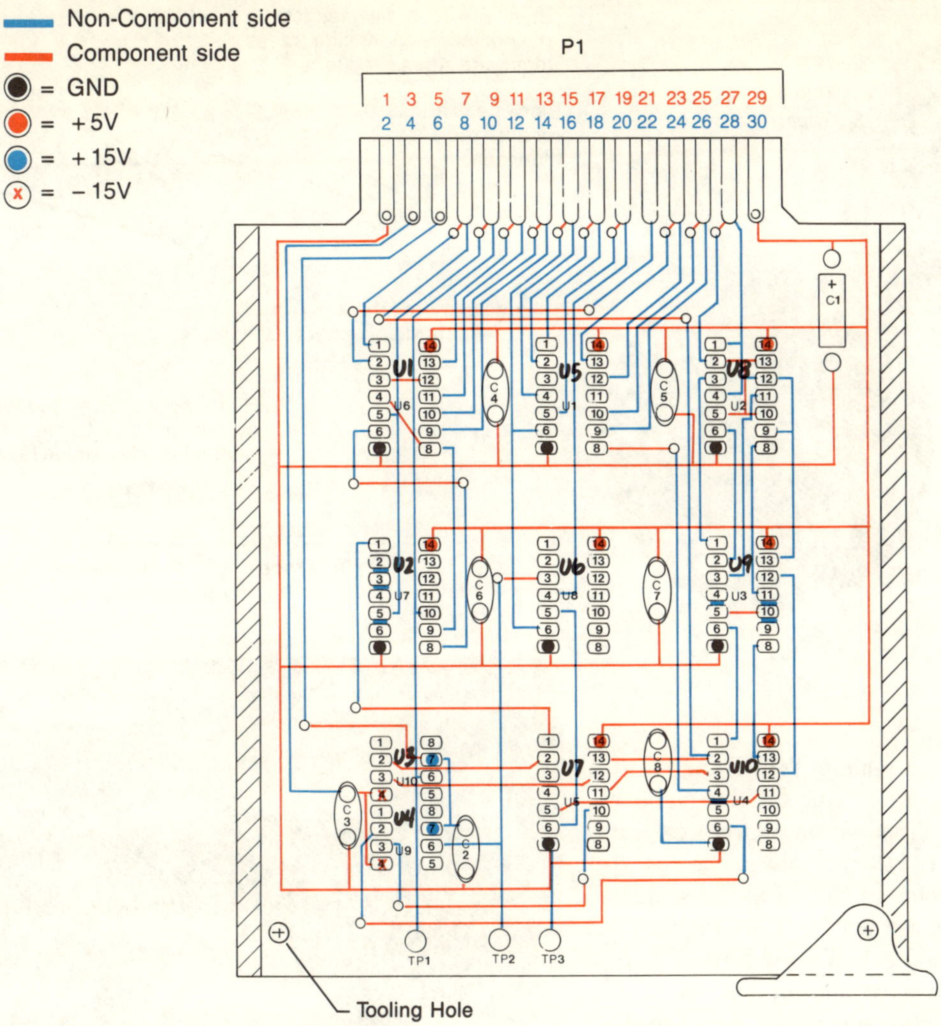

Figure 10-20.

This is the biggest advantage of the red and blue technique. One other advantage is the edge line definition on traces and pads. If you were to magnify the black crepe tape, you might see a saw tooth like trace edge. The red and blue technique gives a much smoother looking edge on the trace. The example in Figure 10-21 is highly magnified.

The red and blue technique is being replaced by an even new system. This is called the three-sheet, or the three layer technique. It is important to understand the evolution of the techniques used today. You have learned the two-sheet method in chapter 9, and the red and blue technique in this chapter. The three-sheet technique will be shown in chapter 11. Figure 10-22 shows the artmaster technique called the one sheet red and blue. One piece of polyester film representing one side uses red traces (perhaps the component side). And the opposite side uses blue traces. The black pads are placed on one, (perhaps the red side), with the connector fingers in crop marks all being black. This is only on one side. All markings that should appear on both red and blue surfaces would be placed in black. The camera will see black and red on one photographic shot. When the filter is inserted it would see blue and black for the next photographic sequence. A tape up using this technique is shown in the layout of Figure 10-4 (the clean up stage).

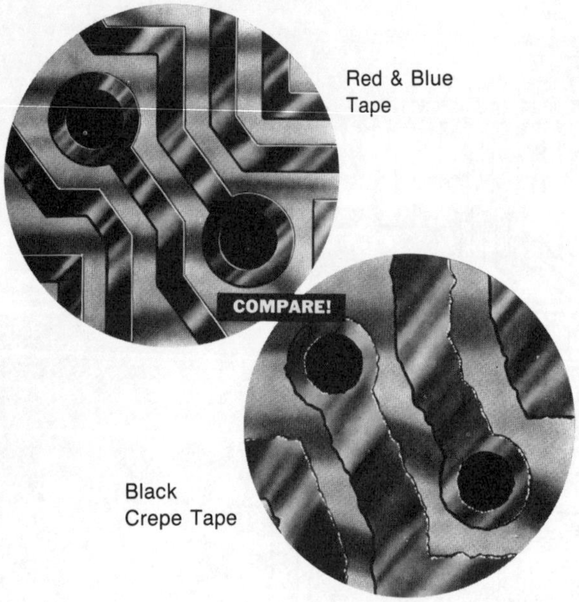

Figure 10-21. Black tape vs. red & blue.

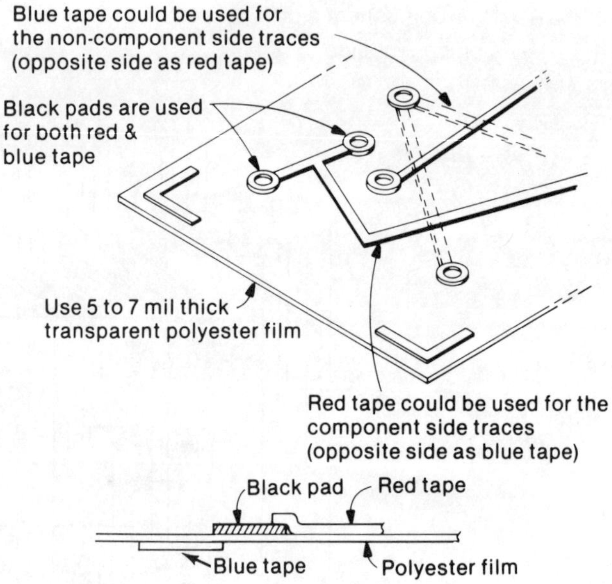

This layout is done at a scale of 2:1. The artmaster that we will be producing for this layout is a tape up (or a tool). Prepared on a large scale, it can be either 2:1 or 4:1. It must always be the same scale as the layout. In this case, the scale is 2:1. The material that we will be using for our tape up is a polyester sheet. We will have a minimum of .005 to a maximum of .007 thick stable base polyester film. Before analyzing the one sheet red and blue technique shown in Figure 10-22 for our layout Figure 10-4, use the following standards. Routing traces should have all turns at 90 degrees or 45 degrees. Because this type of tape does not bend, the free flow style of taping can not be used. Review chapter 6 before taping the PC board. The following is the tape up procedure of the one sheet red and blue technique.

1. Use one sheet of film, usually a clear polyester. Place black pads and all common features that would be registered to both sides. Place all interconnections or circuits paths for red traces to be applied to one side of the clear polyester film. Apply circuitry for the opposite side of the board (noncomponent side), using blue tape. A colored filter is used. The circuit artwork is photographically separated and reduced with the same camera setting under identical environmental conditions. The result is two separate master artwork patterns with a common black feature such as a terminal area (called pad). The reduction targets, connector fingers, and feedthrough pads will be in nearly perfect registration. Cut one sheet of polyester film to overlap the layout. Leave enough room on the outer perimeters of the board outline for title blocks, reduction targets, and tape up identification.

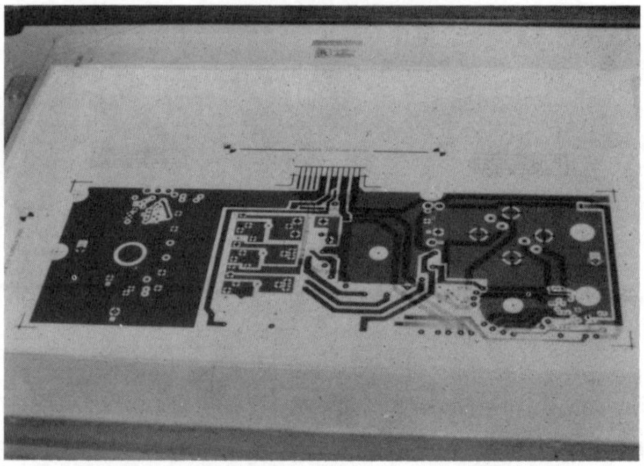

Figure 10-22. One Sheet Red & Blue Method

2. After the grid is placed down, the layout Figure 10-4 is registered to that grid and taped down over it. Overlay the polyester sheet on top

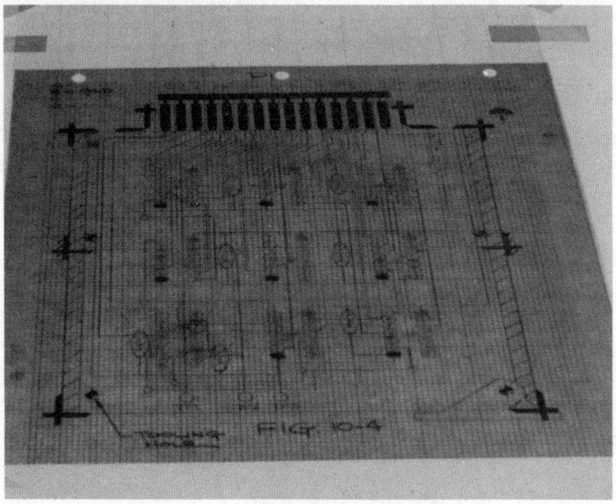

Figure 10-23. Layout Tape-Up Red and Blue

of the layout. Tape this down very securely. See Figure 10-23.

SPECIAL NOTE: Do not tape directly on a blue or a solid black line grid: the lines cannot be eliminated while photographically retaining the blue connector traces. To ensure consistently high quality results, use only one manufacturer's red and blue drafting tape. Do not mix one manufacturer with another. For results, apply red tape traces to one side of the film and the blue tape to the opposite side. This facilitates changes and corrections by not requiring the removal of one color tape to access the opposite color.

3. Make the first taping on this one sheet of polyester film on top of the layout in Figure 10-4. It becomes the board outline or crop marks. Use a .100 wide tape. Place the crop marks (Figure 10-24) only in the corners of the PC board. Do the same for the outline of the P1 connector tongue. Make the black tooling holes. There are two of them on our board outline (Figure 10-4).

4. Place all pads for the components on the side on which we just placed the crop marks. This is the board outline. The pad sizes can be found in Table 10-1.

5. Interconnect or route the traces on the component side. The layout in Figure 10-4, shows all the red traces representing the component side of the board. Use .030 tape at a 2:1 scale for the signal paths. (This is because we are routing through the IC pins). The voltage and ground traces (which are the color coded pads) have .100 tape at a scale of 2:1 in red also. These two different widths are the correct sizes for voltage, ground, and signal paths. Complete all of the red traces on the layout in Figure 10-4. Use red tape on the existing tape up board outline. See Figure 10-25.

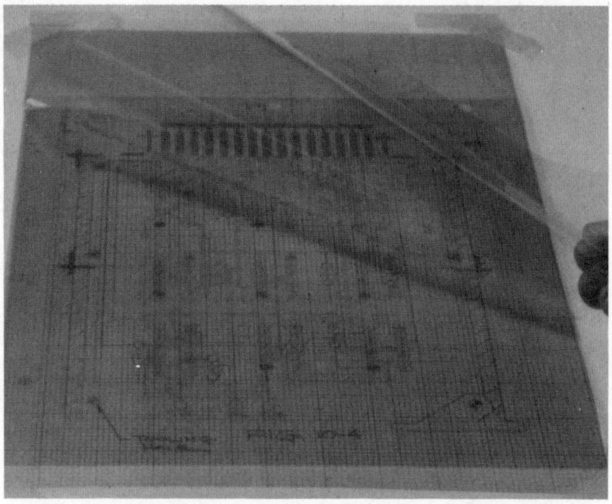

Figure 10-24. Board Outline, Connector, Tooling Holes

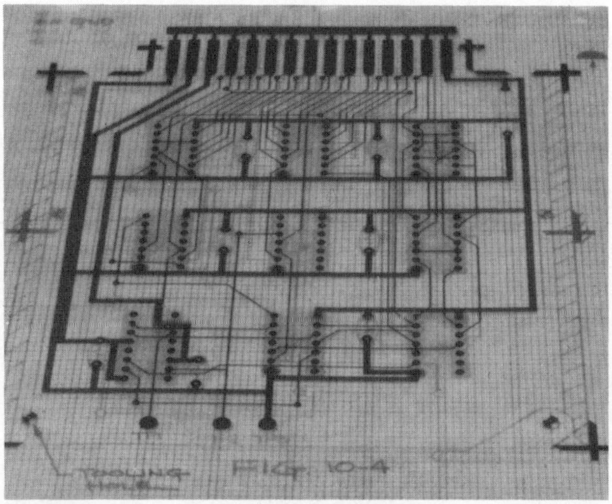

Figure 10-25. Comp Side, Red Traces

6. Remove the tape up sheet from the top of the layout in Figure 10-4. Remove the layout from the grid. Turn over the layout so that the compo-

nent side view is facing down on the grid. Reregister the layout to the grid, and tape it down. Turn over the component side tape up. Place it down on top of the layout so that the noncomponent side of the board is now facing up. Tape the blue traces for the noncomponent side (using blue tape) on the polyester film. Signal traces will be .030 at a scale of 2:1; the voltage and ground traces will be .100 at a scale of 2:1. See Figure 10-26 for an example of a finished noncomponent side tape up (using blue tape) for the layout in Figure 10-4.

7. This is the component designation and marking for the PC board. Identify the location of all the components on the schematic for the layout. Mark the identifying points and reference designations on the component side of the PC board. Etch the designation and markings across the surface (like a trace or pad). Another technique is to silkscreen these markings for the component designations on the surface of the PC board. See Figure 7-12. Silkscreening is the painting of an image on the surface using a type of epoxy paint that is very strong and adherent. On the red and blue artwork, place the reference designations or markings for board identification or part numbers. (Stick-ons are purchased from a drafting supply store). Stick-ons must be the correct color: red for the component side, and blue for the noncomponent side. See the layout in Figure 10-4. These stick-ons are produced from pressure-sensitive or vinyl tape. They come in a variety of sizes and colors. The two-sheet method in chapter 9 would use the opaque black pressure-sensitive stick-ons. Mark these components for Figure 10-4: for the positive end of the input filtering capacitor (C1), the test points (TP1, TP2, and TP3), all IC's (U1 through U10), the pin numbers on P1 (particularly pins 1, 2, 29, and 30) and the bypass capacitors (C2 through C8). This depends on company requirements. Marking requirements vary from company to company. Some companies only require major components like IC's, variable resistors, etc. Other companies might require that all components be identified and marked. The most important identification is the test points or parts that the engineer or technician needs to test adjust the board. See Figure 10-27. An alternate technique is to silkscreen the component outline and designa-

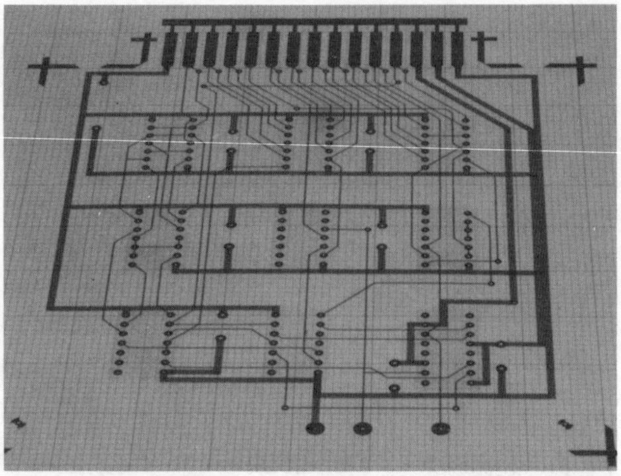

Figure 10-26. Finished Non Component Side Tape Up (Using Blue Tape)

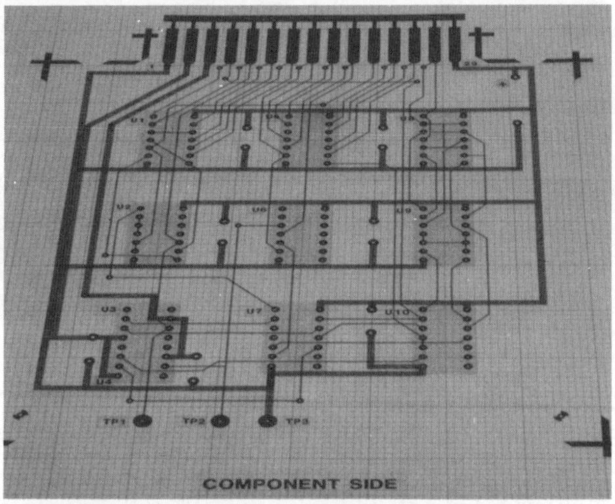

Figure 10-27A. Finished Component Side Artwork. (Using Red Tape)

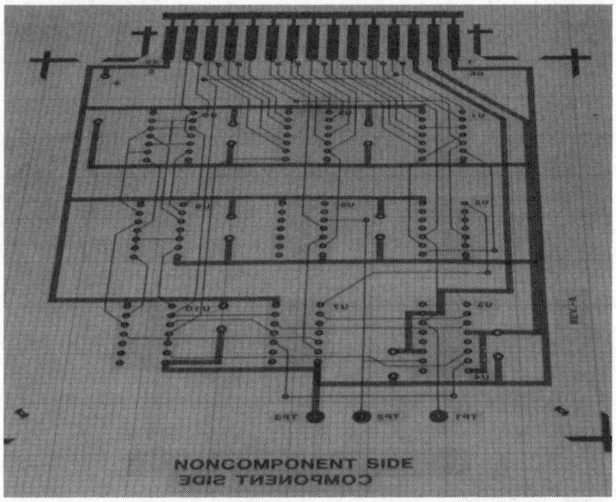

Figure 10-27B. Finished Non-Component Side Artwork (Using Red Tape)

tions on the component side, the noncomponent side, or both. For examples of this technique, see chapter 7.

8. The reduction targets and title blocks are the last steps on the tape-up procedure. They are primarily used by the photographer. Tape the 2:1 tape up. Reduce it to a 1:1 film positive and negative. Tell the photographer how much to reduce the artwork by. In chapter 6, we discussed reduction targets. The larger your artwork is, the larger your reduction target should be. Figure 10-4 shows twelve inch reduction targets. The twelve inches would be reduced to six inches at a 1:1 scale, (as shown in Figure 10-28). We have reduction targets in both the X and Y directions. This is used to check that the film represents a properly reduced artwork, in both horizontal and

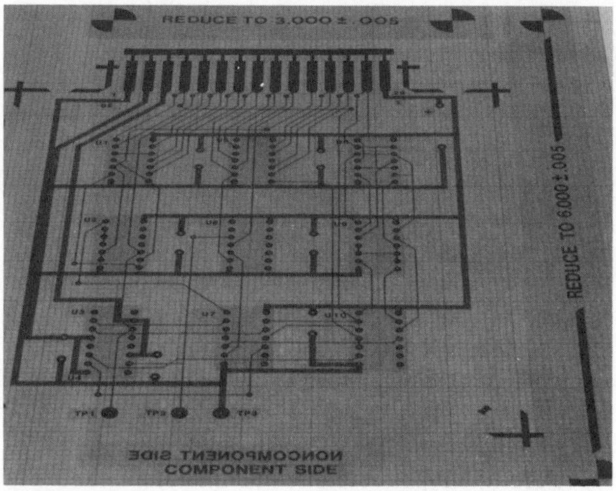

Figure 10-28. Reduction Targets

vertical directions. The title block varies from company to company. Place one title block on each side. Typically, the company's name and address, the part number, and revision level of the PC board is placed in the title block. You are ready to send the 2:1 red and blue artwork to the photographer, request a reduction, and receive one set of positives and one set of negatives. They are used to fabricate the printed circuit board. See Figure 10-29.

Figure 10-29. Example of Pos. & Negs.

SPECIAL NOTE: Tape your own version of the schematic layout in Figure 10-5. Use the red and blue technique. When you are taping the layout, you can move rows of IC's, individual discrete components, and traces. You are not locked into the exact location shown in Figure 10-5.

Step #13—The material that we have used for the first board (3-IC digital board), and the second board (6-IC digital board), and the third board (the 10-IC digital/analog board), becomes the foundation for the remaining boards. At the end of this chapter, there is a summary of all the material that has been added to this foundation.

Summary

Compare the schematics from chapters 8 and 9 to chapter 10. There are more gates and a mixture of open collectors and non-open collectors. We presented the operational amplifier (covered extensively in the analog design book). The operational amplifier requires short lead lengths on the input and output pins. It affects the plotting and grouping of gates on the schematic.

The grouping of the gates was more difficult in this chapter because there were more of them. Specification sheets identified the hole diameter, pad sizes, body case outlines for all the components. We used Table 10-1. We discussed the mechanical interface of this board in much greater detail than in the previous two chapters. We introduced the card ejector.

It is important for the student to layout his/her own version of the 10-IC digital/analog schematic. We grouped, plotted, and layed out a version in the schematic in Figure 10-3. Used as a reference, the student did his/her own version shown in the schematic in Figure 10-5. We did group the schematic in Figure 10-5 to show another grouping combination. You should route the traces through the IC pads. On high density PC boards, this technique would be used more often than not.

We introduced the red and blue method of taping (the single sheet technique). This new technique achieves nearly perfect registration on artwork. Because the one piece of mylar uses the red and blue tape against the black pads, we can filter out the red or the blue while keeping the black for both red and blue sides. This technique is being replaced by a system called the 3-sheet technique (to be demonstrated in chapter 11). But many companies still use the red and blue technique. However, the 3-sheet system has some advantages over the red and blue. These will be itemized in chapter 11.

We combined some of the techniques learned in chapters 8 and 9. We introduced new components and procedures required for the layout and design of digital PC boards. Chapter 11 builds this knowledge to design and layout digital PC boards.

Quiz—Chapter 10

1. This chapter introduces a new component called? ________
2. This P.C. designer is responsible to insure that the schematic has the correct symbols, REF designations & identification. T F
3. This P.C. board design is a digital/analog design because: ____ ________ ________
4. CI input filtering capacitor (polarized) name the composition the component is made of: ________
5. CI input filtering capacitor: using FIG 10-6 find the max lead size? ________
6. C2 & C3 bypass capacitors (non-polarized), name the composition the components are made of: ________
7. The number of bypass capacitors per IC is determinded by the engineer, however the total number of capacitors are not know until:? ________
8. Operational amplifier: using the spec sheet figure 10-8 & 10-9 name the package type we are using on our board layout? ________
9. Schematic figure 10-1 has seven different types of IC spec numbers. Which means that you have a total of seven IC's. T F
10. Schematic figure 10-1 has ________ collector gate, name it? ________
11. For this chapter, the formulas to calculate the hole and pad size are taken from:? ________
12. Component mounting table 10-1 column 8. This column is very important in that the leads must be:? ________
13. Under questions to ask the engineer, we found that all of our components fit a standard case size on our template. T F
14. By pass capacitors: the by-pass capacitors used for the op-amps are for the + & − input pins 2 and 3. T F

15. Operational amplifiers typically would like to have there inputs and outputs relatively close. **T F**
16. The input filtering capacitor is placed to the connector & very close to the +5V pins. **T F**
17. Spare IC gates: for schematic figure 10-3 the engineer does not want any spare gates left over. **T F**
18. It would be correct to assume that most P.C. board's would not have pre-assigned pins. **T F**
19. Test points: just because the schematic figure 10-3, does not show test points, doesn't mean that the engineer does not require them. **T F**
20. The board outline is usally provided by the mechanical designer. **T F**
21. In order to calculate the total number of usable square inches, the designer must define:? ________ ____________________________
22. P.C. board ejector's & card guides typically do not restrict component traces. **T F**
23. The connector specification sheet is very important in setting up the board outline, it will determine the size & number of contacts, & the ______________ of each contact.
24. Grouping: the designer should combine to find the lowest number of IC's and the shortest possible trace connection at the same time. **T F**
25. Grouping: Before grouping & combining the designer must identify the logic using the: ____________ ____________________________
26. Lead Length: digital and analog board's typically have long leads communicating from one IC to the other or component to component. **T F**
27. When "tie off" is required on combining gates, the designer should seek the engineer's permission. **T F**
28. The P.C. designer should change the symbol on the schematic to match the combining changes. **T F**
29. Using the specification sheet the designer can eliminate additional IC's by using the spares the he/she has accumulated. **T F**
30. All digital boards require a minumum of one square inch per I.C. **T F**
31. When caluculating the area required, the designer needs only to add-up the total number on IC's. **T F**
32. Plotting: One of the primary consideraitons in plotting is which IC's interconnect to the connector. **T F**
33. Plotting: The designer should try to assign IC locations so that IC's are communicating in a horizontal or vertical direction X or Y. **T F**
34. Plotting: After the plot is finished the designer will complete a free hand sketch before starting a formal layout. **T F**
35. Formal Layout: The designer must establish the conductor size, air gaps, and pad size before proceeding. **T F**
36. On digital boards color codes are used to identify ______________ and ________________ pads.
37. Unlike analog boards, digital boards have ____________________ and ________________ traces bussed first.
38. Most digital boards establish the X-Y direction color for traces with the bussing of ________________ and ________________________.
39. Review the schematic: it is the designers responsibility to insure that every input and output pin number and their U numbers have been assigned. **T F**
40. Formal layout: One of the last steps is a clean-up. This step is used to clean up all unnecessary ________________ holes.
41. Artmaster: Red and blue tape up technique is the same as black crepe technique except it can be taped on _______ sheet of polyester film.

42. Artmaster: If red & blue is used for traces, which color is used for the pads:? ______________
43. Photographer: When photographing the red side of the board, the photgrapher must use a ______________ to drop out the blue traces.
44. Photographer: The black pads shows up for only blue or red side when photographing that side. T F
45. The black pads for both the component side and the non-component side provide nearly perfect ______________.
46. One of the advantages of red and blue over black crepe tape is the ______________ definition on trace and pads.
47. Red & blue tape: Routed traces should have all turns at 90 degrees or 45 degrees because this type of tape does not bend. T F
48. Red & blue tape: Use one sheet of film pads and all traces should be on the same side. T F
49. Red & blue tape: To ensure consisitently high quality results, use only one manufacturer's red and blue tape. T F
50. Red & blue board identification: Markings for the board is accomplished using stick-ons, and must be the correct color which is: ______________

The answers for each question can be found on the pages listed below:

Question 1. Pg. 308	Question 26. Pg. 325
Question 2. Pg. 308	Question 27. Pg. 325
Question 3. Pg. 308	Question 28. Pg. 328
Question 4. Pg. 309, 313	Question 29. Pg. 328
Question 5. Pg. 313	Question 30. Pg. 328
Question 6. Pg. 309	Question 31. Pg. 329
Question 7. Pg. 309	Question 32. Pg. 329
Question 8. Pg. 316, 317	Question 33. Pg. 330
Question 9. Pg. 317	Question 34. Pg. 330
Question 10. Pg. 317	Question 35. Pg. 330
Question 11. Pg. 320	Question 36. Pg. 332
Question 12. Pg. 322	Question 37. Pg. 332
Question 13. Pg. 322	Question 38. Pg. 332
Question 14. Pg. 323	Question 39. Pg. 333
Question 15. Pg. 323	Question 40. Pg. 333
Question 16. Pg. 323	Question 41. Pg. 334
Question 17. Pg. 323	Question 42. Pg. 334
Question 18. Pg. 323	Question 43. Pg. 334
Question 19. Pg. 323	Question 44. Pg. 334
Question 20. Pg. 324	Question 45. Pg. 334
Question 21. Pg. 324	Question 46. Pg. 335
Question 22. Pg. 324	Question 47. Pg. 336
Question 23. Pg. 325	Question 48. Pg. 336
Question 24. Pg. 325	Question 49. Pg. 337
Question 25. Pg. 325	Question 50. Pg. 338

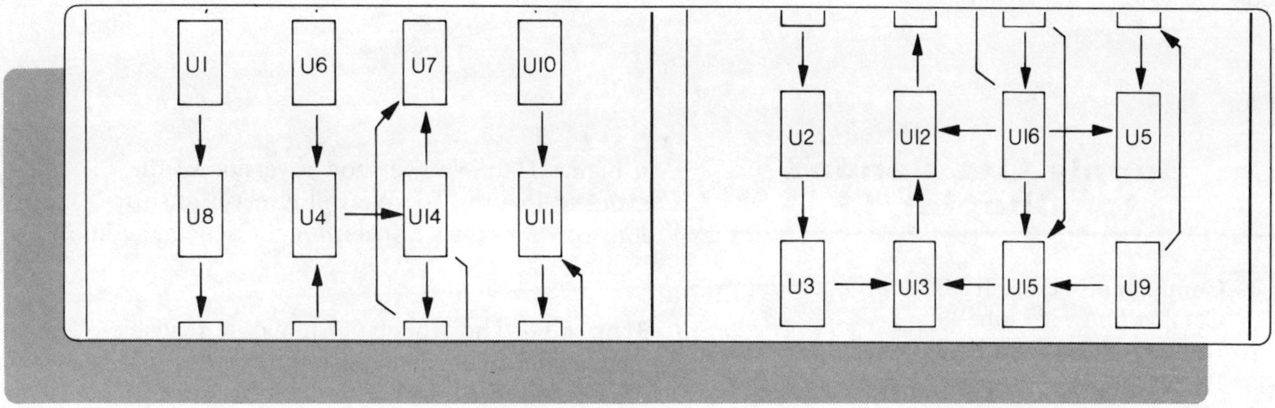

11

16/15-IC DIGITAL BOARD

This chapter illustrates the design of a 16-IC digital printed circuit board, using the X-Y system on a two-layer, double-sided board. The schematic and layout not only introduces new features, but also encourages the student to try his/her own version of the same schematic.

The layout and the student's layout use the same schematic. We will lay it out while using a grouping of the 16-IC's shown in the schematic in Figure 11-1. We will introduce a new taping technique called a three-sheet method for this schematic. You will layout the same schematic, but will group using the 15 IC's shown in the schematic in Figure 11-2. Following the same steps and design rules, you will layout your own version using the three-sheet tape up.

We have grouped, plotted, and drafted a version called the 16-IC layout. There is a second version where we have grouped the schematic and assigned the board outline. We have even placed the IC patterns within a board outline. Your job is to plot the IC's and formally lay out your own version of the schematic. By laying out and showing you the completed drawing (the 16-IC layout), refer to the layout as you design. Follow the same itemized steps that we have used. Observe the same restrictions that the layout has. Ask the same questions of the engineer. The design rules are the same.

Supply List Needed for Chapter 11

1. Component Template: Bishop Graphics Catalog No. 3367 (one)
2. Black Drafting Film Lead Pencil: Mars 100-30 Duralar (one)
3. Colored Pencils (dark colors): Venus Col-erase, or equivalent (red, green, yellow, blue) one each
4. Eraser, Drafting Film: Castell #1954 or equivalent (one)
5. Drafting Dots
6. Knife: X-acto, Bishop Graphics Catalog No. 3141
7. Knife Blades: X-acto, Bishop Graphics Catalog No. 3116
8. Donut Pads: .125", 100, Bishop Graphics Catalog No. D102*
9. Donut Pads: .150", 25, Bishop Graphics Catalog No. D144*
10. Donut Pads: .250", 6, Bishop Graphics Catalog No. D141*
11. Precision Slit Tape: Black Crepe .031", 2 rolls Bishop Graphics Catalog No. 20103111
12. Precision Slit Tape: Black Crepe .100, 1 roll, Bishop Graphics Catalog No. 20110011
13. Connector Pattern: .156" C-C, 15 Fingers, Bishop Graphics Catalog No. 6720*
14. Reduction Mark: 4, Bishop Graphics Catalog No.4052*
15. Datum Target (tooling holes): 2, Bishop Graphics Catalog No.4011*
16. Prepunched Drafting ACCUFILM® : matte one side, .003", 11" x 17", 2 sheets
17. Prepunched Drafting ACCUFILM® : clear taping film, .005", 11" x 17", 3 sheets
18. Dual In Line Pattern: 14-lead, oval pads, 15, Bishop Graphics Catalog No. 6242*
19. Dual In Line Pattern: 16-lead, oval pads, 2, Bishop Graphics Catalog No. 6215*
20. Pin Bar: "A" Size, one

* Bishop catalog number shown for these products represent order numbers for Bishop's standard industrially packaged products. The package quantities for these catalog numbers may be greater than supplied with the "Digital" project kit, #10009C, available from Bishop.

All the steps in chapters 8, 9, and 10 apply to this 16/15-IC digital schematic. We will analyze each step required for the layout of the schematic in Figure 11-1. We will show a version of the layout in Figure 11-3. We will give you instructions of your layout version for the schematic in Figure 11-2.

Step #1—The schematic should be readable and understood: symbols, reference designations, and identification of all logic gates used by the engineer. Review the schematic in Figure 11-1 and your own schematic in Figure 11-2. Make sure that all symbols and markings are understood.

Step #2—Specification sheets identify the case and body size, width, and length of all discrete components. They identify all logic gates or symbols. The engineer typically provides the specification sheets. All discrete components for the schematic in Figures 11-1 and 11-2 are in the notes (listed as a case code number). Usually you would use the specification sheets to find case code numbers. We will provide all necessary information for the discrete components. They will be itemized in the design from schematic Figure 11-1. The IC specification sheets are required to lay out the schematic in Figure 11-2. The IC specifications are shown in Figures 11-4 and 11-5.

Step #3—Determine the component lead spacing and body size. The spacing of all discrete component pads is the location where the leads insert into the boards. This information comes from the specification sheet, and can be reviewed in chapter 3 of the analog design book. Determine the IC body size (such as 14-pin, 16-pin, or 24-pin, etc). Identify each component on the schematic, Figures 11-1 and 11-2. All of this information can be found on Table 11-1. The discrete components, C1 through C6, R1, and R2 are listed on Table 11-1 which gives the case codes, finished hole sizes, and pad sizes. To verify the information in Table 11-2, check the specification sheets in the previous chapters. The IC's, U1 through U16, are 14 or 16-pin dual inline packages. For gate identification, see Figures 11-4, and 11-5. This information has been provided for this layout. In the normal course of business, it is your responsibility to determine all of this information.

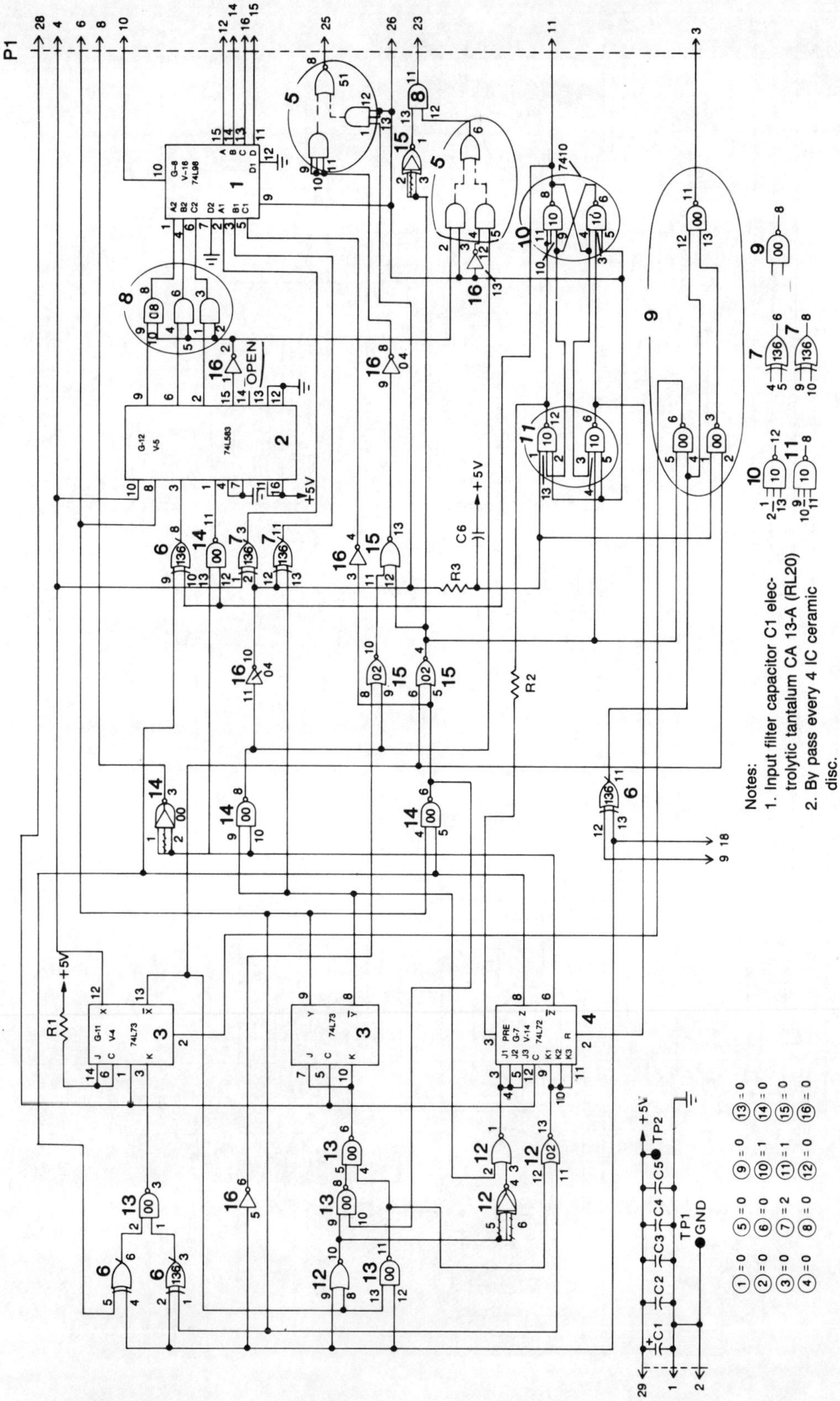

Figure 11-1. 16 IC Grouping

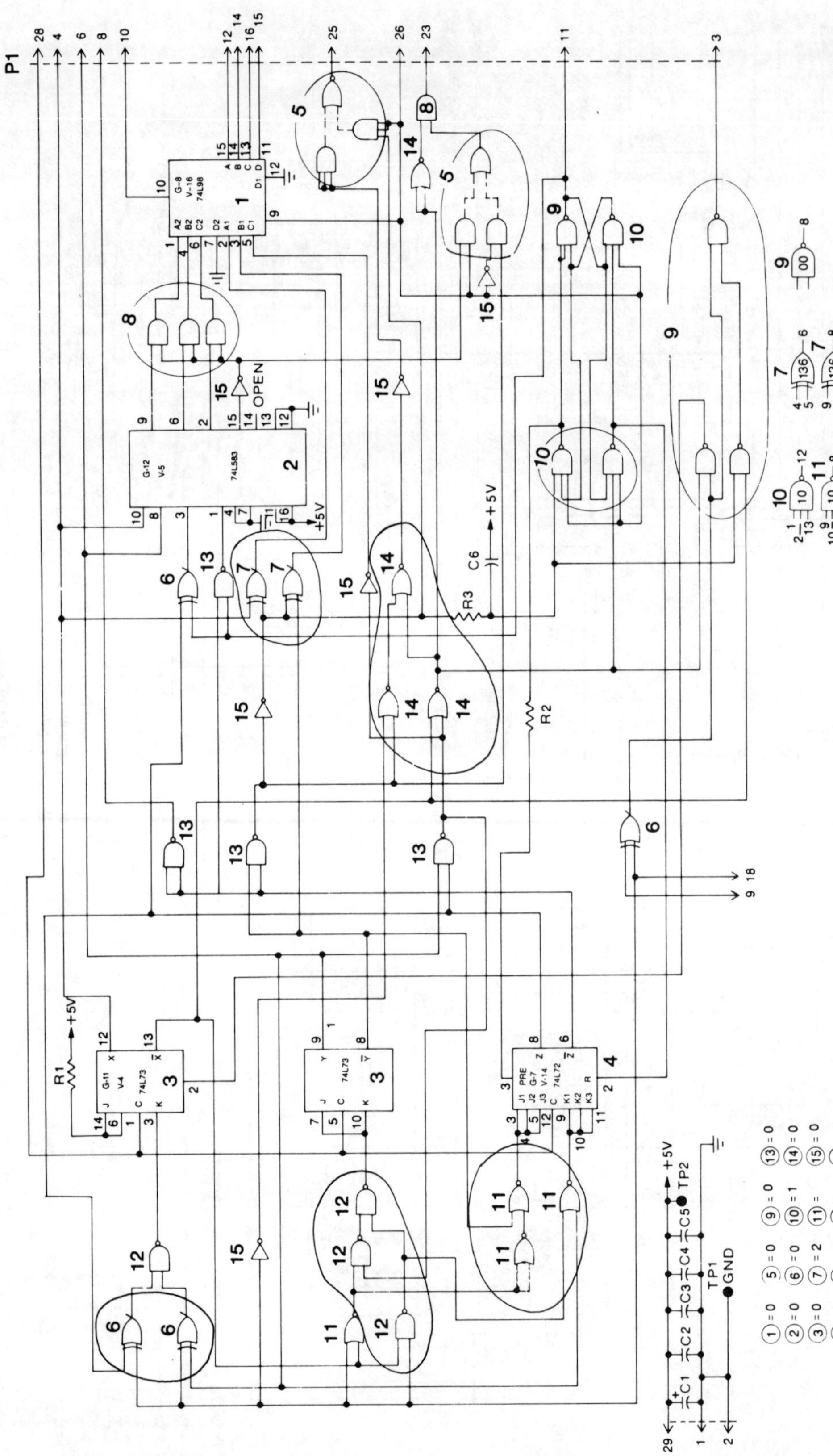

Figure 11-2. Student Schematic (15-IC Grouping)

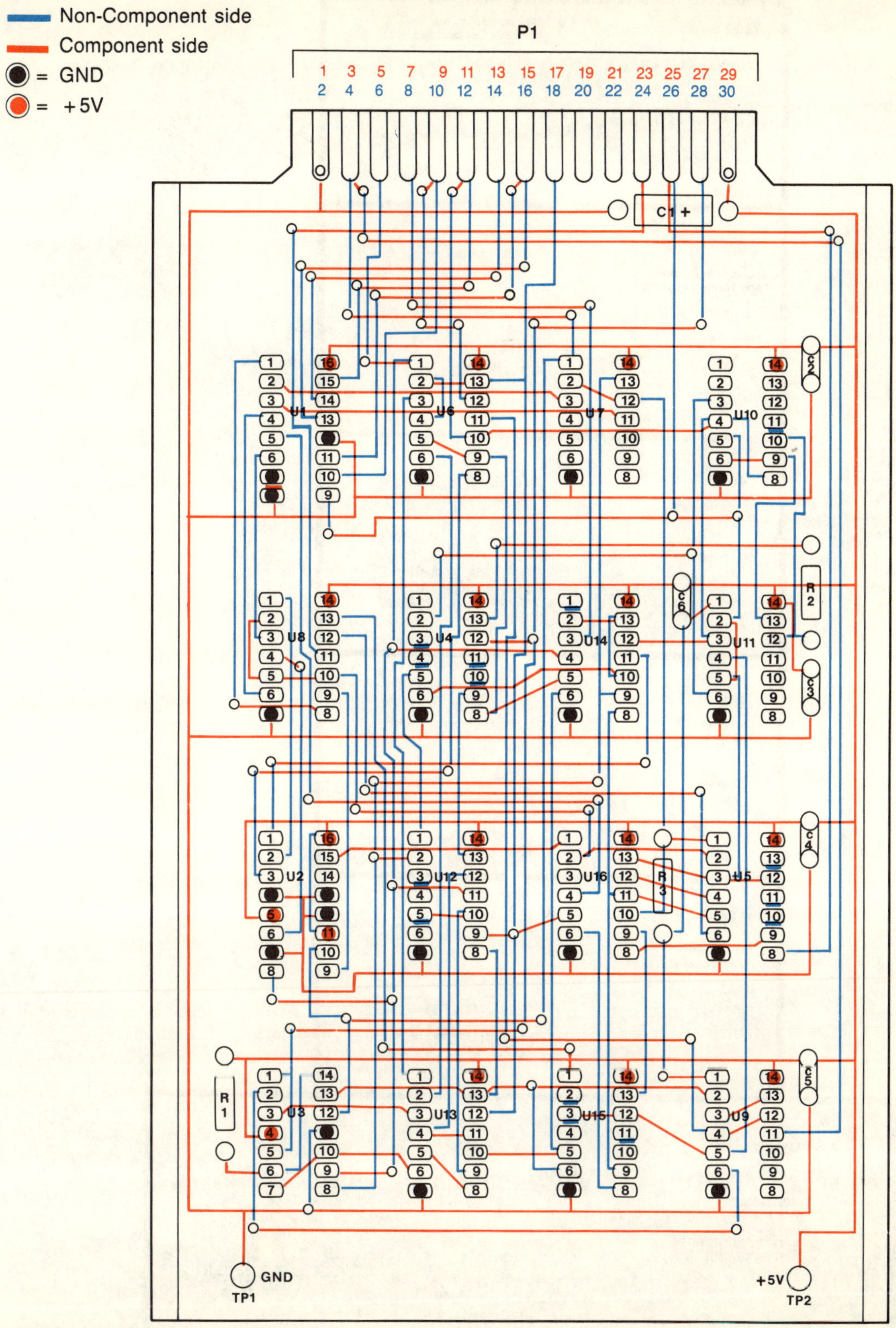

Non-Component side
Component side
= GND
= +5V
P1
1 3 5 7 9 11 13 15 17 19 21 23 25 27 29
2 4 6 8 10 12 14 16 18 20 22 24 26 28 30
C1 +
U1
U6
U7
U10
U8
U4
U14
U11
U2
U12
U16
U5
U3
U13
U15
U9
C2
C3
C4
C5
C6
R1
R2
R3
GND
TP1
+5V
TP2

Figure 11-3.

7400

QUADRUPLE 2-INPUT
POSITIVE-NAND GATES

positive logic:
$Y = \overline{AB}$

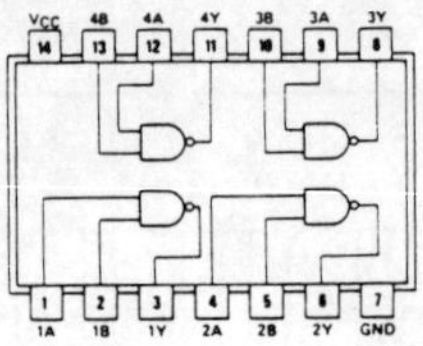

7402

QUADRUPLE 2-INPUT
POSITIVE-NOR GATES

positive logic:
$Y = \overline{A+B}$

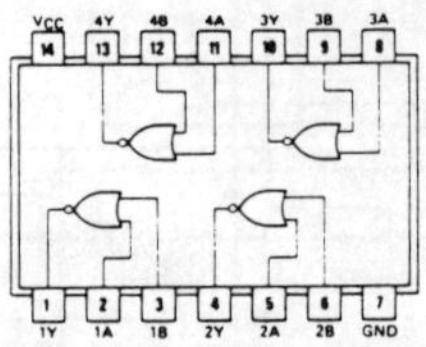

7404

HEX INVERTERS

positive logic:
$Y = \overline{A}$

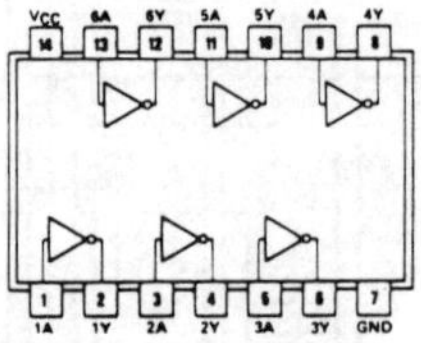

7408

QUADRUPLE 2-INPUT
POSITIVE-AND GATES

positive logic:
Y = AB

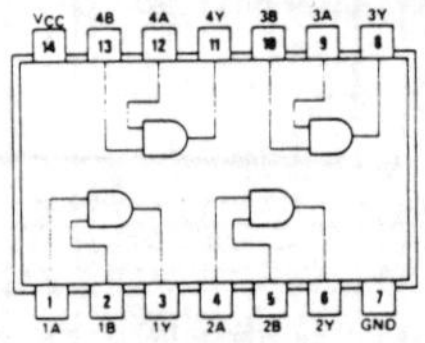

7410

TRIPLE 3-INPUT
POSITIVE-NAND GATES

positive logic:
$Y = \overline{ABC}$

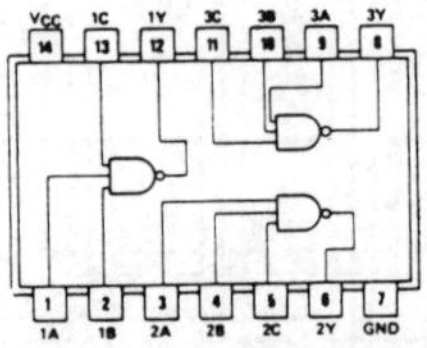

AND-OR-INVERT GATES

51 'L51, 'LS51
2-WIDE 3-INPUT,
2-WIDE 2-INPUT
positive logic:
$1Y = \overline{(1A \cdot 1B \cdot 1C)+(1D \cdot 1E \cdot 1F)}$
$2Y = \overline{(2A \cdot 2B)+(2C \cdot 2D)}$

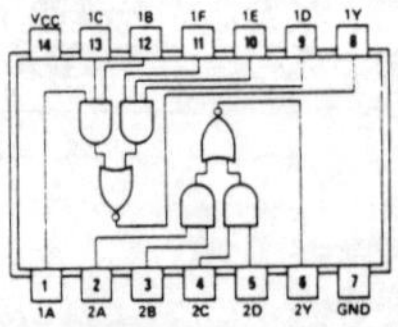

SN54L51 (J) SN74L51 (J, N)
SN54LS51 (J, W) SN74LS51 (J, N)

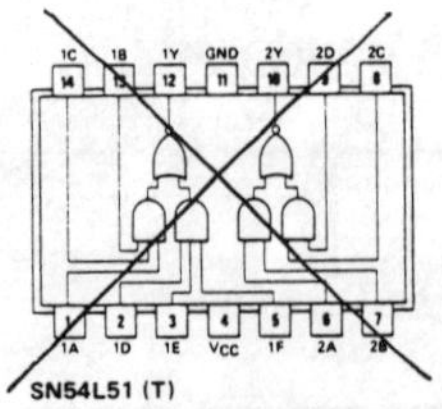

SN54L51 (T)

Figure 11-4. Specification Information

PIN ASSIGNMENTS (TOP VIEWS)

AND-GATED J-K MASTER-SLAVE FLIP-FLOPS WITH PRESET AND CLEAR

72

FUNCTION TABLE

INPUTS					OUTPUTS	
PRESET	CLEAR	CLOCK	J	K	Q	$\bar{Q}$
L	H	X	X	X	H	L
H	L	X	X	X	L	H
L	L	X	X	X	H*	H*
H	H	⎍	L	L	Q_0	$\bar{Q}_0$
H	H	⎍	H	L	H	L
H	H	⎍	L	H	L	H
H	H	⎍	H	H	TOGGLE	

positive logic: J = J1·J2·J3; K1·K2·K3

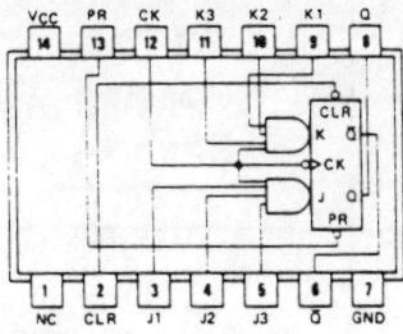

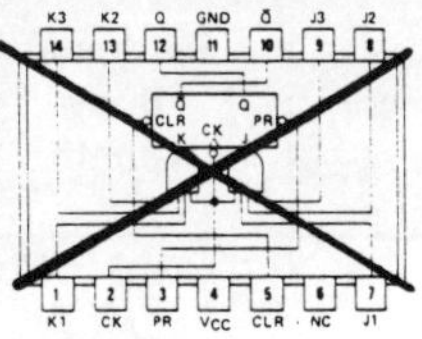

SN5472 (J) SN7472 (J, N)
SN54H72 (J) SN74H72 (J, N)
SN54L72 (J) SN74L72 (J, N)

SN5472 (W)
SN54H72 (W)
SN54L72 (T)

NC–No internal connection

DUAL J-K FLIP-FLOPS WITH CLEAR

73

'73, 'H73, 'L73
FUNCTION TABLE

INPUTS				OUTPUTS	
CLEAR	CLOCK	J	K	Q	$\bar{Q}$
L	X	X	X	L	H
H	⎍	L	L	Q_0	$\bar{Q}_0$
H	⎍	H	L	H	L
H	⎍	L	H	L	H
H	⎍	H	H	TOGGLE	

'LS73
FUNCTION TABLE

INPUTS				OUTPUTS	
CLEAR	CLOCK	J	K	Q	$\bar{Q}$
L	X	X	X	L	H
H	↓	L	L	Q_0	$\bar{Q}_0$
H	↓	H	L	H	L
H	↓	L	H	L	H
H	↓	H	H	TOGGLE	
H	H	X	X	Q_0	$\bar{Q}_0$

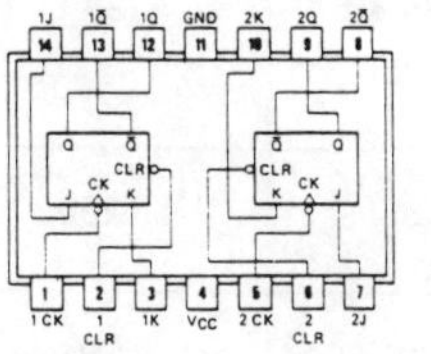

SN5473 (J, W) SN7473 (J, N)
SN54H73 (J, W) SN74H73 (J, N)
SN54L73 (J, T) SN74L73 (J, N)
SN54LS73 (J, W) SN74LS73 (J, N)

4-BIT BINARY FULL ADDERS WITH FAST CARRY

83

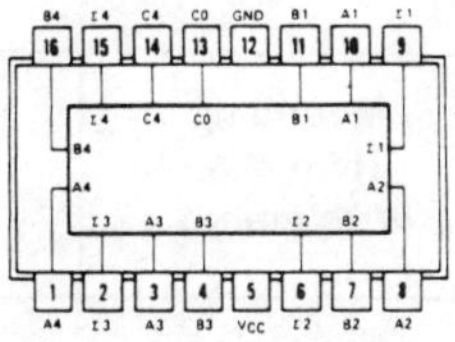

SN5483A (J, W) SN74 83A (J, N)
SN54LS83A (J, W) SN74LS83A (J, N)

4-BIT DATA SELECTOR/STORAGE REGISTERS

98

SELECTS 1 OF 2 4-BIT WORDS
PARALLEL IN/OUT

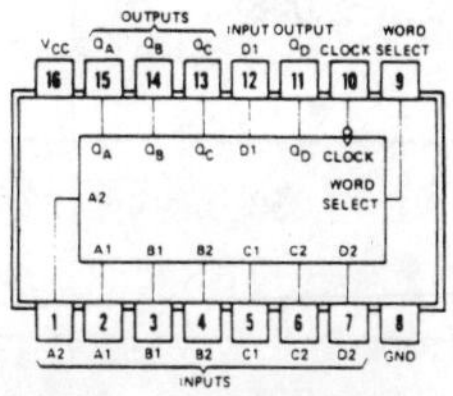

SN54L98 (J) SN74L98 (J, N)

136

QUADRUPLE 2-INPUT EXCLUSIVE-OR GATES
WITH OPEN-COLLECTOR OUTPUTS

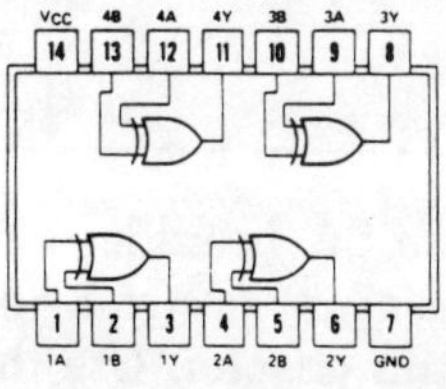

positive logic:

$Y = A \oplus B = \bar{A}B + A\bar{B}$

Figure 11-5. Spec Info con't.

Table 11-1 Component Mounting Table .

Reference Designation	Component Type	Maximum Body Length	1:1 Pad Spacing (inches)	Maximum Lead Diameter	Lead Mounting Hole (Min.)	Lead Mounting Hole (Min.)
1 From Schematic	2 From Specification Sheet	3 From Specification Sheet	4 Column 3+.120 (Round Up)	5 From Specification Sheet	6 Column 5 + .006	7 Column 5 + .020
C1	CS13-A (RL20)	.422	.600	.022	.028	.042
C2 thru C6	CK06	.290	.200	.029	.035	.049
R1 thru R2	RC07	.281	.500	.027	.033	.047
U1 & U16	Dual-in-line 14/16 pin	Per Template	Per Template	.018	.024	.038

With a PCB MFG Tolerance of +.004/-00.2 for the mounting holes, the above components require:

Maximum Lead Diameter	Mounting Hole Min W/Tolerance	Mounting Hole Max W/Tolerance	Hole Call-Out	Min Pad Size	Selected Pad Size
8 Column 5 In Increasing Size	9 Column 6 Plus Tolerance (Add .002)	10 Column 7 Minus Tolerance (Sub .004)	11 From Column 9 and 10	12 Column 11 +.030	13 Standard Pad Size at 2:1
.018 U1 – U16 =	.026	.034	.032 +.004 –.002	0.62	Pattern "D" (Fig. 6-21)
.022 C1 =	.030	.038	.032 +.004 –.002	.062	.150
.027 R1 & R2 =	.035	.043	.040 +.004 –.002	.070	.150
.029 – C6 =	.037	.045	.040 +.004 –.002	.070	.150

Step #4—Establish the board layout design rules. Use these rules for designing and taping the PC board in this chapter. Use the design rules established in chapter 8 in this chapter. Step 4 in chapter 8 establishes the PC board design conventions for this book. It includes the general requirements for the formal layout and the artmasters. Before starting the design, review chapter 8, illustration 2 for the design rules.

Step #5—Refer to the following list of questions to ask the engineer about the 15/16-IC digital schematic. We will borrow the questions to ask the engineer from chapter 10. We will answer only those questions that apply to the schematic in Figures 11-1 and 11-2.

Questions To Ask The Engineer

(1) Are there any special size components or are all the components standard?
Table 11-2 indicates there are standard case code body outlines for all discrete components and IC's.

(2) How many bypass capacitors are there for each IC?
Figures 11-1 and 11-2, note 2 instructs to bypass every four IC's using a ceramic disc, CK06 capacitor. The schematic shows that C2, C3, C4, and C5 are this kind of capacitor. For placement on the board outline, see Figure 11-3.

(3) Are there any critical leads on the board?
There are no critical leads required for schematic Figures 11-1 or 11-2.

(4) Are there any special size filtering capacitors required on the input voltage pins?
The schematics in Figures 11-1 and 11-2 show C1 as a polarized input filtering capacitor. Note 1 on both schematics identifies C1 as an electrolytic tantalum input filtering capacitor with a case code of CS13A (substituted on the template as an RL20). For placement on the board layout, see Figure 11-3.

(5) Should there be any spare IC gates?
The schematic in Figure 11-1 (1 16-IC layout) required two 3-input NAND gates. These are shown as required spares on the schematic. The student's schematic in Figure 11-2 with 15 grouped IC's required no spares.

(6) By using the maximum size of all the components, will all the components fit on the board outline?
The schematic in Figure 11-1 (with the layout shown in Figure 11-3) shows that the layout is not very dense. Calculate the area required for your schematic using the procedures in previous chapters.

(7) Are there any pre-assigned pins on the connector?
The schematics in Figures 11-1 and 11-2 do have pre-assigned pins on the connector. You cannot change them. For an example of routing to pre-assigned pins, see Figure 11-3.

(8) Are there any variable resistors or variable capacitors?
There are no variable components used on the schematic in Figures 11-1 or 11-2.

(9) Should there be any test points?
There are test points (TP1 and TP2) required for the voltage (+5) and the ground shown on the schematic in Figures 11-1 and 11-2. These should be placed on the accessible edge of the board. For an example of placement, see the layout in Figure 11-3.

(10) Do any components require heat sinks?
This question is not applicable to either schematic, Figure 11-1 or 11-2.

(11) Have all package considerations been examined?
This question will be covered in step 6.

Step #6—Determine the board outline. Find out what method of fastening or mounting is being used: card guide, card ejectors, stand-offs, etc. What type of electrical interface is required: (connectors, cables, wires, etc)? The board size must be known to compute the total square inches available for the schematic in Figure 11-2. Figure 11-6 shows the board's overall dimensions. It also establishes the connector size and the tooling holes (sometimes called datum holes). This board size has also been used for the layout shown in Figure 11-3. It will be used for your layout in Figure 11-2. This board will be placed inside a chassis similar to the board shown in Figure 10-13. The difference is that we will not use a card ejector on the layout for the schematic in Figures 11-1 or 11-2.

Step #7—Grouping the logic gates. Compute the total number of IC's required to fit within the assigned board outline. One square for each IC is a minimum. The schematic in Figure 11-1 (which is our version of 16-IC grouping) uses the board outline shown in Figure 11-6. For an approximate estimate of useable area, multiply the width times the length (excluding the connector tongue): 4 inches x 6 inches = 24 square inches. Even though this is not the total useable area, it does give us an approximation. We should be able to place 24 IC's within that board outline. Placing 16-IC's gives us approximately 1.5 square inches per IC. You should be able to fit all of the components and connect all the traces for signal, voltage and ground, with this amount of area for each IC. See Figure 11-3. The schematic in Figure 11-2 has been grouped into a 15-IC position. The IC count was reduced to 15 IC's because we eliminated the spare requirement required on the schematic in Figure 11-1. We changed the grouping in order to organize, plot, and layout your own version of this schematic. With 15 IC's and the same board outline in Figure 11-6, you should be able to fit all of the components and connect them within that board outline electrically.

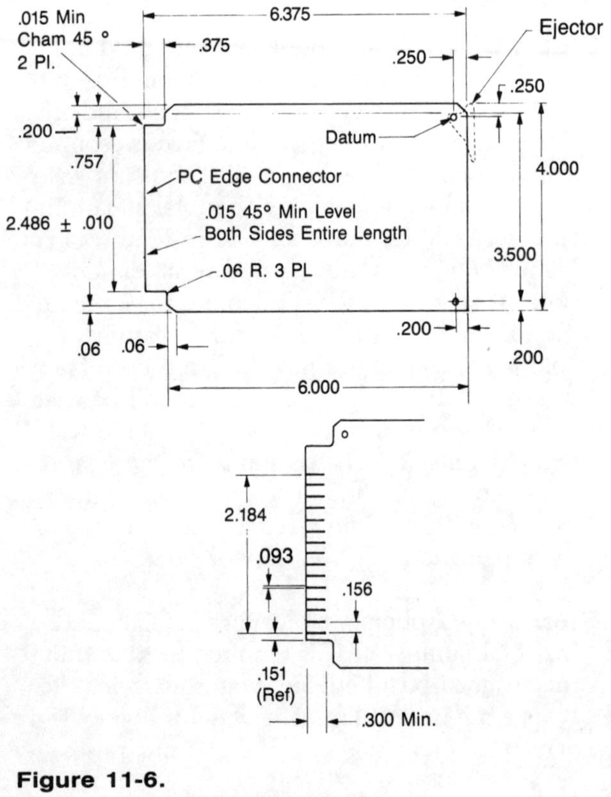

Figure 11-6.

Step #8—Plotting the grouped IC's onto the board is the next step. This is the assignment of the IC numbers to the location on the board. Arrange the IC's (that communicate to one another) in line or in the same rows with each other. See the schematic in Figure 11-1. We have arranged the IC's in a four-by-four pattern shown in Figure 11-7. This is the only logical pattern for the board outline shown in Figure 11-6. The maximum number of IC's in the width of the board is 4 inches, equally spacing the four rows in the length of the board is 6 inches. The key positions for this type of arrangement (4x4) would be the four center positions shown in Figure 11-8. Key position means those IC's that have much communication with other IC's. For example, on our schematic in Figure 11-1, the inverter package (U16) is spread throughout the schematic; it communicates to many different IC's. This should be in one of the four center positions shown in Figure 11-8. See the plotting arrangement in Figure 11-7 for the 16-IC schematic in Figure 11-1. This shows intercommunication from one IC to the other. The flow starts from the P1 connector, goes down the board, and takes advantage of the communication from one IC to another. One of the most obvious plotting arrangements on the schematic in Figure 11-1 was IC 1. It communicates to the connector, particularly to the lower pins on the P1 connector. Also, IC 1 communicates with IC 8 which communicates with IC 2, and IC 2 communicates with IC 3. When grouping the schematic in Figure 11-2, analyze how these components communicate to one another. Figure 11-2 has only 15 IC's grouped. Therefore, the arrangement within the board outline will change (see Figure 11-9). We have arranged a pattern of 3 x 5. This is not the only arrangement for the schematic in Figure 11-2 (3x5). We could use a 4x4, providing one spare position. If we choose not to have a spare IC position, we would use the 3x5 as shown in Figure 11-9. This is the same board outline shown in Figure 11-6. You must use the same pre-assigned pins as in the schematic in Figure 11-1. Therefore, plotting would be similar. Plot your own version of the 15-IC schematic in Figure 11-2 on the pre-arranged IC pattern shown in Figure 11-9.

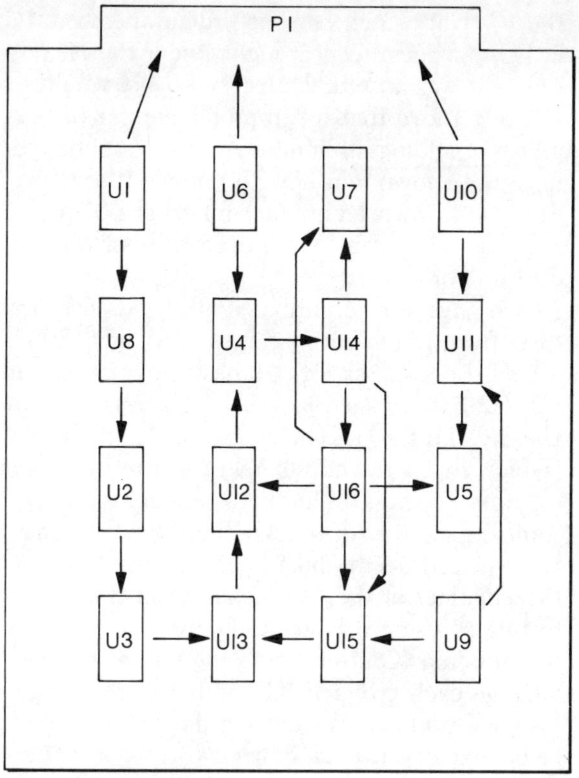

Figure 11-7. IC Plotting

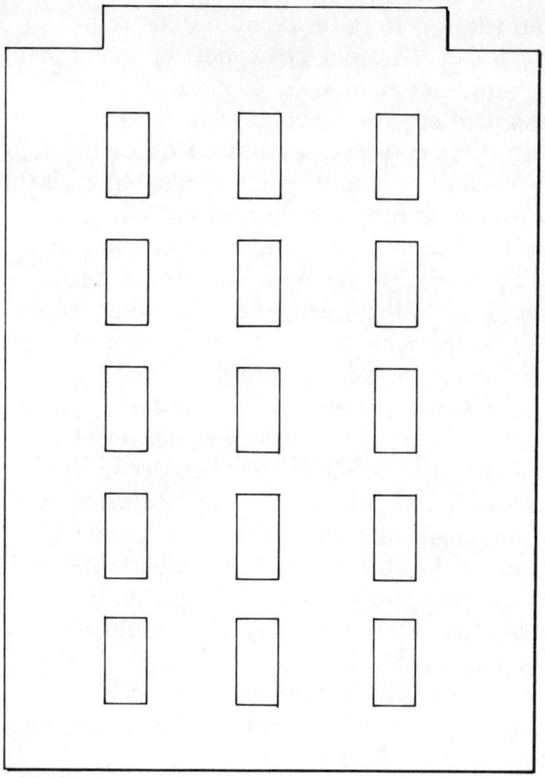
Figure 11-9. IC Plot

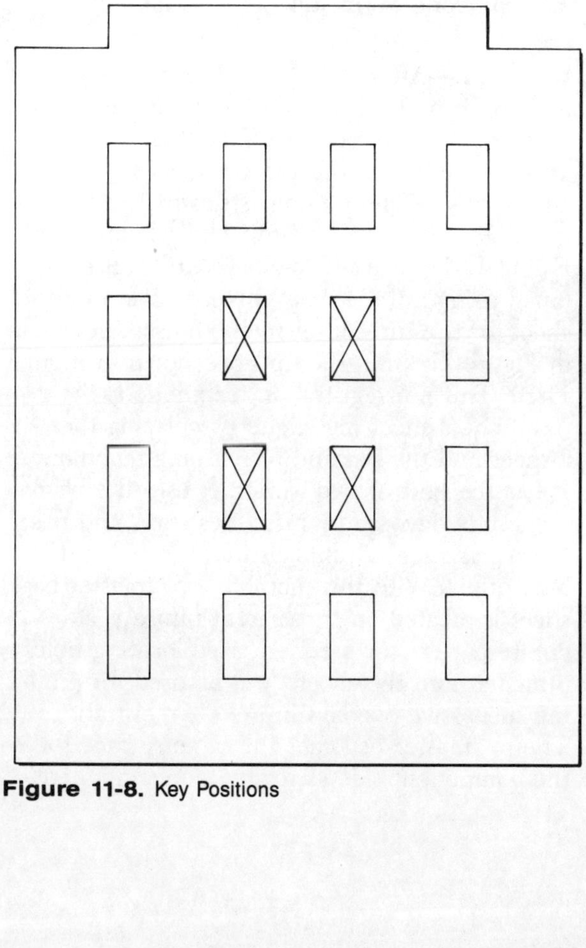
Figure 11-8. Key Positions

Step #9—We will conduct a formal template layout establishing the conductor paths in an X-Y direction. View the layout from the component side of the board. The formal template layout for the schematic in Figure 11-1 is shown in Figure 11-3. Note that the IC positions have been established by the plot that was shown in Figure 11-7. Each IC was formally drawn on a grid pattern that has equal space for each row. Each IC pad was numbered. See the specification sheets in Figures 11-4 and 11-5. The IC patterns were color coded for voltage and ground. All red pads equal +5V, and all black pads equal ground. This information also was taken from the specification sheets shown in Figures 11-4 and 11-5. The first traces to be drafted onto the board were the voltage and ground buses. This establishes the X-Y direction. In this case, red traces equal the component side; blue traces equal the noncomponent side. The P1 connector is numbered 1 to 29 on the component side (in red), and numbered 2 to 30 on the noncomponent side (in blue). This information is from the board outline shown in Figure 11-6. Tooling holes must be established. Locate the tooling holes to avoid placing traces and card edge clearance within that area. Maintain three grids at a scale of 2:1 on each side of the board for card track clearance. Once all this information has been thought out and placed on the board outline, interconnect one gate to another. Assign the input and output pin numbers on the schematic. As we assigned each pin on the schematic and connected each trace on the layout, we yellowed the connections on the schematic as we finished the layout procedures. This keeps track of which traces or lines on the schematic have been connected and which ones have not. Use feedthroughs when changing from the red side (the component side) to the blue side (the noncomponent side). After connecting every trace on the schematic, clean up the layout by removing all unnecessary feedthroughs. For a review of step-by-step interconnections from one gate to another, see the schematic in Figure 11-1 and the layout in Figure 11-3. Make a formal layout using the schematic in Figure 11-2 and the plot from Figure 11-9. Use the following sequence:

A. Draft the board outline (including the connector).
B. Position the IC's.
C. Number the IC pins.
D. Color code all the IC pads.
E. Draft the bus lines for the voltage and ground.
F. Connect all signal lines (yellow out on the schematic).
G. Place the input filtering capacitor.
H. Place the bypass capacitors.
I. Place the test points.
J. Clean up the layout. See step 10.

Step #10—This step cleans up all unnecessary via holes (called feedthroughs). In the formal layout, we placed feedthroughs because the X-Y system was used for routing traces. Remove the vias that were not necessary. Place all the IC's in a numerical order, top-to-bottom, left-to-right, or any practical alphanumerical order. Update the layout and schematic. Figure 11-10 shows the formal layout in Figure 11-3 with the unnecessary feedthroughs or via holes. They can be removed quickly. Circle one via hole under the IC (U8), which has been re-numbered to IC 2. Do not place feedthroughs under the IC's unless it was absolutely necessary. We would like to remove that feedthrough even at the expense of adding an additional feedthrough. We have not shown you how to remove these. Study your own techniques for feedthrough removal. They can be removed without a great deal of work.

We have also renumbered all the IC's. In this case, from top-to-bottom and left-to-right. U1 stayed the same as U1. U8 has been changed to U2. U2 has changed to U3. U3 has changed to U4, etc. This arrangement can be in any logical fashion that you choose. You can use rows such as: A, B, C and D. Number them across. Use some logical way of quickly finding where any IC is placed on the board. This is primarily for the engineer or the technician when they are testing or using the board. Transfer these renumbered IC positions to the schematic. Change each grouped IC number to the new assigned number. We did not do this for our layout, however. Change your own layout and update your own schematic (shown in Figure 11-2). All necessary modifications to the layout must be done at this time (before taping).

Step #11—After completing the layout drawing, an artmaster is taped. The artmaster is made to an enlarged scale (2:1). It is photoreduced to provide a 1:1 scale negative and positive. (The photomaster which is used to manufacture the PC board). The schematic in Figure 11-1 (the 16-IC digital board) will be taped using a three-sheet method. The three-sheet (or the three-layer method) uses the layout in Figure 11-3 and the updates shown in Figure 11-10. This pin registration technique is the best. The 3-sheet technique is replacing the 2-sheet and the red and blue taping techniques. It has the best of two worlds. It uses the black crepe tape favored by many designers and has the registration advantages that the red and blue offers. With this method, a padmaster base sheet is created on transparent film. An accurate pattern (of precisely sized locating holes punched into the sheets) will be used for pinning successive overlays in precise registration. The padmaster becomes the same surface for the component side as for the noncomponent

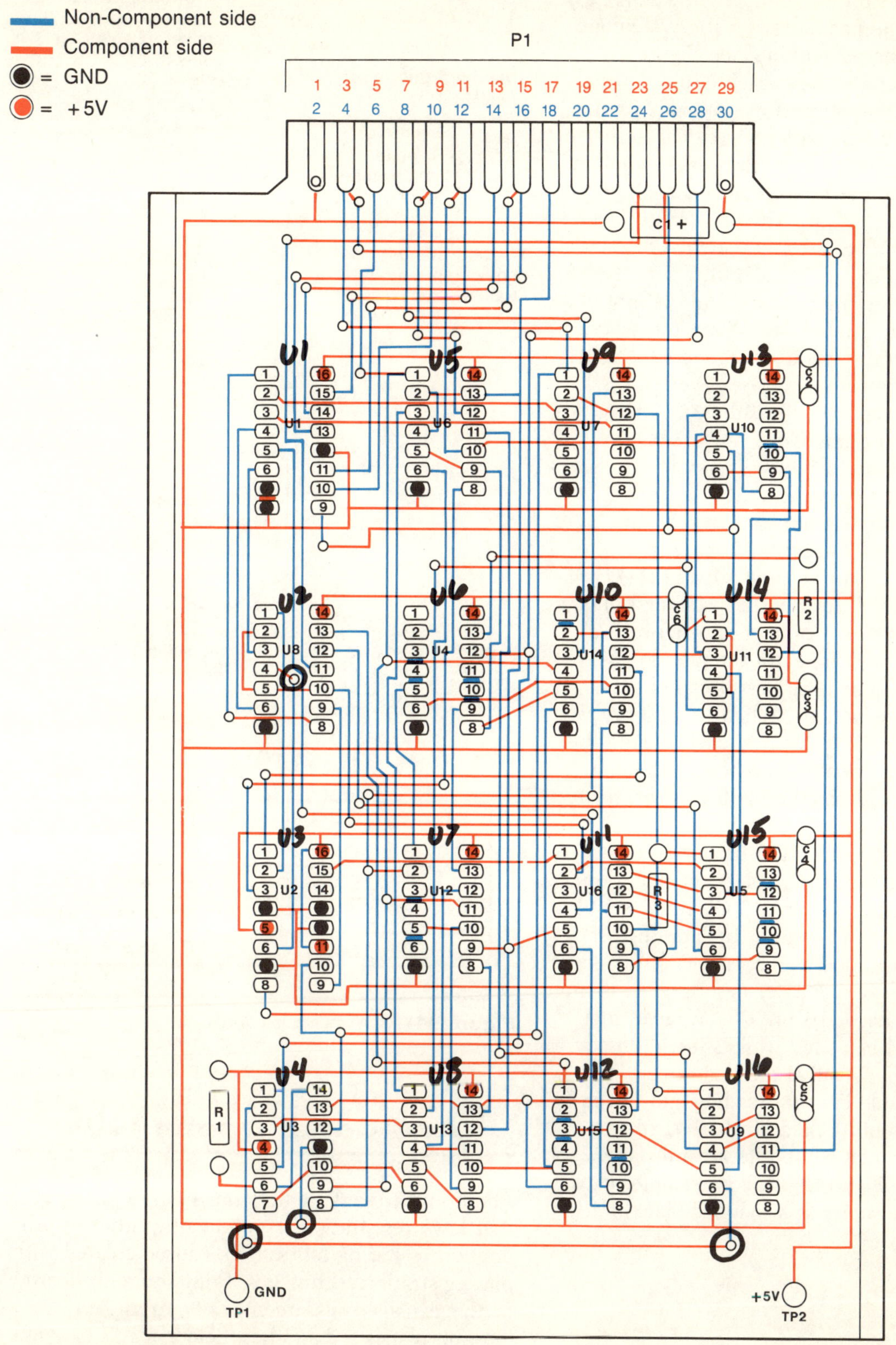

Figure 11-10. Clean up

side. The traces for the component side are overlayed on the padmaster (or the noncomponent side on the padmaster). See Figure 11-11. Overlays are produced for each circuit layout on sheets of transparent film. They are prepunched with a hole pattern identical to that of the padmaster. These artwork overlays contain only the interconnection circuitry and the other features unique to that particular circuit layer or side (in this case, the component side or the noncomponent side). The identification of each overlay and layer description should be noted in a legend on the artwork set. When the artwork is photographically reduced to generate the 1:1 film, the individual circuit layer overlays are placed in contact with the padmaster in a camera vacuum frame (copy holder). Registration of the padmaster to the component versus noncomponent sides is accurately aligned through the precisely punched registration holes. All fabrication register-features are photographed from the same artwork sheet called a padmaster. All padmaster patterned photos then will be in nearly perfect registration. All pads, crop marks, connector tongue, fingers, etc. remain in the same position throughout the photographic reductions. When using a multi-sheet artwork system, identify each sheet. This ensures correct orientation when assembling composite artwork for the photographer, for rework, or for checking your own board.

Place a sheet index chart on the base sheet (in the area that will remain clear on each overlay, beginning with the base sheet). Clearly describe all artwork layers in this chart. Include the artwork and identification numbers, revision codes, sheet numbers, sheet or layer descriptions, or any pertinent data. Each artwork overlay sheet identification should be added while superimposed over the base sheet (called the padmaster). The information should be right reading and not interfere with any other data on these sheets or other overlays (when composite artworks are assembled). See Figure 11-12. For example of positives and negatives at a reduced scale of 1:1, see Figure 11-13.

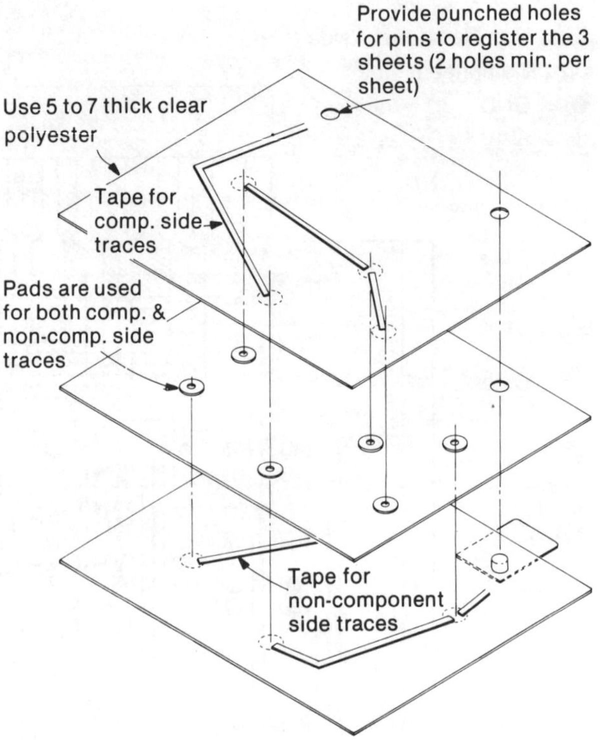

Figure 11-11. Padmaster (3 Sheet Method)

4 O

DWG. NO	SH	REV	DESCRIPTION
350045	1	B	PAD MASTER
350045	2	B	COMP SIDE CKT
350045	3	B	SOLDER SIDE CKT
D 350045	4	B	MARKING ARTW
350045	5	B	SOLDER MASK

Figure 11-12. Identification of Sheets

Additional Documents

Additional artwork and documentation drawings that have registration requirements, either in production or during fabrication, can be created with pin registration techniques. Printed circuit board markings (called silkscreen) is one example. Ground planes are another. Solder masks can also be prepared using this technique of various sheets of pin registered composite artwork overlays.

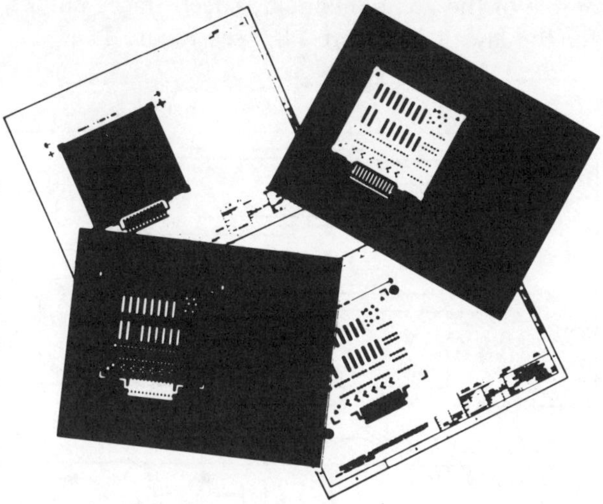

Figure 11-13. Film Pos. & Negs.

Pin Pattern

Hole shapes and locations are very important. Accuracy and consistency of the hole pattern, and the closeness of fit between the pin and the holes is important because film material tends to change size when temperature and humidity changes. Locate a single round hole as close to the artwork center as possible. This round hole is punched to a tolerance that prevents change. When the film is secured on its corresponding round metal pin, almost no change will affect its position or registration. Additional registration holes are necessary to prevent the film from rotating around this center hole. These holes are elongated to allow film expansion and extraction caused by environmental changes. These holes keep misregistration (at the outer edges of the artwork sheets) to a minimum. See Figure 11-14 for an example of punched film. To maintain registration accuracy, all holes in a sheet must be simultaneously punched from the same punching tool. This punch must have absolute repeatability. It must hold the film flat. Fit tolerance of the holes to the pins must be maintained with precision.

Pin registration must incorporate all these requirements. An artwork image registration accuracy of $\pm .002$ should be consistently obtained. For an example of the pin bar technique, see Figure 11-15.

Use the layout in Figure 11-3 and the updates on Figure 11-10. The following procedures show the three-sheet technique.

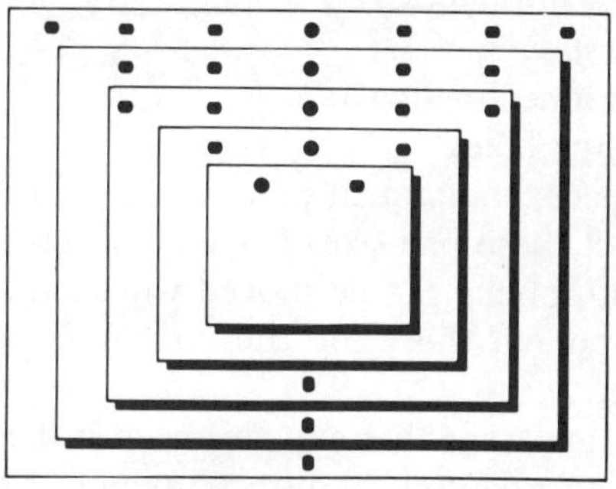

Figure 11-14. Punched Film

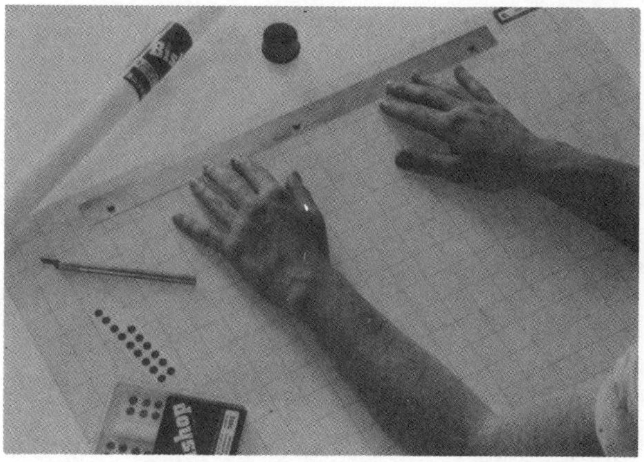

Figure 11-15. Pin Bar

3-Sheet Tape-Up

1. Whenever possible, prepare pin registration artwork on a cold light table to ensure precise registration. Stability problems can be caused by table heat. First, fasten a precision prespaced pin bar to a precision film or glass grid. Use a double StikOn pin fastener or regular masking tape. Be sure to fasten the pin bar to the glass or film grid at a point under the center pin (for maximum stability).

2. Insert the registration pins or the pin bar on the first sheet. This becomes the padmaster sheet, which must have the following information:

A. Board outline (which would be the crop marks).
B. All connector fingers.
C. Tooling holes.
D. All pads, component pads, and feedthroughs. Select these pad sizes from Table 11-1. The feedthroughs can be moved when taping the component sheet and the noncomponent sheet.
E. Any markings that appear on both the component side and the noncomponent side. The padmaster will be used or pinned with each side or layer. For an example of a padmaster for the layout in Figure 11-3, see Figure 11-16.

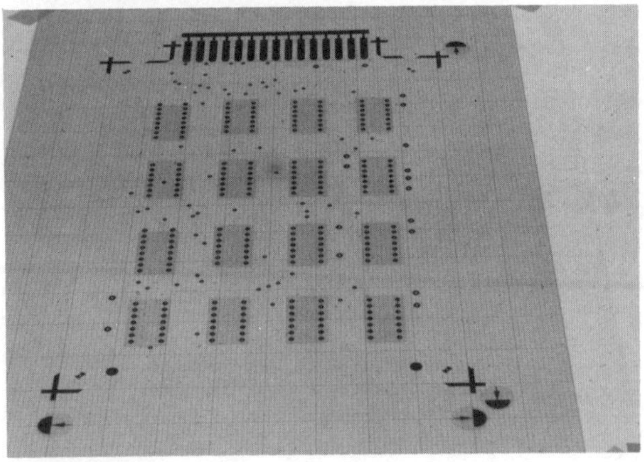

Figure 11-16. Padmaster

3. Using the same pin holes. Insert the component side sheet into place on top of the padmaster. Tape all red traces using .100 at a scale of 2:1 (for all voltages and ground traces). For signal path traces, use .030 tape at a scale of 2:1. The tape must overlap all pads, but do not cover the pad holes. Pads may be adjusted at this time on the padmaster by lifting up the component sheet momentarily and moving the pads on the padmaster. Place all markings that appear on the component side only (such as plus symbols for capacitors or diode symbols and the P1 connector finger numbers, etc). No other markings should appear on this component side sheet. For an example of the component side sheet (traces only) for the layout in Figure 11-3, see Figure 11-17.

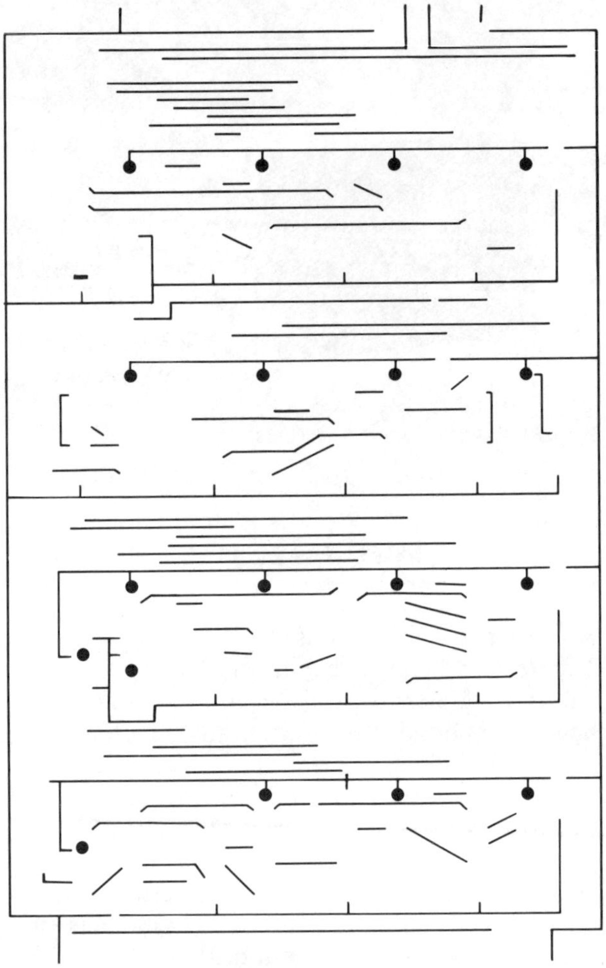

Figure 11-17. Component Side Tape-up

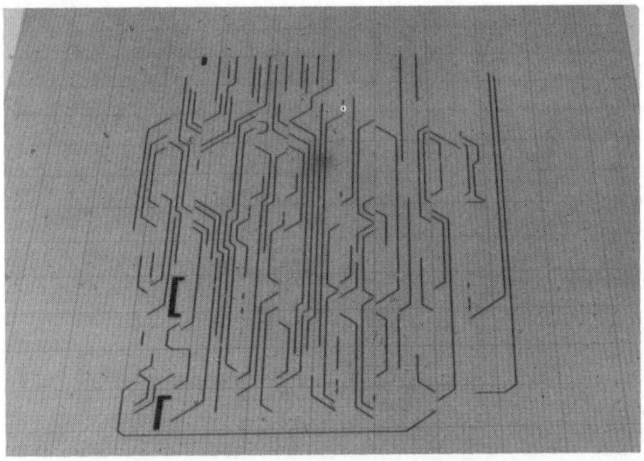

Figure 11-18. Noncomponent Side Tape-up

4. Use the same pin holes. Remove the component side sheet. Replace with noncomponent side sheet, placing it down over the padmaster. Take all blue traces. Use .100 tape for voltage and ground and .030 tape for all signal paths at a scale of 2:1. Traces must overlap all pads, without covering the holes on the component side sheet, and the noncomponent side sheet. All markings that belong only on the noncomponent side must now be placed on the sheet: board identification, company logo, markings for test purposes, etc. For an example of the noncomponent side sheet (traces only) for the layout in Figure 11-3, see Figure 11-18.

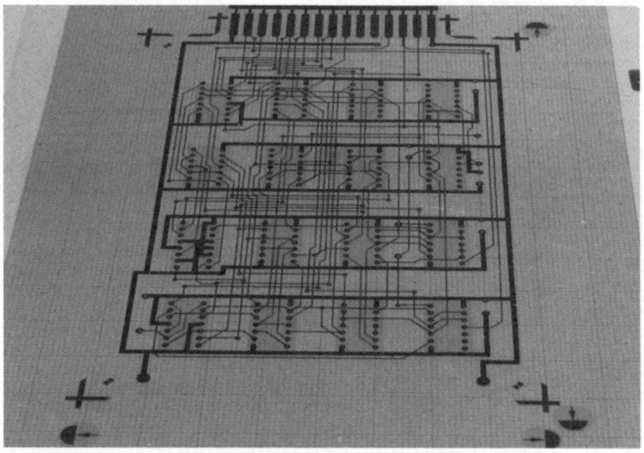

Figure 11-19. All Three Sheets Pinned Together

5. Pin all three sheets together to check pad alignment to traces for both the component side and the noncomponent side. To check that all markings are in the proper position. Check that all traces have been connected (both red and blue). For an example of all three sheets pinned together of the layout in Figure 11-3, see Figure 11-19. Practice this technique using your own layout from the schematic in Figure 11-2. Follow the procedures 1 through 5 for the layout in Figure 11-3. Tape-up using a three-sheet technique (scale of 2:1). Follow the guidelines shown in Figure 11-20. This is the check list for the three-sheet tape-up.

Taping Guidelines

1. Minimize handling of drafting film and tapes.
2. DO NOT fold, crease or tightly roll the artwork.
3. Burnish the tape when complete.
4. Run the tapes to the center of the annular ring of the pad.
5. DO NOT cut through tapes or pads (it will show up on the photograph).
6. DO NOT pull the tape too HARD or it will creep.
7. Tape and place the pads either on the grid coordinate or coincident with the lines.
8. Trim all corners as required.
9. Be neat.

Artmaster Checklist

________ Tape/Pad/Drafting Film Cut
________ Pad Off Grid
________ Corner Trim
________ Air Gap/Tape Width
________ Run Tape to Center of Pad
________ Pad Center Covered
________ Not Tapes at 45/90 Degrees
________ Angle Less Than 90 Degrees
________ Electrical Short, Bridging
________ Ragged Edge, Ragged Cut, Crooked Trace
________ Trace Too Short/Open Circuit

Figure 11-20. Taping Guidelines, Artmaster Checklist

Summary

You were given a completed layout, and tape-up of a 16-IC schematic. You used the same schematic, but with a different grouping. You plotted the layout.

There were pre-assigned pins on the connector. The taping exercise presented a new technique. The three-sheet system achieves the best of both worlds. The padmaster can be overlayed on a sheet (called component side traces), removed, and overlayed on the noncomponent side traces. You use the same pads for both the component side and the noncomponent side. Therefore, you can achieve nearly perfect registration. You can use either the black crepe tape or the red tape for the traces. Red tape has a much finer line definition on its edges than the black. The black is easier to handle than the red. However, many companies still use the 2-sheet and the red and blue systems. But the new 3-sheet system has some advantages over the other two systems.

We identified and marked the individual sheets using the three-sheet technique. It is important to properly identify the film and tape-ups. We combined some of the techniques learned in chapters 8, 9 and 10. We used all these techniques on the design in this chapter.

Quiz—Chapter 11

1. In your own words describe the meaning of the following questions to ask the engineer:

(1.1) Are there any special size components or are all the components standard?

(1.2) Are there any critical leads on the board?

(1.3) Are there any special size filtering capacitor required on the input voltage pins.

(1.4) How many bypass capacitors are there for each IC?

(1.5) Should there be any spare IC gates?

(1.6) By using the maximum size of all the components, will all the components fit on the board?

(1.7) Are there any pre-assigned pins on the connector?

(1.8) Are there any variable resistors or variable capacitors?

(1.9) Should there be any test points?

(1.10) Do any components require head sinks?

(1.11) Have all package considerations been examined?

2. Usable area: When calculating usable board area, should the designer count the connector tongue area.
   Yes ________ No ________
3. Plotting: In your own words describe the meaning of the work plotting.
4. Plotting: In your own words what should the designer think about first?
5. Plotting: IC's that have much communication with other IC's should be placed in a key position, such as the center of the board.
   Yes ________ No ________

6. Formal template layout: When we establish the conductor paths on a formal template layout one color is used for X direction and one color for Y direction. Question: Which side of the board is X and which is Y?

 ____

 ____

7. Formal template layout: To establish the X-Y direction on your layout, which traces were drafted onto the board first?

 ____

 ____

8. Interconnecting signal lines: Once you are ready to start interconnecting signal lines where do you start?

 ____

 ____

 ____

9. Interconnecting: As a means of keeping track of which traces or lines on the schematic have been connected and which ones have not, the designer should ____________ out the schematic as he/she finished each connection.
10. Clean-up: This is the last step in the layout process. In your own words describe the meaning of clean-up?

 ____

 ____

 ____

11. Artmaster - Three sheet method

(11.1) This pin registration technique is replacing the _________ sheet and the _________ and _________ taping technique.

(11.2) The main difference between the three sheet methode and the other techniques is the _________________ sheet.

(11.3) Registration of the overlays is accomplished by punched holes used for ____________________.

(11.4) When using a multisheet artwork system, the designer should ________________ each sheet. This ensures correct orientation.

(11.5) Additional artwork and documentation drawings can be created with pin registration techniques. Name two:

 ____

 ____

(11.6) Artwork stability: Whenever possible, the designer should prepare pin registration artwork on a ________________ light table.

(11.7) Marking: When marking appears on both the component side and the non-component side. The ______________________ will be used with each side or layer.

(11.8) Marking: All marking that belongs only on the non-component side must be place on ____________________________.

(11.9) Marking: Company logo, marking for test purposes, etc. are typically marked on the _____________ side.

(11.0) Final check: A final check on the artwork should be made for markings and proper positioning of tape and pad. The designer should also check for ________________ that have not been connected.

The answers for each question can be found on the pages listed below:

Question 1.1. Pg. 351	Question 6. Pg. 353
Question 1.2. Pg. 351	Question 7. Pg. 353
Question 1.3. Pg. 351	Question 8. Pg. 353
Question 1.4. Pg. 351	Question 9. Pg. 354
Question 1.5. Pg. 351	Question 10. Pg. 354
Question 1.6. Pg. 351	Question 11.1. Pg. 354
Question 1.7. Pg. 351	Question 11.2. Pg. 354
Question 1.8. Pg. 351	Question 11.3. Pg. 356
Question 1.9. Pg. 351	Question 11.4. Pg. 356
Question 1.10. Pg. 351	Question 11.5. Pg. 356
Question 1.11. Pg. 351	Question 11.6. Pg. 357
Question 2. Pg. 352	Question 11.7. Pg. 357
Question 3. Pg. 352	Question 11.8. Pg. 358
Question 4. Pg. 352	Question 11.9. Pg. 358
Question 5. Pg. 352	Question 11.0. Pg. 359

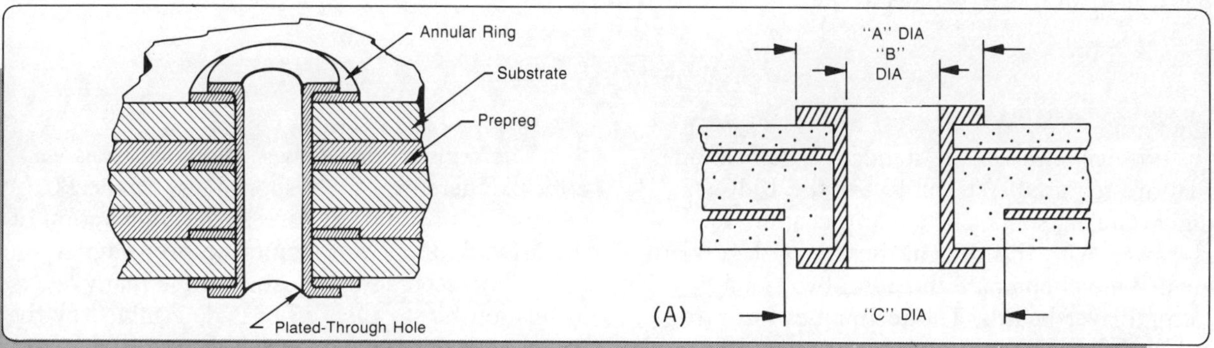

12

10-IC MULTILAYER BOARD

This chapter introduces the design of multilayer PC boards that accommodate up to 2½ times the component density of single layer and double layer boards. Their wiring is much more dense. Multilayer boards are used as fixed position back planes through which component laden PC boards are interconnected. They are also used as plug-in mother boards which hold components and edge connected daughter boards. They are used individually to facilitate highly complex circuits. This chapter defines the construction of the multilayer board. It compares advantages with disadvantages. This chapter is different from the previous chapters. Each individual step is not itemized. It gives an overview of the multilayer design. Relying on the experience of the previous chapters, it discusses the differences between multilayer and two sided boards.

We will also list the standard specifications that are required. We will layout the 10-IC digital/analog schematic used in chapter 10. However, with this design, there is far less board area. We will compare the areas two sided boards to multilayer boards. The techniques for laying out multilayer boards are similar to that of double sided boards, with a few exceptions.

Artmasters and helpful hints for the fabrication drawing will be discussed. Analyze the manufacturing sequence required for multilayer boards. It is quite different from the manufacturing sequence for double sided boards (covered in the analog design book). A check list is provided at the end of this chapter.

Multilayer PC Board Construction

Multilayer PC boards contain from three to as many as twenty or more layers. Each is composed of a single plane of copper. The copper laminated to one side of the substrate is referred to as the layer. Each layer is etched by the board manufacturer from the reduced artwork. These etched layers are bonded together using materials of glass fiber and epoxy resin. These bonding materials are referred to as "prepeg" or "B"-stage (Figure 12-1).

The registration of these many layers is very critical. Therefore, the responsibility of the PC designer is to use great care in the preparation of the artwork; this point cannot be over emphasized. The accuracy in locating these many layers in relationship to each other is determined by the width of the annular rings (which are used to interconnect the various points in each layer). Misalignment of layers could result in either short circuits or open connects within the multilayer board. Interconnections of the many layers are made possible through the use of plated-through holes (Figure 12-2). This technique is the same as that used for conventional double-sided printed circuit boards.

The plating method for making interconnections through holes in the dielectric substrate of PC boards has been established and proved in double-sided PC boards. This appears to be the most widely accepted technique. In practice it must account for 80 to 90 percent of the multilayer PC boards used today. Figure 12-3A, B and C show a typical cross section of a plated-through hole in a multilayer PC board. It connects the inner layers to the feedthrough by means of the plated wall on the hole.

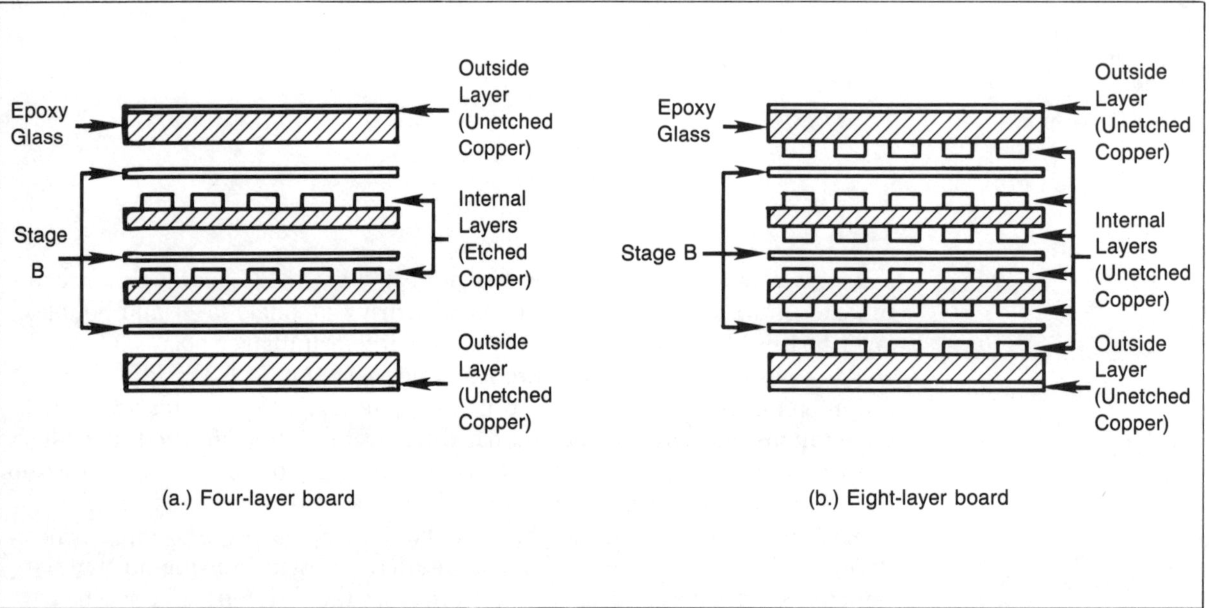

Figure 12-1. Cross section of a multilayer board

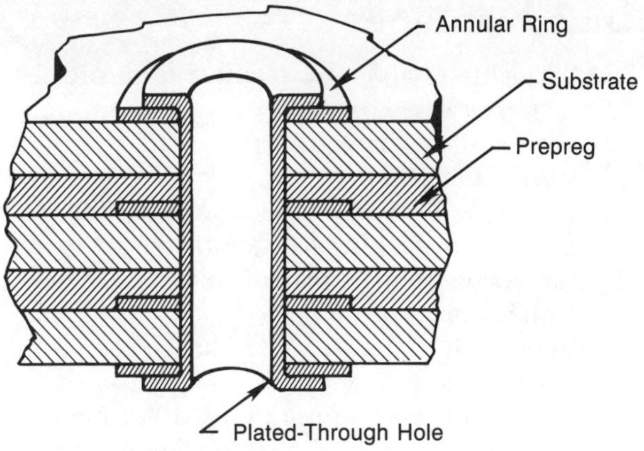

Figure 12-2. Plated-through hole

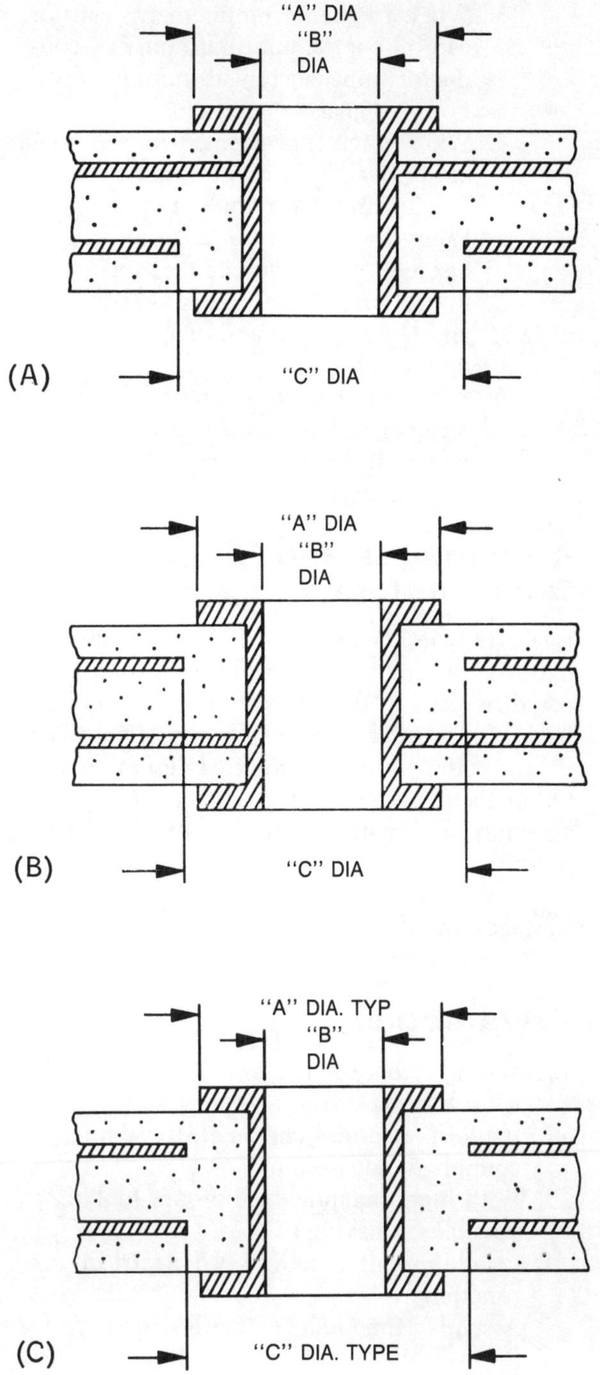

Figure 12-3. (A & B) show typical connection of inner layers to feed-through. (C) shows the clearance around the feed-through.

The plated-through interconnection technique is the most common. Most miniaturized multilayer PC boards benefit from its compactness and its ability to tale from two to twenty layers of interconnections without affecting the space requirement adversely.

The most common use of the plated-through hole is seen in Figure 12-3. The use of this technique is important because of the frequent need to connect only the outer layer of the multilayer PC board. This is accomplished by providing a clearance around the feedthrough pad on the inner layers. This is typical when a signal passes through a ground or voltage plane. While manufacturing the PC board, a concentric circle of copper is etched away, leaving a sufficient diameter around the hole. This minimum clearance assures that any misregistration during fabrication will not result in the shorting of traces or voltage and ground planes.

Need for Multilayer PC Board

Many factors influence the need for a multilayer printed wiring design:

1. When the interconnections in subassemblies are very complex and will require very complicated and costly wiring or harnessing.
2. When a large percentage of the interconnections require coupling or shielding.

3. When frequency requirements call for careful control and uniformity of conductor impedance, with minimal distortion in signals.
4. When interconnections, size, and weight are critical, as in military, airborne, missile, and space application.
5. When the quantity of components, discrete, integrated circuits, and terminals exceed the normal design area and close spacing and high density is a necessity.
6. When the need to connect a large number of interconnections from one PC board to another, as in the backplanes of complex systems.

Advantages and Disadvantages

Evaluate whether to use multilayer boards. Compare the different forms of printed wiring that could be used in the end product. Determine how many double-sided boards versus multilayer boards could fit into the allotted design area. Other factors are cost of manufacturing, assembly, and maintenance time of the end product.

The following is a list of some possible advantages and disadvantages when deciding to use multilayer boards.

Advantages

1. Wiring errors eliminated (compared to wire wrap boards)
2. Uniform electrical characteristics (impedence, coupling, and shielding)
3. With high quantities and proper tooling, a possible cost savings
4. Combining of structural and electrical functions
5. Assembly time reduced (hard wiring)
6. Space saved by tighter packaging
7. Higher reliability of interconnections

Disadvantages

1. Special tools and skills required for board repairs
2. High cost in small quantities
3. Design time is longer
4. Lead time for board fabrication is longer
5. Board warpage (thermal sensitivity)
6. Inspection of boards is difficult and generally requires microsectioning
7. Very costly and difficult to change completed boards
8. Open connections form (SMEAR) - (smearing of the epoxy pre-preg from drilling heat) in plated-through hole.

Packing Density

The use of multilayer boards is particularly important when the maximum component densities possible with double-sided boards must be exceeded. For instance, typical IC densities per square inch on double-sided boards are on the average of one IC per square inch. The multilayer board has a packing density of .6 square inch for each IC. This is also possible when using other components. The double-sided board generally uses four TO-18 case sizes transistors per square inch. This can be increased to over eight TO-18 case sizes per square inch when multilayer designs are used. These densities in component packaging can be increased to twice the amounts when the components are mounted so that the bodies touch each other in final assembly. This method does often require more layers for the interconnecting circuitry.

Specifications and Standards

Many standards and specifications that apply to single and double-sided boards also apply to multilayer PC boards. These are a few of the most commonly used:

A. Institute of Printed Circuits Specifications (IPC)

1. IPC-ML-910: Design and end product specification for Rigid Multilayer boards
2. IPC-ML-950: Performance specifications for Multilayer Printed wiring boards
3. IPC-ML-975: End product documentation specification for Multilayer Printed wiring boards

B. Department of Defense Specifications
1. MIL-STD-1495: Multilayer Printed wiring boards for electronic equipment
2. MIL-P-55617: Plastic sheet, thin laminate, metal clad, for printed wiring, primarily for multilayer
3. MIL-G-55636: Glass cloth, resin preimpregnated, B-stage, for multilayer printed wiring boards
4. MIL-P-55640: Printed wiring boards, multilayer, plated-through holes
5. MIL-STD-275 Printed Wiring For Electronic Equipments

General Design Considerations

When considering the use of multilayer PC boards over the conventional single or double-sided board, a great deal of pre-design time is required. There is an optimal limit to the size and cost of any discrete unit or functional subassembly/integrated circuit. It becomes uneconomical and inefficient to increase the size or complexity beyond that limit, because the unit then becomes too costly, specialized or not easily inter-changeable. Therefore, a great number of packaging requirements must be considered. If the investigation into the various approaches indicates a need for a multilayer PC board to achieve the desired component density, then the following list of design factors should be evaluated for the multilayer board:

A. Design Considerations:
1. Allotted board area
2. Number of layers required
3. Number of hole sizes
4. Total board thickness
   a. Inner layer thickness
5. Width of conductors (internal and external)
6. Terminal areas (internal and external)
7. Spacing considerations
   a. Center-to-center component mounting holes
   b. Terminal area to terminal area
   c. Terminal area to conductor
   d. Conductor to conductor
   e. Terminal area to board edge
   f. Conductor to mounting holes
   g. Conductor to board edge
8. Tolerances (design)
9. Drawing requirements
10. Special components
11. Special design requirements
12. Terms and definitions (unique to design)
13. Plating (type and thickness)

B. Performance and environmental considerations
1. Tensile strength
2. Flexure strength
3. Vibration
4. Mechanical shock
5. Temperature cycling
6. Thermal shock
7. Moisture resistance
8. Fungus resistance
9. Corrosion resistance (salt spray)
10. Twist or warp
11. Solderability and resolderability
12. Insulation resistance
13. Dielectric strength (voltage)
14. Temperature storage (high-low)
15. Conductor temperature rise
16. Conductor resistance
17. Flame resistance
18. Altitude requirement (high-low)
19. Machineability

Board Design Layout

When the complexity of a schematic diagram indicates the need for a multilayer board and when the many design considerations have been weighed, one of the first problems to solve is the distribution of signals, ground, and voltage planes between the number of layers required. There are many possible solutions to this problem. Therefore it becomes a trial-and-error method on the part of the drafter or designer to achieve the design with the minimum number of layers. In some companies, computers help determine the simplest multilayer design. This eliminates some of the trial-and-error method.

Before the actual design layout can be started, ground rules about conductor sizes must be developed. Preparation of the artwork can be accelerated if a uniform conductor width is used throughout he entire board design, although, some narrowing down in critical areas is permitted occasionally. Also, if power or ground distribution buses are used instead of planes, they should be made as wide as possible to minimize voltage drop and maximize shielding.

Once the board size has been determined, all module positions distributed, pad or hole location fixed and input-output methods selected, the formal layout of the board can be started. The layout should contain all the above information as well as mounting holes to secure the board to a frame or holes for mounting other connectors or components to the board. It is also important to indicate any useable areas on the board where no conductor should be placed. Figure 12-4 shows some typical examples of useable area. This will prevent the possible danger of the conductors on the inner layers from being cut through or damaged during hole drilling or final operation, before assembly.

Spacing of conductors is very important. It is best to determine the maximum number of conductors that can be spaced properly between two terminal areas in various regions of the board. Follow those ground rules during the layout. Figure 12-5 gives an example of the maximum number of traces that might run between the pads of an IC.

One person should work on the design from start to finish; the work should not be divided among several designers or drafters. The checking should be done by someone other than the designer.

At this point, the layouts of both a double-sided PC board and a multilayer PC board are compared. The necessary grouping of the logic gates and pins for gates should have been assigned, color codes for the different voltages, grounds and layers selected, and the number of layers determined.

When comparing the multilayer PC layout to the double-sided PC layout, the first things that appear different are that the ground and voltage IC pads on the multilayer design are color-coded and not layed in as bus lines. Figures 12-6 and 12-7 show this difference. The reason for this is that the voltages and ground will be on the inner layers of the multilayer PC board. The connections of these voltages and ground are made possible through feedthroughs which pass through the inner layer of the PC board.

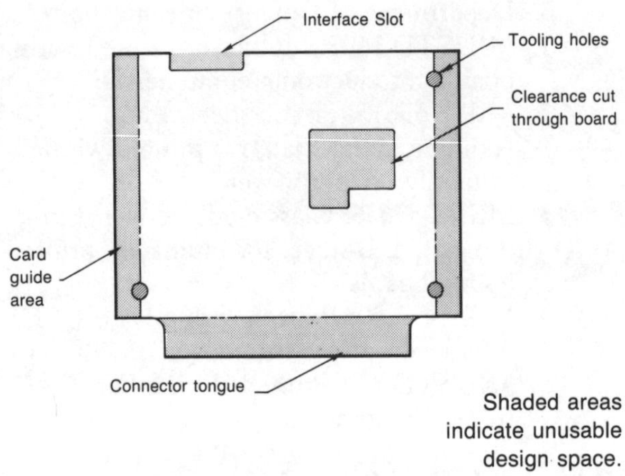

Figure 12-4. Usable area

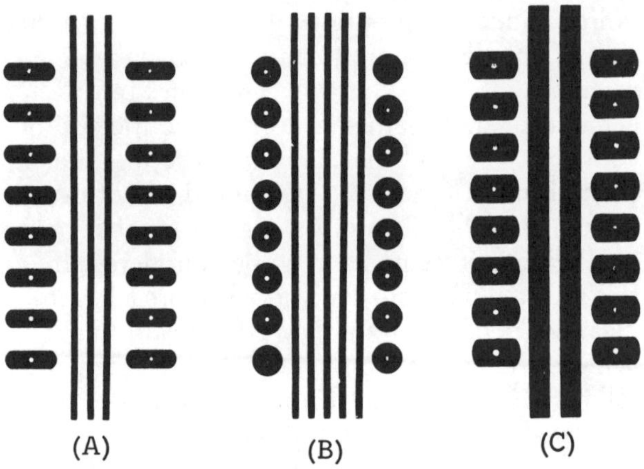

Figure 12-5. Traces between IC's.

The technique for placing these voltages and grounds on the inner layers provides more space for IC's and signal traces. This becomes very apparent when the areas of the two boards are compared. See Figures 12-8A & 12-8B and 12-9 for board dimensions. The amount of area allotted for the double-sided board is two square inches for each IC on Figure 12-8A. The area allotted for the multilayer board is about .6 square inches for each IC (shown in Figure 12-9). This is an increase of more than two and a half times in the multilayer design.

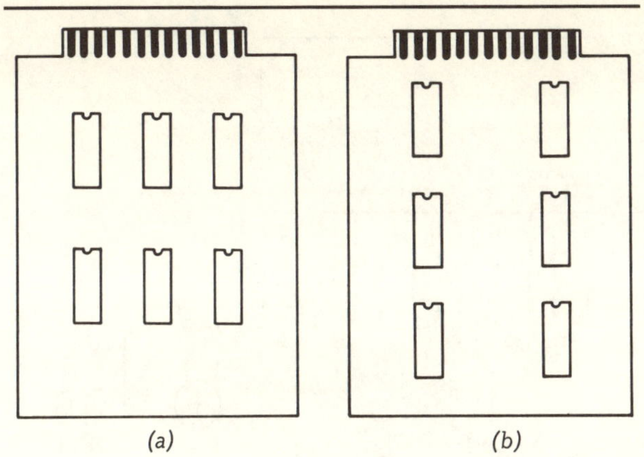

Two possible board layouts.

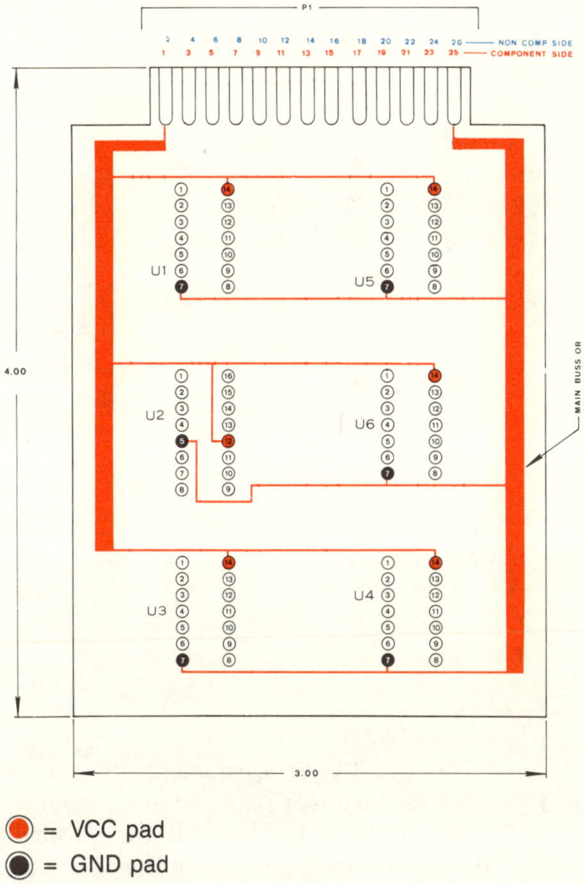

Figure 12-6. Shows the ground and voltage traces layed into the design.

With the IC's placed close together as seen in Figure 12-9, the next thing to notice is that the aisleways are much closer together. Therefore, if the techniques used in the double-sided boards were used, this multilayer design would not be possible.

Therefore, the technique of running traces between IC pads is used. The main idea when running traces between pads is to make these connections as short as possible and not to run a trace from one side of the board to the other. Careful grouping and placement of the IC's will insure the shortest possible routes. If the traces must run a great distance, then use the aisleway if possible. Figure 12-10 shows examples of both techniques. A very important reason for running traces between IC pads is that the amount of feedthroughs is eliminated or at least reduced. This increases the design "real estate". At this time it is also advisable to eliminate all unnecessary feedthroughs.

Figure 12-11 shows some possible feedthroughs that can be removed. Figure 12-12 shows three tooling holes that have been placed within the board outline. These tooling holes are used when the different layers of the PC board are laminated together during final assembly.

These tooling holes are important and should not be overlooked when making the layout. It is a recommended design practice to locate at least two holes so that the board cannot be misaligned during lamination and assembly. Also provide enough space around these tooling holes so that no traces will be cut or destroyed during drilling.

Once the penciled layout of the design is completed and all design parameters are met, prepare the artwork.

Designer Footnote

The above discussion has described the procedures for laying out a four-layer PC board. What must also be considered is how to layout (using color for the signal paths) a multilayer board with more than two signal planes. If the design requires more than two signal planes, it is recommended that a separate sheet of drafting film be used for each additional signal plane. This will eliminate the confusion of having all the signal paths on one sheet of drafting film, and will provide a less confusing layout for the designer when the artmaster is prepared.

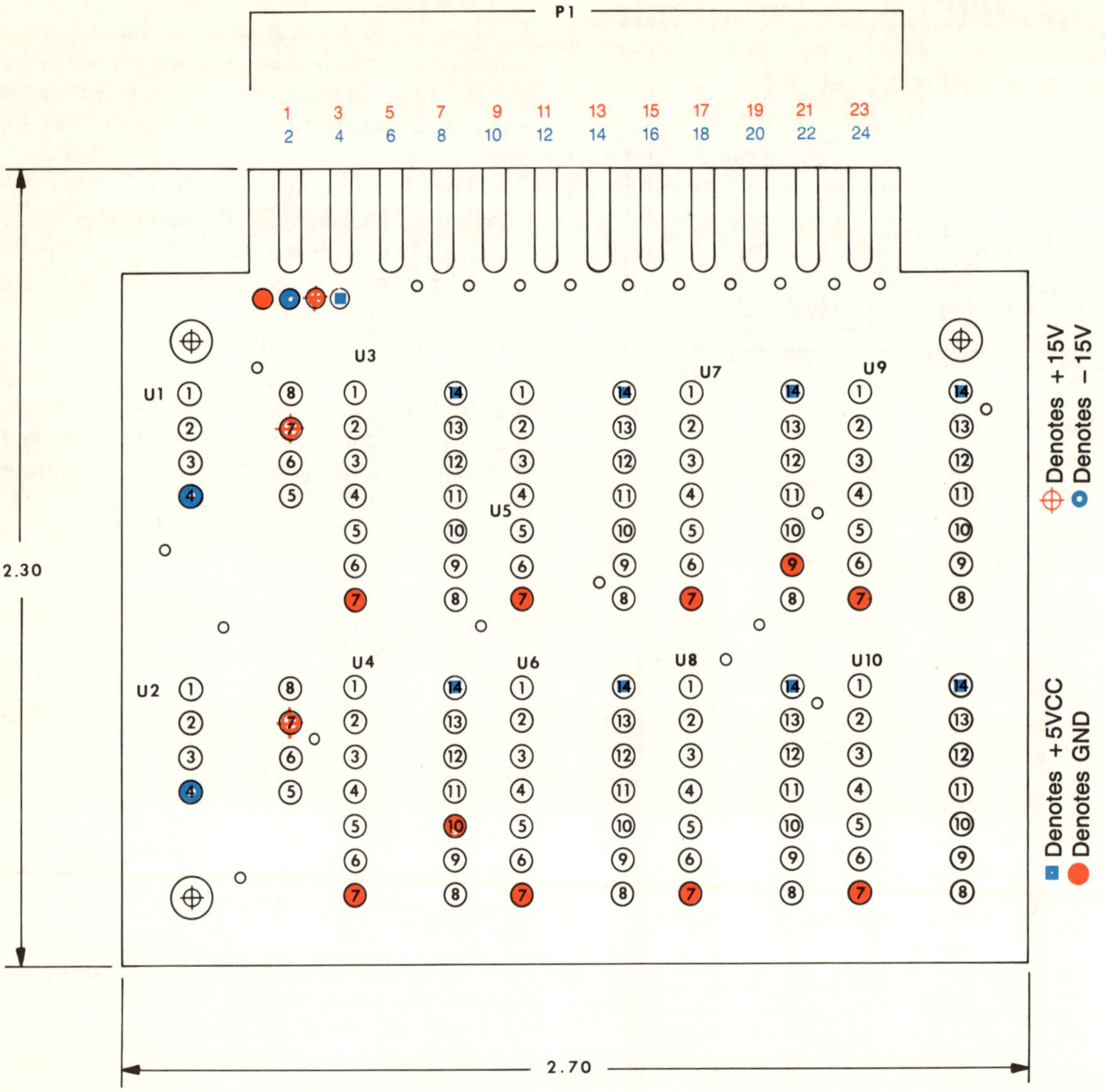

Figure 12-7. Use of color code GND & voltage.

Artwork

The artwork required for a multilayer PC board consists of extremely accurate patterns for each layer of the PC board. To obtain such accuracy, the artwork is made two or four times the finished size of the PC board. The layers of the board require registration or tooling holes, the board outline dimensions, extremely accurate outline marks of the board, and dimensions to which the board artwork is to be reduced.

One of the most critical parameters in producing multilayer boards is registration between layers. Proper registration of layers in the finished product depends on the precision of the artwork or tape-up. This cannot be overly emphasized.

To achieve this, tooling holes, preferably three in number, should be located relative to the outer edges of the finished board (Figure 12-14). These holes should be located close to the edge of the board, and appear on each layer of the art-

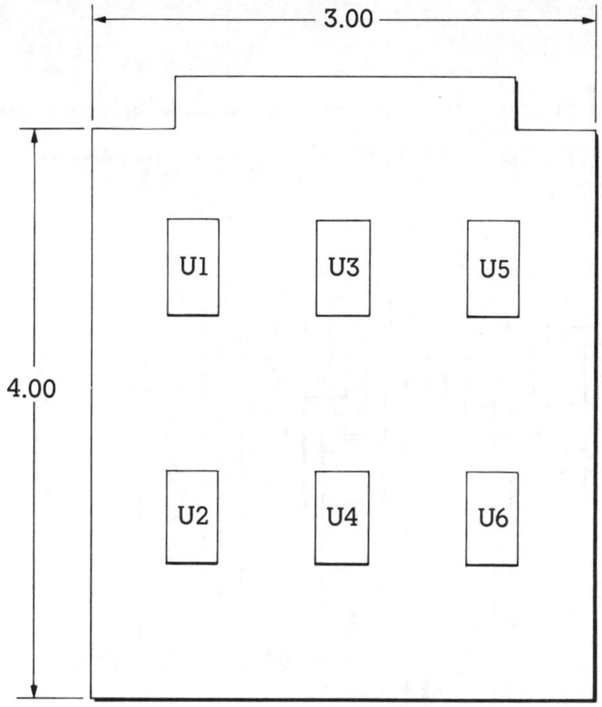

Figure 12-8A. Double-sided Board with 2 Square Inches per IC.

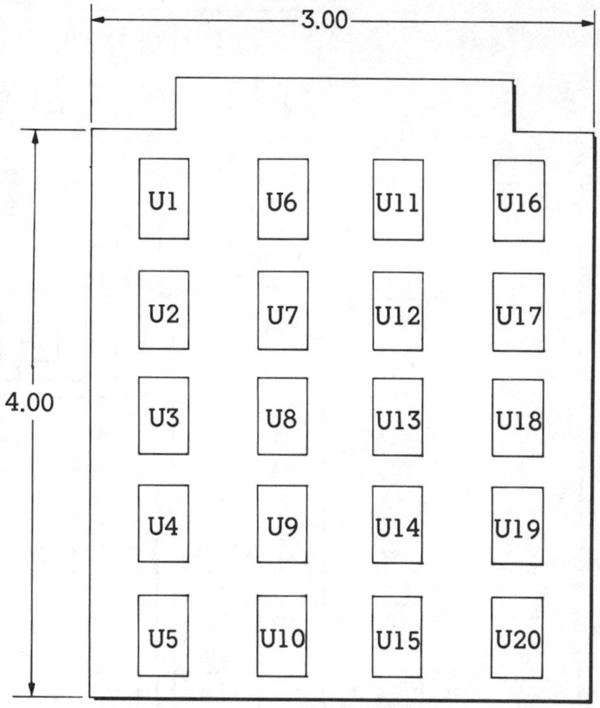

Figure 12-9. Multilayer Board with (.6) Square Inches per IC.

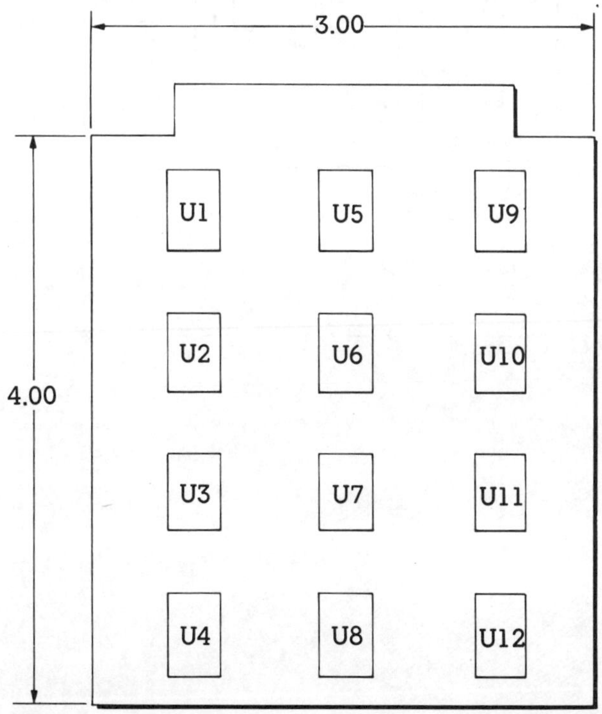

Figure 12-8B. Double-sided Board with 1 Square Inch per IC.

work. The circuit pattern is then located in relationship to these three holes. In Figure 12-4, note that these holes are dimensioned relative to each other rather than from the board edges. This procedure eliminates the tolerance build-up that would otherwise contribute to inaccuracy in the relative placement of the board layers.

In addition to the tooling holes, each layer of the artwork should also include two reduction targets with the exact reduced dimension and tolerance between them. This serves as a guide for the photographer when reducing the artwork (Figure 12-15). In some cases two sets of reduction targets and dimensions are used on the artwork (Figure 12-16). The reason for the two sets of targets and dimensions is that, as the size of the board in one direction becomes greater than 6.00 at 1:1, it becomes necessary for the photographer to check in each direction of measurement for errors in the camera parallax.

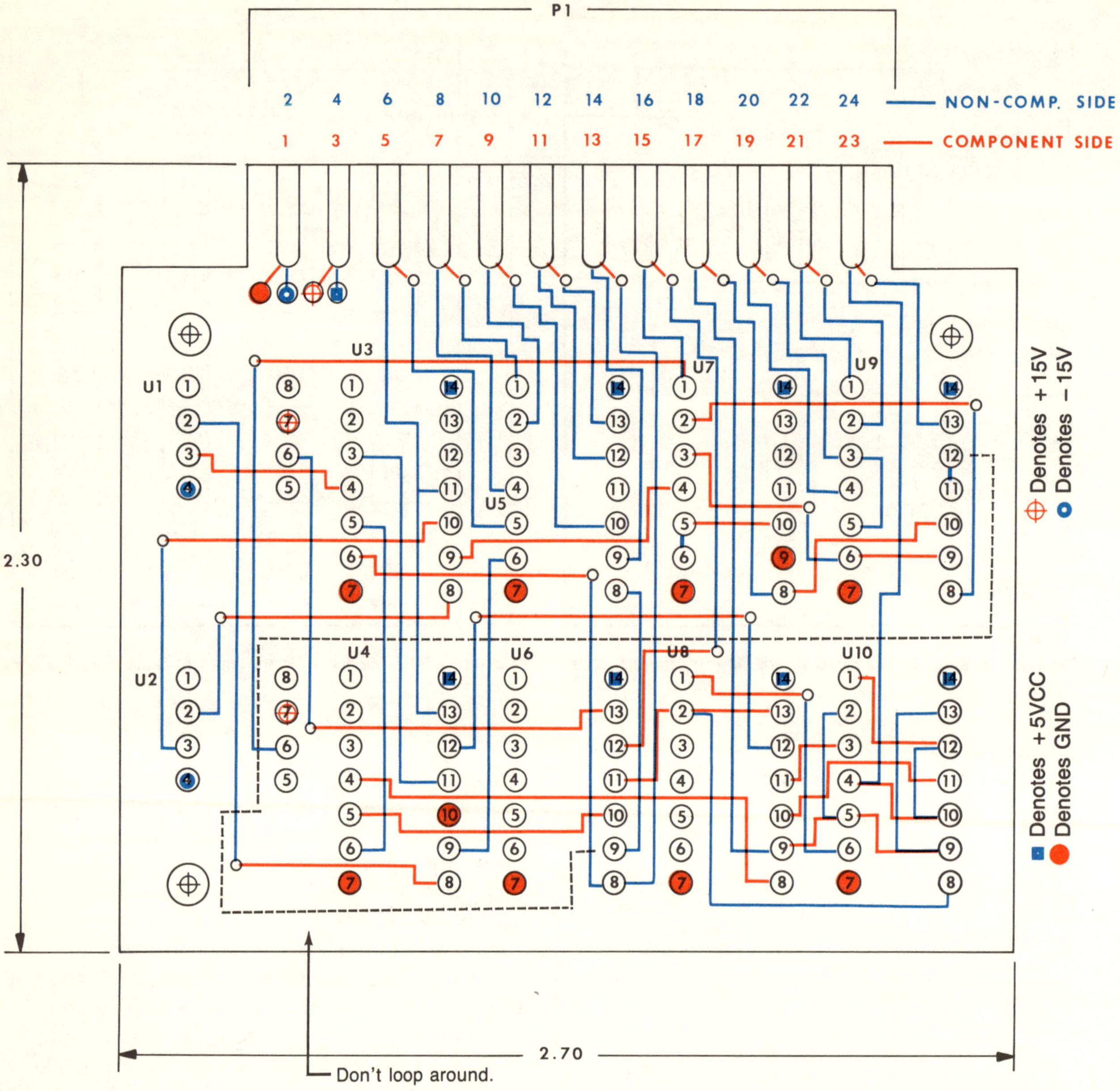

Figure 12-10. Use of Aisleways.

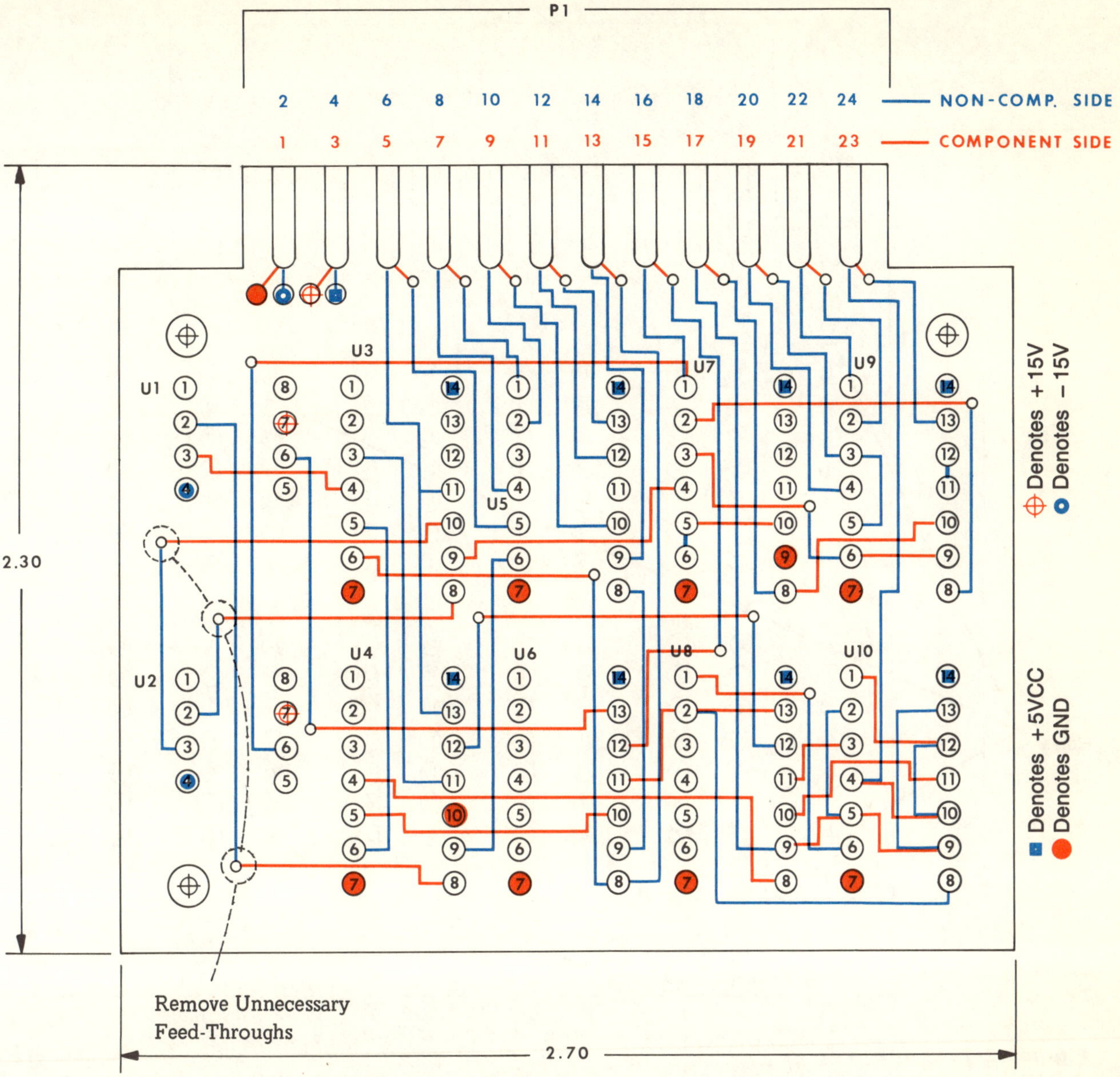
P1
2 4 6 8 10 12 14 16 18 20 22 24 NON-COMP. SIDE
1 3 5 7 9 11 13 15 17 19 21 23 COMPONENT SIDE
U1
U2
U3
U4
U5
U6
U7
U8
U9
U10
2.30
2.70
Remove Unnecessary
Feed-Throughs
Denotes +15V
Denotes −15V
Denotes +5VCC
Denotes GND

Figure 12-11.

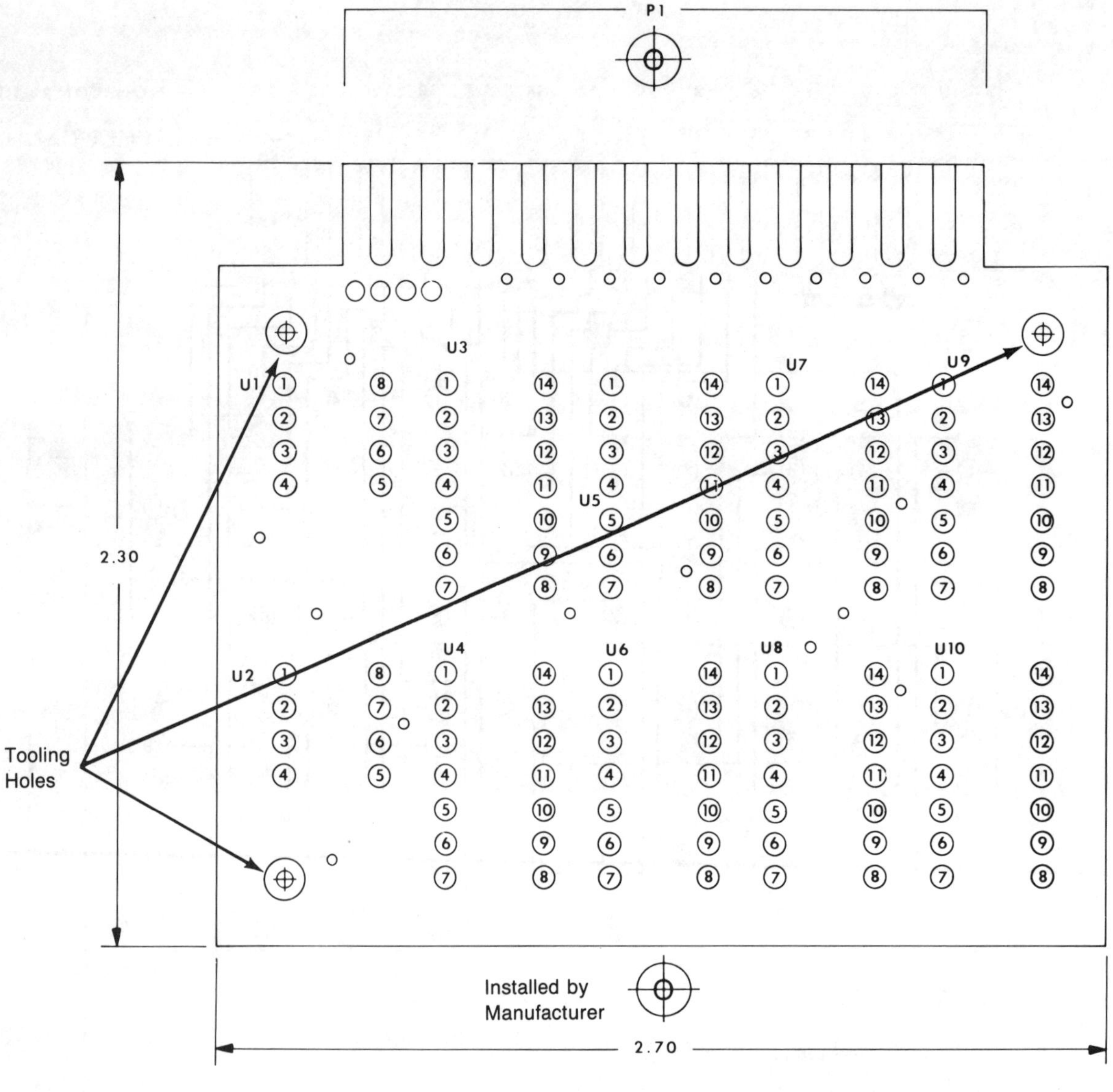
P1
U1
U2
U3
U4
U5
U6
U7
U8
U9
U10
2.30
Tooling
Holes
Installed by
Manufacturer
2.70

Figure 12-12.

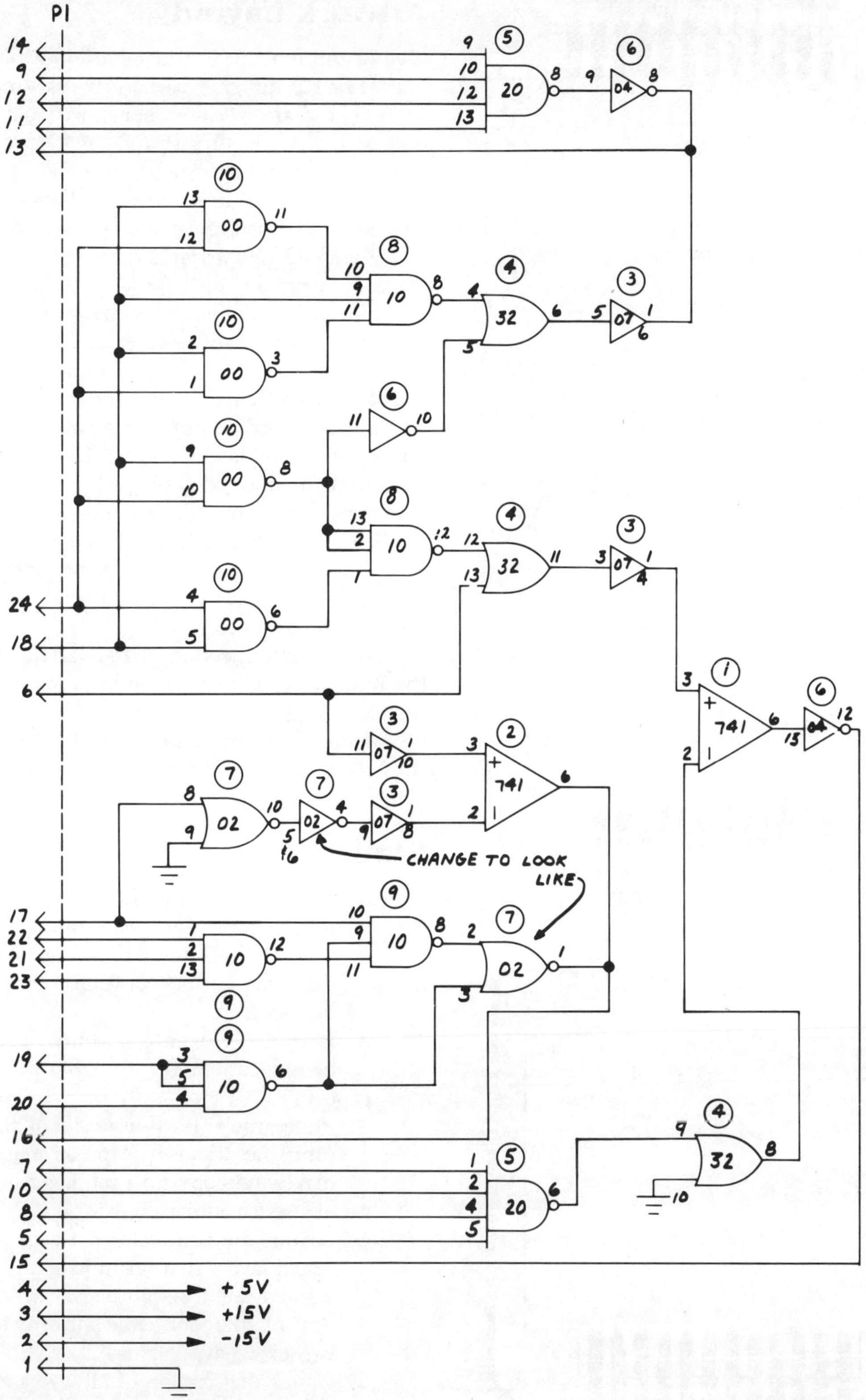

Figure 12-13. Multilayer board schematic shown is for tape-up shown in Figures 12-7, 12-10 12-11, and 12-12

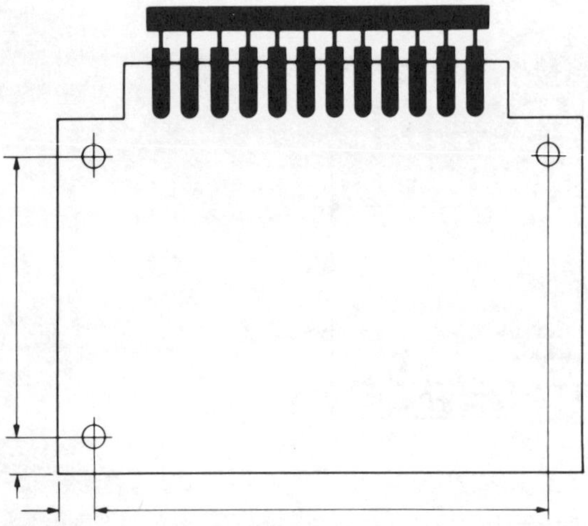

Figure 12-14. Tooling Hole Location

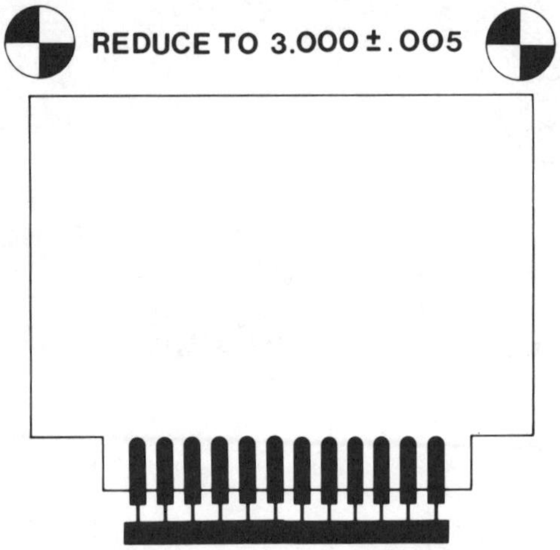

Figure 12-15. One Set of Reducing Targets

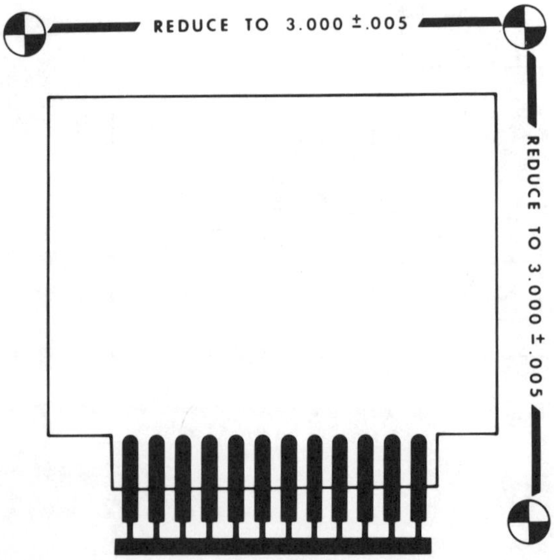

Figure 12-16. Two Sets of Reducing Targets

Artwork Layout

One of the most critical responsibilities of the drafter or PC designer is the artwork preparation itself. The drafter or designer must make sure that each sheet has registration marks or tooling holes, reduction targets and dimensions, and that each layer is identified correctly. These items are very important and must not be overlooked.

Figure 12-17 illustrates the cross section of the four layers on this board.

Figure 12-17A illustrates the four taped layers of a multilayer artwork. Note: The three tooling holes or registration marks, reduction dimension targets, corner marks, connector fingers and overall board outline appear on the padmaster only. Note taped layers 1 and 4 have only traces. The items mentioned above will be combined when the padmaster is overlayed and photographed. See photographed layers 1 and 4, Figure 12-17B. Layer 2 is the ground plane, and layer 3 is the voltage plane. These two layers were photographically produced.

The interconnection of the various layers of the multilayer board is made by using plated-through holes (as done in a double-sided board).

Keep the above information in mind. With the following list of items, a multilayer artwork will easily be completed:

Steps

1. The first step should be to layout the tooling holes and board outline.
2. Two or three reduction targets and the exact dimension between them should be included.
3. Layout the main component pads (or hole) locations for IC patterns, resistors, capacitors, etc.
4. Photograph (1:1 film positive) or contact print the 2:1 or 4:1 tape-up master. Now make one copy for each layer of artwork or use a padmaster.
5. Layout the conductor patterns for the internal layers first. Eliminate any unwanted feedthrough pads.
6. Layout the conductor patterns for the two external layers last.
7. Check registration of all layers and clearance for all conductor patterns.
8. The following considerations must also

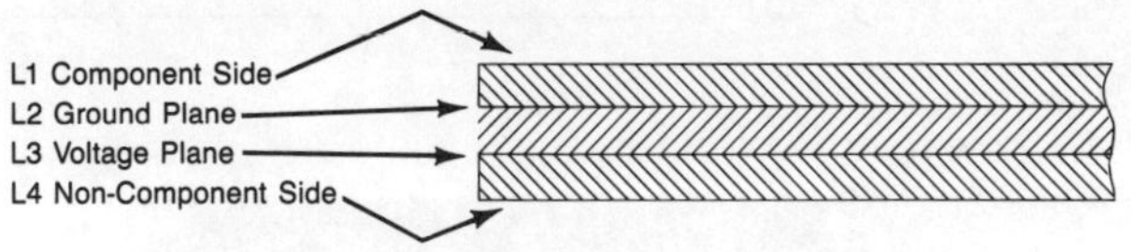

Figure 12-17. Cross Section

be given careful attention to insure a satisfying end product:

a. Conductor layout - tape of proper widths should be selected to produce the desired conductor width after reduction.
b. Pad and hole layout - pads are usually round, although square and specially shaped pads are sometimes used (see hole formula, for selecting proper pad size). Where ground and voltage planes appear as inner layers, the clearance hole required to eliminate unwanted connections to the plated- through holes must be clearly marked. This is essential as these are fabricated in the plane prior to laminating.

9. Identify each layer of artwork starting with sheet one component side, sheet two voltage, etc. This identification is to be placed outside the board outline.
10. See Figure for coupled schematic for layout and tape-up Figure 12-17A and 12-17B.

Corner Markings

Corner marks and part outlines are useful features for the finished products, especially when hand tooling does not exist. As a minimum, corner marks should appear on each layer of the artwork; these simplify the checking processes.

Identification Markings

Markings such as nomenclature, part number, and component symbols are frequently desired on multilayer PC boards. Such markings can be made either by etching simultaneously with the conductor pattern or by printing in ink.

Etched markings are usually lower in cost, provided the marks are large enough. Care must be taken when locating these marks so as to prevent any possible shorts between traces or pads on the final board. Markings printed by using epoxy-based ink may be smaller than etched markings. Small printed markings will be easier to read, and the ink is as good an insulator as the base laminate. Also, ink markings remain readable after flow soldering, since solder will not adhere to the ink.

Whatever method of identification is used, the marks should be placed on the outer layers of the final product.

Hole Sizes

A plated-through hole must be sufficiently large enough to allow the required deposit of metal on the wall. (See the standard hole formula, chapter 8). As a rule of thumb, minimum hole diameters should be approximately ½ of the board thickness after plating. This will allow the deposition of approximately .0005-in. to .0002-in. of metal on the wall of each hole. The metal deposited upon the surface will be up to two times as much as that deposited in the hole. Very small holes can be plated-through. However, the ratio increases of metal deposited on the surface to that deposited within the holes.

Round holes should be used in multilayer boards unless a round hole is completely unacceptable because of the shape of some components to be mounted on the PC board. Slots, square holes, etc., add significantly to the cost of any PC board because of the complexity of the operation. Therefore, try to avoid slots or square holes. Figures 12-18A and 12-18B illustrate the cross section of a plated-through hole of a multilayer PC board.

Hole Location

There are three methods of locating holes in multilayer boards:

1. Hole location by pad center on the reduced artmaster (or film): Using this method, all holes are located with reference to pad centers. This method permits the use of optical drilling with holes located directly from the artmaster (or film). Figure 12-19.
2. Hole location by dimensions from the tooling holes: This method of locating hole centers is by dimension from datum

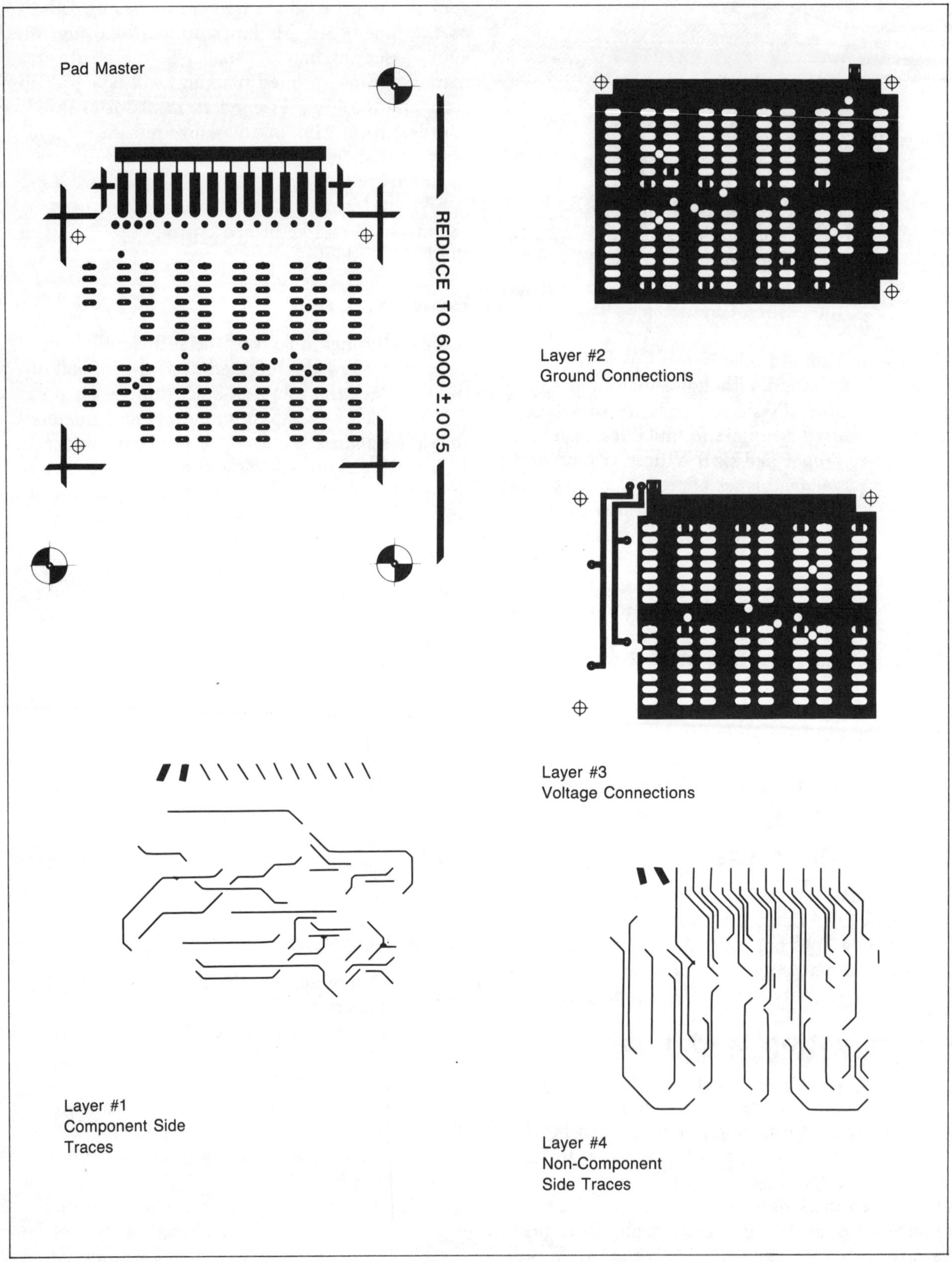

Figure 12-17A. Tape-up Layers

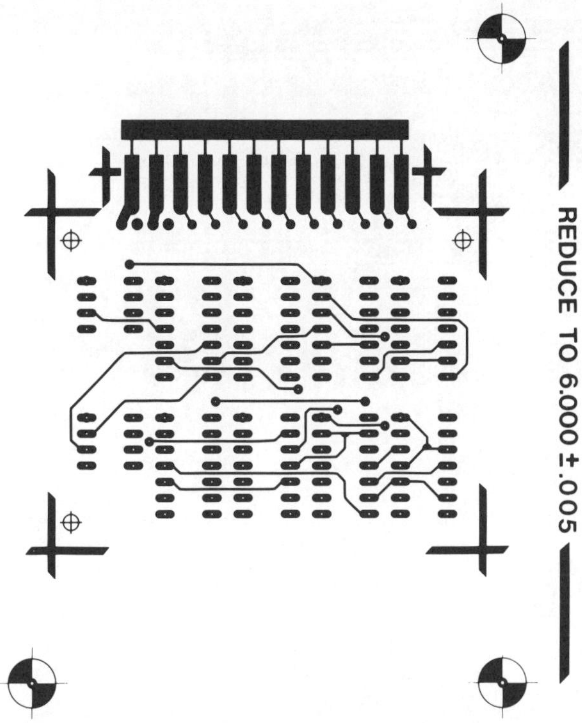

Figure 12-18A. Cross Section

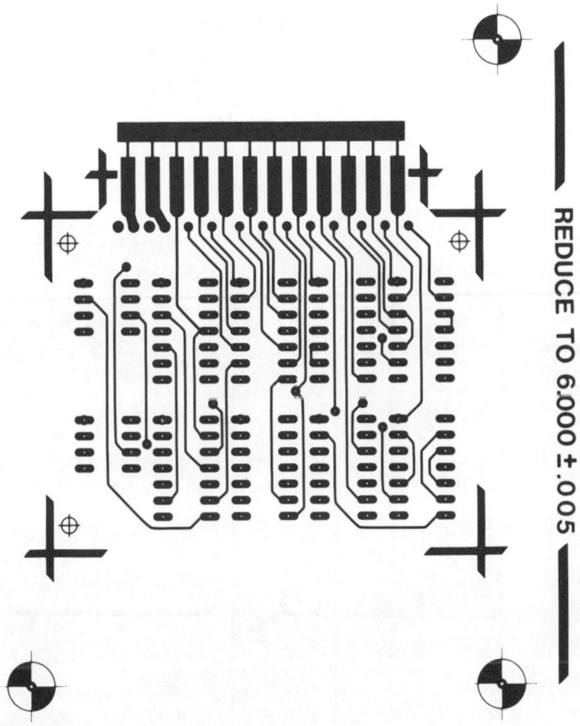

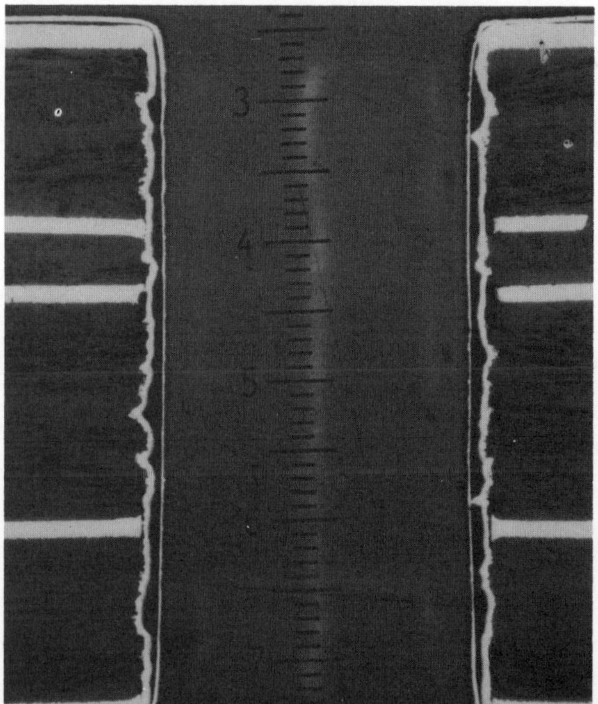

Figure 12-18B. Cross Section of Plated Thru Hole

Figure 12-17B.

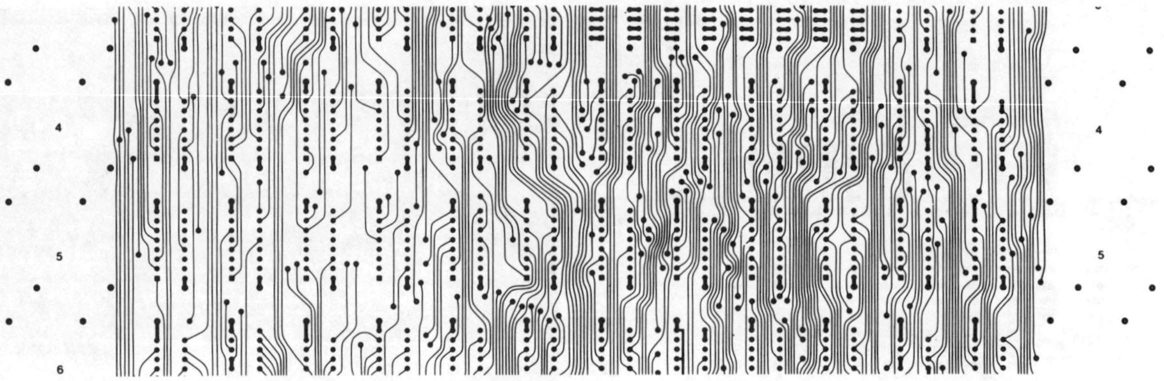

Figure 12-19. Artmaster

lines or X and Y coordinates. When this method is used, the artwork must be carefully layed out to match dimensions which identify hole locations. Figure 12-20. Groups of holes may be required to match a specific component pattern such as an IC, TO-5 case or transistor pattern (Figure 12-21). These holes may be located dimensionally. One hole within the group should be designated as a reference hole and located dimensionally to the tooling hole. The remainder of the holes within the group should be located dimensionally to the reference hole.

3. Hole location by grid pattern: Most designers locate hole centers on a predetermined grid pattern. Grid patterns containing increments of .100-in., and .050-in. are commonly used. Listed below are some of the reasons for using a grid.
   a. Provides a standard base for spacing and locating component mounting holes, resulting in a simplification of component lead bending when machine bending and trimming is used.
   b. Using a grid pattern on the taping is helpful in precisely locating pads on a "pattern to hole" and layer to layer dimensional system (see Figure 12-22).
   c. A grid system is often used only for design standardization. Unless a good reason for locating holes to a grid exists, all holes should be located to pad centers. Multilayer boards or any type of PC board with holes located to pad center will have a better registration and cost less to produce, as one potential set of tolerance build-up is eliminated.

Materials

The base material of the multilayer PC board should be specified as GE or GF grade glass-laminated epoxy, polyester film, polyimide film, or teflon. The use of epoxy glass as a base material assures high stability and minimum warpage. This

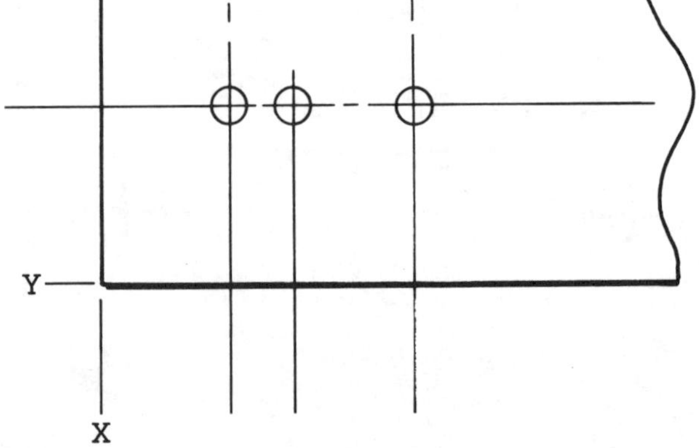

Figure 12-20. X-Y Coordinate Locating

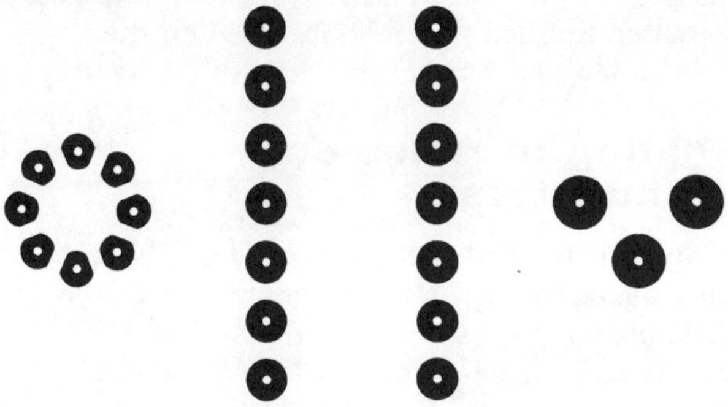

Figure 12-21. Specific Component Hole Patterns

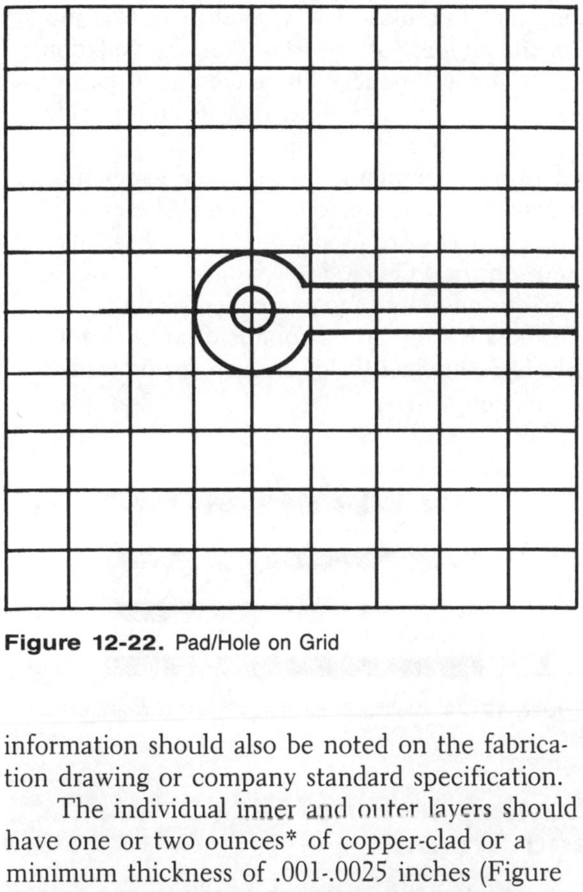

Figure 12-22. Pad/Hole on Grid

information should also be noted on the fabrication drawing or company standard specification.

The individual inner and outer layers should have one or two ounces* of copper-clad or a minimum thickness of .001-.0025 inches (Figure 12-23). Outside layers should be one ounce copper since the plating adds to the external copper and makes etching more difficult. Photo 1 shows individual inner and outer layers.

The etched layers will be laminated together using prepreg or "B" stage epoxy glass cloth. Prepreg or "B" stage epoxy glass is a single layer of glass cloth which has been saturated with a specific amount of epoxy resin. The resin is uncured or partially cured, and will become plastic and flow under the application of heat and pressure. Photo 2 shows "B" stage material.

The prepreg or "B" stage material will range in thickness from .0025 to .006 of an inch before lamination. The thickness used depends on the thickness of the copper on each layer and the number of layers to be laminated together. Therefore, it is recommended that the final cross sectional thickness of the multilayer board be specified on the fabrication drawing (Figure 12-24). The manufacturer selects laminating materials which produce the desired structure.

Lamination

Laminating is one of the most critical processes in producing a multilayer PC board. A multilayer PC board is produced from a combination of etched internal layers and prepreg or "B" stage material. Figure 12-25 shows a typical multilayer board lamination lay-up process. Lamination is performed by using fixtures with registration pins to maintain etched layer relationships to one another. Once the layers are registered, the multilayer assembly is then placed under high pressure and heat, usually 300 PSI and 350°F for about 55 minutes. The application of heat and pressure to the laminate fixture package converts "B" stage semi-cured material between etched layers to fully cured "C" stage epoxy. During the process, the "B" stage resin becomes a liquid adhesive which fills all etched circuit pattern voids and bonds the layers together. The structure must be totally free from all voids and this can be done only by a complete adherence of the epoxy resin throughout the conductor patterns and glass fibers on the inner layers. Two ounces of copper, for example, requires a larger amount of resin to fill the voids around the conductor than does one ounce of copper. Similarly, three ounces of copper require more.

The bonding materials, which are obtained from different suppliers, differ considerably from one another in thickness. This affects the end product thickness. Therefore it is recommended that the manufacturer of the multilayer PC board be given the freedom to select the base and

* Ounces of copper in PC industries mean the weight of 1 sq. ft. of copper (the thickness is usually .0014 for 1 ounce.)

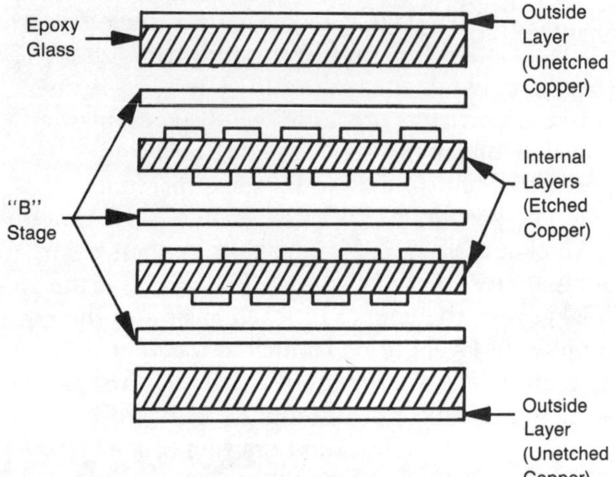

Figure 12-23. Individual Layers

Photo 1. Inner & outer layers

Photo 2. "B" stage material

prepreg or "B" stage materials that will meet the required finished thickness specified on the fabrication drawing.

Plating of Holes and Conductors

Copper is the first metal deposited on the walls of a plated-through hole in a multilayer PC board. The plating process used for the multilayer PC board is the same as for the double-sided PC board.

Although some PC boards are produced with only copper-plated circuit patterns, the majority of PC boards have conductors plated with a second metal or alloy. The second metal will protect the copper pattern from erosion, oxidation, etc. In many instances, these conductor patterns are plated with tin-lead for easy assembly soldering. Gold is plated on some conductor patterns, but is most commonly used for connector fingers or conductor terminations. Tin-nickel and nickel-rhodium alloys are used when much wear may occur on the PC board.

The minimum recommended plating thickness is from 80 millioniths of an inch for gold to 300 millionths of an inch for tin-lead, tin-nickel, and others.

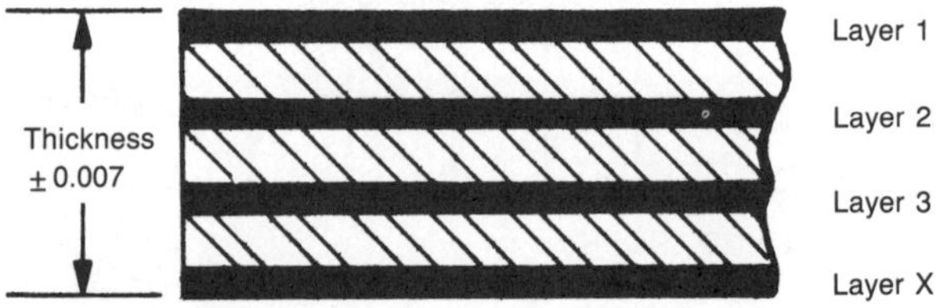

Figure 12-24. Multilayer Final Thickness (Cross Section)

Cutting to Final Size and Shape

To facilitate production and handling, one sheet of copper may actually contain many final etched conductor patterns. Figure 12-27 shows a typical multi-etched sheet.

Various methods are used to cut multilayer PC boards to their final size and shape. For small to medium quantities, routing to size and shape is the most common. Parts will be within ±.010 inch of the specified dimension. If large quan-

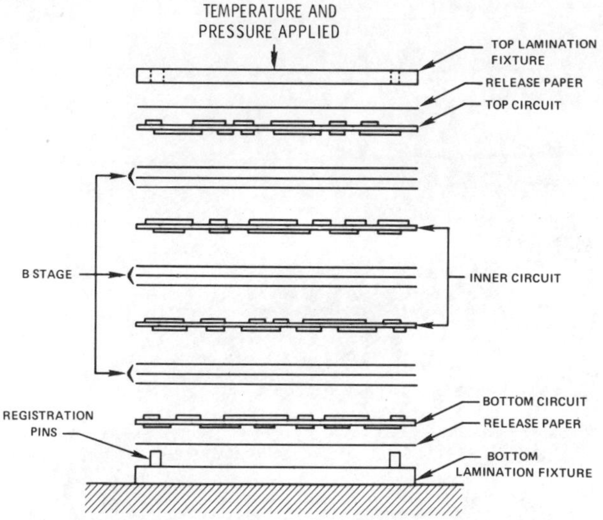

Figure 12-25. Multilayer Board Lay-up Process

Figure 12-26. Lay-up Process (Steps 1 thrugh 7). The following photos show the board lamination lay-up process.

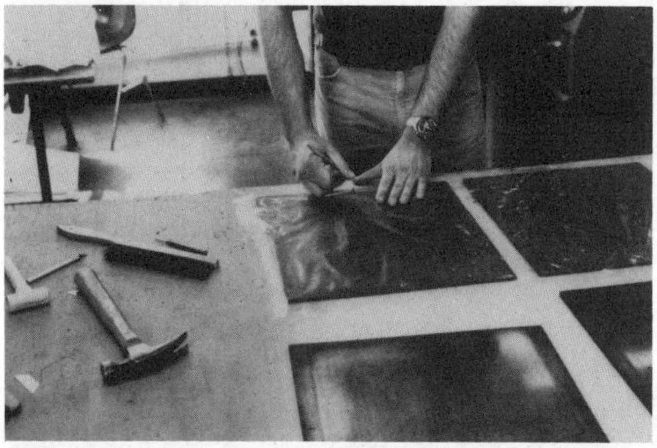

Step 1 shows bottom lamination fixture with release paper.

Step 2 shows registration pins, bottom circuit, and "B" stage material.

Step 3 shows stop circuit and top lamination fixture.

Step 4 shows hot press for the application of heat and pressure to laminate the board.

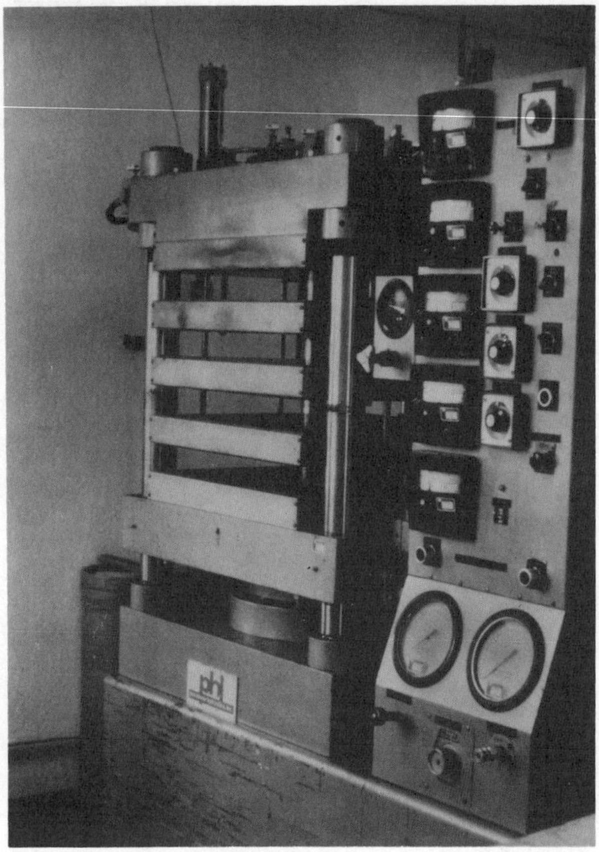

Step 5 shows cold press. After the board is unloaded from the hot press, it may be desirable to quickly transfer the board to another press for cooling under pressure. Cooling under pressure is a good practice to minimize warping of the finished laminate.

tities are required (for example, a PC board used in the common calculator or digital watches), die and special tooling are used to produce the final shape. The technique is more expensive, although the final dimension is very accurate and consistent. Cost is then justified by quantity.

PC Board Thickness

On multilayer boards designed with fingers or edges intended for insertion into connectors, the board thickness is important. To meet the requirements of most commercial connectors, the PC board thickness typically is .060 or $\pm$.007-in of the nominal thickness specified. This is standard tolerancing. See Figure 12-24.

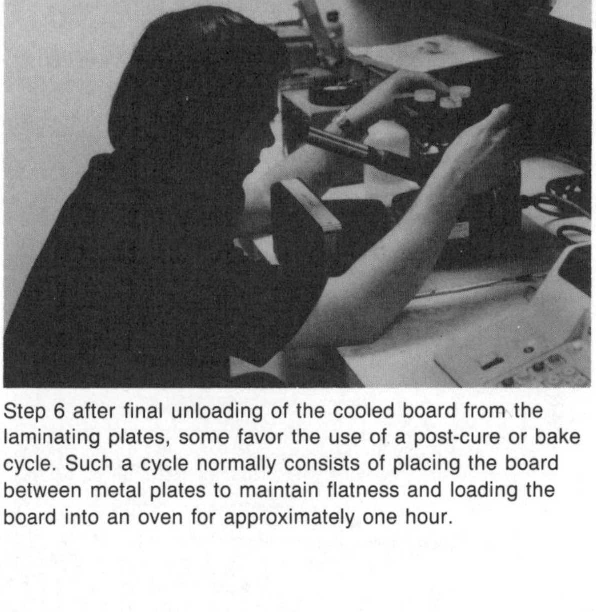

Step 6 after final unloading of the cooled board from the laminating plates, some favor the use of a post-cure or bake cycle. Such a cycle normally consists of placing the board between metal plates to maintain flatness and loading the board into an oven for approximately one hour.

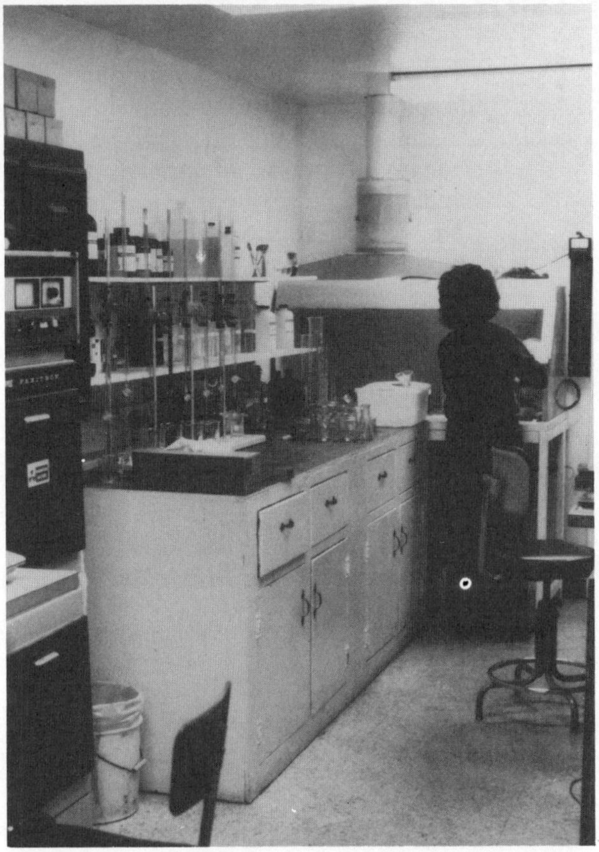

Step 7 most multilayer board manufacturers check the completed board for material movement, shrinkage, growth, misregistration, voids, delamination, blistering, and a host of other flaws.

Figure 12-27. Typical Multi-Etched Sheet

Fabrication Drawing

Possibly the most important document required for an accurate and economic production of a multilayer PC board is the initial preparation of a correct and complete fabrication drawing.

The following list of items should be required on any fabrication:

1. An outline drawing of the finished multilayer PC board that shows all the necessary dimensions together with their tolerance.
2. A cross-sectional view showing the desired end product, indicating all layers (by number), and critical dimensions (Figure 12-24).
3. Conductor pattern images for the non-component side is identified by layer number (this is optional, but very desirable). Also, each conductor pattern layer should be identified numerically.
4. A description of the base material such as; G10 or FR4 or other selected materials.
5. A table showing hole diameters, tolerances, and number of holes, and which holes are clearance or plated-through holes. Also, a coding system that shows which holes in the table can be identified on the multilayer circuit board. See Figure 12-28 for a typical hole table.
6. A system to locate holes either by standard grid pattern, or by X-Y coordinate dimensions. Use only one corner when specifying coordinate dimensions.
7. Show any markings such as reference designations, nomenclature, part number of the fabricated PC board, part number of the board assembly, and so forth. These markings can either be printed on the board with special ink or etched

simultaneously with the conductor pattern. In the latter case, these markings will appear on the artwork.

8. Name and number of any other notes and specifications that may apply to the end product.

Hole	Quantity	Diameter	Tolerance
A	2	.375	± 005
B	13	.093	± 002
C	7	.125	± 003
D	All other	.040	± 003

Figure 12-28. Typical Hole Table

PC Board Checklist

One of the last things done before a multilayer PC board is released for fabrication is the checking of all required design and artwork parameters. The following is a typical list that could be used for this important final step in the multilayer design:

1. Part numbers and logo.
2. Required reference designation. All IC's, transistors, pots, connector numbers, TP, E numbers.
3. Rev. letter
4. Title Block - complete.
5. Reduction targets.
6. All pin 1 on IC's are marked, positive end of capacitors marked, cathode end of diodes marked.
7. Board thickness.
8. Environmental and performance requirements.
9. Made in USA label.
10. Size of board and its relationship with mating structures. .125 min. clearance on two sides from all components.
11. Connector location on the PC board and its relationship with mating connector.
12. Operating temperature.
13. Conductor width: The standard width at 1:1 shall be .015 for signal lines and .050 for voltage, (or company standards).
14. Conductor spacing at 1:1 shall be .015 standard, .012 min., or company standards.
15. Component lead spacing.
16. All pads must be on grid intersections except feedthrough pads.
17. Rough assembly drawing.
18. Parts list.
19. Ground stakes should be placed on the PC board approximately every 16 square inches.

Quiz—Chapter 12

1. The techiques for laying out multilayer boards are similar to that of double sided boards, with a few exceptions. T F
2. The fabrication sequence for multiplayer boards is quite different from the manufacturing sequence for double sided boards. T F
3. Multilayer boards contain from three to as many as four layers of single plane copper. T F
4. Registration: The registration of multilayer board in most cases is not critical. T F
5. The interconnection method is the same as the double sided P.C. board for plated through holes. T F
6. Many factors influence the need for multilayer boards such as? List 3:
   1. ________________
   2. ________________
   3. ________________
7. Multilayer boards in some cases have advantages over a double sided board? List 3.
   1. ________________
   2. ________________
   3. ________________
8. Multilayer boards in some cases have disadvantages compared to double sided boards. List 3.
   1. ________________
   2. ________________
   3. ________________
9. The use of multilayer boards is particulary important when the maximum component densities possible with double sided boards must be exceeded. T F
10. When comparing multiplayer boards & double sided boards on pre-design time there is very little difference. T F
11. When you compare multilayer P.C. layout to the double sided layout, one of the first things that appears different is that the ground and voltage pins are not bused with a trace. They are only color coded. T F
12. Multilayer boards typically place the ground and voltage on the outside layers. T F
13. The area allotted for IC's on multilayer boards is .5 to .6 typically. T F
14. On multilayer boards it is very important to spend time on grouping and placement of the IC's. T F
15. When drafting the layout the designer should use a separate sheet of mylar for every two layers of signal lines. T F
16. Proper registration of layers in the finished product depends on the precesion of the __________.
17. Artwork: Corner marks should appear on each layer of the artwork. T F
18. Markings: Marking the multilayer board and the double sided board is basically the same T F
19. Multilayer boards: As a rule of thumb, minimum hole diameters should be approximately _______ of the board thickness.
20. Materials: The use of epoxy glass as a base material assures high stability a minimum ________.
21. The etched layers will be laminated together using prepreg or ______ epoxy glass cloth.
22. Lamination: Laminating is one of the most __________ processes in producing a multilayer board.
23. Plating: Although some P.C. boards are produced with only copper-plated circuit patterns, the majority of P.C. boards have conductors plated with a second metal or alloy except multilayer boards. T F
24. Multilayer board thickness: To meet the requirements of most commercial connectors, the P.C. board thickness should be ______ thick.
25. Fabrication drawing: A cross-sectional view showing the desired end product indicating all layers by number is required for both multilayer & double sided boards. T F

The answers for each question can be found on the pages listed below:

Question 1. Pg. 354
Question 2. Pg. 354
Question 3. Pg. 354
Question 4. Pg. 354
Question 5. Pg. 354
Question 6. Pg. 355, 357
Question 7. Pg. 356
Question 8. Pg. 356
Question 9. Pg. 356
Question 10. Pg. 357
Question 11. Pg. 358
Question 12. Pg. 358
Question 13. Pg. 358
Question 14. Pg. 359
Question 15. Pg. 359
Question 16. Pg. 360
Question 17. Pg. 367
Question 18. Pg. 367
Question 19. Pg. 367
Question 20. Pg. 370
Question 21. Pg. 371
Question 22. Pg. 371
Question 23. Pg. 372
Question 24. Pg. 374
Question 25. Pg. 375

APPENDIX

The appendix provides supplemental information for general charts and tables, schematic symbols, component data sheets, packaging considerations, layout considerations such as current wire charts, artmaster pad sizes, materials for taping and recommendations, general drawing requirements and terms and definitions. It also includes standard notes and grids required for PC design.

Printed Circuit Documentation

Whether a PC board is to be a one-of-a-kind prototype, or a high-volume production article, it should have some informative type of documentation describing the means to the end. A printed circuit drawing package should include the drawings shown below. Exactly *how much documentation,* and how it should be prepared will vary with budget and schedule as well as application.

Adequate documentation conveys to the user the basic electromechanical design concept, the type and quantity of parts and materials required, special manufacturing instructions, and up-to-date revisions.

Too little documentation results in misinterpretation, information gaps, and loss of uniform configuration. Manufacturing becomes dependent on individuals rather than documentation, often resulting in expensive rework and valuable lost time.

Too much documentation can result in increased drafting costs, and decreased manufacturing productivity due to time-consuming interpretation of overly complicated and confusing drawings.

Printed Circu

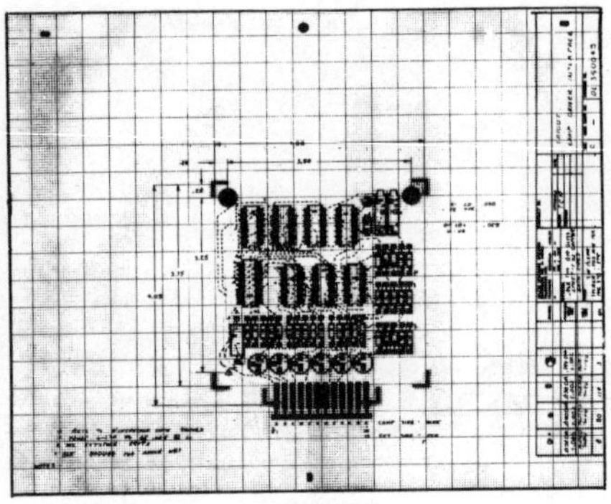

LAYOUT: The conceptual intermediate link between the schematic diagram and the master pattern.

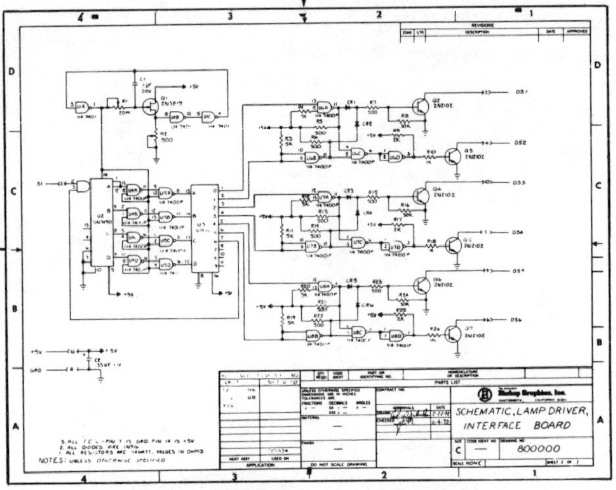

SCHEMATIC DIAGRAM: Consists of graphic symbols indicating the interconnections and functions of an electronic circuit. It is the basis for a printed circuit design. It is also used to test, evaluate and troubleshoot the completed circuit board.

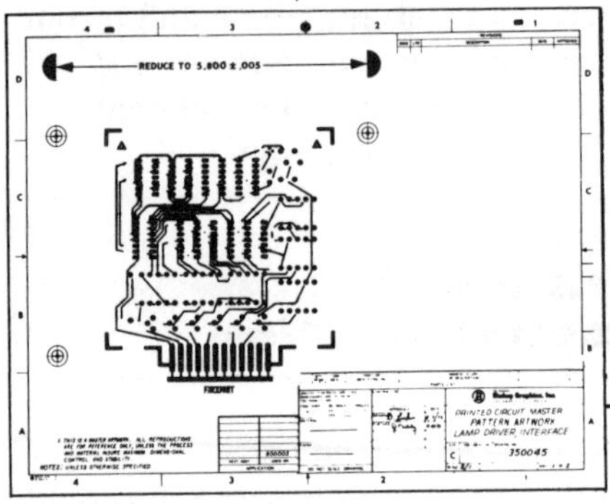

ARTWORK: Accurately scaled configuration of the printed circuit pattern from which the master pattern is photographically produced.

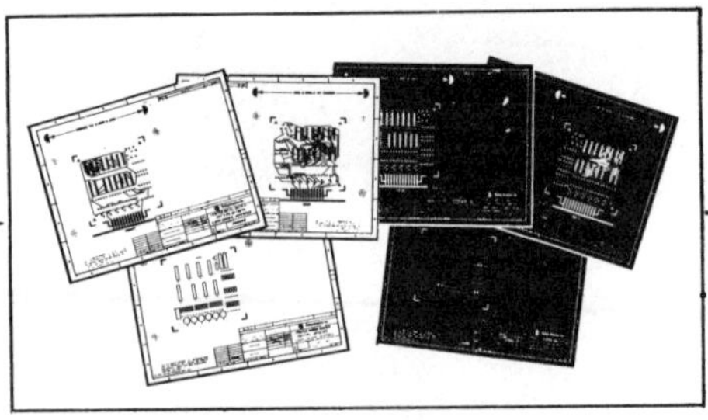

MASTER PATTERN: The actual size photo tool used to produce the printed circuit.

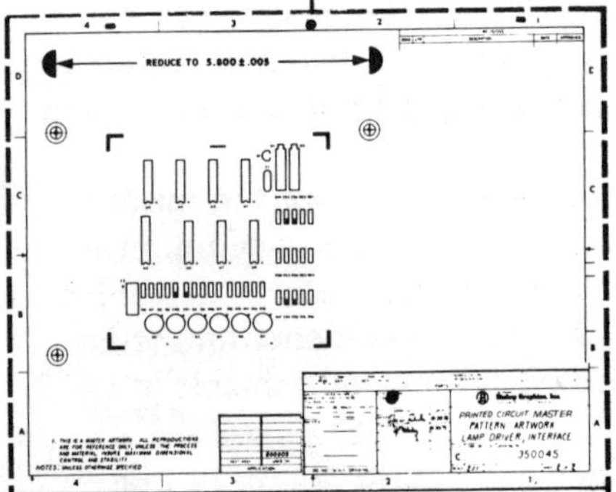

MARKING ARTWORK: Accurately scaled marking configurations to be printed on the printed circuit board.

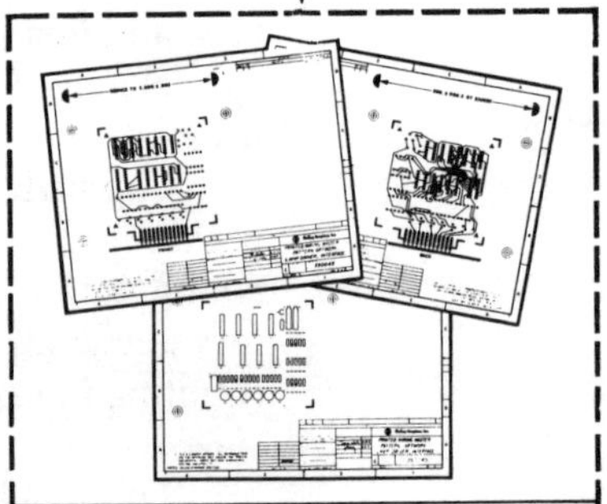

FILE MASTER ARTWORK (OPTIONAL): Photocopies of the original master artwork that prevent damage during handling and storage.

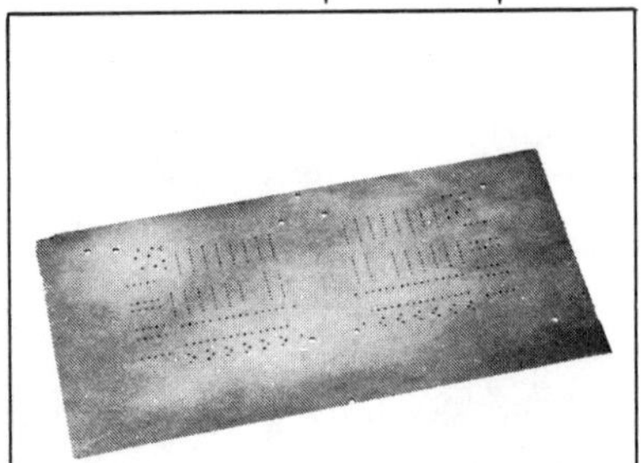

Two-Sided Drilled Panel (Two Boards) with Plated Through Holes

ocumentation

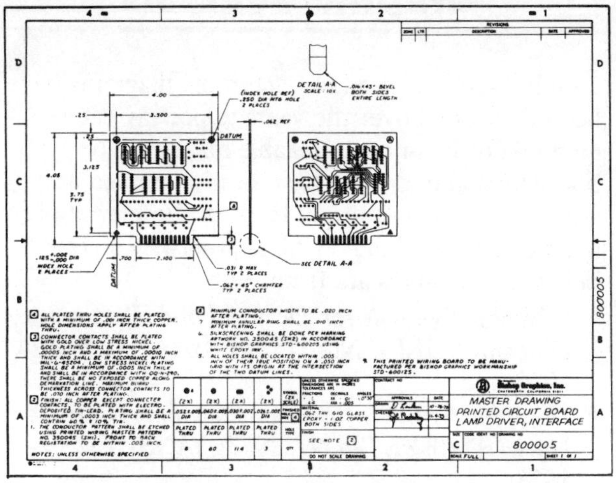

MASTER DRAWING: A complete engineering description of all design features essential to printed circuit board fabrication.

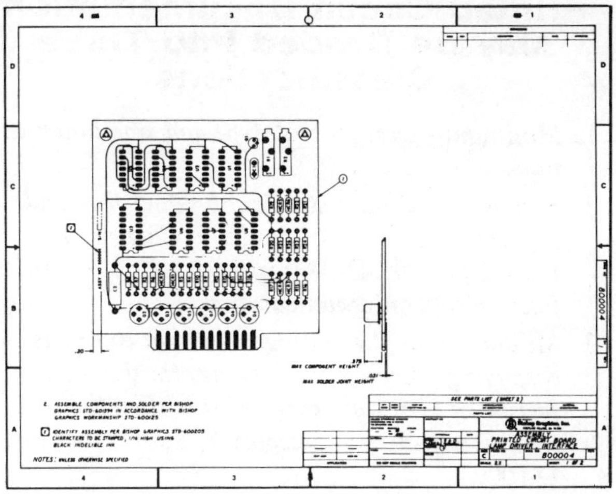

ASSEMBLY DRAWING: A pictorial representation of the finished, assembled board that lists all items and processes required for complete assembly.

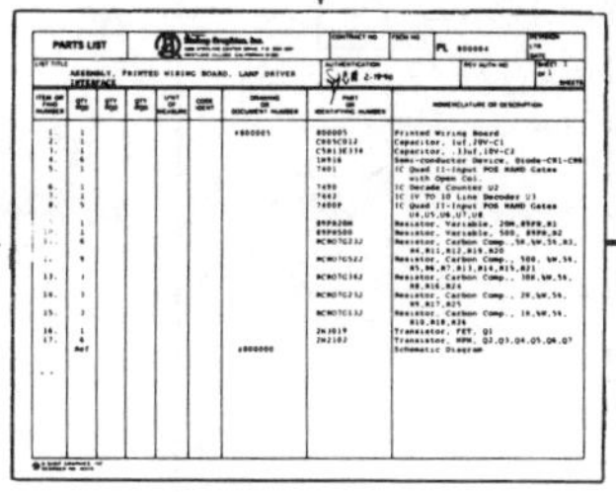

PARTS LIST: Description, quantity and reference data for items in assembly. May be included on assembly drawing.

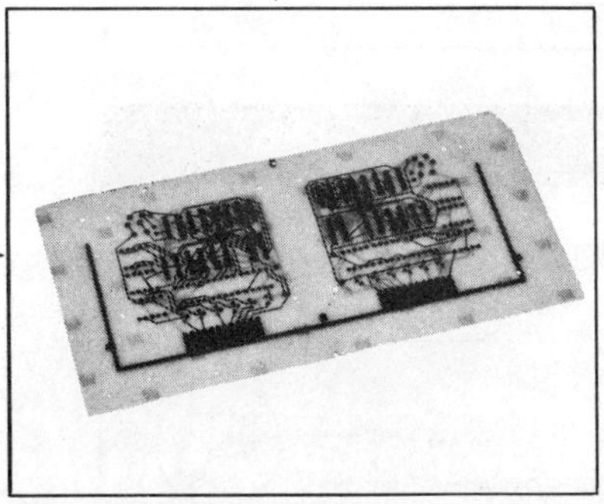

Panel (Two Boards) with Resist and Etched Prior to Plating

Fabricated Printed Circuit Board

Completed Printed Circuit Board Assembly

Printed Circuit Documentation May Be Divided Into Three Classifications

1. *Minimum—used for prototype and small quantity runs.*
2. *Formal—used for a standard product line and/or production quantities. Similar to Category E, Form 2 per MIL-D-1000 without source or specification support documentation.*
3. *Military—complies with government contracts specifying procurement drawings for the manufacture of identical items by other than the original manufacturer. (Category E, Form 1, per MIL-D-1000.)*

Standard format and overlay drafting techniques should be used to improve drawing readability and reduce costs.

Graphic Symbols

Graphic symbols used for electronic diagrams have been developed over the years to portray each component in an unmistakable manner. The symbols which appear here have been compiled and edited from many sources to show those symbols most often used today. Unless otherwise stated below, the symbols are from:

- *"Graphic Symbols for Electrical and Electronics Diagrams"* (IEEE No. 315, 1971)/(ANSI Y32.2-1970) and
- *"Graphic Symbols for Logic Diagrams (Two-State Devices)"* (IEEE Std. 91, 1973)/(ANSI Y32.14-1973)

These documents have been approved by the Institute of Electrical and Electronic Engineers

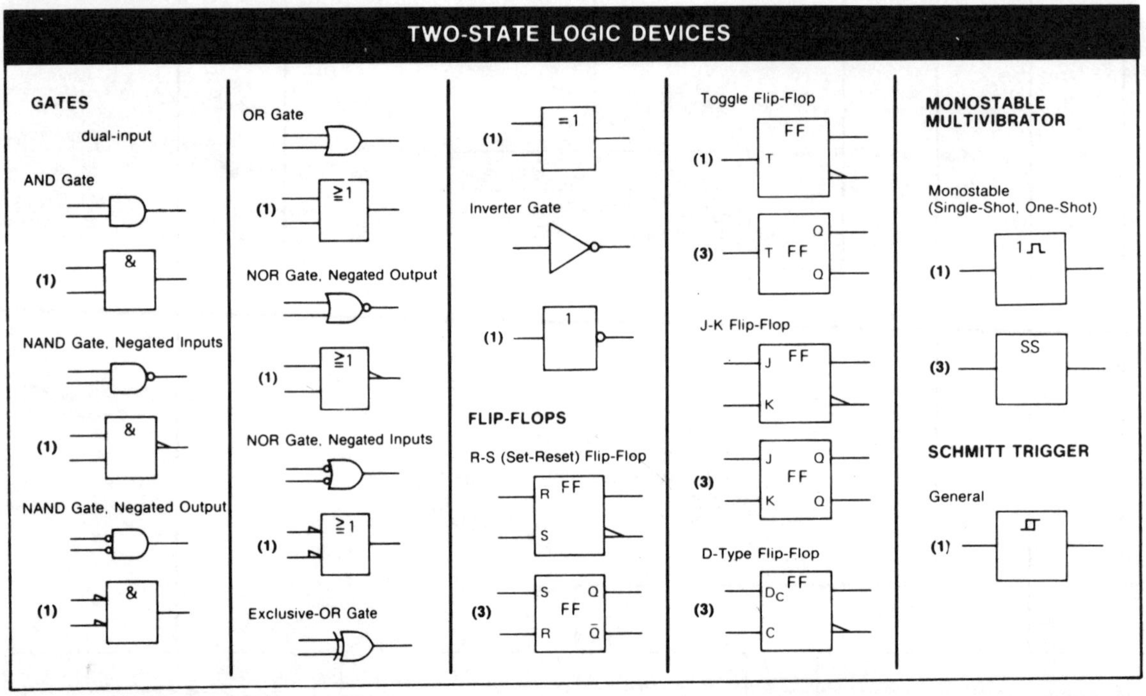

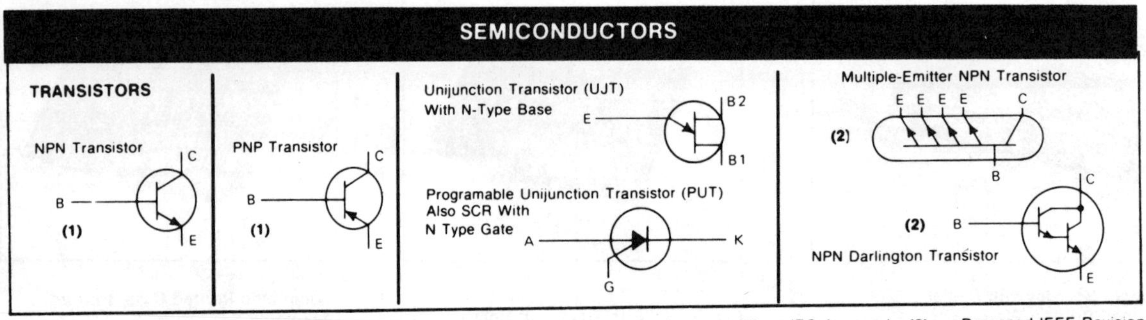

(1) = IEC Approval (2) = Proposed IEEE Revision

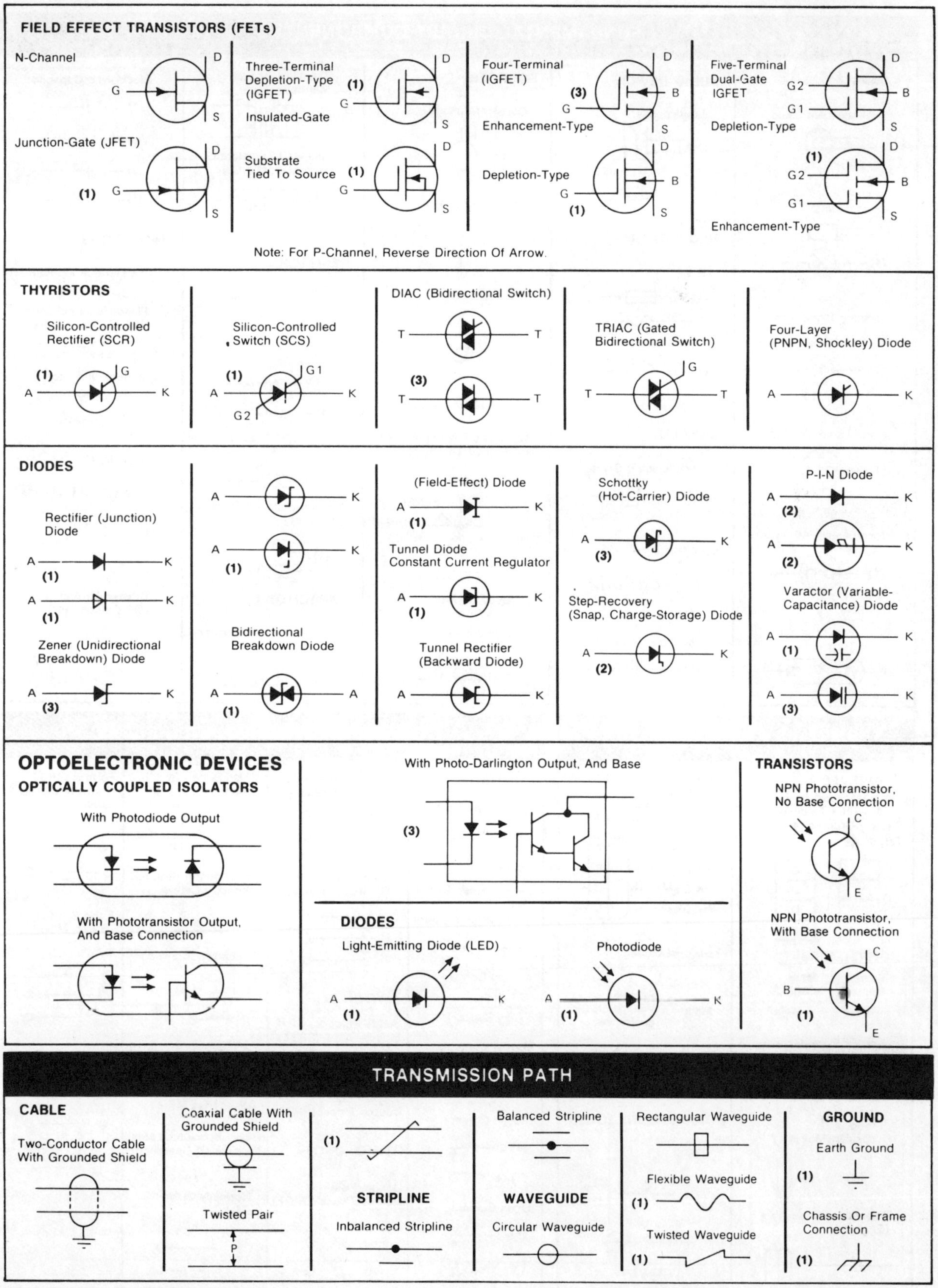

(3) = Popular Industry Usage

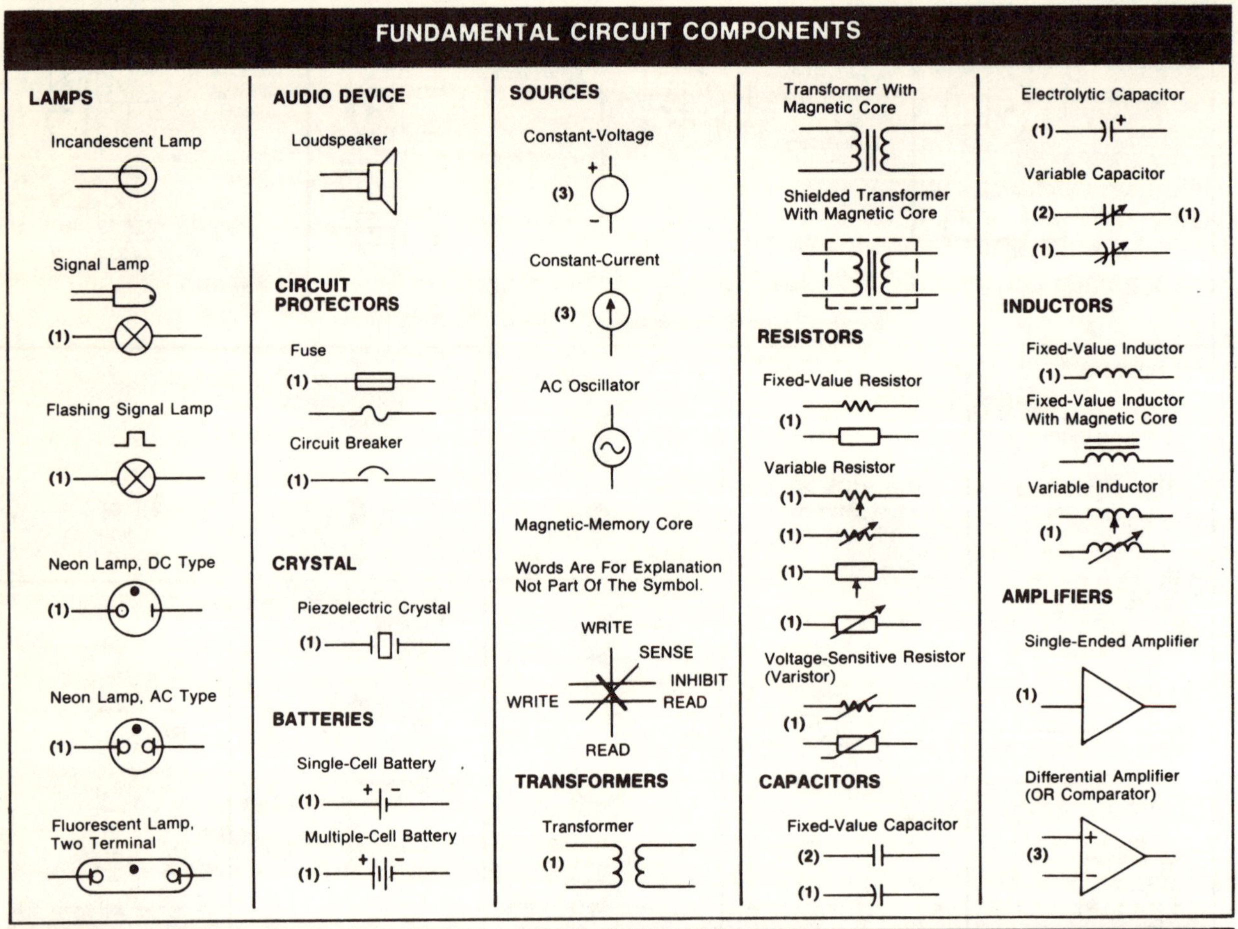

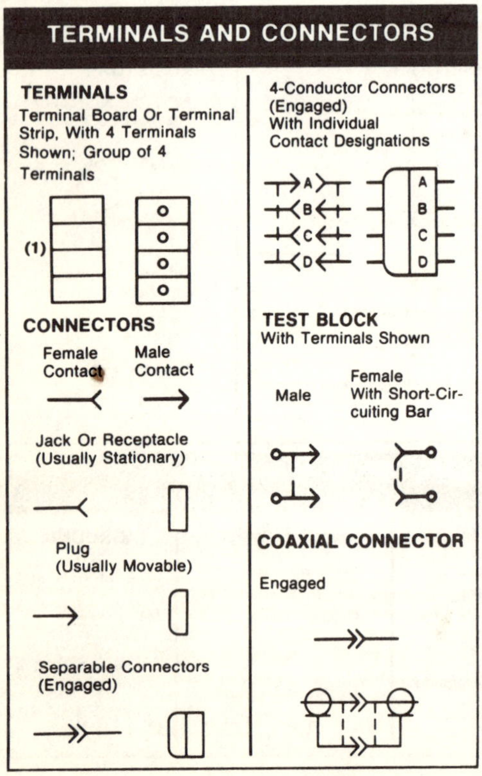

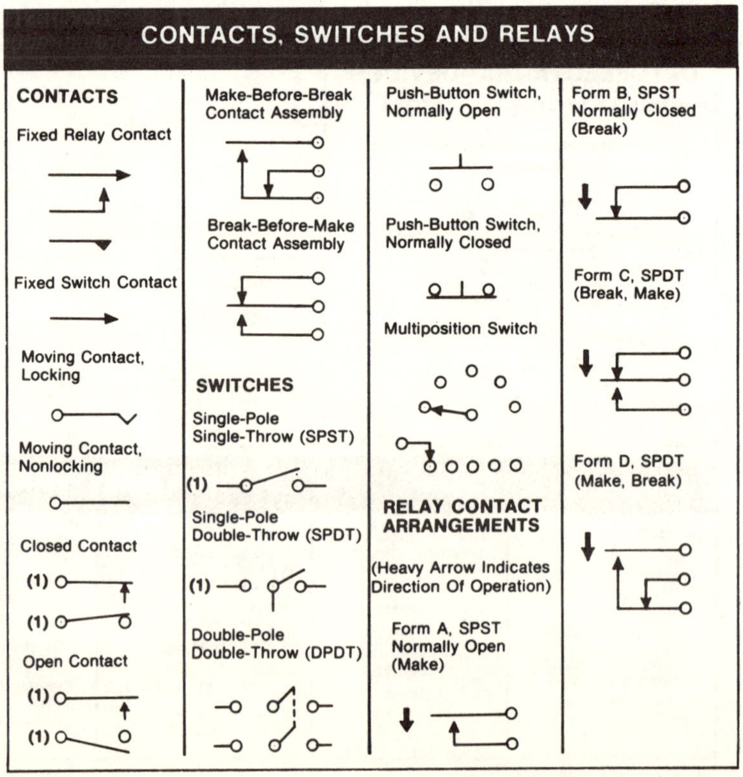

(1) = IEC Approval (2) = Proposed IEEE Revision (3) = Popular Industry Usage

(IEEE), the American National Standards Institute (ANSI), the American Society of Mechanical Engineers (ASME) and the Department of Defense (DOD). In addition, IEEE #315 has been approved b the Canadian Standards Association (CSA).

These standards are updated at least every five years by standing committees of the IEEE and ANSI.

Grid System

The use of a grid system is essential in laying out and preparing the master pattern artwork. It facilitates the placement of components, terminal areas and conductor traces on the PC layout. A grid system is mandatory if automated circuit board artwork preparation, manufacturing or assembly techniques are being considered because the machines that perform these functions are coordinated to basic grid location systems. An increasing number of multiple lead components such as dual in-line packages conform to standard grid increments.

A grid is a two-dimensional rectangular network consisting of a set of equidistant parallel lines superimposed upon another set of equidistant parallel lines with one set of lines perpendicular to the other. The line intersections provide the basis for an incremental location system. The standard increments most commonly used are .100" *(2,54mm)* .050" *(1,27mm)* and .025" *(0,635mm)* in order of preference. If more finite increments are needed, MIL-STD-275D specifies any multiple of .005" *(0,127mm)* while the International Electrotechnical Commission (IEC) specifies *0,5mm* (.0197") and *0,1mm* (.0039").

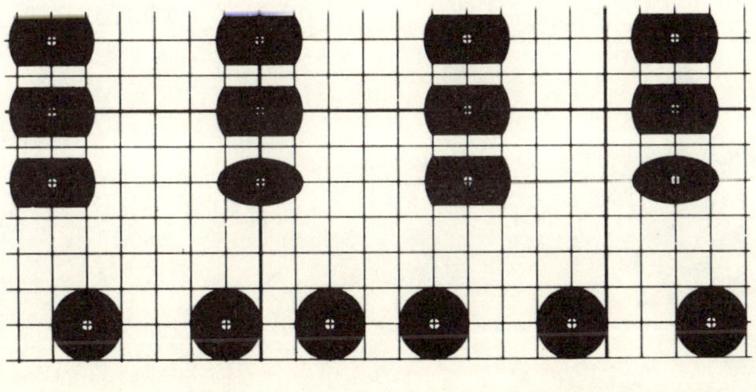

Grid layout sheets are available on many different types of materials. Their accuracy falls into three main classifications: Precision, Semi-precision and Non-precision.

Always use precision grids for manual printed circuit layout and artwork preparation to insure maximum accuracy on the final circuit board. Precision grids are usually photographically prepared on dimensionally stable materials such as glass or polyester film with accuracy tolerances of ±.002" *(0,051mm)* over 36" *(91,4cm)* at specific environmental conditions. Grids produced on Glass are the most stable, therefore the most preferable. However, they are more expensive and fragile to handle than film.

Effects of Temperature on Grid Materials

Semi-precision grids may be used during layout of circuit boards when the master pattern is to be generated by computer controlled equipment which can compensate for minor inaccuracies in the grid pattern. They can also be used for layout and artwork preparation of single sided circuit boards where close tolerances are unnecessary. They are generally manufactured by printing or photo techniques on polyester or acetate film with accuracy tolerances of ±.010" *(0,254mm)* to ± .020" *(0,508mm)* over 36" *(91,4cm)*.

EFFECT OF TEMPERATURE

MATERIAL	AMOUNT OF CHANGE	EFFECT
GLASS	.0027 - .0033 in. (0.0686 - 0.0838 mm)	Average size changes in a 30 in. (76.2 cm) length of base material with a temperature increase of 20° F. (6.7° C.)
* POLYESTER	.008 - .010 in. (0.203 - 0.254 mm)	
TRI-ACETATE	.018 - .022 in. (0.457 - 0.559 mm)	
VINYL OR POLYSTYRENE	.016 - .020 in. (0.406 - 0.508 mm)	

EFFECT OF HUMIDITY

MATERIAL	AMOUNT OF CHANGE	EFFECT
GLASS	.000	Average size changes in a 30 in. (76.2 cm) length of base material with a 20% increase in relative humidity.
* POLYESTER	.007 - .010 in. (0.178 - 0.254 mm)	
TRI-ACETATE	.007 - .010 in. (0.178 - 0.254 mm)	
VINYL OR POLYSTYRENE	.033 - .040 in. (0.838 - 1.016 mm)	

* KODAK Estar base films have same properties as polyester material.

Non-precision grids are *not* recommended to be used for printed circuit master artwork layouts. However non-precision grids may be used economically for sketches or drawings with less critical applications. They are manufactured by printing on vellum or drafting ACCUFILM®, usually in roll form. Their lack of precision may cause a severe dimensional disparity between the X and Y axis coordinates. Tolerances are usually not stated, *but* variations of ± .062" (*1,57mm*) over 36" (*91,4cm*) are typical with non-precision grids.

Scale

All printed circuit board artwork and layout should be prepared at an enlarged scale. Drafting errors, such as terminal pad misalignment with grid centers, conductor spacing variances and imperfections in drafting aids, are reduced proportionately with the reduction of the artwork to actual finished board size.

The scales most universally used are evenly multiplied factors of 2X, 4X and 1X, in order of preference.

Selection of scale should be based on the relationship of artwork drafting error to finished board tolerance requirements within practical limitations. Scale can be estimated by using the following formula:

$$\text{Artwork Scale} = \frac{DE}{FL}$$

DE = Drafting Error
usually .015" (*0,381mm*) diametric true position for manually prepared artwork and .0025" (*0,064mm*) diametric true position for artwork prepared by automated techniques

FL = Tolerance required for location of circuitry with respect to true position on the finished board.

$$\text{Example: } \frac{.015\text{" } (0{,}38mm)}{.005\text{" } (0{,}127mm)} = 3X \text{ Scale}$$

The reason for using standard scales (2X, 4X, 1X) is the ready availability of artwork and layout drafting aids and templates in these scales. Other limiting factors in scale selection are ease of artwork handling, availability of standard film sizes and camera copy board capacity [usually 30" x 40" (*76,2 x 101,6cm*) or 30" x 44" (*76,2 x 111,8cm*)].

Avoid using 1X scale whenever possible since there can be no reduction in drafting error. 1X scale artwork should be used only:

1. When there is absolutely no reduction facility available;
2. There are no limiting tolerances on the circuit board (e.e., large conductor spacings and oversize pads);
3. For prototype work where speed and cost are important
4. When the master pattern is prepared with automated photo-techniques insuring a minimal drafting error.

Choice of 4X scale greatly reduces drafting errors and should be used only when extremely accurate dimensions are required on the printed circuit board. However, due to size limitations of light tables, the artwork itself, camera copyboards, and material costs, 4X scale often is not the practical choice.

With the advent of precision prespaced artwork symbols, a 2X scale can produce the exacting dimensional requirements, yet maintain lower artwork material costs and ease of handling. 2X scale is commonly preferred for all-around use.

Conductor Width Processing Tolerances

Final Conductor Width Dimensions after Etching Depends on the Following:

1. Photographic opacity of tape
2. Tape width tolerance
3. Over or underdevelopment in photo reduction
4. Copper laminate thickness
5. Manufacturing process (photo direct or silk-screened, and intensity of etch)

The following tolerances can generally be maintained (dimensions based upon reduced 1:1 artwork):

PLATING	CONDUCTIVE MATERIAL	CONDUCTOR WIDTH in.	CONDUCTOR WIDTH mm
unplated	2 oz. copper	±.005	±0.127
	1 oz. copper	±.003	±0.076
	½ oz. copper	±.001	±0.025
panel plated copper .001 in. (0,025 mm) minimum	2 oz. copper	+.005 −.008	+0.127 −0.203
	1 oz. copper	+.003 −.005	+0.076 −0.127
	½ oz. copper	+.001 −.003	+0.025 −0.076
pattern plated copper .001 in. (0,025 mm) minimum	2 oz. copper	±.005	±0.127
	1 oz. copper	±.003	±0.076
	½ oz. copper	±.0015	±0.038

FIGURE I **CONDUCTOR THICKNESS AND WIDTH**

(For use in determining current carrying capacity and sizes of etched copper conductors for various temperature rises above ambient)

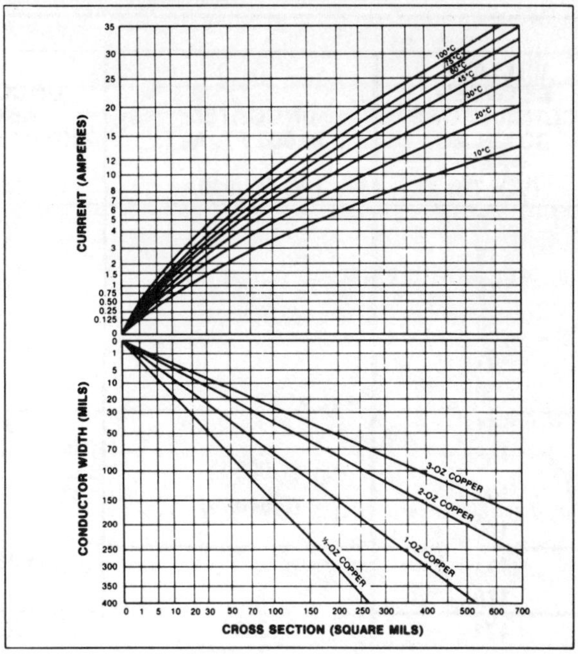

Conductor Spacing Minimums

<table>
<tr><th>VOLTAGE BETWEEN CONDUCTORS DC OR AC PEAK (VOLTS)</th><th>UNCOATED 0-10,000 FT. ALT.
IPC-ML-910A</th><th>UNCOATED ABOVE 10,000 FT. ALT.
IPC-ML-910A</th><th>COATED (1) & INTERNAL LAYERS MIL-STD-275D & IPC-ML-910A</th><th>VOLTAGE BETWEEN CONDUCTORS DC OR AC PEAK (VOLTS)</th></tr>
<tr><td>0
15</td><td rowspan="3">.015″
(0,38mm)</td><td rowspan="3">.025″
(0,64mm)</td><td>.005″
(0,13mm)</td><td>0
15</td></tr>
<tr><td>16
30</td><td>.010″
(0,25mm)</td><td>16
30</td></tr>
<tr><td>31
50</td><td>.015″
(0,38mm)</td><td>31
50</td></tr>
<tr><td>51
100</td><td rowspan="2">.025″
(0,64mm)</td><td>.060″
(1,52mm)</td><td>.020″
(0,51mm)</td><td>51
100</td></tr>
<tr><td>101
150</td><td rowspan="2">.125″
(3,18mm)</td><td rowspan="4">.030″
(0,76mm)</td><td>101
150</td></tr>
<tr><td>151
170</td><td rowspan="3">.050″
(1,27mm)</td><td>151
170</td></tr>
<tr><td>171
250</td><td>.250″
(6,35mm)</td><td>171
250</td></tr>
<tr><td>251
300</td><td rowspan="2">.500″
(12,70mm)</td><td>251
300</td></tr>
<tr><td>301
500</td><td>.100″
(2,54mm)</td><td>.060
(1,52mm)</td><td>301
500</td></tr>
<tr><td>500 +</td><td>.0002 in/volt
(0,0051mm/volt)</td><td>.0010 in/volt
(0,0030mm/volt)</td><td>.00012 in/volt
(0,00305mm/volt)</td><td>500 +</td></tr>
</table>

(1) COATING PER MIL-I-46058

Ground Planes

A ground plane is a continuous conductive area used as a common reference point for circuit returns, signal potentials, shielding or as a heat sink.

Ground planes larger than a .50" *(12,7mm)* circle should be broken up (relieved) into a striped or checkered pattern (see below) with the non-conductive area equal to approximately 50% of the conductive area. This prevents blistering and warping during soldering operations.

Adequate clearance should be provided around non-functional terminal areas in ground planes on external board layers. On internal layers of multilayer sandwich boards, use of a nonfunctional terminal area is not necessary if a diameter (usually the size of the terminal area) is left clear around the hole. (See opposite.)

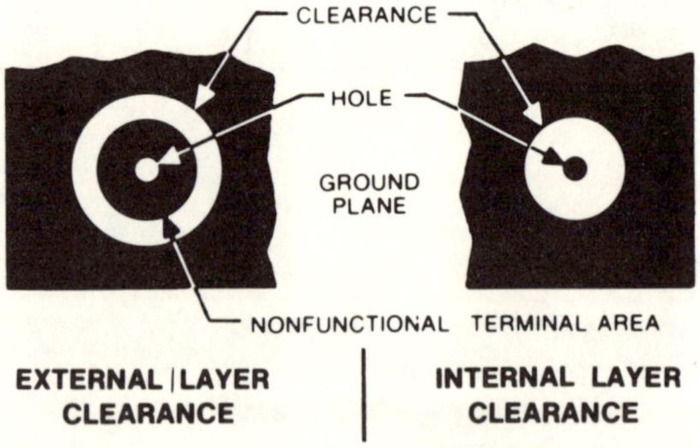

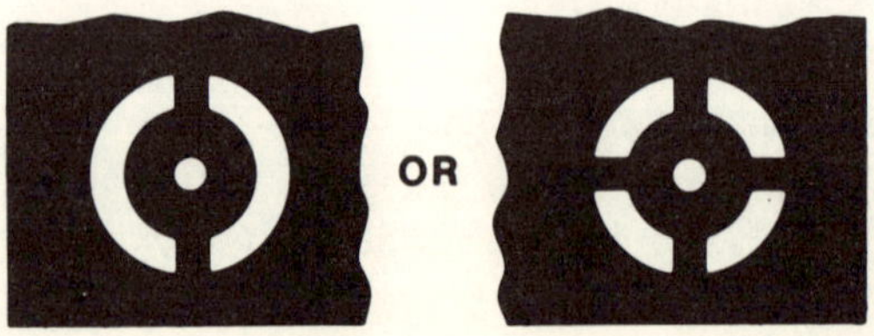

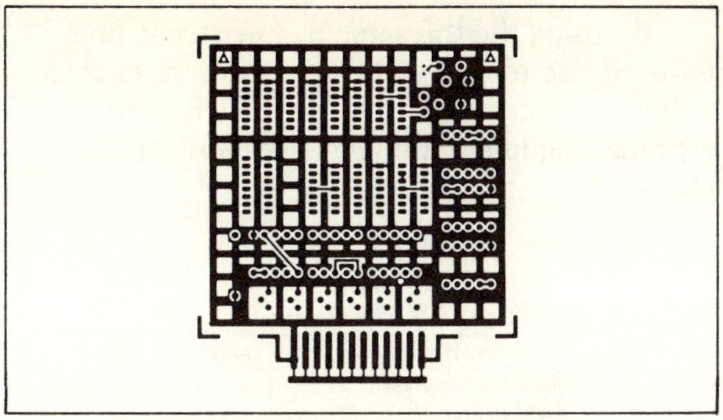

When a hole terminates in a ground plane and electrical continuity is required, a terminal area should be used. Clearance should be provided between the terminal area and ground plane for two to four connections to preserve circuit continuity. (See above) This prevents "heat sinking" of the terminal area during soldering operations which can result in an inferior solder joint.

Registration Techniques

The relative position of one or more printed circuit patterns, or portions thereof, with respect to their desired location on the printed circuit board is called register. The relationships of the elements of register and the techniques used to achieve register are called registration:

Printed circuit board register consists of three basic elements.

1. *Overall pattern alignment from layer to layer*
2. *Alignment of individual terminal areas (pads) from layer to layer*
3. *Alignment of drilled holes to the circuit pattern*

The first and third elements are primarily manufacturing registration concerns. If the photo reduction of the master pattern is not carefully controlled, size differences between artwork layers can contribute to layer-to-layer misregistration of the circuit patterns. The same condition can occur if master artwork sheets are exposed to different environmental conditions. If the mechanical alignment of the master patterns to the drilled hole pattern, and to each other, is not accomplished accurately during fabrication, holes will not align with terminal areas. See above.

The second element is an artwork function. Precise control of photography and fabrication alignment cannot compensate for terminal areas individually misaligned on the artwork.

Artwork Preparation

Four major criteria should be taken into consideration prior to artwork generation: 1. end-product board requirements; 2. material considerations; 3. photographic considerations; and 4. manufacturing capabilities. Consider these factors carefully when deciding which method of artwork generation is required to produce a satisfactory printed circuit board for various categories of board complexity. (End-product requirements are defined in IPC-D-320 for single and double-sided boards and in IPC-ML-910 for multi-layer printed circuit boards.)

Proper preparation of the artwork is crucial because of the many board fabrication processes necessary to complete the final PC board. The dimensional tolerances required in the end-product board, plus the ability of the manufacturing and artwork generation techniques to hold these tolerances, determines the method used to achieve the desired results.

The use of pressure-sensitive drafting materials has greatly improved and facilitated printed circuit artwork generation. With pressure-sensitive symbols, changes or corrections can be made quickly and easily, artwork preparation is simplified, and drafting speed and accuracy are increased. Registration has also been greatly improved through the use of pre-spaced multiple pad appliques and systematic approaches to PC artwork preparation and drafting.

Drafting Media

Stable-based drafting film has the dimensional stability required for the base material of printed circuit artworks, artwork masters, and related products. This stable-based material is available in several convenient forms, including precision grids, drafting films, scales, and photographic and diazo film products.

The use of stable-based drafting film throughout the artwork generation process will minimize the effect of temperature and humidity on your artwork (see chart above), and help insure the maximum accuracy of your finished printed circuit board.

Pad Master

The pad master concept eliminates artwork misregistration of individual terminal pads and holes. There are several methods of implementing this concept, but all employ the use of a single base artwork sheet containing terminal area pads with hole and circuitry reference markings common to all circuit layers.

Effects of Temperature and Humidity on Grid Materials

The pad master artwork is prepared with terminal areas, registration, indexing hole, board outline and reduction marks. This single sheet is then utilized to create a master artwork by serving as the base sheet of the Red and Blue PC Tape Method, or as the base layer of an overlay drafting system with circuit pattern overlays. Stable contact duplicate prints of the pad master can also be made for direct taping of each layer. However, with this duplicate print method, dimensional variations in the artwork sheets—due to environmental changes, processing conditions, or basic film stability—can cause misregistration and create a recurring added expense.

By using the this same pad master at final reduced size to digitize hole locations or to create a drill template, you insure accurate hole-to-terminal alignment during board fabrication and checking.

EFFECT OF TEMPERATURE

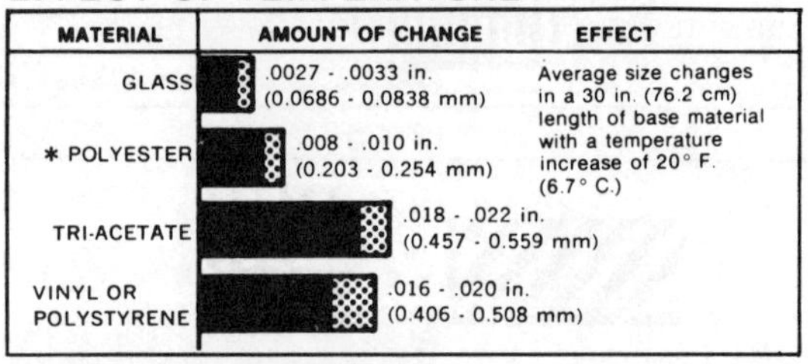

EFFECT OF HUMIDITY

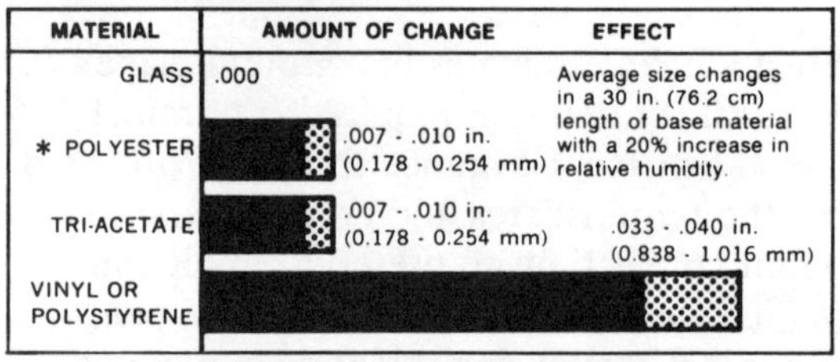

* KODAK Estar base films have same properties as polyester material.

Artwork Markings

Printed circuit board artwork markings relating to registration include datum points, register marks, indexing holes, and photo reduction points. (Other types of board markings are part numbers, reference designations, corner markers, etc.)

Datum Points are points, assumed to be exact, from which the location of printed wiring board features may be established by dimensional computation. Three datum points establish the perpendicular Cartesian coordinate system of grid dimensioning. Datum points are normally indicated by distinctively shaped pads when they are located directly on the printed circuit board. All mechanical dimensioning of the master pattern should be made from datum points.

TYPICAL PADS USED FOR DATUM POINTS OR INDEXING HOLES

Register Marks are bold geometric shapes which are used to establish registration. Register marks are affixed to the artwork on all layers so, when the layers are overlaid, the patterns will be superimposed. Register marks may be located inside or outside of the board outline as appropriate. When located inside the board area, they may be drilled for mounting holes after they have served their original purpose.

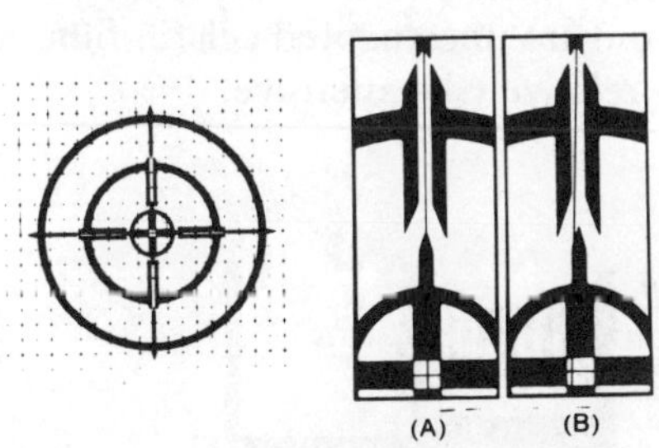

Bishop Graphics' Universal Target is designed to achieve optimum visual register. Lines sighted through the apertures in figure (A) above are in a tightly controlled position and are in proper alignment. The line sighted through the apertures in figure (B) are obviously out of alignment.

Indexing or Tooling Holes are holes placed in a printed circuit board to accurately position the board during manufacturing processes. They may or may not appear within the finished board area, but they must appear on the master pattern.

A minimum of two indexing holes are required, and they should be diagonally located to encompass the greatest area. A third hole is sometimes used, and should form a right triangle with the first two. They should always be located on grid intersections.

Datum points, register marks, and indexing holes may be combined in the same locations, or they may be completely independent of each other. A registration target such as Bishop Graphics' Universal Target or Modified Crosshair may be used as a combination datum point symbol, indexing hole symbol and register mark.

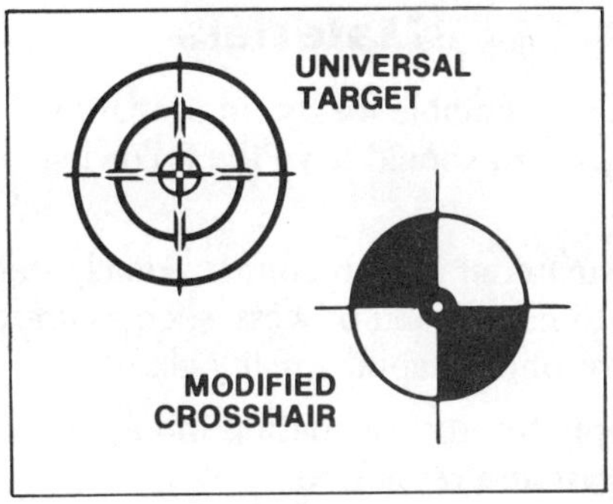

Photo Reduction Points are two points spaced an exact distance apart to visually indicate the extent to which printed circuit artwork is to be photographically reduced. Care should be used in selecting a mark which clearly indicates the precise measurement point.

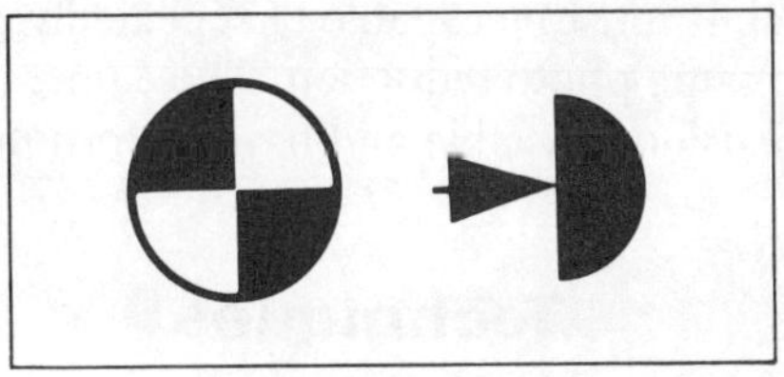

Use reduction marks or simple quadrants since the exact point of measurement cannot be confused—it must be taken from the edge. In addition, the points of intersection indicate proper photo exposure and development.

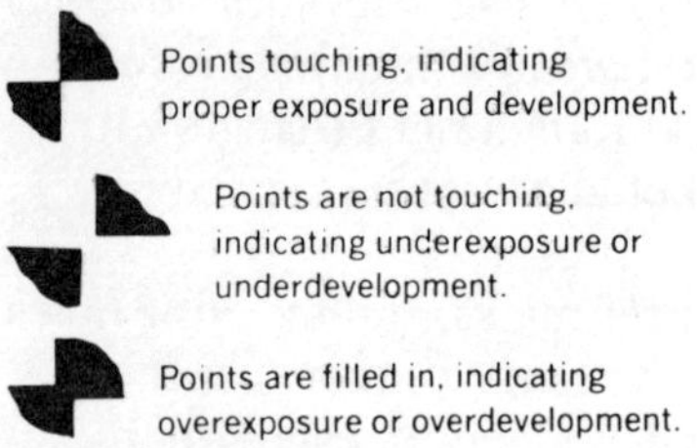

DO NOT USE Crosshairs for reduction points since the width of the line makes the exact measurement point ambiguous.

Materials

Materials acceptable for use in preparing the master pattern should have the following characteristics:

1. Base material must be dimensionally stable and easy to handle and process, such as drafting film or photographic quality glass.
2. The photosensitive coating should have high contrast and resolution.
3. The photosensitive coating should be of the stabilized type.

Camera

PC artwork is photographically reduced using a graphic arts process camera. This type of camera is explicitly designed for copying line or half-tone material. It must be capable of repeatedly producing the same photo reduction. The camera must be dimensionally stable and have distortion-free optics.

Techniques

Back light the copy board to avoid fuzzy edges, particularly on taped-up master artwork. Back lighting provides contrast of 1,000:1, compared with approximately 50:1 for front lighting.

A green light should be used to achieve the optimum results *except* when the Red and Blue separation method is used.

Always specify the side of the film on which you want the emulsion (image). The film should always have the emulsion in direct contact with the emulsion of that which it is being contacted to. Normally, negatives are right reading (R/R), emulsion down and positives, right reading, emulsion up.

Applying the images to the printed circuit board in register can be accomplished by many methods (see Registration Techniques). One of the best uses identically pre-punched boards and master patterns. The working copies of the master patterns are made using the PC board indexing hole system to expose pre-punched photo film.

Another method of applying the image is to make the master pattern visually transparent by "bleaching the photo." This allows the operator to actually see through the circuit pattern, and shift the photo around to obtain the best registration. Diazo coated film is available which is ideally suited to this method.

Photo Separation of Red and Blue Artwork

Properly prepared Red and Blue master artwork will produce two negatives in perfect register by following these instructions:

- Camera lens must be color corrected for lateral color aberration. Backlighting from a constant white light source is essential for color separation.
- Proper filters are vital—use only Kodak's Wratten Gelatin filters, PM47 Blue and PM29 Red. These are uncemented gelatin film filters and are relatively inexpensive.

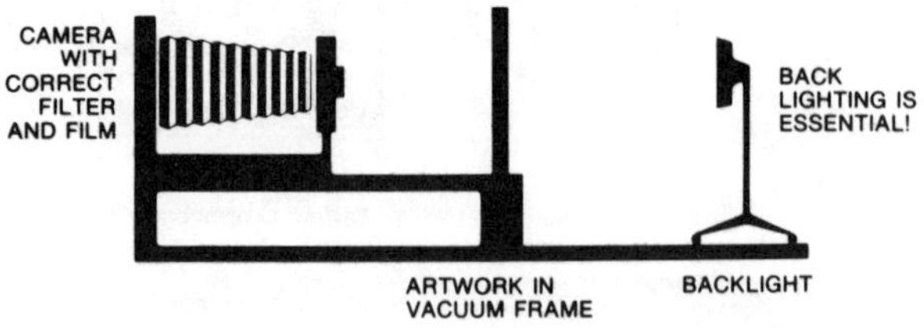

Note: The blue filter is often not necessary because this film is not sensitive to Blue light. This allows shorter exposure times and less chance of overexposure.

(Environmental, material, and equipment variations may necessitate adjustments to the data below.)

To Drop Red and Hold Blue

Film:	Kodalith Estar Base Panchromatic #2568 (check exposure data packed with film) If required, .007" (0,18mm) base pan film is available from DuPont. For test trial: camera setting same size.
Filter:	#29 (Kodak-Wratten)
Exposure:	f22 for 7 seconds
Develop:	2½ minutes at 68°F (20°C) Kodak Kodalith Super Developer (powder), develop with continuous agitation.

To Drop Blue and Hold Red

Film:	Kodalith Type 3, Estar Base. For test trial: camera setting same size.
Filter:	#47 (Kodak-Wratten)*
Exposure:	f16 for 17 seconds
Develop:	2¾ minutes at 68°F (20°C) Kodak Kodalitih Super Developer (powder), develop with continuous agitation.
Light Source:	2 - 1,000 watt pulsed xenon lamps.

*Filter may not be necessary to effectively drop blue and hold red when using suggested film.

Bishop Graphics has published a separate handbook (T.B. 1001R) fully describing the Red and Blue Artwork Separation Drafting Techniques including complete photography instructions.

Master Pattern Inspection Checklist

The master pattern is a production tool. It should be inspected and controlled in the same manner as other precision tooling.

1. SIZE—Each master pattern should be measured at the photographic reduction marks, and at least one other dimension at right angles to the first measurement to insure proper reduction and size.
2. REGISTRATION—All patterns required to produce a printed circuit board must be inspected together to insure that registration as stated on the master drawing has been achieved.
3. SHARPNESS—Image sharpness and resolution are necessary to prevent ragged conductor edge definition on the etched printed circuit board. Sample conductor line width and spacing measurements must be made to verify proper exposure and development.
4. DENSITY—The image on the master pattern must be opaque enough to insure that proper exposure of the photosensitive coating can be made.
5. WORKMANSHIP—All master patterns should be inspected for pinholes, scratches, fingerprints and general cleanliness. All touch-up work should be made on the non-emulsion side of the film.

Handling and Storage of PC Artwork

The completed PC artwork is the master tool, affecting all subsequent stages through production of the printed circuit board. Every way in which tolerance buildup can be reduced or eliminated is invaluable.

The proper handling of all of the materials which are used to create the master artwork will produce the optimum results.

Drafting Media

Only drafting film and glass have the dimensional stability required for the development of printed circuit artworks. This includes precision grids and scales, drafting films and photographic and diazo films.

The following procedures are recommended to safeguard all drafting film products.

Storage

- Store cartons in an area where the temperature and humidity remain relatively constant.
- DO NOT expose to temperatures in excess of 120°F (40°C) or below 33°F (1°C).
- Films should be stored and transported flat in a closed container.

Master Drawing

A printed circuit board master fabrication drawing presents a complete engineering description of all design features.

Information recurring from board to board should always be placed in the same location on the drawing to facilitate communications between departments, and between the company and its outside vendors.

Contents

The printed circuit board master drawing should always include:

1. A reproduction of the conductor pattern image for the solder side of the board. It may also include the circuit patterns images of all other layers.
2. The delineation of the complete board outline with all dimensions and tolerances necessary for fabrication, including any structural requirements.
3. The complete designation of the material, process and workmanship specifications that apply to the board.
4. All hole identification information, including type, finished size and location.

Hole Location

There are three fundamental methods for locating holes on printed circuit boards.

Location by a *grid system* is often used to meet military requirements or to have all features on the board located to the grid pattern. (See the "Grid System" section of this Appendix.) This method is essential for automated drilling and component assembly techniques since digitized grid coordinates are used for locating components. The drawback to this system is the precise accuracy required in:

1. Layout Grid;
2. Location of terminal areas to the grid pattern;
3. Photo reduction of the artwork;
4. Layer-to-layer registration.

Inaccuracy in any of these areas can result in serious pattern-to-hole misalignment.

Hole location by *dimension* from datum lines or "X" and "Y" axis coordinates is similar to the grid system method because terminal areas must be precisely placed in reference to the hole location dimensions. When locating a multi-lead component pattern, only one hole in the group should be located from the standard datum points or indexing holes. The other holes should be located dimensionally from this reference hole to preserve accuracy within the pattern. Besides the high risk of pattern-to-hole misalignment, this method can be very time-consuming for the draftsman and the PC board manufacturer.

The simplest method is to locate holes optically *from the terminal area centers* on the master pattern or actual PC board. This method is usually least expensive and produces the most accurate pattern-to-hole alignment. It should be used whenever precise alignment to external parts or tooling is not a requirement.

Master Drawing Check List

1. Has the circuit image been properly shown from the solder side?
2. Have hole types, sizes and tolerances been properly indicated and referenced to the circuit image?
3. Has a hole location method been clearly defined?
4. Is the complete board outline clearly indicated and properly dimensioned?
5. Have all non-standard holes (i.e., board mounting holes, component hardware holes, etc.) been properly dimensioned?
6. Are all inside corners of notches, cutouts, etc. specified as maximum radii?
7. Has the base material been correctly specified?

8. Have all reference drawings such as master pattern, marking drawing, etc., been properly referenced?
9. Are all applicable manufacturing and workmanship specifications and standards referenced in the notes?
10. Has all plating been correctly specified, and toleranced for circuitry, holes, and connector contacts?
11. Have processing variance allowances been indicated for registration, conductor width and minimum annular ring? (See "Conductor Width and Spacing" section of this manual.)
12. Has the allowable amount and direction of base material warp been specified if critical?
13. Is the drawing scale clearly indicated?
14. Has the board identification number been specified as either being etched or printed on the board?
15. Can the printed circuit board be completely and correctly fabricated from the master drawing without question?

Assembly Drawing

The printed circuit board assembly drawing is a complete engineering description of a printed circuit board assembly including all components and parts mounted by fastening, soldering or bolting.

Contents

The drawing includes, as a minimum:

1. A view of the component side of the completed board;
2. A complete parts list or bill of materials (often a separate document);
3. All general and specific notes to enable the board to be fabricated, inspected and tested.

The printed circuit board assembly drawing is usually drawn at the scale used in creating the master artwork. However, any easily readable scale may be used. The scale must be shown.

The parts list should have the following information for each entry: part number, item description, number of parts required and the parts list item numbers. For electronic components, the reference designation(s) are used as find numbers and should be included. Other information that sometimes is helpful or required by contract is the Federal Code Identification and drawing zone numbers indicating location.

Techniques

Printed circuit board assembly drawings are easily prepared by using a number of contact reproduction methods utilizing either wash-off or diazo photo-sensitive drafting films.

Using the artwork or master pattern of the component side of the board and a half-tone screen (85 line, 40% for example), a screened copy of the artwork configuration is produced within a standard drawing format. The component outlines and reference designations are then added in solid black, using a pencil and template or preprinted pressure-sensitive StikOns. The screened artwork pattern provides a reference which allows components to be easily located during assembly and testing. The same basic technique may use an outline artwork pattern instead of the screened pattern. This technique uses a spread negative made by placing a clear film spacer between the master negative and duplicating film. The spread negative and the original artwork then are used to produce the assembly drawing. An assembly drawing made using the outline method is easier to change or revise.

Minimum documentation requires only that the component outlines are placed in their proper location within a blank board outline.

The quickest, easiest method of preparing assembly drawings is to make a reproducible copy of the component portion of a layout made with Bishop PUPPETS™, then draw in the board outline and add the appropriate notes and details.

Assembly Drawing Checklist

1. Printed circuit board assembly drawings should be drawn viewed from the component side. If components are mounted on both sides, then both front and back views should be shown.
2. Use Auxiliary views to clarify any details such as the precise location and orientation of mechanical parts.
3. Maximum component height or overall board thickness should be included.

4. Electronic components should be identified within the component outline with the reference designation used on the schematic diagram.
5. All parts, other than electronic components, should be identified by ballooned item numbers.
6. The orientation of components (pin number, polarity, etc.), when applicable, should be clearly shown.
7. The assembly part number, revision level, and serial number (if applicable) should be clearly marked on the assembled board.

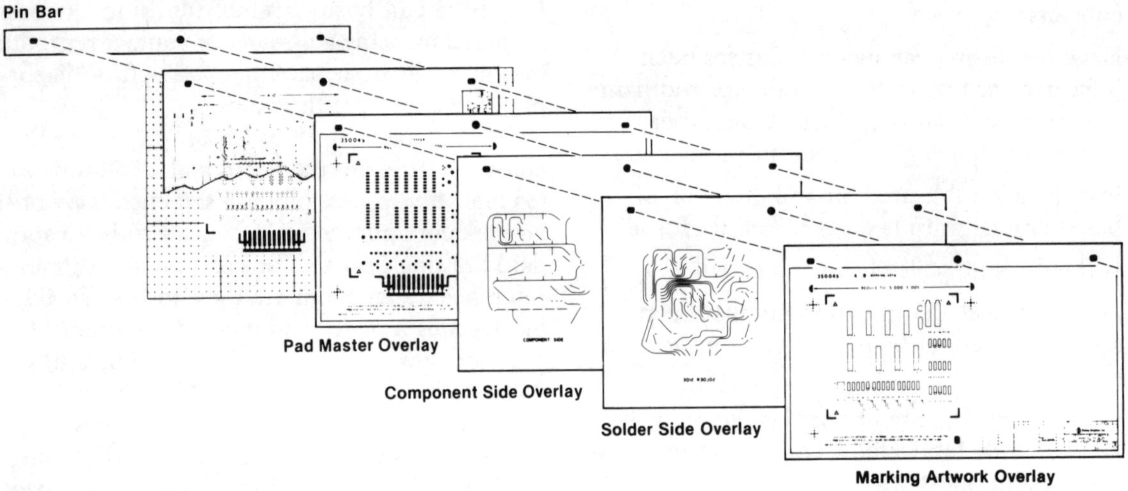

Donut Pads

Bishop all-purpose tape donut pads serve as the basic termination area for a variety of prototyping and PC artwork functions. These photographically opaque terminal areas are accurate to ±.002" (0,05mm). They reproduce with sharp, clear edges and *no* distortion. A special formula, pressure-sensitive backing allows adjustment before application of pressure. Pads will not shift or creep after they are permanently affixed.

STANDARD INCH

The basic all-purpose PC terminal area for termination of wire or component leads extending through the PC board. ID aids in accurate pad location to grid centers.

OD		ID		Cat. No.
in.	mm	in.	mm	
.050	1,27	.015	0,38	D135
.062	1,57	.025	0,64	D136
▶ .070	1,78	.025	0,64	D370
.075	1,91	.025	0,64	D311
.080	2,03	.031	0,79	D216
▶ .085	2,16	.031	0,79	D371
▶ .090	2,29	.031	0,79	D372
.093	2,36	.031	0,79	D137
.100	2,54	.025	0,64	D312
		.031	0,79	D101
		.040	1,02	D207
		.050	1,27	D211
.103	2,62	.039	0,99	D238
.104	2,64	.032	0,81	D263
.106	2,69	.032	0,81	D277
▶ .110	2,79	.031	0,79	D373
.111	2,82	.032	0,81	D279
.113	2,87	.040	1,02	D245
.115	2,92	.031	0,79	D265
.120	3,05	.031	0,79	D239
.125	3,18	.025	0,64	D180
		.031	0,79	D102
		.040	1,02	D179
		.050	1,27	D230
		.062	1,57	D169
.130	3,30	.032	0,81	D276
.140	3,56	.031	0,79	D280
		.040	1,02	D246
.150	3,81	.031	0,79	D144
		.040	1,02	D203

OD		ID		Cat. No.
in.	mm	in.	mm	
.150 (cont'd)	3,81	.050	1,27	D164
		.062	1,57	D231
		.080	2,03	D165
.156	3,96	.031	0,79	D103
		.040	1,02	D313
		.062	1,57	D240
▶ .160		.031	0,79	D374
	4,06	.040	1,02	D247
.170	4,32	.031	0,79	D307
.175	4,45	.040	1,02	D331
.180	4,57	.031	0,79	D306
		.040	1,02	D248
.187	4,75	.025	0,64	D181
		.031	0,79	D138
		.040	1,02	D104
		.050	1,27	D150
		.062	1,57	D105
		.093	2,36	D182
.190	4,83	.135	3,43	D264
.200	5,08	.025	0,64	D183
		.031	0,79	D139
		.040	1,02	D204
		.050	1,27	D158
		.062	1,57	D147
		.080	2,03	D106
		.093	2,36	D184
▶		.125	3,18	D345
.208	5,28	.062	1,57	D168
.218	5,54	.031	0,79	D140
		.040	1,02	D143
		.050	1,27	D155
		.062	1,57	D107
.220	5,59	.093	2,36	D185
.227	5,76	.062	1,57	D215
.230	5,84	.031	0,79	D298
.240	6,10	.040	1,02	D156
		.062	1,57	D170
.245	6,22	.062	1,57	D242
.250	6,35	.025	0,64	D201
		.031	0,79	D141
		.040	1,02	D108
		.050	1,27	D152
		.062	1,57	D109

OD		ID		Cat. No.
in.	mm	in.	mm	
.250 (cont'd)		.080	2,03	D186
		.093	2,36	D176
		.125	3,18	D166
▶ .260	6,60	.062	1,57	D358
.270	6,86	.062	1,57	D275
.275	6,99	.040	1,02	D314
▶ .280	7,11	.050	1,27	D375
.281	7,14	.031	0,79	D145
		.062	1,57	D110
.300	7,62	.031	0,79	D187
		.040	1,02	D293
		.050	1,27	D332
		.062	1,57	D111
		.080	2,03	D172
		.093	2,36	D177
		.100	2,54	D174
.312	7,92	.031	0,79	D315
		.040	1,02	D112
		.050	1,27	D161
		.062	1,57	D113
		.080	2,03	D171
		.093	2,36	D173
.320	8,13	.062	1,57	D159
▶ .340	8,64	.062	1,57	D376
.343	8,71	.062	1,57	D205
.350	8,89	.062	1,57	D146
		.180	4,57	D241
▶ .360	9,14	.062	1,57	D377
.375	9,53	.031	0,79	D210
		.040	1,02	D294
		.050	1,27	D162
		.062	1,57	D114
		.080	2,03	D178
		.093	2,36	D142
		.100	2,54	D160
		.125	3,18	D333
		.187	4,75	D167
.400	10,16	.031	0,79	D243
		.040	1,02	D189
		.050	1,27	D157
		.062	1,57	D115
		.080	2,03	D116

TO PATTERNS

(CONT'D)

8 & 10 LEAD

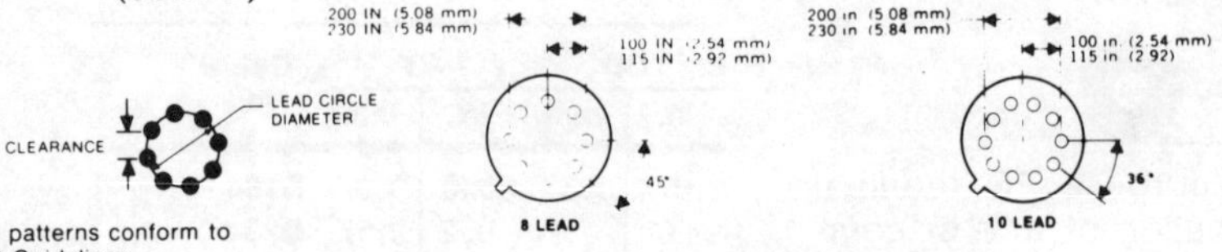

Pattern Tolerance: .002 in. (0.05 mm) All patterns conform to MIL-STD-275 and Institute of Printed Circuit Guidelines

Dimensions Below Are For 1X Scale (1:1). For 2X And 4X Scale Dimensions, Multiply Values By 2 Or 4, Respectively.

Typical Mounting Configurations	All Patterns Are Shown 2X Scale (2:1)	Pattern Type	No. of Leads	OD in.	OD mm	ID in.	ID mm	Clearance in.	Clearance mm	Lead Cir. Dia. in.	Lead Cir. Dia. mm	Qty. Per Pkg	Scale	Cat. No.
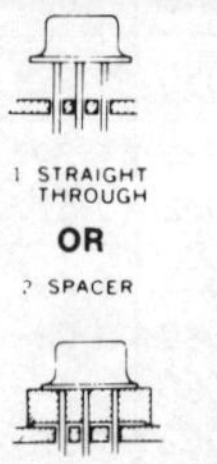 1 STRAIGHT THROUGH OR 2 SPACER	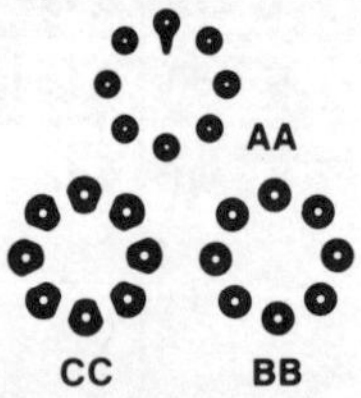 AA CC BB	AA		.046	1,17	.010	0,25	.031	0,79	.200	5,08		1X 2X 4X	**6032** **6090** **6005**
		BB		.054	1,37	.015	0,38	.024	0,61	.200	5,08	250	1X 2X 4X	**6026** **6092** **6094**
		CC CUT PAD		.062	1,57	.015	0,38	.025	0,63	.200	5,08		1X 2X 4X	**6021** **6197** **6199**
				.060	1,52	.015	0,38	.032	0,81	.230	5,84		1X 2X 4X	**6436** **6205** **6206**
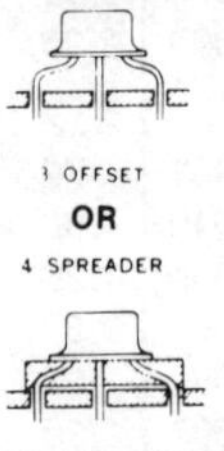 3 OFFSET OR 4 SPREADER	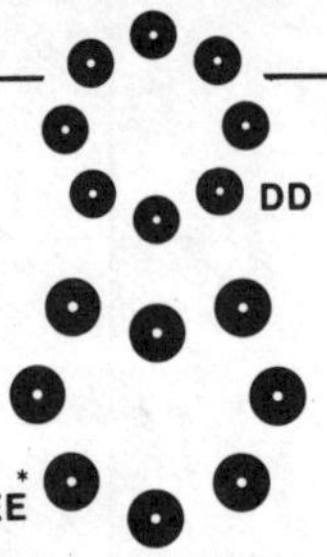 DD EE*	DD		.078	1,98	.015	0,38	.037	0,94	.300	7,62	250 250 100	1X 2X 4X	**6437** **6107** **6105**
				.093	2,36	.015	0,38	.040	1,02	.350	8,89	250 250 100	1X 2X 4X	**6304** **6300** **6301**
				.100	2,54	.015	0,38	.050	1,27	.400	10,16	250 250 100	1X 2X 4X	**6305** **6302** **6303**
		EE*		.093	2,36	.015	0,38	.057	1,45	.400	10,16	250 250 100	1X 2X 4X	**6438** **6211** **6212**
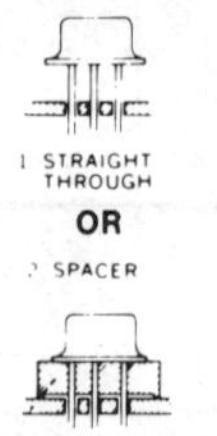 1 STRAIGHT THROUGH OR 2 SPACER	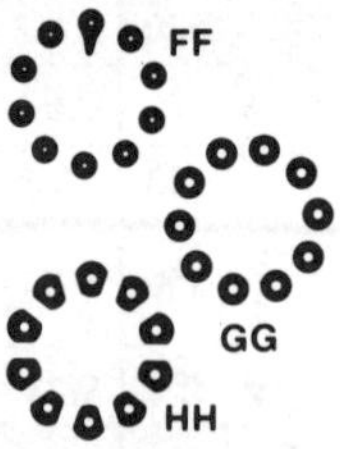 FF GG HH	FF		.046	1,17	.010	0,25	.025	0,64	.230	5,84		1X 2X 4X	**6033** **6091** **6006**
		GG		.046	1,17	.015	0,38	.015	0,38	.200	5,08	250	1X 2X 4X	**6439** **6357** **6358**
				.054	1,37	.015	0,38	.016	0,41	.230	5,84		1X 2X 4X	**6027** **6093** **6095**
		HH CUT PAD		.062	1,57	.015	0,38	.025	0,64	.230	5,84		1X 2X 4X	**6022** **6198** **6200**
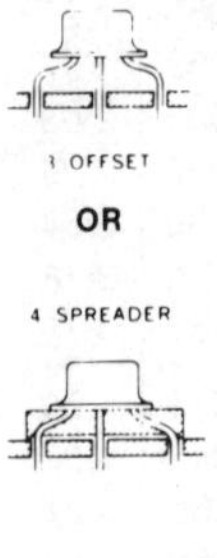 3 OFFSET OR 4 SPREADER	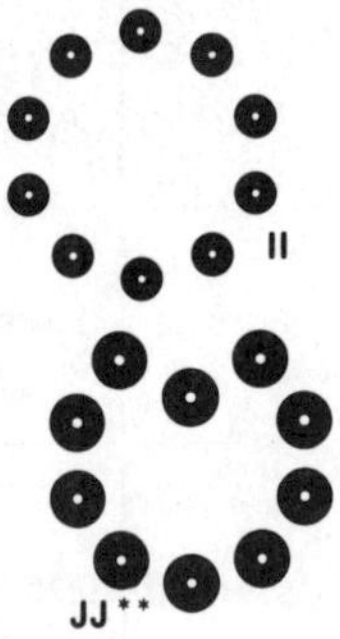 II JJ**	II		.070	1,78	.015	0,38	.040	1,02	.350	8,89	250 250 100	1X 2X 4X	**6440** **6306** **6307**
				.070	1,78	.015	0,38	.055	1,40	.400	10,16	250 250 100	1X 2X 4X	**6381** **6308** **6309**
		JJ**		.093	2,36	.015	0,38	.030	0,76	.400	10,16	250 250 100	1X 2X 4X	**6441** **6221** **6222**

*Pattern fits Thermalloy Spreader # 7717-19

**Pattern fits Thermalloy Spreader # 7717-21

Resistor and Capacitor Standard Color Code

Standard Color Code for Resistors and Capacitors

The standard color code provides the necessary information required to properly identify color coded resistors and capacitors. Refer to the color code for numerical values and the number of zeros (or multiplier) assigned to the colors used. A fourth color band on resistors determines the tolerance rating. Absence of the fourth band indicates a 20% tolerance rating. (RED MIL-STD-221)

AXIAL LEAD RESISTOR

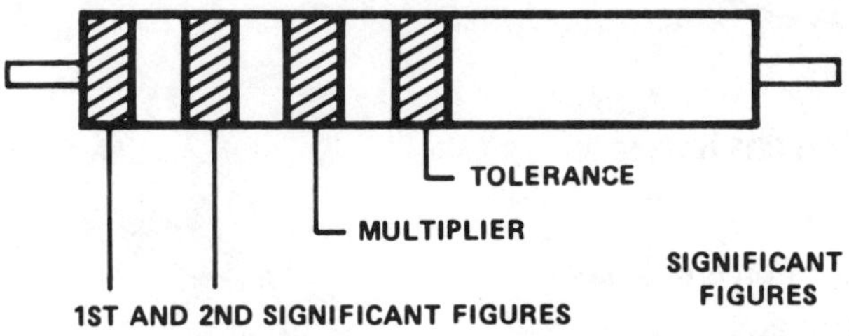

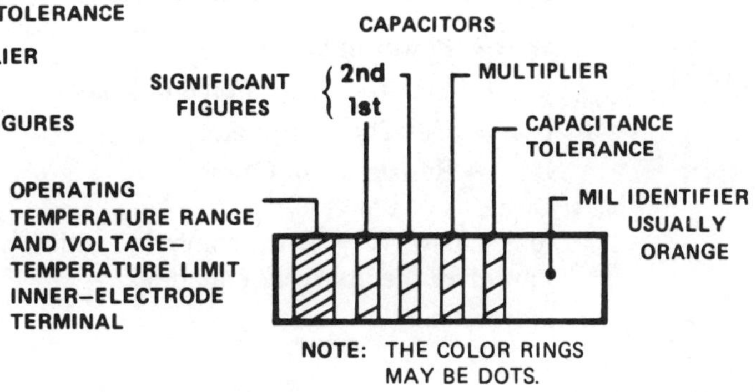

WIRE-WOUND RESISTORS HAVE FIRST DIGIT BAND DOUBLE WIDTH

COLOR CODE

COLOR	FIRST FIGURE	SECOND FIGURE	MULTIPLIER	TOLERANCE	
				%	LTR
BLACK	0	0	1		
BROWN	1	1	10		
RED	2	2	100	± 2%	G
ORANGE	3	3	1,000		
YELLOW	4	4	10,000		
GREEN	5	5	100,000		
BLUE	6	6	1,000,000		
PURPLE (VIOLET)	7	7			
GRAY	8	8			
WHITE	9	9			
SILVER			0.01	± 10%	K
GOLD			0.1	± 5%	J
				± 20%	M
				± 1%	F

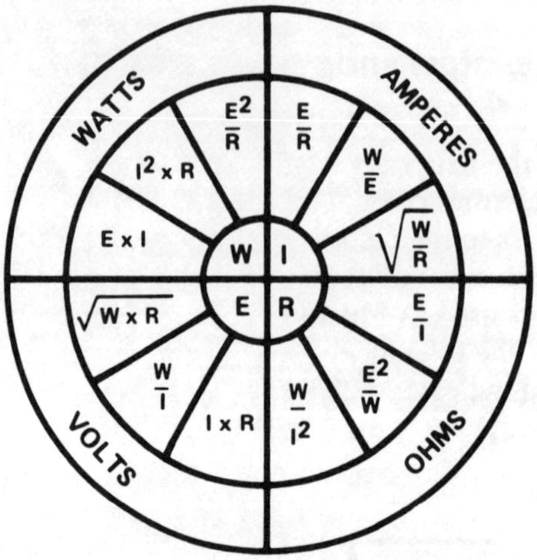

Ohm's Law Equations

W = Power in Watts
E = Electro-Motive Force in Volts
I = Current in Amperes
R = Resistance in Ohms

Joule's Law, W = E x I, combined with Ohm's Law gives the above 12 equations.

Conversion of Electrical Component Values

CAPACITORS	
1,000,000 Microfarads (UF) = 1 Farad	
1,000,000 Picofarads (PF) = 1 Microfarad (UF)	
RESISTORS	
1,000 ohms = 1 k	
1,000,000 ohms = 1 Megohm (Meg)	
VOLTS & AMPERES	
1000 Millivolts = 1 Volt	1000 Milliamps = Amp
1000 Microvolts = 1 Millivolt	1000 Microamps = 1 Milliamp
INDUCTANCE	
1000 Millihenries = 1 Henry	
1000 Microhenries = 1 Millihenry	

Solid Bare Wire Gauge Diameters And Printed Circuit Drill Sizes

AMER WIRE GAGE (AWG)	NOM DIA	MM DIA	AREA CIRCULAR MILLS (sq. in.)	OHMS PER 1000 FT at 20°C	FEET PER POUND	PC BOARD HOLE DIA
36	.005	.127	25.0	415.	13,210.	.016 (#78)
35	.0056	.142	31.5	331.	10,481.	.016 (#78)
34	.006	.152	39.8	261.	8,310.	.016 (#78)
33	.007	.178	50.1	206.	6.591.	.016 (#78)
32	.008	.208	63.2	162.	5,227.	.016 (#78)
31	.009	.228	79.7	131.	4,145.	.016 (#78)
30	.010	.254	101.0	104.	3,287.	.016 (#78)
29	.011	.280	127.0	81.2	2,607.	.016 (#78)
28	.013	.330	160.0	65.3	2,067.	.016 (#78)
27	.014	.356	202.0	51.4	1,639.	.026 (#71)
26	.016	.406	254.0	41.0	1,300.	.026 (#71)
25	.018	.457	320.0	32.4	1,031.	.026 (#71)
24	.020	.508	404.0	25.7	817.7	.031 (#68)
23	.023	.584	509.0	20.3	648.4	.031 (#68)
22	.025	.635	642.0	16.2	514.2	.031 (#68)
21	.028	.711	810.0	12.8	407.8	.040 (#60)
20	.032	.813	1,020.0	10.1	323.4	.040 (#60)
19	.036	.914	1,290.0	8,04	256.5	.052 (#55)
18	.040	1.016	1,620.0	6.38	203.4	.052 (#55)
17	.045	1.143	2,050.0	5.04	161.3	.052 (#55)
16	.050	1.270	2,580.0	4.01	127.9	
15	.057	1.448	3,260.0	3.18	101.4	
14	.064	1.626	4,110.0	2.52	80.44	
13	.072	1.829	5,180.0	2.00	63.80	
12	.081	2.057	6,530.0	1.59	50.59	
11	.090	2.286	8,230.0	1.26	40.12	
10	.102	2.590	10,400.0	.999	31.82	
9	.114	2.896	13,100.0	.792	25.23	
8	.128	3.251	16,500.0	.628	20.01	
7	.144	3.658	20,800.0			
6	.162	4.115	26,300.0			
5	.182	4.623	33,100.0			
4	.204	5.182	41,700.0			
3	.229	5.817	52,600.0			
2	.258	6.553	66,400.0			
1	.289	7.340	83,700.0			
0	.325	8.255	106,000.0			

NOTE: PC Board Hole Diameters are for manual insertion and NOT plated through holes.

Wire Current Capabilities Chart

AWG	FUSING CURRENT (AMPS)	NORMAL LOAD (AMPS)
26	20	0.6
24	29	1.0
22	41	1.6
20	58	2.5
18	83	4.0
16	117	6.0
14	166	15.0
12	235	20.0

1. THE FUSING CURRENT is the current at which the wire will melt.
2. THE NORMAL LOAD (wire rating) is based on 400 circular mils per ampere.
3. Stranded wire specifications allow strands to run from 7 to 19 strands but manufacturers usually use 7 or 19.

How To Use This Chart: Locate the conductor's desired AWG size on the chart and trace it vertically. The number and size of strands needed to make the stranded conductor will be indicated by the horizontal line (strand number) and diagonal line (strand size) respectively.

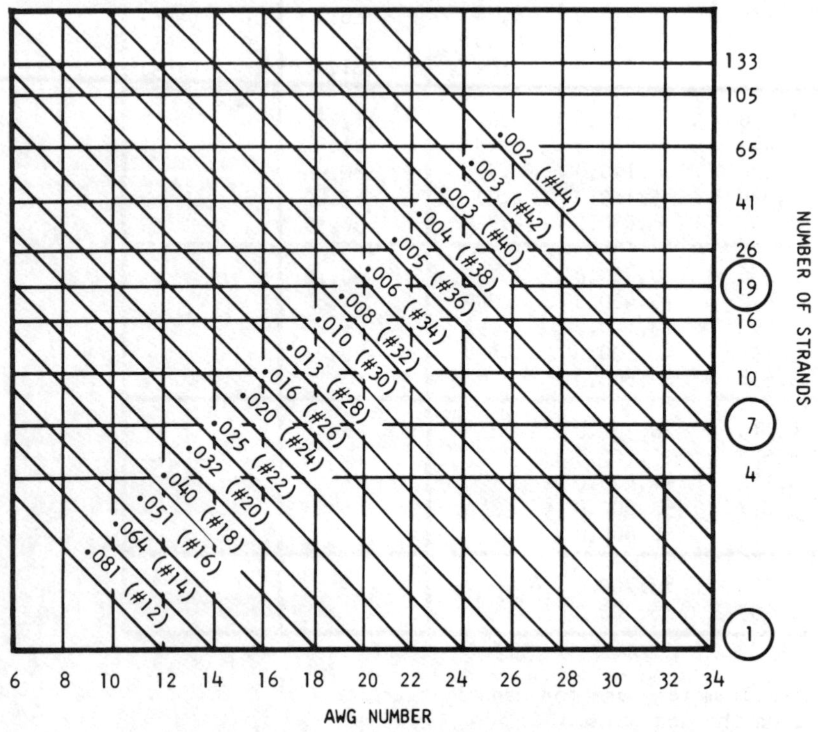

Drill Sizes—Decimal and Metric Equivalents

SIZE	DECIMAL	MM
85	.011	.2794
84	.0115	.2921
83	.012	.3048
82	.0125	.3175
81	.013	.3302
80	.0135	.3429
79	.0145	.3683
1/64	.0156	.3962
78	.016	.4064
77	.018	.4572
76	.020	.5080
75	.021	.5334
74	.0225	.5715
73	.024	.6096
72	.025	.6350
71	.026	.6604
70	.028	.7112
69	.0292	.7417
68	.031	.7874
1/32	.0312	.7925
67	.032	.8128
66	.033	.8382
65	.035	.8890
64	.036	.9144
63	.037	.9398
62	.038	.9652
61	.039	.9906
60	.040	1.0160
59	.041	1.0414
58	.042	1.0668
57	.043	1.0922
56	.0465	1.1811
3/64	.0469	1.1912
55	.052	1.3208
54	.055	1.3970
53	.0595	1.5113
1/16	.0625	1.5875
52	.0635	1.6129
51	.067	1.7018
50	.070	1.7780
49	.073	1.8542
48	.076	1.9304
5/64	.0781	1.9837
47	.0785	1.9939
46	.081	2.0574
45	.082	2.0828
44	.086	2.1844
43	.089	2.2606
42	.0935	2.3749
3/32	.0938	2.3825

SIZE	DECIMAL	MM
41	.096	2.4384
40	.098	2.4892
39	.0995	2.5273
38	.1015	2.5780
37	.104	2.6415
36	.1065	2.7050
7/64	.1094	2.7787
35	.110	2.7939
34	.111	2.8193
33	.113	2.8701
32	.116	2.9463
31	.120	3.0480
1/8	.125	3.1749
30	.1285	3.2638
29	.136	3.4543
28	.1405	3.5686
9/64	.1406	3.5712
27	.144	3.6575
26	.147	3.7337
25	.1495	3.7972
24	.152	3.8607
23	.154	3.9115
5/32	.1562	3.9674
22	.157	3.9877
21	.159	4.0385
20	.161	4.0893
19	.166	4.2163
18	.1695	4.3052
11/64	.1719	4.3662
17	.173	4.3941
16	.177	4.4957
15	.180	4.5719
14	.182	4.6227
13	.185	4.6989
3/16	.1875	4.7624
12	.189	4.8005
11	.191	4.8513
10	.1935	4.9148
9	.196	4.9783
8	.199	5.0545
7	.201	5.1053
13/64	.2031	5.1586
6	.204	5.1815
5	.2055	5.2196
4	.209	5.3085
3	.213	5.4101
7/32	.2188	5.5574
2	.221	5.6133
1	.228	5.7911

Drill Sizes—Decimal and Metric Equivalents (Continued)

SIZE	DECIMAL	MM
A	.234	5.943
15/64	.2344	5.954
B	.238	6.045
C	.242	6.147
D	.246	6.248
E & ¼	.250	6.350
F	.257	6.528
G	.261	6.629
17/64	.2656	6.746
H	.266	6.756
I	.272	6.909
J	.277	7.036
K	.281	7.137
9/32	.2812	7.142
L	.290	7.366
M	.295	7.493
19/64	.2969	7.541
N	.302	7.671
5/16	.3125	7.937
O	.316	8.026
P	.323	8.204
21/64	.3281	8.334
Q	.332	8.433
R	.339	8.610
11/32	.3438	8.732
S	.348	8.839
T	.358	9.093
23/64	.3594	9.129
U	.368	9.347
3/8	.375	9.525
V	.377	9.576
W	.386	9.804
25/64	.3906	9.921
X	.397	10.084
Y	.404	10.261
13/32	.4062	10.317
Z	.413	10.490

SIZE	DECIMAL	MM
27/64	.4219	10.716
7/16	.4375	11.112
29/64	.4531	11.509
15/32	.4688	11.907
31/64	.4844	12.304
1/2	.5000	12.700
33/64	.5156	13.096
17/32	.5312	13.492
35/64	.5469	13.891
9/16	.5625	14.287
37/64	.5781	14.683
19/32	.5938	15.082
39/64	.6094	15.478
5/8	.625	15.875
41/64	.6406	16.271
21/32	.6562	16.667
43/64	.6719	17.066
11/16	.6875	17.462
45/64	.7031	17.858
23/32	.7188	18.257
47/64	.7344	18.653
3/4	.750	19.050
49/64	.7656	19.446
25/32	.7812	19.842
51/64	.7969	20.241
13/16	.8125	20.637
53/64	.8281	21.033
27/32	.8438	21.432
55/64	.8594	21.828
7/8	.875	22.225
57/64	.8906	22.621
29/32	.9062	23.017
59/64	.9219	23.416
15/16	.9375	23.812
61/64	.9531	24.208
31/32	.9688	24.607
63/64	.9844	25.003
1	1.000	25.400

NOTES: METRIC EQUIVALENTS ARE BASED UPON 1 INCH = 25.3995mm.

TO FIND CENTIMETERS, DIVIDE MILLIMETERS BY 10.

Temperature Conversion

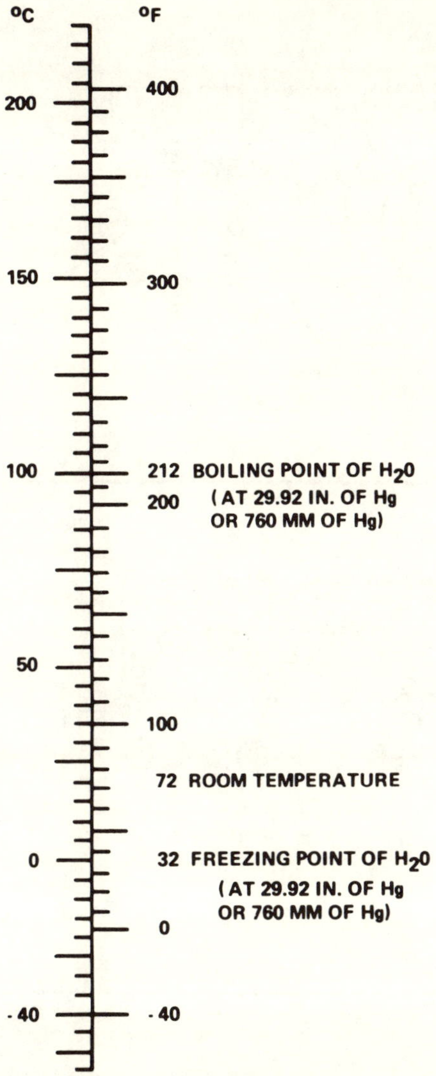

°C	°F	°C	°F
-70	-94	110	230
-60	-76	120	248
-50	-58	130	266
-40	-40	140	284
-30	-22	150	302
-20	- 4	160	320
-10	14	170	338
0	32	180	356
10	50	190	374
20	68	200	392
30	86	210	410
40	104	220	428
50	122	230	446
60	140	240	464
70	158	250	482
80	176	260	500
100	212	300	572

VISCOSITY INTERPRETATION

Listing viscosities by centipoise is sometimes confusing unless one is thoroughly familiar with this particular system. The following practical interpretation will be of assistance. The viscosity readings are for mixed compounds.

The readings were taken at 25°C with a Brookfield viscosimeter.

1 CENTIPOISE (C.P.S.) - WATER
400 CENTIPOISE (C.P.S.) - #10 MOTOR OIL
1,000 CENTIPOISE (C.P.S.) - CASTOR OIL
3,500 CENTIPOISE (C.P.S.) - KARO SYRUP
4,500 CENTIPOISE (C.P.S.) - #40 MOTOR OIL
25,000 CENTIPOISE (C.P.S.) - HERSHEY CHOCOLATE SYRUP

To convert temperatures use following formulas:

$^{\circ}C = 5/9\ (^{\circ}F - 32)$ (Celsius) Formerly Centigrade

$^{\circ}F = 9/5\ (^{\circ}C + 32)$ (Fahrenheit)

$^{\circ}K = {}^{\circ}C + 273.15$ (Kelvin)

$^{\circ}R = {}^{\circ}F + 459.67$ (Rankin)

TRANSISTORS

OVERALL DIMENSIONS ARE MAXIMUM

TO-76

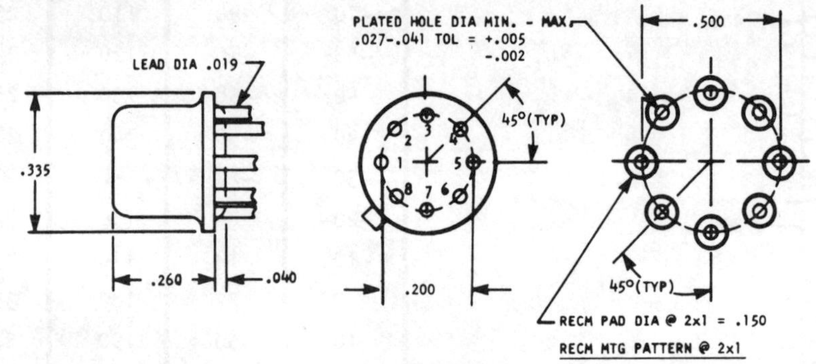

TO-100

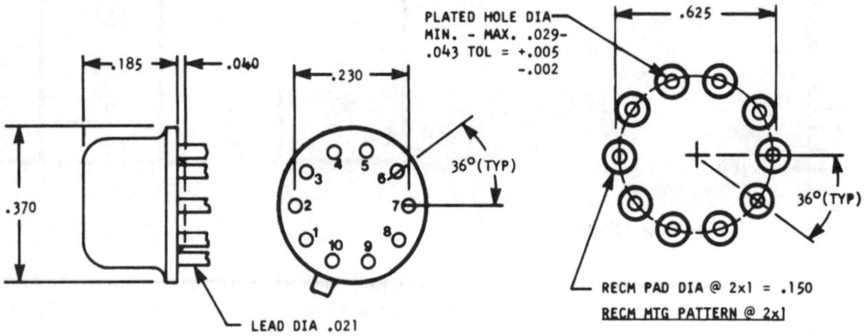

TO-220

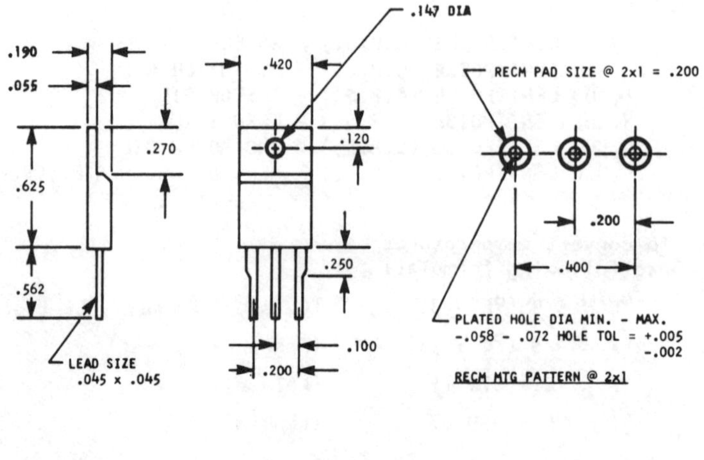

TRANSISTORS
OVERALL DIMENSIONS ARE MAXIMUM

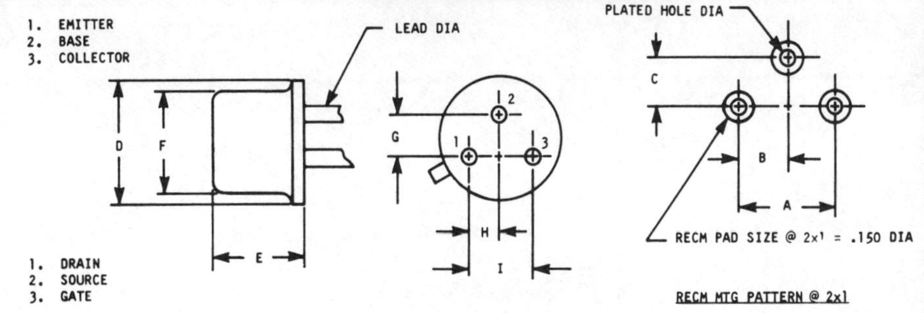

STYLE	A 2x1	B 2x1	C 2x1	MIN. - MAX. PLATED HOLE DIA TOL = + .005 - .002	LEAD DIA	D	E	F	G NOM	H NOM	I NOM
TO-5	.400	.200	.200	.029 - .043	.021	.370	.260	.335	.100	.100	.200
TO-12	.400	.200	.200	.029 - .043	.021	.370	.260	.335	.100	.100	.200
TO-18	.350	.175	.150	.029 - .043	.021	.230	.210	.195	.050	.050	.100
TO-39	.400	.200	.200	.029 - .043	.021	.370	.260	.335	.100	.100	.200
TO-46	.350	.175	.150	.029 - .043	.021	.230	.085	.195	.050	.050	.100
TO-52	.350	.175	.150	.029 - .043	.021	.230	.150	.195	.050	.050	.100
TO-107	.350	.175	.150	.029 - .043	.021	.350	.195	.215	.050	.050	.100

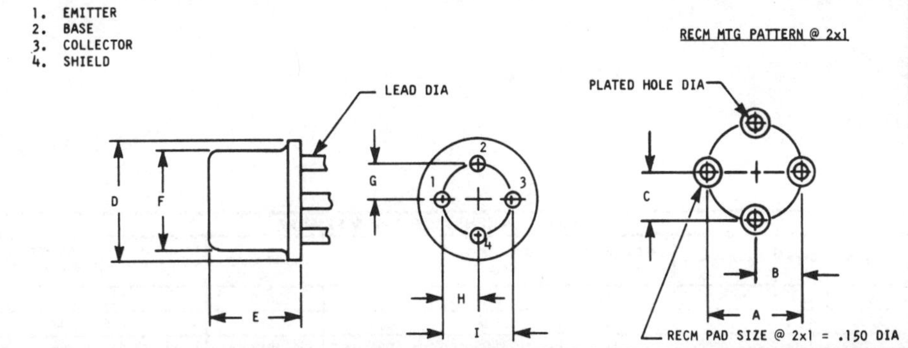

STYLE	A 2x1	B 2x1	C 2x1	MIN. - MAX. PLATED HOLE DIA TOL = + .005 - .002	LEAD DIA	D	E	F	G	H	I
TO-72	.350	.175	.175	.029 - .043	.021	.230	.210	.195	.050	.050	.100
TO-17	.350	.175	.175	.029 - .043	.021	.215	.210	.168	.036	.036	.071

TO-92

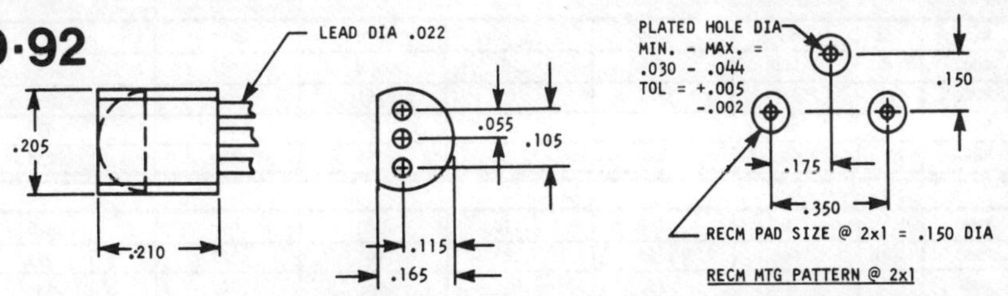

RESISTORS

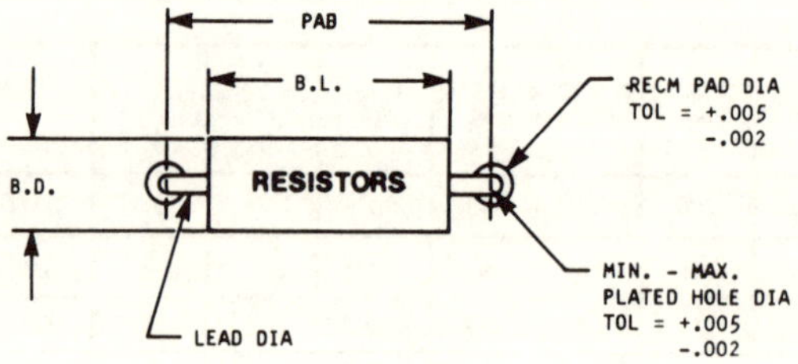

OVERALL DIMENSIONS ARE MAXIMUM

STYLE	WATT	PAD @ 2 x 1	RECM PAD DIA @ 2 x 1	BODY LENGTH (B.L.)	BODY DIA (B.D.)	LEAD DIA	MIN. - MAX. PLATED HOLE DIA TOL = +.005 / -.002
RC05	1/8	.800	.150	.160	.066	.018	.026 - .040
RC06	1/10	.800	.150	.160	.075	.018	.026 - .040
RC07	1/4	1.000	.150	.281	.098	.027	.035 - .049
RC08	1/10	.800	.150	.225	.115	.018	.026 - .040
RC12	1/4	1.200	.150	.390	.148	.028	.036 - .050
RC20	1/2	1.200	.187	.416	.161	.036	.044 - .058
RC22	1/2	1.600	.187	.575	.233	.036	.044 - .058
RC32	1	1.600	.187	.593	.240	.042	.050 - .064
RC42	2	2.000	.200	.728	.336	.048	.056 - .070
RL05	1/8	.800	.150	.170	.070	.017	.025 - .040
RL07	1/4	1.000	.150	.281	.098	.027	.035 - .049
RL20	1/2	1.200	.187	.416	.161	.036	.044 - .058
RL32	1	1.600	.187	.593	.205	.042	.050 - .064
RL42	2	2.000	.200	.728	.336	.050	.058 - .072
RN50	1/10	.800	.150	.170	.080	.018	.026 - .040
RN55	1/10	1.000	.150	.281	.141	.027	.035 - .049
RN60	1/8	1.200	.150	.437	.165	.027	.035 - .049
RN65	1/4	1.800	.150	.656	.250	.027	.035 - .049
RN70	1/2	2.200	.187	.875	.328	.034	.042 - .056
RN75	1	2.800	.187	1.125	.437	.034	.042 - .056
RN80	2	5.000	.187	2.281	.437	.034	.042 - .056
RWP20	3	1.600	.187	.563	.250	.036	.044 - .058

DIODES

OVERALL DIMENSIONS ARE MAXIMUM

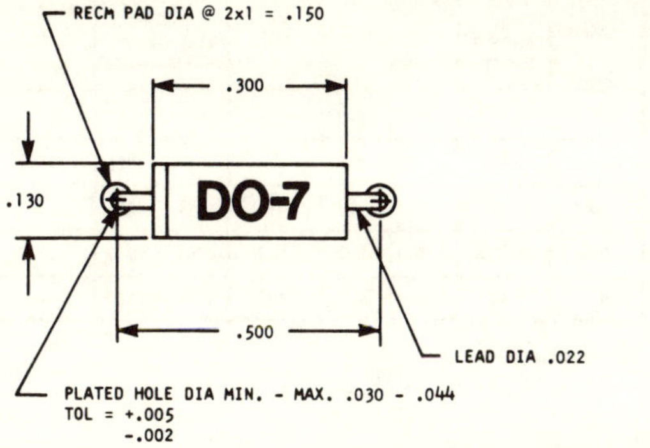

(RECM PAD SPACING @ 2x1 = 1.000)

REFERENCE JAN 1N NUMBERS	
1N 270	THRU
1N 276	1N 939B
1N 277	1N 941
1N 457	1N 943B
1N 458	1N 944B
1N 459	1N 945B
1N 483B	1N 692B
1N 485B	THRU
1N 486B	1N 992B
1N 645	1N 3595
1N 647	1N 4245
1N 649	THRU
1N 653	1N 4249
1N 746A	1N 4370A
THRU	1N 4371A
1N 759A	1N 4372A
1N 821	1N 4942
1N 823	1N 4944
1N 825	1N 4946
1N 827	1N 4947
1N 829	1N 4948
1N 935B	

JAN 1N 3611, THRU 1N 3614, 1N 3957

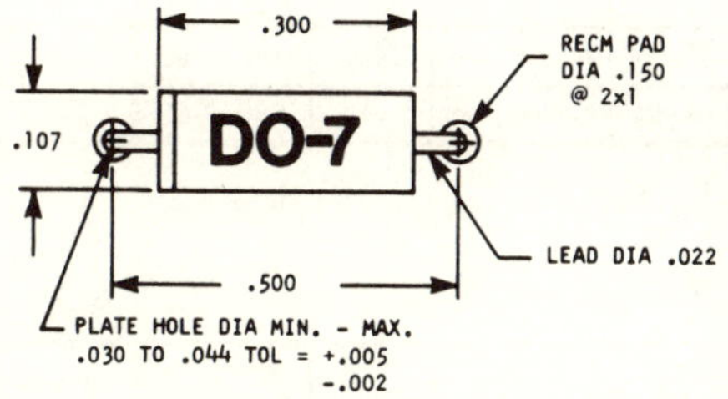

(RECM PAD SPACING @ 2x1 = 1.000)

JAN 1N 914, 1N 3064, 1N 3070, 1N 3600

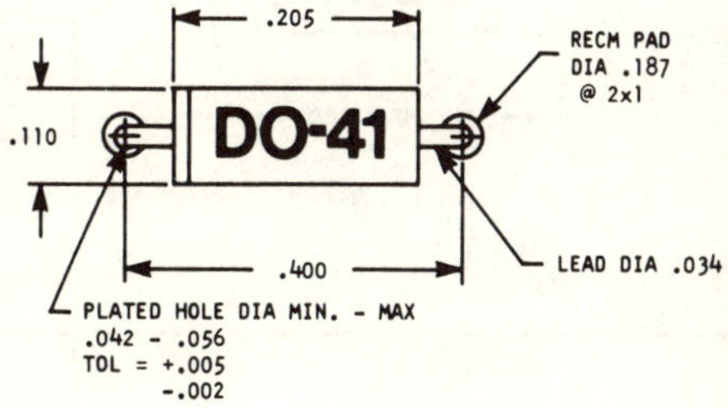

(RECM PAD SPACING @ 2x1 = .800)

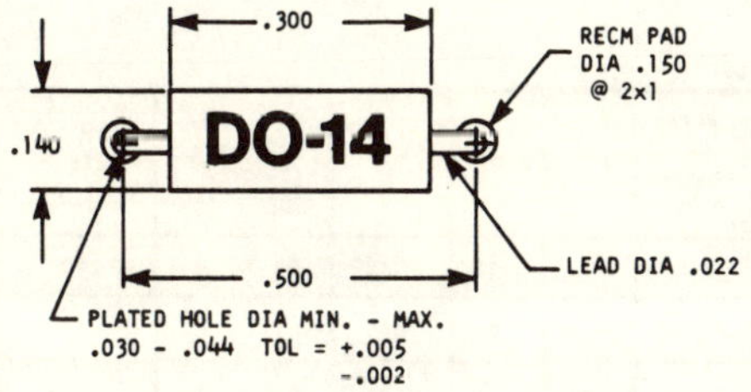

(RECM PAD SPACING @ 2x1 = 1.000)

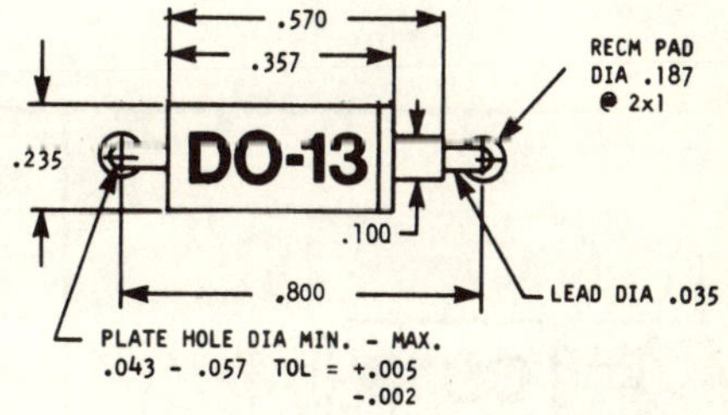

(RECM PAD SPACING @ 2x1 = 1.600)

CAPACITORS

OVERALL DIMENSIONS ARE MAXIMUM

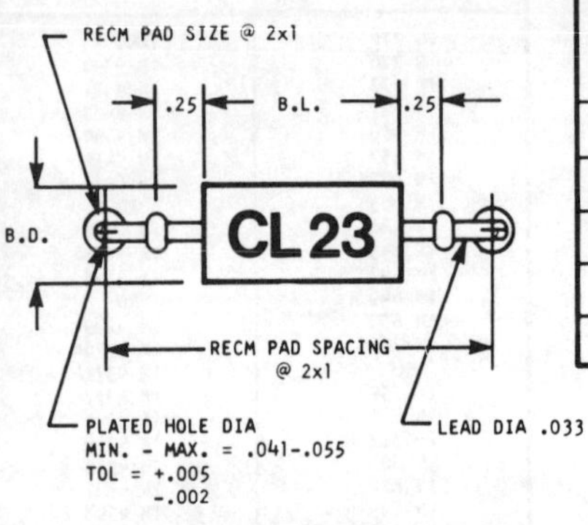

CASE	RECM PAD SPACING @ 2x1	B.L.	B.D.	LEAD DIA	MIN - MAX PLATED HOLE SIZE TOL = +.005, -.002	RECM PAD SIZE @ 2x1
D1	3.200	.844	.234	.033	.041 - .055	.187
D2	3.800	1.125	.328	.033	.041 - .055	.187
D3	4.800	1.656	.422	.033	.041 - .055	.187
D4	6.200	2.344	.422	.033	.041 - .055	.187
D5	7.400	2.969	.422	.033	.041 - .055	.187

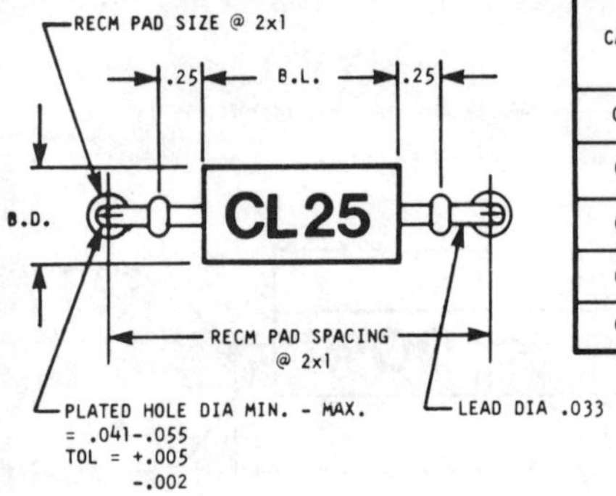

CASE	RECM PAD SPACING @ 2x1	B.L.	B.D.	LEAD DIA	MIN - MAX PLATED HOLE SIZE TOL = +.005, -.002	RECM PAD SIZE @ 2x1
C1	3.200	.875	.219	.033	.041 - .055	.187
C2	3.600	1.063	.313	.033	.041 - .055	.187
C3	4.800	1.625	.406	.033	.041 - .055	.187
C4	6.200	2.313	.406	.033	.041 - .055	.187
C5	7.400	2.933	.406	.033	.041 - .055	.187

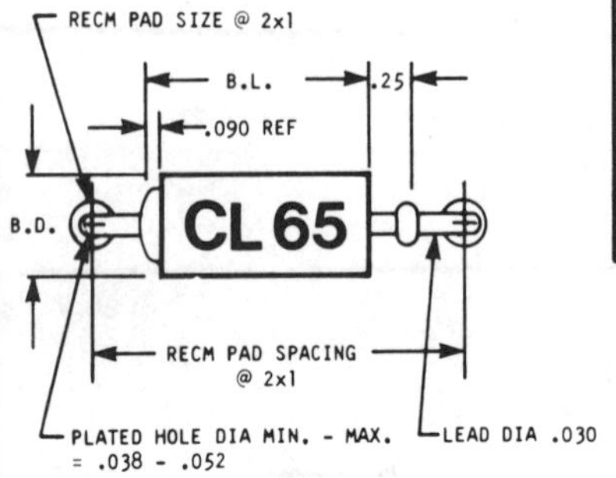

CASE	RECM PAD SPACING @ 2x1	B.L.	B.D.	LEAD DIA	MIN - MAX PLATED HOLE SIZE TOL = +.005, -.002	RECM PAD SIZE @ 2x1
T1	2.400	.668	.219	.030	.038 - .052	.150
T2	2.800	.856	.313	.030	.038 - .052	.150
T3	3.000	.981	.406	.030	.038 - .052	.150

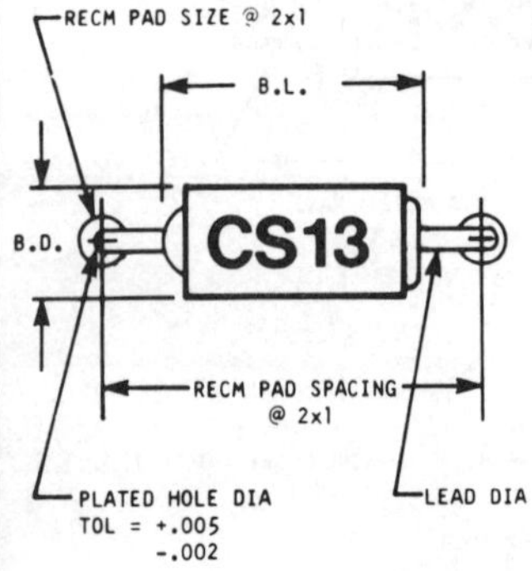

CASE	RECM PAD SPACING @ 2x1	B.L.	B.D.	LEAD DIA	MIN - MAX PLATED HOLE SIZE TOL = +.005, -.002	RECM PAD SIZE @ 2x1
A	1.400	.422	.151	.025	.033 = .047	.150
B	1.800	.610	.201	.025	.033 - .047	.150
C	2.200	.822	.305	.030	.038 - .052	.150
D	2.400	.922	.367	.030	.038 - .052	.150

CAPACITORS

OVERALL DIMENSIONS ARE MAXIMUM

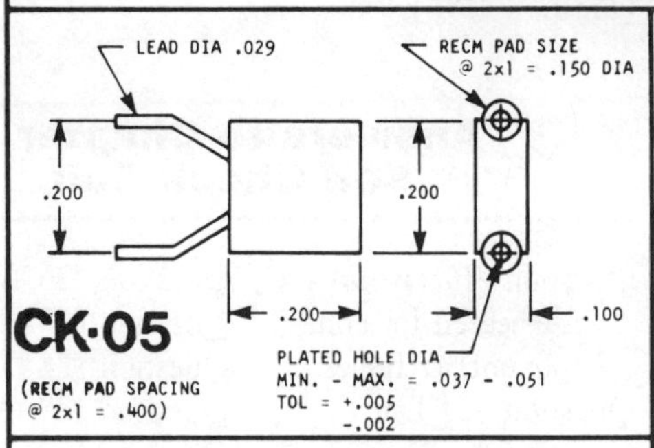

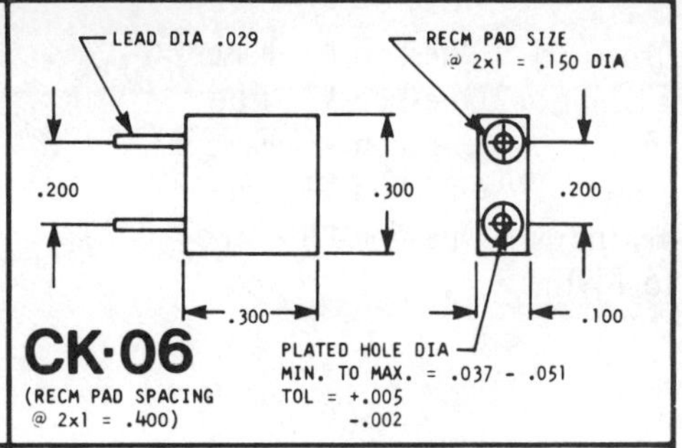

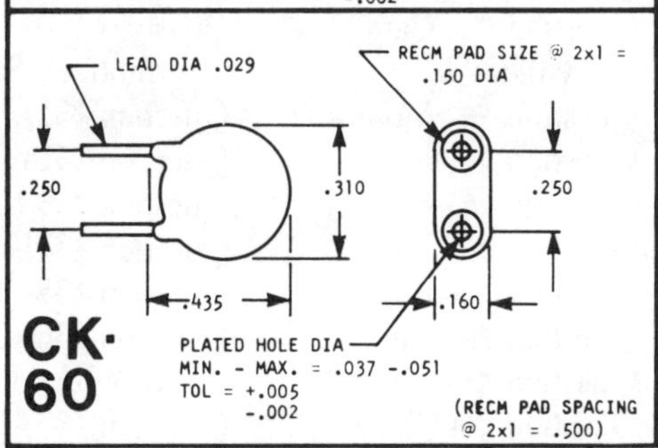

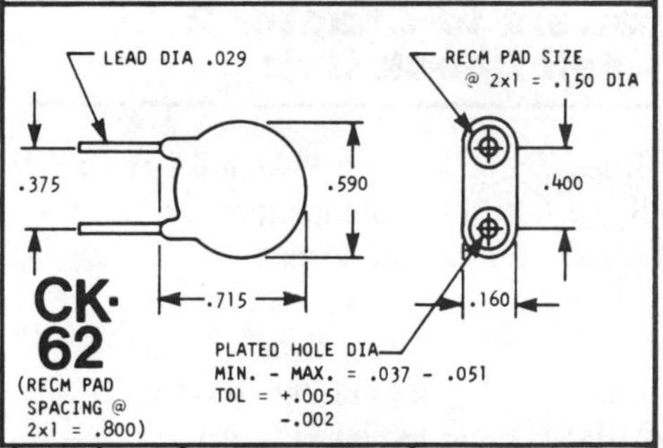

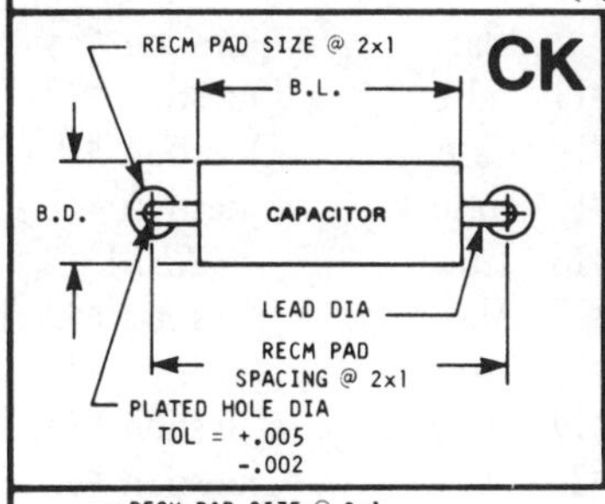

STYLE	RECM PAD SPACING @ 2x1	BODY LENGTH (B.L.)	BODY DIA (B.D.)	9,999 Pf & UNDER			10,000 Pf & OVER		
				LEAD DIA	MIN-MAX PLATED HOLE DIA	RECM PAD SIZE @2x1	LEAD DIA	MIN-MAX PLATED HOLE DIA	RECM PAD SIZE @2x1
CK12	.800	.220	.100	.020	.028-.042	.150	.029	.037-.051	.187
CK13	1.400	.500	.120	.020	.028-.042	.150	.029	.037-.051	.187
CK14	1.600	.500	.180	.020	.028-.042	.150	.029	.037-.051	.187
CK15	1.600	.500	.200	.020	.028-.042	.150	.029	.037-.051	.187
CK16	1.800	.650	.240	.020	.028-.042	.150	.029	.037-.051	.187
CK17	2.000	.750	.310	.020	.028-.042	.150	.029	.037-.051	.187

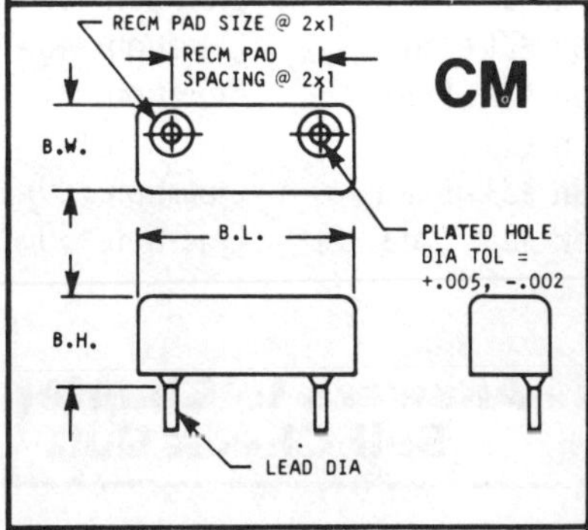

CM05, CM06, CM07, CM08

CAPACITANCE IN Pf	RECM PAD SPACING @2x1	B.L.	B.H.	B.W.	LEAD DIA	PLATED HOLE DIA TOL=+.005 -.002	RECM PAD SIZE @2x1
5-390	.450	.480	.473	.220	.027	.035-.049	.150
430-3300	.700	.720	.591	.260	.034	.042-.056	.150
3600-4700	.700	.730	.625	.340	.034	.042-.056	.150
5100-13K	.850	.830	.962	.360	.042	.050-.064	.187
15K-20K	.850	.830	1.000	.450	.042	.050-.064	.187
22K-30K	2.100	1.426	.960	.342	.042	.050-.064	.187
33K-39K	2.100	1.444	.960	.393	.042	.050-.064	.187
43K-47K	2.100	1.463	1.000	.443	.042	.050-.064	.187
51K-62K	2.100	1.400	1.000	.443	.042	.050-.064	.187

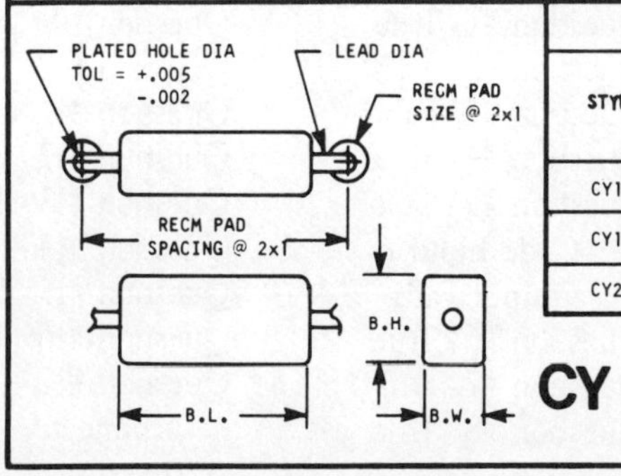

CY-CASE SIZES

STYLE	RECM PAD SPACING @ 2 x 1	B.L.	B.H.	B.W.	LEAD DIA	PLATED HOLE DIA TOL=+.005 -.002	RECM PAD SIZE @2x1
CY10	1.200	.391	.203	.109	.024	.032-.046	.150
CY15	1.400	.516	.297	.156	.029	.037-.051	.150
CY20	2.000	.797	.469	.188	.036	.044-.058	.187

Answers to Chapter 1 Self Check Quiz

Question #1 - True
Question #2 - Etching
Question #3 - True
Question #4 - False
Question #5 - By the question to ask the Engineer
Question #6 - False
Question #7 - True
Question #8 - False
Question #9 - True
Question #10 - True

Answers to Chapter 2 Self Check Quiz

Question #1 - True
Question #2 - R
C
Q
CR
Question #3 - False
Question #4 - MIL-STD 275 Y14.5
Question #5 - IEEE & ANSI
Question #6 - Metal or Paper
Question #7 - Carbon, Wire-wound or Metal Film
Question #8 - Transistor
Question #9 - True
Question #10 - False
Question #11 - False
Question #12 - True
Question #13 - Small Scale Intergration
Medium "
Large "
Very large"
Question #14 - Inverter
Question #15 - True
Question #16 - False
Question #17 - True
Question #18 - True
Question #19 - High Low
Question #20 - True
Question #21 - True
Question #22 - True
Question #23 - True
Question #24 - False
Question #25 - True
Question #26 - True
Question #27 - False
Question #28 - True

Answers to Chapter 3 Self Check Quiz

Question #1 - True
Question #2 - MOS
Question #3 - Transistor Transistor Logic
Emitter Couple Logic
Complementry Metal Oxide Semi-Conductor
Question #4 - True
Question #5 - False
Question #6 - False
Question #7 - True
Question #8 - True
Question #9 - False
Question #10 - True
Question #11 - False
Question #12 - True
Question #13 - True
Question #14 - True
Question #15 - VDD
VEE
Question #16 - True
Question #17 - 20, 22, 24, 28 & 40
Question #18 - False
Question #19 - True
Question #20 - False

Answers to Chapter 4 Self Check Quiz

Question #1 - The area dedicated for components & traces.
Question #2 - True
Question #3 - Card Guides
Question #4 - Mounting
Question #5 - False
Question #6 - False
Question #7 - True
Question #8 - Edge
Question #9 - False
Question #10 - True
Question #11 - True
Question #12 - Flat
Question #13 - False
Question #14 - Ease
Question #15 - False
Question #16 - True
Question #17 - Absorbing
Question #18 - True
Question #19 - False
Question #20 - False
Question #21 - True
Question #22 - False
Question #23 - True
Question #24 - False
Question #25 - True
Question #26 - Vibration
Question #27 - True
Question #28 - .005 & .010
Question #29 - True
Question #30 - Subtract, Subtract, Subtract
Question #31 - Depth
Question #32 - False
Question #33 - True
Question #34 - True
Question #35 - True
Question #36 - Temperature & Vibration
Question #37 - Conformal
Question #38 - False
Question #39 - False
Question #40 - True
Question #41 - I.P.C. ANSI
Question #42 - False
Question #43 - True
Question #44 - True
Question #45 - False
Question #46 - True
Question #47 - False
Question #48 - Uniformly
Question #49 - Bend
Question #50 - Pads

Answers to Chapter 5 Self Check Quiz

Question #1 - True
Question #2 - .030
Question #3 - False
Question #4 - True
Question #5 - Color Code Input & Output Pads
Question #6 - True
Question #7 - .010
Question #8 - True
Question #9 - .100
Question #10 - False
Question #11 - False
Question #12 - Lead
Question #13 - True
Question #14 - False
Question #15 - False
Question #16 - True
Question #17 - Interconnect

Question #18 - False
Question #19 - Components
Question #20 - False
Question #21 - True
Question #22 - Twice
Question #23 - When the quantity of comp's exceed the normal 2 sided design area.
2. when a large number of traces or interconnections are needed
Question #24 - 2 x 1
Question #25 - True
Question #26 - False
Question #27 - False
Question #28 - True
Question #29 - Shock
Question #30 - Clamping and Adhesives
Question #31 - True
Question #32 - .100
Question #33 - True
Question #34 - Connector
Question #35 - False
Question #36 - True
Question #37 - False
Question #38 - True
Question #39 - True
Question #40 - True
Question #41 - False
Question #42 - False
Question #43 - True
Question #44 - Fabrication
Question #45 - True

Answers to Chapter 6 Self Check Quiz

Question #1 - 2 x 1
Question #2 - .005
Question #3 - True
Question #4 - One layer is used for a pad master
Question #5 - False
Question #6 - False
Question #7 - True
Question #8 - 45°
Question #9 - True
Question #10 - False
Question #11 - True
Question #12 - True
Question #13 - Step around the feed through pads
Question #14 - False
Question #15 - True
Question #16 - True
Question #17 - any component that requires access
Question #18 - False
Question #19 - True
Question #20 - True
Question #21 - Same
Question #22 - False
Question #23 - True
Question #24 - True
Question #25 - Platemaker
Question #26 - registration
Question #27 - traces
Question #28 - False
Question #29 - True
Question #30 - multipoint
Question #31 - photography
Question #32 - True
Question #33 - bus
Question #34 - True
Question #35 - True

Answers to Chapter 7 Self Check Quiz

Question #1 - True
Question #2 - False
Question #3 - True
Question #4 - False
Question #5 - False
Question #6 - False
Question #7 - True
Question #8 - True
Question #9 - True
Question #10 - False
Question #11 - noncomponent
Question #12 - False
Question #13 - False
Question #14 - False
Question #15 - True
Question #16 - parts list
Question #17 - artwork
Question #18 - component
Question #19 - True
Question #20 - True
Question #21 - printed circuit
Question #22 - False
Question #23 - printed circuit
Question #24 - False
Question #25 - maintenance
Question #26 - stamping
Question #27 - True
Question #28 - True
Question #29 - epoxy
Question #30 - True
Question #31 - False
Question #32 - False
Question #33 - True
Question #34 - humidity
Question #35 - reduction

Answers to Chapter 8 Self Check Quiz

Question #1 - Engineer
Question #2 - Inputs
Question #3 - Pin
Question #4 - Designer
Question #5 - Never
Question #6 - Specification
Question #7 - Dual Inline
Question #8 - Dual Inline
Question #9 - Pin
Question #10 - Shape
Question #11 - Lead
Question #12 - Length Diameter Diameter
Question #13 - Rating Tolerance Composition
Question #14 - .060
Question #15 - Up
Question #16 - C
Question #17 - Close
Question #18 - Guide
Question #19 - Pad Size
Question #20 - Lead Size
Question #21 - Hole & Pad
Question #22 - .006 .004
Question #23 - .028
Question #24 - Because it would save money on fabricating the board
Question #25 - .010
Question #26 - Stocked
Question #27 - Military ANSI IPC Commercial
Question #28 - Possible
Question #29 - Clamping Bolting Adhesive's
Question #30 - Accessible
Question #31 - .050
Question #32 - Common Hole
Question #33 - Spread more than the

components
Pre-bent leads
Question #34 - The company could have a standard parts book
Question #35 - Pertains to: signal paths or traces
Question #36 - Input voltage
Question #37 - P.C. Board
Question #38 - Connector
Question #39 - .150
Question #40 - The lowest number of IC's required
Question #41 - Communicate
Question #42 - One
Question #43 - Pads
Question #44 - In line Row
Question #45 - Voltage pads
Ground pads
Component side traces
Non-component side traces
Question #46 - Bus lines for voltage & ground
Question #47 - Opposite
Question #48 - This keeps track of completed and incomplete traces
Question #49 - Feedthroughs
Question #50 - It is better to have one continuous connection on the same side of the P.C. board.

Answers to Chapter 9 Self Check Quiz

Question #1 - C
Question #2 - C
Question #3 - B
Question #4 - A
Question #5 - B
Question #6 - C
Question #7 - D
Question #8 - D
Question #9 - B
Question #10 - A
Question #11 - B
Question #12 - A
Question #13 - C
Question #14 - A
Question #15 - C
Question #16 - D
Question #17 - D
Question #18 - B
Question #19 - E
Question #20 - B
Question #21 - D
Question #22 - A
Question #23 - C
Question #24 - B
Question #25 - B
Question #26 - D
Question #27 - C
Question #28 - A
Question #29 - E
Question #30 - A
Question #31 - B
Question #32 - C
Question #33 - A
Question #34 - D
Question #35 - C
Question #36 - B
Question #37 - A
Question #38 - D
Question #39 - D
Question #40 - C
Question #41 - B
Question #42 - C
Question #43 - D
Question #44 - A
Question #45 - B
Question #46 - C
Question #47 - C
Question #48 - C
Question #49 - B
Question #50 - D

Answers to Chapter 10 Self Check Quiz

Question #1 - Operational Amplifier
Question #2 - T
Question #3 - The operational amplifier shares characteristics of both digital and analog worlds
Question #4 - Electrolyte
Question #5 - .022
Question #6 - Ceramic disc
Question #7 - The layout configuration is established
Question #8 - P Dual-in-line
Question #9 - F
Question #10 - One 07, Buffer
Question #11 - Good commercial practices
Question #12 - Increasing size, starting from the smallest to the largest
Question #13 - T
Question #14 - F
Question #15 - T
Question #16 - T
Question #17 - F
Question #18 - F
Question #19 - T
Question #20 - T
Question #21 - What interfaces to the P.C. board
Question #22 - F
Question #23 - Numbering
Question #24 - T
Question #25 - Manufacturing specification number: 00, 04 etc.
Question #26 - F
Question #27 - T
Question #28 - T
Question #29 - T
Question #30 - F
Question #31 - F
Question #32 - T
Question #33 - T
Question #34 - F
Question #35 - T
Question #36 - Voltage ground
Question #37 - Voltage ground
Question #38 - Voltage ground
Question #39 - T
Question #40 - Via
Question #41 - One
Question #42 - Black
Question #43 - Filter
Question #44 - T
Question #45 - Registration
Question #46 - Edge line
Question #47 - T
Question #48 - F
Question #49 - T
Question #50 - Red

Answers to Chapter 11 Self Check Quiz

Question #1.1- Most companies have a standard parts or components book they would use to purchase most of their components from. And the engineer drafting dept. would list all of these parts with respective body size, lead size, pad size, etc. Occasionally special parts must be purchased non-standard, and all of the data for body size, lead size, pad size must be computed.

1.2- There are some types of logic that require short traces such as ECL, and Schottky logic. Unlike the more common logic such as TTL, and C-MOS. Another example might be P.C. boards that have both analog & digital on the same board.

1.3- Both analog & digital P.C. boards require input filtering capacitors or input filtering circuits. Typically this is a protection from the input source.

1.4- Bypass capacitors are used on digital boards, and are connected between the power bus & ground bus. These capacitors are spaced out in the rows of IC's to remove the AC of the power bus and terminate it to the ground.

1.5- Typically most digital boards will end up with some spare gates. However, occasionally the project engineer will request a specific type and quantity of spare gates.

1.6- The designer would like to know with a high degree of confidence that all the components, traces, via's test points will fit on the board before he/she starts.

1.7- Pre-assigned pins can effect the way you plot your IC's on a digital board. All pre-assigned pin require more board area.

1.8- The main concern above variable components on the P.C. board is placement or pre-assigned location.

1.9- Analog and digital boards typically require test points. These are to be used by the test technicians, field repair or th project engineer.

1.10- Most analog or digital boards do not have components that require head sinks, unless it is a power supply board, if heat sinks are required the most likely components are transistors or large resistors.

1.11- Package considerations are typically concerned with the chassis which houses the P.C. board. Method of fastening, cards guides, card ejectors, stand ffs, mother board interface etc.

Question #2- No

Question #3- This assignment of the IC numbers to the location on the board.

Question #4- The main concern is the arrangement is IC's that communicate to one another in line or in the same rows with each other.

Question #5- Yes

Question #6- Does not matter unless you consider wave soldering: If you are wave soldering the board the traces should be perpendicular to the wave.

Question #7- The first traces to be drafted onto the board first is the voltage and ground bases.

Question #8- Any gate—interconnect one gate to another. Assign the input and output pin numbers on the schematic.

Question #9- Yellow

Question #10- This steps cleans up all unnecessary via holes, and places the IC's in an alphanumerical order.

Question #11.1- Two, red, blue

11.2- Pad master

11.3- Pinning

11.4- Identify

11.5- Silkscreen is one, ground planes are another, solder masks can also be used

11.6- Cold

11.7- Padmaster

11.8- Noncomponent trace sheet

11.9- component

11.0- traces

Answers to Chapter 12 Self Check Quiz

Question #1 - True

Question #2 - True

Question #3 - False

Question #4 - False

Question #5 - True
Question #6 - When the need to connect a large number of interconnections.
When frequency requirements call for careful control.
When the quantity of components, exceed the normal design area.
Question #7 - Wiring errors eliminated compared to wire wrap boards.
Uniform electrical characteristics, coupling and shielding.
Space saved by tighter packaging.
Question #8 - Special tools and skills required for board repairs.
High cost in small quantities
Very costly and difficult to change.
Question #9 - True
Question #10 - False
Question #11 - True
Question #12 - False
Question #13 - True
Question #14 - True
Question #15 - True
Question #16 - Artwork
Question #17 - True
Question #18 - True
Question #19 - 1/2
Question #20 - Warpage
Question #21 - "B" stage
Question #22 - Critical
Question #23 - False
Question #24 - .060
Question #25 - False

Index